Advancement in Propagation, Breeding, Cultivation and Marketing of Ornamentals

Advancement in Propagation, Breeding, Cultivation and Marketing of Ornamentals

Editors

Margherita Beruto
Emmy Dhooghe
Bruce Dunn

MDPI

Basel • Beijing • Wuhan • Barcelona • Belgrade • Novi Sad • Cluj • Manchester

Editors

Margherita Beruto
International Society for
Horticultural Science (ISHS)
Sanremo (IM)
Italy

Emmy Dhooghe
Ghent University
Ghent
Belgium

Bruce Dunn
Oklahoma State University
Stillwater
USA

Editorial Office
MDPI AG
Grosspeteranlage 5
4052 Basel, Switzerland

This is a reprint of articles from the Special Issue published online in the open access journal *Horticulturae* (ISSN 2311-7524) (available at: https://www.mdpi.com/journal/horticulturae/special_issues/propagation_marketing).

For citation purposes, cite each article independently as indicated on the article page online and as indicated below:

Lastname, A.A.; Lastname, B.B. Article Title. *Journal Name* **Year**, *Volume Number*, Page Range.

ISBN 978-3-7258-1811-2 (Hbk)
ISBN 978-3-7258-1812-9 (PDF)
doi.org/10.3390/books978-3-7258-1812-9

Cover image courtesy of Stefano Maritano

Contents

About the Editors

Margherita Beruto

Margherita Beruto is the Chair of ISHS Division Ornamental Plants. She obtained her PhD on the topic of Agricultural Sciences and Applied Biology: Cell and Gene Biotechnology in 1997 from the University of Ghent (Belgium). From 2009 to 2021, Margherita Beruto, as Director of the Regional Institute for Floriculture (IRF), Sanremo, Italy, coordinated the research activities of four research units (breeding; in vivo propagation and crop management; pathology and crop protection and tissue culture) and developed specialized services to support breeders and nurserymen with the initial stock of in vitro plant material, which was subsequently transferred to commercial laboratories and/or to companies for industrial application. She is an Associate Editor of scientific journals and has acted as a Guest Editor for Special Issues of Ornamentals.

Emmy Dhooghe

Emmy Dhooghe is a professor in the Department of Plants and Crops at Ghent University. She has over 20 years of research experience. In 2009, she obtained her PhD (Ghent University) on intergeneric crosses in the *Ranunculaceae* family. After her PhD, she worked as a senior researcher at ILVO (Belgium, Flanders Research Institute for Agriculture, Fisheries and Food). Since 2022, she has been working as a professor in High-Technology Horticulture, where her group is conducting research in the field of vertical farming, soilless culture, LED lights, plant breeding, horticulture, abiotic stress and plant physiology.

Bruce Dunn

Bruce Dunn is a Professor in the Department of Horticulture and Landscape Architecture at Oklahoma State University. He obtained his PhD in Plant Sciences (with an emphasis on plant breeding) in 2008 from the University of Arkansas. He currently has a split research and teaching appointment and has taught principles of horticulture, plant propagation, greenhouse management, flower and fruit crop physiology, urban horticulture production, soilless production, and sustainable nursery production. Bruce Dunn received the 2024 Ferguson College Teaching Excellence award and the Southern Region ASHS Teaching Excellence award. He has published over 80 peer reviewed research articles with research focused on new plant development and improvement of production practices in controlled environments, focusing on use of optical sensors, irrigation, hydroponics, LED lighting, plant nutrition, plant evaluation, and product trials.

horticulturae

Editorial

Advancement in Propagation, Breeding, Cultivation, and Marketing of Ornamentals

Margherita Beruto [1,*], Emmy Dhooghe [2,*] and Bruce Dunn [3,*]

1 International Society for Horticultural Science (ISHS), 18038 Sanremo, IM, Italy
2 Department Plants and Crops, Faculty of Bioscience Engineering, Ghent University, 9000 Ghent, Belgium
3 Department of Horticulture and Landscape Architecture, Oklahoma State University, Stillwater, OK 74078, USA
* Correspondence: margheberuto@gmail.com (M.B.); emmy.dhooghe@ugent.be (E.D.); bruce.dunn@okstate.edu (B.D.)

Citation: Beruto, M.; Dhooghe, E.; Dunn, B. Advancement in Propagation, Breeding, Cultivation, and Marketing of Ornamentals. *Horticulturae* **2024**, *10*, 747. https://doi.org/10.3390/horticulturae10070747

Received: 17 June 2024
Accepted: 4 July 2024
Published: 16 July 2024

Ornamental horticulture is an important branch of horticulture addressed to grow and market plants or cut flowers used for decorative purposes and landscape design. Accordingly, the ornamental industry appears to be segmented and diversified. Moreover, the industry should meet the fast-changing needs of consumers, and, in consideration of the aesthetic value of its final product, the achievement of high-quality standards is mandatory.

In 2021, the flower and ornamental plants market was globally estimated at about 27 billion USD, and the sector is forecast to increase its value more than one and a half times by 2029 (source: Global Flowers and Ornamental Plants Market—Industry Trends and Forecast to 2029 Report; https://www.databridgemarketresearch.com/reports/global-flowers-and-ornamental-plants-market, accessed on 7 June 2024). The European market has a great impact on the ornamental industry and represents about 47% of the global market. The ornamental industry is fast changing; new producer and consumer countries are entering the global market [1], and consequently, in the coming years, the ornamental market and the dynamics of trade at an international level could be modified.

Several factors are driving this industry, linked to the increasing demand for cut flowers and ornamental plants used for floral arrangement gardening and landscaping. On one side, the expected increase in demand can be related to the increases in urban consumers, a greater spending capacity, and the growing population of retiring and elderly people. On the other side, technological advancements and innovative approaches enhanced the number of high-standard products and efficient crop management.

Therefore, the question is whether the ornamental industry can satisfy this request and, at the same time, maintain high standards of innovation and quality, an essential prerequisite of this sector. Indeed, several challenges must be overcome to satisfy the market trend for innovative, sustainable, and eco-friendly products and to enhance product quality and efficiency. In this light, the major issues the ornamental sector must meet are related to the following:

- Limited resources availability in term of natural resources, as well as labor and capital to maintain the operations required for crop management.
- Climate change, which can greatly affect the plant growth and health.
- Necessity to implement sustainable and effective methods for pest and disease management, minimizing the use of chemicals and enhancing an environmentally friendly approach.
- Diversification of supply through products meeting the criteria of innovation and sustainability in a broad sense (environmental, economic, and social).
- Marketing and distribution optimization.

Breeding is a key step in the obtention of new and performant cultivars, and through the classical methods (selection, hybridization, and mutation breeding), new flower colors and shapes and new cultivars resistant to biotic and abiotic stresses have been introduced into the ornamental industry [2]. However, conventional breeding of ornamentals can be a long and costly process; modern methods, including in vitro techniques and/or genomic tools, can represent a valuable aid to speed up and make the breeding process more efficient [3]. In recent years, the novel tools of genomics, nanotechnology, and gene editing have been shown to be useful in developing new traits of flowers to meet the current requirements of the global market [4,5]. However, although advancements in breeding strategies plays a pivotal role, a new product can reach the market only if efficient systems for propagation, cultivation, and marketing are put into action. The ornamental industry has been constantly improved thanks to the availability of new propagation and cultivation systems, which enable the production of uniform and high-quality plant material. Innovative technologies are needed to reduce production costs and satisfy the need for more efficient and sustainable cultivation [6]. In addition, to face the global production, the ornamental sector could benefit from new networking and new logistics which can bring innovation from the individual farm level to the system level [7].

This Special Issue (SI) was addressed to present the advancement in the ornamental sector, considering the entire productive flow from breeding to marketing. In more detail this SI wanted to focus on the innovative approaches that can meet the above-mentioned challenges in the present ornamental industry. In total, 20 papers, of which 3 are review papers and 17 are scientific papers, were gathered in this SI.

Contributions (1, 2) addressed molecular aspects, (9, 11, 14) breeding, (5, 8, 16) propagation, (4, 13, 17) growth regulators, (6, 12) physiology, (3, 7, 10, 15) production, and (18, 19, 20) literature review.

Contribution 1 looked to determine the role of the *PEBP* gene family in *Cymbidium* species. Researchers mapped multiple *PEBP* genes to the chromosomes of *C. ensifolium*, *C. sinense*, and *C. goeringii*. In addition, the physiochemical characteristics of the protein encoded by 40 *PEBP* genes were analyzed in *Cymbidium*, *Phalaenopsis*, and *Arabidopsis* using a phylogenetic tree. The results indicated that *PEBP* genes play a role in growth and development, flowering, vegetative and reproductive growth, and root development.

Contribution 2 aimed to add to the understanding of molecular mechanisms regulating lily (*Lilium* sp.) stem bulblet formation as a result of hormonal regulation. Researchers used RNA sequencing in lily 'Aladdin' to characterize the transcriptional response of 6-BA. The results indicated that sucrose and genes associated with auxin and cytokinin transport signaling aid in triggering bulblet formation, emergence, and growth.

Contribution 3 investigated a way to cultivate potted chrysanthemums suitable for outdoor production in Northern China. Researchers investigated different substrates, water, and fertilizer ratios to reduce cost and improve efficiency. The results indicated that a combination of coir, peat moss, perlite, and pine needle mulch for the substrate, along with 336 mg/L nitrogen, 93 mg/L^{-1} phosphorus, and 273 mg/L^{-1} potassium for nutrient fertility and 40% water capacity, reduced production by 16%.

Contribution 4 examined the effects of growth regulators on cyclamen 'Halios F1 Salmon Rose' to correct stem elongation due to low light conditions from February to March in Eastern Europe. Researchers applied daminozide, paclobutrazol, gibberellic acid, and benzyladenine. The results indicated that daminozide and paclobutrazol did control plant elongation, while GA_3 and GA_{4+7} + BAP increased the number of flower buds, number of flowers, and inflorescence diameter.

Contribution 5 focused on developing an efficient micropropagation protocol using a new bioreactor system (SETIS™) and biostimulant, IQ Forte, on *Monstera deliciosa* Liebm 'Thai Constellation'. Researchers studied three different immersion durations and aeration frequencies using the bioreactor plus semi-solid medium as a control and IQ Forte at three different rates plus a control. The results indicated that bioreactor treatments improved multiplication rates, while the biostimulant did not improve growth during acclimation.

Contribution 6 identified peony (*Paeonia lactiflora* Pall.) cultivars with strong photosynthetic productivity using a portable photosynthesis machine. Researchers evaluated 20 different cultivars for photosynthetic rate, stomatal conductance, intracellular carbon dioxide concentration, and transpiration rate. The results indicated that 'Xueyuanhonghua', 'Qingwen', 'Taihuafeixue', 'Chifen', and 'Qihualushuang' all had high photosynthetic productivity.

Contribution 7 evaluated extending the economic productivity of *Amaryllis belladonna* L. as a specialty cut flower and potted plant. Researchers put bulbs into six different warm (23 °C) storage regimes over different weekly intervals. The results indicated that warm temperature storage affected flower production, flowering time, and quality attributes, with 8–12 weeks of warm bulb storage being recommended.

Contribution 8 wanted to introduce efficient propagation methods for *Dactylorhiza umberosa* and *Epipactis veratifolia*, which are endangered orchid species. Researchers explored the effects of different light spectrums, explant types, wounding, and plant growth regulators on direct somatic embryogenesis. The results indicated that wounding the protocorm, adding 3 mg/L^{-1} of TDZ, and white light for *D. umberosa* and red light for *E. veratrifolia* increased production by 94% and 99%, respectively.

Contribution 9 aimed to improve the success of interspecific hybridization in hardy geranium (*Geranium* sp.). Researchers made crosses among 42 different genotypes and studied pollen tube growth, seed development, seed set, and parental differences. The results showed that interspecific hybridization can be predicted by parental distance and logistic regression models, which can predict the number of crosses needed to achieve 10 successful products.

Contribution 10 assessed if biostimulants could help mitigate drought stress under continuous or periodic irrigation deficits on container-grown *Hydrangea paniculata*. Researchers used seaweed extracts (*Ascophyllum nodosum, Soliera chordalis, Ecklonia maxima,* and *Saccharina latissimi*) and a microbial biostimulant. The results found that biostimulants had limited effects, but *A. nodosum* accelerated flowering and *E. maxima* reduced branching under repeated drying and wet cycles.

Contribution 11 looked at the genetic diversity of *Ilex* germplasm for breeding purposes. Researchers determined the genome size, chromosome number, and genetic fingerprints of mostly *I. crenata* accessions. The results indicated wide intra- and interspecific genetic diversity among accessions that will be useful to ornamental breeding programs looking to develop newly improved hybrids.

Contribution 12 determined the morphological, physiological, anatomical, and biochemical responses of zinnia (*Zinnia* sp.) to salinity stress. Researchers used *Z. elegans* Jacq. 'Zinnita Scarlet' (sensitive) and *Z. marylandica* D.m. Spooner et al. 'Double Zhara Fire Improved' under salt concentrations ranging from 1–200 mM NaCl at intervals of 50 mM NaCl. The results showed that 'Zinnita Scarlet' had high Na content, high ion leakage, slow stomatal closure, reduced photosynthetic pigments, and decreased stomatal number under salt stress, while 'Double Zhara Fire Improved' showed quicker stomatal closure, early proline synthesis, maintained photosynthetic pigments, and had low ion leakage.

Contribution 13 analyzed the effects of thidiazuron (TDZ) on the flowering and physiological metabolism of potted *Dendrobium nobile*. Researchers applied 0, 200, 500, and 1000 mg/L^{-1} concentrations to the plant roots. The results found that TDZ greatly influenced flowering and increased relative membrane permeability with 500 mg/L^{-1}, leading to the greatest amount of flowering and highest quality morphological flower features.

Contribution 14 clarified how parental pigment composition and content influence flower color separation in Asiatic hybrid lilies (*Lilium* sp.). Researchers made crosses among 'Easy Waltz', 'Red Life', 'Tresor', and 'Pearl Loraine', which have different pigment compositions and contents. The results revealed that carotenoid content was highly heritable, generally leading to orange offspring, while parents with varying levels of carotenoids and anthocyanins produced more extensive color segregation progeny.

Contribution 15 established case studies using RGB visual imagery from a UAV to assist in the breeding and selection of nursery plants. Researchers used sweet box (*Sarcococca* Lindl.) and garden rose (*Rosa* L.) to compare UAV-based measurements to on-ground measurements. The results showed that plant architecture, flowering, and disease resistance can be assessed faster and more objectively using UAV imaging.

Contribution 16 found that air-layering could be a new technique for tree peony (*Paeonia suffruticosa* Andr.) propagation. Researchers used tree peony cultivars 'Baoqing Hong', Quehao', and 'Xishi' and looked at propagation during mid-May, June, and July along with different rates of NAA and IBA plant growth regulators. The results indicated that the rooting rate was greatest when air-laying was performed in mid-June, while the type and concentration of growth regulators used were cultivar-dependent.

Contribution 17 studied the plant defense-related hormone responses of jasmonic acid and salicylic acid to broad mite presence on azalea (*Rhododendron simsii* Planch.). Researchers looked at both short-term and long-term responses to artificial and natural infestations. The results indicated that the primary plant response to broad mite feeding is through the jasmonic pathway but later switches to the salicylic acid pathway, and variation in broad mite susceptibility occurs within azalea germplasm.

Contribution 18 reviewed the literature on tissue culture methods for propagating bulb-type geophytes. Researchers pooled the literature to explain the history, economy, techniques, and stages of micropropagation, challenges, and opportunities for geophyte tissue culture production. Findings indicate that tissue culture techniques are an important propagation method for geophytes and that future advances in protocols may require insight from multiple disciplines or artificial intelligence.

Contribution 19 reviewed the literature on how plant genetic considerations can contribute to urban planning. Researchers discussed recent genomic sources, defining urban areas, the role of genetics in plant sustainability and adaptability, and the services different genetic traits provide for urban environments and the human population. Findings indicate that identification of genetic resources for urban spaces is fundamental for future breeding programs to develop plants with resilient traits.

Contribution 20 reviewed the literature on emerging and evolving consumption trends, marketing strategies, and governance settings of the ornamental plant market. Researchers reviewed integrated data sources and the structured grey literature. Finding outlined the European ornamental market and the need for public and private stakeholder collaboration to advance research for greater social and environmental goals.

As a general comment, we would like to highlight that the contributions of this SI came from a wide geographical area (China, Romania, the United States, South Africa, Nigeria, Iran, Belgium, Turkey, and Italy) and covered a diverse range of research in the field of the ornamental industry. This highlights the relevance of this SI by presenting different methodologies and case studies that testify to the richness of the research in the ornamental field. In addition, this SI wanted to consider the research related to the entire productive flow, allowing the reader to find a wide overview of the challenges and opportunities in the ornamental sector. We hope that the reader finds inspiration for further research in the worldwide ornamental industry.

Author Contributions: Writing, review, and editing, M.B., E.D. and B.D. All authors have read and agreed to the published version of the manuscript.

Funding: This research received no external funding.

Conflicts of Interest: The authors declare no conflicts of interest.

List of Contributions

1. Chen, J.; Wang, F.; Zhang, Y.; Zheng, R.; Wu, X.; Ai, Y.; Ahmad, S.; Liu, Z.; Peng, D. Genome-Wide Identification of the PEBP Family Gene in Three *Cymbidium* Species and Its Expression Patterns Analysis in *C. ensifolium*. *Horticulturae* **2024**, *10*, 252. https://doi.org/10.3390/horticulturae10030252.
2. Zhang, K.; Lyu, T.; Lyu, Y. Transcriptional Insights into Lily Stem Bulblet Formation: Hormonal Regulation, Sugar Metabolism, and Transcriptional Networks in LA Lily 'Aladdin'. *Horticulturae* **2024**, *10*, 171. https://doi.org/10.3390/horticulturae10020171.

3 Lu, Y.; Zhou, T.; Guo, J.; Zhong, J.; Li, D.; Shi, H.; Liu, C.; Xiang, R.; Sun, M. Effects of Substitute Substrate, Water, and Fertilizer Management on the Growth of Potted Chrysanthemums. *Horticulturae* **2024**, *10*, 138. https://doi.org/10.3390/horticulturae10020138.
4. Bálint, J.; Benedek, K.; Csorba, A.B. Assessing the Effect of Plant Growth Stimulants and Retardants on Cyclamen "Halios F1 Salmon Rose" Cultivar. *Horticulturae* **2023**, *10*, 53. https://doi.org/10.3390/horticulturae10010053.
5. Jing, Y.; Beleski, D.; Vendrame, W. Micropropagation and Acclimatization of *Monstera deliciosa* Liebm. 'Thai Constellation'. *Horticulturae* **2023**, *10*, 1. https://doi.org/10.3390/horticulturae10010001.
6. Xie, A.; Wu, J.; Shi, Y.; Lei, F.; Dong, L.; Zhang, D.; Sun, L.; Sun, X. Photosynthetic Characteristics of 20 Herbaceous Peony Cultivars. *Horticulturae* **2023**, *9*, 1331. https://doi.org/10.3390/horticulturae9121331.
7. Wilmot, C.M.; Jimoh, M.O.; Laubscher, C.P. Warm Bulb Storage Optimises Flowering Attributes and Foliage Characteristics in *Amaryllis belladonna* L. *Horticulturae* **2023**, *9*, 1271. https://doi.org/10.3390/horticulturae9121271.
8. Naderi Boldaji, H.; Dianati Daylami, S.; Vahdati, K. Use of Light Spectra for Efficient Production of PLBs in Temperate Terrestrial Orchids. *Horticulturae* **2023**, *9*, 1007. https://doi.org/10.3390/horticulturae9091007.
9. Akbarzadeh, M.; Quataert, P.; Van Huylenbroeck, J.; Werbrouck, S.P.; Dhooghe, E. Prediction Model for Breeding Hardy Geraniums. *Horticulturae* **2023**, *9*, 617. https://doi.org/10.3390/horticulturae9060617.
10. De Clercq, P.; Pauwels, E.; Top, S.; Steppe, K.; Van Labeke, M. Effect of Seaweed-Based Biostimulants on Growth and Development of *Hydrangea paniculata* under Continuous or Periodic Drought Stress. *Horticulturae* **2023**, *9*, 509. https://doi.org/10.3390/horticulturae9040509.
11. Geukens, E.; Haegeman, A.; Van Meulder, J.; Van Laere, K.; Smolders, E.; Ruttink, T.; Leus, L. Exploring Genetic Diversity in an *Ilex crenata* Breeding Germplasm. *Horticulturae* **2023**, *9*, 485. https://doi.org/10.3390/horticulturae9040485.
12. Yasemin, S.; Koksal, N. Comparative Analysis of Morphological, Physiological, Anatomic and Biochemical Responses in Relatively Sensitive *Zinnia elegans* 'Zinnita Scarlet' and Relatively Tolerant *Zinnia marylandica* 'Double Zahara Fire Improved' under Saline Conditions. *Horticulturae* **2023**, *9*, 247. https://doi.org/10.3390/horticulturae9020247.
13. Ren, S.; Hu, M.; Wu, Q.; Wang, L.; Gu, H.; Chen, Z.; Ming, Z.; Li, Z. Flowering Time and Physiological Reaction of *Dendrobium nobile* Lindl in Response to TDZ Application. *Horticulturae* **2023**, *9*, 129. https://doi.org/10.3390/horticulturae9020129.
14. Li, J.; Chen, J.; Zhang, Q.; Yu, P.; Zhou, Y.; Jia, G. The Composition of Anthocyanins and Carotenoids Influenced the Flower Color Heredity in Asiatic Hybrid Lilies. *Horticulturae* **2022**, *8*, 1206. https://doi.org/10.3390/horticulturae8121206.
15. Van Laere, K.; Lootens, P.; Leus, L. Breeding and Selection of Nursery Plants Assisted by High-Throughput Field Phenotyping Using UAV Imagery: Case Studies with Sweet Box (*Sarcococca*) and Garden Rose (*Rosa*). *Horticulturae* **2022**, *8*, 1186. https://doi.org/10.3390/horticulturae8121186.
16. Zhang, Y.; Yu, S.; Hu, Y. Air Layering Improves Rooting in Tree Peony Cultivars from the Jiangnan Group. *Horticulturae* **2022**, *8*, 941. https://doi.org/10.3390/horticulturae8100941.
17. Leus, L.; Luypaert, G.; Dhooghe, E.; Witters, J.; Pauwels, E.; Van Poucke, C.; Van Pamel, E.; Van Huylenbroeck, J.; Audenaert, J. Jasmonic Acid and Salicylic Acid Levels in Defense Response of Azalea (*Rhododendron simsii* Hybrid) to Broad Mite (*Polyphagotarsonemus latus*). *Horticulturae* **2022**, *8*, 840. https://doi.org/10.3390/horticulturae8090840.
18. Yasemin, S.; Beruto, M. A Review on Flower Bulb Micropropagation: Challenges and Opportunities. *Horticulturae* **2024**, *10*, 284. https://doi.org/10.3390/horticulturae10030284.
19. Farinati, S.; Betto, A.; Palumbo, F.; Scariolo, F.; Vannozzi, A.; Barcaccia, G. The New Green Challenge in Urban Planning: The Right Genetics in the Right Place. *Horticulturae* **2022**, *8*, 761. https://doi.org/10.3390/horticulturae8090761.
20. Gabellini, S.; Scaramuzzi, S. Evolving Consumption Trends, Marketing Strategies, and Governance Settings in Ornamental Horticulture: A Grey Literature Review. *Horticulturae* **2022**, *8*, 234. https://doi.org/10.3390/horticulturae8030234.

References

1. International Association of Horticultural Producers (AIPH). *International Statistics Flowers and Plants Yearbook 2019*; AIPH and Union Fleurs Publishers: Oxfordshire, UK, 2019; Volume 67, ISBN 978-1-9164807-1-1.
2. Kim, J.B. Current status on applications of conventional breeding techniques and biotechnological system in ornamentals. *J. Plant Biotechnol.* **2020**, *47*, 107–117. [CrossRef]
3. Van Huylenbroeck, J.; Bhattarai, K. Ornamental plant breeding: Entering a new era? *Ornam. Hortic.* **2022**, *28*, 297–305. [CrossRef]
4. Ricroch, A.; Martin-Laffon, J.; Rault, B.; Pallares, V.C.; Kuntz, M. Next Biotech Plants: New Traits, Crops, Developers, and Technologies for Addressing Global Challenges. *New Biotechnol.* **2021**, *37*, 613–620. [CrossRef] [PubMed]
5. Penna, S.; Jain, S.M. Innovative technologies and advancements in designing custom-made ornamental plants. *Front. Plant Sci.* **2023**, *14*, 1348949. [CrossRef]
6. Cardoso, J.C.; Vendrame, W.A. Innovation in Propagation and Cultivation of Ornamental. *Horticulturae* **2022**, *8*, 229. [CrossRef]
7. Lambrecht, E.; Taragola, N.; Kühne, B.; Crivits, M.; Gellync, X. Networking and innovation within the ornamental plant sector. *Agric. Food Econ.* **2015**, 3, 10. [CrossRef]

horticulturae

MDPI

Article

Genome-Wide Identification of the *PEBP* Family Gene in Three *Cymbidium* Species and Its Expression Patterns Analysis in *C. ensifolium*

Jinliao Chen †, Fei Wang †, Yangting Zhang, Ruiyue Zheng, Xiaopei Wu, Ye Ai, Sagheer Ahmad, Zhongjian Liu * and Donghui Peng *

Key Laboratory of National Forestry and Grassland Administration for Orchid Conservation and Utilization at Landscape Architecture and Art, Fujian Agriculture and Forestry University, Fuzhou 350002, China; fjchenjl@126.com (J.C.); 19559119005@163.com (Y.Z.); wuxp7872@163.com (X.W.); sagheerhortii@gmail.com (S.A.)

* Correspondence: zjliu@fafu.edu.cn (Z.L.); fjpdh@fafu.edu.cn (D.P.)

† These authors contributed equally to this work.

Abstract: The *PEBP* gene family is involved in many biological processes in plants, including plant growth and development, flowering regulation, light response, and abiotic stress response. But there is little information about the role of the *PEBP* gene family in *Cymbidium* species. In this study, we identified 11, 9, and 7 *PEBP* genes in *C. ensifolium*, *C. sinense*, and *C. goeringii*, respectively, and mapped them to the chromosomes. We also studied the physicochemical characteristics of the proteins encoded by these *PEBP*s and analyzed their intra-species collinearity, gene structure, conserved motifs, and cis-acting elements. Furthermore, a total of forty PEBP genes from *C. sinense*, *C. ensifolium*, *C. goeringii*, *Phalaenopsis*, and *Arabidopsis* were divided into three clades based on the phylogenetic tree. The expression patterns of 11 *PEBP* genes in different tissues and organs of *C. ensifolium* were analyzed based on transcriptome data, indicating that the *CePEBP*s might play an important role in the growth and development, especially in the flower bud organs (1–5 mm). *CePEBP5* plays an indispensable role in both the vegetative and reproductive growth cycles of *C. ensifolium*. *CePEBP1* is essential for root development, while *CePEBP1*, *CePEBP3*, *CePEBP5*, and *CePEBP10* regulate the growth and development of different floral organ tissues at various stages. The findings of this study can do a great deal to understand the roles of the *PEBP* gene family in *Cymbidium*.

Keywords: *PEBP* gene family; gene family analysis; expression pattern; *Cymbidium* species

Citation: Chen, J.; Wang, F.; Zhang, Y.; Zheng, R.; Wu, X.; Ai, Y.; Ahmad, S.; Liu, Z.; Peng, D. Genome-Wide Identification of the *PEBP* Family Gene in Three *Cymbidium* Species and Its Expression Patterns Analysis in *C. ensifolium*. *Horticulturae* **2024**, *10*, 252. https://doi.org/10.3390/horticulturae10030252

Academic Editors: Margherita Beruto, Emmy Dhooghe and Bruce Dunn

Received: 1 February 2024
Revised: 29 February 2024
Accepted: 5 March 2024
Published: 6 March 2024

1. Introduction

Phosphatidylethanolamine-binding proteins (PEBPs) contain a conserved PEBP structural domain and exhibit a strong affinity for phosphatidylethanolamine [1]. They play an important role in regulating flowering, seed development, and germination in plants [1–4]. The PEBP gene family can be divided into three clades in angiosperms: MOTHER OF FT AND TFL1-like (MFT-like), FLOWERING LOCUS T-like (FT-like), and TERMINAL FLOWERING 1-like (TFL1-like) [5]. FT-like and TFL1-like genes are reported only in gymnosperms and angiosperms, whereas MFT-like genes can be traced back to the origin of land plants. Therefore, MFT-like genes are the common ancestor of FT-like and TFL1-like genes [4–6].

Despite the high degree of sequence similarity between members of the PEBP gene family, their functions are not identical. MFT-like genes play an important role in the regulation of the development, germination, and dormancy of seeds [7–9]. It has been demonstrated that the regulation of flowering time and morphogenesis can be controlled by most members of the FT-like gene and TFL1-like genes [10]. The proteins encoded by FT-like genes act as flowering promoters in plants [8,11,12]. In *Arabidopsis thaliana*, FT acts as a floral signal transducer, moving from leaves, passing through the phloem to the shoot apical meristem, and binding to Floring Locus D (FD) proteins. It promotes the expressio of downstream flowering-related genes (such as AP1), thereby regulating the flowerin process in plants. The TFL1 subfamily consists of TFL1, CENTRORADIALIS (CEN), an BROTHER OF FT AND TFL1 (BFT). TFL1-like genes have high sequence similarity t FT-like genes, but they have opposite functions. They inhibit flowering by binding the bZIP-type transcription factor FD and maintain the infinite growth of inflorescenc meristematic tissue [13–17]. There is 60% homology in the amino acid sequences betwee FT and TFL1, but only a few amino acids need to be changed to convert FT from a flora promoter to TFL1, a floral repressor. This is mainly due to inconsistencies in two key amin acid sites, Tyr85 in FT and His88 in TFL1 [18,19]. In addition, the 14 amino acid fragmen LGRQTVYAFGWRQN and the triplet LYN in exon 4 of FT/TFL1-like also play importar roles in the opposite function of FT/TFL1-like [20].

C. ensifolium, *C. goeringii*, and *C. sinense* are the most significant ornamental orchid because of their beautiful and unique flowers. They have a long history of cultivation an are loved by consumers in China [21]. The *PEBP* gene family has not yet been systematicall analyzed by bioinformatics, although some members have been identified and studie in *Cymbidium*. Given the considerable role of *PEBP* genes in regulating plant flowering seed development, and germination, this study utilized bioinformatics methods to perforn genome-wide identification of three *Cymbidium* species in Orchidaceae. In this study, w identified twenty-seven members of the PEBP gene family in three *Cymbidium* species, deter mined their chromosomal localization, constructed phylogenetic trees, and analyzed the gen structure, conserved motifs, and cis-acting element types. Additionally, we analyzed 11 mem bers of the *CePEBP* genes in different tissues of *C. ensifolium*. Our findings further elucidat the functions of *PEBP* genes in the flowering and vegetative development of three *Cymbidiun* species and provide suggestions for improvement and the creation of new varieties.

2. Materials and Methods

2.1. Data Sources

The genome sequences and annotation information of *C. ensifolium*, *C. goeringii*, and *C. sinense* were retrieved from their whole-genome sequencing data [22–24]. The protein sequence of PEBP gene family of *A. thaliana* was retrieved from the Arabidopsis Infor mation Resource (TAIR, https://www.arabidopsis.org/, accessed on 20 August 2023). The *Phalaenopsis* 'Little Gem Stripes' data (PhFT1 (Peq009747), PhFT2 (Peq006920), PhFT3 (Peq017805), PhFT4 (Peq012163), PhFT5 (Peq006349), PhFT6 (Peq009750), and PhMFT (Peq004653)) were downloaded from National Center for Biotechnology Information (NCBI, https://www.ncbi.nlm.nih.gov/, accessed on 20 August 2023) [25].

2.2. Identification and Physicochemical Properties of PEBP Genes from Three Cymbidium Species

The conserved domain of PEBP (PF01161) was downloaded from Pfam. Using the BLAST and Simple HMM Search functions of TBtools (version 1.132). The PEBP family members of three *Cymbidium* species were identified from the genome databases [26,27]. The screening parameter had an E-value lower than 1×10^{-5}. Then, NCBI CD-Search (https://www.ncbi.nlm.nih.gov/Structure/cdd/wrpsb.cgi, accessed on 20 August 2023) [28–30] and SMART (http://smart.embl.de/, accessed on 20 August 2023) [31] websites were used to analyze the structure of candidate PEBP proteins and eliminate incomplete and redundant protein sequences. The PEBP genes were named and classified according to the naming rules of *A. thaliana*. Finally, the ExPASy website (https://www.expasy.org/, accessed on 20 August 2023) was used to calculate the amino acid (aa), isoelectric point (pI), molecular weight (MW), grand average hydrophilicity (GRAVY), instability index (II), and lipid index (AI) of the PEBP proteins [32,33].

2.3. Phylogenetic Analysis of PEBP Genes

The protein sequences of 7 *PEBPs* from *C. goeringii*, 11 *PEBPs* from *C. ensifolium*, 9 *PEBPs* from *C. sinense*, 6 *PEBPs* from *A. thaliana*, and 7 *PEBPs* from *Phalaenopsis* 'Little Gem Stripes' were imported in MEGA 7.0 software [34]. A total of 40 protein sequences were aligned using the MUSCLE program with default parameters. The phylogenetic tree of *PEBPs* was constructed based on maximum likelihood (ML), with Bootstrap parameters set to 1000 and partial deletion to 75% [34]. For better visualization, the phylogenetic tree was processed using the online software iTOL 6.8.2 (https://itol.embl.de/itol.cgi, accessed on 20 August 2023) [35].

2.4. Chromosome Distribution and Collinear Correlation of PEBP Genes in Three Cymbidium Species

The visualization and analysis of chromosomal localization of *PEBP* genes in three *Cymbidium* were conducted using Tbtools software, utilizing the genome and annotation files of *C. ensifolium*, *C. goeringii*, and *C. sinense*. In addition, the genomic data of the three *Cymbidium* species were analyzed in collinearity analysis using the One-Step MCScanx program in Tbtools [27]. In Tbtools, the replication patterns of three *Cymbidium* species were visualized using Advance Circos [27].

2.5. Gene Structure and Conserved Motif Analysis of PEBP Gene

The conserved domains of the *PEBP* genes of three *Cymbidium* species were predicted using the CDD website (https://www.ncbi.nlm.nih.gov/cdd, accessed on 20 August 2023) [29,30], and the motifs of these *PEBP* genes were analyzed using the MEME website (https://meme-suite.org/meme/tools/meme, accessed on 20 August 2023) [36]. Ten motifs were followed, and the other was the default value.

2.6. Sequence Analysis of PEBP Gene Promoter

The sequence of 2000 bp upstream of the transcription start site was extracted using Tbtools as the promoter sequence of the *PEBP* genes in three *Cymbidium* species [26,27]. Additionally, potential cis-acting elements on the promoter sequences were predicted using PlantCARE website (https://bioinformatics.psb.ugent.be/webtools/plantcare/html/, accessed on 20 August 2023) [37]. Then, the prediction results were classified and analyzed using Excel 2019 and Tbtools software.

2.7. Expression Pattern and qRT-PCR Analysis

The expression patterns of 11 *CePEBP* genes were analyzed to investigate the potential impact of *PEBP* genes across different organs of *C. ensifolium*. The sampled organs for investigation included the root, leaf, buds of various sizes (1–5 mm, 6–10 mm, and 11–15 mm), petal, lip, sepal, pedicel, and gynostemium. Three biological replicates were analyzed, each of which was a pooled sample from five plants. We conducted transcriptome analysis on all ten samples, calculating the fragments per kilobase of transcript per million mapped reads (FPKM) [22]. The heatmap showed the patterns of expression using TBtools (version 1.132) [27].

Quantitative real-time PCR (qRT-PCR) was used to further analyze the expression patterns of the *CePEBPs*. The root, leaf, buds of various sizes (1–5 mm, 6–10 mm, and 11–15 mm), petals, lip, sepal, pedicel, and gynostemium at blooming period were sampled from *C. ensifolium* 'Longyansu' planted at the Fujian Agriculture and Forestry University. Primer Premier 5.0 software was used to design primers. The details of the primers and reference genes are listed in Supplementary Table S3. Total RNA was extracted by the TIANGEN DP441 Reagent Kit (Tiangen, Beijing, China). A HiScript III 1st Strand cDNA Synthesis Kit (+gDNA wiper; Vazyme, Nanjing, China) was used to reverse-transcribe RNA to cDNA. Based on the Taq Pro Universal SYBR qPCR Master Mix kit (Vazyme, Nanjing, China), the ABI 7500 Real-Time System (Applied Biosystems, Foster City, CA, USA) was used to analyze the RT-qPCR. Finally, the $2^{-\Delta\Delta CT}$ method was used to calculate the expression level [26,33,38].

3. Results

3.1. Identification and Sequence Analysis of PEBP Genes in Three Cymbidium Species

The basic information and physicochemical properties of the *PEBP* genes for three *Cymbidium* species are shown in Table 1. A total of 11, 9, and 7 *PEBP* genes were found in *C. ensifolium*, *C. sinense*, and *C. goeringii*, respectively. Based on the sequentia distribution on chromosomes, these 27 *PEBP* genes were named *CePEBP1-11*, *CsPEBP1-9* and *CgPEBP1-7*, respectively. A sequence analysis of the encoded proteins showed tha the physicochemical properties of amino acids, isoelectric point, molecular weight, granc average of hydropathicity, aliphatic index, and instability index of the *PEBP* genes in three *Cymbidium* species differed significantly (Table 1). The deduced protein length (AA) o *PEBP* genes ranged from 66 (*CgPEBP3*) to 379 (*CsPEBP9*) amino acids. The isoelectric point (pI) values of the 27 *PEBP* genes in *Cymbidium* ranged from 5.13 (*CgPEBP1*) to 10.75 (*CgPEBP3*). Among them, 6 PEBP proteins had an acidic pI below seven, while the 21 PEBP proteins with a pI higher than seven were alkaline. The grand average of hydropathicity (GRAVY) values of *PEBP* genes were ranged from −0.577 (*CgPEBP3* to −0.118 (*CePEBP2*), with all GRAVY values of less than 0, indicating that they were hydrophilic. The molecular weight (Mw) ranged from 7806.99 kD (*CgPEBP3*) to 42330.61 kD (*CsPEBP9*), with the aliphatic index (AI) between 66.52 (*CgPEBP3*) and 89.15 (*CePEBP4*) The maximum instability index (II) value was 58.26 (*CePEBP2*), and the minimum value was 30.19 (*CgPEBP4*).

Table 1. PEBP gene family protein properties table from three *Cymbidium* species.

Gene Name	Gene ID	Protein Length (AA)	Isoelectric Point (pI)	Molecular Weight (Mw)	Grand Average of Hydropathicity (GRAVY)	Aliphatic Index (AI)	Instability Index (II)
CePEBP1	JL006795	176	6.42	19,848.39	−0.311	80.74	43.37
CePEBP2	JL020923	173	7.74	19,256.21	−0.118	83.29	58.26
CePEBP3	JL010014	173	9.06	19,607.53	−0.202	82.14	45.42
CePEBP4	JL026838	189	6.73	21,260.35	−0.151	89.15	46.08
CePEBP5	JL020421	174	6.42	19,926.58	−0.355	75.52	35.95
CePEBP6	JL027939	101	5.62	10,995.31	−0.265	73.27	41.88
CePEBP7	JL002228	174	9.03	19,523.09	−0.375	77.82	48.29
CePEBP8	JL001165	177	9.18	20,180.03	−0.227	83.11	45.55
CePEBP9	JL013430	178	8.48	20,082.78	−0.39	74.44	42.64
CePEBP10	JL027407	112	5.27	12,308.11	−0.14	86.88	52.12
CePEBP11	JL028740	183	6.12	20,580.41	−0.248	81.97	48.35
CgPEBP1	GL13937	236	5.13	26,017.24	−0.323	80.04	48.92
CgPEBP2	GL01335	176	6.42	19,848.39	−0.311	80.74	43.37
CgPEBP3	GL28974	66	10.75	7806.99	−0.577	66.52	54.87
CgPEBP4	GL14129	125	6.83	14,228.19	−0.331	82.56	30.19
CgPEBP5	GL07645	174	9.03	19,522.14	−0.333	80.06	48.67
CgPEBP6	GL00658	201	7.8	22,704.79	−0.341	77.06	38.75
CgPEBP7	GL09595	181	6.73	20,394.24	−0.229	82.87	49.19
CsPEBP1	cymsin_Mol026710	118	5.34	13,110.05	−0.251	85.85	51.29
CsPEBP2	cymsin_Mol020839	189	6.42	21,659.53	−0.355	79.31	37.43
CsPEBP3	cymsin_Mol012759	173	9.06	19,598.52	−0.203	82.14	45.59
CsPEBP4	cymsin_Mol020552	187	7	20,394.34	−0.184	73.48	57.35
CsPEBP5	cymsin_Mol006878	190	6.08	21,404.16	−0.297	80.95	41.78
CsPEBP6	cymsin_Mol006013	243	6.08	27,504.27	−0.363	80.16	42.56
CsPEBP7	cymsin_Mol018868	174	9.03	19,523.09	−0.375	77.82	48.29
CsPEBP8	cymsin_Mol003216	178	9.18	20,292.16	−0.23	83.2	42.1
CsPEBP9	cymsin_Mol017371	379	9.57	42,330.61	−0.181	84.62	52.69

Note: Ce: *Cymbidium ensifolium*; Cg: *C. goeringii*; Cs: *C. sinense*.

3.2. Chromosome Localization and Collinearity Analysis of PEBP Genes in Three Cymbidium Species

The relationship between the location of the *PEBP* genes on the chromosomes and the collinearity within three *Cymbidium* species is shown in Figure 1. The *PEBPs* are exclusively located on the partial chromosomes of three *Cymbidium* species and are dispersed. Ten *CePEBPs* were unevenly distributed on seven chromosomes (Chr02, 05, 06, 09, 11, 14, and 17). Chromosome 02,05,09 contained two genes, while each of the remaining chromosomes contained one gene. *CePEBP9* was localized to the unanchored scaffold, named Scaffold5149. Chromosome localization analysis in *C. sinense* showed that nine *CsPEBPs* were unevenly distributed on seven chromosomes (Chr02, 06, 07, 08, 11, 13, and 17). The chromosome 08 contained three genes (*CsPEBP4*, *CsPEBP5*, and *CsPEBP6*), while the other chromosomes contained one gene each. Six *CgPEBPs* were unevenly distributed on four chromosomes (Chr01, 05, 11, and 16). Chromosome 01 contained three genes (*CgPEBP1*, *CgPEBP2*, and *CgPEBP3*), whereas the other chromosomes contained one gene each. *CgPEBP7* was localized to the unanchored scaffold, named Scaffold10 (Figure 1A).

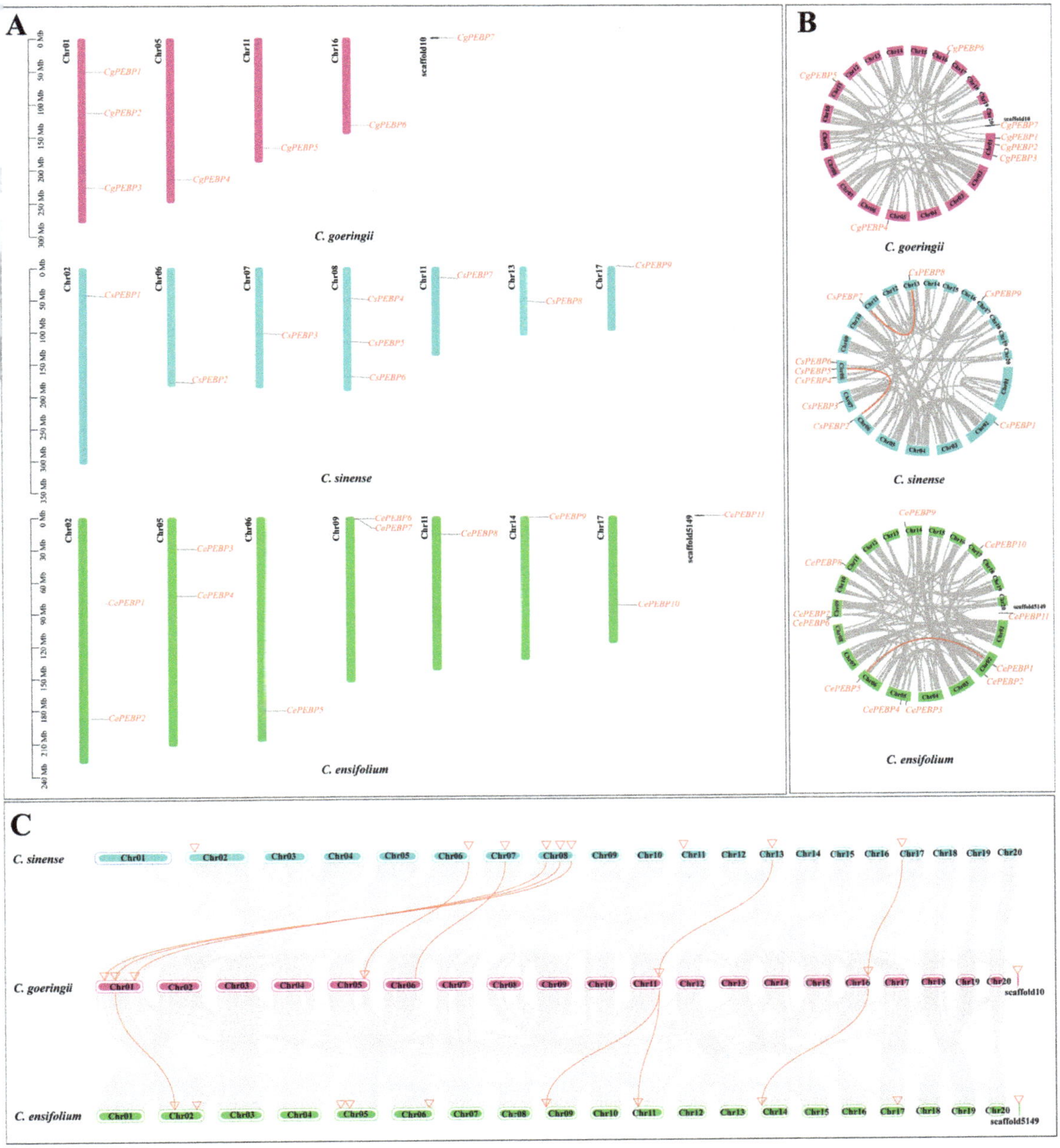

Figure 1. (**A**) The position of PEBP gene family members on chromosomes in three *Cymbidium* species. The left-hand scale is used to estimate the length of chromosomes. (**B**) Intraspecific collinearity of the *PEBP* genes in the three *Cymbidium* species. (**C**) Interspecific collinearity relationship between PEBP gene family members and *C. goeringii*, *C. ensifolium*, and *C. sinense*. The chromosomes of *C. goeringii*, *C. ensifolium*, and *C. sinense* are marked with different colors. Red lines connect the collinear relationship between PEBP gene family members of different species, and the location of *PEBPs* is represented by the red triangle. Circles of different colors represent different *Cymbidium* species.

The analysis of collinearity between genes revealed replication relationships between them. Within *C. ensifolium*, the *PEBP* genes had one collinear relationship, which was between *CePEBP5* on Chr06 and *CePEBP1* on Chr02 (Figure 1B). There were two collinea relationships in *C. sinense*, which were *CsPEBP2* on Chr06 and *CsPEBP5* on Chr08, an *CsPEBP7* on Chr011 and *CsPEBP8* on Chr11, respectively. They exhibited similar conserve motifs and gene arrangements. No collinear relationship was detected in *C. goering* (Figure 1B). Figure 1C shows the collinear relationships of *PEBP* genes in three *Cymbidiu* species. The analysis results indicated that *C. goeringii* shared seven collinearities wit *C. sinense* and four collinearities with *C. ensifolium*. Among the three species, *CgPEBP* wa most closely related to *CsPEBP*.

3.3. Phylogenetic Relationship Analysis of PEBP Genes

To analyze the phylogenetic relationships of *PEBPs* and other homologous gene in three *Cymbidium* species, we constructed a maximum likelihood (ML) phylogeneti tree using the amino acid sequences of 6 *AtPEBP* proteins, 7 *PhPEBP* proteins, 7 *CgPEB* proteins, 9 *CsPEBP* proteins, and 11 *CePEBP* proteins (Figure 2). The phylogenetic tree showed that the forty PEBP proteins were categorized into three subfamilies: FT, TFL1 and MFT. Among these, twenty-two *PEBP* genes belonged to the FT subfamily for thre *Cymbidium* species (six *CgPEBP*, seven *CsPEBP*, and nine *CePEBP*, respectively), and th number of members of the TFL1 subfamily was the lowest (two members, *CePEBP3* an *CsPEBP3*, respectively). Moreover, the TFL1 gene was absent in *C. goeringii*. The *PEBPs* o the three *Cymbidium* species were classified into four types (I-IV) based on the topology o the phylogenetic tree. The like-I clade consists of genes that are very similar to the FT-lik genes in *A. thaliana*.

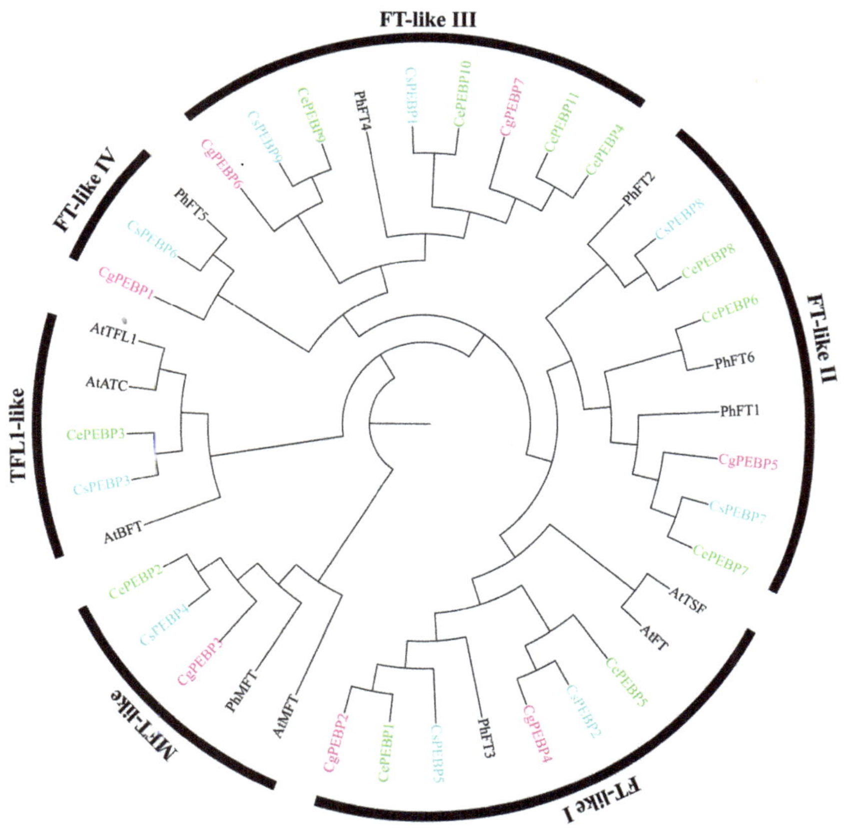

Figure 2. Phylogenetic tree of the 40 PEBP proteins from *Cymbidium goeringii*, *C. ensifolium*, *C. sinense*, *A. thaliana*, and *Phalaenopsis hybrid*. Circles of different colors represent different *Cymbidium* species.

3.4. Conserved Motif and Gene Structure Analysis of PEBP Gene Family

To understand the gene structure of *PEBPs* in three *Cymbidium* species, we predicted ten conserved motifs of *PEBP* genes by MEME and demonstrated the exon intron structure using Tbtools. The result showed that the majority of *PEBP* genes exhibited six conserved motifs, numbered one to six. Members of the same subfamily within the PEBP family share similar conserved motifs. Additionally, motifs 7 and 9 were exclusively observed in MFT-like genes. Except for *CePEBP10* and *CsPEBP1*, almost all *PEBP* genes contained at least four conserved motifs (Figure 3A). Motif 1 (e-value = 2.0e − 723) contained a conserved motif D-P-D-X-P and its critical AA84 amino acid residue (Y). Motif 2 (e-value = 3.6e − 708) contained the critical AA139 amino acid residue (Q). Motif 4 (e-value = 8.3e − 285) contained a conserved motif G-X-H-R, and motif G-X-H-R had a strong effect on the Ile (I) residue with a preference (Figure 3B, 3C). The gene structure analysis indicated that 17 *PEBPs* (accounting for 63%) contained four exons and three introns, and four *PEBPs* (accounting for 15%) contained three exons and two introns. Three *PEBPs* (accounting for 11%) contained five exons and four introns, and three *PEBPs* (accounting for 11%) contained six exons and five introns. All *PEBP* genes had between one and five introns, with *CsPEBP9* having the longest intron and *CgPEBP3* having only one intron (Figure 3B).

The PEBP gene family had two key amino acid (AA) residues at the AA85 (Tyr, Y) and AA140 (Gln, Q) positions in *Arabidopsis* [18,20]. We performed the protein alignment of *PEBP* homologs from three *Cymbidium* species (Figure 3D). In these three *Cymbidium* species, Tyr (Y) at AA85 was replaced by Cys (C) and His (H) in MFT-like (*CgPEBP3, CsPEBP4,* and *CePEBP2*) and TFL1-like (*CsPEBP3, CePEBP3*), respectively. In the FT subfamily, Tyr (Y) at AA85 was replaced by His (H) and Leu (L) in three genes (*CgPEBP6, CsPEBP9, CePEBP9*) and five genes (*CgPEBP7, CePEBP10, CePEBP11, CePEBP4, CsPEBP1*) of FT-like III, respectively. The key amino acid residues of other PEBP genes in the FT subfamily were highly conserved at AA85. In addition, another key amino acid residue at the AA140 (Gln, Q) positions of *PEBP* genes was replaced by Asp (D) in the TFL1-like subfamily and replaced by Glu (E) and His (H) in the FT-like II tapy, respectively, and in *CgPEBP4* by Lys (K). Key amino acid residues of other *PEBP* genes were highly conserved at AA140. The amino acid comparison showed that the functions of these *PEBP* genes might be largely conserved (Figure 3D).

3.5. Cis-Element Analysis of Three Cymbidium Species

We extracted the 2000 bp sequence upstream region of each gene in the *PEBP* genes for three *Cymbidium* species and predicted cis-acting elements using the PlantCARE databases. There were 690 predicted cis-acting elements in three *Cymbidium* species, and *C. ensifolium* had the most cis-elements (273/690), followed by *C. sinense* (239/690) and *C. goeringii* (178/690). These were categorized into six groups: light-responsive elements (317), hormone-responsive elements (213), developmental-associated elements (43), environmental stress-related elements (88), site-binding-associated elements (22), and promoter-associated elements (7). Among them, the maximum number of light-responsive elements were Box 4 (94/317, 29.65%) and G-box (62/317, 19.55%), followed by TCT-motifs (24/317) and GT1-motifs (223/317). Among the phytohormone response elements, a higher number of ABRE (58/213, 27.23%), CGTCA-motif (39/213, 7.8%), and TGACG-motifs (39/213, 7.8%), were associated with abscisic acid response and MeJA, respectively. The remaining elements were associated with the salicylic acid response. Among the plant growth- and development-related response elements, GCN4-motif (14/43, 32.55%) and O2-site (12/43, 27.91%) were associated with circadian rhythm control and arginine metabolism, respectively. In contrast, other elements such as CAT-box motifs, circadian, and RY-elements were associated with phloem tissue expression and seed development. Thus, the *PEBP* genes of the three orchids were mainly associated with light response and the regulation of tissue metabolism (Figure 4, Supplementary Table S1).

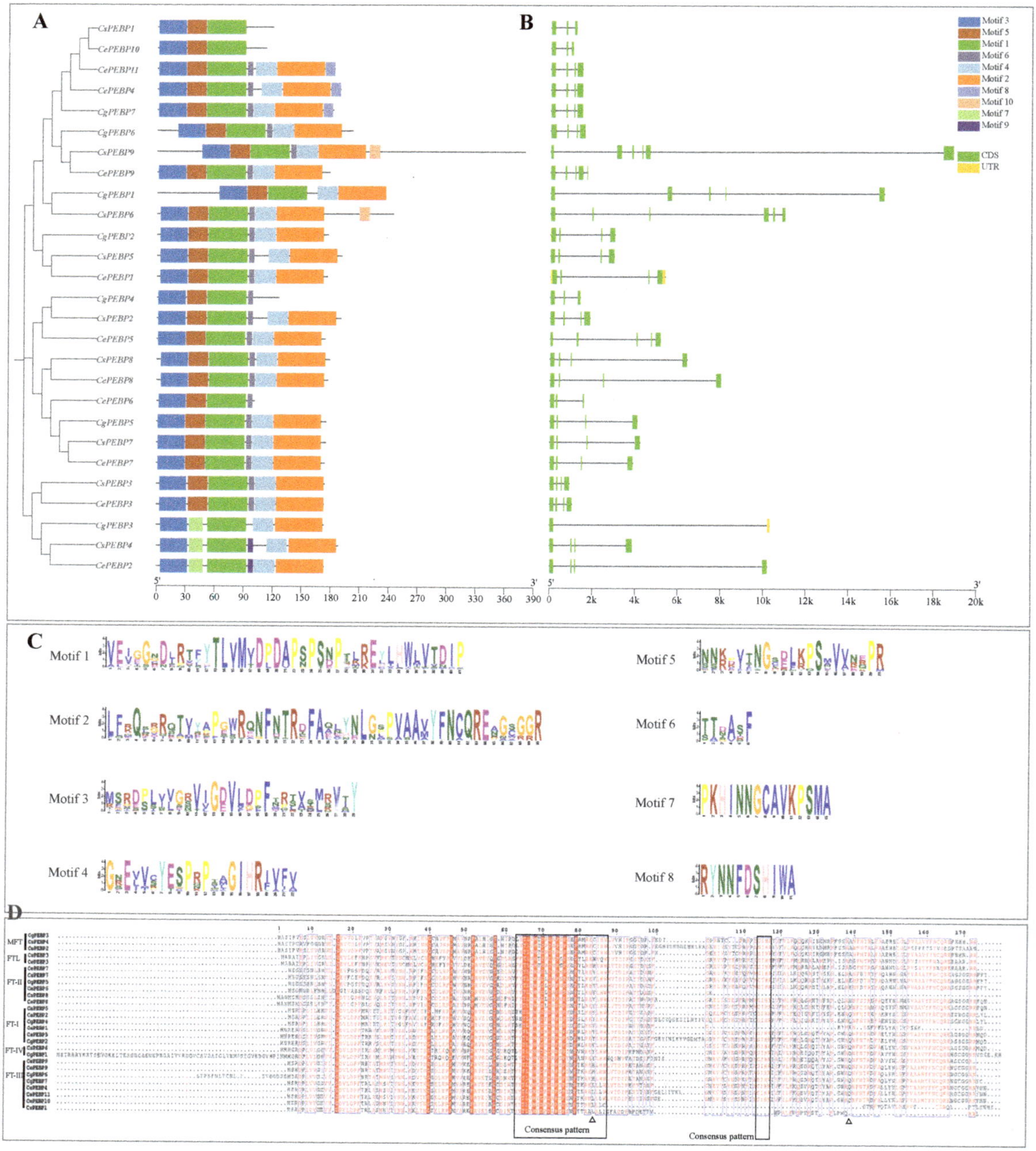

Figure 3. (**A**) The comparative map of *PEBPs* is based on the phylogenetic tree and conserved protein motifs of the three *Cymbidium* species. (**B**) Distribution of UTRs and CDSs of PEBP gene family members of the three *Cymbidium* species. Green represents CDS s and yellow represents UTRs. The scale at the bottom is used to compare the lengths of different genes and proteins. (**C**) Conserved domains of the three *Cymbidium* species protein sequences. The overall height of each stack indicates the sequence conservation at that position. (**D**) The *PEBP* homeodomain sequence alignment analysis of three *Cymbidium* species. The red blocks represent highly conserved residues. The red blocks represent highly conserved residues. Ce: *Cymbidium ensifolium*; Cg: *C. goeringii*; Cs: *C. sinense*.

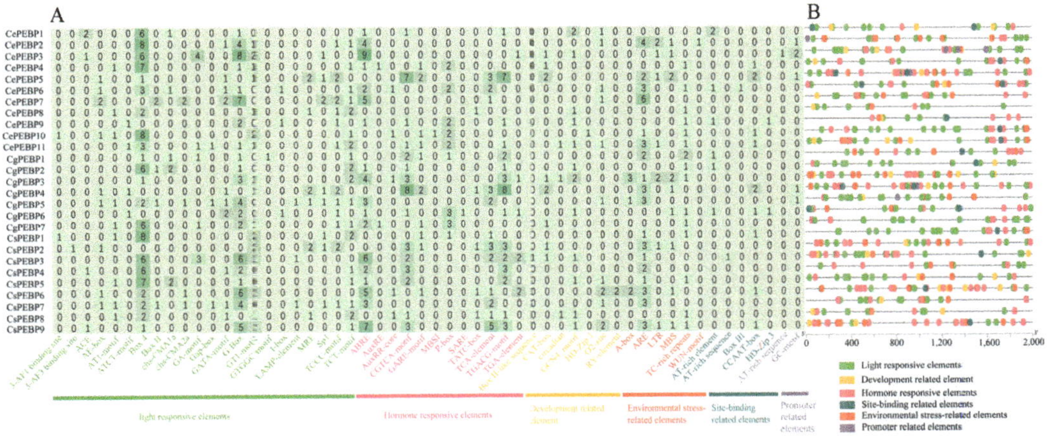

Figure 4. (**A**) Classification and statistics of cis-acting elements of three *Cymbidium* species. The numbers in the grid represent the number of elements, with darker colors indicating a larger number and lighter colors indicating a smaller number. (**B**) The promoter region's distribution of cis-acting elements of three *Cymbidium* species. The various types of cis-acting elements are represented by different colors and shapes. The sequence direction and length are indicated by the ruler at the bottom. Ce: *Cymbidium ensifolium*; Cg: *C. goeringii*; Cs: *C. sinense*.

3.5. Expression Patterns of PEBP Genes in C. ensifolium

Based on the transcriptome data, three *PEBP* genes were significantly expressed in the buds and flowers of *C. ensifolium*, and two genes were significantly expressed in the leaves and roots (Figure 5). Some genes (*CePEBP5*, *CePEBP1*) were expressed in several tissues, whereas some genes (*CePEBP8*, *CePEBP9*, and *CePEBP4*) showed little or no expression in the tissues. The average FPKM values of the transcriptome of *CePEBP* genes in different tissues of *C. ensifolium* are shown in Supplementary Table S2.

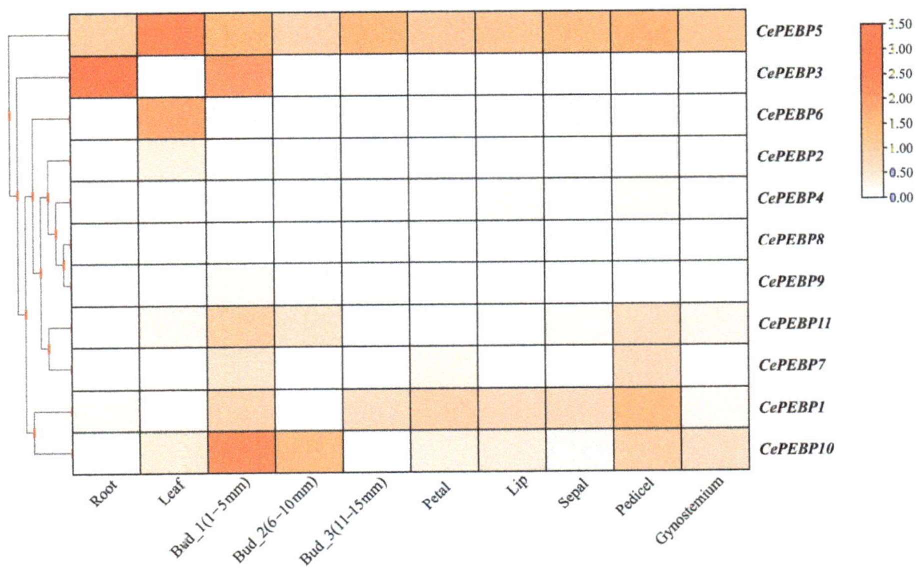

Figure 5. Expression of *CePEBPs* in various tissues of *Cymbidium ensifolium*. The dendrogram on the left displays the results of the inter-gene clustering analysis.

We selected four genes, *CePEBP1*, *CePEBP3*, *CePEBP5*, and *CePEBP10*, for RT-qPC experiments according to the transcriptome data (primer sequence information is show in Table S3). The RT-qPCR results showed that they were expressed highly in the bu (1–5 mm), leaves, roots, and pedicels, suggesting their vital role in multiple developmenta stages of *C. ensifolium* (Figure 6). The transcriptome data and RT-qPCR results for *CePEBP* and *CePEBP3* were basically the same. Transcriptome data indicated that *CePEBP5* wa expressed in all tissues and had the highest expression in leaves. The RT-qPCR result indicated that it had the highest expression in the gynostemium. *CePEBP10* exhibited th highest expression in the bud (1–5 mm), while the RT-qPCR results showed the highes expression in leaves.

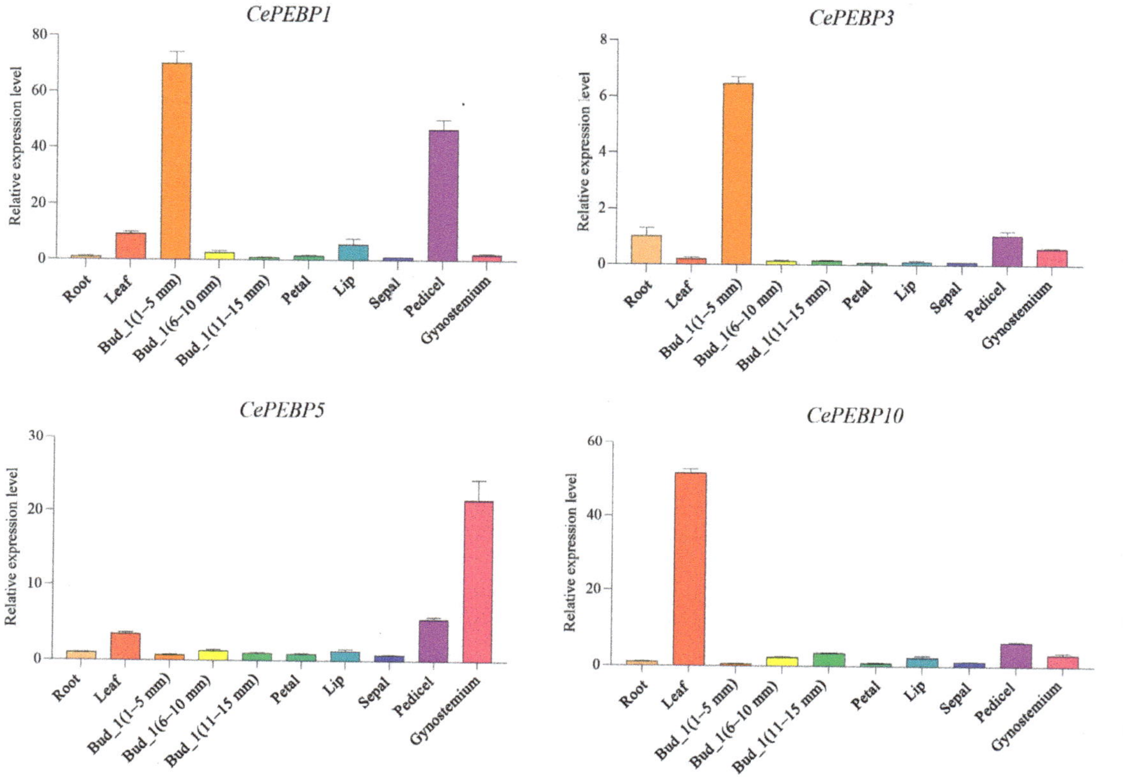

Figure 6. Analysis of gene expression of four *CePEBPs* in *Cymbidium ensifolium* at ten different organ materials.

4. Discussion

The growth habit, flowering time, flower number, and floral organ development of *Cymbidium* species are closely related to their ornamental value. Previous studies have cloned one PEBP homologous gene in each of the three *Cymbidium* species and found that *C. goeringii* might be primarily regulated by low temperatures, while *C. ensifolium* and *C. sinense* are regulated by the photoperiod [39]. This suggests that the study of the PEBP gene family may contribute to an improvement in ornamental traits in *Cymbidium* species. However, the role of the PEBP gene family in *Cymbidium* species has not been systematically studied to examine the common characteristics of its members.

In angiosperms, the PEBP gene family has undergone two ancient duplications, giving rise to three types: FT-like, TFL1-like, and MFT-like [4–6]. In this study, twenty-seven *PEBP* genes from three *Cymbidium* species were classified into three subfamilies (FT, TFL1, and

MFT) through phylogenetic analysis, which was consistent with other species [4–6]. The FT-like genes were the most diversified in terms of copy number among the three PEBP subfamilies. In a study on the evolution of FT/TFL1 in tropical new orchids, the FT-like genes in monocotyledonous plants were divided into two branches: MonFT1 and MonFT2. Genes in the MonFT1 sub-branch might have played a role in delaying flowering, while genes in the MonFT2 sub-branch could have retained the function of promoting flowering [40]. Similarly to other monocots, three *Cymbidium* species carried more FT-like homologous sequences than TFL1-like and MFT-like, which could be further divided into four types (like I–IV). Among them, type I belongs to the MonFT2 sub-branch and promotes flowering through genes closely related to the *AtFT* gene, while types II, III, and IV belong to the MonFT1 sub-branch and inhibit flowering [40]. Additionally, previous studies have reported that the TFL1-like genes have undergone duplication during evolution in dicotyledons, followed by functional divergence from the TFL1 and CEN gene lineages [1,41]. In contrast to the FT genes, the TFL1 subfamily is more diverse in dicotyledons than in monocotyledons [40]. TFL1-like genes are either completely absent or very few are present in orchids, frequently as single copies. Only homology between *Oncidium* 'Gower Ramsey' and *Vanilla planifolia* has been reported [40,42,43]. This may be caused by the progressive loss of function of TFL1-like genes together with functional compensation by FT-like copies, but it still needs to be tested in the necessary experiments [40]. Among the three *Cymbidium* species, the absence of TFL1-like in *C. goeringii*, while *C. ensifolium* and *C. sinense* had a single copy, agrees with the results of previous studies [25,40]. These results reveal the functional differentiation and diversity in the PEBP gene family of three *Cymbidium* species.

PEBP protein has highly conserved D-P-D-x-P and G-x-H-R motifs in plants, and the binding of these motifs to anions is important for the conformation of the ligand-binding site of the PEBP protein [1,44]. Mutations close to this region may affect the binding of the PEBP protein with phosphate ions, thereby altering its interaction with FD [1,45]. Previous studies have indicated that a single amino acid determines the antagonistic activity of the floral regulators, including FT-like and TFL1-like. The residues Tyr85/His88 and Gln140/Asp144 in the FT-like and TFL1-like proteins may be the key residues that distinguish FT-like and TFL1-like activity, where they form hydrogen bonds in TFL1-like but not in FT-like [18,20]. For example, one amino acid substitution (replacing His-88 with Tyr in TFL1-like) can convert TFL1-like into FT-like, which promotes flowering [18]. In another study, specific mutations at the Glu-109, Trp-138, Gln-140, and Asn-152 sites can convert the FT-like into the TFL1-like, which inhibits flowering [19]. In this study, 27 *PEBPs* were identified in three *Cymbidium* species, and the results of the conserved motifs of all *PEBPs* indicated that these genes contained not only key amino acid residues but also two conserved motifs (D-P-D-X-P and G-X-H-R) [44,46]. Among them, the aa85 position of the FT-like III branch (*CgPEBP6, CsPEBP9, CePEBP9*) and five genes (*CePEBP4, CePEBP10, CePEBP11, CgPEBP7, CsPEBP1*) were replaced by His (H) and Leu (L) instead of Tyr (Y). Moreover, the aa140 (Gln, Q) of the FT-like II members was replaced by (Glu, E) and (His, H). FT-like II and III belong to the MonFT1 subbranch, which inhibits flowering in plants [40]. This indicates that the changes in Tyr85/His88 and Gln140/Asp144 residues of the PEBP gene family of three *Cymbidium* species can determine the functional conversion of FT/TFL1, which is similar to the results of previous researches [18,19,40].

Flowering is a key developmental process for environmental adaptation and reproduction in higher plants and requires a complex network of signaling pathways, which has been studied in many plants [1]. PEBP functions as a gene hub, integrating the photoperiodic pathway, vernalization pathway, autonomous pathway, gibberellin pathway, and age pathway in major floral induction pathways [1,47–49]. Investigating the transcriptional regulation of gene expression at the level of promoters by cis-acting elements has advanced our basic understanding of gene regulation and enriched the arsenal of readily available promoters [50]. In this study, a series of functional regulatory elements in the promoter region of the *PEBPs* were identified in three *Cymbidium* species, including growth and development factors, stress response factors, and plant hormone response factors. Among them, the light-responsive elements had the maximum number, indicating that the *PEBPs* might be regulated by light signals and growth and development (Supplementary Table S1). Previous studies also found that *C. ensifolium* and *C. sinense* were regulated by photoperiod, while *C. goeringii* was regulated by other factors, such as low temperatures. The research results also indirectly confirmed this point [39].

The PEBP transcripts are abundant in numerous organs during the growth and development of orchids [25,42,51]. In *Phalaenopsis* 'Little Gem stripes', transcription of the *PEBP* genes among the various organs is detected. *PhFT1* is mainly expressed in vegetative buds, *PhFT2* is specifically expressed in leaves, and the expression level of *PhFT3* is highest in inflorescence [25]. But the FT homologous gene is highly expressed during flower organ development and growth processes in *Dendrobium* 'Chao Praya Smile' [51]. In *Oncidium* 'Gower Ramsey', the *OnFT* mRNA is widely detected in different organs at different growth stages and had the highest level in tender flower buds (2 mm) [42]. In this study, we observed that certain *CePEBP* genes exhibited tissue-specific expression (Figure 6), with *CePEBP3*, *CePEBP5*, and *CePEBP6* being specifically expressed in the developing roots and leaves of *C. ensifolium*, independently. *CePEBP1*, *CePEBP3*, *CePEBP5*, and *CePEBP10* were expressed specifically in buds and flowers, which might be related to flower differentiation and development. None of *CePEBP8*, *CePEBP9*, and *CePEBP4* were expressed in any of the tested tissues or organs, indicating that they were not expressed in *C. ensifolium*. The result of RT-qPCR analysis showed that *CePEBP1* and *CePEBP3* had high expression in flower buds (1–5 mm) and pedicels; *CePEBP5* exhibited high expression in the gynostemium and pedicel; and *CePEBP10* showed high expression in the leaves of *C. ensifolium* 'Longyan Su'. This may be due to an incomplete correlation between sequencing and RT-qPCR samples. These findings also suggest that the various expression patterns of the PEBP gene family may contribute to further research on functional differentiation of the FT-like branch in orchids.

5. Conclusions

In this study, seven *CgPEBP*, nine *CsPEBP*, and eleven *CePEBP* were identified in three *Cymbidium* species, which were classified into three clades. The *PEBP* genes of *C. ensifolium* can play a significant role in the development and growth of the plant, particularly in the bud (1–5 mm). It was noteworthy that *CePEBP5* also played an indispensable role in both the vegetative and reproductive growth of *C. ensifolium*. The *CePEBP1* gene was crucial for root development, while *CePEBP1*, *CePEBP3*, *CePEBP5*, and *CePEBP10* might be involved in the growth and development of multiple floral organ tissues. These findings can provide possible directions for further investigations on the regulation of *PEBPs* on the flowering times of *C. ensifolium*.

Supplementary Materials: The following supporting information can be downloaded at: https://www.mdpi.com/article/10.3390/horticulturae10030252/s1. Table S1: Details of cis-acting elements in the promoter region of *PEBP* genes from three *Cymbidium* species (2000 bp upstream of the initiation codon); Table S2: Average FPKM value of *CePEBP* genes transcriptome in different tissues of *C. ensifolium*; Table S3: Primers used in this study.

Author Contributions: Conceptualization, methodology, J.C., F.W., Z.L. and D.P.; Software, formal analysis, visualization, writing—original draft preparation, J.C. and F.W.; Investigation, resources and data curation, validation, J.C., F.W., R.Z., Y.Z. and X.W.; Writing—review and editing, S.A., Y.A., Z.L. and D.P.; Supervision, J.C., F.W., Z.L. and D.P.; Project administration, D.P.; Funding acquisition D.P. All authors have read and agreed to the published version of the manuscript.

Funding: This research was funded by the National Natural Science Foundation of China (32071815).

Data Availability Statement: All data generated or analyzed during this study are included in this published article (Supplementary Files) and are also available from the corresponding author on reasonable request.

Conflicts of Interest: The authors declare no conflicts of interest.

References

1. Zhang, M.; Li, P.; Yan, X.; Wang, J.; Cheng, T.; Zhang, Q. Genome-wide characterization of PEBP family genes in nine Rosaceae tree species and their expression analysis in *P. mume*. *BMC Ecol. Evol.* **2021**, *21*, 32. [CrossRef] [PubMed]
2. Yu, X.; Liu, H.; Sang, N.; Li, Y.; Zhang, T.; Sun, J.; Huang, X. Identification of cotton MOTHER OF FT AND TFL1 homologs, GhMFT1 and GhMFT2, involved in seed germination. *PLoS ONE* **2019**, *14*, e0215771. [CrossRef] [PubMed]
3. Cheng, X.; Li, G.; Krom, N.; Tang, Y.; Wen, J. Genetic regulation of flowering time and inflorescence architecture by MtFDa and MtFTa1 in *Medicago truncatula*. *Plant Physiol.* **2021**, *185*, 161–178. [CrossRef] [PubMed]
4. Xu, H.; Guo, X.; Hao, Y.; Lu, G.; Li, D.; Lu, J.; Zhang, T. Genome-wide characterization of PEBP gene family in *Perilla frutescens* and PfFT1 promotes flowering time in *Arabidopsis thaliana*. *Front. Plant Sci.* **2022**, *13*, 1026696. [CrossRef]
5. Zheng, X.-M.; Wu, F.-Q.; Zhang, X.; Lin, Q.-B.; Wang, J.; Guo, X.-P.; Lei, C.-L.; Cheng, Z.-J.; Zou, C.; Wan, J.-M. Evolution of the PEBP gene family and selective signature on *FT*-like clade. *J. Syst. Evol.* **2016**, *54*, 502–510. [CrossRef]
6. Hedman, H.; Källman, T.; Lagercrantz, U. Early evolution of the MFT-like gene family in plants. *Plant Mol. Biol.* **2009**, *70*, 359–369. [CrossRef] [PubMed]
7. Xi, W.; Liu, C.; Hou, X.; Yu, H. MOTHER OF FT AND TFL1 regulates seed germination through a negative feedback loop modulating ABA signaling in *Arabidopsis*. *Plant Cell* **2010**, *22*, 1733–1748. [CrossRef]
8. Nakamura, S.; Abe, F.; Kawahigashi, H.; Nakazono, K.; Tagiri, A.; Matsumoto, T.; Utsugi, S.; Ogawa, T.; Handa, H.; Ishida, H.; et al. A wheat homolog of MOTHER OF FT AND TFL1 acts in the regulation of germination. *Plant Cell* **2011**, *23*, 3215–3229. [CrossRef]
9. Li, Q.; Fan, C.; Zhang, X.; Wang, X.; Wu, F.; Hu, R.; Fu, Y. Identification of a soybean MOTHER OF FT AND TFL1 homolog involved in regulation of seed germination. *PLoS ONE* **2014**, *9*, e99642. [CrossRef]
10. Eshed, Y.; Lippman, Z.B. Revolutions in agriculture chart a course for targeted breeding of old and new crops. *Science* **2019**, *366*, eaax0025. [CrossRef]
11. Karlgren, A.; Gyllenstrand, N.; Kallman, T.; Sundstrom, J.F.; Moore, D.; Lascoux, M.; Lagercrantz, U. Evolution of the PEBP gene family in plants: Functional diversification in seed plant evolution. *Plant Physiol.* **2011**, *156*, 1967–1977. [CrossRef] [PubMed]
12. Corbesier, L.; Vincent, C.; Jang, S.; Fornara, F.; Fan, Q.; Searle, I.; Giakountis, A.; Farrona, S.; Gissot, L.; Turnbull, C.; et al. FT protein movement contributes to long-distance signaling in floral induction of *Arabidopsis*. *Science* **2007**, *316*, 1030–1033. [CrossRef] [PubMed]
13. Bradley, D.; Carpenter, R.; Copsey, L.; Vincent, C.; Rothstein, S.; Coen, E. Control of inflorescence architecture in *Antirrhinum*. *Nature* **1996**, *379*, 791–797. [CrossRef] [PubMed]
14. Bradley, D.; Ratcliffe, O.; Vincent, C.; Carpenter, R.; Coen, E. Inflorescence commitment and architecture in *Arabidopsis*. *Science* **1997**, *275*, 80–83. [CrossRef]
15. Kaneko-Suzuki, M.; Kurihara-Ishikawa, R.; Okushita-Terakawa, C.; Kojima, C.; Nagano-Fujiwara, M.; Ohki, I.; Tsuji, H.; Shimamoto, K.; Taoka, K.I. TFL1-Like Proteins in Rice Antagonize Rice FT-Like Protein in Inflorescence Development by Competition for Complex Formation with 14-3-3 and FD. *Plant Cell Physiol.* **2018**, *59*, 458–468. [CrossRef] [PubMed]
16. Freytes, S.N.; Canelo, M.; Cerdan, P.D. Regulation of Flowering Time: When and Where? *Curr. Opin. Plant Biol.* **2021**, *63*, 102049. [CrossRef] [PubMed]
17. Zuo, X.; Xiang, W.; Zhang, L.; Gao, C.; An, N.; Xing, L.; Ma, J.; Zhao, C.; Zhang, D. Identification of apple TFL1-interacting proteins uncovers an expanded flowering network. *Plant Cell Rep.* **2021**, *40*, 2325–2340. [CrossRef] [PubMed]
18. Hanzawa, Y.; Money, T.; Bradley, D. A single amino acid converts a repressor to an activator of flowering. *Proc. Natl. Acad. Sci. USA* **2005**, *102*, 7748–7753. [CrossRef]
19. Ho, W.W.; Weigel, D. Structural features determining flower-promoting activity of Arabidopsis FLOWERING LOCUS T. *Plant Cell* **2014**, *26*, 552–564. [CrossRef]
20. Ahn, J.H.; Miller, D.; Winter, V.J.; Banfield, M.J.; Lee, J.H.; Yoo, S.Y.; Henz, S.R.; Brady, R.L.; Weigel, D. A divergent external loop confers antagonistic activity on floral regulators FT and TFL1. *EMBO J.* **2006**, *25*, 605–614. [CrossRef]
21. Ramya, M.; Park, P.H.; Chuang, Y.C.; Kwon, O.K.; An, H.R.; Park, P.M.; Baek, Y.S.; Kang, B.C.; Tsai, W.C.; Chen, H.H. RNA sequencing analysis of *Cymbidium goeringii* identifies floral scent biosynthesis related genes. *BMC Plant Biol.* **2019**, *19*, 337. [CrossRef] [PubMed]
22. Ai, Y.; Li, Z.; Sun, W.H.; Chen, J.; Zhang, D.; Ma, L.; Zhang, Q.H.; Chen, M.K.; Zheng, Q.D.; Liu, J.F.; et al. Correction: The Cymbidium genome reveals the evolution of unique morphological traits. *Hortic. Res.* **2021**, *8*, 264. [CrossRef] [PubMed]
23. Yang, F.X.; Gao, J.; Wei, Y.L.; Ren, R.; Zhang, G.Q.; Lu, C.Q.; Jin, J.P.; Ai, Y.; Wang, Y.Q.; Chen, L.J.; et al. The genome of *Cymbidium sinense* revealed the evolution of orchid traits. *Plant Biotechnol. J.* **2021**, *19*, 2501–2516. [CrossRef] [PubMed]
24. Sun, Y.; Chen, G.-Z.; Huang, J.; Liu, D.-K.; Xue, F.; Chen, X.-L.; Chen, S.-Q.; Liu, C.-G.; Liu, H.; Ma, H.; et al. The *Cymbidium goeringii* genome provides insight into organ development and adaptive evolution in orchids. *Ornam. Plant Res.* **2021**, *1*, 1–13. [CrossRef]
25. Jiang, L.; Jiang, X.; Li, Y.; Gao, Y.; Wang, S.; Ma, Y.; Wang, G. FT-like paralogs are repressed by an SVP protein during the floral transition in *Phalaenopsis* orchid. *Plant Cell Rep.* **2022**, *41*, 233–248. [CrossRef] [PubMed]
26. Zheng, R.; Peng, Y.; Chen, J.; Zhu, X.; Xie, K.; Ahmad, S.; Zhao, K.; Peng, D.; Liu, Z.J.; Zhou, Y. The Genome-Level Survey of the WOX Gene Family in *Melastoma dodecandrum* Lour. *Int. J. Mol. Sci.* **2023**, *24*, 17349. [CrossRef] [PubMed]

27. Chen, C.; Chen, H.; Zhang, Y.; Thomas, H.R.; Frank, M.H.; He, Y.; Xia, R. TBtools: An Integrative Toolkit Developed for Interactiv Analyses of Big Biological Data. *Mol. Plant* **2020**, *13*, 1194–1202. [CrossRef]
28. Wang, J.; Chitsaz, F.; Derbyshire, M.K.; Gonzales, N.R.; Gwadz, M.; Lu, S.; Marchler, G.H.; Song, J.S.; Thanki, N.; Yamashita, R.A et al. The conserved domain database in 2023. *Nucleic Acids Res.* **2023**, *51*, D384–D388. [CrossRef]
29. Marchler-Bauer, A.; Bo, Y.; Han, L.; He, J.; Lanczycki, C.J.; Lu, S.; Chitsaz, F.; Derbyshire, M.K.; Geer, R.C.; Gonzales, N.R.; et a CDD/SPARCLE: Functional classification of proteins via subfamily domain architectures. *Nucleic Acids Res.* **2017**, *45*, D200–D20 [CrossRef]
30. Lu, S.; Wang, J.; Chitsaz, F.; Derbyshire, M.K.; Geer, R.C.; Gonzales, N.R.; Gwadz, M.; Hurwitz, D.I.; Marchler, G.H.; Song, J.S et al. CDD/SPARCLE: The conserved domain database in 2020. *Nucleic Acids Res.* **2020**, *48*, D265–D268. [CrossRef]
31. Letunic, I.; Khedkar, S.; Bork, P. SMART: Recent updates, new developments and status in 2020. *Nucleic Acids Res.* **2021**, *4* D458–D460. [CrossRef] [PubMed]
32. Duvaud, S.; Gabella, C.; Lisacek, F.; Stockinger, H.; Ioannidis, V.; Durinx, C. Expasy, the Swiss Bioinformatics Resource Portal, a designed by its users. *Nucleic Acids Res.* **2021**, *49*, W216–W227. [CrossRef] [PubMed]
33. Zheng, Q.; Zhao, X.; Huang, Y.; Zhang, M.M.; He, X.; Ke, S.; Li, Y.; Zhang, C.; Ahmad, S.; Lan, S.; et al. Genome-Wide Identificatio of the *YABBY* Gene Family in *Dendrobium* Orchids and Its Expression Patterns in *Dendrobium chrysotoxum*. *Int. J. Mol. Sci.* **202** *24*, 10165. [CrossRef] [PubMed]
34. Kumar, S.; Stecher, G.; Tamura, K. MEGA7: Molecular Evolutionary Genetics Analysis Version 7.0 for Bigger Datasets. *Mol. Bio Evol.* **2016**, *33*, 1870–1874. [CrossRef] [PubMed]
35. Letunic, I.; Bork, P. Interactive tree of life (iTOL) v3: An online tool for the display and annotation of phylogenetic and other trees *Nucleic Acids Res.* **2016**, *44*, W242–W245. [CrossRef] [PubMed]
36. Bailey, T.L.; Williams, N.; Misleh, C.; Li, W.W. MEME: Discovering and analyzing DNA and protein sequence motifs. *Nuclei Acids Res.* **2006**, *34*, W369–W373. [CrossRef] [PubMed]
37. Rombauts, S.; Déhais, P.; Montagu, M.V.; Rouzé, P. PlantCARE, a plant cis-acting regulatory element database. *Nucleic Acids Re* **1999**, *27*, 295–296. [CrossRef]
38. Chen, G.Z.; Huang, J.; Zhou, X.Q.; Hao, Y.; Chen, J.L.; Zhou, Y.Z.; Ahmad, S.; Lan, S.; Liu, Z.J.; Peng, D.H. Comprehensiv Analysis for GRF Transcription Factors in Sacred Lotus (*Nelumbo nucifera*). *Int. J. Mol. Sci.* **2022**, *23*, 6673. [CrossRef]
39. Huang, W.; Fang, Z.; Zeng, S.; Zhang, J.; Wu, K.; Chen, Z.; Teixeira da Silva, J.A.; Duan, J. Molecular cloning and functiona analysis of Three FLOWERING LOCUS T (FT) homologous genes from Chinese *Cymbidium*. *Int. J. Mol. Sci.* **2012**, *13*, 11385–11398 [CrossRef]
40. Ospina-Zapata, D.A.; Madrigal, Y.; Alzate, J.F.; Pabon-Mora, N. Evolution and Expression of Reproductive Transition Regulatory Genes *FT/TFL1* With Emphasis in Selected Neotropical Orchids. *Front. Plant Sci.* **2020**, *11*, 469. [CrossRef]
41. Gao, J.; Huang, B.H.; Wan, Y.T.; Chang, J.; Li, J.Q.; Liao, P.C. Functional divergence and intron variability during evolution of angiosperm TERMINAL FLOWER1 (TFL1) genes. *Sci. Rep.* **2017**, *7*, 14830. [CrossRef]
42. Hou, C.J.; Yang, C.H. Functional analysis of FT and TFL1 orthologs from orchid (*Oncidium* Gower Ramsey) that regulate the vegetative to reproductive transition. *Plant Cell Physiol.* **2009**, *50*, 1544–1557. [CrossRef] [PubMed]
43. Cai, J.; Liu, X.; Vanneste, K.; Proost, S.; Tsai, W.C.; Liu, K.W.; Chen, L.J.; He, Y.; Xu, Q.; Bian, C.; et al. The genome sequence of the orchid *Phalaenopsis equestris*. *Nat. Genet.* **2015**, *47*, 65–72. [CrossRef] [PubMed]
44. Banfield, M.J.; Brady, R.L. The structure of *Antirrhinum* centroradialis protein (CEN) suggests a role as a kinase regulator. *J. Mol. Biol.* **2000**, *297*, 1159–1170. [CrossRef] [PubMed]
45. Si, Z.; Liu, H.; Zhu, J.; Chen, J.; Wang, Q.; Fang, L.; Gao, F.; Tian, Y.; Chen, Y.; Chang, L.; et al. Mutation of SELF-PRUNING homologs in cotton promotes short-branching plant architecture. *J. Exp. Bot.* **2018**, *69*, 2543–2553. [CrossRef]
46. Zhou, S.; Jiang, L.; Guan, S.; Gao, Y.; Gao, Q.; Wang, G.; Duan, K. Expression profiles of five FT-like genes and functional analysis of PhFT-1 in a *Phalaenopsis* hybrid. *Electron. J. Biotechnol.* **2018**, *31*, 75–83. [CrossRef]
47. Amasino, R. Seasonal and developmental timing of flowering. *Plant J.* **2010**, *61*, 1001–1013. [CrossRef]
48. Teotia, S.; Tang, G. To bloom or not to bloom: Role of microRNAs in plant flowering. *Mol. Plant* **2015**, *8*, 359–377. [CrossRef]
49. Song, Y.H.; Shim, J.S.; Kinmonth-Schultz, H.A.; Imaizumi, T. Photoperiodic flowering: Time measurement mechanisms in leaves. *Annu. Rev. Plant Biol.* **2015**, *66*, 441–464. [CrossRef]
50. Hernandez-Garcia, C.M.; Finer, J.J. Identification and validation of promoters and cis-acting regulatory elements. *Plant Sci.* **2014**, *217–218*, 109–119. [CrossRef]
51. Wang, Y.; Liu, L.; Song, S.; Li, Y.; Shen, L.; Yu, H. DOFT and DOFTIP1 affect reproductive development in the orchid *Dendrobium* Chao Praya Smile. *J. Exp. Bot.* **2017**, *68*, 5759–5772. [CrossRef] [PubMed]

horticulturae

Article

Transcriptional Insights into Lily Stem Bulblet Formation: Hormonal Regulation, Sugar Metabolism, and Transcriptional Networks in LA Lily 'Aladdin'

Kewen Zhang [1,†], Tong Lyu [2,†] and Yingmin Lyu [1,*]

1 Beijing Key Laboratory of Ornamental Plants Germplasm Innovation & Molecular Breeding, China National Engineering Research Center for Floriculture, College of Landscape Architecture, Beijing Forestry University, Beijing 100083, China
2 Beijing Flower Engineering Technology Research Center, Plant Institute, China Botanical Garden North Garden, Beijing 100093, China
* Correspondence: luyingmin@bjfu.edu.cn
† These authors contributed equally to this work.

Abstract: Bulblets, as the main reproductive organ of lilies, have a tremendous impact on the reproductive efficiency of lilies. Cytokinin is known to promote the formation of lily bulblets, but little is known about the mechanisms involved. In this study, a combination of full-length transcriptome and high-throughput RNA sequencing (RNA-Seq) was performed at the leaf axils of LA lily 'Aladdin' to characterize the transcriptional response to 6-BA treatment during the critical period of stem-to-bulblet transition. Kyoto Encyclopedia of Genes and Genomes (KEGG) analysis showed that 6-BA treatment caused significant changes in starch and sucrose metabolism and plant hormone balance. In particular, the high expression of *SUS1* and *TPS6* in the 6-BA-treated group suggests that sucrose may act as a key signal to promote bulblet initiation. Furthermore, the induction of elevated expression of genes associated with cytokinin and auxin transport and signaling is crucial for initiating bulblet emergence and stimulating growth. WGCNA analysis revealed that hub TFs such as BLHs, ARFs, HD-ZIPs, AP2/ERFs, and SBPs were significantly overexpressed with genes involved in carbohydrate metabolism and phytohormone signaling, which warranted more in-depth functional studies. This study enriches the understanding of plant hormone-related genes, sugar metabolism-related genes and various transcription factors in the regulation of plant organ development, and lays the foundation for further studies on the molecular mechanisms of lily stem bulblet formation.

Keywords: LA hybrid lily; stem bulblet formation; 6-benzylaminopurine (6-BA); phytohormone biosynthesis and signaling; starch and sucrose metabolism

Citation: Zhang, K.; Lyu, T.; Lyu, Y. Transcriptional Insights into Lily Stem Bulblet Formation: Hormonal Regulation, Sugar Metabolism, and Transcriptional Networks in LA Lily 'Aladdin'. *Horticulturae* **2024**, *10*, 171. https://doi.org/10.3390/horticulturae10020171

Academic Editors: Margherita Beruto, Emmy Dhooghe and Bruce Dunn

Received: 15 December 2023
Revised: 4 February 2024
Accepted: 11 February 2024
Published: 14 February 2024

1. Introduction

Lily (*Lilium* spp.), an old worldwide flower, has high ornamental and edible value. Bulblets are the reproductive organ of lilies, and the quantity and quality of bulblets have a great influence on the yield of lilies. Previous studies mainly focused on the formation of lily bulbs under in vitro conditions, and very few studies have been carried out on their natural proliferation, especially in the absence of a clear molecular regulatory mechanism [1]. LA hybrid lily 'Aladdin' has a very strong self-propagation ability under natural conditions, and can spontaneously form a large number of good quality underground-stem bulblets on underground stems, which is an excellent test material to study the mechanism of lily bulb formation under natural conditions [2,3].

The synthesis and signaling of several plant hormones have been reported in the literature to act on the formation of bulblets, like auxin (AUX), cytokinin (CK), ethylene (ETH), gibberellin (GA), abscisic acid (ABA), etc. [4–6]. Among them, cytokinin and auxin are considered to be important hormones regulating bulb formation in plants [7,8]. Previous studies found that applying exogenous cytokinin (6-BA) to LA hybrid lily 'Aladdin' can accelerate the speed of underground stem-bulblet initiation and increase the number of bulblet occurrence [9]. In *Lilium lancifolium*, 6-BA promotes bulbil initiation while lovastatin delays it, and genes involved in cytokinin metabolism and signaling undergo opposite changes [10]. According to previous studies, auxin may have a dual role in lily bulb organ formation, with auxin acting as a promoter during the initiation phase of organ induction but inhibiting the response to auxin at a later stage to maintain proper hormonal homeostasis within the newly formed meristematic tissue [11–13]. The exogenous application of IAA promotes the formation of bulbils, but the inhibitors of auxin transport, NPA, TIBA, and the inhibitor of auxin function, PCIB, inhibited the formation of bulbils in in vitro stem segments of *Lilium lancifolium* to different extents [14]. In addition, it was found that the application of IAA induces aerial bulbils on the cut surface of the upper stem through a process similar to callus formation [15]. In a recent study, 2,4-dichlorophenoxyacetic acid (2,4-D) significantly increased the regeneration rate of in vitro bulblets in *L. brownii* and *L. brownii* var. giganteum [16].

Recent studies showed that carbohydrate metabolism was also associated with bulb formation [17,18]. In *Lilium davidii* var. unicolor, genes involved in starch synthesis (ADPG pyrophosphorylase gene *AGP*, starch branching enzyme gene *SBE*, soluble starch synthase gene *SSS*, and bound starch synthase gene *GBSS*) were mainly highly expressed at the emergence of small bulbs and during later bulb development; therefore, it is speculated that starch may play a key role in later bulb development [18]. A previous study demonstrated that bulbing involves a shoot-to-bulblet transition and the higher expression of the sucrose synthase gene (*LohSuSy*) could accelerate this transformation process [19]. In *Lycoris sprengeri*, a study indicated a sugar-mediated model of the regulation of vegetative propagation in which high cell wall invertase (*CWIN*) expression or activity could promote bulblet initiation via enhancing apoplasmic unloading of sucrose or sugar signals [20].

Stem bulblets, formed in the leaf axils of the above- and below-ground stems of lilies, are axillary organs and originate from axillary meristematic tissue (AM) [2,21]. In *Arabidopsis thaliana*, numerous studies have revealed that several transcription factors affect AM initiation [22,23]. However, regarding the molecular regulation of bulblet formation, only a small number of genes have been validated so far, and their regulatory mechanisms are still unclear. It was demonstrated that HOMEOBOX PROTEIN KNOTTED-1-LIKE (LaKNOX1) interacts with LaKNOX2 and BEL1-LIKE HOMEBOX (LaBEL1) to simultaneously regulate multiple plant hormones to achieve proper hormone homeostasis, thereby exerting a positive effect in stem bulblet formation in *L.* 'Aladdin' [9]. In *Agave tequilana*, *AtqKNOX1* and *AtqKNOX2* genes were induced at bulbil initiation and their expression increased with bulbil development [24]. In *Lilium lancifolium*, the *Argonaute 1* (*LlAGO1*) was suggested to play an positive role in the formation of bulbils [25]. In other studies, the *LoAGO1* gene was also found to play a role in meristem induction along with other genes regulating growth hormone signaling such as *TOPLESS* (*LoTPL*) and *AUXIN SIGNALING F-BOX 3* (*LoAFB3*) [4]. According to a previous study, cytokinin type-B response regulators *LlRRs* could bind to the promoters of *LlWOX9* and *LlWOX11* and encourage their transcription to promote bulbil generation [26]. In addition to TFs associated with AM initiation, TFs associated with organ boundaries were also shown to affect bulb production. In the oriental lily 'Siberia', a *LoLOB18* gene belonging to the LOB (lateral organ boundaries domain) transcription factor family was demonstrated to have a positive regulatory role in bulb formation by VIGS experiments [27].

In this study, we applied 6-BA to the underground stems of lilies, 40 days after planting, and analyzed RNA sequences of treated and control samples on days 1, 10, and 20 thereafter. We investigated the patterns of gene expression changes during lily bulb development related to starch and sucrose metabolism, hormone synthesis and signaling, and some transcription factors related to axillary bud initiation, and then constructed the network relationships of these genes by WGCNA analysis. In conclusion, our results provide insight into the molecular regulatory mechanisms regulating lily bulblet formation, which will contribute to future studies to improve the reproductive efficiency of lily plants.

2. Materials and Methods

2.1. Plant Materials

At Beijing Forestry University (BJFU) (116.3E, 40.0N), the study was carried out in the identical climate-controlled facility. Three-year-old LA (*Lilium longiflorum* × *Lilium asiatic*) hybrid cultivars 'Aladdin' with similar diameters were selected as experimental materials. The planted lilies were divided into two groups, and 100 mg/L of 6-BA (total 25 mg) and water (as control) were applied to the underground stems 40 days after planting. The second application was repeated after 48 h. We collected leaf axils of underground stems at 1, 10, 20, and 25 days after application, as with the control group. The leaf axils (including leaf axil tissues and part of the stem bode) were cut with a double-sided razor blade at about 0.2 mm thickness. Samples of each period and group were mixed from three lily samples, each containing 3–5 leaf axils, and three biological replicates were performed. The control samples were named C1, C10, C20, and C25 while the 6-BA treated groups were named B1, B10, B20, and B25. The collected leaf axil samples were wrapped in tin foil bags and quickly frozen in liquid nitrogen for 30 min, and then transferred to a −80 °C refrigerator for storage until further use.

2.2. Morphological and Histological Observation

Morphological changes of bulblet initiation in *L.* 'Aladdin' were observed under a stereomicroscope (SP8 SR, Lecia, Wetzlar, Germany). The induction rate of bulblets was calculated as follows: induction rate = number of axils that produced axillary meristems/total number of underground axils. Samples from leaf axils with partial leaves were fixed with FAA (70% ethanol: glacial acetic acid: formalin, 90:5:5) at 4 °C for 24 h. Fixed samples were dehydrated and embedded in paraffin wax through an ethanol gradient, and longitudinal sections were prepared using a sectioning machine (Leica HistoCore BIOCUT, Wetzlar, Germany). The resulting sections were stained with safranin–alcian green, and the morphological and structural characteristics of the samples were subsequently observed under a light microscope and recorded.

2.3. RNA Extraction and Quality Assessment

Total RNA was extracted using a Plant RNA Extraction Kit (RN53, Aidlab, Beijing, China) according to the manufacturer's instructions, and treated with RNase-free DNase I to remove residual DNA contamination. The integrity and purity of the total RNA were determined using a NanoDrop-1000 UV spectrophotometer (Thermo Fisher Scientific, Waltham, MA, USA) and 1.2% agarose gel electrophoresis.

2.4. PacBio Iso-Seq Library Preparation, Sequencing, and Analysis

Total RNA from above-ground stem axils of C1, C10, C20, B1, B10, and B20 were mixed in equal amounts and sequencing libraries were constructed through the PacBio platform to obtain the full-length transcriptome sequences of *L.* 'Aladdin'. First, Clontech SMARTer PCR cDNA Synthesis Kit was used to reverse translate total RNA to first-strand cDNA, and then PCR amplification was performed to synthesize double-strand cDNA. Second, SMRTbell library construction was performed after the purification of the products. Finally, the SMRTbell library was annealed with primers and polymerase and sequenced using MagBead Loading.

The raw reads were analyzed, and circular consensus sequence (CCS) sequences were extracted using SMRT Link v6.0 [28]. The primers, barcodes, poly(A) tails, and tandems were then removed to obtain full-length non-chimeric sequences (FlNC). Subsequently, the full-length non-chimeric (FLNC) reads were clustered by error-corrected iterative clustering (ICE) to generate clustered consensus heterodimers. Next, low quality isomers were further corrected by the LoRDEC tool [29] using Illumina short reads obtained from the same lily-leaf axil samples. CD-HIT-v4.6.7 software was used to remove redundant sequence using a threshold of 0.99 identity to obtain the final transcriptome isoform sequence.

2.5. Illumina Transcriptome Library Preparation, Sequencing, and Analysis

To characterize the expression of transcriptomic differential gene sets during L. 'Aladdin' stem bulblet initiation, total RNA from aboveground stem axils of C1, C10, C20, B1, B10, and B20 was extracted for RNA-Seq with two replications for each sample. First, after the RNA test was passed, the mRNA with poly(A) structure was enriched. Then, the mRNA was broken into short fragments, and the RNA was used as a template to synthesize double-stranded cDNA and purified. The purified double-stranded cDNA was then subjected to end repair, Then, the purified double-stranded cDNA was end-repaired, sequencing junction and A-tail structure were added, fragment size was selected, with PCR amplification and purification of PCR products. The final PCR products were sequenced using Illumina HiSeqTM 2000 (Illumina, San Diego, CA, USA).

Using in-house Perl scripts with default parameters, we filtered the raw data (raw reads) generated from the Illumina sequencing platform, including reads containing adapters, reads containing more than 10% unknown nucleotides (N), and low-quality reads containing more than 50% low-quality (Q-value $\leq$ 10) bases. The rRNA removed high-quality clean reads were mapped to the reference transcriptome using the short reads alignment tool Bowtie2 [30] by default parameters. These clean reads were then mapped to the full-length transcriptome sequences of L. 'Aladdin' obtained by PacBio Iso-Seq using HISAT2 [31].

2.6. Functional Annotation of Tanscripts, Identification of Differentially Expressed Genes (DEGs) and Functional Enrichment

We performed BLASTx searches of all the identified transcripts against seven public databases, including the NR, NT, Pfam, KOG, Swiss-Prot, KEGG, and GO databases, with a threshold E-value of $\leq 10^{-5}$. Gene expression levels were calculated as fragments per kilobase of transcript per million mapped reads (FPKM) [32]. DESeq2 was utilized to conduct a differential expression study between two groups. Based on the Benjamini and Hochberg method, differential expression was defined as a fold change of more than 1 with an adjusted e-value (false discovery rate, FDR) of less than 0.05.

2.7. Identification of Co-Expression Network Modules

Gene co-expression modules were created using the weighted gene co-expression network analysis program for R (WGCNA, version 1.47) [33]. To construct co-expression modules, the WGCNA package used the block-wise module construction feature with detailed settings to import the transcript expression values. This TOMType was unsigned, the mergeCutHeight was 0.8, and the minModuleSize was 50. The networks were visualized using Cytoscape v3.9.1 [34].

2.8. Quantitative Real-Time PCR Validation

To validate the RNA-Seq data, 12 significant DEGs involved in the plant hormone signal transduction pathway, starch and sucrose metabolism, and transcription factors were selected randomly, and qRT–PCR assays were performed. cDNA was synthesized from total RNA using Prime Script II 1st strand cDNA Synthesis Kit (Takara, Shiga, Japan). According to the manufacturer's protocol, the cDNA template was fully mixed with SYBR R® qPCR mix (Takara, Japan) and used for qRT–PCR in a Bio-Rad/CFX ConnectTM Real-Time PCR Detection System (Bio-Rad, Irvine, CA, USA). Gene expression levels were analyzed with the $2^{-\Delta\Delta CT}$ method [35] using the *TIP41* gene of *Lilium* [36] as internal standards. All analytical procedures were repeated on three biological replicates. Specific quantitative primers were designed using NCBI Primer-BLAST (http://www.ncbi.nlm.nih.gov) (accessed on 12 July 2022) (Table S1).

3. Results

3.1. Morphological and Histological Observations on Bulblet Formation after Application of 6-BA

To understand the formation of stem bulblet initiation in *L.* 'Aladdin', samples were collected from underground leaf axils at 1d, 10d, 20d, and 25d after 6-BA was applied to observe it morphologically and histologically (Figure 1). Based on our observations, there were no apparent changes in the axillary region of the leaves (Figure 1a,e) one day after treatment with 6-BA (B1). Through paraffin section imaging, in comparison to the B1 stage, cells in the axils at the B10 stage exhibited a more closely arranged pattern, with increased cell numbers, and more prominent cell nuclei (Figure 1e,f). At the B20 stage, the distinct band-like protrusions in the axillary region became clearly visible (Figure 1c), and the dense cells had formed axillary meristematic tissue (Figure 1g). By the B25 stage, these structures developed into bulbous formations (Figure 1d), with the axils distinctly showing the initiation of bulblet primordia, including axillary meristems and scale primordia (Figure 1h). Thus, we classified the process facilitated by 6-BA into the axillary meristem activation stage (B1–B10), bulblet initiation stage (axillary meristem formation stage) (B20), and bulblet primordia formation stage (B25). In contrast, morphological and histological observations in the control group treated with distilled water during the corresponding time periods did not reveal the formation of bulbs (Figure S1). During the bulblet initiation stage (B20), we observed the formation of axillary meristems in both the control and treatment groups. The results revealed a significantly higher induction rate in the 6-BA treatment group compared to the control group ($p < 0.0001$), with no axillary bud formation observed in the control group (Figure S2). Therefore, we hypothesize that the use of 6-BA plays a very important role in the germination of underground stem bulblets.

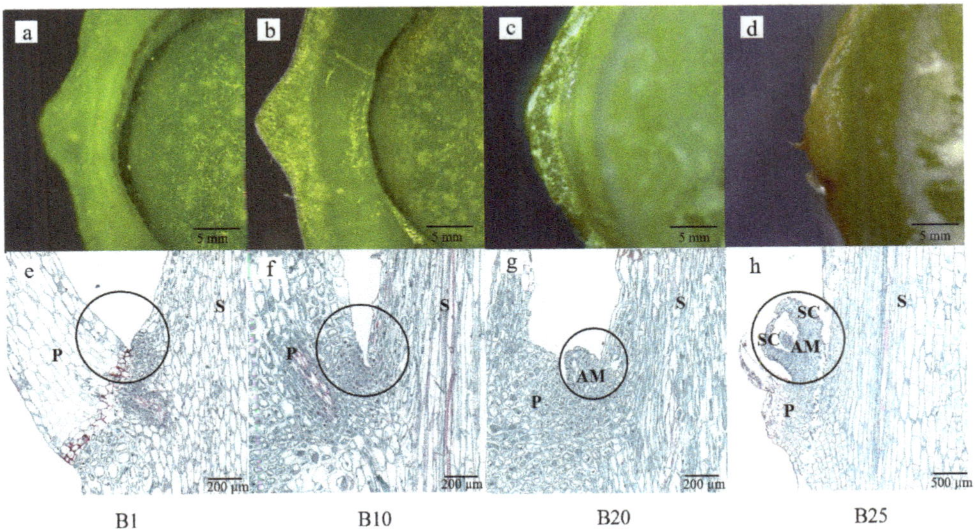

Figure 1. Morphology and anatomy of stem bulblet formation under the influence of 6-BA. Scale bar for sample images (**a–d**) = 5 mm; scale bar for microscope images (**e–g**) = 200 and (**h**) = 500 μm. P, Petiole; S, stem; SC, scale; AM, axillary meristem; black circle, leaf axil.

3.2. RNA-Seq Analysis of *L.* 'Aladdin'

For the purpose of obtaining a comprehensive annotation of the genes of *L.* 'Aladdin', a total of 32,400 full-length non-redundant transcripts were obtained using PacBio SMRT sequencing. The transcript sequences ranged from 54 to 7596 base pairs (bp) in length, with an average transcript length of 1362.86 bp and an N50 of 1783 bp (Figure S3a). Of the clean reads obtained from Illumina RNA-Seq, 80.33% was mapped to the reference transcripts (Table S2), and the number of transcripts annotated in the four major protein databases, Nr,

SwissProt, KEGG and KOG, was 29,244, 16,404, 18,679, and 25,547, respectively. Additionally, out of the 32,400 transcripts, 12,975 were annotated in all four major databases, while 3011 transcripts were not annotated (Figure S3b).

The differentially expressed genes (DEGs) were determined by the *L.* 'Aladdin' 6-BA treated and control groups at three stages (1d, 10d, 20d). We screened for expression levels $|\log2(FC)| > 1$ and $FDR < 0.05$ as DEGs by comparing each comparison group. First, three comparison groups were obtained by comparing samples from different treatments during the same time period. In total, 2275, 4704, and 4378 of the DEGs were differentially expressed between B1 versus (vs.) C1, B10 vs. C10, and B20 vs. C20, respectively. Meanwhile, 755, 2196, and 1999 DEGs were specific for B1 vs. C1, B10 vs. C10, and B20 vs. C20, respectively (Figure S3e). Among the three comparison groups, B10 vs. C10 had the highest number of DEGs with 4704, including 2519 up-regulated genes and 2185 down-regulated genes. This was followed by B20 vs. C20 with 4378 DEGs, including 2614 up-regulated and 1764 down-regulated genes. The least number of differential genes was in the B1 vs. C1 comparison group with only 2275, including 1393 up-regulated and 882 down-regulated genes (Figure S3c). Meanwhile, we compared samples at different time points after the application of 6-BA and obtained three comparison groups: B10 vs. B1, B20 vs. B1, and B20 vs. B10. Among these three comparison groups, the highest number of commonly expressed DEGs was observed, totaling 2505 (Figure S3f). Additionally, all three comparison groups showed a higher number of upregulated genes than downregulated genes, and the number of upregulated DEGs increased with the number of days of application (Figure S3d).

To validate the reliability of the RNA-Seq data, 12 DEGs were selected for qRT–PCR analysis. We found that the RNA-seq and RT-qPCR data were similar and yielded consistent expression patterns. Therefore, the experimental results confirmed the accuracy and reliability of the sequencing results (Figure S4).

3.3. KEGG Pathway Enrichment Analysis of DEGs

We performed KEGG pathway enrichment analysis for DEGs obtained from B1 vs. C1, B10 vs. C10, B20 vs. C20, B10 vs. B1, B20 vs. B1, and B20 vs. B10 comparisons, respectively (Figure 2). There was a significant difference in KEGG enrichment between B1 and C1 DEGs, with "plant hormone signal transduction" the most significant, followed by "biosynthesis of secondary metabolites", "biosynthesis of amino acids", "plant circadian rhythm", "linoleic acid metabolism", and "carbon metabolism" (Figure 2a). For B10 vs. C10, DEGs were mainly enriched in "biosynthesis of secondary metabolites", "carbon fixation in photosynthetic organisms", "metabolic pathways", "starch and sucrose metabolism", and "plant hormone signal transduction pathway" (Figure 2b). Additionally, the DEGs for B20 vs. C20 were mainly enriched in the "biosynthesis of secondary metabolites", "starch and sucrose metabolism", "glyoxylate and dicarboxylate metabolism", "garbon metabolism", "metabolic pathways" and "plant hormone signal transduction" (Figure 2c). In the comparison of B10 vs. B1, DEGs were significantly enriched in the pathways of "plant hormone signal transduction", "plant-pathogen interaction", "biosynthesis of secondary metabolites", and "MAPK signaling pathway-plant" (Figure 2d). Similar to B10 vs. B1, the two most significantly enriched pathways for DEGs in the B10 vs. B1 comparison were "plant hormone signal transduction" and "plant-pathogen interaction". Additionally, DEGs were also enriched in the "starch and sucrose metabolism" pathway (Figure 2e). Furthermore, in the comparison B20 vs. B10, DEGs were significantly enriched in the pathways of "plant hormone signal transduction" and "starch and sucrose metabolism" (Figure 2f). In all six comparative groups, DEGs were enriched in the pathway of "plant hormone signal transduction". In addition, except for the two comparative groups B1 vs. C1 and B10 vs. B1, all other comparative groups were enriched in the pathway of "starch and sucrose metabolism". These results suggest that 6-BA promotes phytohormone responses throughout bulblet emergence and promotes bulblet formation through signal transduction. Moreover, starch and sucrose metabolism as well as sugar signaling may play an important role in the bulblet process.

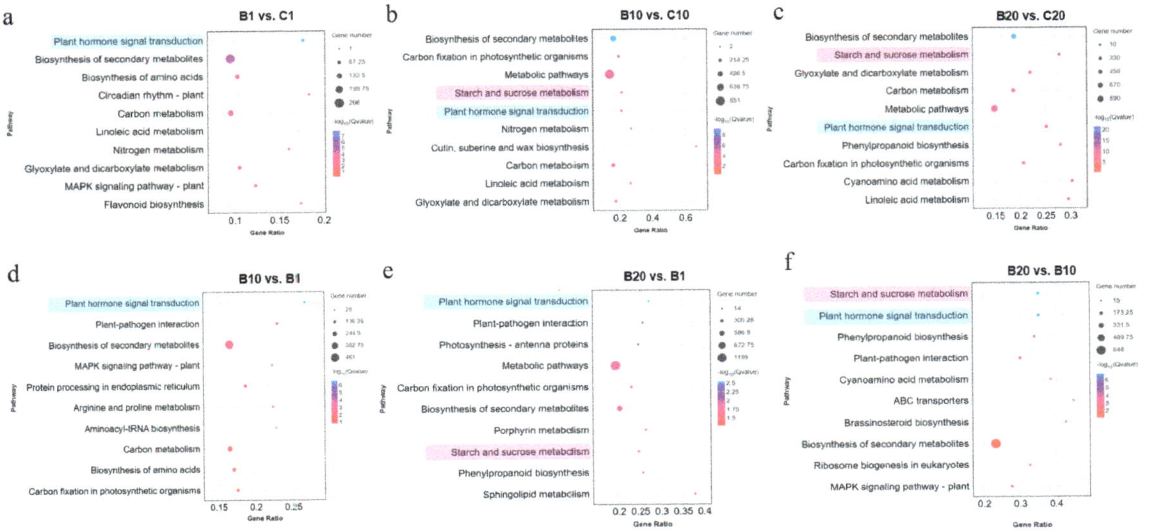

Figure 2. KEGG enrichment analyses of stem bulblet initiation. The top 10 KEGG pathways of the DEGs in each comparison: (**a**) B1 vs. C1, (**b**) B10 vs. C1, (**c**) B20 vs. C20, (**d**) B10 vs. B1, (**e**) B20 vs. B1, (**f**) B20 vs. B10. The phytohormone signaling pathway is highlighted in blue, while the starch and sucrose pathway is highlighted in red.

3.4. Expression of Genes Involved in Starch and Sucrose Metabolism Pathways under 6-BA Treatment

In our research, according to the KEGG enrichment, the 6-BA treatment induced genes related to starch and sucrose metabolism. Therefore, we discovered 25 genes associated with this pathway that showed differences in expression in at least one comparison (Figure 3). Seven *SUS1* genes encoding sucrose synthases were upregulated at B10 compared to C10. Compared to the control group, seven genes encoding sucrose synthases (*SUSs*) in the 6-BA-treated group were upregulated at different time points. Among them, two *SUS1* genes (Isoform0001718 and Isoform0028472) exhibited an upregulation pattern in the B10 vs. C10 comparison, with *SUS1* (Isoform0001718) showing a particularly remarkable upregulation fold change of 14.60 (Table S3). Sucrose may be a signal that promotes bulb growth by upregulating genes in the T6P pathway. In our data, the gene *TPS* regarding trehalose-phosphate synthase and the gene *TPP* regarding trehalose-phosphate phosphatase presented distinct expression patterns in the 6-BA treatment and control groups, respectively. Among them, a higher level of expression was observed in B20 than in C20 for seven *TPS* isoforms. However, another two *TPP* transcripts (*TPPG* and *TPP6*) exhibited an opposite expression profile by the 6-BA treatment. Particularly, 6-BA consistently decreased *TPP6* expression in all three periods. Genes encoding sucrose phosphate synthase, including *SPS2* (Isoform0009502) and *SPS4* (Isoform0009583), were up-regulated in the B20 vs. C20 comparison. In addition, *INV* genes (Isoform0004628 and Isoform0031226), which are related to fructose and glucose synthesis, showed an up-regulated expression pattern after 10 d and 20 d of 6-BA application compared to the control group. One *AMY3* gene encoding alpha-amylase, which degrades starch to glucose, is highly expressed at B10. Moreover, *PGMP* (phosphoglucomutase) and *AGPS1* (ADP-glucose pyrophosphorylase catalytic subunit) genes related to glucose metabolism were upregulated at B10. Additionally, one *AGPS1* (Isoform0030445) was also expressed upregulated at B20.

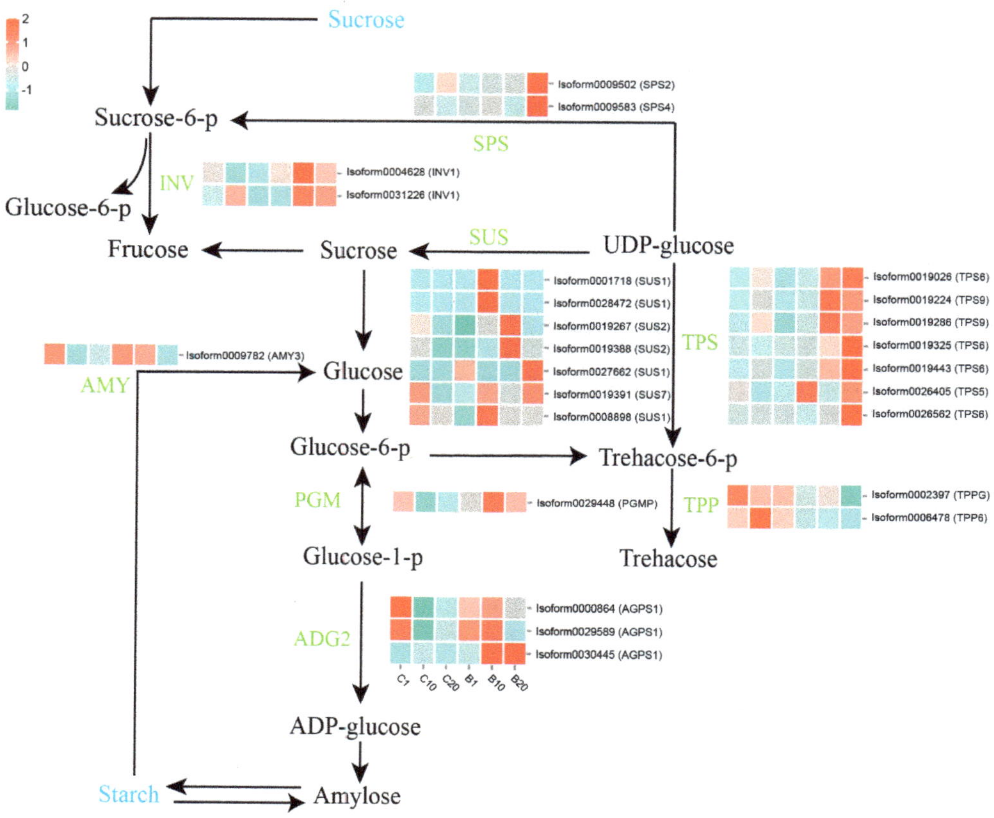

Figure 3. Expression of DEGs related to starch and sucrose metabolism pathways in 6-BA treated and control groups. The gene expression levels are standardized into Z−scores, with red indicating upregulation and purple indicating downregulation.

3.5. Analysis of DEGs Involved in Phytohormone Biosynthesis and Signal Transduction Pathway during Stem Bulblet Initiation

Studies have shown that plant hormones such as auxin (AUX), cytokinin (CK), abscisic acid (ABA), gibberellin (GA), brassinosteroid (BR), ethylene (ETH), and jasmonic acid (JA) regulate the formation of lily bulbs. We filtered a large number of DEGs involved in phytohormone synthesis and the signal transduction pathway from all pairwise comparisons to analyze their effects on the development of lily stem bulblets.

In this study, we hypothesized that genes related to cytokinin biosynthesis and signaling would undergo larger changes in expression levels compared to the control groups due to the application of 6-BA. We identified one *IPT3* (Isoform0002852) gene encoding isopentenyl transferase, which expression is up-regulated at B1 and down-regulated at the B10 and B20 stages. In contrast, most of the genes related to cytokinin degradation (*CKXs*) were down-regulated in expression at B1, but showed an up-regulated expression pattern at B10, including *CKX3* (Isoform0002852), *CKX5* (Isoform0002852), *CKX8* (Isoform0002852), *CKX9* (Isoform0002852), and *CKX11* (Isoform0002852) (Table S4). We also identified genes related to cytokinin signaling, such as cytokinin receptors (*AHKs*), phosphotransmitters (*AHPs*), and cytokinin response regulators (*RRs*). Among them, the *AHK3* gene was down-regulated at B1 and up-regulated at B10 and B20 stages. However, the *AHP4* gene was induced to be up-regulated by CK at B1 and down-regulated in expression at the following two periods (B10 and B20) as compared to control groups. Meanwhile, the response regulator *ARR12* (Isoform0019327) was significantly upregulated in 6-BA treated groups (Figure 4a).

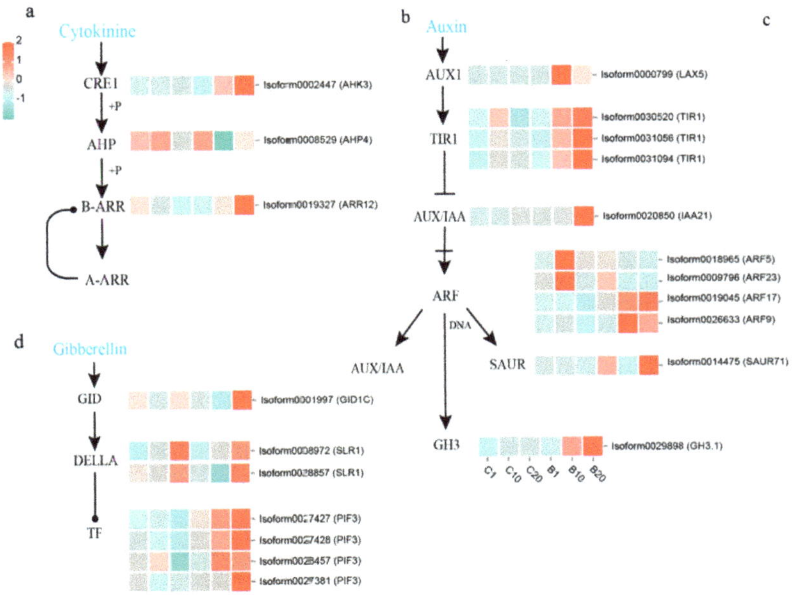

Figure 4. (**a–d**) Heatmap of the expression of DEGs related to plant hormone signal transduction pathway in the indicated groups during bulblet initiation, respectively. The gene expression levels are standardized into Z−scores, with red indicating upregulation and blue indicating downregulation.

In the auxin signaling pathway, the expression of the auxin transporter-like protein (*LAX5*) (Isoform0000799) increased in response to the 6-BA treatment after 10 and 20 days (Table S4). Similarly, we identified an auxin efflux carrier, *PIN1* (Isoform0012372), which was expressed at higher levels at B10 and B20 (Table S4). The transport inhibitor response 1 genes (*TIR1s*) did not show significant changes in gene expression at B1 and B10, but exhibited up-regulation at B20 vs. C20. In our study, the indole-3-acetic acid inducible gene (*IAA21*) was up-regulated at B20. We observed that several auxin response factor genes were induced by 6-BA, including *ARF5*, *ARF9*, *ARF17*, and *ARF23*. Among them, *ARF5* (Isoform0018965) and *ARF23* (Isoform0009796) exhibited an up-regulation pattern at B1. Meanwhile, the expression of *ARF17* (Isoform0019045) and *ARF9* (Isoform0026633) was significantly higher at 10 and 20 days in samples treated with 6-BA. In addition, the small auxin upregulated RNA gene (*SAUR71*) (Isoform0014475) was also upregulated in response to 6-BA after 20 days. At B10 and B20, there was an upregulation of the *GH3* gene (Isoform0029898), which encodes the enzyme IAA-amido synthetase, responsible for converting IAA to amino acids (Figure 4b).

In the ABA signaling pathway, abscisic acid receptor *PYL* genes (Isoform0010687, Isoform0012614 and Isoform0009072) were upregulated at B10 and B20. In contrast, two genes encoding *PP2C* (Isoform0005538 and Isoform0006273), a negative regulator of ABA, had a down-regulation at B20. The expression of serine/threonine-protein kinase (*SAPK*) genes, which contribute to ABA signaling, was mostly down-regulated (Figure 4c). As for GA metabolism, two *GA2OX* genes encoding GA2-beta-dioxygenase showed a downregulation at B10 and B20 (Table S4). Moreover, in the GA signal pathway, one GA receptor *GID1C* (Isoform001997) and one *SLR* gene (Isoform0028857) encoding DELLA protein were downregulated at B10 compared to C10. In contrast, one *SLR* gene (Isoform0008972) and four GA-regulated transcription factors *PIF3* (Isoform0027427, Isoform0027428 and Isoform0028457) showed different degrees of up-regulation in the three periods after 6-BA application (Figure 4d).

Additionally, numerous genes were related to brassinosteroid, ethylene, and jasmonic acid pathways. As for BR signaling, five *CYP* genes involved in BR biosynthesis were significantly upregulated in B10 and B20 compared to C10 and C20. Consistent herewith, *CURL3* (brassinosteroid insensitive 1 protein) and *BSK3* (brassinosteroid insensitive 2 protein) were upregulated during three periods under the 6-BA treatment. The 6-BA stress also led to up-regulation of several *XTH* genes encoding xyloglucan endotransglucosylase/hydrolase proteins. The ethylene signaling transduction pathway contains eight genes, among which three ethylene insensitive 3 (*EIN3*) genes and two *EBF1* genes encoding EIN3-binding F-box protein were downregulated at B10, while ethylene receptor (*ETR3*) was upregulated at B10. In terms of JA, three JAZ genes, which encode a protein with a ZIM domain, were primarily downregulated over the course of three treated periods. However, except for one *JAZ* gene (Isoform0016877), which was significantly upregulated by 6.04-fold at B10 vs. C10. Meanwhile, jasmonic acid-amino synthetase (*GH3.5*) and coronatine-insensitive protein 1 (*COI1*) were identified (Table S4).

3.6. Co-Expression Network Construction and Identification of WGCNA Modules

WGCNA allows clustering genes with similar expression patterns and analyzing the association between modules and specific traits or phenotypes. Based on the resemblance of expression profiles in our investigation, 16 modules were found (Figure S5a). Genes in the modules were evaluated using cluster analysis, which is shown using a heat map (Figure S5b). Furthermore, correlations between modules and samples were established to discern modules that exhibited significant associations with different treatments and stages (Figure S5c). Black module was significantly correlated with sample B1 (0.91), ivory module was strongly correlated with C10 (0.81), darkgreen module was positively correlated with samples B10 and B20 (0.47 and 0.66), and darkgreenolive2 module was most strongly correlated with C20 (0.89).

To identify the core modules, the expression profiles of the four obtained modules may be evaluated, as the eigengene expression patterns describe the changes in gene expression for the whole module. We found that eigengenes in the black module were up-regulated at B1 and down-regulated in all other samples (Figure S5d). Furthermore, the eigengene expression of the ivory module was down-regulated at all three periods under 6-BA treatment, but the eigengenes were significantly up-regulated at both C10 and C20 (Figure S5e). The eigengenes in the darkgreen module showed up-regulated expression at the B10 and B20 periods, while the genes in the control group showed a down-regulated pattern at all three periods (Figure S5f). The eigengenes of darkgreenolive2 module were also highly expressed only in C20. Subsequently, we annotated the four modules by KEGG analysis. According to the enrichment of KEGG, there were only three pathways in the black module with $p < 0.05$, which were "linoleic acid metabolism", "proteasome" and "peroxisome". The ivory module was clustered to two pathways regulated to hormone biosynthesis and signaling transduction (Figure S6). Similarly, the darkgreen module was enriched for three pathways related to phytohormones, which are "plant hormone signal transduction", "brassinosteroid biosynthesis" and "zeatin biosynthesis". Meanwhile, the darkgreenolive2 module genes were significantly enriched in the "starch and sucrose metabolism" pathway (Figure S6). It has been shown that the genes related to lily bulblet development are mainly enriched in two KEGG pathways, the plant hormone metabolism and the starch and sucrose pathways, respectively [3,18,19,37,38]. Therefore, the initial response of lily bulbs to 6-BA treatment may be strongly affected by the darkgreen, ivory and darkgreenolive2 module genes, so we focused on the in-depth analysis of the gene regulatory relationships of these three modules.

3.7. Identification of Hub TFs and Network Construction

To elucidate the regulatory role played by transcription factors on the formation of stem bulblets, we constructed network relationships between transcription factors in each module and genes significantly enriched in the hormone signaling transduction and starch and sucrose metabolism pathways We analyzed the DEGs of each module and screened the TFs with high connectivity in the network relationships as hub genes. According to previous studies, the regulation of TFs such as HD-ZIP, TALE, MYB, BHLH, and others plays a pivotal role in the formation of new organs in plants [39].

In the darkgreen module, we found seven highly connected transcription factors as hub genes, which are HOX20 (Isoform0008631), AP2 (Isoform0002884), BLH4 (Isoform 0002562), LRP1 (Isoform0004786), ARR1 (Isoform0027655), and ARR12 (Isoform0000808). Through the gene expression heatmap, we found that except for BLH4 whose expression was down-regulated in B10 and B20 compared to the control group, the rest of the hub TFs expressed were up-regulated (Figure 5a). In the ivory module, we found many ERFs, SBPs, WRKYs, MYBs, and TALEs were identified as hub TFs. Among them, WRKYs and ERFs were the two most abundant TFs, with 15 and 10, respectively. Interestingly, BLH9 and the two SPL15 TFs expressed extremely high rates of crosstalk, and we found that all three genes were significantly down-regulated after 10 days of treatment by gene expression analysis (Figure 5b). Based on the gene connectivity relationships in the darkolivegreen2 module, a large number of TFs were found to be enriched in the module, among which were MYBs, GRAs, TALEs, and BHLHs. In addition, the two genes encoding ARF5 and HOX32 showed a high degree of connectivity in darkolivegreen2, and the expression of HOX32 was greater than that of C10 during the B10 period, whereas the expression of ARF5 was greater than that of C20 at the B20 stage (Figure 5c). From the above analysis, we know that the formation of stem bulblets is inextricably related to the collaborative regulation of multiple transcription factors.

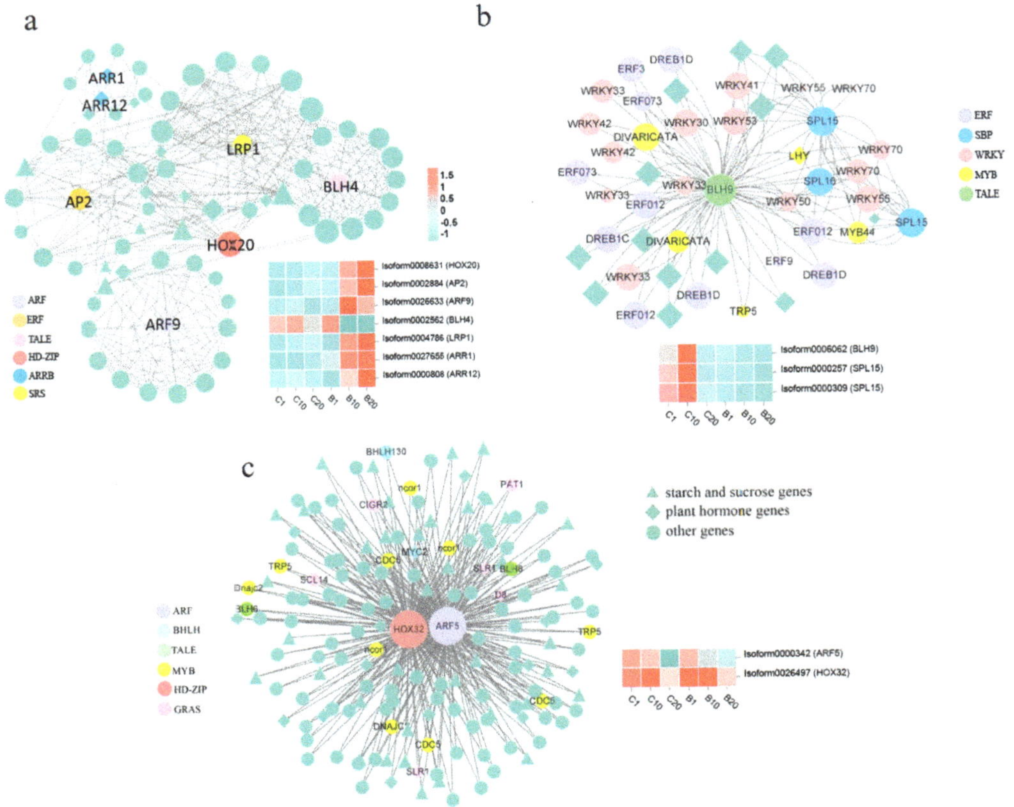

Figure 5. Identification and selection of vital transcription factors in four modules. (**a**) Network analysis of TFs in darkgreen module. (**b**) Network of the related genes in ivory module. (**c**) Network analysis of hub TFs in darkolivegreen2 module. The gene expression levels are standardized into Z−scores, with red indicating upregulation and green indicating downregulation.

4. Discussion

4.1. The Starch and Sucrose Metabolism Mediates Lily Stem Bulblet Initiation

As shown by the KEGG enrichment data, a significant portion of the genes in our study were enriched for the starch and sucrose metabolic pathways, demonstrating that carbohydrate metabolism affected bulblet formation in response to CK signaling under the application of 6-BA. Vacuolar convertase (INV) and sucrose synthase (SUS) are key components of the response to CK [40]. In lilies, previous studies demonstrated that high expression of *SUSs* may induce bulblet initiation in scales or leaf axils [3,16,18,19,41,42]. In our study, all six *SUS1* genes were up-regulated at B10 vs. C10, suggesting that *SUS1* can positively regulate bulblet formation and that the period of 10 days after 6-BA application may be the critical time point for bulblet initiation. In *Lycoris*, the expression of *INVs* was dramatically up-regulated to promote bulbogenesis, while leading to a decrease in sucrose content in maternal scales [43]. Similarly, the invertase (INV) activity in leaf axils was significantly increased, leading to the fact that sucrose can exert its signaling function by rapidly breaking down hexose to promote bulbil formation [42]. Here, the expression levels of *INVs* were significantly increased by 6-BA, showing a possible role of *INVs* in increasing bulblet regeneration ability.

The synthesis of starch is widely recognized as a pivotal process in the initiation of bulblets. Many investigations have suggested that the ability of a meristem to generate bublet primordia is contingent upon its aptitude to amass starch [44]. In *L. radiata*, AGPase as a major starch synthase was shown to be relevant in bulb expansion [43]. In this study the *AGPS1* genes encoding AGPase exhibited an upregulated expression pattern in B10 vs. C10. Trehalose 6-phosphate (Tre6P) is an intermediate in trehalose biosynthesis, and it has been suggested that Tre6P can act as a signal for sugar supply that may be specific to the sucrose state [45]. Enhanced sucrose signaling during yam bulbil initiation leads to upregulation of TPS genes as well as downregulation of TPP genes. These up-regulated TPS genes can increase T6P levels, which can promote bulbil growth [46]. In our study most of the *TPS* genes were up-regulated in 6-BA-treated groups compared to the control groups, especially at the time of 20 days of treatment. So, it may be speculated that genes related to sucrose synthesis such as *SUSs* and *INVs* were triggered to be up-regulated at B10 and thus positively affected *TPS* expression. All these results suggest that 6-BA as a CK signal can influence the metabolism of starch and sucrose and thus play a key positive regulatory role in the initiation of lily stem bulblets.

4.2. The Involvement of 6-BA in Plant Hormone Signal Transduction in Stem Bulblet Formation

In lily, CK was found to stimulate the outgrowth of bulblets or bulbils when directly applied to the scale or axillary part of the stem [9,10,47]. However, transcriptome analysis related to lily bulbogenesis after 6-BA is currently imposed. In transcriptome analyses of axillary bud formation in apple, samples treated with 6-BA show differences in the expression of genes related to CK synthesis and signal transduction. Among them, the expression of *IPT* genes related to cytokinin synthesis as well as CK degradation gene *CKXs* were highly expressed after 6-BA treatment [48]. In the present study, the expression of *IPT3* and several *CKXs* increased at B1 and B10, respectively. Furthermore, the upregulation of the cytokinin receptor gene *AHK3* on B10 and B20 is observed. In the case of *L. lancifolium*, the application of 6-BA on scales also enhances the expression of the *AHKs* gene, which is consistent with our experimental results and further validates the reliability of our data. In *L. lancifolium*, the level of type-B *LlRRs* expression was found to be more pronounced in the axillary meristematic (AM) tissue compared to other tissues in lilies. It was also verified by the VIGs experiment that silencing *LlRRs* decreases the induction rate of bulbils. Moreover, *LlRRs* could bind to the promoters of *LlWOX9* and *LlWOX11* and promote their transcription to regulate the bulbil formation [26]. In this study, the B-type *ARR* gene (*ARR12*) was up-regulated in expression at all 6-BA samples. Taken as a whole, we conclude that CK signaling encourages cell proliferation and is crucial for bulb formation.

Research conducted on *L. lancifolium* revealed that the phenomenon of bulb formation in this particular plant serves as a noteworthy illustration of axillary organogenesis [10]. Simultaneously, auxins play a substantial role in fostering the growth of axillary meristematic tissue cells and the initiation of lateral organs [49]. In *yam*, high expression of auxin transporter proteins, including PINs and LAXs, triggered bulbil growth [46]. Research conducted on *Agave tequilana* has demonstrated that the upregulation of *AtqSoPIN1* expression can be triggered by the elimination of flower buds, resulting in the accumulation of auxin [50]. Consequently, this process stimulates the formation of bulbils. In our investigation, it was observed that the genes *PIN1* and *LAX5* exhibited a significant increase in expression levels at 10 and 20 days subsequent to the application of 6-BA. This observation leads to the proposition that the augmented expression of these genes, which are related to the transport of auxin, might play a crucial role in facilitating the process of lily stem bulblet formation. In our study, it was discovered that the four *ARF* genes displayed distinct patterns of expression at various stages following the induction by 6-BA. Specifically, *ARF5* and *ARF23* exhibited an increase in expression levels at B10, whereas *ARF17* and *ARF9* demonstrated an up-regulation at both B10 and B20. Genes related to the auxin signaling were significantly up-regulated in the sample group after treatment with 2,4-D, especially *ARFs*, which were expressed in an up-regulated pattern in the treated group, similar to our findings [16]. Therefore, we concluded that both 6-BA and 2,4-D could promote auxin signaling transduction.

In the current study, 6-BA administration upregulated *PYL* expression while down-regulating *PP2C* and *SAPK7* expression. As a result, 6-BA administration increased the expression of downstream genes involved in ABA anabolism, which prevented the growth of bulbs. In a prior study, the same findings showed that ABA had an inhibiting influence on bulb growth [43]. Our transcriptome results demonstrated that most BR and JA signaling pathway-related genes were expressed at higher levels in the 6-BA-treated group compared to the control group. In *P. ternate*, BR application can increase the content of GA and BR in bulbils, and decreased the ABA content [51]. A previous study showed that JA regulates tulip bulbs, and ectopic overexpression of *TgLOX4* and *TgLOX5* in Arabidopsis raised endogenous JA levels, promoted plant development, and multiplied the number of lateral roots [52]. Overall, we hypothesize that BR and JA can positively regulate the formation of bulbs. In addition, 6-BA treatment up-regulated *ETR* and down-regulated *EIN* and *EBF* in the ethylene signaling pathway. Together, 6-BA treatment may control the shift in plant hormone signal transduction to promote the production of stem bulblets.

4.3. TFs Involved in Bulblet Formation

In this study, the WGCNA and transcriptional regulatory modules were analyzed to screen out the highly connected transcription factors in the modules, and specific analysis was performed on the hub genes.

The TALE superfamily of genes is ubiquitous in plants and is involved in meristem development, leaf structure, and tuber formation [53]. It includes two subfamily members, KNOTTED1-like homeobox (KNOX) and BEL1-like homeobox (BELL or BLH) [54]. In the model plant Arabidopsis (*A. thaliana*), the BEL1 transcription factor regulates the production of lateral progenitors during ovule development and maintenance of the undifferentiated state of the inflorescence meristem by interacting with SHOOT MERISTEMLESS (STM) [55,56]. PENNYWISE(PNY)/BLH9 and POUND-FOOLISH(PNF)/BLH8 transcription factors promote normal meristematic structure both by interacting with BREVIPEDICELLU(BP)/KNAT1 and STM transcription factors in the class I KNOX subfamily [57–59]. In addition, BLH6 can negatively regulate the expression of secondary cell wall-related genes, and the secondary cell wall of the mutant *blh6* is thicker than that of the wild type [60]. In our study, all three modules were enriched for TALE family transcription factors, and BLH4 and BLH9 showed high connectivity in the darkgreen and ivory modules, respectively (Figure 5a,b). These two genes showed genetically down-regulated expression at the critical stages of bulb initiation and formation, i.e., B10 and B20, and, therefore, we hypothesized that they may exercise a negative regulatory role in bulblet development.

The APETALA2/Ethylene Responsive Factor (AP2/ERF) transcription factors have been shown to integrate multiple stress-related hormones during development to activate cell division and lead to organogenesis [61]. In *Lilium*, it was found that external wounding first activates LoERF109, which is closely related to the auxin metabolic pathway, and *LoERF115*, which confers the ability to divide cells during organ regeneration *LoERF115* [62–64]. According to a study, EBE has an impact on Arabidopsis cell proliferation, axillary bud outgrowth, and shoot branching [65]. The *AP2* gene in the darkgreen module maintains high expression at the B10 versus B20 stage and may promote bulb formation (Figure 5a).

An earlier investigation found that the transcription factor AUXIN RESPONSE FACTOR 5 (ARF5) facilitated the initiation of auxin-dependent organs [49]. In old leaf axils of Arabidopsis, *AtARF5* up-regulates the expression of *ARGONAUTE10* (*AtAGO10*) [66], which can regulate the expression of HD-ZIP III as a target by competing with *AtAGO1* and specifically binding to miR166/165, to regulate AM [67]. HD-Zip transcription factors participate in plant organ and vascular development, meristem maintenance, regulation of hormones, and participation in responses to environmental conditions [68]. Among them, PHABULOSA (PHB), PHAVOLUTA (PHV), and REVOLUT (REV) belong to the HD-ZIP III subfamily, which have overlapping roles in embryogenesis and leaf polarity determination and are involved in the establishment of root tip bilateral symmetry and shoot apical meristem [69]. In the analysis, the HOX32 transcription factor in darkolivegreen2 module shared the same structural domain as the transcription factor of HD-Zip III. At the same time, we were surprised to find that HOX32 and ARF5 showed extremely high connectivity in the module, with which we boldly predicted that the interaction between these two genes might be important for bulblet formation, but the exact function needs further experimental validation (Figure 5c).

SQUAMOSA PROMOTER BINDING PROTEIN LIKE (SPL) controls many aspects of plant development and physiology, including the vegetative phase change, flowering time, leaf initiation, shoot and inflorescence branching [70]. In a previous study, miRNA156 could promote shoot branching and the vegetative to reproductive phase transition by regulating SPL9 and SPL15 [71]. In *lilium*, *LbrSPL9* and *LbrSPL15* could regulate phase transition [72]. In the ivory module, genes encoding SPL15 demonstrated extreme connectivity and we speculated that the initiation of bulbs may be facilitated (Figure 5b). Through WGCNA analysis, we similarly screened for MYBs, WRKYs, NACs, GRAs, BHLHs, ARRs, and other transcription factors, suggesting that the regulation of bulblet initiation is subject to the synergistic cooperation of multiple transcription factors and that the specific functions require further analysis.

5. Conclusions

The information gathered in this study provides a fresh perspective for a comprehensive understanding of the molecular processes, revealing the molecular mechanisms through which 6-BA promotes the formation of lily stem bulblets. Based on morphological and histological analyses, we divided the stimulation of underground stem bulblets by 6-BA into three stages: (1) axillary meristem activation stage (B1–B10), (2) bulblet initiation stage (axillary meristem formation stage) (B20), and (3) bulblet primordia formation stage (B25). Transcriptome sequencing was performed on leaf axil samples during the first two stages, covering three time points (1, 10, 20 days). Through the enrichment of KEGG pathways based on differentially expressed genes in various comparison groups, it was observed that 6-BA treatment had a significant impact on "plant hormone transduction" and "carbohydrate metabolism" pathways. Therefore, we propose a schematic model based on our findings (Figure 6). The *SUS1* gene, which is involved in sucrose synthesis, was up-regulated at the critical stage, so we hypothesized that sucrose may act as a signal to promote bulb growth by up-regulating a gene involved in the T6P pathway (*TPS6*). In addition, auxin is consumed and transported through up-regulation of auxin conjugated gene (*GH3.1*) and auxin efflux proteins (*PIN1* and *LAX5*) to maintain bulb growth. *ARF9*, a gene related to growth hormone signaling, was up-regulated in expression relative to the control groups during the bulb initiation and formation period (B10 and B20). CK promotes its growth by rapidly activating the CK receptor (*AHK3*) gene while raising the cytokinin type-B response regulators (*ARR12*). By WGCNA analysis, we obtained the core transcription factors. It is possible that TFs such as BLHs, ARFs, HD-ZIPs, AP2/ERFs, and SBPs play a role in mediating cytokinin signaling and promoting stem bulblet formation.

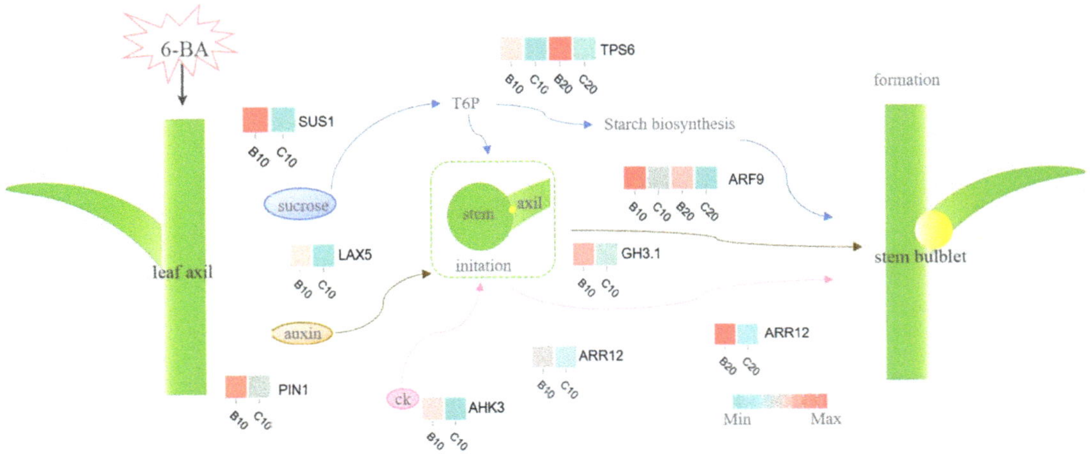

Figure 6. Schematic modeling of gene regulation by auxin, cytokinin, and sucrose during bulblet initiation and formation. The gene expression levels are standardized into Z-scores, with red indicating upregulation and green indicating downregulation.

In this study, after 6-BA treatment, genes involved in the synthesis and signal transduction of different phytohormones and starch and sucrose metabolism were differentially expressed, forming a comprehensive regulatory network to promote bulb formation. To establish the groundwork for the cultivation of lily cultivars with notable rejuvenation rates, it is imperative to prioritize additional investigations concerning the functional authentication of pivotal TFs. Furthermore, a more comprehensive exploration of the precise mechanisms underlying bulb development is warranted.

Supplementary Materials: The following supporting information can be downloaded at: https://www.mdpi.com/article/10.3390/horticulturae10020171/s1. Figure S1: Morphology and anatomy of control samples. Figure S2: Induction rate of the control group and 6-BA treated group during in stem bulblet initiation. Figure S3: PacBio Iso-Seq and expression profiles of *L.* 'Aladdin'. Figure S4: Expression of representative genes determined by RNA-seq and qRT-PCR. Figure S5: WGCNA network of axil samples. Figure S6: Top 20 KEGG pathways in each relevant module. Table S1: The primer sequences required for qRT-PCR. Table S2: Statistical summary of the reference transcriptome. Table S3: Starch and sucrose metabolism-related genes that were differentially expressed in response to 6-BA treatment. Table S4: Differently-expressed related to hormone metabolism and signaling in response to 6-BA treatment.

Author Contributions: K.Z., T.L. and Y.L. conceptualized and designed the experiment; K.Z. and T.L. performed the laboratory experiments and statical analysis; K.Z. and T.L. wrote and revised the manuscript; Y.L. funding acquisition and supervision. All authors have read and agreed to the published version of the manuscript.

Funding: This work was supported by the National Key R&D Program of China (grant No. 2019YFD1000400) and the China National Natural Science Foundation (grant No. 31872138).

Institutional Review Board Statement: Not applicable.

Informed Consent Statement: Not applicable.

Data Availability Statement: Data are contained within the article and Supplementary Materials.

Conflicts of Interest: The authors declare no conflict of interest.

References

1. Nhut, D.T. Micropropagation of lily (*Lilium longiflorum*) via in vitro stem node and pseudo-bulblet culture. *Plant Cell Rep.* **1998**, *17*, 913–916. [CrossRef]
2. Zhang, Y.; Yong, Y.B.; Wang, Q.; Lu, Y.M. Physiological and Molecular Changes during Lily Underground Stem Axillary Bulbils Formation. *Russ. J. Plant Physiol.* **2018**, *65*, 372–383. [CrossRef]
3. Yang, P.; Xu, L.; Xu, H.; Tang, Y.; He, G.; Cao, Y.; Feng, Y.; Yuan, S.; Ming, J. Histological and Transcriptomic Analysis during Bulbil Formation in *Lilium lancifolium*. *Front. Plant Sci.* **2017**, *8*, 1508. [CrossRef] [PubMed]
4. Moreno-Pachón, N. Mechanisms of Vegetative Propagation in Bulbs: A Molecular Approach. Ph.D. Dissertation, Wageningen University, Wageningen, The Netherlands, 2017.
5. Chatfield, S.P.; Raizada, M.N. Ethylene and shoot regeneration: Hookless1 modulates de novo shoot organogenesis in *Arabidopsis thaliana*. *Plant Cell Rep.* **2008**, *27*, 655–666. [CrossRef] [PubMed]
6. Motte, H.; Vereecke, D.; Geelen, D.; Werbrouck, S. The molecular path to in vitro shoot regeneration. *Biotechnol. Adv.* **2014**, *32*, 107–121. [CrossRef]
7. Sang, Y.L.; Cheng, Z.J.; Zhang, X.S. Plant stem cells and de novo organogenesis. *New Phytol.* **2018**, *218*, 1334–1339. [CrossRef]
8. Greb, T.; Clarenz, O.; Schafer, E.; Muller, D.; Herrero, R.; Schmitz, G.; Theres, K. Molecular analysis of the LATERAL SUPPRESSOR gene in *Arabidopsis* reveals a conserved control mechanism for axillary meristem formation. *Genes. Dev.* **2003**, *17*, 1175–1187. [CrossRef]
9. Zhang, Y.; Zeng, Z.; Yong, Y.B.; Lyu, Y.M. Hormonal Regulatory Patterns of LaKNOXs and LaBEL1 Transcription Factors Reveal Their Potential Role in Stem Bulblet Formation in LA Hybrid Lily. *Int. J. Mol. Sci.* **2021**, *22*, 13502. [CrossRef]
10. He, G.; Yang, P.; Tang, Y.; Cao, Y.; Qi, X.; Xu, L.; Ming, J. Mechanism of exogenous cytokinins inducing bulbil formation in *Lilium lancifolium* in vitro. *Plant Cell Rep.* **2020**, *39*, 861–872. [CrossRef]
11. Rosspopoff, O.; Chelysheva, L.; Saffar, J.; Lecorgne, L.; Gey, D.; Caillieux, E.; Colot, V.; Roudier, F.; Hilson, P.; Berthomé, R.; et al. Direct conversion of root primordium into shoot meristem relies on timing of stem cell niche development. *Development* **2017**, *144*, 1187–1200. [CrossRef]
12. de Reuille, P.B.; Bohn-Courseau, I.; Ljung, K.; Morin, H.; Carraro, N.; Godin, C.; Traas, J. Computer simulations reveal properties of the cell-cell signaling network at the shoot apex in *Arabidopsis*. *Proc. Natl. Acad. Sci. USA* **2006**, *103*, 1627–1632. [CrossRef] [PubMed]
13. Heisler, M.G.; Ohno, C.; Das, P.; Sieber, P.; Reddy, G.V.; Long, J.A.; Meyerowitz, E.M. Patterns of auxin transport and gene expression during primordium development revealed by live imaging of the *Arabidopsis* inflorescence meristem. *Curr. Biol.* **2005**, *15*, 1899–1911. [CrossRef] [PubMed]
14. Tang, Y. *The Mechanism of Auxin Involving in Bulbil Formation of Lilium lancifolium*; Chinese Academy of Agricultural Sciences: Beijing, China, 2019.
15. Li, J.; Sun, M.; Li, H.; Ling, Z.; Wang, D.; Zhang, J.; Shi, L. Full-length transcriptome-referenced analysis reveals crucial roles of hormone and wounding during induction of aerial bulbils in lily. *BMC Plant Biol.* **2022**, *22*, 415. [CrossRef]
16. Gao, C.; Zhang, L.; Xu, Y.; Liu, Y.; Xiao, X.; Cui, L.; Xia, Y.; Wu, Y.; Ren, Z. Full-length transcriptome analysis revealed that 2,4-dichlorophenoxyacetic acid promoted in vitro bulblet initiation in lily by affecting carbohydrate metabolism and auxin signaling. *Front. Plant Sci.* **2023**, *14*, 1236315. [CrossRef]
17. Gao, M.; Zhang, S.; Luo, C.; He, X.; Wei, S.; Jiang, W.; He, F.; Lin, Z.; Yan, M.; Dong, W. Transcriptome analysis of starch and sucrose metabolism across bulb development in *Sagittaria sagittifolia*. *Gene* **2018**, *649*, 99–112. [CrossRef]
18. Li, X.Y.; Wang, C.X.; Cheng, J.Y.; Zhang, J.; da Silva, J.A.T.; Liu, X.Y.; Duan, X.; Li, T.L.; Sun, H.M. Transcriptome analysis of carbohydrate metabolism during bulblet formation and development in *Lilium davidii* var. *unicolor*. *BMC Plant Biol.* **2014**, *14*, 358. [CrossRef]
19. Wu, Y.; Ren, Z.; Gao, C.; Sun, M.; Li, S.; Min, R.; Wu, J.; Li, D.; Wang, X.; Wei, Y.; et al. Change in Sucrose Cleavage Pattern and Rapid Starch Accumulation Govern Lily Shoot-to-Bulblet Transition In Vitro. *Front. Plant Sci.* **2020**, *11*, 564713. [CrossRef]
20. Ren, Z.M.; Zhang, D.; Jiao, C.; Li, D.Q.; Wu, Y.; Wang, X.Y.; Gao, C.; Lin, Y.F.; Ruan, Y.L.; Xia, Y.P. Comparative transcriptome and metabolome analyses identified the mode of sucrose degradation as a metabolic marker for early vegetative propagation in bulbs of Lycoris. *Plant J.* **2022**, *112*, 115–134. [CrossRef]
21. Wang, Q.; Hasson, A.; Rossmann, S.; Theres, K. Divide et impera: Boundaries shape the plant body and initiate new meristems. *New Phytol.* **2016**, *209*, 485–498. [CrossRef] [PubMed]
22. Tian, C.; Zhang, X.; He, J.; Yu, H.; Wang, Y.; Shi, B.; Han, Y.; Wang, G.; Feng, X.; Zhang, C.; et al. An organ boundary-enriched gene regulatory network uncovers regulatory hierarchies underlying axillary meristem initiation. *Mol. Syst. Biol.* **2014**, *10*, 755. [CrossRef] [PubMed]
23. Yang, M.; Jiao, Y. Regulation of Axillary Meristem Initiation by Transcription Factors and Plant Hormones. *Front. Plant Sci.* **2016**, *7*, 183. [CrossRef] [PubMed]

24. Abraham-Juárez, M.J.; Martínez-Hernández, A.; Leyva-González, M.A.; Herrera-Estrella, L.; Simpson, J. Class I KNOX genes are associated with organogenesis during bulbil formation in Agave tequilana. *J. Exp. Bot.* **2010**, *61*, 4055–4067. [CrossRef] [PubMed]
25. Yang, P.; Xu, H.; Xu, L.; Tang, Y.; He, G.; Cao, Y.; Yuan, S.; Ren, J.; Ming, J. Cloning and expression analysis of LlAGO1 in *Lilium lancifolium*. *Acta Hortic. Sin.* **2018**, *45*, 784–794. [CrossRef]
26. He, G.; Cao, Y.; Wang, J.; Song, M.; Bi, M.; Tang, Y.; Xu, L.; Ming, J.; Yang, P. WUSCHEL-Related Homeobox Genes Cooperate with Cytokinin to Promote Bulbil Formation in *Lilium lancifolium*. *Plant Physiol.* **2022**, *190*, 387–402. [CrossRef] [PubMed]
27. Fang, S.; Yang, C.; Ali, M.M.; Lin, M.; Tian, S.; Zhang, L.; Chen, F.; Lin, Z. Transcriptome Analysis Reveals the Molecular Regularity Mechanism Underlying Stem Bulblet Formation in Oriental Lily 'Siberia'; Functional Characterization of the LoLOB18 Gene. *Int. J. Mol. Sci.* **2022**, *23*, 15246. [CrossRef] [PubMed]
28. Gordon, S.P.; Tseng, E.; Salamov, A.; Zhang, J.; Meng, X.; Zhao, Z.; Kang, D.; Underwood, J.; Grigoriev, I.V.; Figueroa, M.; et al. Widespread Polycistronic Transcripts in Fungi Revealed by Single-Molecule mRNA Sequencing. *PLoS ONE* **2015**, *10*, e0132628. [CrossRef] [PubMed]
29. Salmela, L.; Rivals, E. LoRDEC: Accurate and efficient long read error correction. *Bioinformatics* **2014**, *30*, 3506–3514. [CrossRef] [PubMed]
30. Li, R.; Yu, C.; Li, Y.; Lam, T.W.; Yiu, S.M.; Kristiansen, K.; Wang, J. SOAP2: An improved ultrafast tool for short read alignment. *Bioinformatics* **2009**, *25*, 1966–1967. [CrossRef]
31. Kim, D.; Langmead, B.; Salzberg, S.L. HISAT: A fast spliced aligner with low memory requirements. *Nat. Methods* **2015**, *12*, 357–360. [CrossRef]
32. Trapnell, C.; Williams, B.A.; Pertea, G.; Mortazavi, A.; Kwan, G.; van Baren, M.J.; Salzberg, S.L.; Wold, B.J.; Pachter, L. Transcript assembly and quantification by RNA-Seq reveals unannotated transcripts and isoform switching during cell differentiation. *Nat. Biotechnol.* **2010**, *28*, 511–515. [CrossRef]
33. Zhang, B.; Horvath, S. A general framework for weighted gene co-expression network analysis. *Stat. Appl. Genet. Mol. Biol.* **2005**, *4*, 17. [CrossRef]
34. Shannon, P.; Markiel, A.; Ozier, O.; Baliga, N.S.; Wang, J.T.; Ramage, D.; Amin, N.; Schwikowski, B.; Ideker, T. Cytoscape: A software environment for integrated models of biomolecular interaction networks. *Genome Res.* **2003**, *13*, 2498–2504. [CrossRef]
35. Livak, K.J.; Schmittgen, T.D. Analysis of relative gene expression data using real-time quantitative PCR and the 2(-Delta Delta C(T)) Method. *Methods* **2001**, *25*, 402–408. [CrossRef]
36. Liu, Q.; Wei, C.; Zhang, M.F.; Jia, G.X. Evaluation of putative reference genes for quantitative real-time PCR normalization in *Lilium regale* during development and under stress. *Peerj* **2016**, *4*, e1837. [CrossRef]
37. Qi, N.N.; Hou, X.M.; Wang, C.L.; Li, C.X.; Huang, D.J.; Li, Y.H.; Wang, N.; Liao, W.B.A. Methane-rich water induces bulblet formation of scale cuttings in *Lilium davidii* var. *unicolor* by regulating the signal transduction of phytohormones and their levels. *Physiol. Plant.* **2021**, *172*, 1919–1930. [CrossRef] [PubMed]
38. Hou, X.; Qi, N.; Wang, C.; Li, C.; Huang, D.; Li, Y.; Wang, N.; Liao, W. Hydrogen-rich water promotes the formation of bulblets in *Lilium davidii* var. *unicolor* through regulating sucrose and starch metabolism. *Planta* **2021**, *254*, 106. [CrossRef]
39. Schmitz, G.; Theres, K. Shoot and inflorescence branching. *Curr. Opin. Plant Biol.* **2005**, *8*, 506–511. [CrossRef] [PubMed]
40. Roman, H.; Girault, T.; Barbier, F.; Péron, T.; Brouard, N.; Pěnčík, A.; Novák, O.; Vian, A.; Sakr, S.; Lothier, J.; et al. Cytokinins Are Initial Targets of Light in the Control of Bud Outgrowth. *Plant Physiol.* **2016**, *172*, 489–509. [CrossRef] [PubMed]
41. Wang, F.; Sanz, A.; Brenner, M.L.; Smith, A. Sucrose Synthase, Starch Accumulation, and Tomato Fruit Sink Strength. *Plant Physiol.* **1993**, *101*, 321–327. [CrossRef] [PubMed]
42. Hao, C.; Yang, P.; Qu, Y.; Hao, Z.; Yin, X.; Tang, Y.; Bi, M.; Xu, L.; Hu, F.; Ming, J. Sucrose function on the bulbil formation of *Lilium lancifolium*. *Sci. Hortic.* **2024**, *323*, 112538. [CrossRef]
43. Xu, J.X.; Li, Q.Z.; Yang, L.Y.; Li, X.; Wang, Z.; Zhang, Y.C. Changes in carbohydrate metabolism and endogenous hormone regulation during bulblet initiation and development in *Lycoris radiata*. *BMC Plant Biol.* **2020**, *20*, 180. [CrossRef]
44. Wu, Y.; Sun, M.; Li, S.; Min, R.; Gao, C.; Lyu, Q.; Ren, Z.; Xia, Y. Molecular cloning, characterization and expression analysis of three key starch synthesis-related genes from the bulb of a rare lily germplasm, *Lilium brownii* var. *giganteum*. *J. Zhejiang Univ. Sci. B* **2021**, *22*, 476–491. [CrossRef]
45. Yadav, U.P.; Ivakov, A.; Feil, R.; Duan, G.Y.; Walther, D.; Giavalisco, P.; Piques, M.; Carillo, P.; Hubberten, H.M.; Stitt, M.; et al. The sucrose-trehalose 6-phosphate (Tre6P) nexus: Specificity and mechanisms of sucrose signalling by Tre6P. *J. Exp. Bot.* **2014**, *65*, 1051–1068. [CrossRef]
46. Wu, Z.G.; Jiang, W.; Tao, Z.M.; Pan, X.J.; Yu, W.H.; Huang, H.L. Morphological and stage-specific transcriptome analyses reveal distinct regulatory programs underlying yam (*Dioscorea alata* L.) bulbil growth. *J. Exp. Bot.* **2020**, *71*, 1899–1914. [CrossRef]
47. Mo, J.L.; Qu, Y.X.; He, G.R.; Yang, P.P.; Wang, L.H.; Zhang, L.N.; Wu, X.W.; Zhang, D.; Li, L.L.; Ming, J. Effect of exogenous 6-BA induced bulblets formation in aerial cultivation. *Sci. Hortic.* **2023**, *309*, 111644. [CrossRef]
48. Li, G.; Tan, M.; Cheng, F.; Liu, X.; Qi, S.; Chen, H.; Zhang, D.; Zhao, C.; Han, M.; Ma, J. Molecular role of cytokinin in bud activation and outgrowth in apple branching based on transcriptomic analysis. *Plant Mol. Biol.* **2018**, *98*, 261–274. [CrossRef]
49. Luo, L.; Zeng, J.; Wu, H.; Tian, Z.; Zhao, Z. A Molecular Framework for Auxin-Controlled Homeostasis of Shoot Stem Cells in *Arabidopsis*. *Mol. Plant* **2018**, *11*, 899–913. [CrossRef] [PubMed]

50. Abraham Juárez, M.J.; Hernández Cárdenas, R.; Santoyo Villa, J.N.; O'Connor, D.; Sluis, A.; Hake, S.; Ordaz-Ortiz, J.; Terry, L.; Simpson, J. Functionally different PIN proteins control auxin flux during bulbil development in *Agave tequilana*. *J. Exp. Bot.* **2015**, *66*, 3893–3905. [CrossRef] [PubMed]
51. Guo, C.; Li, J.; Li, M.; Xu, X.; Chen, Y.; Chu, J.; Yao, X. Regulation Mechanism of Exogenous Brassinolide on Bulbil Formation and Development in *Pinellia ternata*. *Front. Plant Sci.* **2021**, *12*, 809769. [CrossRef]
52. Ravnikar, M.; El, J.; Plaper, I.; Pacapan, A. Jasmonic acid stimulates shoot and bulb formation of garlic in vitro. *J. Plant Growth Regul.* **1993**, *12*, 73–77. [CrossRef]
53. Hamant, O.; Pautot, V. Plant development: A TALE story. *Comptes Rendus Biol.* **2010**, *333*, 371–381. [CrossRef]
54. Bürglin, T.R. Analysis of TALE superclass homeobox genes (MEIS, PBC, KNOX, Iroquois, TGIF) reveals a novel domain conserved between plants and animals. *Nucleic Acids Res.* **1997**, *25*, 4173–4180. [CrossRef]
55. Bellaoui, M.; Pidkowich, M.S.; Samach, A.; Kushalappa, K.; Kohalmi, S.E.; Modrusan, Z.; Crosby, W.L.; Haughn, G.W. The *Arabidopsis* BELL1 and KNOX TALE homeodomain proteins interact through a domain conserved between plants and animals. *Plant Cell* **2001**, *13*, 2455–2470. [CrossRef] [PubMed]
56. Robinson-Beers, K.; Pruitt, R.E.; Gasser, C.S. Ovule Development in Wild-Type *Arabidopsis* and Two Female-Sterile Mutants. *Plant Cell* **1992**, *4*, 1237–1249. [CrossRef] [PubMed]
57. Smith, H.M.; Hake, S. The interaction of two homeobox genes, BREVIPEDICELLUS and PENNYWISE, regulates internode patterning in the *Arabidopsis* inflorescence. *Plant Cell* **2003**, *15*, 1717–1727. [CrossRef]
58. Kanrar, S.; Onguka, O.; Smith, H.M. *Arabidopsis* inflorescence architecture requires the activities of KNOX-BELL homeodomain heterodimers. *Planta* **2006**, *224*, 1163–1173. [CrossRef] [PubMed]
59. Bhatt, A.M.; Etchells, J.P.; Canales, C.; Lagodienko, A.; Dickinson, H. VAAMANA—A BEL1-like homeodomain protein, interacts with KNOX proteins BP and STM and regulates inflorescence stem growth in *Arabidopsis*. *Gene* **2004**, *328*, 103–111. [CrossRef] [PubMed]
60. Liu, Y.; You, S.; Taylor-Teeples, M.; Li, W.L.; Schuetz, M.; Brady, S.M.; Douglas, C.J. BEL1-LIKE HOMEODOMAIN6 and KNOTTED *ARABIDOPSIS THALIANA7* interact and regulate secondary cell wall formation via repression of REVOLUTA. *Plant Cell* **2014**, *26*, 4843–4861. [CrossRef]
61. Licausi, F.; Ohme-Takagi, M.; Perata, P. APETALA2/Ethylene Responsive Factor (AP2/ERF) transcription factors: Mediators of stress responses and developmental programs. *New Phytol.* **2013**, *199*, 639–649. [CrossRef]
62. Mao, J.L.; Miao, Z.Q.; Wang, Z.; Yu, L.H.; Cai, X.T.; Xiang, C.B. *Arabidopsis* ERF1 Mediates Cross-Talk between Ethylene and Auxin Biosynthesis during Primary Root Elongation by Regulating ASA1 Expression. *PLoS Genet.* **2016**, *12*, e1005760. [CrossRef]
63. Cai, X.T.; Xu, P.; Zhao, P.X.; Liu, R.; Yu, L.H.; Xiang, C.B. *Arabidopsis* ERF109 mediates cross-talk between jasmonic acid and auxin biosynthesis during lateral root formation. *Nat. Commun.* **2014**, *5*, 5833. [CrossRef]
64. Heyman, J.; Cools, T.; Canher, B.; Shavialenka, S.; Traas, J.; Vercauteren, I.; Van den Daele, H.; Persiau, G.; De Jaeger, G.; Sugimoto, K.; et al. The heterodimeric transcription factor complex ERF115-PAT1 grants regeneration competence. *Nat. Plants* **2016**, *2*, 16165. [CrossRef]
65. Mehrnia, M.; Balazadeh, S.; Zanor, M.I.; Mueller-Roeber, B. EBE, an AP2/ERF transcription factor highly expressed in proliferating cells, affects shoot architecture in *Arabidopsis*. *Plant Physiol.* **2013**, *162*, 842–857. [CrossRef]
66. Zhang, C.; Fan, L.; Le, B.H.; Ye, P.; Mo, B.; Chen, X. Regulation of ARGONAUTE10 Expression Enables Temporal and Spatial Precision in Axillary Meristem Initiation in *Arabidopsis*. *Dev. Cell* **2020**, *55*, 603–616.e605. [CrossRef]
67. Zhu, H.L.; Hu, F.Q.; Wang, R.H.; Zhou, X.; Sze, S.H.; Liou, L.W.; Barefoot, A.; Dickman, M.; Zhang, X.R. *Arabidopsis* Argonaute10 Specifically Sequesters miR166/165 to Regulate Shoot Apical Meristem Development. *Cell* **2011**, *145*, 242–256. [CrossRef]
68. Ariel, F.D.; Manavella, P.A.; Dezar, C.A.; Chan, R.L. The true story of the HD-Zip family. *Trends Plant Sci.* **2007**, *12*, 419–426. [CrossRef] [PubMed]
69. Prigge, M.J.; Clark, S.E. Evolution of the class III HD-Zip gene family in land plants. *Evol. Dev.* **2006**, *8*, 350–361. [CrossRef] [PubMed]
70. Preston, J.C.; Hileman, L.C. Functional Evolution in the Plant SQUAMOSA-PROMOTER BINDING PROTEIN-LIKE (SPL) Gene Family. *Front. Plant Sci.* **2013**, *4*, 80. [CrossRef] [PubMed]
71. Schwarz, S.; Grande, A.V.; Bujdoso, N.; Saedler, H.; Huijser, P. The microRNA regulated SBP-box genes SPL9 and SPL15 control shoot maturation in *Arabidopsis*. *Plant Mol. Biol.* **2008**, *67*, 183–195. [CrossRef] [PubMed]
72. Zhao, M.; Liu, R.; Chen, Y.; Cui, J.; Ge, W.; Zhang, K. Molecular identification and functional verification of SPL9 and SPL15 of *Lilium*. *Mol. Genet. Genom.* **2022**, *297*, 63–74. [CrossRef] [PubMed]

horticulturae

Article

Effects of Substitute Substrate, Water, and Fertilizer Management on the Growth of Potted Chrysanthemums

Yufan Lu [1,†], Tongjun Zhou [1,†], Jinqing Guo [1,2], Jian Zhong [1], Dawei Li [1], Huajin Shi [1], Chang Liu [1], Ruizi Xiang [1] and Ming Sun [1,*]

1 State Key Laboratory of Efficient Production of Forest Resources, Beijing Key Laboratory of Ornamental Plants Germplasm Innovation and Molecular Breeding, National Engineering Research Center for Floriculture, Beijing Laboratory of Urban and Rural Ecological Environment, Key Laboratory of Genetics and Breeding in Forest Trees and Ornamental Plants of Ministry of Education, School of Landscape Architecture, Beijing Forestry University, Beijing 100083, China; lyf1997@bjfu.edu.cn (Y.L.); zhoutj000@bjfu.edu.cn (T.Z.); guojinqing163@163.com (J.G.); zhongjianbjfu@163.com (J.Z.); lidawei929@163.com (D.L.); shihuajinbjfu@163.com (H.S.); liuchangbjfu@163.com (C.L.); xiangruizibjfu@163.com (R.X.)

2 Institute of Leisure Agriculture, Shandong Academy of Agricultural Sciences, Jinan 250100, China

* Correspondence: sunmingbjfu@163.com

† These authors contributed equally to this work.

Abstract: The chrysanthemum is a perennial herbaceous flower with a long history of cultivation dating back over 3000 years. The potted chrysanthemum is a significant type and is widely used in landscaping. Expensive substrate costs, complicated management of water and fertilizer, and uneven product quality currently plague the potted chrysanthemum industry. This study systematically investigated the growth status of potted chrysanthemums under different substrates, water, and fertilizer ratios and established a simplified cultivation system for potted chrysanthemums. The substitute substrate experiment demonstrated that coir: moss peat: perlite: pine needle mulch = 2:4:2:2 is the most suitable substitute substrate. Research on fertilizer ratios found that chrysanthemums' best growth and flowering characteristics were achieved with nitrogen, phosphorus, and potassium concentrations of 336 mg/L, 93 mg/L, and 273 mg/L, respectively. A comprehensive, simplified cultivation system was established when utilizing T4 substitute substrate (2:4:2:2 ratios of coir, moss peat, perlite, and pine needle mulch), 40% water capacity, and F9 fertilizer (336 mg/L nitrogen, 93 mg/L phosphorus, and 273 mg/L potassium). This study comprehensively and systematically explored the cultivation and maintenance schemes in the production of potted chrysanthemums and built a light, simple, and efficient production technology system of potted chrysanthemums in the open field suitable for the climatic characteristics of northern China, which provides feasible technical specifications and a theoretical basis for the refinement and large-scale management of potted chrysanthemums. This experiment lays the foundations for cost reduction and efficiency in the potted chrysanthemum industry.

Keywords: potted chrysanthemum; substitute substrates; fertilizers; water

Citation: Lu, Y.; Zhou, T.; Guo, J.; Zhong, J.; Li, D.; Shi, H.; Liu, C.; Xiang, R.; Sun, M. Effects of Substitute Substrate, Water, and Fertilizer Management on the Growth of Potted Chrysanthemums. *Horticulturae* **2024**, *10*, 138. https://doi.org/10.3390/horticulturae10020138

Academic Editors: Margherita Beruto, Emmy Dhooghe and Bruce Dunn

Received: 20 December 2023
Revised: 21 January 2024
Accepted: 24 January 2024
Published: 31 January 2024

1. Introduction

Chrysanthemum is widely cultivated for ornamental purposes worldwide and has a long cultivation history [1]. The potted chrysanthemum is popular in horticulture mainly because of its vivid colors, long flowering duration, and attractive rounded form [2]. Traditional cultivation technologies of potted chrysanthemums lack optimal dosages and ratios for potted chrysanthemums' unique needs [3]. They also lack precision in resource management, leading to overuse or underutilization of inputs such as water, peat, and fertilizers. These problems seriously restrict the sustainability of potted chrysanthemum production [4,5].

Peat is the most commonly utilized substrate component in horticulture out of all the organic substrates [6]. It features a unique blend of high air space and drainage qualities, low nutrient content, low pH, high water holding capacity, and excellent chemical, biological, and physical attributes [7,8]. However, drainage of peatlands for extraction is one of the significant causes of CO_2 and CH_4 emissions [9,10]. Harvesting moss peat has led to numerous environmental concerns. The large-scale drainage of peatlands has resulted in annual emissions exceeding 2 Gt CO_2-eq, with CO_2 emissions alone estimated at 1.3 Gt annually. This accounts for approximately 5.6% of the global anthropogenic CO_2 emissions [11]. On the other hand, peat is a non-renewable resource, and the ecology needs to improve its use of peat bogs. Therefore, there is an urgent need to find other organic substrates to entirely or partially replace peat. Some organic resources are already being utilized commercially as alternatives to peat, including compost, coir, bark, and wood fiber [12,13]. Research is needed to determine optimal organic waste use and mixing ratios [14,15].

Fertilizer research plays a crucial role in streamlining agricultural production for crops. Crop growth relies on essential elements nitrogen, phosphorus, and potassium [16]. Understanding the optimal nutrient requirements and application methods enhances crop yields and promotes resource efficiency. By tailoring fertilization practices to specific crop needs, researchers contribute to simplifying production processes. This precision in nutrient management boosts agricultural productivity and minimizes environmental impact by reducing excess nutrient runoff. Moreover, precision fertilization is paramount in realizing substantial cost savings in agricultural production. Skillful NPK fertilizer application can foster plant development, increase dry matter, leaf photosynthetic efficiency, and root strength [17–19]. Nutrient ratio customization affects traits like germination, flower pedicel length, and flower amount [20,21]. However, excessive fertilization will lead to soil compaction and low soil nutrient use efficiency, which is not conducive to the sustainable and healthy development of crop production, and will also cause profound nitrate accumulation in soil and crops, which is not conducive to human health [22,23]. Chrysanthemum production and flower cultivations can benefit from precise nitrogen, phosphorus, and potassium fertilization [24].

Water is vital in crop production, serving as a critical resource that directly influences plant growth, development, and overall yield. However, water scarcity and extreme weather events are likely to intensify, which may lead to lower yields and increased yield volatility [25]. Agriculture is recognized as a substantial consumer of water resources, utilizing 70% of the world's freshwater to irrigate 25% of its croplands [26]. Therefore, the significance of precision irrigation lies in its capacity to address water scarcity challenges, particularly in regions prone to drought or facing limited water resources. Higher crop yields with lower water inputs could be achieved by maximizing the efficiency of water delivery. Moreover, precision irrigation contributes to environmental sustainability by minimizing the potential negative impacts of excess water usage, such as soil erosion and nutrient leaching [27–29].

To ensure cost-effectiveness and bloom quality, careful studies about simplified cultivation technologies of potted chrysanthemum are required for large-scale chrysanthemum production [30]. This study systematically investigated the growth status of potted chrysanthemums under different substrates, water, and fertilizer ratios and laid essential theoretical and practical foundations for improving the future quality and efficiency of potted chrysanthemum production.

2. Materials and Methods

2.1. Experimental Site and Materials

C. morifolium 'Hanluqiushi' was provided by Beijing Florascape Co., Ltd. The experiment was conducted from July to October 2021 at the Beijing Florascape Co. in Shunyi District, Beijing, China (40°21′ N, 116°51′ E). The average local temperature from July to October is 24.5 °C, with an average of 13.6 h of sunshine. The seedlings were potted on

July 1st in plastic pots (diameter = 13.5 cm, height = 15.5 cm). A controlled-release fertilizer with a nitrogen, phosphorus, and potassium ratio of 14:14:14 was mixed in the substrate at a dosage of 3 kg/m^3 with one seedling per pot. Xinyang Guotong Co., Ltd. (Xinyang, China) provided the coir, and Beijing Florascape Co. supplied the moss peat, pine needle mulch, and perlite. Urea (containing N: 46%) was used as the nitrogen fertilizer, calcium superphosphate (containing P_2O_5: 16%) was used as the phosphate fertilizer, and potassium sulfate (containing K_2O: 52%) was used as the potassium fertilizer. Beijing Florascape Co., Ltd. supplied the recovered substrate from the 2021 and 2022 potted chrysanthemum substrate that was recycled through open-air composting and primary crushing.

2.2. Experiments on Substitute Substrates

The fermented peanut shell, coir, bark, fungal residue, and rice husk were used to gradually replace the peat, which was used in the traditional potted substrate (peat: perlite: pine needle soil = 6:2:2), and 15 composite substrates were formed (T1–T15), peat: perlite: pine needle soil = 6:2:2 was used as the control group, and T16 and T17 were the recovered substrate group. The specific substitute substrate ratio is shown in Table 1. A total of 18 experimental groups were set up, including one control group (CK) and 17 treatment groups (T1–T17), each repeated ten times. Each treatment group was treated with ten pots, one seedling per pot.

Table 1. Cultivation experiment plan with different substrates.

Treatment	Component of Substrate (*v*/*v*)	The Proportion of Moss Peat (*v*/*v*%)
CK	Moss peat: Perlite: Pine Needle Soil = 6:2:2	60
T1	Peanut Shells: Moss peat: Perlite: Pine Needle Soil = 2:4:2:2	40
T2	Peanut Shells: Moss peat: Perlite: Pine Needle Soil = 4:2:2:2	20
T3	Peanut Shells: Perlite: Pine Needle Soil = 6:2:2	0
T4	Coir: Moss peat: Perlite: Pine Needle Soil = 2:4:2:2	40
T5	Coir: Moss peat: Perlite: Pine Needle Soil = 4:2:2:2	20
T6	Coir: Perlite: Pine Needle Soil = 6:2:2	0
T7	Bark: Moss peat: Perlite: Pine Needle Soil = 2:4:2:2	40
T8	Bark: Moss peat: Perlite: Pine Needle Soil = 4:2:2:2	20
T9	Bark: Perlite: Pine Needle Soil = 6:2:2	0
T10	Rice Husk: Moss peat: Perlite: Pine Needle Soil = 2:4:2:2	40
T11	Rice Husk: Moss peat: Perlite: Pine Needle Soil = 4:2:2:2	20
T12	Rice Husk: Perlite: Pine Needle Soil = 6:2:2	0
T13	Fungal Residue: Moss peat: Perlite: Pine Needle Soil = 2:4:2:2	40
T14	Fungal Residue: Moss peat: Perlite: Pine Needle Soil = 4:2:2:2	20
T15	Fungal Residue: Perlite: Pine Needle Soil = 6:2:2	0
T16	Recovered substrate (heap rot for one year)	100
T17	Recovered substrate (heap rot for two years)	100

Watering during cultivation was conducted before 10:00 every day according to weather conditions. Other cultivation management measures were treated the same except for the test conditions.

After potting, plants were watered once a day, and water-soluble fertilizer was applied once a week. We used LARRY (Wuhan Greencare Fertilizers Co., Ltd., Wuhan, China), a water-soluble fertilizer with a ratio of nitrogen, phosphorus, and potassium 20:10:20 in the growth period.

2.3. Experiments on Nitrogen–Phosphorus–Potassium Rationing

The fertilizer program including a control group (Clear water treatment) and LARRY fertilizer (water-soluble fertilizer commonly used in floriculture production, N: P: K = 20:10:20, Wuhan Greencare Fertilizers Co., Ltd., Wuhan, China) is shown in Table 2. It was applied using a three-factor, three-level, completely randomized group orthogonal experimental design. A total of 100 mL of the fertilizer solution was applied to each flower container once the fertilizer had been equally distributed and dissolved in the water. The concentration of the fertilizer was ascertained and converted into the appropriate mass. During fertilization, fertilizer was dissolved in water and placed evenly in pots, and 100 mL of fertilizer solution was added to each pot each time. Five pots, one seedling per pot, were treated by each treatment group.

Table 2. Fertilization plan.

Treatment	N ($mmol \cdot L^{-1}$)	P ($mmol \cdot L^{-1}$)	K ($mmol \cdot L^{-1}$)
LR *	3	1.5	3
CK *	0	0	0
F1	N1 (8)	P1 (1)	K1 (3.5)
F2	N1 (8)	P2 (2)	K2 (7.0)
F3	N1 (8)	P3 (3)	K3 (10.5)
F4	N2 (16)	P1 (1)	K2 (7.0)
F5	N2 (16)	P2 (2)	K3 (10.5)
F6	N2 (16)	P3 (3)	K1 (3.5)
F7	N3 (24)	P1 (1)	K3 (10.5)
F8	N3 (24)	P2 (2)	K1 (3.5)
F9	N3 (24)	P3 (3)	K2 (7.0)

* Note: LR: treated with LARRY fertilizer, a water-soluble fertilizer commonly used in floriculture production, N:P:K = 20:10:20; CK: control group treatment.

2.4. *Experiments on Substrate and Water-Fertilizer Rationing*

The N3P3K2 (24 mmol/L N, 3.0 mmol/L P, and 7.0 mmol/L K, which was tested as the best NPK ratio through the experiment above) was chosen as the reference for this experiment. Three gradients of substrate treatments were referenced from our previous study, including the poor substrate combination (coir: perlite: pine needle mulch = 6:2:2), the best substrate combination (coir: moss peat: perlite: pine needle mulch = 2:4:2:2), and the substrate combination typically used in production. The percentage of saturated water capacity was used to express the three moisture treatments (W1, W2, and W3), and the weighing method was used to manage the moisture levels. Three concentration gradients of N3P3K2 (0.5×, 1×, and 1.5×) were established for fertilizer. The detailed experimental plan is shown in Table 3. The concentration of the fertilizer was ascertained and converted into the appropriate mass. During fertilization, fertilizer was dissolved in water and placed evenly in pots, and 100 mL of fertilizer solution was added to each pot each time. Three pots, one seedling per pot, were treated by each treatment group.

Table 3. Substrate and water fertilizer combination test plan.

Treatment	Water Capacity	Fertilizer Concentration	Substrate (*v*/*v*)
X1	40%	0.5	Coir: Perlite: Pine Needle Soil = 6:2:2
X2	40%	1.0	Moss peat: Perlite: Pine Needle Soil = 6:2:2
X3	40%	1.5	Coir: Moss peat: Perlite: Pine Needle Soil = 2:4:2:2
X4	60%	0.5	Moss peat: Perlite: Pine Needle Soil =6:2:2
X5	60%	1.0	Coir: Moss peat: Perlite: Pine Needle Soil = 2:4:2:2
X6	60%	1.5	Coir: Perlite: Pine Needle Soil = 6:2:2
X7	80%	0.5	Coir: Moss peat: Perlite: Pine Needle Soil = 2:4:2:2
X8	80%	1.0	Coir: Perlite: Pine Needle Soil = 6:2:2
X9	80%	1.5	Moss peat: Perlite: Pine Needle Soil = 6:2:2

2.5. *Measurement of Growth and Physiological Indices*

After 80 days of cultivation, growth parameters including height, crown diameter, stem diameter, flower numbers, flower diameter, chlorophyll content, photosynthetic index, shoot dry weight, shoot fresh weight, total N, total P, and total K were measured. The detailed methods were adopted from a previous study [31–33].

2.5.1. Determination of Physical Properties of Substrates

The extracts were obtained by saturation extraction, pH was determined by a pH acidity meter (Potentiometric method), and EC values were determined by a DDS-307 conductivity meter.

A ring knife with a volume of 200 mL was used; the ring knife was weighed to W_0 mass, filled with a naturally dried mixed matrix, weighed as W_1, and soaked in water for 24 h, weighed as W_2. Soil bulk density and total porosity are calculated according to the following formula:

$$Bulk\ density = (W_1 - W_0) / 200$$

$$Total\ porosity = (W_1 - W_0) / 200 \times 100\%$$

2.5.2. Plant Height and Crown Width

A ruler was used to measure the height from the base of the chrysanthemum rhizome to the highest branch (accurate to 0.01 cm).

2.5.3. Photosynthetic Index

Measurements were made with a portable plant photosynthetic instrument (LI-6400, LI-COR Biosciences, Beijing, China). Three plants were selected for each treatment, three mature leaves in the middle and periphery of each plant were selected for measurement, and the average value was calculated. Assay times were 9–11 a.m. The net photosynthetic rate (Pn) (accurate to 0.001 μmol $CO_2/(m^2{\cdot}s)$), transpiration rate (Tr) (accurate to 0.001 mmol $H_2O/(m^2{\cdot}s)$), intercellular CO_2 concentration (Ci) (accurate to 0.001 μmol/mol), and stomatal conductance (Gs) (accurate to 0.001 mol $H_2O/(m^2{\cdot}s)$) were measured.

2.5.4. Flower Diameter

When the potted chrysanthemum was in full bloom, the largest flower diameter was measured using vernier calipers; 3 plants were selected for each treatment, five flowers were taken from each plant to measure, and the average value was calculated (accurate to 0.01 mm).

2.5.5. Determination of Fresh and Dry Weight of Shoots

The chrysanthemums were cut from the rhizomes, and the above-ground parts were kept and weighed with an electronic balance to determine the fresh weight (accurate to 0.01 g). Then, they were put in an oven at 120 °C for 30 min, dried at 80 °C to constant weight, and weighed with an electronic balance to determine the dry weight (accurate to 0.01 g).

2.5.6. Analysis of Plant Nutrients

Plant samples were crushed into a powder in a grinder for biomass analysis, and the powder was subsequently filtered through a sieve with a mesh size of 100. H_2SO_4–H_2O_2 was used to filter the abrasive material (0.2 g). The total nitrogen concentration was ascertained using the Kjeldahl nitrogen determination method and the KDY-9820 automatic nitrogen ion meter. The molybdenum phosphate yellow technique and the UV-2550 UV-Vis spectrophotometer were used to calculate the total phosphorus concentrations. SpectraAA220 atomic absorption spectrometry was used to measure total K concentrations.

2.6. Statistical Analysis

One-way analysis of variance (ANOVA) and principal component analysis (PCA) were performed using SPSS 21.0, containing all the indices, and multiple comparisons were performed using Duncan's test for significant differences ($p < 0.05$). The results were plotted using GraphPad Prism 8.

3. Results

3.1. Effects of Substitute Substrates on the Growth and Physiological Indices of Potted Chrysanthemum

3.1.1. Analysis of Physicochemical Properties of Different Substrates

Table 4 displays the physicochemical characteristics of several substitute substrates. The recovery matrix T17 group had the highest bulk density, while the coir matrix T6 group had the lowest. In contrast to the CK group, the bulk density of the substrate entirely substituted with bark (T9), rice husk (T12), and fungal residue (T15) increased, whereas the bulk density of the substrate substituted with peanut shell (T3) and coir (T6) decreased. Regarding total porosity, coir substrate T6 had the most extensive total porosity, and recovered matrix T16 had the most minor total porosity. The substitute substrate group of peanut shell (T1, T2, T3) and coir (T4, T5, T6) increased compared with the CK group. In contrast, the substitute substrate group of bark (T7, T8, T9), rice husk (T10, T11, T12), and fungal residue (T13, T14, T15) decreased compared with the control group. Regarding the EC value, the EC value of the T12 group was the highest, while that of the T17 group was the most minor. The recovery substrates (T16, T17) decreased compared with the control group, and the rest increased. In terms of pH value, the pH value of T10 was the smallest while that of T17 was the largest, the substitute substrate group of rice husk (T10, T11, T12) was acidic, and the other groups were neutral or weakly alkaline. The T17 treatment group was alkaline, and the other groups were within the appropriate range.

Table 4. Physicochemical properties of different substrates.

Treatment	Bulk Density (g/cm^3)	Total Porosity(%)	EC(mS/cm)	pH	Total N (g/100 g)	Total P (g/100 g)	Total K (g/kg)
CK	0.25	70.3	0.45	7.11	1.84	0.19	11.28
T1	0.24	73.2	0.74	7.68	1.79	0.24	11.88
T2	0.22	77.6	0.94	7.76	1.53	0.20	7.95
T3	0.18	81.4	1.04	7.85	1.52	0.50	20.18
T4	0.21	77.4	0.51	7.21	1.39	0.14	13.70
T5	0.18	79.1	0.65	7.35	1.33	0.21	14.83
T6	0.15	82.6	1.21	7.68	0.72	0.17	20.46
T7	0.26	68.9	0.52	7.27	1.26	0.15	11.88
T8	0.29	65.3	0.71	7.33	0.96	0.13	10.79
T9	0.43	60.7	0.97	7.57	0.76	0.16	8.02
T10	0.27	53.7	0.76	6.73	1.69	0.48	16.82
T11	0.35	64.9	1.21	6.81	1.64	0.46	17.71
T12	0.52	60.3	1.42	6.85	1.39	0.56	16.55
T13	0.38	61.3	0.93	7.39	1.31	0.33	14.53
T14	0.43	66.7	1.14	7.42	1.12	0.28	13.28
T15	0.63	70.4	1.52	7.53	1.00	0.48	16.39
T16	0.72	40.1	0.37	7.71	/	/	/
T17	0.78	48.3	0.21	8.12	/	/	/

The total nitrogen content was the highest in CK, followed by the T3 substitute substrate, and the lowest in the T6 substitute substrate. The entire nitrogen content of the substitute substrate group of peanut shells (T1, T2, T3) and rice husks (T11, T12, T13) was higher than the rest. The total nitrogen content decreased with the decrease in peat in all substitute substrate groups. The entire phosphorus content was the highest in the T12 substitute substrate and the lowest in the T8 substitute substrate. The total potassium content was the highest in the T6 substitute substrate and the weakest in T3. The total phosphorus content increased as the peat decreased in all substitute substrate groups. The organic carbon content was the highest in CK and the lowest in T15. As the moss peat content decreased, the organic carbon content in all substrates decreased.

3.1.2. Effects of Substitute Substrates on Growth Indices of *C. morifolium* 'Hanluqiushi'

The plants' growth indices of 'Hanluqiushi' under various circumstances are shown in Table 5. Gradually rising fertility was accompanied by progressive plant height and crown diameter increases.

Table 5. Effects of substitute substrates on the height and crown diameter of potted chrysanthemum.

No.	Height (cm)				Crown Diameter (cm)			
	20 d	40 d	60 d	80 d	20 d	40 d	60 d	80 d
CK	4.8 ± 1.0 bcdef *	9.4 ± 0.8 abc	12.9 ± 1.1 abc	16.5 ± 1.0 cd	7.0 ± 1.1 ab	17.9 ± 1.9 a	24.7 ± 1.9 a	26.5 ± 1.3 abc
T1	5.6 ± 0.5 abc	9.6 ± 1.5 abc	13.0 ± 0.8 abc	15.9 ± 0.8 cde	5.7 ± 1.0 cdefghi	16.3 ± 1.1 abc	22.2 ± 0.9 bc	27.5 ± 1.5 a
T2	5.2 ± 0.9 abcde	9.7 ± 0.5 abc	11.7 ± 1.1 cde	14.3 ± 0.3 g	4.7 ± 0.5 ij	14.9 ± 1.1 cde	21.0 ± 0.5 cd	22.8 ± 0.3 d
T3	4.4 ± 0.3 ef	8.5 ± 0.1 cd	10.8 ± 1.2 ef	13.3 ± 0.3 h	5.3 ± 0.9 abcde	10.9 ± 1.3 gh	17.0 ± 0.6 f	17.8 ± 0.8 fg
T4	5.7 ± 0.4 ab	10.5 ± 1.3 a	14.2 ± 1.1 a	18.9 ± 0.7 a	6.5 ± 0.5 abcdef	14.8 ± 1.8 cde	20.9 ± 1.1 cd	27.7 ± 0.9 a
T5	5.7 ± 0.9 abc	9.7 ± 1.2 abc	12.1 ± 0.7 bcde	15.3 ± 1.1 efg	6.6 ± 0.4 abcde	15.6 ± 1.0 bcd	23.0 ± 0.5 ab	25.1 ± 1.1 c
T6	4.7 ± 0.4 cdef	8.3 ± 0.4 cd	9.7 ± 0.8 f	11.3 ± 1.4 j	5.1 ± 0.2 hij	12.5 ± 0.7 fg	15.2 ± 0.8 g	17.8 ± 0.8 fg
T7	4.9 ± 0.8 abcdef	8.9 ± 0.3 bcd	12.2 ± 1.2 bcde	15.3 ± 0.4 efg	6.3 ± 0.6 abcdefg	16.0 ± 1.3 abcd	19.8 ± 2.9 de	25.1 ± 0.8 c
T8	5.5 ± 0.6 abcd	10.5 ± 0.5 a	12.6 ± 0.3 bc	16.7 ± 0.3 bc	6.8 ± 0.9 abc	16.4 ± 1.1 abc	21.7 ± 1.9 bc	26.8 ± 0.8 ab
T9	5.9 ± 0.4 a	9.4 ± 1.0 abc	12.8 ± 1.7 abc	15.0 ± 0.4 efg	5.8 ± 0.3 cdefgh	17.3 ± 0.8 ab	20.8 ± 0.7 cd	22.5 ± 1.3 d
T10	5.4 ± 0.5 abcd	9.1 ± 0.6 bcd	12.8 ± 1.2 abc	17.5 ± 0.5 b	6.7 ± 0.4 abcd	17.7 ± 1.0 ab	20.6 ± 1.0 cd	26.0 ± 0.5 bc
T11	5.8 ± 1.0 ab	10.1 ± 0.9 ab	12.3 ± 0.3 bcd	16.3 ± 0.6 cd	7.4 ± 1.4 a	16.4 ± 1.2 abc	20.9 ± 0.6 cd	25.3 ± 0.6 c
T12	4.5 ± 0.2 def	7.9 ± 0.4 de	11.6 ± 0.7 cde	15.0 ± 0.3 efg	5.4 ± 0.5 fghi	12.9 ± 1.0 efg	17.3 ± 0.3 f	19.5 ± 0.9 e
T13	5.4 ± 0.5 abcd	9.0 ± 1.0 bcd	13.5 ± 0.6 ab	15.7 ± 0.4 def	6.2 ± 0.1 abcdefgh	15.9 ± 0.7 abcd	21.0 ± 1.2 cd	22.3 ± 1.3 d
T14	2.9 ± 0.2 g	7.8 ± 0.4 de	11.1 ± 1.4 def	14.8 ± 0.2 fg	4.1 ± 0.4 j	14.2 ± 2.0 cdef	18.4 ± 1.1 ef	22.2 ± 0.3 d
T15	4.1 ± 1.0 f	8.0 ± 0.4 de	10.9 ± 0.7 def	15.0 ± 0.2 efg	5.6 ± 0.7 defghi	14.0 ± 0.8 def	18.2 ± 0.7 ef	18.2 ± 1.1 ef
T16	4.2 ± 0.4 ef	6.9 ± 0.7 e	8.0 ± 0.5 g	12.7 ± 0.6 hi	6.0 ± 0.5 bcdefgh	10.1 ± 0.8 h	12.3 ± 0.6 h	19.0 ± 0.9 ef
T17	4.4 ± 0.1 def	5.0 ± 0.6 f	7.4 ± 0.1 g	12.0 ± 1.0 ig	5.5 ± 0.6 efghi	7.6 ± 1.0 i	10.0 ± 1.1 i	16.7 ± 1.5 g

* Note: Data are mean ± standard error ($n = 3$); different lowercase letters indicate significant differences ($p < 0.05$).

At 80 days of cultivation, the plant height of the potted chrysanthemum treated with traditional cultivation substrate CK was 16.47 cm, and the highest plant height was 18.87 cm in the T4 treatment group, which was significantly 114.6% higher than that of the CK group. The lowest was 8.87 cm in the T6 treatment group, significantly different from the CK group. In terms of crown diameter, the crown width of the CK group was 26.5 cm. The crown width of the T4 treatment group was the largest, 27.73 cm, 104.6% larger than that of the CK group. The T17 treatment group had the smallest crown width, 16.67 cm, significantly smaller than the CK group.

3.1.3. Effects of Substitute Substrates on Physiological Indices of *C. morifolium* 'Hanluqiushi'

As is shown in Figure 1a–d, different treatments affected the photosynthetic indexes of 'Hanluqiushi'. Regarding the net photosynthetic rate, the CK group had the highest photosynthetic rate, while the T17 treatment group had the lowest, significantly different from the CK group. Moreover, the stomatal conductance was the highest in the T2 treatment group and the smallest in the T10 treatment group. Regarding intercellular carbon dioxide concentration, the T17 treatment group was the highest and the T8 treatment group was the lowest, while the transpiration rate was the highest in the T10 treatment group and the lowest in the T6 treatment group.

As is shown in Figure 2, significant differences existed in the fresh and dry weight of shoots of chrysanthemum cultivated in different substrate treatment groups ($p < 0.05$). The fresh weight of the shoots in the CK treatment group was 241.9 g. The T8 treatment group had the most significant fresh weight, which was 284.48 g, an increase of 117.6% compared with the CK group, while the T17 group, which was the most minor (48.21 g) was 80.1% lower than that of the CK group. The fresh weight of the T8 and T4 treatment groups was significantly higher than that of the CK group. Regarding dry weight, the dry weight of the shoots of the CK treatment group was 52.7 g. Among the treatment groups, the dry weight of the T7 treatment group was the largest, reaching 54.68 g, which was 103.7% higher than that of the CK group, while the dry weight of the T17 treatment group was the smallest at 8.63 g, 83.6% lower than that of the CK group.

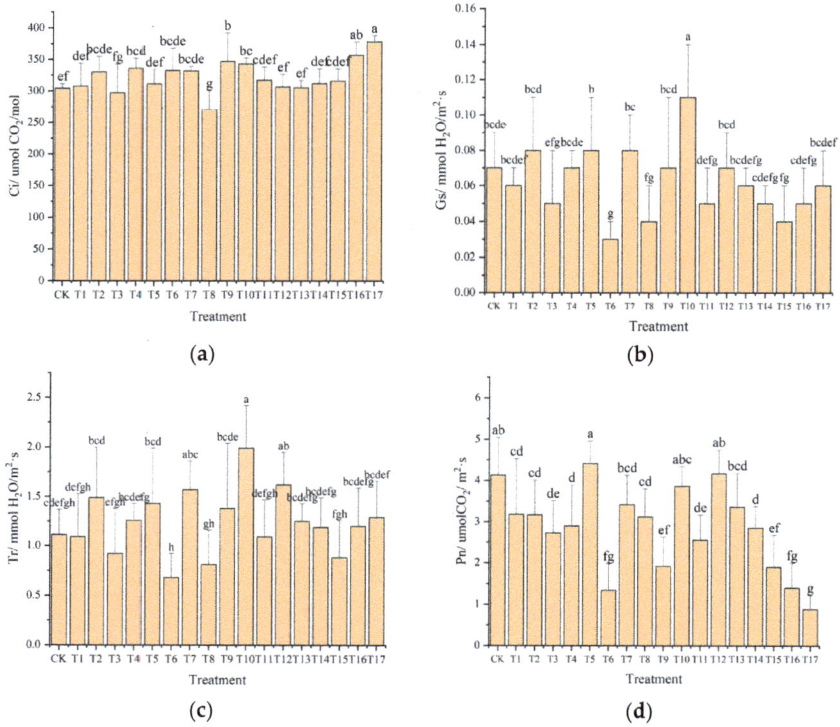

Figure 1. Effect of substitute substrates on the physiological indices of potted chrysanthemums: (**a**) Intercellular CO_2 concentration (vertical coordinate: Ci; mmol $H_2O/m^2{\cdot}s$); (**b**) Stomatal conductance (vertical coordinate: Gs; μmol $CO_2/m^2{\cdot}s$); (**c**) Transpiration rate (vertical coordinate: Tr; mmol $H_2O/m^2{\cdot}s$); (**d**) Net photosynthetic rate (vertical coordinate: Pn; μmol $CO_2/m^2{\cdot}s$). The horizontal coordinates represent the 18 treatment groups CK, T1-T17. Data are mean ± standard error (n = 3). Different lowercase letters indicate significant differences ($p < 0.05$), as determined by Duncan's multiple range test.

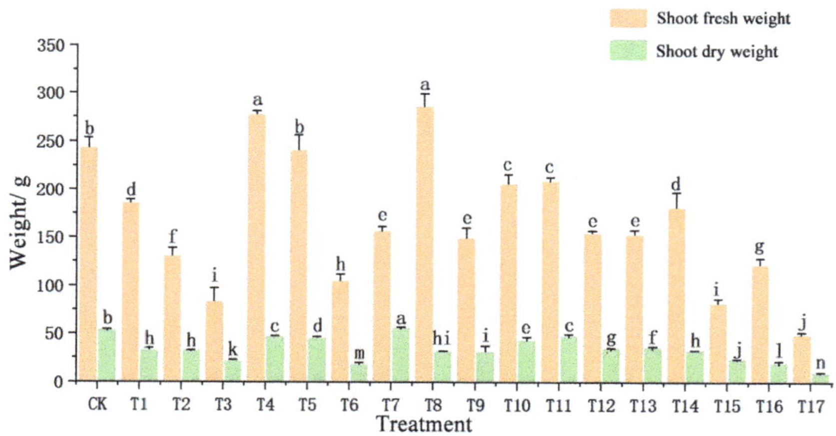

Figure 2. Effect of substitute substrates on the biomass of potted chrysanthemums. The vertical coordinate represents the weight (g). The horizontal coordinates represent the 18 treatment groups CK, T1–T17. Data are mean ± standard error (n = 3). Different lowercase letters indicate significant differences ($p < 0.05$), as determined by Duncan's multiple range test.

3.1.4. Effects of Substitute Substrates on Blooming Indices of *C. morifolium* 'Hanluqiushi'

As is shown in Figures 3 and 4, during the entire flowering period of 'Hanluqiushi', the chrysanthemum growth of the T6 and T17 treatment groups was significantly weaker than that of the CK group, and the T17 treatment group was the weakest. Significant differences existed in the 'Hanluqiushi' flower diameter cultivated with the different substrates ($p < 0.05$). The flower diameter of the CK group was 48.88 mm, while that of the T4 treatment group was the largest (52.37 mm), which was 7.1% higher than that of the control group. The flower diameter of the T16 treatment group was the smallest (36.38 mm), which was 25.6% smaller than that of the control group.

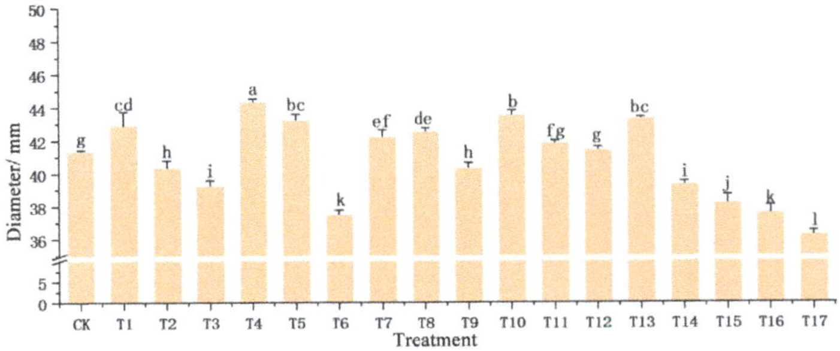

Figure 3. Effect of substitute substrates on the flower diameter of potted chrysanthemums. The vertical coordinate represents the flower diameter (mm). The horizontal coordinates represent the 18 treatment groups CK, T1–T17. Different lowercase letters indicate significant differences ($p < 0.05$), as determined by Duncan's multiple range test.

Figure 4. Effect of substitute substrates on the flower quality of potted chrysanthemums.

3.1.5. Comprehensive Evaluation of Substitute Substrates

According to the Figure 5, nine indicators for potted chrysanthemums with various nitrogen, phosphorus, and potassium ratios were evaluated using principal component analysis (PCA). Three significant components, accounting for 89.26% of the total variance, were retrieved. Principal component scores (F) and growth indicators were shown to be related. Table 6 presents the final scoring formulas for each principal component. Chrysanthemum growth was evaluated thoroughly using composite ratings (F). The score for treatment group T4 (coir: moss peat: perlite: pine needle soil = 2:4:2:2) was 1.27, followed by T5 (coir: moss peat: perlite: pine needle soil = 4:2:2:2) with 1.20, while the control group CK came in seventh overall with a score of 0.23. Using this analysis, the best substitute substrate can be chosen for improved chrysanthemum production, shown in Table 7. According to the integrated evaluation of the principal component analysis, the combined performance of the T4 treatment group was significantly better than that of the CK and the T4 treatment groups (coir: moss peat: perlite: pine needle soil = 2:4:2:2) was the best substitute substrate under the traditional cultivation substrate conditions.

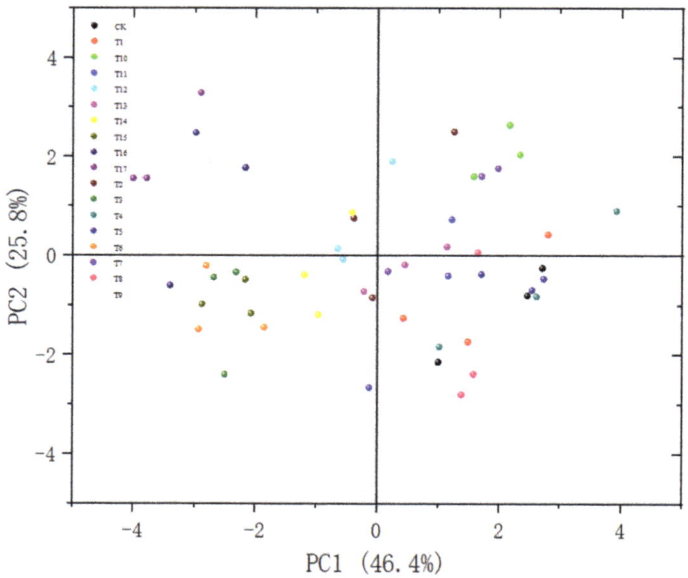

Figure 5. Principal component analysis for substitute substrates.

Table 6. Final score formula for the three principal components of substitute substrates.

Principal Component	Score Formula
1	$Y = 0.189X_1 + 0.185X_2 + 0.197X_3 + 0.197X_4 + 0.174X_5 + 0.004X_6 - 0.105X_7 + 0.028X_8 + 0.124X_9$
2	$Y = -0.124X_1 + 0.017X_2 + 0.020X_3 - 0.025X_4 + 0.073X_5 + 0.313X_6 + 0.290X_7 + 0.349X_8 + 0.222X_9$
3	$Y = -0.040X_1 + 0.489X_2 + 0.312X_3 + 0.326X_4 - 0.837X_5 + 0.611X_6 + 0.278X_7 - 0.526X_8 - 0.172X_9$

Table 7. Comprehensive evaluation of substitute substrates.

No.	F 1	F 2	F 3	F	Rank
CK	0.428	0.107	−1.004	0.228	7
T1	0.623	0.616	0.622	0.621	3
T2	0.290	1.355	−0.547	0.587	4
T3	−0.934	−0.348	−0.225	−0.694	16
T4	1.094	1.870	−0.164	1.269	1
T5	1.364	0.949	0.923	1.198	2
T6	−2.055	1.995	−0.416	−0.608	15
T7	0.532	−0.599	1.506	0.222	8
T8	0.422	0.078	0.227	0.295	6
T9	−0.319	−1.290	0.462	−0.589	14
T10	0.739	−1.091	1.178	0.163	9
T11	0.352	−0.401	−0.070	0.075	10
T12	0.203	−1.045	−2.621	−0.395	12
T13	0.684	0.050	−0.285	0.411	5
T14	0.271	−0.861	−0.392	−0.146	11
T15	−0.175	−0.937	−0.999	−0.481	13
T16	−1.479	−0.424	1.424	−0.940	17
T17	−2.038	−0.026	0.381	−1.214	18

3.2. *Effects of Nitrogen, Phosphorus, and Potassium Ratios on the Growth and Physiological Indices of Potted Chrysanthemum*

3.2.1. Effects of Nitrogen, Phosphorus, and Potassium Ratios on *C. morifolium* 'Hanluqiushi' Growth Indices

The plants' growth indices of 'Hanluqiushi' under various circumstances are shown in Table 8. Gradually rising fertility was accompanied by progressive increases in plant height, crown diameter, and stem diameter. The fertilizer level influenced the substantial difference in plant height across treatments ($p < 0.05$). At day 80, the plants in the LR, F3, and F6 treatment groups were the tallest, the LR and F9 treatment groups had the largest crown diameter, and the F9 treatment group had the thickest stems.

3.2.2. Effects of Nitrogen, Phosphorus, and Potassium Ratios on Physiological Indices of *C. morifolium* 'Hanluqiushi'

To assess the effects of different fertilizer treatments on potted chrysanthemums, we focused on comparing the biomass, chlorophyll content, and nutrient content of the whole plant at the 80 days of cultivation (Figures 6 and 7). Significant interactions between different ratio treatments and organs were observed, greatly influencing the nutrient content ($p < 0.05$). The shoot fresh weight of the aerial parts in the CK treatment group was 14.76 g. The F9 treatment group had the highest fresh weight of 52.29 g, 354.27% higher than the control group. The dry weight of the aerial parts of the CK treatment group was 3.38 g. The maximum dry weight of the F4 treatment group was 12.60 g, which was 372.78% higher than that of the control group, and the CK dry weight was the lowest in the control group. The F9 treatment group had the most incredible chlorophyll content (118.28% of CK).

Regarding the total nitrogen (TN) content and total phosphorus (TP) content, the TN content ranged between 2.99 to 44.37 g·100 g^{-1}, while the F9 treatment group had the highest TN content. Although variations in the TP content were negligible, fertilization significantly increased the TP content of all plants. The F6 treatment group exhibited the highest increase in TP content (222.48% of CK). For the total potassium (TK) content, all treatment groups showed a significant deviation from the CK. The LR and F9 treatment groups had the highest TK content (221.74% and 216.60% of CK, respectively).

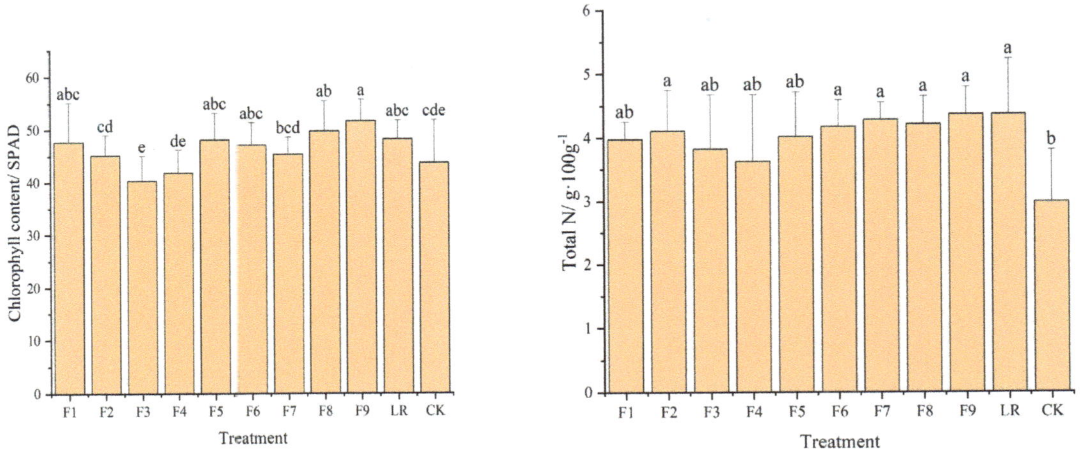

(a) (b)

Figure 6. *Cont.*

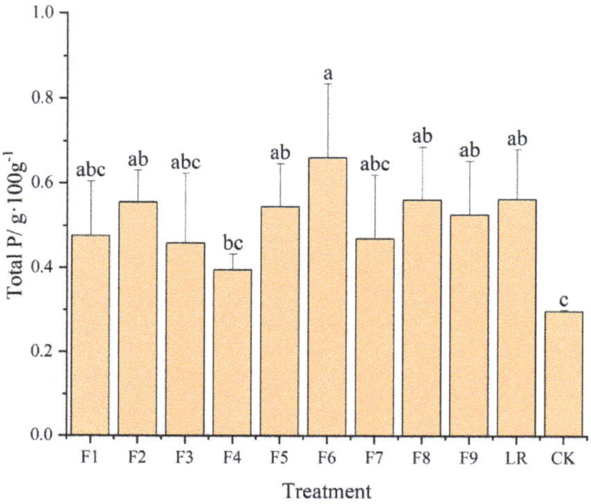

(**c**)

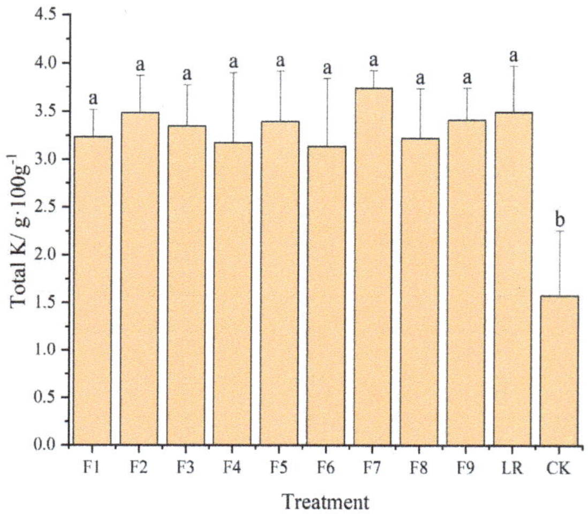

(**d**)

Figure 6. Effect of chlorophyll content, nitrogen, phosphorus, and potassium ratio on the total chlorophyll content and N/P/K of potted chrysanthemums: (**a**) The total chlorophyll content (SPAD); (**b**) Total N (g/100 g); (**c**) Total P (g/100 g); (**d**) Total K (g/100 g). The horizontal coordinates represent the 11 treatment groups CK, LR, and F1-F9. Data are mean ± standard error (n = 5). Different lower-case letters indicate significant differences ($p < 0.05$), as determined by Duncan's multiple range test.

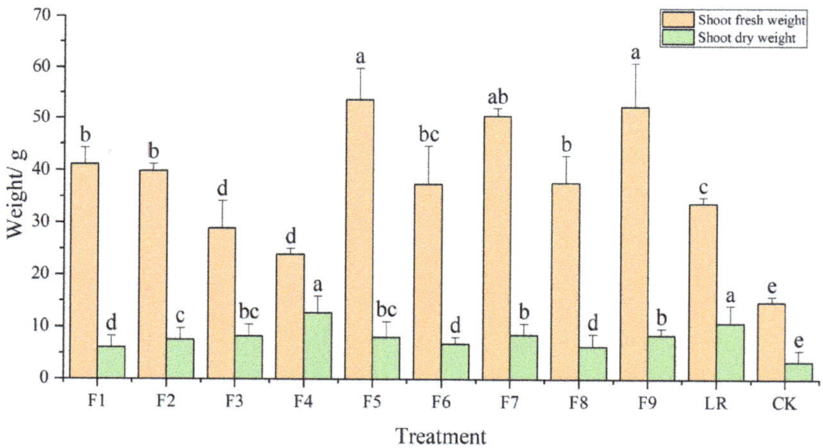

Figure 7. Effect of nitrogen, phosphorus, and potassium ratio on the biomass of potted chrysanthemums. The horizontal coordinates represent the 11 treatment groups (CK, LR, and F1–F9). Data are mean ± standard error (n = 5). Different lowercase letters indicate significant differences ($p < 0.05$), as determined by Duncan's multiple range test.

Table 8. Effects of nitrogen, phosphorus, and potassium ratios on the growth indices of potted chrysanthemum.

No.	Height (cm)				Crown Diameter (cm)				Stem Diameter (mm)			
	20 d	40 d	60 d	80 d	20 d	40 d	60 d	80 d	20 d	40 d	60 d	80 d
LR *	8.5 ± 0.2 a *	12.8 ± 0.3 a	14.3 ± 0.56 a	20.5 ± 1.6 a	13.7 ± 0.7 ab	16.0 ± 2.3 a	20.0 ± 0.9 a	22.8 ± 0.2 a	4.9 ± 0.2 a	5.1 ± 0.2 a	5.3 ± 0.8 a	5.9 ± 0.7 bcd
CK	8.5 ± 0.9 a	12.2 ± 0.4 a	12.4 ± 0.6 bc	14.0 ± 0.4 e	15.7 ± 0.6 a	15.3 ± 1.2 ab	16.4 ± 1.5 bc	15.4 ± 1.0 e	4.1 ± 0.1 b	4.3 ± 0.4 b	4.6 ± 0.5 ab	5.0 ± 0.7 d
F1	8.8 ± 0.7 a	11.8 ± 0.4 b	12.4 ± 0.6 b	14.5 ± 0.3 e	11.1 ± 1.0 c	14.0 ± 0.8 ab	15.8 ± 0.6 bc	17.1 ± 0.8 d	3.0 ± 0.3 c	3.4 ± 0.4 c	4.0 ± 0.5 bc	5.9 ± 1.1 bcd
F2	7.9 ± 0.6 b	12.5 ± 1.4 a	13.3 ± 0.5 ab	15.8 ± 0.6 bcd	10.5 ± 1.8 c	11.8 ± 1.5 b	15.8 ± 1.2 bc	19.5 ± 0.3 bc	3.6 ± 0.1 b	3.8 ± 0.3 c	4.2 ± 0.9 bc	4.9 ± 0.5 d
F3	8.3 ± 0.9 ab	13.7 ± 0.7 a	13.3 ± 0.4 ab	17.3 ± 0.5 b	11.9 ± 0.5 bc	14.3 ± 0.8 ab	16.4 ± 0.6 bc	20.0 ± 0.3 b	4.0 ± 0.1 b	4.3 ± 0.3 bc	4.0 ± 0.3 abc	6.6 ± 0.9 abcd
F4	8.0 ± 0.6 b	11.9 ± 0.4 ab	13.3 ± 0.6 ab	16.6 ± 0.2 bc	11.6 ± 0.8 bc	14.0 ± 0.7 ab	15.9 ± 0.4 bc	18.5 ± 0.3 cd	3.9 ± 0.1 b	4.3 ± 0.4 bc	4.6 ± 0.6 abc	6.6 ± 0.4 abcd
F5	8.6 ± 0.5 a	13.0 ± 0.5 a	12.9 ± 0.2 abc	15.6 ± 0.2 bcd	12.7 ± 1.0 b	16.1 ± 1.5 a	17.1 ± 1.1 bc	18.9 ± 0.2 c	3.0 ± 0.1 c	3.4 ± 0.4 c	3.7 ± 0.3 c	5.7 ± 0.4 bcd
F6	8.5 ± 1.1 a	12.5 ± 0.8 a	13.6 ± 0.7 ab	17.3 ± 0.4 b	11.8 ± 1.0 bc	14.1 ± 1.0 ab	17.0 ± 0.4 bc	19.6 ± 0.6 bc	3.9 ± 0.2 b	4.4 ± 0.3 b	4.5 ± 0.9 abc	5.4 ± 0.9 cd
F7	8.4 ± 1.1 a	12.5 ± 0.7 a	12.4 ± 0.7 bc	15.8 ± 0.5 c	11.5 ± 1.0 bc	13.6 ± 0.9 ab	16.5 ± 0.9 bc	18.3 ± 0.6 cd	4.0 ± 0.1 b	4.4 ± 0.4 b	4.7 ± 1.4 ab	6.0 ± 0.3 bcd
F8	8.6 ± 0.6 a	12.5 ± 0.4 a	12.3 ± 0.3 bc	14.6 ± 0.2 d	11.2 ± 0.4 c	13.6 ± 1.0 ab	14.5 ± 0.3 c	17.8 ± 0.3 d	3.8 ± 0.1 b	4.1 ± 0.4 bc	4.2 ± 0.8 bc	5.6 ± 0.5 cd
F9	8.0 ± 0.9 ab	13.2 ± 0.6 a	11.5 ± 0.2 c	16.8 ± 0.6 bc	11.9 ± 0.9 bc	17.1 ± 1.5 a	17.9 ± 1.4 ab	21.6 ± 0.6 a	3.7 ± 0.1 b	4.3 ± 0.4 b	4.2 ± 0.7 bc	7.1 ± 1.0 a

* Note: LR: treated with LARRY fertilizer, a water-soluble fertilizer commonly used in floriculture production, N: P: K = 20: 10: 20; CK: control group treatment. Data are mean ± standard error (n = 5); different lowercase letters indicate significant differences ($p < 0.05$).

3.2.3. Effects of Nitrogen, Phosphorus, and Potassium Ratios on Flower Qualities of *C. morifolium* 'Hanluqiushi'

As is shown in Figures 8 and 9, regarding the number of flowers, the X9 group had the most, 172.80, while the CK group had the fewest, 41.80. Concerning flower diameter, the X5 group had the biggest (55.16 mm), while the CK group had the smallest (43.94 mm).

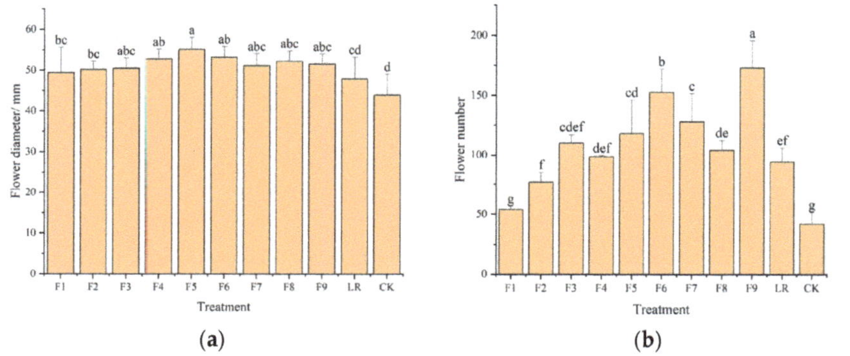

Figure 8. Effect of nitrogen, phosphorus, and potassium ratio on the flowering number and diameter of potted chrysanthemums: (**a**) Flower diameter (mm); (**b**) Flower number. The horizontal coordinates represent the 11 treatment groups (CK, LR, and F1–F9). Data are mean ± standard error (n = 5). Different lowercase letters indicate significant differences (p< 0.05), as determined by Duncan's multiple range test.

Figure 9. Effect of nitrogen, phosphorus, and potassium ratio on the flowering quality of potted chrysanthemums.

3.2.4. Comprehensive Evaluation of Nitrogen, Phosphorus, and Potassium Fertilizer Rationing

As shown in Figure 10, eleven indicators for potted chrysanthemums with various nitrogen, phosphorus, and potassium ratios were evaluated using principal component analysis (PCA). Four significant components, accounting for 85.25% of the total variance, were retrieved. Principal component scores (F) and growth indicators were shown to be related. Table 9 presents the final scoring formulas for each principal component. Chrysanthemum growth was evaluated thoroughly using composite ratings (F). The score for treatment group X9 (N3 P3 K2) was 0.83, was followed by LR with 0.72, while the control group CK came in eleventh overall with a score of 0.19. With use of this analysis, the best nutrient ratios can be chosen for improved chrysanthemum production, as shown in Table 10. According to the integrated evaluation of the principal component analysis, the combined performance of the X9 treatment group was significantly better than that of the

CK and LR, and the X9 treatment group (N3 P3 K2) was the best NPK fertilizer program under the traditional cultivation substrate conditions.

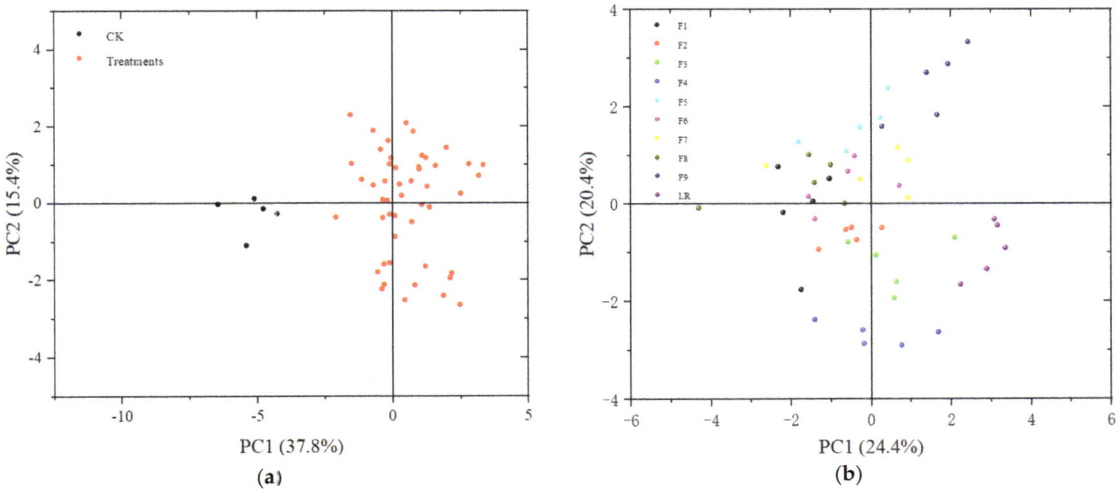

Figure 10. Principal component analysis for nitrogen, phosphorus, and potassium fertilizer ration-ing. (**a**) PCA between CK and other treatments; (**b**) PCA between different treatments.

Table 9. Final score formula for the three principal components of nitrogen, phosphorus, and potassium fertilizer rationing.

Principal Component	Score Formula
1	$Y = 0.104X_1 + 0.142X_2 + 0.083X_3 + 0.08X_4 + 0.139X_5 + 0.121X_6 + 0.103X_7 + 0.13X_8 + 0.161X_9 + 0.136X_{10} + 0.154X_{11}$
2	$Y = 0.288X_1 + 0.177X_2 + 0.237X_3 - 0.3X_4 - 0.015X_5 - 0.081X_6 + 0.334X_7 - 0.244X_8 - 0.109X_9 - 0.183X_{10} + 0.02X_{11}$
3	$Y = -0.407X_1 - 0.306X_2 + 0.414X_3 - 0.161X_4 + 0.177X_5 + 0.406X_6 + 0.115X_7 + 0.182X_8 - 0.135X_9 - 0.274X_{10} + 0.097X_{11}$
4	$Y = 0.016X_1 + 0.184X_2 + 0.583X_3 + 0.57X_4 + 0.261X_5 - 0.446X_6 - 0.207X_7 + 0.085X_8 - 0.014X_9 - 0.322X_{10} - 0.341X_{11}$

Table 10. Comprehensive evaluation of nitrogen, phosphorus, and potassium fertilizer rationing.

No.	F 1	F 2	F 3	F 4	F	Rank
LR	0.79608	1.17602	−2.23594	0.36924	0.721133	2
CK	−2.62065	−0.11042	−0.49501	0.96326	0.188883	11
X1	−0.50485	−0.77516	0.30873	0.17846	0.476229	10
X2	−0.05415	−0.46167	−0.81387	−1.38358	0.492573	9
X3	−0.10074	1.36793	0.39266	−0.29265	0.679201	4
X4	−0.22755	1.82108	1.36269	−0.62273	0.719399	3
X5	0.51626	−0.83428	0.78928	−0.55996	0.628747	6
X6	0.55854	−0.62173	−0.6976	−0.7094	0.593011	7
X7	0.39555	−0.20306	0.77807	−0.22384	0.660954	5
X8	0.04062	−1.2068	0.09926	−0.09379	0.515717	8
X9	1.20088	−0.15191	0.51173	2.37498	0.832422	1

3.3. *Effects of Substrate, Water, and Fertilizer Ratio on the Growth and Physiological Indices of Potted Chrysanthemum*

3.3.1. Effect of Substrate, Water, and Fertilizer Ratio on the Growth Indices of *C. morifolium* 'Hanluqiushi'

The plants' growth indices of 'Hanluqiushi' under various circumstances are shown in Table 11. Gradually rising fertility was accompanied by progressive plant height, crown diameter, and stem diameter increases. The fertilizer level influenced the substantial difference in plant height across treatments ($p < 0.05$). At day 80, the plants in the W1, W5, and W6 treatment groups were the tallest, the W6 treatment group had the largest crown diameter, and the W2, W4, and W6 treatment groups had the thickest stems.

3.3.2. Effect of Substrate, Water, and Fertilizer Ratio on the Physiological Indices of *C. morifolium* 'Hanluqiushi'

To assess the effects of different fertilizer treatments on potted chrysanthemums, we focused on comparing the biomass, chlorophyll content, and nutrient content of the whole plant at the 80 days of cultivation (Figures 11 and 12). Significant interactions between different ratio treatments and organs were observed, greatly influencing the nutrient content ($p < 0.05$). The shoot fresh weight of the aerial parts in the W1 treatment group was 24.55 g. The W4 treatment group had the highest fresh weight of 43.01 g, 175.19% higher than the control group. The dry weight of the aerial parts of the W1 treatment group was 5.78 g. The maximum dry weight of the W6 treatment group was 7.56 g, which was 130.80% higher than that of the control group, and the CK dry weight was the lowest in the control group. The W6 treatment group had the most incredible chlorophyll content (126.84% of CK).

Regarding the total nitrogen (TN) content and total phosphorus (TP) content, the TN content ranged between 2.95 to 4.10 g·100 g^{-1}, while the W2 treatment group had the highest TN content. Although variations in the TP content were negligible, fertilization significantly increased the TP content of all plants. The W7 treatment group exhibited the highest increase in TP content (119.72% of W1). For the total potassium (TK) content, the W9 treatment group had the highest TK content (101.62% of W1).

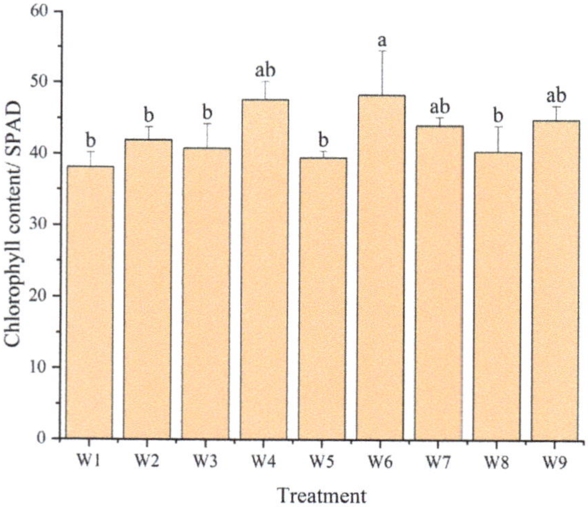

(a)

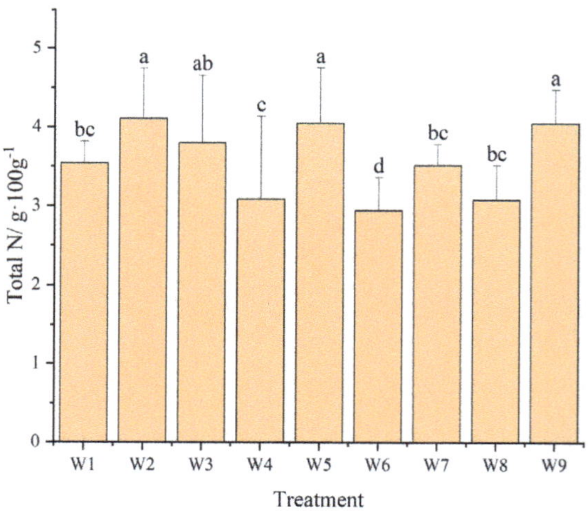

(b)

Figure 11. *Cont.*

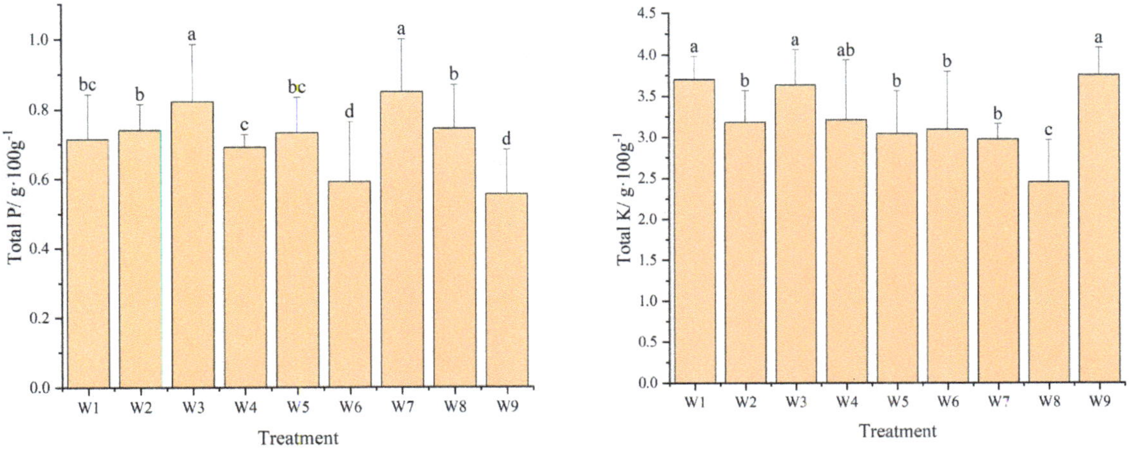

(c) (d)

Figure 11. Effect of substrate, water, and fertilizer on the total chlorophyll content and N/P/K of potted chrysanthemums: (**a**) Total chlorophyll content (SPAD); (**b**) Total N (g/100 g); (**c**) Total P (g/100 g); (**d**) Total K (g/100 g). The horizontal coordinates represent the 9 treatment groups (W1–W9). Data are mean ± standard error ($n = 5$). Different lowercase letters indicate significant differ-ences ($p < 0.05$), as determined by Duncan's multiple range test.

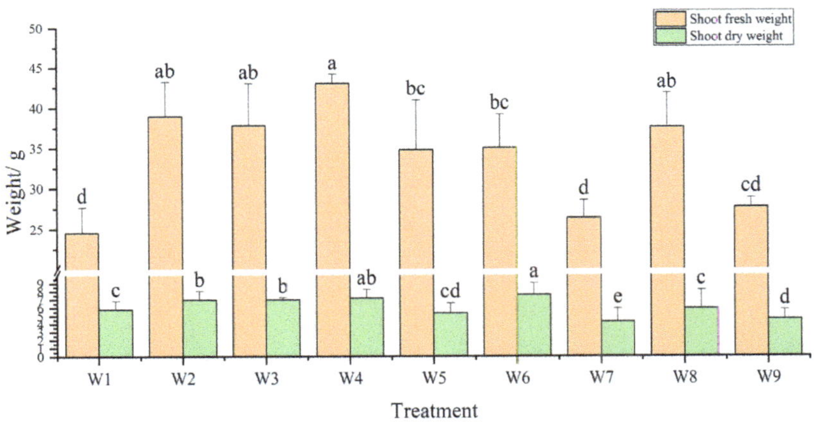

Figure 12. Effect of substrate, water, and fertilizer on the biomass of potted chrysanthemums. The horizontal coordinate represents the 9 treatment groups (W1–W9). Data are mean ± standard error ($n = 3$). Different lowercase letters indicate significant differences ($p < 0.05$), as determined by Duncan's multiple range test.

Table 11. Effects of substrate, water, and fertilizer on the growth indices of potted chrysanthemum.

No.	Height (cm)				Crown Diameter (cm)				Stem Diameter (mm)			
	20 d	40 d	60 d	80 d	20 d	40 d	60 d	80 d	20 d	40 d	60 d	80 d
W1	7.3 ± 0.5 a *	10.4 ± 1.6 ab	10.5 ± 1.5 ab	17.3 ± 0.3 a	8.3 ± 0.6 ab	9.4 ± 2.1 b	11.2 ± 0.4 b	18.1 ± 1.4 abc	3.9 ± 0.1 ab	5.1 ± 0.1 b	5.4 ± 0.2 ab	5.5 ± 1.2 abc
W2	5.6 ± 1.0 bc	10.2 ± 0.9 ab	10.8 ± 2.0 ab	14.7 ± 2.1 ab	7.4 ± 0.7 abc	10.3 ± 2.0 ab	12.8 ± 2.5 ab	18.4 ± 2.8 abc	4.0 ± 0.2 a	5.1 ± 0.2 a	5.2 ± 1.0 abc	6.1 ± 0.8 a
W3	6.9 ± 0.8 ab	11.5 ± 1.6 a	11.6 ± 0.9 ab	16.0 ± 1.6 ab	7.3 ± 0.4 bc	10.2 ± 1.2 ab	13.8 ± 0.5 ab	18.5 ± 1.4 abc	4.0 ± 0.1 a	4.5 ± 0.1 c	4.7 ± 0.2 cd	5.0 ± 0.4 abc
W4	5.4 ± 0.2 bc	10.8 ± 1.2 ab	10.4 ± 1.3 ab	15.7 ± 1.8 ab	7.2 ± 1.3 bc	11.9 ± 1.5 a	15.1 ± 1.4 a	19.3 ± 1.2 ab	3.9 ± 0.1 ab	5.4 ± 0.2 a	5.8 ± 0.2 a	5.8 ± 0.4 ab
W5	7.3 ± 0.8 a	12.1 ± 2.6 a	11.9 ± 2.0 a	17.0 ± 1.1 a	7.6 ± 1.2 abc	11.7 ± 1.0 ab	13.2 ± 0.6 ab	18.1 ± 0.7 abc	3.9 ± 0.2 ab	4.5 ± 0.2 c	4.7 ± 0.1 bcd	5.1 ± 0.4 abc
W6	5.4 ± 1.0 bc	10.3 ± 1.0 ab	10.3 ± 0.3 ab	16.2 ± 1.0 ab	9.1 ± 1.5 a	12.3 ± 0.2 a	15.3 ± 2.1 a	21.2 ± 1.1 a	4.0 ± 0.2 ab	5.1 ± 0.2 b	5.3 ± 0.2 ab	5.5 ± 0.5 abc
W7	6.4 ± 1.0 ab	11.0 ± 0.9 ab	10.7 ± 0.3 ab	14.6 ± 1.3 ab	7.1 ± 0.7 bc	10.5 ± 0.2 ab	11.2 ± 1.0 b	15.2 ± 0.1 c	3.7 ± 0.2 b	3.7 ± 0.2 d	4.1 ± 0.4 d	4.3 ± 0.5 c
W8	5.8 ± 0.8 abc	10.8 ± 1.0 ab	9.8 ± 1.5 ab	15.2 ± 1.0 ab	7.8 ± 0.3 abc	10.5 ± 0.7 ab	12.9 ± 0.4 ab	18.1 ± 1.4 abc	3.9 ± 0.2 ab	4.6 ± 0.1 c	4.7 ± 0.2 bcd	5.1 ± 0.8 abc
W9	4.4 ± 1.7 c	8.5 ± 2.7 b	9.1 ± 2.4 b	13.5 ± 4.2 b	6.2 ± 1.7 c	9.3 ± 2.5 b	11.2 ± 4.1 b	17.8 ± 4.0 bc	3.8 ± 0.2 ab	3.9 ± 0.2 d	4.1 ± 0.1 d	4.7 ± 1.2 bc

* Note: Data are mean ± standard error (n = 3); different lowercase letters indicate significant differences ($p < 0.05$).

3.3.3. Effects of Substrate, Water, and Fertilizer Ratio on Flower Qualities of *C. morifolium* 'Hanluqiushi'

As is shown in Figures 13 and 14, regarding the number of flowers, the W4 group had the most (159.30), while the W7 group had the fewest (56.33). Concerning flower diameter, the W6 group had the biggest (52.67 mm), while the W8 group had the smallest (38.73 mm).

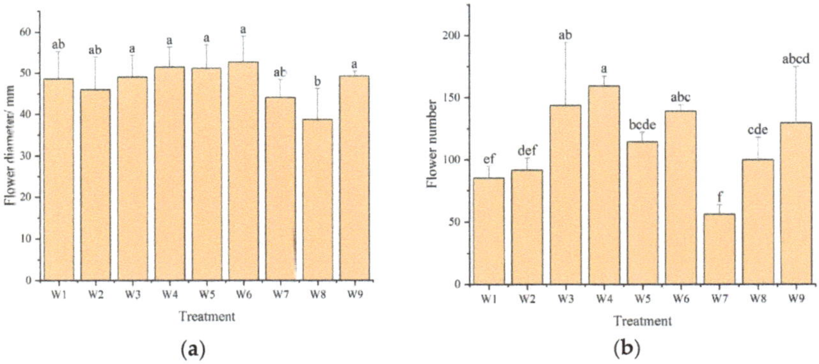

Figure 13. Effect of substrate, water, and fertilizer on the flower diameter and number of potted chrysanthemums: (**a**) Flower diameter (mm); (**b**) Flower number. The horizontal coordinates represent the 9 treatment groups (W1–W9). Data are mean ± standard error (n = 3). Different lowercase letters indicate significant differences ($p < 0.05$), as determined by Duncan's multiple range test.

Figure 14. Effect of substrate, water, and fertilizer on the flower quality and number of potted chrysanthemums.

3.3.4. Integrated Evaluation of Substrate and Water Fertilization

As is shown in Figure 15, eleven indicators for potted chrysanthemums with various nitrogen, phosphorus, and potassium ratios were evaluated using principal component analysis (PCA). Four significant components, accounting for 89.25% of the total variance, were retrieved. Principal component scores (F) and growth indicators were shown to be related. Table 12 presents the final scoring formulas for each principal component. Chrysanthemum growth was evaluated thoroughly using composite ratings (F). The score for treatment group W6 was 0.69, followed by W4 with 0.66, while W7 came in ninth with a score of 0.21. With use of this analysis, the best substrate and water fertilization can be chosen for improved chrysanthemum production, as shown in Table 13. According to the integrated evaluation of the principal component analysis, the combined performance of the W6 treatment group was significantly better than that of the others, and the W6 treatment group was the best substrate and water fertilization under the traditional cultivation substrate conditions.

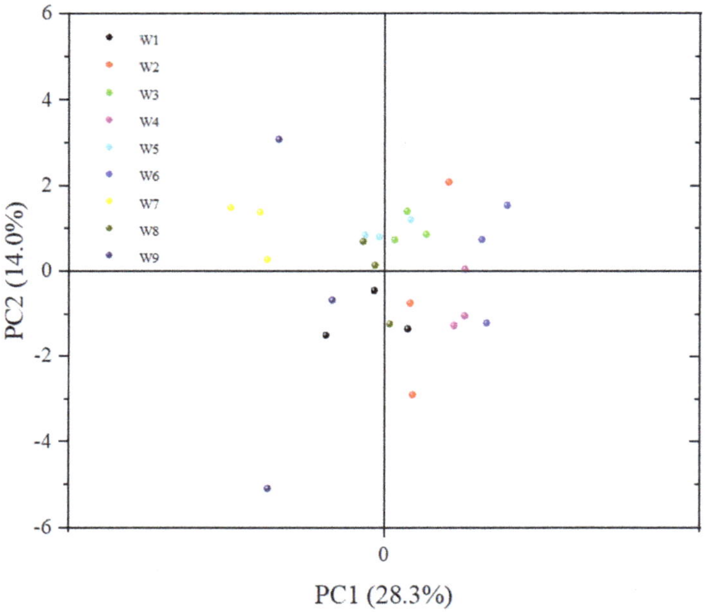

Figure 15. Principal component analysis for substrate and water fertilization.

Table 12. Final score formula for the four principal components of substrate and water fertiliza-tion.

Principal Component	Score Formula
1	$Y = 0.079X_1 + 0.214X_2 + 0.15X_3 + 0.144X_4 + 0.173X_5 + 0.135X_6 + 0.192X_7 + 0.134X_8 - 0.092X_9 - 0.114X_{10} + 0.016X_{11}$
2	$Y = -0.031X_1 + 0.012X_2 - 0.088X_3 + 0.012X_4 + 0.103X_5 + 0.301X_6 - 0.132X_7 - 0.243X_8 + 0.243X_9 - 0.234X_{10} + 0.406X_{11}$
3	$Y = 0.219X_1 - 0.062X_2 + 0.286X_3 - 0.422X_4 + 0.042X_5 + 0.065X_6 + 0.167X_7 + 0.234X_8 - 0.362X_9 + 0.313X_{10} + 0.193X_{11}$
4	$Y = 0.713X_1 + 0.014X_2 - 0.037X_3 + 0.139X_4 - 0.26X_5 - 0.177X_6 - 0.001X_7 - 0.354X_8 - 0.228X_9 + 0.231X_{10} - 0.034X_{11}$

Table 13. Comprehensive evaluation of substrate and water fertilization.

No.	F 1	F 2	F 3	F 4	F	Rank
W1	−0.26155	0.70391	0.45811	1.75321	0.613235	4
W2	0.08254	−0.43436	1.17124	−1.00389	0.530772	6
W3	0.19618	0.20796	1.15125	−0.01159	0.621736	3
W4	1.14453	−0.37846	0.42559	−0.52124	0.657696	2
W5	−0.15548	0.28797	0.71599	0.65183	0.576752	5
W6	1.70391	0.04471	−1.6102	0.66549	0.688633	1
W7	−1.79657	−0.40817	−0.82109	0.47642	0.209011	9
W8	−0.41279	−1.86618	−0.64326	−0.48966	0.287113	8
W9	−0.50078	1.84261	−0.84763	−1.52057	0.455252	7

4. Discussion

Simplified cultivation in ornamental plants, including *Pelargonium hortorum*, *Cyclamen persicum* [34], *Euphorbia pulcherrima* [35], etc., has been reported in the literature. *Chrysanthemum morifolium*, one of the top ten cut flowers and one of the most popular potted flowers in the international market, is widely used in landscaping [21]. Currently, China's potted chrysanthemum industry suffers from problems of high substrate cost, sloppy water and fertilizer management, uneven quality of potted flowers, a severe mismatch between the scale of the industry and existing industrial or technological innovations, etc. Therefore, applying simplified cultivation in the potted chrysanthemum industry is necessary.

In soilless culture, the substrate is the medium for plant growth, providing water, fertilizer, and an excellent inter-root environment; its physicochemical properties are closely related to plant growth and development, and it helps to reduce soil-related problems in conventional crop cultivation [22]. Moss peat is currently the most commonly used substrate for soilless culture. However, due to the high environmental damage associated with its acquisition, there is a need to develop new and excellent soilless substrates to replace moss peat in its application. A few of the more prevalent alternatives to substrates are rice husk, bark [36], coir [37], and compost [38]. It has been discovered that most of the impact requirements for *Pelargonium hortorum* could be met by raising the compost content to 40%, reducing the need for chemical fertilizers [34]. To grow *Gerbera jamesonii*, compost and coir can be mixed together as a culture substrate instead of peat [39]. Compared to regular imported peat, *Anthurium andraeanum* can flower at its best when a portion of moss peat is replaced with rotten rice husks and agroforestry wastes [40], which can also be applied as a substitute substrate in production [41]. In this experiment, nine indicators of potted chrysanthemums were combined by principal component analysis to evaluate 18 substrates. The results showed that the T4 treatment group (coir: moss peat: perlite: pine needle soil = 2:4:2:2) was the best, which was consistent with the results of Xiong et al. [38] and Riaz et al. [39].

Different N, P, and K fertilizer applications have a significant impact on the physiological indices of *Machilus chinensis* seedlings [42]; the use of N: P: K = 2: 3: 2 fertilizers in chrysanthemum cultivation resulted in the best performance of chrysanthemums in photosynthetic shape and flowering traits [43]. Appropriate N, P, and K fertilization increased plant height, crown spread, diameter, and total biomass of seedlings [41], which agrees with the present study's findings. The present study found that chrysanthemum growth was significantly better than CK after fertilizer application. The F9 treatment group was superior to CK and LR used in general flower production with an F value of 0.83, which was ranked first overall. Nitrogen is an essential nutrient for plant growth and photosynthesis [43]. In this experiment, plant height, crown width, and stem thickness increased significantly with nitrogen concentration. Consistent with the previous study's findings, potash is an essential nutrient for plant protein, transpiration, respiration, and chlorophyll production, and is also necessary to improve the flower traits of flower crops [42]. In the present study, the plants' chlorophyll content and flower size also increased with the increase in potassium content, which agrees with previous studies findings. Therefore, using 336 mg/L nitrogen, 93 mg/L phosphorus, and 273 mg/L potassium, as seen in the F9 treatment group, can be considered an ideal NPK fertilizer in conventional cultivation substrates.

In this experiment, the S6 treatment group using 40% water holding capacity, N3P3K2 fertilizer had the most significant plant height and crown diameter, the largest flower diameter, and the highest number of flower heads, which indicates that 40% water holding capacity is beneficial to the growth of plant height and crown size of potted chrysanthemum. On the other hand, S4 treated with N3P3K2 fertilizer had the highest fresh and dry weight, while plants treated with 1.5 times N3P3K2 fertilizer concentration showed a decrease in all indicators. This indicates that the appropriate increase in fertilizer concentration can achieve the effect of increased yield, and too high fertilizer concentration will instead inhibit the growth of plants, which is consistent with the results of previous studies [44].

Through the screening of substitute substrates, the screening of nitrogen, phosphorus, and potassium ratios, and the screening of substrate water and fertilizer formulas, this study comprehensively and systematically explored the cultivation and maintenance schemes in the production of potted chrysanthemums and built a light, simple, and efficient production technology system for potted chrysanthemums in an open field that is suitable for the climatic characteristics of northern China, providing feasible technical specifications and a theoretical basis for the refinement and large-scale management of potted chrysanthemums. However, this study only used the chrysanthemum quality 'Hanluqiushi' as the experimental material; thus, more data are needed for the whole chrysanthemum cultivation industry. In addition, the cultivation time was half a month later than the critical time of chrysanthemum production, which resulted in the overall growth of the subsequent chrysanthemums being weaker than that of actual production. Therefore, it is recommended that future trials are conducted for more cultivars at the appropriate production time to verify the cultivation effect of the substitute substrate and water and fertilizer formulations.

Last but not least, the production of chrysanthemums in pots with the ideal ratios of water holding capacity (40%), fertilizer (N3P3K2), and substrate (coir: moss peat: perlite: pine needle soil = 2:4:2:2) produced a production cost of RMB 2.64/m^3, 16.4% less than that of the CK group (with a production cost of RMB 3.28/m^3).

5. Conclusions

This study aimed to enhance the quality and efficiency of potted chrysanthemum production in China. The substitute substrate experiment demonstrated that coir: moss peat: perlite: pine needle mulch = 2:4:2:2 is the most suitable substitute substrate among others. Through research on fertilizer ratios, it was found that the best growth and flowering characteristics of chrysanthemums were achieved with nitrogen, phosphorus, and potassium concentrations of 336 mg/L, 93 mg/L, and 273 mg/L, respectively. The substrate–water–fertilizer coupling experiment demonstrated that utilizing a 2:4:2:2 ratio of coir, moss peat, perlite, and pine needle mulch, along with 40% water capacity, and 336 mg/L nitrogen, 93 mg/L phosphorus, and 273 mg/L potassium led to improved chrysanthemum behavior. In addition, production costs are reduced by 16.4% with the use of new water, fertilizer and substrate management. These findings provide valuable insights for optimizing potted chrysanthemum cultivation practices.

Author Contributions: Conceptualization, Y.L. and J.G. methodology, Y.L., T.Z. and J.G.; software, T.Z., H.S. and Y.L.; validation, Y.L., T.Z. and J.G.; formal analysis, D.L.; investigation, Y.L., C.L. and R.X.; resources, M.S., J.Z. and D.L.; data curation, T.Z., C.L. and R.X.; writing—original draft preparation, Y.L., T.Z. and J.G.; writing—review and editing, Y.L. and J.Z.; visualization, Y.L. and H.S.; supervision, M.S.; project administration, M.S.; funding acquisition, M.S. All authors have read and agreed to the published version of the manuscript.

Funding: This research was funded by the National Natural Science Foundation of China, grant number 32271947, and the Science, Technology and Innovation Project of Xiongan New Area, grant number 2022XAGG0100.

Data Availability Statement: Data are contained within the article.

Conflicts of Interest: The authors declare no conflicts of interest.

References

1. Zhou, X.; Chen, X.; Wu, X.; Cao, G.; Zhang, J. Characterization of the chemical composition of white chrysanthemum flowers of Hangzhou by using high-performance ion trap mass spectrometry. *J. Sep. Sci.* **2016**, *39*, 1218–1222. [CrossRef] [PubMed]
2. Herrera-Estrella, L.; Lopez-Arredondo, D. Phosphorus: The underrated element for feeding the world. *Trends Plant Sci.* **2016**, *21*, 461–463. [CrossRef] [PubMed]
3. Ribeiro, H.M.; Vasconcelos, E.; Santos, J. Fertilisation of potted geranium with a municipal solid waste compost. *Bioresour. Technol.* **2000**, *73*, 247–249. [CrossRef]
4. Jia, Y.; Li, F.; Wang, X.; Yang, S.-M. Soil water and alfalfa yields as affected by alternating ridges and furrows in rainfall harvest in a semiarid environment. *Field Crops Res.* **2006**, *97*, 167–175. [CrossRef]
5. Time, A.; Ponce, C.; Kuhn, N.; Arellano, M.; Sagredo, B.; Donoso, J.M.; Meisel, L.A. Canopy spraying of abscisic acid to improve fruit quality of different sweet cherry cultivars. *Agronomy* **2021**, *11*, 1947. [CrossRef]
6. Gruda, N.; Bisbis, M.B.; Tanny, J. Impacts of protected vegetable cultivation on climate change and adaptation strategies for cleaner production—A review. *J. Clean. Prod.* **2019**, *225*, 324–339. [CrossRef]
7. Gruda, N.; Caron, J.; Prasad, M.; Maher, M.J. Growing media. In *Encyclopedia of Soil Sciences*, 3rd ed.; Lal, R., Ed.; CRC Press Taylor & Francis Group: Boca Raton, FL, USA, 2016; pp. 1053–1058.
8. Savvas, D.; Gruda, N. Application of soilless culture technologies in the modern greenhouse industry—A review. *Eur. J. Hortic. Sci.* **2018**, *83*, 280–293. [CrossRef]

9. Lloyd, I.L.; Thomas, V.; Ofoegbu, C.; Bradley, A.V.; Bullard, P.; D'Acunha, B.; Delaney, B.; Driver, H.; Evans, C.D.; Faulkner, K.J.; et al. State of Knowledge on UK Agricultural Peatlands for Food Production and the Net Zero Transition. *Sustainability* **2023**, *15*, 16347. [CrossRef]
10. Felix, E.; Benedikt, J.; Jessica, S.; Garske, B. Peatland Governance: The Problem of Depicting in Sustainability Governance, Regulatory Law, and Economic Instruments. *Land* **2020**, *9*, 83.
11. Reumer, M.; Harnisz, M.; Lee, H.J.; Reim, A.; Grunert, O.; Putkinen, A.; Fritze, H.; Bodelier, P.L.; Ho, A. Impact of peat mining and restoration on methane turnover potential and methane cycling microorganisms in a northern bog. *Appl. Environ. Microbiol.* **2018**, *84*, e02218-17. [CrossRef]
12. Papafotiou, M.; Phsyhalou, M.; Kargas, G.; Chatzipavlidis, I.; Chronopoulos, J. Olive-mill wastes compost as growing medium component for the production of poinsettia. *Sci. Hortic.* **2004**, *102*, 167–175. [CrossRef]
13. Altieri, R.; Esposito, A.; Baruzzi, G. Use of olive mill waste mix as peat surrogate in substrate for strawberry soilless cultivation. *Int. Biodeterior. Biodegrad.* **2010**, *64*, 670–675. [CrossRef]
14. El-Mola, M.; El-Behairy, U.A. Effect of different organic media on growth of cucumber seedlings. *J. Appl. Sci. Res.* **2012**, *3*, 1430–1440.
15. Polat, E.; Uzun, H.I.; Topçuoglu, B.; Önal, K.; Onus, A.N.; Karaca, M. Effects of spent mushroom compost on quality and productivity of cucumber (*Cucumis sativus* L.) grown in greenhouses. *Afr. J. Biomed. Res.* **2009**, *8*, 176–180.
16. Djidonou, D.; Zhao, X.; Koch, K.E.; Zotarelli, L. Nitrogen accumulation and root distribution of grafted tomato plants as affected by nitrogen fertilization. *HortScience* **2019**, *54*, 1907–1914. [CrossRef]
17. Jia, X.; Wang, L.; Zeng, H.; Yi, K. Insights of intracellular/intercellular phosphate transport and signaling in unicellular green algae and multicellular land plants. *New Phytol.* **2021**, *232*, 1566–1571. [CrossRef] [PubMed]
18. Johnson, R.; Vishwakarma, K.; Hossen, M.S.; Kumar, V.; Shackira, A.M.; Puthur, J.T.; Abdi, G.; Sarraf, M.; Hasanuzzaman, M. Potassium in plants: Growth regulation, signaling, and environmental stress tolerance. *Plant Physiol. Biochem.* **2022**, *172*, 56–69. [CrossRef] [PubMed]
19. Dana, S.; Herdean, A.; Lundin, B.R.; Spetea, C. Retracted: Each of the chloroplast potassium efflux antiporters affects photosynthesis and growth of fully developed Arabidopsis rosettes under short-day photoperiod. *Physiol. Plant.* **2016**, *158*, 483–491. [CrossRef]
20. Sainju, U.M.; Singh, B.P.; Whitehead, W.F. Comparison of the effects of cover crops and nitrogen fertilization on tomato yield, root growth, and soil properties. *Sci. Hortic.* **2001**, *91*, 201–214. [CrossRef]
21. Thakur, T.; Grewal, H.S. Influence of potting media compositions on flower production of chrysanthemum (*Chrysanthemum morifolium* Ramat) cultivar Kikiobiory. *J. Plant Nutr.* **2019**, *42*, 1861–1867. [CrossRef]
22. Gruda, N. Increasing Sustainability of Growing Media Constituents and Stand-Alone Substrates in Soilless Culture Systems. *Agronomy* **2019**, *9*, 298. [CrossRef]
23. Meharg, A. *Marschners Mineral Nutrition of Higher Plants*, 3rd ed.; Marschner, P., Ed.; Elsevier: Amsterdam, The Netherlands; Academic Press: Cambridge, MA, USA, 2011; p. 684. 124.95. ISBN 978-0-12-384905-2. *Exp. Agr.* **2012**, *48*, 305. [CrossRef]
24. Elser, J.J.; Bracken, M.E.; Cleland, E.E.; Gruner, D.S.; Harpole, W.S.; Hillebrand, H.; Ngai, J.T.; Seabloom, E.W.; Shurin, J.B.; Smith, J.E. Global analysis of nitrogen and phosphorus limitation of primary producers in freshwater, marine and terrestrial ecosystems. *Ecol. Lett.* **2007**, *10*, 1135–1142. [CrossRef] [PubMed]
25. FAO. *The state of food and agriculture 2020. Overcoming Water Challenges in Agriculture*; FAO: Rome, Italy, 2020.
26. Aerts, R.; Chapin, F.S. The Mineral Nutrition of Wild Plants Revisited: A Re-evaluation of Processes and Patterns. In *Advances in Ecological Research*; Academic Press: Cambridge, MA, USA, 1999; Volume 30, pp. 1–67.
27. Ohmiya, A. Molecular mechanisms underlying the diverse array of petal colors in chrysanthemum flowers. *Breed. Sci.* **2018**, *68*, 119–127. [CrossRef] [PubMed]
28. Yin, H.; Cao, Y.; Marelli, B.; Zeng, X.; Mason, A.J.; Cao, C. Soil sensors and plant wearables for smart and precision agriculture. *Adv. Mater.* **2021**, *33*, 2007764. [CrossRef]
29. Chen, M.; Zhu, K.; Tan, P.; Liu, J.; Xie, J.; Yao, X.; Chu, G.; Peng, F. Ammonia–nitrate mixture dominated by NH_4^+–N promoted growth, photosynthesis and nutrient accumulation in pecan (*Carya illinoinensis*). *Forests* **2021**, *12*, 1808. [CrossRef]
30. Zou, N.; Huang, L.; Chen, H.; Huang, X.; Song, Q.; Yang, Q.; Wang, T. Nitrogen form plays an important role in the growth of moso bamboo (*Phyllostachys edulis*) seedlings. *PeerJ* **2020**, *8*, e9938. [CrossRef] [PubMed]
31. Wu, Y.J.; Lu, Q.Y.; Gong, Y.; Zhang, Y.; Xu, Y.; Cai, M.; Pan, H.; Zhang, Q. Optimizing nitrogen, phosphorus, and potassium fertilization levels for container plants of *Lagerstroemia Indica* 'Whit III' based on the comprehensive quality evaluation. *HortScience* **2023**, *58*, 222–230. [CrossRef]
32. Wang, Y.; Wu, W.H. Potassium transport and signaling in higher plants. *Annu. Rev. Plant Biol.* **2013**, *64*, 451–476. [CrossRef]
33. Raviv, M. Can compost improve sustainability of plant production in growing media? *Acta Hortic.* **2017**, *1168*, 119–134.
34. Bonaguro, J.; Coletto, L.; Sambo, P.; Nicoletto, C.; Zanin, G. Environmental Analysis of Sustainable Production Practices Applied to Cyclamen and Zonal Geranium. *Horticulturae* **2021**, *7*, 8. [CrossRef]
35. Zanin, G.; Coletto, L.; Passoni, M.; Nicoletto, C.; Bonato, S.; Ponchia, G.; Sambo, P. Organic by-product substrate components and biodegradable pots in the production ofPelargonium × hortorum Bailey and *Euphorbia pulcherrima* L. *Acta Hortic.* **2016**, *1112*, 371–378. [CrossRef]

36. Horie, K.; Baron, M.; Fox, R.B.; He, J.; Hess, M.; Kahovec, J.; Kitayama, T.; Kubisa, P.; Maréchal, E.; Mormann, W.; et al. Definitions of terms relating to reactions of polymers and to functional polymeric materials. *Pure Appl. Chem.* **2006**, *76*, 889–906. [CrossRef]
37. Preeti, G.; Mulji, G.; Jignasa, R.; Halepotara, F.; Solanki, M.; Malam, V.R.; Barad, R. Role of Growing Media for Ornamental Pot Plants. *Int. J. Pure App. Biosci.* **2018**, *6*, 1219–1224.
38. Xiong, J.; Tian, Y.; Wang, J.; Liu, W.; Chen, Q. Comparison of Coconut Coir, Rockwool, and Peat Cultivations for Tomato Production: Nutrient Balance, Plant Growth and Fruit Quality. *Front. Recent. Dev. Plant Sci.* **2017**, *8*, 1327. [CrossRef]
39. Riaz, A.; Younis, A.; Ghani, I.; Tariq, U.; Ahsan, M. Agricultural waste as growing media component for the growth and flowering of *Gerbera jamesonii* cv. hybrid mix. *Int. J. Recycl. Org. Waste Agric.* **2015**, *4*, 197–204.
40. Suo, L.N.; Sun, X.Y.; Li, S.Y. Use of organic agricultural wastes as growing media for the production of *Anthurium andraeanum* 'Pink Lady'. *J. Hortic. Sci. Biotechnol.* **2011**, *86*, 366–370. [CrossRef]
41. Lin, W.; Zhu, X.; Wu, Y.; Li, L.; Huang, H. Effects of Ratio Fertilization of N, P, K on Seedling Growth of Machilus chinensis. *J. For. Environ. Sci.* **2021**, *12*, 142–147.
42. Choudhary, A.; Kumar, A.; Kumar, U.; Choudhary, R.; Kumar, R.; Jat, R.; Nidhibahen, P.; Hatamleh, A.A.; Al-Dosary, M.A.; Al-Wasel, Y.A.; et al. Various Fertilization Managements Influence the Flowering Attributes, Yield Response, Biochemical Activity and SoilNutrient Status of Chrysanthemum (Chrysanthemum morifolium Ramat.). *Sustainability* **2022**, *14*, 4561. [CrossRef]
43. Abioye, O.F.; Nwoke, O.C. Phosphate rock: Origin, importance, environmental impacts and future roles. *Environ. Rev.* **2019**, *24*, 403–415.
44. Luo, Z.; Song, H.; Huang, M.; Zhang, Z.; Peng, Z.; Zi, T.; Tian, C.; Eissa, M.A. Nitrogen-Reduction in Intensive Cultivation Improved Nitrogen Fertilizer Utilization Efficiency and Soil Nitrogen Mineralization of Double-Cropped Rice. *Agronomy* **2022**, *12*, 1103. [CrossRef]

horticulturae

MDPI

Article

Assessing the Effect of Plant Growth Stimulants and Retardants on Cyclamen "Halios F1 Salmon Rose" Cultivar

János Bálint [1], Klára Benedek [1,*] and Artúr Botond Csorba [1,2,*]

1 Department of Horticulture, Faculty of Technical and Human Sciences, Sapientia Hungarian University of Transylvania, Calea Sighișoarei nr. 2, 540485 Târgu-Mureș, Romania; balintjanos@ms.sapientia.ro
2 Doctoral School in Biology and Sport Biology, University of Pécs, Faculty of Sciences, Ifjúság útja 6., 7624 Pécs, Hungary
* Correspondence: benedekklara@ms.sapientia.ro (K.B.); csorba.artur@ms.sapientia.ro (A.B.C.)

Abstract: In Eastern Europe, the traditional marketing of cyclamen is in the period from the middle of February to the middle of March. The poor light of winter months and the higher number of plants (because of the reduction in heating costs) often result in elongated plants. To avoid this, it is recommended to use plant growth retardants. At the same time, another problem is that flowers do not rise from the level of rosette due to the unfavorable cultivation conditions. This can be solved with growth stimulants. In the experiment, we tested the effect of growth regulators on the growth of a frequently used variety of cyclamen. We used daminozide and paclobutrazol plant growth retardants for height control and gibberellic acid (GA) and benzyladenine (BAP) as growth promoters for increasing the number of flowers and the length of stems. The results show that daminozide and paclobutrazol are both effective for height control in cyclamen production.

Keywords: cyclamen; cytokinin; daminozide; gibberellic acid; paclobutrazol

Citation: Bálint, J.; Benedek, K.; Csorba, A.B. Assessing the Effect of Plant Growth Stimulants and Retardants on Cyclamen "Halios F1 Salmon Rose" Cultivar. *Horticulturae* **2024**, *10*, 53. https://doi.org/10.3390/horticulturae10010053

Academic Editors: Margherita Beruto, Emmy Dhooghe and Bruce Dunn

Received: 12 December 2023
Revised: 28 December 2023
Accepted: 29 December 2023
Published: 5 January 2024

1. Introduction

The cultivation of ornamental plants, in addition to the cultivation of potted ornamental plants, is gaining more and more use worldwide [1,2]. Cyclamens are cultivated in many regions around Europe and are currently considered one of the most popular potted ornamental plants, even though their production can be very expensive due to heating requirements, especially in Middle and Eastern European countries [3,4]. Annually, nearly 200 million plants are produced and sold worldwide, out of which 140–150 million are grown in Europe [5].

On the other hand, some cyclamen species, like *C. purpurascens* or *C. coum var. coum*, are used as medicinal plants [6–8]. *C. persicum* is traditionally propagated by seeds at high costs despite its lack of homogeneity of cultivars [5,9,10]. The cyclamens taxonomically belong to the family *Primulaceae* and the genus *Cyclamen* with about 21 species [11,12]. It is native to Cyprus, Turkey, Crete, Libya, and Syria [13,14]. The ornamental value is given by its flowers, and the colors range from white to red, pink, and purple, and these flowers usually appear between December and early May [15,16]. In Western European countries, the cyclamen plants are most likely sold in the Christmas period before New Year's Eve [17,18]. The cultivation conditions are favorable for the plants in the period from June to November and, in general, the usage of growth regulators is not necessary. In Romania, the traditional marketing of cyclamen is in the period from the middle of February to International Women's Day in March [18]. In this region, the poor light of winter months and the higher number of plants (because of the reduction in heating costs) often result in elongated plants. To avoid this situation, it is recommended to use plant growth retardants. At the same time, another problem is that the flowers do not rise from the level of rosette due to the unfavorable cultivation conditions. A solution for this problem can be the usage of growth regulators. The role of plant growth retardants and regulators has increased since the spread of flowering timing and the influence of stem elongation [19]. This can be critical to a successful cultivation because of the market demands [19,20].

Plant growth regulators (PGRs) are defined as natural or synthetic compounds that affect the developmental and metabolic processes in plants [21]. These compounds influence the yield, the number, and quality of flowers, and they are widely used in ornamental plant production [22]. A high number of growth regulators are used to adjust the size of the plants, and it can help if an excessive amount of growth retardants are applied [23].

Cytokinins promote cell division, and these phytohormones play an important role in regulating plant growth and differentiation [24,25]. They also play a crucial role in the regulation of plant responses to environmental stresses such as drought and salinity. Moreover, cytokinins have been identified as key players in the intricate network of plant stress responses. They exhibit a remarkable ability to modulate and mediate plant reactions to various stress conditions, including challenging situations such as drought, salinity, and pathogen attacks. Cytokinins contribute to the maintenance of cellular viability and play a pivotal role in enhancing stress tolerance in plants. These hormones actively engage in activating defense mechanisms, enabling plants to effectively combat and mitigate the detrimental effects of stressors. Through their multifaceted actions, cytokinins serve as critical components in the intricate machinery that orchestrates plant responses to stress [26].

6-Benzylaminopurine (BAP) was the first synthetic cytokinin used in potted plant production, and previous research has shown that the application of BAP successfully delayed senescence in plants [27]. Gibberellins are primarily involved in promoting seed germination and flowering. They also regulate various developmental processes such as leaf expansion, fruit development, and the transition from vegetative to reproductive growth [28]. The effects of gibberellic acid (GA3) have been proven and documented in numerous crops and ornamental plants, as well as cyclamens, as they have the remarkable ability to significantly improve stem elongation and early flowering [29,30]. Gibberellins are also involved in regulating plant responses to environmental cues such as light and temperature [30].

Another way to control the development of potted ornamental plants is to use plant growth retardants. Within the realm of potted cyclamen production, the utilization of growth retardants to regulate plant height is deemed to have limited practicality; however, the literature on this subject matter is rather scarce with only a few papers documenting such findings [31]. Most of the plant growth retardants inhibit the biosynthesis of gibberellin (GA), which is the hormone that is responsible for plant growth [32,33]. It is already known that GA metabolism is sensitive to irradiance, and in Eastern European countries, the low irradiance increases GA biosynthesis [34], which results in shoot elongation, and this effect has been demonstrated for a large number of plant species in previous research [35]. A popular plant growth retardant used by ornamental plant growers is daminozide. This chemical compound reduces the plant cell size, shortens the length of the stem, and inhibits the growth of the apical meristem [36]. Paclobutrazol is another effective and widely used growth retardant, which shortens the length of stems but can also reduce the size of bracts in the case of excessive use in ornamental plants [37]. The effect of this retardant can be described by a compact growth plant, which also accelerates the inflorescence and reduces the height of the potted ornamental plants [38].

Our aim was to investigate and evaluate the effects of plant growth regulators and retardants used in Romania in potted *Cyclamen persicum* cultivation. In addition, another aim was to develop a cultivation technology that results in compact, healthy, and attractive plants to fulfill consumer expectations and meet market demands.

2. Materials and Methods

The experiment was performed in a greenhouse with a double-inflated foil cover in Miercurea Nirajului, Romania. The tables were 2 m wide and 22 m long on which the potted cyclamens were grown, and the irrigation was constant during the experiment under controlled humidity and temperature. In summer, the average temperature was 20 °C, while in autumn/winter, it was 15 °C (80% relative humidity average, and 17–18 °C temperature average under the whole experiment) (Figure 1). The humidity and temperature were measured with a data logger (LASCAR EL-USB-2), and the heating was operated by a central heating system.

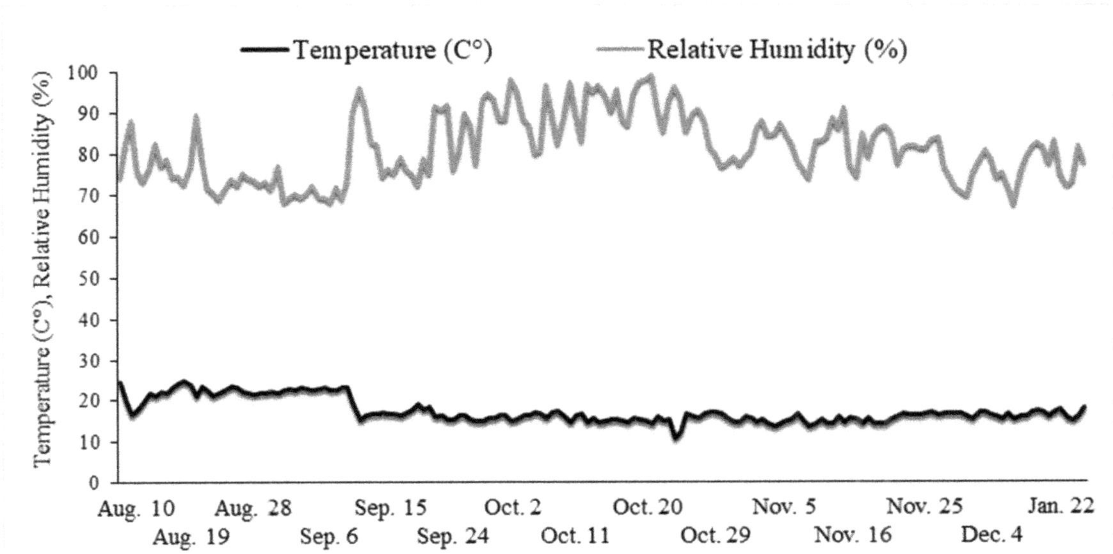

Figure 1. Temperature (°C) and relative humidity (%) during the experiment.

The cyclamen cultivar Halios F1 Salmon Rose with pink flowers was used in the experiment. The morphological description of the "Halios F1 Salmon Rose" is the following: large flowers suitable for 13–20 cm pots, flowering from October to March, good shelf life and a sturdy plant habit, tolerant of high temperatures.

The plant material was bought from S.A.S. Morel Diffusion, France. The cyclamens were grown in propagation trays (Xtray 128), and we planted them into pots with a 12 cm diameter (700 cm^3 volume) on 10 August, in five rows, each row containing six plants in four replications. The first row from each replication did not receive any treatment; they were kept as a control (plants were sprayed only with clean water) (Table 1).

Table 1. Different treatments used in the experiment.

Experimental Variants	Commercial Name	Active Ingredient (According to Label)	Dosage	Observations
T1 (control)	-	-	-	-
T2	Bonzi (Syngenta, Ontario, USA)	paclobutrazol (PBZ)	30 mg/L PBZ	used two times
T3	Alar 85 (Arysta LifeScience, Tokyo, Japan)	daminozide (DAM)	2500 mg/L DAM	used two times
T4	Florgib (Fine Americas, California, USA)	gibberellic acid (GA)	20 mg/L GA3	used two times
T5	Fascination (Nufarm, Illinois, USA)	gibberellins A_4A_7 + 6-benzylaminopurine (GA4+7 + BAP)	10 mg/L GA4+7 + 10 mg/L BAP	used two times

A total of 120 plants were used in the experiment. The substrate used for pots was a combination of peat and perlite with 75% and 25% proportions. On 5 October, the cyclamen cultivars were sprayed for the first time with the selected plant growth regulators and retardants. The second treatment was made on 28 October 2016, and the first cyclamens with full bloom were obtained in early January 2017 (Figure 2). The time of the treatments was selected based on our applied technological experience. During the experiment, five measurements on 120 potted cyclamens were performed at four different treatments. The measured plant characteristics during the experiment were the following: rosette diameter, rosette height, peduncle height, peduncle diameter, plant height, number of flower buds, number of flowers, and inflorescence diameter. In the case of rosette, peduncle, and inflorescence diameter and height, we used a tape measure and a digital caliper. The first measurement took place on 15 November, and the following measurements were on 30 November, 14 December, 4 January, and 22 January (Table 2). The first cyclamens in full bloom appeared in early January.

Weeks	1	2	3	4	5	6	7	8	9	10	11	12	13	14	15	16	17	18	19	20	21	22	23	24	25	26	27
Month	August			September				October				November				December					January				February		
Week number	35	36	37	38	39	40	41	42	43	44	45	46	47	48	49	50	51	52	1	2	3	4	5	6	7	8	9

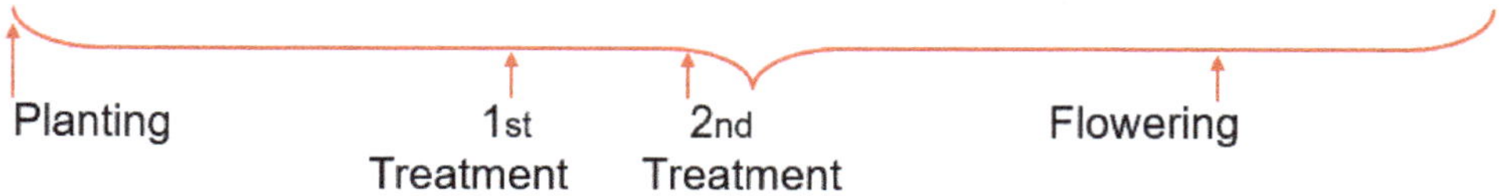

Figure 2. Date of actions during the experiment in chronological order.

Data Analyses

The distribution of the data was examined with the Kolmogorov–Smirnov test. The original data were not distributed normally; therefore, the nonparametric Kruskal–Wallis was used, and afterwards, Mann–Whitney U tests were used to compare the treatments. Means with different letters on figures represent statistically significant differences.

Table 2. Date and time with the measured characteristics.

	15 November				30 November						14 December						4 January							
	Rosette Diameter		Rosette Height		Rosette Diameter		Rosette Height		Number of Buds		Rosette Diameter		Rosette Height		Number of Buds		Rosette Diameter		Rosette Height		Number of Buds		Number of Flowers	
	Mean	*a*	*Mean*	*a*	*Mean*	*a*	*Mean*	*a*	*Mean*	*a*	*Mean*	*a*	*Mean*	*a*	*Mean*	*a*	*Mean*	*a*	*Mean*	*a*	*Mean*	*a*	*Mean*	*a*
Control	19.48	*b*	8.27	*a*	21.33	*a*	8.58	*bc*	0.42	*ab*	22.77	*ab*	8.79	*b*	1.71	*b*	22.90	*b*	8.96	*c*	11.08	*b*	3.92	*ab*
DAM	19.68	*b*	8.16	*a*	21.25	*a*	8.42	*c*	0.58	*a*	22.38	*bc*	8.83	*b*	1.70	*bc*	22.79	*b*	9.38	*b*	8.58	*c*	3.25	*bc*
PBZ	19.98	*ab*	8.13	*a*	20.71	*b*	8.25	*c*	0.04	*c*	22.10	*c*	8.81	*b*	0.79	*c*	22.90	*c*	9.19	*bc*	7.08	*c*	2.29	*c*
GA4+7 + BAP	19.63	*b*	8.35	*a*	21.60	*a*	9.25	*a*	0.29	*abc*	23.02	*a*	9.48	*a*	2.83	*a*	23.74	*a*	10.33	*a*	23.09	*a*	5.13	*a*
GA3	20.46	*a*	8.18	*a*	20.54	*b*	8.92	*b*	0.25	*bc*	21.02	*d*	8.90	*b*	2.67	*ab*	21.71	*d*	9.08	*bc*	25.22	*a*	5.13	*a*

Significance: Means with different letters are significantly different from each other according to the applied statistical test ($p \leq 0.05$, Mann–Whitney U test).

3. Results

3.1. Rosette Diameter under the Effect of PGRs and Plant Growth Regulators

According to our results, the growth retardants and regulators influenced the rosette diameter significantly. The smallest diameter was obtained in cyclamens sprayed with paclobutrazol (PBZ), which significantly differed from the control treatment and from the plants treated with growth regulators (GA3, GA3 + BAP). Conversely, no noteworthy variation was observed between the control treatment and the plants that were sprayed with a combination of growth regulators (GA4+7 + BAP). (Figure 3).

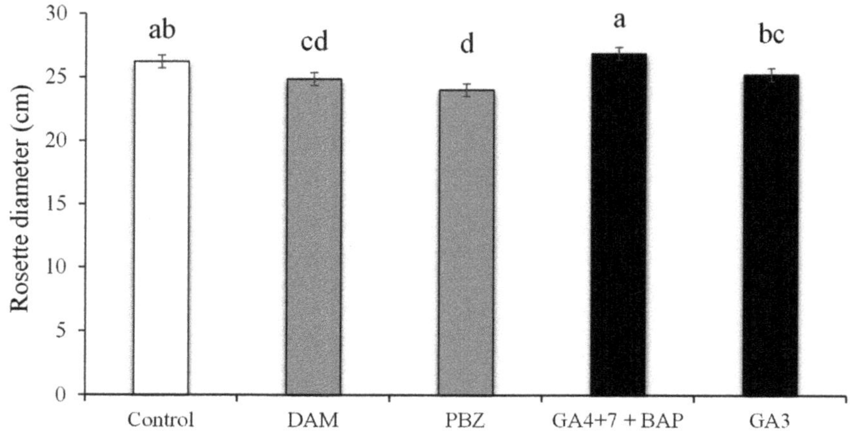

Figure 3. Rosette diameter under different plant growth retardant and plant growth regulator treatments in *C. persicum* "Halios F1 Salmone Rose" variety (Mann–Whitney test $p < 0.05$). Different letters mean statistically significant differences.

3.2. Rosette Height Influenced by PGRs and Plant Growth Regulators

Retardant treatments significantly reduced the height of the rosettes of potted cyclamens. The plants sprayed with paclobutrazol (PBZ) and daminozide (DAM) produced smaller rosettes than plants under plant growth regulator treatments (GA3, GA4+7 + BAP). However, it is worth noting that there was no significant disparity observed between the control treatment and the plants treated with plant growth regulators (Figure 4).

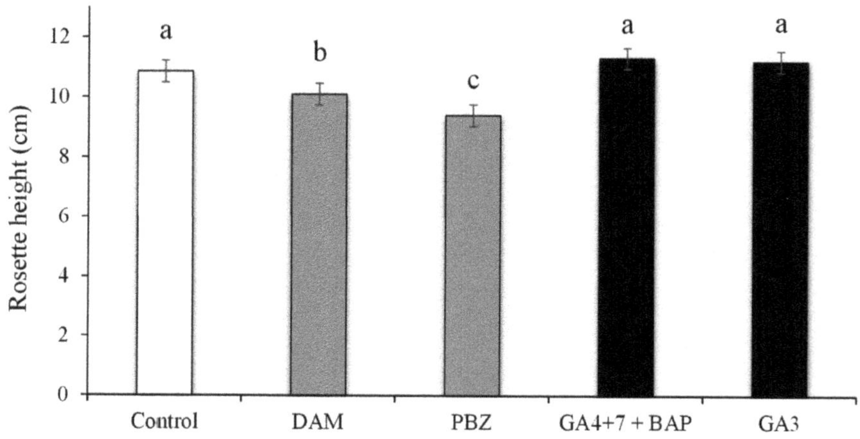

Figure 4. Rosette height under different plant growth retardant and plant growth regulator treatments in *C. persicum* "Halios F1 Salmone Rose" variety (Mann–Whitney test $p < 0.05$). Different letters mean statistically significant differences.

3.3. Peduncle Height under the Effect of PGRs and Plant Growth Regulators

The length (height) of the peduncles was profoundly impacted by the application of both plant growth retardants and plant growth regulators. Notably, there were significant enhancements observed in the peduncle length of the plants treated with plant growth regulators (GA3) when compared to the control treatment. Additionally, substantial disparities were found between the cyclamens subjected to plant growth regulators (GA3, GA4+7 + BAP) and those treated with plant growth retardants (PBZ, DAM). It is worth highlighting that the shortest peduncle length was recorded in the plants treated with paclobutrazol (PBZ). (Figure 5).

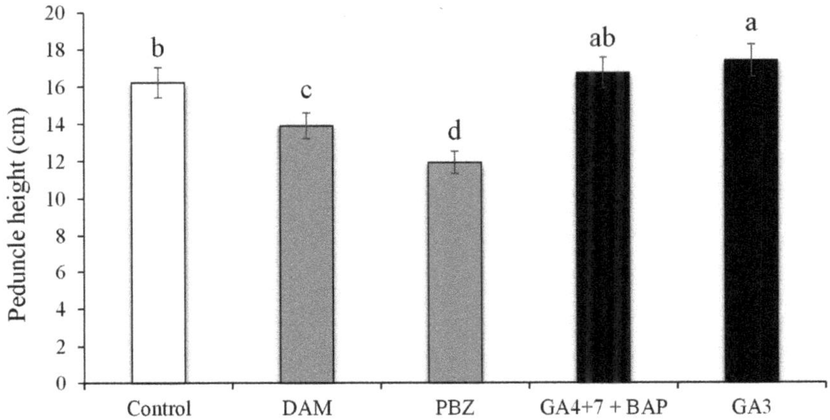

Figure 5. Peduncle height under different plant growth retardant and plant growth regulator treatments in *C. persicum* "Halios F1 Salmone Rose" variety (Mann–Whitney test $p < 0.05$). Different letters mean statistically significant differences.

3.4. Peduncle Diameter Influenced by PGR's and Plant Growth Regulator

Based on our findings, it is evident that the application of growth regulators and growth retardants had a significant impact on the diameter of peduncles in cyclamen plants. Specifically, the cyclamen cultivars that were treated with paclobutrazol (PBZ) and daminozide (DAM) exhibited thicker peduncles compared to the plants treated with plant growth regulators (GA3, GA4+7 + BAP) or the control treatment of water (Figure 6).

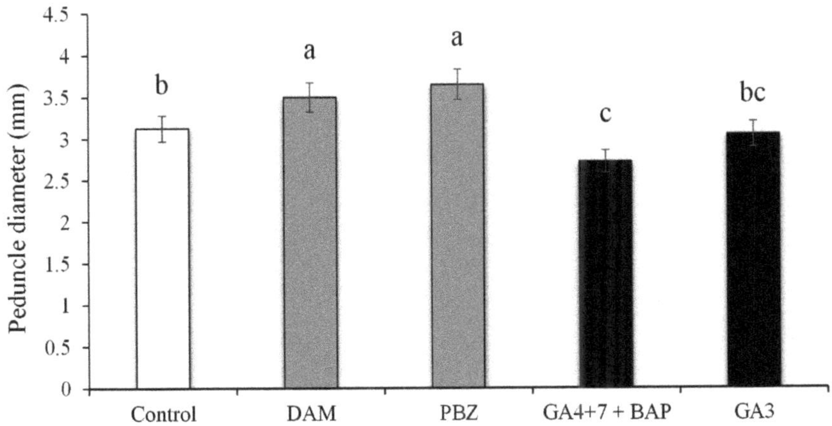

Figure 6. Peduncle diameter under different plant growth retardant and plant growth regulator treatments in *C. persicum* "Halios F1 Salmone Rose" variety (Mann–Whitney test $p < 0.05$). Different letters mean statistically significant differences.

3.5. Plant Height under the Effect of PGRs and Plant Growth Regulators

The plant heights were substantially affected by the application of both plant growth retardants and plant growth regulators. Notably, the plants treated with paclobutrazol (PBZ) and daminozide (DAM) were considerably shorter compared to the plants treated with plant growth regulators (GA3, GA4+7 + BAP). The tallest plants were observed after the treatment with plant growth regulators, and this difference was statistically significant when compared to all other treatments (Figure 7).

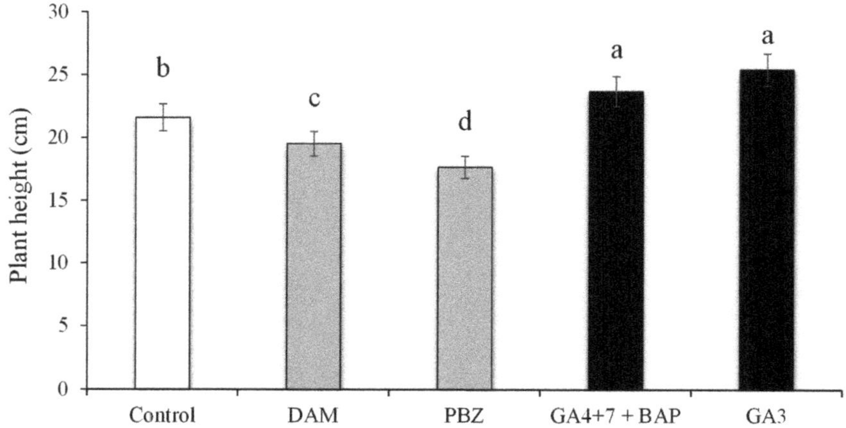

Figure 7. Plant height under different plant growth retardant and plant growth regulator treatments in *C. persicum* "Halios F1 Salmone Rose" variety (Mann–Whitney test $p < 0.05$). Different letters mean statistically significant differences.

3.6. Number of Flower Buds Influenced by PGRs and Plant Growth Regulators

The number of flower buds was significantly higher on cyclamen cultivars treated with plant growth regulators (GA3, GA4+7 + BAP) compared to the cultivars treated with plant growth retardants (DAM, PBZ) or to control treatment. In addition, a higher number of flower buds were recorded in the control group compared to the cyclamens treated with PBZ (Figure 8).

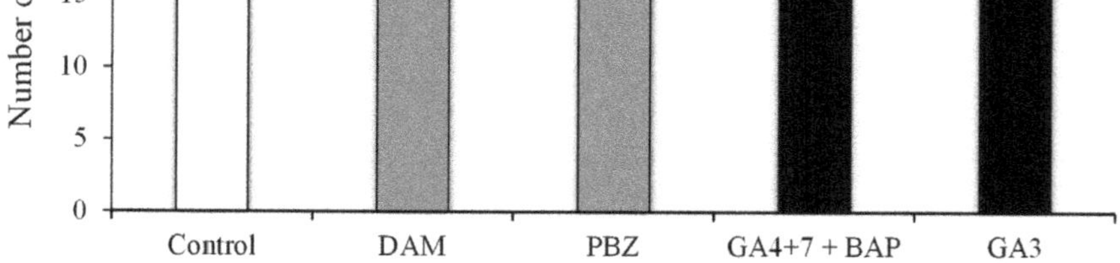

Figure 8. Number of flower buds under different plant growth retardant and plant growth regulator treatments in *C. persicum* "Halios F1 Salmone Rose" variety (Mann–Whitney test $p < 0.05$). Different letters mean statistically significant differences.

3.7. Number of Flowers under the Effect of PGRs and Plant Growth Regulators

According to our results, the plant growth regulators and growth retardants significantly influenced the number of flowers in cyclamen cultivars. The best results were obtained after the treatments with plant growth regulators (GA3, GA4+7 + BAP), which significantly differed from all treatments. The weakest result in the number of flowers was obtained after the PBZ treatment, which was significantly different from the control group, too (Figure 9).

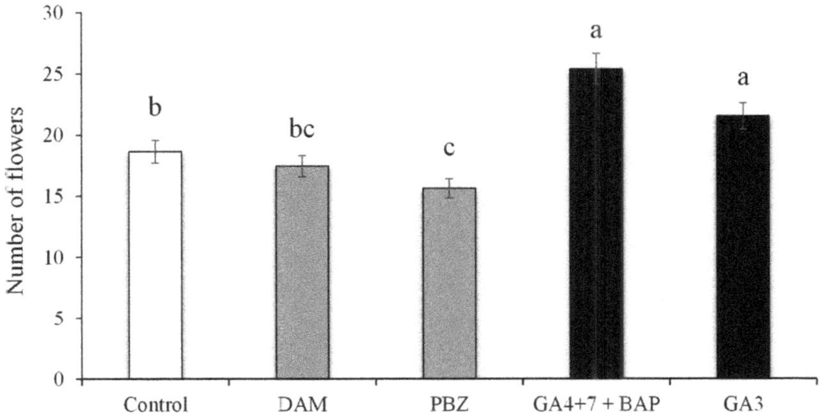

Figure 9. Number of flowers under different plant growth retardant and plant growth regulator treatments in *C. persicum* "Halios F1 Salmone Rose" variety (Mann–Whitney test $p < 0.05$). Different letters mean statistically significant differences.

3.8. Inflorescence Diameter Influenced by PGRs and Plant Growth Regulators

The inflorescence diameters were influenced after the plant growth regulator and plant growth retardant treatments. Significantly higher inflorescence diameters were found between the GA3, GA4+7 + BAP-treated cyclamen cultivars compared to plants treated with plant growth retardants (DAM, PBZ) and control. No significant differences were obtained between the control and plants treated with DAM or PBZ regarding inflorescence diameter (Figure 10). At the end of the experiment, the differences between the different treatments were clearly visible (Figure 11).

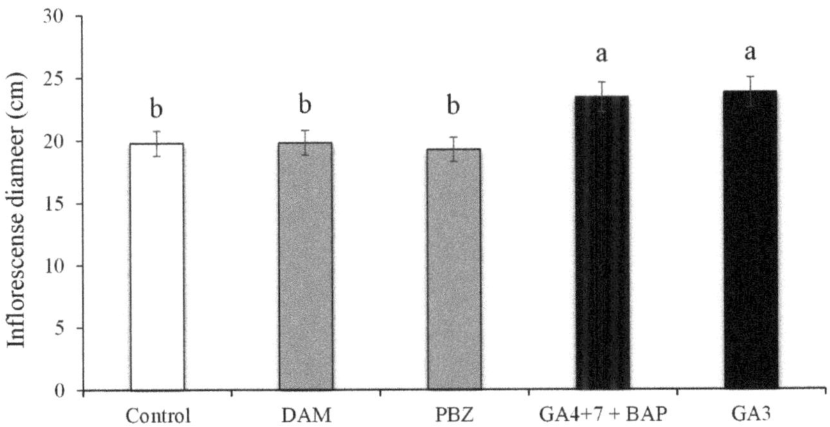

Figure 10. Inflorescence diameter under different plant growth retardant and plant growth regulator treatments in *C. persicum* "Halios F1 Salmone Rose" variety (Mann–Whitney test $p < 0.05$). Different letters mean statistically significant differences.

Figure 11. Differently treated plants at the end of the experiment.

4. Discussion

In our experiment, we have demonstrated that the proper use of plant growth regulators can favorably influence the growth, flower, and quality parameters of the selected cyclamen cultivar, and these plants can also fulfill the customer demands. In a two-year experiment in Jordan, it was found that treatments with GA3 and gibberellic acid A4+A7 significantly hastened the flower development and increased the peduncle height [29]. The treatment with 6-benzylaminopurine (BAP) was ineffective regarding peduncle length [29]. Our results are in agreement with the results obtained in the experiment. However, in the case of peduncle length, we observed the opposite because cyclamens treated with BAP provided one of the best results for us. In contrast, in the case of peduncle diameter, GA3 and the combination of BAP and GA4+7 treatment provided the lowest results.

Plants treated with GA3 exhibited higher plant height, leaf area and root length in a research study performed in Romania with different cyclamen cultivars [39]. Research investigations have revealed that the application of exogenous gibberellic acids (Gas), as well as the utilization of GA biosynthesis inhibitors, have demonstrated an impact on the elongation of petioles. According to previous reports by Oh et al. [40], it was observed that *Cyclamen persium* cv. 'Metis Scarlet Red' plants exhibited leaf expansion and petiole elongation when subjected to short-day conditions. These findings suggest that endogenous GA3 and GA4 may play a significant role in the process of petiole elongation in *C. persicum* Mill [39]. In another experiment, it was demonstrated that plant growth retardants inhibit chrysanthemum growth in a significant way but had a negative effect on inflorescence in *Chrysanthemum indicum* and other ornamental plants [41,42]. GA3 has been extensively studied in cyclamen cultivation, and alongside its positive effects, there can also be negative effects from incorrect usage. However, recently, the combined use of GA4+7 + BAP growth stimulants is being applied more frequently in various ornamental plants, which can help promote flowering and reduce negative effects when used in appropriate doses. In Eastern European countries, a unique situation and tradition have developed where the sale of cyclamen peaks on Valentine's Day (February 14) and International Women's Day (March 8). As a result, a significant portion of cultivation falls in the winter months. One prerequisite for successful cultivation is raising the tuberous rosette utilizing the warmth of autumn, which then necessitates the use of growth regulators during the winter months to prevent the excessive elongation of the plants.

Another solution is to provide higher heating during winter while attempting to achieve denser spacing to reduce heating costs per plant. In such cases, the application of plant regulators and retardants may be necessary to counteract elongation. However, after winter cultivation, if the plants cannot be adequately raised under suitable conditions, there may be a need to stimulate flowering to produce high-quality and marketable plants.

In their research, Osterc et al. [43] discovered that *Cyclamen* rosettes of plants exposed to long-day conditions were found to be larger in size compared to those exposed to short-day conditions. Due to this cultivation characteristic, *Botrytis cinerea* also poses a major problem [44]. In the mentioned research, Csorba et al. (2023) [44] used the same Halios F1 Salmon Rose cultivar. Plant hormones like cytokinin and auxin treatments are increasing the resistance against the *Botrytis cinerea* infection, and this effect of the resistance is based on delaying senescence [45]. Bosch et al. 2016 [46] demonstrated that a high concentration of PBZ application can reduce the plant height to a desired level. Similar results were obtained in two different studies: containerized *A. graminifolia* [47] and *R. tingitana* [48]. Our results are in agreement with the published research.

Irradiance is an important factor in the development of *Cyclamen* cultivars. Ravnjak et al. (2019) [49] found that cyclamen species are capable of surviving in forests because of the increased anthocyanin content, which protects them against solar radiation. Furthermore, it was demonstrated in another research study that cyclamen rosettes of long-day exposed plants are higher compared to their short-day exposed peers [43]. Growth retardants can increase the number of flowers, but they also can reduce it or have no influence on them. Alshakhaly and Qrunfleh 2019 [29] showed in their studies that GA3 increased the number of flowers and peduncle length on five different *C. persicum* hybrids. In a similar research study from Poland, the authors demonstrated that spraying the leaves with GA3 at a lower concentration not only initiates the growth of flower stalks but also significantly accelerates flowering and increases the number of flowers [50]. These results are in agreement with our results. In the case of the flower buds and the number of flowers, the best results were obtained after the treatments with plant growth regulators (GA3, GA4+7 + BAP) in our experiment.

In *Chrisanthemum* cultivars, these plant growth regulators had the same positive effect on flowering time and flower size [51]. Malik et al. (2017) [52] described that in each concentration used, plant growth regulators increased the flower number per plant in *Dahlia* variabilis. In another research study with tomato plants (*Lycopersicon esculentum*), a couple of parameters were measured and described under different plant growth regulator treatments. The highest number of flower clusters per plant and the highest number of flowers were recorded after GA3 treatment [53].

The treatments with plant growth retardants negatively influenced the rosette diameter and height in the selected cyclamen cultivar. In the case of PBZ treatment, we measured the lowest values. Another research study conducted with *Reichardia tingitana* showed that none of the PBZ doses altered the inflorescence diameter [48]. In contrast, we measured the lowest inflorescence diameter in control and PBZ-treated cyclamens.

5. Conclusions

Our study provides experimental data on the *C. persicum* "Halios F1 Salmone Rose" variety and the effect of plant growth regulators and plant growth retardants on the measured parameters mentioned above.

Long-term experiments are highly important in ornamental horticulture as well as in other industries. By observing and analyzing the growth, development, and reactions of plants over an extended period of time under various conditions, researchers and growers can gain a better understanding of their needs, optimal conditions, and best cultivation methods. Similar experiments provide opportunities for the development of new varieties, the formulation of nutrient and water management strategies, and the improvement of plants' stress tolerance. As a result, they contribute to the development of more efficient and sustainable practices in ornamental horticulture.

The use of plant growth stimulants and retardants can bring several advantages in the case of cyclamen cultivation. They allow for control of plant size and shape. This is particularly important for products where uniform size and shape are important for marketability. Growth regulators enable plants to become more compact, making them easier to handle and shape. By using growth stimulants and retardants, growers are able to harvest the crop in a timely and efficient manner. According to our results, active ingredients have delaying effects on flowering, and they slightly reduce the number of flower buds and flowers. We also found a positive effect of these products on the flowering time and number of flowers.

Our results and findings can help the cyclamen growers plan the most appropriate time to intervene and use the plant growth regulators and retardants, especially in Eastern Europe, in order to achieve the best ornamental quality.

To conclude, PGRs have a positive impact on product quality and contribute to the adoption of economical and sustainable cultivation practices.

Author Contributions: Conceptualization, J.B., K.B. and A.B.C.; methodology, J.B. and A.B.C., formal analysis, A.B.C.; investigation, J.B. and A.B.C.; resources, J.B.; writing—original draft preparation, A.B.C.; writing—review and editing; J.B., K.B. and A.B.C.; visualization, J.B. and A.B.C.; All authors have read and agreed to the published version of the manuscript.

Funding: This research received no external funding.

Data Availability Statement: Data associated with the current study are presented in the manuscript.

Conflicts of Interest: The authors declare no conflicts of interest.

References

1. Turiné Farkas, Z.; Palkovics, A.; Jóljárt, F. Analysis of Growing *Cyclamen persicum*. *Gradus* **2020**, *7*, 43–49. [CrossRef]
2. Van Huylenbroeck, J. Status of Floriculture in Europe. In *Protocols for In Vitro Propagation of Ornamental Plants*; Jain, S.M., Ochatt, S.J., Eds.; Methods in Molecular Biology; Humana Press: Totowa, NJ, USA, 2010; Volume 589, pp. 365–376. ISBN 978-1-60327-390-9.
3. Darras, A.I. Implementation of Sustainable Practices to Ornamental Plant Cultivation Worldwide: A Critical Review. *Agronomy* **2020**, *10*, 1570. [CrossRef]
4. Karlsson, M.; Werner, J. Temperature Affects Leaf Unfolding Rate and Flowering of *Cyclamen*. *HortScience* **2001**, *36*, 292–294. [CrossRef]
5. Winkelmann, T. Clonal Propagation of *Cyclamen persicum* via Momatic Embryogenesis. In *Protocols for In Vitro Propagation of Ornamental Plants*; Jain, S.M., Ochatt, S.J., Eds.; Methods in Molecular Biology; Humana Press: Totowa, NJ, USA, 2010; Volume 589, pp. 281–290. ISBN 978-1-60327-390-9.
6. Seyring, M.; Ewald, A.; Mueller, A.; Haensch, K.-T. Screening for Propagation Suitability in Vitro of Different *Cyclamen* Species. *Electron. J. Biotechnol.* **2009**, *12*, 1–11. [CrossRef]
7. Foubert, K.; Theunis, M.; Apers, S.; Vlietinck, A.; Pieters, L. Chemistry, Distribution and Biological Activities of 13,28-Epoxy-Oleanane Saponins from the Plant Families Myrsinaceae and Primulaceae. *Curr. Org. Chem.* **2008**, *12*, 629–642. [CrossRef]
8. Quave, C.L.; Plano, L.R.W.; Pantuso, T.; Bennett, B.C. Effects of Extracts from Italian Medicinal Plants on Planktonic Growth, Biofilm Formation and Adherence of Methicillin-Resistant *Staphylococcus aureus*. *J. Ethnopharmacol.* **2008**, *118*, 418–428. [CrossRef]
9. Rode, C.; Gallien, S.; Heintz, D.; Van Dorsselaer, A.; Braun, H.-P.; Winkelmann, T. Enolases: Storage Compounds in Seeds? Evidence from a Proteomic Comparison of Zygotic and Somatic Embryos of *Cyclamen persicum* Mill. *Plant Mol. Biol.* **2011**, *75*, 305–319. [CrossRef]
10. Takamura, T. Cyclamen. In *Flower Breeding and Genetics*; Anderson, N.O., Ed.; Springer: Dordrecht, The Netherlands, 2006; pp. 459–478. ISBN 978-1-4020-4427-4.
11. Compton, J.A.; Clennett, J.C.B.; Culham, A. Nomenclature in the Dock. Overclassification Leads to Instability: A Case Study in the Horticulturally Important Genus *Cyclamen* (Myrsinaceae). *Bot. J. Linn. Soc.* **2004**, *146*, 339–349. [CrossRef]
12. Yesson, C.; Culham, A. A Phyloclimatic Study of *Cyclamen*. *BMC Ecol. Evol.* **2006**, *6*, 72. [CrossRef]
13. Jalali, N.; Naderi, R.; Shahi-Gharahlar, A.; da Silva, J.A.T. Tissue Culture of *Cyclamen* spp. *Sci. Hortic.* **2012**, *137*, 11–19. [CrossRef]
14. Schwartz-Tzachor, R.; Eisikowitch, D.; Dafni, A. Flower Characteristics and Breeding System of Two Phenological Ecotypes of *Cyclamen persicum* Mill. (Myrsinaceae) in Israel. *Plant Syst. Evol.* **2008**, *274*, 127–134. [CrossRef]
15. Terakawa, T.; Yamamura, T.; Murayama, T. Improvement of Regeneration and Transformation Systems for *Cyclamen persicum* Using Somatic Embryo Culture. *Plant Biotechnol.* **2008**, *25*, 77–80. [CrossRef]
16. Pavlíček, T.; Bureš, P.; Horová, L.; Raskina, O.; Nevo, E. Genome Size Microscale Divergence of *Cyclamen persicum* in Evolution Canyon, Israel. *Open Life Sci.* **2008**, *3*, 83–90. [CrossRef]
17. Savona, M.; Ruffoni, B.; Giovannini, A.; Altamura, M.M. *Cyclamen persicum* Mill Cv. "Halios": Somatic Embryogenesis and Phenotypic Analysis of Somatic Embryo-Derived Plants. *Acta Hortic.* **2007**, 91–97. [CrossRef]
18. Bala, M.; Berecici, D. The Behaviour of Some Old *Cyclamen persicum* Varieties in Growing Conditions of a Modern Greenhouse from the Didactic Base of the Faculty of Horticulture and Forestry in Timişoara. *Bull. Univ. Agric. Sci. Vet.* **2010**, *67*, 298–301.
19. Lewis, K.P.; Faust, J.E.; Sparkman, J.D.; Grimes, L.W. The Effect of Daminozide and Chlormequat on the Growth and Flowering of Poinsettia and Pansy. *HortScience* **2004**, *39*, 1315–1318. [CrossRef]

20. Batelja Lodeta, K.; Goreta Ban, S.; Perica, S.; Dumicic, G.; Bucan, L. Response of Poinsettia to Drench Application of Growth Regulators. *J. Food Agric. Environ.* **2010**, *8*, 297–301.
21. Rademacher, W. Plant Growth Regulators: Backgrounds and Uses in Plant Production. *J. Plant Growth Regul.* **2015**, *34*, 845–872. [CrossRef]
22. Bons, H.K.; Kaur, M. Role of Plant Growth Regulators in Improving Fruit Set, Quality and Yield of Fruit Crops: A Review. *J. Hortic. Sci. Biotechnol.* **2020**, *95*, 137–146. [CrossRef]
23. Runkle, E.; Blancard, M.; Olrich, M. Fascination on Poinsettia. Available online: https://www.canr.msu.edu/resources/fascination-on-poinsettia (accessed on 17 February 2022).
24. Hwang, I.; Sheen, J.; Müller, B. Cytokinin Signaling Networks. *Annu. Rev. Plant Biol.* **2012**, *63*, 353–380. [CrossRef]
25. Markovich, O.; Steiner, E.; Kouřil, Š.; Tarkowski, P.; Aharoni, A.; Elbaum, R. Silicon Promotes Cytokinin Biosynthesis and Delays Senescence in *Arabidopsis* and *Sorghum*: Silicon and Cytokinin. *Plant Cell Environ.* **2017**, *40*, 1189–1196. [CrossRef] [PubMed]
26. Waadt, R.; Seller, C.A.; Hsu, P.-K.; Takahashi, Y.; Munemasa, S.; Schroeder, J.I. Plant Hormone Regulation of Abiotic Stress Responses. *Nat. Rev. Mol. Cell Biol.* **2022**, *23*, 680–694. [CrossRef]
27. Massolo, J.F.; Lemoine, M.L.; Chaves, A.R.; Concellón, A.; Vicente, A.R. Benzyl-Aminopurine (BAP) Treatments Delay Cell Wall Degradation and Softening, Improving Quality Maintenance of Refrigerated Summer Squash. *Postharvest Biol. Technol.* **2014**, *93*, 122–129. [CrossRef]
28. Lee, J.W.; Kim, Y.C.; Kim, J.U.; Jo, I.H.; Kim, K.H.; Kim, D.H. Effects of gibberellic acid and alternating temperature on breaking seed dormancy of Panax ginseng CA Meyer. *Korean J. Med. Crop. Sci.* **2016**, *24*, 284–293. [CrossRef]
29. Oh, W.; Kim, K.S. Light Intensity and Temperature Regulate Petiole Elongation by Controlling the Content of and Sensitivity to Gibberellin in *Cyclamen persicum*. *Hortic. Environ. Biotechnol.* **2014**, *55*, 175–182. [CrossRef]
30. Alshakhaly, Z.M.; Qrunfleh, M.M. Effect of Plant Growth Regulators on Flower Development and Quality of Five *Cyclamen persicum* Hybrids. *Acta Hortic.* **2019**, 215–222. [CrossRef]
31. Hedden, P.; Thomas, S.G. Gibberellin Biosynthesis and Its Regulation. *Biochem. J.* **2012**, *444*, 11–25. [CrossRef] [PubMed]
32. An, K.B.; Lee, Y.S.; Son, K.C. Effect of Plant Growth Retardants on Producing High Quality of *Cyclamen*. *Korean J. Hortic. Sci.* **2004**, *22*, 65.
33. Fletcher, R.A.; Gilley, A.; Sankhla, N.; Davis, T.D. Triazoles as Plant Growth Regulators and Stress Protectants. In *Horticultural Reviews*; Janick, J., Ed.; John Wiley & Sons, Inc.: Oxford, UK, 2010; pp. 55–138. ISBN 978-0-470-65077-6.
34. Rostami, S.; Azhdarpoor, A. The Application of Plant Growth Regulators to Improve Phytoremediation of Contaminated Soils: A Review. *Chemosphere* **2019**, *220*, 818–827. [CrossRef]
35. García-Martinez, J.L.; Gil, J. Light Regulation of Gibberellin Biosynthesis and Mode of Action. *J. Plant Growth Regul.* **2001**, *20*, 354–368. [CrossRef]
36. Tan, Z.G.; Qian, Y.L. Light Intensity Affects Gibberellic Acid Content in Kentucky Bluegrass. *HortScience* **2003**, *38*, 113–116. [CrossRef]
37. Nishijima, T. Use of Plant Growth Regulators for Floriculture in Japan. *Sci. Hortic.* **2023**, *309*, 111630. [CrossRef]
38. Niu, G.; Heins, R.; Carlson, W. Using Paclobutrazol to Control Height of Poinsettia 'Freedom'. *HortTechnology* **2002**, *12*, 232–236. [CrossRef]
39. Elisheba, B.P.; Sudghar, R. Growth Manipulation in Ornamental Sunflower (*Helianthus annuus*) Cv. Ring of Fire as a Bedding Plant. *Crop Res.* **2021**, *56*, 30–36. [CrossRef]
40. Cornea-Cipcigan, M.; Pamfil, D.; Sisea, C.R.; Mărgăoan, R. Gibberellic Acid Can Improve Seed Germination and Ornamental Quality of Selected *Cyclamen* Species Grown under Short and Long Days. *Agronomy* **2020**, *10*, 516. [CrossRef]
41. Kentelky, E.; Szekely-Varga, Z.; Bálint, J.; Balog, A. Enhance Growth and Flower Quality of Chrysanthemum Indicum l. with Application of Plant Growth Retardants. *Horticulturae* **2021**, *7*, 532. [CrossRef]
42. Hg, M.; Dt, L.; Dt, B. Effects of Plant Growth Retardants and Pot Sizes on the Height of Potting Ornamental Plants: A Short Review. *J. Hortic.* **2018**, *5*. [CrossRef]
43. Osterc, G.; Mikulic Petkovsek, M.; Stampar, F.; Ravnjak, B.; Bavcon, J. Impact of Specific Environmental Characteristics of the Site of Origin (Shady, Sunny) on Anthocyanin and Flavonol Contents of Replanted Plants at Common *Cyclamen* (*Cyclamen purpurascens* Mill.). *Acta Physiol. Plant* **2017**, *39*, 64. [CrossRef]
44. Csorba, A.B.; Kentelky, E.; Szabó, M.-E.; Jakab, M.; Nyárádi, I.-I.; Bálint, J. Controlling Grey Mold (*Botrytis cinerea*) in Flowering *Cyclamen* Production. *Europ. J. Hortic. Sci.* **2023**, *88*, 1–8. [CrossRef] [PubMed]
45. Shaul, O.; Elad, Y.; Kirshner, B.; Volpin, H.; Zielsin, N. Control of Botrytis Cinerea in Cut Rose Flowers by Gibberellic Acid, Ethylene Inhibitors and Calcium. In *Recent Advances in Botrytis Research*; Verhoeff, K., Malathrakis, N.E., Williamson, B., Eds.; Pudoc Scientific Publishers Wageningen, The Netherlands, 1992; pp. 257–261.
46. Bosch, E.; Cuquel, F.L.; Tognon, G.B. Physalis Size Reduction for Potted Ornamental Plant Use. *Ciência Agrotecnologia* **2016**, *40*, 555–564. [CrossRef]
47. Wanderley, C.D.S.; de Faria, R.T.; Ventura, M.U.; Vendrame, W. The Effect of Plant Growth Regulators on Height Control in Potted Arundina Graminifolia Orchids (Growth Regulators in Arundina Graminifolia). *Acta Sci. Agron.* **2014**, *36*, 489. [CrossRef]
48. Bañón, S.; Ochoa, J.; Fernández, J.A.; González, A. Plant Growth Retardants for Introduction of Native Reichardia Tingitana. *Acta Hortic.* **2003**, 271–277. [CrossRef]

49. Ravnjak, B.; Bavcon, J.; Osterc, G. Physiological Response of Local Populations of Species *Cyclamen purpurascens* Mill. to Forest Gaps. *Appl. Ecol. Env. Res.* **2019**, *17*, 11489–11508. [CrossRef]
50. Nowak, J. Wplyw roznych preparatow zawierajacych kwas giberelinowy na wzrost i kwitnienie cyklamena i gerbery. *Zesz. Nauk. Inst. Sadow. I Kwiaciarstwa W Skiern.* **2000**, *7*, 259–263.
51. Sahu, J.K.; Tamrakar, S.K.; Lakpale, R.; Tirkey, T. Effect of Planting Geometry and Plant Growth Regulators on Growth and Flowering of Chrysanthemum. *Progress. Hortic.* **2021**, *53*, 105–108. [CrossRef]
52. Malik, S.; Rather, Z.; Wani, M.; Din, A.; Nazki, I. Effect of Growth Regulators on Plant Growth and Flowering in Dahlia (*Dahlia variabilis*) Cv. Charmit. *J. Exp. Agric. Int.* **2017**, *15*, 1–7. [CrossRef]
53. Choudhury, S.; Islam, N.; Sarkar, M.D.; Ali, M.A. Growth and Yield of Summer Tomato as Influenced by Plant Growth Regulators. *Int. J. Sustain. Agric.* **2013**, *5*, 25–28. [CrossRef]

horticulturae

MDPI

Article

Micropropagation and Acclimatization of *Monstera deliciosa* Liebm. 'Thai Constellation'

Yifan Jing, David Beleski and Wagner Vendrame *

Environmental Horticulture Department, Institute of Food and Agricultural Sciences, University of Florida, Gainesville, FL 32611, USA; yifan.jing@ufl.edu (Y.J.); dbeleski@ufl.edu (D.B.)
* Correspondence: vendrame@ufl.edu

Abstract: *Monstera deliciosa* Liebm. 'Thai Constellation' is a variegated variety of *M. deliciosa* belonging to the family Araceae, which has become a new favorite in the foliage plant market. However, limited studies exist on its propagation, and growers have difficulties in achieving large-scale production. This study aimed at developing an efficient protocol for the micropropagation of *M. deliciosa* using SETIS™ temporary immersion bioreactors. Furthermore, we aimed at evaluating the role of a novel biostimulant (IQ Forte) in the acclimatization of *M. deliciosa*. Significant differences were observed among the different treatments, showing higher multiplication rates under TIS conditions as compared to the semi-solid control. Adjusting immersion parameters also showed benefits in improving multiplication rates. The novel biostimulant (IQ Forte) did not provide significant gains in growth of *M. deliciosa* 'Thai Constellation' during acclimatization.

Keywords: acclimatization; bioreactor; biostimulant; in vitro; IQ Forte; micropropagation; *Monstera deliciosa*; semi-solid; TIS

Citation: Jing, Y.; Beleski, D.; Vendrame, W. Micropropagation and Acclimatization of *Monstera deliciosa* Liebm. 'Thai Constellation'. *Horticulturae* **2024**, *10*, 1. https://doi.org/10.3390/horticulturae10010001

Academic Editors: Margherita Beruto, Emmy Dhooghe and Bruce Dunn

Received: 28 November 2023
Revised: 11 December 2023
Accepted: 14 December 2023
Published: 19 December 2023

1. Introduction

Monstera deliciosa Liebm., also known as ceriman (Trinidad) or piñanona (Mexico), is a climbing vine (growing to a height of 30 ft or more) that belongs to the family Araceae [1,2]. It has been spread around the world as an ornamental foliage plant that can be used indoors or outdoors [3]. *M. deliciosa* 'Thai Constellation' is a new variant of *M. deliciosa*, possessing variegated leaves and being relatively genetically stable. Unfortunately, the plant grows slowly, and cuttings can be a challenge to root, driving up the price. The rapid propagation of *M. deliciosa* by tissue culture technology not only is conducive to the maintenance of excellent characters but also provides a large number of plantlets for plant cultivation [4].

In vitro clonal propagation of plants is popularly called micropropagation because of the miniaturization of the process [5]. Temporary immersion system (TIS) is one of the most advanced tools for commercial micropropagation. It works by using a semi-automated bioreactor to immerse cells, tissues, or organs in a liquid culture medium for a period of time. Hyperhydricity can be remedied by TIS, which exposes explants to liquid media intermittently rather than continuously [6]. Two variants of this system have been developed and are currently on the market: the Recipient for Automated Temporary Immersion system (RITA®) and the twin-flasks system (BIT®) [7]. More recently, a new bioreactor system has been developed (SETIS™, Vervit, Lochristi, Belgium), which allows for increased light irradiation, automation of cultivation systems, easy handling, and large culture medium capacity, and successful protocols have been reported for the micropropagation of banana and vanilla [8–10].

The term acclimatization is defined as the climatic adaptation of an organism, especially a plant, that has been moved to a new environment [11,12]. Acclimatization is an important step in micropropagation. Biostimulants are defined as materials that contain substance(s) and/or microorganisms, whose function when applied to plants or the rhizosphere is to stimulate natural processes to enhance/benefit nutrient uptake, nutrient efficiency, tolerance to abiotic stress, and/or crop quality, independent of its nutrient content [13,14]. Protein hydrolysates (PHs) are a category of plant biostimulants defined as mixtures of polypeptides, oligopeptides, and amino acids that are manufactured from protein sources using partial hydrolysis [15,16]. Plant-derived PHs are gaining greater acceptance by farmers due to their richness in bioactive compounds and their great efficacy in enhancing crop performance [17]. The use of protein hydrolysis products as stimulants to assist plants in acclimatization seems to be a viable option at present.

The goal of this study was to evaluate the efficiency of the SETIS™ bioreactor for the micropropagation of *M. deliciosa* 'Thai Constellation' as compared to a conventional semi-solid culture system. The development of a new micropropagation system for *M. deliciosa* may contribute to its large-scale commercialization. In view of the absence of data regarding the acclimatization of *M. deliciosa* 'Thai Constellation', the present study also had the objective of evaluating the effects of IQ Forte as a biostimulant on the acclimatization of in vitro-derived plantlets of *M. deliciosa* 'Thai Constellation'.

2. Materials and Methods

2.1. Plant Material and Culture Establishment

Plants of *M. deliciosa* 'Thai Constellation' were obtained from an ornamental nursery in Homestead, Florida, and in vitro cultures were established in the micropropagation laboratory of the Environmental Horticulture Dept., University of Florida, in Gainesville, FL. The axillary buds were collected and washed in the sink with soap and water, surface sterilized using 2% sodium hypochlorite (NaClO) for 20 min, and then rinsed in dH2O (distilled water). Subsequently, the outer leaf layer and tissue were removed. The buds again were put in a fresh solution of 2% sodium hypochlorite for another 20 min and then rinsed 3 times. Explants were initially propagated in agar-based medium.

In vitro shoots were multiplied through axillary shoot multiplication in semi-solid modified MS medium [18] including B5 vitamins supplemented with 30 g/L sucrose, 2.5 g/L gellan gum, 7.5 mg/L BAP (6-benzylaminopurine), and 0.5 mg/L NAA (naphthalene acetic acid). The medium pH was adjusted to 5.6–5.8 before autoclaving at 121 °C for 25 min at 1.2 kg cm^{-2}. About 40 mL of medium was dispensed into baby food jars. This allowed the development of multiple axillary shoots and roots.

After 60 days, prior to subculture, shoots formed directly from axillary buds showed some roots and a small callus formation at the base of the shoot. In the subculture process, a small number of roots were trimmed, and the shoots were divided into single shoots containing a portion of the callus.

2.2. Acclimatization and Biostimulant

In vitro-derived shoots generated from bioreactors were used for this study. The acclimatization of in vitro-derived plantlets was conducted in a greenhouse.

Shoots were separated into single shoots under aseptic conditions and transplanted into baby food jars containing rooting medium. The rooting medium was a modified MS medium supplemented with 30 g/L sucrose, 2.5 g/L gellan gum, and 0.5 mg/L NAA (naphthalene acetic acid). The medium pH was adjusted to 5.6–5.8 before autoclaving at 121 °C for 20 min at 15 lbs pressure. About 40 mL of medium was dispensed into baby food jars. Cultures were maintained under fluorescent lighting (General Electric fluorescent bulbs 59W) at 50 µmol m^{-2} s^{-1} (FL), and the photoperiod was 16/8 h (light/dark). The temperature of the culturing room was set to 25 °C.

After 60 days, the plantlets with roots were transplanted to the greenhouse.

IQ Forte (IQBiotech, Miami, FL, USA), which is a natural fertilizer based on oligopeptides and free amino acids obtained by enzymatic hydrolysis of proteins from seeds, was used as the biostimulants.

2.3. In Vitro Multiplication Using Bioreactors

Young shoots (4–6 cm) collected from in vitro seedlings were used as explants for shoot proliferation.

Shoots were grown in four different culture systems: (1) in baby food jars containing 40 mL semi-solid medium (2.5 g/L gellan gum) as a control, (2) in temporary immersion bioreactors (SETIS™, VERVIT, Zelzate, Belgium) with an immersion and aeration frequency of 1.5 h and immersion duration of 1 min (Treatment 1), (3) in temporary immersion bioreactors with an immersion and aeration frequency of 1.5 h and immersion duration of 2 min (Treatment 2), (4) in temporary immersion bioreactors with an immersion and aeration frequency of 3 h and immersion duration of 2 min (Treatment 3) (Figure 1).

Figure 1. Micropropagation of *M. deliciosa* 'Thai Constellation' under different culture systems. (**A**,**B**) Semi-solid medium in baby food jar. (**C**,**D**) Liquid medium in SETIS™ temporary immersion bioreactors.

For the control group, two trays were prepared, each with 30 baby food jars (Figure 1). The 60 baby food jars in the two trays had identical experimental treatments. Cultures were maintained under fluorescent lighting (General Electric fluorescent bulbs 59W) at 50 μmol m^{-2} s^{-1} (FL), and the photoperiod was 16/8 h (light/dark). The temperature of the culturing room was 25 $\pm$ 2 °C. At the end of the experiment (90 days), 30 baby food jars were taken out to collect the experimental data.

The treatments in TIS were comprised of 4 replicates per treatment, each replicate containing 10 explants. The cultures were maintained under controlled environment conditions with continuous light (Valoya's L-Series LED grow light, model L35-144, 35W,

AP67 spectrum; 14% Blue, 16% Green, 53% Red, and 17% Far Red) and light intensity of 79 μmol m^{-2} s^{-1}, temperature of 25 ± 2 °C, and photoperiod of 16/8 h (light/dark) After 90 days of culture establishment, the initial 10 explants turned into 10 clusters with multiple shoots. Data were provided by 7 randomly selected clusters out of the 10. Plant height (height of the tallest shoot), fresh weight, dry weight, chlorophyll content, and shoot multiplication rate were evaluated. To obtain the dry weight, the plant material was placed in an oven and dried at 80 °C for 48 h. For the shoot multiplication rate, shoots that were clearly differentiated were counted, the total number was calculated, and they were classified into small, medium, and large grades based on their length (small: ≤2 cm, medium: 2–4 cm, large: >4 cm). Due to the low survival rate of small shoots in the acclimatization process, the medium and large shoots were summed up and referred to as 'effective shoots'.

It is important to note that in addition to the differences in the type of media used and the method of cultivation, the light source used in the bioreactor system is different from that of a traditional semi-solid system. The traditional semi-solid systems use regular fluorescent lights, as they are the traditional systems used in our lab for many different cultures, whereas for the bioreactor, we use the new LED system, as recommended for optimal growth, propagation, and development.

2.4. Relative Chlorophyll Content Analysis

Fully expanded leaves were selected from 5 plants per treatment for chlorophyll analysis. Chlorophyll relative content was measured as SPAD values by placing the third leaf of each plantlet, counted from top downwards, in a portable SPAD-502 chlorophyll meter (SPAD-502, Minolta Co., Ltd., Osaka, Japan).

2.5. Statistical Analysis

A completely randomized experimental design was used for all experiments. Data were collected and submitted to analysis of variance (ANOVA) using the software R version 4.1.3 (Copyright © 2022 The R Foundation for Statistical Computing). Tukey's post hoc multiple comparison adjustment ($\alpha = 0.05$) was used for all pairwise comparisons of means.

2.6. Effect of Concentration of Biostimulant on Plantlet Growth

The experiment was completely randomized with 3 treatments (3 concentrations of biostimulant; ml IQ Forte/L water) and one control group: (1) Treatment 1: 2.0 mL/L, (2) Treatment 2: 3.0 mL/L, (3) Treatment 3: 4.0 mL/L, (4) Control: water without biostimulant. The concentration of biostimulant was determined based on the manufacturer's label recommendations. The water in all treatments came from the irrigation system of the greenhouse with a water-soluble fertilizer (Peter's 20-10-20 Peat Lite at 150 ppm). There were 20 repetitions for each treatment, where one repetition means one plant in a pot. At the end of the experiment, the following parameters were evaluated: (1) survival: % of plants that survived after transplant to greenhouse, (2) growth: shoot and root development (length, fresh and dry weight), and (3) chlorophyll relative content (SPAD).

The biostimulant IQ Forte was applied at two-week intervals during the course of the experiment (according to the instructions provided by the manufacturer), 50 mL per pot per time. Meanwhile, foliar phosphite fertilizer was applied every two weeks to ensure that there was no microbial infection. The amount applied was 50 mL per pot of an aqueous solution containing 1.6% of the original solution. The interval between the application of these two products was one week.

After 100 days of the plants growing in the greenhouse, data were collected. Survival rates were recorded first. Other data were provided by 10 randomly selected plantlets out of plants that survived. Plant height (height of the tallest shoot), fresh weight, dry weight, and chlorophyll content were evaluated.

3. Results

3.1. Micropropagation Using TIS Bioreactors

Significant differences were observed among culture systems for plant height. The highest plant height occurred in treatment 1 (1 min,1.5 h; 5.2cm), followed by treatment 2 (2 min, 1.5 h; 4.2 cm) and treatment 3 (2 min, 3 h; 4.0 cm). The control group in semi-solid medium had a plant height of 4.1cm. The plant height of treatment 1 was significantly higher than the height in the other groups (Figure 2).

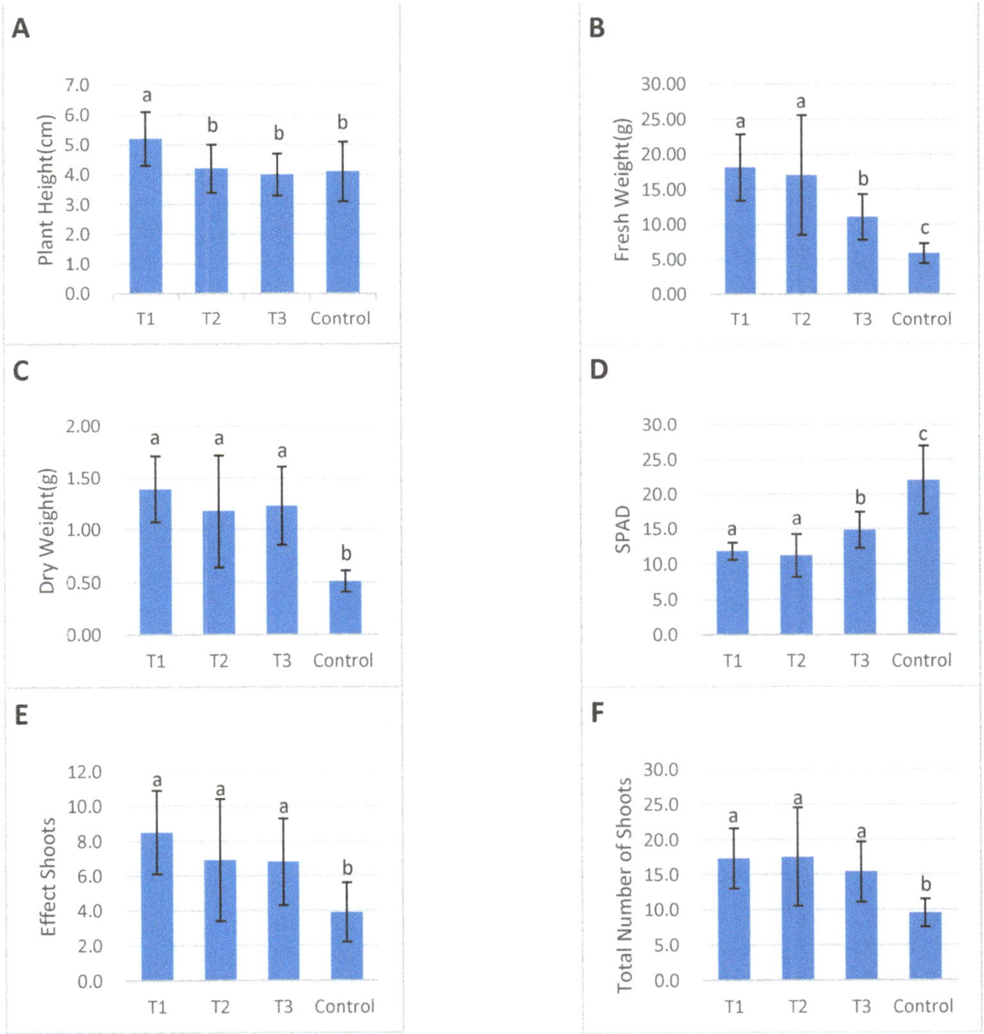

Figure 2. Effect of different in vitro culture systems on plants' development of *M. deliciosa* 'Thai Constellation' after 90 days culturing. (**A**) Plant height, (**B**) Fresh weight, (**C**) Dry weight, (**D**) SPAD, (**E**) Effective shoots, and (**F**) Total number of shoots. Treatment: Liquid medium in SETIS™ bioreactor; treatment 1: immersion frequency of 1.5 h, immersion duration of 1 min; treatment 2: immersion frequency of 1.5 h, immersion duration of 2 min; treatment 3: immersion frequency of 3 h, immersion duration of 2 min. Control: Semi-solid medium in baby food jar. Bars indicate mean ± SE. Different letters indicate significant differences by Tukey's test at $p \leq 0.05$.

For fresh weight, treatments 1 and 2 ($p < 0.001$) (17.02 g and 18.13 g, respectively) were significantly higher than treatment 3 (11.05 g), and all three treatments in bioreactors were significantly higher than the explants cultured in a semi-solid medium, with 5.85 g. Similarly, for dry weight, all three treatments in bioreactors ($p < 0.001$) (1.18 g to 1.39g) were significantly higher than the control cultured in a semi-solid medium (0.51 g); however, there were no significant differences among the dry weights of the three treatments.

Total numbers of shoots ($p < 0.004$) (15.4 to 17.5 shoots per explant) in bioreactors were significantly higher than those in the control group using a semi-solid medium, with 9.5 shoots per explant. The number of effective shoots ($p < 0.007$) (6.8 to 8.5 shoots per explant) in bioreactors was significantly higher than the control, with 3.9 shoots per explant.

The highest chlorophyll relative content was recorded in the control group in a semi-solid medium with an SPAD value of 22.0 ($p < 0.001$), followed by treatment 3 with an SPAD value of 14.8 ($p < 0.007$) and then treatment 1 and treatment 2 with SPAD values of 11.8 and 11.2, respectively (Figure 2).

3.2. Acclimatization Using IQ Forte as a Biostimulant

Survival rates varied greatly among the four groups. Treatments 1 (2.0 mL/L; 85%) and 2 (3.0 mL/L; 90%), where lower concentrations of the biostimulant were applied, had higher survival rates, while treatment 3 (4.0 mL/L; 70%) and the control group (75%) had lower survival rates (Figure 3).

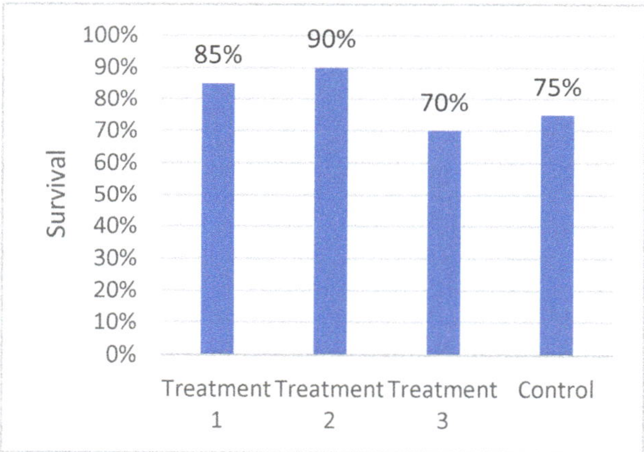

Figure 3. Survival of *M. deliciosa* 'Thai Constellation' after 100 days of growing in the greenhouse. Treatment 1: 2.0 mL/L biostimulant; treatment 2: 3.0 mL/L biostimulant; treatment 3: 4.0 mL/L biostimulant; control: water without biostimulant.

For the plant growth data (including plant height, fresh weight, and dry weight), the differences between the three groups were not significant at the 0.05 level of significance according to ANOVA. At the same time, treatment 3 was significantly lower than the other two treatments, but it was not significantly different from the control group.

The highest plant height (16.1 cm) occurred in treatment 2, followed by treatment 1 (15.8 cm) and the control group (15.2 cm). Treatment 3 had a plant height of 12.9 cm, significantly lower than treatments 1 and 2 ($p < 0.02$). For fresh weight, treatments 1 and 2 (22.21 g to 20.48 g, respectively) were significantly higher than treatment 3 (11.17 g) ($p < 0.03$). There was no significant difference between the control group (16.24 g) and the other treatments. Similarly, for dry weight, treatment 3 (0.87 g) was significantly lower than treatments 1 (1.79 g) and 2 (1.61 g) ($p < 0.05$), with no significant difference between the control group (1.37 g) and the other groups (Figure 4).

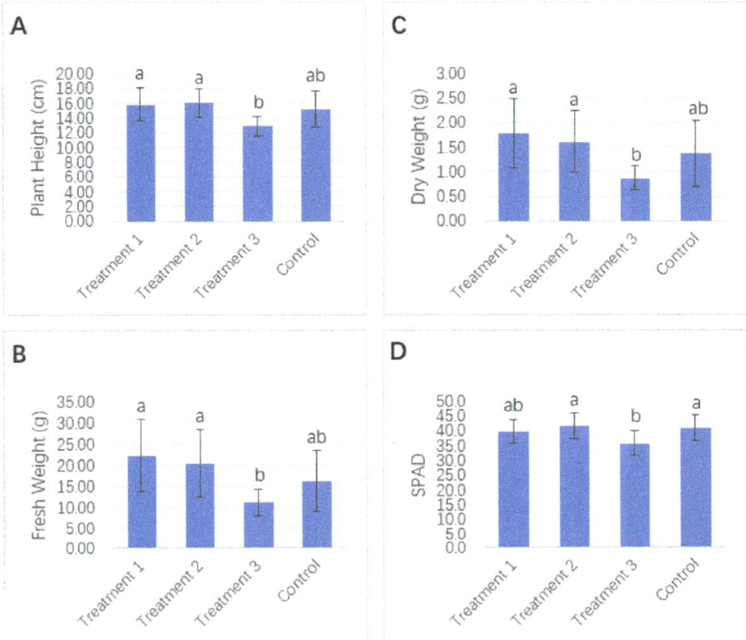

Figure 4. Effect of different concentrations of biostimulant on plants' development of *M. deliciosa* 'Thai Constellation' after 100 days of growing in the greenhouse. (**A**) Plant height, (**B**) Fresh weight, (**C**) Dry weight, (**D**) SPAD; treatment 1: 2.0 mL/L biostimulant; treatment 2: 3.0 mL/L biostimulant; treatment 3: 4.0 mL/L biostimulant; control: water without biostimulant. Bars indicate mean ± SE. Different letters indicate significant differences by Tukey's test at $p \leq 0.05$.

For the relative chlorophyll content, results were slightly different. Treatment 3 was significantly lower than treatment 2 and the control, while treatment 1 was not significantly different from the other groups (Figure 4).

4. Discussion

The evaluation of the different in vitro culture systems demonstrated the usefulness of TI (temporary immersion) to increase the quantity and quality of shoots produced with respect to the conventional micropropagation system in a semi-solid medium [19]. Similarly, *Anthurium andreanum* Lind. and malanga (*Colocasia esculenta* L. Schott) multiplication can also be greatly improved in efficiency by TIS [19,20]. In studies with sugarcane and different types of TIS bioreactors, the efficiency of using temporary immersion systems for direct organogenesis has also been shown [21].

The two species mentioned above belong to the Araceae family, like *M. deliciosa*. In our study, *M. deliciosa* 'Thai Constellation' propagated in SETIS™ bioreactors achieved higher multiplication rates and increased fresh/dry weight as compared to those in a semi-solid culture medium. For the shoot multiplication rate, not only was the total number of new shoots significantly greater in temporary immersion bioreactors than in a semi-solid culture medium, but the number of effective shoots produced by it was likewise much greater, which is of great significance for commercial production practice. Because each effective shoot that emerges can be used as a new single explant to enter the next round of multiplication, the effective shoot yield improvement during each round of multiplication will greatly affect the final production efficiency. The increase in fresh/dry weight in the bioreactor is also important. Shoots that accumulate more nutrients have a higher adaptation and survival rate during subsequent multiplication or acclimatization.

One of the main advantages of using TIS is that it promotes ventilation within the culture vessel, allowing the removal of volatile compounds such as ethylene [22], recirculating the carbon dioxide necessary for photosynthesis [23], and increasing stomatal functionality in the leaves compared to those obtained in semi-solid environments [24]. These factors could explain the increased efficiency of *M. deliciosa* multiplication in TIS.

Unlike the above parameters, however, for plant height, the data in the temporary immersion bioreactors were not fully ahead of the traditional semi-solid medium culture method. The average plant height in the control was not significantly different from either treatment 2 or treatment 3.

For the relative chlorophyll content, the experimental data collected contradicts many experiments. An increase in the content of photosynthetic pigments in TIS compared to a semi-solid system has been observed in malanga (*Colocasia esculenta*) [20]. In addition, an increase in chlorophyll synthesis has also been reported by using TIS in the micropropagation of *A. andreanum*, and the relative chlorophyll content was significantly higher for *Brassavola nodosa* (L.) Lindl. plantlets produced in TIS bioreactors as compared to those from a semi-solid medium [25]. In this experiment, however, in temporary immersion bioreactors, the relative chlorophyll content of the leaves of the explants was significantly lower than that of the control using the traditional method. The control group was significantly higher than all other treatments. Irradiance is an important factor in physiological processes that affects the synthesis of chlorophyll [26], leading to potentially two main reasons for our results:

1. The light system in the immersion bioreactors in this experiment was different from the lighting used in the control group. It has been shown that different light sources result in significant differences in relative chlorophyll content among banana varieties [10]. The effects of LED and fluorescent lamps were compared as sources of light on the growth and development of sweet basil (*Ocimum basilicum* L.) and lemon balm (*Melissa officinalis* L.), indicating that the response of plants to the applied light is individual and depends on the species [27].
2. For micropropagated plants in the lab, photosynthesis is not really useful for their growth since we have already added a sufficient amount of sucrose to the medium as their nutrient source. Plant materials in TIS may instead accumulate relatively little chlorophyll due to higher growth rates.

Immersion and aeration frequency and duration in TIS bioreactors are factors that require adjustments for each species, and they play an important role in plant in vitro multiplication and plant development [28]. The optimization of such parameters is important to prevent physiological disorders such as hyperhydricity. Generally, the incidence of hyperhydricity increases due to the over-accumulation of water in plant tissues [29]. Two different immersion frequencies (1.5 h and 3 h) and two immersion durations (1 min and 2 min) were set to form three experimental groups in this experiment. For several parameters, such as dry weight, total new shoots produced, and number of effective shoots, the results of the three treatments were not significantly different. This is the first study on variegated *M. deliciosa* using TIS, and there were no previous reports to suggest optimum conditions for *Monstera* in general. Thus, this initial study explored some basic parameters for immersion using TIS bioreactors, warranting further studies to optimize the system, such as using shorter immersion duration and lower immersion frequency.

For plant height, treatment 1 with an immersion duration of 1 min was significantly higher than treatments 2 and 3 with an immersion duration of 2 min. At the time of experimental data collection, no significant hyperhydricity was observed. This suggests that a shorter immersion duration helps to increase the plant height of variegated *M. deliciosa*.

While the dry weights of the three treatments were not significantly different, the fresh weights of treatments 1 and 2 were significantly greater than those of treatment three. This implies that the water content of the plant material may be related to the frequency of immersion. A higher frequency of immersion would increase the water content of the culture. A similar trend has been described, where shorter immersion intervals gave rise to a higher number of shoots and a greater ratio of hyperhydric carnation plantlets cultured in the TIS [30].

The relative chlorophyll content of treatment 3 was significantly higher than that of treatments 1 and 2. This may indicate that in TIS, the relative chlorophyll content of the plant material is related to its water content.

The roles played by accumulated amino acids in plants vary, from acting as an osmolyte, regulating ion transport, modulating stomatal opening, and detoxifying heavy metals. Amino acids also affect the synthesis and activity of some enzymes, gene expression, and redox homeostasis [31]. By providing exogenous amino acids to maize seedlings, it was observed that in addition to proline, alanine, serine, and asparagine also delayed wilting under stress conditions [32]. In wheat, a foliar spray of methionine in a mixture increased drought tolerance and was associated with an increase in plant–water relations, physio-biochemical attributes, yield attributes, and nutrition quality when applied at 0.2 mg/mL [33]. Using two levels of methionine, i.e., 100 mg/L and 200 mg/L, on green parasol plants improved the absorption of N, P and K, which was associated with increases in dry weight of shoots, leaf area, and leaf chlorophyll contents [34]. In cowpea, a 0.4 mM concentration of methionine was effective in enhancing stress tolerance and improving growth, yield characteristics, contents of chlorophyll, carotenoids, shoot and seed nutrients, and other components [35].

Several studies mentioned above have demonstrated that a wide range of amino acid compositions have a beneficial effect on many species of plants, especially in terms of nutrient uptake and increasing their resistance to stress. The main components of the biostimulant used in this experiment are amino acids obtained from the hydrolysis of plant proteins, and we expected that they would play a significant role in the acclimatization of variegated *M. deliciosa*.

The results show that application of low concentrations of a biostimulant returned higher survival rates for variegated *M. deliciosa* seedlings, while high concentrations of a biostimulant had an opposite effect, leading to reduced survival rates.

In the experiment, the growth data of the plantlets (plant height, fresh weight, and dry weight) treated with low concentrations of biostimulants were not significantly different from the control. The reasons for this were analyzed as follows:

1. Insufficient sample number. This is mainly due to the high value of the material (variegated *M. deliciosa*) in this experiment, the high cost of expanding the sample numbers, and the lack of experience with its acclimatization process (death of plant material due to excessive sunlight and high temperatures). Subsequent experiments could increase the number of samples based on more mature acclimatization experience.
2. It is true that a proportion of plants are insensitive to biostimulants by means of increasing exogenous amino acids.

It is evident from the data that the use of low concentrations of a biostimulant did not significantly affect the relative chlorophyll content of leaves of variegated *M. deliciosa*. It is noteworthy that the SPAD values of treatment 3 were significantly lower than those of treatment 2 and the control, indicating that higher concentrations of amino acid solutions may, on the contrary, hinder the development of chloroplasts in the green parts of the leaves of variegated *M. deliciosa*.

5. Conclusions

This is the first study reporting the use of the SETIS™ bioreactor for the micropropagation of *M. deliciosa* 'Thai Constellation'. We determined that this variegated *M. deliciosa* under TIS bioreactors provided higher multiplication rates and increased fresh/dry weight compared to conventional culture systems using a semi-solid agar-based medium. Different immersion and aeration frequencies and duration in TIS bioreactors do affect the growth of the material in a number of ways, such as plant height, fresh weight, and chlorophyll content. Although the plant material in TIS has grown favorably under the conditions we set up, we still need to follow up with more experiments to find out if there are more suitable culture conditions for *M. deliciosa*. This is also the first study reporting the use of IQ Forte as a biostimulant for the acclimatization of *M. deliciosa* 'Thai Constellation'. We found that this biostimulant at low concentrations enhanced the acclimatization survival of variegated *M. deliciosa*. The gain in plantlets' growth was not significant, while higher concentrations of the biostimulant may have a negative effect on the growth of variegated *M. deliciosa*.

For future research, expanded experimentation is necessary. The first step could be to set up more different culture conditions, mainly different immersion and aeration intervals and durations, under conditions of more adequate plant material and more bioreactors. This setup of culture conditions could be based on this study to increase the differences in conditions to get more significant results. This study only proved the advantages of the TIS bioreactor in the multiplication of variegated *M. deliciosa*, and experiments on its rooting efficiency under the new system would be a good addition. This experiment did not demonstrate the effect of IQ Forte, the new biostimulant, well. The following directions can be taken for the subsequent improvement of the experiment: (1) Consider the effect of seasonal climate to provide more stable and safe acclimatization conditions for variegated *M. deliciosa* to avoid the occurrence of stress. (2) Prepare sufficient experimental materials before the experiment. (3) Other cheaper varieties of *Monstera* can be used to test this novel biostimulant to reduce the cost of testing.

For commercial production, this study suggests the use of SETIS™ bioreactors instead of the system using a semi-solid agar-based medium at the stage of multiplication of *M. deliciosa* 'Thai Constellation'. The new system not only saves production space and labor but also expands the micropropagation of *M. deliciosa* at a higher efficiency.

Author Contributions: Conceptualization, W.V.; methodology, W.V. and Y.J.; validation, W.V., Y.J. and D.B.; formal analysis, Y.J.; investigation, Y.J.; resources, W.V. and D.B.; data curation, Y.J.; writing—original draft preparation, Y.J.; writing—review and editing, W.V.; visualization, Y.J.; supervision, W.V. and D.B.; project administration, W.V. and D.B.; funding acquisition, W.V. All authors have read and agreed to the published version of the manuscript.

Funding: This research received no external funding.

Institutional Review Board Statement: Not applicable.

Informed Consent Statement: Not applicable.

Data Availability Statement: Data is contained within the article.

Acknowledgments: This work is supported by the USDA National Institute of Food and Agriculture, Hatch project 1012202.

Conflicts of Interest: The authors declare no conflict of interest.

References

1. Bailey, L.H.; Bailey, E.Z. *Hortus Third: A Concise Dictionary of Plants Cultivated in the United States and Canada*; MacMillan: New York, NY, USA, 1976; pp. 739–740.
2. Peppard, T.L. Volatile flavor constituents of *Monstera deliciosa*. *J. Agric. Food Chem.* **1992**, *40*, 257–262. [CrossRef]
3. Spínola, V.; Perestrelo, R.; Câmara, J.S.; Castilho, P.C. Establishment of *Monstera deliciosa* fruit volatile metabolomic profile at different ripening stages using solid-phase microextraction combined with gas chromatography–mass spectrometry. *Food Res. Int.* **2015**, *67*, 409–417. [CrossRef]
4. Dewei, H.A.N. Tissue Culture and Rapid Propagation of *Monstera deliciosa*. *J. Anhui Agric. Sci.* **2010**, *38*, 14270–14271.
5. Bhojwani, S.S.; Dantu, P.K.; Bhojwani, S.S.; Dantu, P.K. Micropropagation. In *Plant Tissue Culture: An Introductory Text*; Springer: New Delhi, India, 2013; pp. 245–274.
6. Berthouly, M.; Etienne, H. Temporary immersion system: A new concept for use liquid medium in mass propagation. In *Liquid Culture Systems for In Vitro Plant Propagation*; Springer: Dordrecht, The Netherlands, 2005; pp. 165–195.
7. Preil, W. General introduction: A personal reflection on the use of liquid media for in vitro culture. In *Liquid Culture Systems for In Vitro Plant Propagation*; Springer: Dordrecht, The Netherlands, 2005; pp. 1–18.
8. Bello-Bello, J.J.; Cruz-Cruz, C.A.; Pérez-Guerra, J.C. A new temporary immersion system for commercial micropropagation of banana (*Musa* AAA cv. Grand Naine). *Vitr. Cell. Dev. Biol.-Plant* **2019**, *55*, 313–320. [CrossRef]

9. Ramírez-Mosqueda, M.A.; Bello-Bello, J.J. SETIS™ bioreactor increases in vitro multiplication and shoot length in vanilla (*Vanilla planifolia* Jacks. Ex Andrews) *Acta Phys. Plant* **2021**, *43*, 52. [CrossRef]
10. Vendrame, W.A.; Feuille, C.; Beleski, D.; Bueno, P.M.C. In vitro growth responses of ornamental Bananas (*Musa* sp.) as affected by light sources. *Horticulturae* **2022**, *8*, 92. [CrossRef]
11. Conover, C.A.; Poole, R.T. Acclimatization of indoor foliage plants. *Hort. Rev.* **1984**, *6*, 119–154.
12. Kozai, T. Acclimatization of micropropagated plants. *Biotech. Agric For.* **1991**, *17*, 127–141.
13. European Biostimulant Industry Council (EBIC). 2016. Available online: http://www.biostimulants.eu/ (accessed on 23 October 2020).
14. Mrid, R.B.; Benmrid, B.; Hafsa, J.; Boukcim, H.; Sobeh, M.; Yasri, A. Secondary metabolites as biostimulant and bioprotectant agents: A review. *Sci. Total Environ.* **2021**, *777*, 146204. [CrossRef]
15. Schaafsma, G. Safety of protein hydrolysates, fractions thereof and bioactive peptides in human nutrition. *Eur. J. Clin. Nutr.* **2009**, *63*, 1161–1168. [CrossRef]
16. Colla, G.; Nardi, S.; Cardarelli, M.; Ertani, A.; Lucini, L.; Canaguier, R.; Rouphael, Y. Protein hydrolysates as biostimulants in horticulture. *Sci. Hort.* **2015**, *196*, 28–38. [CrossRef]
17. Colla, G.; Hoagland, L.; Ruzzi, M.; Cardarelli, M.; Bonini, P.; Canaguier, R.; Rouphael, Y. Biostimulant action of protein hydrolysates: Unraveling their effects on plant physiology and microbiome. *Front. Plant Sci.* **2017**, *8*, 2202. [CrossRef] [PubMed]
18. Murashige, T.; Skoog, F. A revised medium for rapid growth and bioassays with tobacco tissue cultures. *Physiol. Plant* **1962**, *15*, 473–497. [CrossRef]
19. Martínez-Estrada, E.; Islas-Luna, B.; Pérez-Sato, J.A.; Bello-Bello, J.J. Temporary immersion improves in vitro multiplication and acclimatization of *Anthurium andreanum* Lind. *Sci. Hort.* **2019**, *249*, 185–191. [CrossRef]
20. Arano-Avalos, S.; Gómez-Merino, F.C.; Mancilla-Álvarez, E.; Sánchez-Páez, R.; Bello-Bello, J.J. An efficient protocol for commercial micropropagation of malanga (*Colocasia esculenta* L. Schott) using temporary immersion. *Sci. Hort.* **2020**, *261*, 108998. [CrossRef]
21. Orozco-Ortiz, C.; Sanchez, L.; Araya-Mattey, J.; Vargas-Solorzano, I.; Araya-Valverde, E. BIT biorreactor increases in vitro multiplication of quality shoots in sugarcane (*Saccharum* spp. variety LAICA 04-809). *Plant Cell Tissue Org. Cult.* **2022**, *152*, 115–128. [CrossRef]
22. Roels, S.; Noceda, C.; Escalona, M.; Sandoval, J.; Canal, M.J.; Rodriguez, R.; Debergh, P. The effect of headspace renewal in a temporary immersion bioreactor on plantain (*Musa* AAB) shoot proliferation and quality. *Plant Cell Tissue Org. Cult.* **2006**, *84*, 155–163. [CrossRef]
23. Aragón, C.E.; Sánchez, C.; Gonzalez-Olmedo, J.; Escalona, M.; Carvalho, L.; Amâncio, S. Comparison of plantain plantlets propagated in temporary immersion bioreactors and gelled medium during in vitro growth and acclimatization. *Biol. Plant* **2014**, *58*, 29–38. [CrossRef]
24. Afreen, F. Temporary immersion bioreactor. *Plant Tissue Cult. Eng.* **2008**, *2008*, 187–201.
25. Vendrame, W.A.; Xu, J.; Beleski, D.G. Micropropagation of *Brassavola nodosa* (L.) Lindl. using SETIS™ bioreactor. *Plant Cell Tissue Org. Cult.* **2023**, *153*, 67–76. [CrossRef]
26. Mancilla-Álvarez, E.; Pérez-Sato, J.A.; Núñez-Pastrana, R.; Spinoso-Castillo, J.L.; Bello, J.J.B. Comparison of Different Semi-Automated Bioreactors for In vitro Propagation of Taro (*Colocasia esculenta* L. Schott). *Plants* **2021**, *10*, 1010.
27. Frąszczak, B.; Golcz, A.; Zawirska-Wojtasiak, R.; Janowska, B. Growth rate of sweet basil and lemon balm plants grown under fluorescent lamps and LED modules. *Acta Sci. Pol. Hortorum Cultus* **2014**, *13*, 3–13.
28. Georgiev, V.; Schumann, A.; Pavlov, A.; Bley, T. Temporary immersion systems in plant biotechnology. *Eng. Life Sci.* **2014**, *14*, 607–621. [CrossRef]
29. Saher, S.; Piqueras, A.; Hellin, E.; Olmos, E. Hyperhydricity in micropropagated carnation shoots: The role of oxidative stress. *Physiol. Plant.* **2004**, *120*, 152–161. [CrossRef] [PubMed]
30. Ahmadian, M.; Babaei, A.; Shokri, S.; Hessami, S. Micropropagation of carnation (*Dianthus caryophyllus* L.) in liquid medium by temporary immersion bioreactor in comparison with solid culture. *J. Gen. Eng. Biotechnol.* **2017**, *15*, 309–315. [CrossRef] [PubMed]
31. Rai, V.K. Role of amino acids in plant responses to stresses. *Biol. Plant.* **2002**, *45*, 481–487. [CrossRef]
32. Thakur, P.S.; Rai, V.K. Exogenously supplied amino acids and water deficits in *Zea mays* cultivars. *Biol. Plant* **1985**, *27*, 458–461. [CrossRef]
33. Hammad, S.A.; Ali, O.A. Physiological and biochemical studies on drought tolerance of wheat plants by application of amino acids and yeast extract. *Ann. Agric. Sci.* **2014**, *59*, 133–145. [CrossRef]
34. Shekari, G.; Javanmardi, J. Effects of foliar application pure amino acid and amino acid containing fertilizer on broccoli (*Brassica oleracea* L. var. *italica*) transplants. *Adv. Crop Sci. Technol.* **2017**, *5*, 280–287. [CrossRef]
35. Merwad, A.R.M.; Desoky, E.S.M.; Rady, M.M. Response of water deficit-stressed *Vigna unguiculata* performances to silicon, proline or methionine foliar application. *Sci. Hort.* **2018**, *228*, 132–144. [CrossRef]

horticulturae

MDPI

Article

Photosynthetic Characteristics of 20 Herbaceous Peony Cultivars

Anqi Xie [1], Jingyue Wu [1], Yajie Shi [1], Fuling Lei [1], Lingling Dong [1], Dongliang Zhang [1], Limin Sun [2,*] and Xia Sun [1,*]

[1] College of Horticulture Science and Engineering, Shandong Agricultural University, Tai'an 271018, China; leifuling1213@163.com (F.L.); dll1204@163.com (L.D.); z1399391909@163.com (D.Z.)
[2] College of Forestry, Shandong Agricultural University, Tai'an 271018, China
* Correspondence: sunlimin06@163.com (L.S.); unsiax@sdau.edu.cn (X.S.); Tel.: +86-538-8242238 (L.S. & X.S.)

Abstract: To identify herbaceous peony cultivars with strong photosynthetic productivity, we compared the photosynthetic characteristics of 20 herbaceous peony cultivars based on four photosynthetic characteristics parameters and established light–response curves under a light intensity gradient, using CIRAS-3 portable photosynthetic dynamic monitoring. The net photosynthetic rate (Pn) showed a "unimodal" diurnal variation pattern, with a peak around 12:00. The diurnal pattern of the transpiration rate was the same as that of Pn. Stomatal conductance values (Gs) showed similar patterns among the cultivars, with only small differences. The daily variation in intracellular CO_2 concentration (Ci) showed an opposite trend to that of Pn. When the photosynthetically active radiation was 0–400 $\mu mol \cdot m^{-2} s^{-1}$, Pn increased linearly and gradually with increasing light intensity. 'Xueyuanhonghua', 'Qingwen', 'Taohuafeixue', 'Chifen', and 'Qihualushuang' showed high photosynthetic productivity. 'Xueyuanhonghua', 'Fushi', 'Qingwen', 'Tianshanhongxing', 'Qingtianlan', 'Dafugui', and 'Hongfushi' had high light saturation points and the highest light resistance. 'Xueyuanhonghua', 'Qingwen', 'Taohuafeixue', 'Tianshanhongxing', 'Qingtianlan', 'Guifeichacui', 'Chifen', and 'Hongxiuqiu' had low light compensation points. Thus, two cultivars with strong photosynthetic productivity, 'Xueyuanhonghua' and 'Qingwen', can be cross-bred to obtain both light- and shade-tolerant plants. This study provides a theoretical basis for breeding new cultivars with high photosynthetic productivity.

Keywords: herbaceous peony; photosynthetic characteristics; light compensation point; light saturation point

Citation: Xie, A.; Wu, J.; Shi, Y.; Lei, F.; Dong, L.; Zhang, D.; Sun, L.; Sun, X. Photosynthetic Characteristics of 20 Herbaceous Peony Cultivars. *Horticulturae* **2023**, *9*, 1331. https://doi.org/10.3390/horticulturae9121331

Academic Editors: Margherita Beruto, Emmy Dhooghe and Bruce Dunn

Received: 30 October 2023
Revised: 4 December 2023
Accepted: 7 December 2023
Published: 12 December 2023

1. Introduction

The herbaceous peony (*Paeonia lactiflora* Pall.) is a perennial herb belonging to the Paeoniaceae family that is widely cultivated worldwide. Currently, cultivated herbaceous peony cultivars are divided into three groups: lactiflora, hybrid, and Itoh. Herbaceous peony is a typical temperate long-day plant that prefers warmth; however, it is resistant to cold [1]. Numerous herbaceous peony cultivars, with various attractive colors and patterns, exist and are widely used, for example, in landscaping and as fresh-cut flowers and potted plants. Herbaceous peony not only has high ornamental value, but also medicinal, edible, and health-care values. Its roots contain pharmacologically active ingredients, such as paeoniflorin, and its petals contain various nutrients, active ingredients, and minerals [2]. Organ growth and development in herbaceous peony is affected by climatic conditions. Flower bud differentiation, a critical step in the flowering process, occurs during the autumn–winter (short-day season), and flower bud development and flowering occur during the long days of summer. In conditions of insufficient light or during short days, mixed bud germination occurs and may result in plants having long leaves without flowers or with abnormal flowers [3], seriously affecting the ornamental and medicinal value of herbaceous peony.

Plant photosynthetic characteristics play a decisive role in plant productivity [4,5]. Studies have shown that improved photosynthetic capacity increases the quantity and quality of herbaceous peony flowers, extends their flower life, and enhances their ornamental value [6]. The photosynthetic process is sensitive to the environment [7,8]. Environmental factors such as light and humidity play important roles in regulating carbohydrate exchange in plant leaves. Stomatal opening and closing are regulated by the activation of photosensitive ion channels. When light intensity is high, the ion channels are activated, resulting in stomatal opening. In contrast, when light is weak or absent, these channels are not activated and the stomata remain closed [9]. The state of stomatal opening directly affects the transpiration rate of plant leaves. When the humidity around the leaves or in the surrounding environment is too high during the day, the plant controls the transpiration rate by adjusting stomatal closure [10]. When plants adapt to changes in the external environment, photosynthetic characteristics change correspondingly, leading to certain differences in the photosynthetic performance of different germplasm of the same plant.

When cultivated for medicinal purposes, herbaceous peony is generally propagated from seeds sown into a field, and plants can be harvested after 3–5 years of vegetative growth. When cultivated for ornamental purposes, they are generally propagated using division propagation, and daughter plants require 3–4 years of vegetative growth before they can be divided again. Furthermore, the low reproduction coefficient limits the development of high-value cultivars [11]. The more suitable the seasonal light conditions with more energy available, the more vigorous the vegetative growth and the higher the organic matter accumulation [12]. The suitability of the light intensity for herbaceous peony plants is related to the growth of the vegetative organs, photosynthetic production, storage of nutrients in the root system, development of scale buds in the winter, and consumption of accumulated resources for flowering in the spring of the following year [13].

The slow propagation and cultivation of new cultivars restrict the development of the herbaceous peony industry, which requires modernization and industrialization to improve the quantity and quality of herbaceous peony cultivars. A previous study showed clear and stable genetic differences in the photosynthetic rate among peony cultivars and among individuals of the same cultivar, indicating the possibility of breeding crop cultivars with high photosynthetic rates [14]. Hence, high-light-efficiency breeding has become a hot topic in plant genetic breeding. In this study, we aimed to generate a theoretical basis for the breeding of new varieties with high light efficiency. Toward this aim, we dynamically monitored the typical photosynthetic metrics—net photosynthetic rate (Pn), intercellular carbon dioxide concentration (Ci), transpiration rate (Tr), and stomatal conductance (Gs)—of different herbaceous peony cultivars, and light–response curves under various light intensities were dynamically monitored. We compared the photosynthetic characteristics of the different cultivars according to the light saturation and compensation points. Our findings provide technical support for the cultivation and further industrialization of herbaceous peony.

2. Materials and Methods

2.1. Test Site and Materials

The experiment was completed in early May 2022 at the Herbaceous Peony Resource Nursery and Horticultural Experiment Center of Shandong Agricultural University, located in the south of Tai'an City (35°38′ N 116°02′ E), China. The center is located in a temperate semi-humid continental monsoon climate zone with four distinct seasons: summer, winter, autumn, and spring. Summers and winters are long, whereas spring and autumn seasons are short. Rainfall is concentrated during the summer. The mean annual temperature ranges between 11 and 14 °C, with occasionally extremely low temperatures around −20 °C and extremely high temperatures around 41 °C. The region has a frost-free period of approximately 187 days, and annual precipitation of 600–700 mm. The mean annual relative humidity is 65%. The 20 experimental herbaceous peony cultivars analyzed in this study were 'Xuanyuanhonghua', 'Fushi', 'Yangfeichuyu', 'Huguangshanse', 'Fenchijinyu', 'Gaoganhong', 'Bingshan', 'Xuefeng', 'Qingwen', 'Taohuafeixue', 'Tianshanhongxing', 'Qingtianlan', 'Guifeichacui', 'Dafugui', 'Hongfeng', 'Chifen', 'Hongxiuqiu', 'Qihualushuang', 'Dongjingnvlang', and 'Hongfushi' (Figure 1). All cultivars were planted, cultivated, and managed under the same natural conditions. The planting density was 80 cm × 80 cm. The plants were planted in loam soil, pH 7.6, with an organic matter content of 23.32 g kg^{-1}, ammonium nitrogen content of 25.72 g kg^{-1}, nitrate nitrogen content of 1.70 g kg^{-1}, available phosphorus content of 47.63 g kg^{-1}, and available potassium content of 79.00 g kg^{-1}.

Figure 1. Herbaceous peony cultivars tested in this study. Note: 1, 'Xueyuanhonghua'; 2, 'Fushi'; 3, 'Yangfeichuyu'; 4, 'Huguangshanse'; 5, 'Fenchijinyu'; 6, 'Gaoganhong'; 7, 'Bingshan'; 8, 'Xuefeng'; 9, 'Qingwen'; 10, 'Taohuafeixue'; 11, 'Tianshanhongxing'; 12, 'Qingtianlan'; 13, 'Guifeichacui'; 14, 'Dafugui'; 15, 'Hongfeng'; 16, 'Chifen'; 17, 'Hongxiuqiu'; 18, 'Qihualushuang'; 19, 'Dongjingnvlang'; and 20, 'Hongfushi'.

2.2. *Measurement of Daily Variation in Net Photosynthesis*

According to the method of Du [15], net Pn, Ci, Tr, and Gs were measured on a sunny day with full sunshine and few clouds, using a CIRAS-3 portable photosynthesis analyzer (PP Systems; Boston, MA, USA). Measurements were obtained from healthy, functional leaves with robust, consistent growth and at the same plant position. Three herbaceous peony plants were selected for each cultivar, three functional leaves were measured for each plant, and the measurements were repeated thrice for each leaf. The leaf chamber conditions were consistent with the natural environment during the measurement. Daily photosynthesis metrics were measured at 08:00, 10:00, 12:00, 14:00, 16:00, and 18:00.

2.3. Establishment of Photosynthetic Light-Response Curves

According to Du [15], data for photosynthetic light-response curves were collected using the CIRAS-3 instrument. The measurement conditions were set as follows: CO_2 concentration, 360 mg L^{-1}; relative humidity, 60%; and light intensity 2000, 1600, 1200, 800, 400, 200, 100 and 0 μmol·m^{-2}s^{-1}. Light–response curves were obtained on sunny days from 10:00 to 11:30.

2.4. Data Analysis

Microsoft Excel software (Office 2016, Microsoft, Redmond, WA, USA) was used to analyze the data and create graphs. SPSS Statistics V22.0 (IBM Corp., Armonk, NY, USA) was used to analyze the errors, significance of differences, and principal components. The least significant difference and Duncan's new multiple-range tests were used to compare the differences among samples. The Kaiser–Meyer–Olkin test and Bartlett sphere test were used to test the applicability of factor analysis to the original data.

3. Results

3.1. Diurnal Variation in Photosynthetic Parameters in 20 Herbaceous Peony Cultivars

3.1.1. Diurnal Variation in Pn

The Pn of the 20 herbaceous peony cultivars showed a "unimodal" trend (Figure 2). In the morning, with increasing light intensity and temperature, Pn continued to increase, and it peaked around 12:00 for all cultivars. In the afternoon, with the weakening of light intensity, the Pn values of the cultivars continually declined until 18:00, when the lowest values were reached. In the afternoon, 'Dongjingnvlang' exhibited the largest decline in Pn of approximately 10 μmol·m^{-2}s^{-1}, whereas 'Guifeichacui' and 'Dafugui' exhibited the smallest Pn reductions of approximately 4 μmol·m^{-2}s^{-1}.

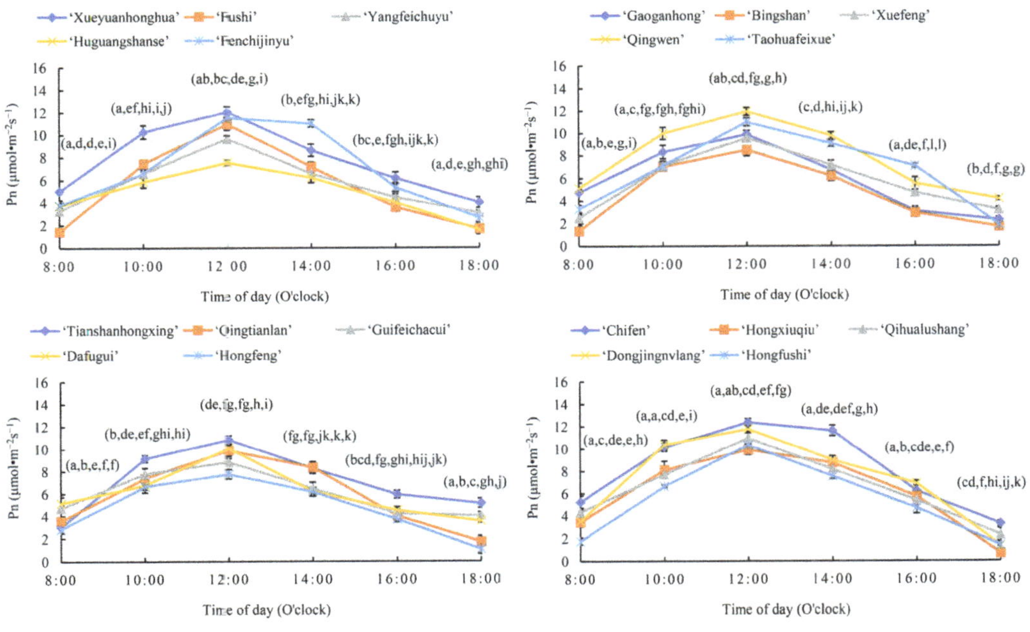

Figure 2. Diurnal changes in the net photosynthesis rate (Pn) of the 20 herbaceous peony cultivars. Different lowercase letters indicate differences among the 20 herbaceous peony cultivars at the same time at $p < 0.05$; error lines indicate standard errors.

Seven cultivars had a Pn peak value between 11 and 13 μmol·m^{-2}s^{-1}: 'Xuanyuan-honghua', 'Fenchijinyu', 'Qingwen', 'Taohuafeixue', 'Chifen', 'Qihualushuang', and

'Dongjingnvlang'. For 13 cultivars, the Pn peak was between 6 and 11 μmol·m^{-2}s^{-1}: 'Fushi', 'Yangfeichuyu', 'Huguangshanse', 'Gaoganhong', 'Bingshan', 'Xuefeng', 'Qingwen', 'Tianshanhongxing', 'Guifeichacui', 'Dafugui', 'Hongfeng', 'Hongxiuqiu', and 'Hongfushi'

3.1.2. Diurnal Variation in Gs

The Gs showed similar patterns, with only small differences noted among different cultivars (Figure 3). The Gs of each cultivar exhibited a variation trend similar to that of Pn. From 08:00 onwards, Gs showed an upward trend and peaked around 12:00 for all cultivars. In the afternoon, the Gs continually declined until 18:00, when the lowest values were observed.

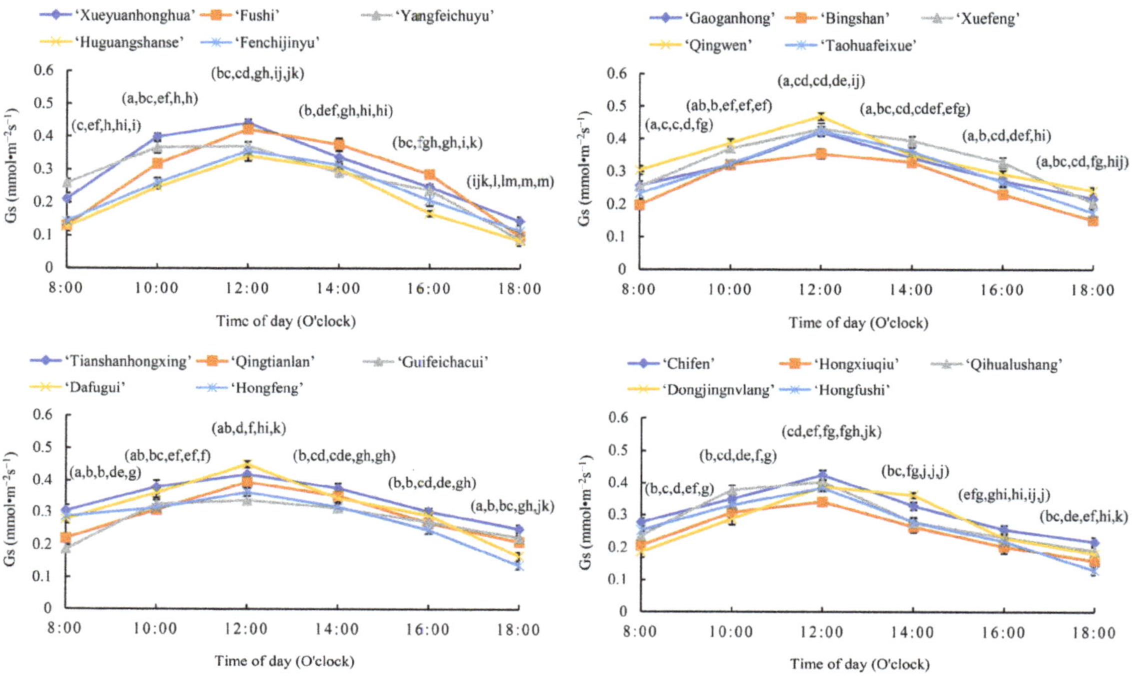

Figure 3. Diurnal changes in the stomatal conductance (Gs) of the 20 herbaceous peony cultivars. Different lowercase letters indicate differences among the 20 herbaceous peony cultivars at the same time at $p < 0.05$; error lines indicate standard errors.

Ten cultivars, 'Yangfeichuyu', 'Huguangshanse', Fenchijinyu', 'Bingshan', 'Qingtianlan', 'Guifeichacui', 'Hongfeng', 'Hongxiuqiu', 'Dongjingnvlang', and 'Hongfushi' had a Gs peak of 0.3–0.4 mmol·m^{-2}s^{-1}, whereas the other 10 cultivars, 'Xuanyuanhonghua', 'Fushi', 'Gaoganhong', 'Xuefeng', 'Qingwen', 'Taohuafeixue', 'Tianshanhongxing', 'Dafugui', 'Chifen', and 'Qihualushuang' had a Gs peak of 0.4–0.5 mmol·m^{-2}s^{-1}.

3.1.3. Diurnal Variation in Ci

The average daily Ci showed similar trends among the tested cultivars. The highest values were observed in the morning, after which they decreased by 12:00, and increased again in the evening. This "high–low–high" trend of Ci contrasted with that of the daily variation in Pn (Figure 4).

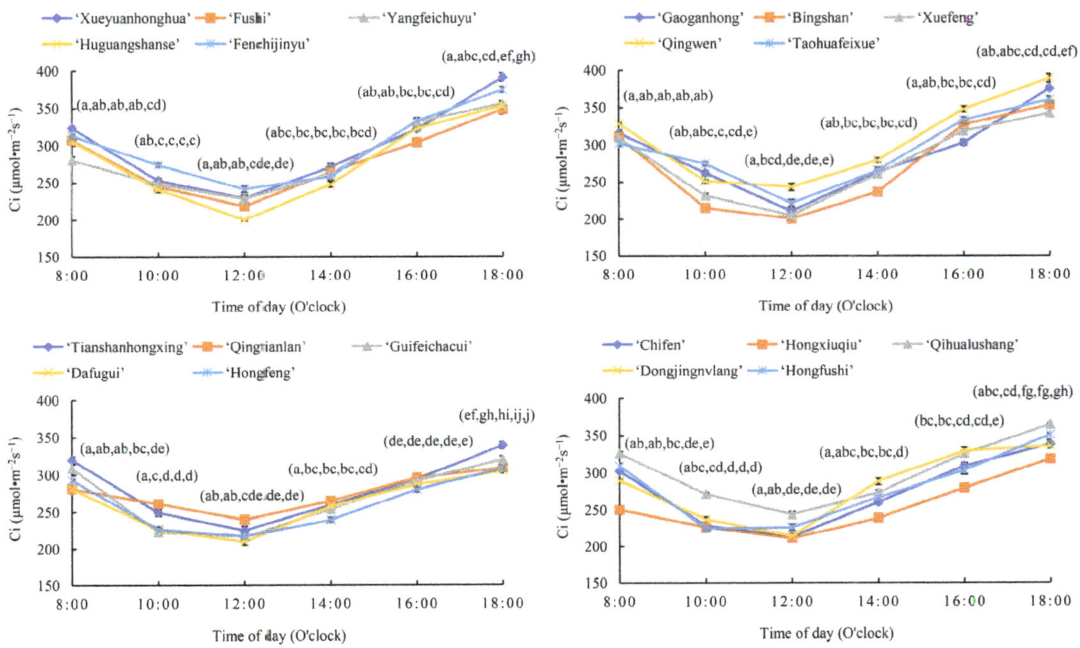

Figure 4. Diurnal changes in the intracellular CO_2 concentration (Ci) of the 20 herbaceous peony cultivars. Different lowercase letters indicate differences among the 20 herbaceous peony cultivars at the same time at $p < 0.05$; error lines indicate standard errors.

From 08:00 to 12:00, the Ci values of all cultivars decreased; the Ci values of 'Yangfeichuyu', 'Qingtianlan', and 'Hongxiuqiu' decreased the least, by approximately 40–50 μmol·m^{-2}s^{-1}. However, the Ci values of 'Huguangshanse', 'Gaoganhong', 'Xuefeng', and 'Qingwen' exhibited the largest decrease of approximately 100 μmol·m^{-2}s^{-1}. The decrease in Ci of the remaining 13 cultivars varied from 70 to 100 μmol·m^{-2}s^{-1}.

From 12:00 to 18:00, the Ci values of all tested cultivars rebounded, and those of 'Xueyuanhonghua', 'Huguangshanse', 'Gaoganhong', and 'Bingshan' increased the most, by 150–160 μmol·m^{-2}s^{-1}. The Ci value of 'Qingtianlan' increased the least, by approximately 70 μmol·m^{-2}s^{-1}. The Ci value of the remaining 15 cultivars increased by 90–140 μmol·m^{-2}s^{-1}.

3.1.4. Diurnal Variation in Tr

The daily variation trend of the Tr of the tested cultivars was similar to that of Pn, both of which were unimodal, and the peak values were observed around 12:00 (Figure 5). From 08:00 to 12:00, the Tr increased the most in 'Yangfeichuyu', by approximately 9 mmol·m^{-2}s^{-1}, followed by 'Tianshanhongxing', 'Guifeichacui', 'Dafugui', and 'Qihualushuang', for which the Tr increased by approximately 8 mmol·m^{-2}s^{-1}. The rate of increase in Tr for 'Fenchijinyu' and 'Bingshan' was approximately 5 mmol·m^{-2}s^{-1}. The Tr of the remaining 13 cultivars increased by 6–7 mmol·m^{-2}s^{-1}.

From 12:00 to 18:00, the Tr values decreased. The largest decrease of 10.2 mmol·m^{-2}s^{-1} was observed for 'Guifeichacui', followed by 'Fenchijinyu', 'Tianshanhongxing', and 'Hongxiuqiu', for which the Tr decreased by approximately 9.3 mmol·m^{-2}s^{-1}. The smallest decrease in Tr was observed for 'Dongjingnvlang' (5.8 mmol·m^{-2}s^{-1}), followed by 'Huguangshanse', 'Xueyuanhonghua', 'Fushi', and 'Bingshan'. The Tr value of the remaining 11 cultivars decreased by 7–8 mmol·m^{-2}s^{-1}.

Ten cultivars, 'Xuanyuanhonghua', 'Fushi', 'Yangfeichuyu', 'Qingwen', 'Taohuafeixue', 'Tianshanhongxing', 'Guifeichacui', 'Dafugui', 'Chifen', and 'Qihualushuang', had a Tr peak of 10–13 mmol·m^{-2}s^{-1}, whereas the other 10 cultivars, 'Huguangshanse', 'Fenchijinyu',

'Gaoganhong', 'Bingshan', 'Xuefeng', 'Qingtianlan', 'Hongfeng', 'Hongxiuqiu', 'Dongjingnvlang', and 'Hongfushi', had a Tr peak of 7–10 mmol·m^{-2}s^{-1}.

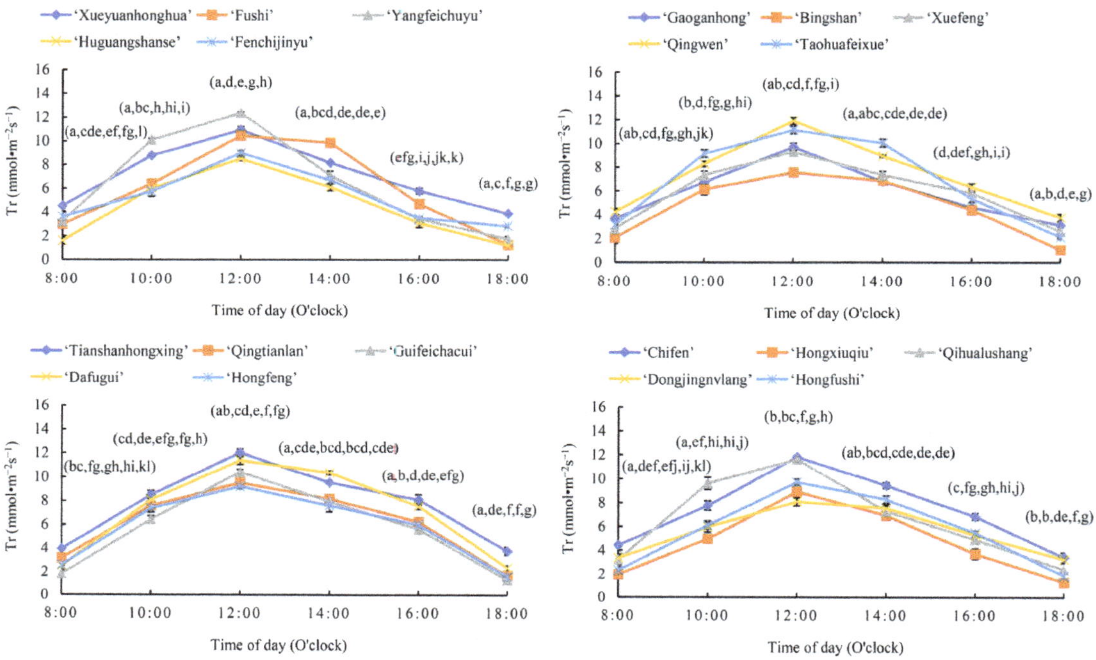

Figure 5. Diurnal changes in the transpiration rate (Tr) of the 20 herbaceous peony cultivars. Different lowercase letters indicate differences among the 20 herbaceous peony cultivars at the same time at $p < 0.05$; error lines indicate standard errors.

3.2. Light–Response Curves for the 20 Herbaceous Peony Cultivars

When the light intensity was less than 400 μmol·m^{-2}s^{-1}, the Pn showed a linear trend and increased rapidly with the increase in photosynthetically active radiation (PAR). When PAR exceeded 400 μmol·m^{-2}s^{-1}, the increase in Pn gradually decreased (Figure 6). When PAR was greater than 1200 μmol·m^{-2}s^{-1}, the Pn value tended to become stable; thus, the maximum Pn (Pn_{max}) was reached. When PAR ranged between 1200 and 2000 μmol·m^{-2}s^{-1} with an increase in light intensity, the Pn_{max} remained stable: it reached the light saturation point (LSP), and the photosynthetic rate at this time reflected the maximum photosynthetic capacity of plant leaves. The Pn_{max} values corresponding to the LSPs of all cultivars were concentrated in the range of 8–13 μmol·m^{-2}s^{-1}. Eight cultivars, 'Huguangshanse', 'Bingshan', 'Xuefeng', 'Guifeichacui', 'Hongfeng', 'Chifen' 'Qihualushuang', and 'Dongjingnvlang' had a Pn_{max} value between 8 and 10 μmol·m^{-2}s^{-1}. Five cultivars, 'Yangfeichuyu', 'Fenchijinyu', 'Gaoganhong', 'Taohuafeixue', and 'Hongxiuqiu' had a Pn_{max} between 10 and 11 μmol·m^{-2}s^{-1}. Seven cultivars, 'Xuanyuanhonghua', 'Fushi', 'Qingwen', 'Tianshanhongxing', 'Qingtianlan', 'Dafugui', and 'Hongfushi' had a Pn_{max} between 11 and 13 μmol·m^{-2}s^{-1}.

When Pn is zero, the nutrients produced by the leaves and the nutrients consumed during respiration reach a dynamic equilibrium, and the corresponding light intensity is termed the light compensation point (LCP). The LCPs of the 20 cultivars varied in the range of 10–70 μmol·m^{-2}s^{-1}. Eight cultivars, 'Xuanyuanhonghua', 'Qingwen', 'Taohuafeixue', 'Tianshanhongxing', 'Qingtianlan', 'Guifeichacui', 'Chifen', and 'Hongxiuqiu' had an LCP between 10 and 30 μmol·m^{-2}s^{-1}. Seven cultivars, 'Fushi', 'Yangfeichuyu', 'Huguangshanse', 'Gaoganhong', 'Bingshan', 'Dafugui', and 'Hongfeng' had an LCP between 10 and 30 μmol·m^{-2}s^{-1}. Five cultivars, 'Fenchijinyu', 'Xuefeng', 'Qihualushuang', 'Dongjingnvlang', and 'Hongfushi' had an LCP between 10 and 30 μmol·m^{-2}s^{-1}.

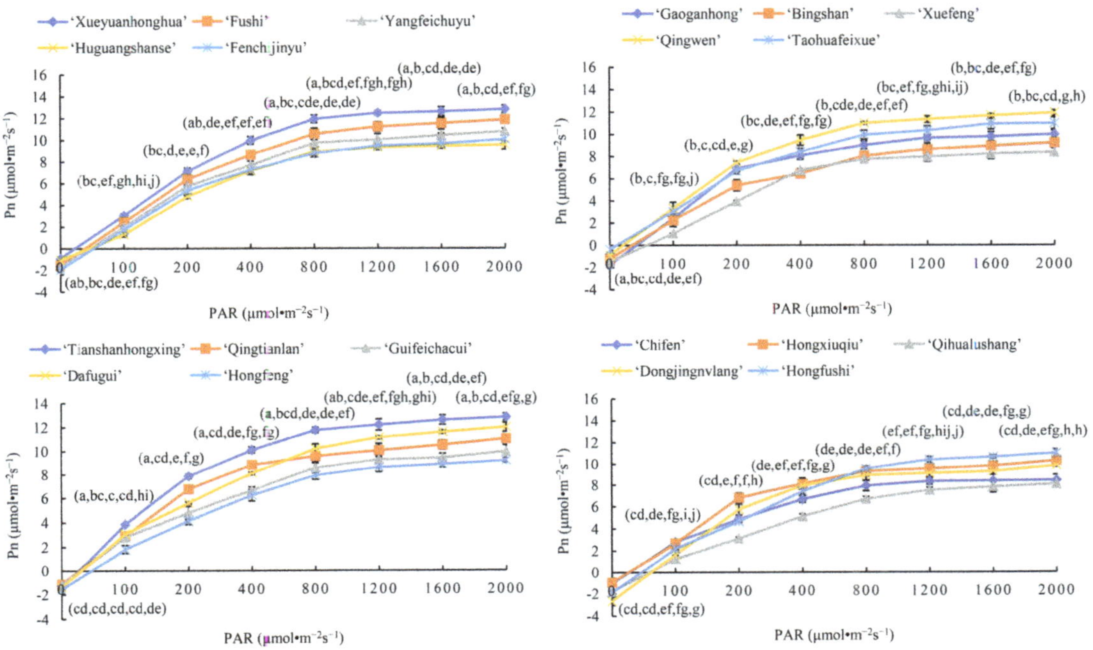

Figure 6. Light–response curves for the 20 herbaceous peony cultivars. Different lowercase letters indicate differences among the 20 herbaceous peony cultivars at the same light intensity at $p < 0.05$; error lines indicate standard errors.

3.3. Comprehensive Evaluation of the Photosynthetic Characteristics of the 20 Herbaceous Peony Cultivars

Principal component analysis of the various indices measured at various times in the 20 cultivars tested revealed that the variance contribution rate of the first six principal components was 39.428%, 16.775%, 11.586%, 6.849%, 6.354%, and 4.667%, respectively, and the cumulative variance contribution rate was 85.659%. All eigenvalues were greater than one, indicating that the principal components well represented the information contained in the photosynthetic characteristics of herbaceous peony. Therefore, these six principal components can be used as comprehensive indicators for evaluating the photosynthetic characteristics of herbaceous peony (Table 1).

Table 1. Factor analysis of photosynthetic parameters for the 20 herbaceous peony cultivars.

Principal Component	Eigenvalue	Variance Contribution Rate %	Cumulative Contribution Rate %
1	12.617	39.428	39.428
2	5.368	16.775	56.204
3	3.707	11.586	67.789
4	2.192	6.849	74.638
5	2.033	6.354	80.992
6	1.493	4.667	85.659

The functional expression of the six principal components was obtained based on the eigenvalues and load values of the principal components. By substituting standardized data into the principal component function expression, F1, F2, F3, F4, F5, and F6 scores were obtained. A comprehensive score function for each cultivar was obtained by summing the product of the score value of each principal component and the variance contribution rate of the corresponding characteristic value: F = 0.394 × F1 + 0.168 × F2 + 0.116 × F3 + 0.685 × F4 + 0.635 × F5 + 0.467 × F6. The composite scores of the different cultivars were calculated and sorted; the results are shown in Table 2.

Table 2. Factor scores and ranking of photosynthetic parameters for the 20 herbaceous peony cultivars.

Cultivar	Score of Each Principal Component						Total Score F	Order
	F1	F2	F3	F4	F5	F6		
'Qingwen'	6.69	1.29	0.42	0.05	1.17	0.52	3.93	1
'Xueyuanhonghua'	6.07	0.55	3.39	0.13	0.58	0.56	3.60	2
'Yangfeichuyu'	−0.18	0.10	1.68	2.85	−0.79	0.20	1.68	3
'Xuefeng'	−0.78	1.90	−1.89	0.86	2.79	−1.52	1.44	4
'Chifen'	2.98	2.90	−3.54	−1.38	−0.05	2.41	1.40	5
'Gaoganhong'	0.50	0.83	0.00	0.01	2.20	−0.76	1.39	6
'Tianshanhongxing'	6.96	−2.87	−0.80	0.32	−1.28	−0.67	1.27	7
'Qihualushuang'	−0.79	4.65	−1.35	2.96	−2.29	0.72	1.22	8
'Taohuafeixue'	1.33	−0.90	0.87	−0.49	1.16	0.56	1.13	9
'Guifeichacui'	−1.59	−1.75	−1.75	0.62	0.13	0.47	−0.39	10
'Bingshan'	−4.40	−0.41	−0.37	0.31	2.14	−0.42	−0.47	11
'Dafugui'	2.27	−4.14	−1.49	−0.24	−0.12	−1.11	−0.73	12
'Huguangshanse'	−5.58	0.27	2.59	0.49	0.76	0.43	−0.84	13
'Fenchijinyu'	−1.97	2.68	2.21	−1.06	−0.11	−0.15	−0.94	14
'Fushi'	−0.49	−1.26	3.19	−0.86	−0.66	−0.23	−1.15	15
'Qingtianlan'	−0.94	−2.17	−1.36	−0.94	−0.39	0.17	−1.70	16
'Hongfeng'	−3.49	−2.08	−2.38	1.39	−0.57	−0.69	−1.74	17
'Hongfushi'	−1.19	−1.19	1.26	0.49	−2.38	−0.74	−2.04	18
'Hongxiuqiu'	−4.45	−2.01	−0.01	−2.12	−0.44	2.84	−2.50	19
'Dongjingnvlang'	−0.96	3.61	−0.66	−3.39	−1.85	−2.60	−4.56	20

3.4. Cluster Analysis of Photosynthetic Metrics among the 20 Herbaceous Peony Cultivars

A systematic cluster dendrogram of photosynthetic metrics directly reflects the similarity of photosynthetic productivity of different cultivars; the earlier they are aggregated, the more similar their photosynthetic characteristics. The average distance clustering method was used to cluster 20 herbaceous peony cultivars with six factors score as variables. (Figure 7). Based on the similarity of their photosynthetic characteristics, the cultivars were divided into three major groups: Group I, whose Pn, Gs, and Tr values were low indicating weak photosynthetic productivity, comprised five cultivars, 'Qingtianlan', 'Guifeichacui', 'Dafugui', 'Hongfeng', and 'Hongxiuqiu'. Group II, with medium photosynthetic productivity, comprised 12 cultivars, 'Fushi', 'Tianshanhongxing', 'Chifen', 'Hongfushi', 'Yangfeichuyu', 'Taohuafeixue', 'Huguangshanse', 'Fenchijinyu', 'Gaoganhong', 'Xuefeng', 'Bingshan', and 'Dongjingnvlang'. And Group III, whose Pn, Gs, and Tr values were high indicating high photosynthetic productivity, comprised three cultivars, 'Xuanyuanhonghua', 'Qingwen', and 'Qihualushuang'. According to the cluster analysis, the cultivars 'Qingwen', 'Xuanyuanhonghua', and 'Qihualushuang' had the best photosynthetic characteristics, and they also ranked high (first, second, and eighth, respectively) in terms of the comprehensive score in the principal component analysis. Thus, the results of the cluster and principal component analyses were highly consistent, suggesting that this approach can be used to effectively screen high-value germplasm resources.

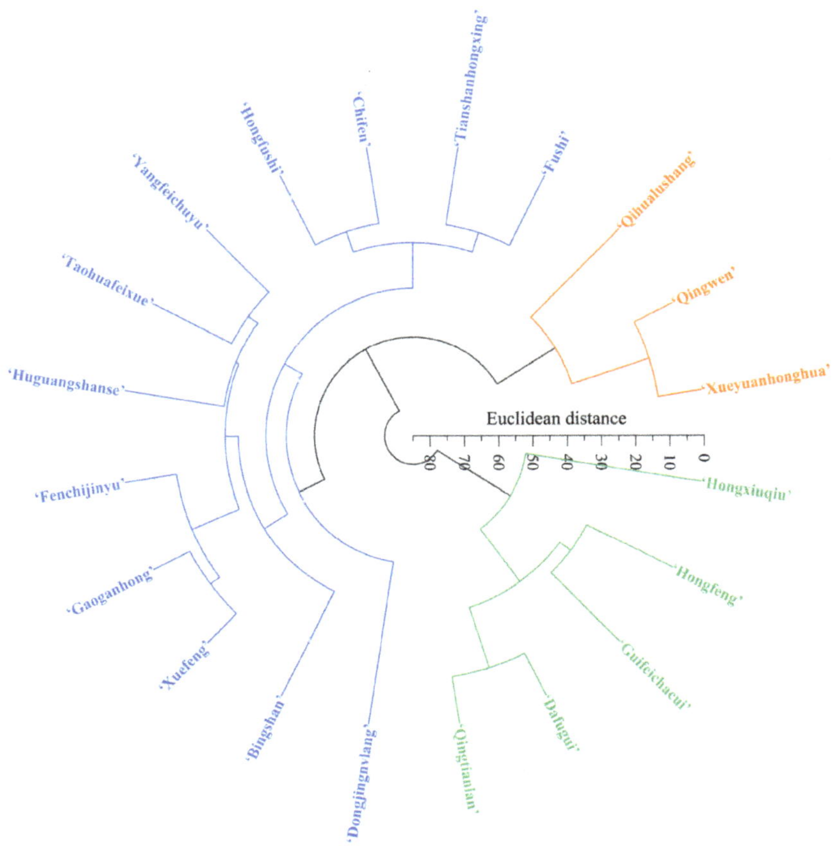

Figure 7. Clustering tree of 20 herbaceous peony cultivars.

4. Discussion

Photosynthesis accounts for 90–95% of dry matter production in plants. As such, photosynthesis is closely related to plant growth and an important parameter in cultivar selection and breeding [16,17]. Li et al. [18] studied the photosynthetic characteristics of four wild oil peony species from 11 sources in different provinces of China and identified cultivars with strong photosynthetic capacity as excellent sources. Chen et al. [13] compared the photosynthetic characteristics of 10 types of *Cypripedium* plants introduced from Changbai Mountain, providing a theoretical basis for the artificial introduction and cultivation of wild resources. The photosynthetic capacity differs among plant species and cultivars, and ecological factors, such as light, temperature, water, and air, affect plant photosynthesis [19]. Differences in the photosynthetic characteristics of different herbaceous peony cultivars have been attributed to a combination of parental differences and long-term adaptation to environmental conditions [20]. Photosynthetic products play an important role in maintaining the high ornamental value of herbaceous peony, and high-light-efficiency breeding is an important method for improving the overall photosynthetic capacity of herbaceous peony cultivars [21].

The photosynthetic characteristics of herbaceous peony leaves change in correspondence with diurnal changes in light intensity, temperature, and humidity. The Pn, which is mainly influenced by Ci, Gs, and Tr, determines the photosynthetic capacity of plants [22–24]. In herbaceous peony, the diurnal variation in Pn reportedly is "bimodal" under strong summer light, with a "photosynthetic siesta phenomenon," and "unimodal" under light shade [15]. In our study, the diurnal variation in Pn measured in early May followed a unimodal pattern in all herbaceous peony cultivars tested, and the peak values were observed around 12:00. Stomatal conductance is readily affected by the environment, with high Gs being conducive to gas exchange, allowing the entry of additional CO_2 from the air [9]. The differences in diurnal Gs values among the cultivars were small, and the curves showed similar trends: the highest Gs occurred around 12:00, and the lowest values were observed in the morning and evening. There was a negative feedback adjustment between Gs and Ci [25]. In our study, the diurnal variation in Ci was the highest in the morning for all cultivars. By 12:00, it had decreased, and in the evening, it increased again. Tr daily variation showed a similar "low–high–low" trend as Pn daily variation, both of which followed single-peak curves, with the peaks occurring around 12:00. Principal component analysis is a relatively mature comprehensive evaluation method widely used in germplasm resource evaluation and screening [26]. According to the comprehensive score ranking, 'Xueyuanhonghua', 'Qingwen', 'Taohuafeixue', 'Chifen', and 'Qihualushuang' had relatively high photosynthetic productivity. Cluster analysis results corroborated that these cultivars had strong photosynthetic capacity, and screening by principal component analysis and cluster analysis validated mutual verification.

Herbaceous peony is a light-loving plant that requires sufficient light during growth; however, some cultivars can grow and develop normally under light shade [12]. Light–response curves reflect plant adaptability to and utilization ability of different light intensities [27]. Studies have reported that moderate shade during the flowering period of herbaceous peony in the summer, which increases the environmental humidity and decreases the temperature, can extend the flowering period and prevent sunburn [28,29]. In this study, when the PAR was 0–400 $\mu mol{\cdot}m^{-2}s^{-1}$, the Pn of the herbaceous peony cultivars increased linearly, and with subsequent increases in light intensity, the Pn gradually reached a saturation level. The LCP and LSP represent the light demand characteristics of plants [30,31]. The LCP represents the ability of plants to adapt to low light [32]. The eight cultivars with the lowest LCPs, 'Xueyuanhonghua', 'Qingwen', 'Taohuafeixue', 'Tianshanhongxing', 'Qingtianlan', 'Guifeichacui', 'Chifen', and 'Hongxiuqiu', were more shade tolerant than the other cultivars and, therefore, can be planted in low-light environments, such as forests and shaded slopes in garden landscapes. The seven cultivars with the highest LSPs, 'Xueyuanhonghua', 'Fushi', 'Qingwen', 'Tianshanhongxing', 'Qingtianlan', 'Dafugui', and 'Hongfushi' were more resistant to bright light [33]. According to the light demand characteristics of different herbaceous peony cultivars, plant allocation can be a reasonable approach to achieving the purposes of obtaining high yields and creating rich landscapes.

5. Conclusions

In this study, the photosynthetic characteristics and light responses of 20 herbaceous peony cultivars were dynamically monitored. Four cultivars with strong shade tolerance and high photosynthetic productivity were 'Xueyuanhonghua', 'Qingwen', 'Taohuafeixue', and 'Chifen', which can be cross-bred to create new cultivars of shade-tolerant flowers. For breeding new cultivars with strong light resistance and high photosynthetic productivity, two cultivars 'Xueyuanhonghua' and 'Qingwen' were identified as potential parents. Two cultivars, 'Xueyuanhonghua', and 'Qingwen', were both light and shade resistant and had strong photosynthetic productivity and can be cross-bred to obtain both light- and shade-resistant flower cultivars. This study provides a theoretical basis for breeding of highly efficient new varieties.

Author Contributions: A.X. designed and largely conducted the study, analyzed the data, and wrote the paper. J.W., Y.S., F.L., L.D. and D.Z. participated in the tests and data analysis. L.S. and X.S. co-designed the study and revised manuscript drafts. All authors contributed to the writing of the manuscript. All authors have read and agreed to the published version of the manuscript.

Funding: This work was funded by the Shandong Province Improved Seed Project (No. 2021S230304-02583).

Data Availability Statement: The datasets used and/or analyzed during this study are available from the corresponding author upon reasonable request.

Conflicts of Interest: The authors declare no conflict of interest.

References

1. Zhang, D.L.; Xie, A.Q.; Yang, X.; Yang, L.J.; Shi, Y.J.; Dong, L.L.; Lei, F.L.; Sun, L.M.; Bao, M.Y.; Sun, X. Analysis of physiological and biochemical factors affecting flower color of herbaceous peony in different flowering periods. *Horticulturae* **2023**, *9*, 502. [CrossRef]
2. Wang, X.X.; Xiong, H.R.; Wang, S.L.; Zhang, Y.L.; Song, Z.H.; Zhang, X.X. Physicochemical analysis, sensorial evaluation, astringent component identification and aroma-active compounds of herbaceous peony (*Paeonia lactiflora* Pall) black tea. *Ind. Crop Prod.* **2023**, *193*, 116159. [CrossRef]
3. Sun, X. *Chinese Herbaceous Peony Varieties Appreciation*; China Forestry Publishing House: Beijing, China, 2019; pp. 10–11. (In Chinese)
4. Ma, Y.; Ding, Y.; Lu, J. Photosynthetic physiological characteristics of four medicinal plants. *Appl. Ecol. Environ. Res.* **2019**, *17*, 12807–12826. [CrossRef]
5. Noor, H.; Wang, Q.; Islam, M.A.; Sun, M.; Lin, W.; Ren, A.X.; Feng, Y.; Yu, S.B.; Fida, N.; Dong, S.F.; et al. Effects of sowing methods and nitrogen rates on photosynthetic characteristics, yield, and quality of winter wheat. *Photosynthetica* **2021**, *59*, 277–285. [CrossRef]
6. Xia, X.; Tang, Y.; Wei, M.G.; Zhao, D.Q. Effect of paclobutrazol application on plant photosynthetic performance and leaf greenness of herbaceous peony. *Horticulturae* **2018**, *4*, 5. [CrossRef]
7. Liu, J.L.; Liu, L.; Fu, Q.; Zhang, L.L.; Li, J.W.; Yu, K.; Xu, Q. Effects of straw mulching and tillage measures on the photosynthetic characteristics of maize leaves. *Transact. ASABE* **2019**, *62*, 851–858. [CrossRef]
8. Li, Z.K.; Gong, X.W.; Wang, J.L.; Cheng, Y.D.; Liu, F.Y.; Li, H.P.; Lü, G.H. Foliar water uptake improves branch water potential and photosynthetic capacity in *Calligonum mongolicum*. *Ecol. Indic.* **2023**, *146*, 109825. [CrossRef]
9. Han, T.; Feng, Q.; Yu, T.F.; Yang, X.M.; Zhang, X.F.; Li, K. Characteristic of stomatal conductance and optimal stomatal behaviour in an arid oasis of northwestern China. *Sustainability* **2022**, *14*, 968. [CrossRef]
10. Cao, Y.B. *Study on Leaf Morphological Structure and Photosynthetic Characteristics of Wild European Plum*; Xinjiang Agricultural University: Wulumuqi, China, 2022. (In Chinese)
11. Xie, A.Q.; Sun, L.M.; Zhang, D.L.; Li, Y.; Liu, Z.M.; Li, X.; Sun, X. Changes in the root system of the herbaceous peony and soil properties under different years of continuous planting and replanting. *Hortic. Plant J.* **2023**, *9*, 801–810. [CrossRef]
12. Yang, Y.L.; Chen, M.Z.; Tian, J.S.; Xiao, F.; Xu, S.Z.; Zuo, W.Q.; Zhang, W.F. Improved photosynthetic capacity during the mid- and late reproductive stages contributed to increased cotton yield across four breeding eras in Xinjiang, China. *Field Crops Res.* **2019**, *240*, 177–184. [CrossRef]
13. Du, J.; Xu, J.G.; Lv, M.W.; Gao, C.R.; Lu, J.; Zhang, Q.X.; Wang, T.L.; Sun, X. Study on photosynthetic mechanism of summer shading slowing aging of herbaceous peony (*Paeonia lactiflora*) leaves. *Plant Physiol. J.* **2018**, *54*, 773–782. (In Chinese)
14. Chen, L.F.; Li, S.; Li, Y.; Zhang, Y.Q.; Bai, Y.; Cong, H.; Liu, W.; Zhou, Y.W. Comparative study of Cypripedium plant photosynthetic characteristics from Changbai Mountain. *Horticulturae* **2023**, *9*, 358. [CrossRef]
15. Du, J. *Effects of Summer Strong Light and Slight Shading on Ultrastructure and Physiological Characteristics in the Herbaceous Peony (Paeonia lactiflora Pall.) Leaf*; Shandong Agricultural University: Taian, China, 2016. (In Chinese)
16. Bai, Y.C.; Zhang, J.P.; Wu, Y.; Huang, R.M.; Chang, Y.Y.; Lei, X.S.; Song, X.B.; Pei, D. Possibility of increasing the growth and photosynthetic properties of precocious walnut by grafting. *Sustainability* **2020**, *12*, 5178. [CrossRef]
17. Kim, D.H.; Son, S.; Jung, J.Y.; Lee, J.C.; Kim, P.G. Photosynthetic characteristics and chlorophyll content of *Cypripedium japonicum* in its natural habitat. *For. Sci. Technol.* **2022**, *18*, 160–171. [CrossRef]
18. Li, M.; Li, W.Y.; Zhang, Y.L.; Zhang, H.; Ji, W.L.; Du, M.Y.; Zhao, X. Comparative photosynthetic characteristics and chlorophyll content of four oil peony species from 11 provenances in China. *Pak. J. Bot.* **2018**, *50*, 143–150.
19. Zhang, J.Y.; Liao, M.; Cheng, Y.H.; Feng, Y.; Ju, W.B.; Deng, H.N.; Li, X.; Plenković-Moraj, A.; Xu, B. Comparative chloroplast genomics of seven endangered Cypripedium species and phylogenetic relationships of Orchidaceae. *Front. Plant Sci.* **2022**, *13*, 911702. [CrossRef]
20. Li, L.L. Photosynthetic Physiology of Peony under Different Light Conditions. Master's Thesis, Shandong Agricultural University, Tai'an, China, 2016. (In Chinese).
21. Zhao, D.Q.; Hao, Z.J.; Tao, J. Effects of shade on plant growth and flower quality in the herbaceous peony (*Paeonia lactiflora* Pall.). *Plant Physiol. Biochem.* **2012**, *61*, 187–196. [CrossRef]
22. Schwinghamer, T.; Souleimanov, A.; Dutilleul, P.; Smith, D. The plant growth regulator lipo-chitooligosaccharide (LCO) enhances the germination of canola (*Brassica napus* [L.]). *J. Plant Growth Regul.* **2015**, *34*, 183–195. [CrossRef]
23. Fang, X.M.; Li, Y.S.; Nie, J.; Wang, C.; Huang, K.H.; Zhang, Y.K.; Zhang, Y.L.; She, H.Z.; Liu, X.B.; Ruan, R.W.; et al. Effects of nitrogen fertilizer and planting density on the leaf photosynthetic characteristics, agronomic traits and grain yield in common buckwheat (*Fagopyrum esculentum* M.). *Field Crops Res.* **2018**, *219*, 160–168. [CrossRef]

24. Fang, Z.W.; Wang, X.X.; Zhang, X.Y.; Zhao, D.Q.; Tao, J. Effects of fulvic acid on the photosynthetic and physiological characteristics of *Paeonia ostii* under drought stress. *Plant Signal. Behav.* **2020**, *15*, 1774714. [CrossRef]
25. Favaretto, V.F.; Martinez, C.A.; Soriani, H.H.; Furriel, R.P.M. Differential responses of antioxidant enzymes in pioneer and late-successional tropical tree species grown under sun and shade conditions. *Environ. Exp. Bot.* **2011**, *70*, 20–28. [CrossRef]
26. Wei, C.B.; Li, M.M.; Ma, Z.L.; Qiao, J.; Sun, G.M. Vitamin diversity in pineapple varieties based on cluster and principal component analysis. *Bangl. J. Bot.* **2018**, *47*, 799–803.
27. Li, C.F.; Dong, S.T.; Liu, R.X.; Ren, H.; Ding, Z.S.; Zhao, M. Diurnal variation of gas exchange, chlorophyll fluorescence, and photosynthetic response of six parental lines of maize released in three eras. *J. Integr. Agric.* **2019**, *18*, 2732–2741. [CrossRef]
28. Wang, X.X.; Fang, Z.W.; Zhao, D.Q.; Tao, J. Effects of high-temperature stress on photosynthetic characteristics and antioxidant enzyme system of *Paeonia ostii*. *Phyton* **2022**, *91*, 599–615. [CrossRef]
29. Tang, Y.H.; Shi, W.B.; Xia, X.; Zhao, D.Q.; Wu, Y.Q.; Tao, J. Morphological, microstructural and lignin-related responses of herbaceous peony stem to shading. *Sci. Hortic.* **2022**, *293*, 110734. [CrossRef]
30. Wang, J.Y.; Shen, J.S.; Gu, M.; Wang, J.; Cheng, T.; Pan, H.; Zhang, Q. Leaf coloration and photosynthetic characteristics of hybrids between *Forsythia 'Courtaneur'* and *Forsythia koreana* 'Suwon gold'. *HortScience* **2017**, *52*, 1661–1667. [CrossRef]
31. Wei, S.Y.; Lei, Q.H.; Ji, W.L.J.R.; Fan, R.; Yang, D.Y.; Zhang, Y.L. Differences in light response characteristics of oil peonies from eight provenances in China. *Photosynthetica* **2019**, *57*, 788–795. [CrossRef]
32. Song, H.Z. *Study on Photosynthetic Characteristics and Carbohydrate of Peony*; Beijing Forestry University: Beijing, China, 2012. (In Chinese)
33. Kong, F. *Studies on Photosynthetic Characteristics of Herbaceous Peony*; Yangzhou University: Yangzhou, China, 2010. (In Chinese)

horticulturae

MDPI

Article

Warm Bulb Storage Optimises Flowering Attributes and Foliage Characteristics in *Amaryllis belladonna* L.

Carolyn Margaret Wilmot [1], Muhali Olaide Jimoh [1,2] and Charles Petrus Laubscher [1,*]

1 Department of Horticultural Sciences, Faculty of Applied Sciences, Cape Peninsula University of Technology, P.O. Box 1906, Bellville 7535, South Africa; wilmotc@cput.ac.za (C.M.W.); muhali.jimoh@oouagoiwoye.edu.ng (M.O.J.)
2 Department of Plant Science, Olabisi Onabanjo University, P.O. Box 2002, Ago-Iwoye 120107, Nigeria
* Correspondence: laubscherc@cput.ac.za

Abstract: *Amaryllis belladonna* is an autumn-flowering bulbous geophyte endemic to the Western Cape, South Africa. The species' erratic flowering disposition and brief flowering period upon maturity limit its economic productivity and competitiveness within the traditional genera of cut flowers and potted plants. However, it can be an attractive, eco-friendly, seasonal addition to the specialty floriculture market. A 10-month study evaluated the effects of a warm storage period on *A. belladonna* bulbs' flowering yield, flowering time, quality characteristics, and foliage growth. The experiment comprised dormant flower-sized bulbs randomly assigned to one of six storage regimes of either a 0- (no storage control), 4-, 6-, 8-, 10-, or 12-week interval periods at a continuous warm temperature of 23 ± 1 °C before planting into pots between mid-November 2021 and mid-February 2022 in the greenhouse. The results showed that flowering production (64.3% flowering after the 12-week storage), flowering time (anthesis occurring 9 days after the 10- and 12-week storage), and quality attributes (number of florets in the inflorescence, scape diameter, inflorescence fullness ratio, and pot longevity) of *A. belladonna* scapes were significantly impacted by warm bulb storage, but not foliage growth. Irrespective of bulb storage, inflorescence abortion occurred. An extended bulb storage did not advance the flowering time despite a greater harvest and shorter cultivation periods after planting. This study established that a cumulative temperature range during bulb dormancy is crucial for supporting the *A. belladonna* inflorescence maturity's energetic demands and the opening of floret buds. Bulbs should be stored at elevated temperatures for at least 8–10 weeks to attain the best floret-quality attributes and longevity. However, for an economical and sustainable greenhouse and specialty cut flower production, 12-week warm bulb storage is recommended to achieve the optimal anthesis in the shortest interval for this seasonal single-harvest species after planting.

Keywords: Amaryllidaceae; anthesis; dormancy; specialty cut flower; temperature duration; traditional cut flower

Citation: Wilmot, C.M.; Jimoh, M.O.; Laubscher, C.P. Warm Bulb Storage Optimises Flowering Attributes and Foliage Characteristics in *Amaryllis belladonna* L. *Horticulturae* **2023**, *9*, 1271. https://doi.org/10.3390/horticulturae9121271

Academic Editors: Margherita Beruto, Emmy Dhooghe and Bruce Dunn

Received: 28 October 2023
Revised: 20 November 2023
Accepted: 24 November 2023
Published: 27 November 2023

1. Introduction

Cut flowers and potted plants are among the most extensively produced and marketed ornamentals in the leading and competitive traditional floriculture sector and a multibillion-dollar international export-oriented industry [1,2]. Successful cultivation and production of these ornamental crops have generally emphasised the species appeal, aesthetic traits, and flowering time variation, with significant efforts to attain these targets to meet timeous demands [3,4]. However, a recent awareness of ornamental horticulture has rekindled the interest in features of production sustainability as a selection criterion [5,6]. This movement has been propelled by the realisation of the high levels of resource consumption, energy-intensive production techniques, rapid distribution channels, and profitability to preserve and maintain the traditional standards of ornamental crops at the expense of biodiversity and contribution to the values of the local ecosystems, culture, and societal well-being [5–7].

Due to the significant sustainability issues associated with ornamental plants, producers, and sellers must analyse and mitigate the species' life cycle assessment (LCA) through sustainable and integrated production techniques for future cultivation [5,6]. Among them, given the high cost of greenhouse production and maintenance, spending less time in the greenhouse would alleviate the load on the LCA and minimise production costs [5,8,9]. As a result, speciality cut flower (SCF) production and sales have increased over the last 15 years, drawing attention to the critical role they serve in the worldwide floriculture industry as a viable and sustainable substitute to traditional cut flower (TCF) crops [4]. The SCF market objectives promote various aesthetic features and are strengthened by the preservation of indigenous and underutilised flora grown locally, seasonally, and sustainably produced [4,6,10].

Amaryllis belladonna L. is an inimitable autumn-flowering bulbous geophyte endemic to the botanically diverse fynbos biome of South Africa's Cape Floristic Region [11–14]. Formerly, *A. belladonna* was classified as a monotypic genus of the Amaryllidaceae family [15]. The species is cultivated and naturalised in Mediterranean regions worldwide, where hot, dry summers alternate with mild, wet winters [11,13,14]. *Amaryllis belladonna* owes its numerous vernacular names, "Belladonna Lily", "March Lily", and "Naked Lady", to its hysteranthous habit, which displays a solitary inflorescence on a naked stem in late summer to early autumn. The umbellate floral arrangement consists of fragrant, trumpet-shaped florets that open in coordinated succession to exhibit a range of iridescent pale ivory to deep pink-shaded tepals (Figure 1A). Its strap-shaped leaves unfurl actively at the onset of winter rains and cooler autumn temperatures. When late spring transcends and seasonal temperatures rise, the leaves wither, initiating the bulb's summer dormancy [11–14].

Figure 1. *Amaryllis belladonna* in full anthesis as a cut flower (**A**) or potted plant (**B**) (photos: C. Wilmot).

In a variety of floriculture, landscape, and ornamental industry settings, the perennial bulb's seasonal adaptability, aesthetic splendour, drought tolerance, and minimal maintenance needs, along with its closely related and well-known species, *Clivia*, *Narcissus*, and *Nerine*, have brought conscious recognition to their valuable and desirable commodities [3,4,16–18]. The comparatively short flowering period of *A. belladonna*, which predictably lasts between six and nine weeks, is a significant shortcoming despite the species' promising allure. In addition, once a critical bulb size is attained, each bulb produces only a single inflorescence per season and often encounters a capricious flowering disposition [12,14,19], limiting its marketable application as an influential cut flower and potted plant under the tight parameters of the traditional floriculture trade (Figure 1A,B).

Flowering is a multifaceted complexity of sequential events governed by endogenous mechanisms and extrinsic environmental signals [20–22]. According to several authors, temperature is a critical factor influencing the numerous physiological and morphological processes involved in the induction and differentiation of floral organs and flowering time [21,23–28]. Temperature regulation during dormancy is crucial to regulate bulbs' reproductive forcing, qualities and timing intended to produce timely high-quality flowers [5,29,30]. Although organogenesis of bulb dormancy is an ecological adaptation that allows plant species to survive unfavourable climatic conditions below the soil surface, it provides a favourable stage for handling, treatment, and shipping [31]. Depending on their natural phenological growth conditions, different taxa require different successive temperature regimes to pass through certain developmental stages to flowering [12,25,32]. Mediterranean geophytes' annual rhythmic cycles typically follow a warm–cold–warm cycle.

According to [33], there are many possibilities to increase flowering in *A. belladonna* bulbs, including physical bulb traits (age, size, and dormancy), bulb disturbance and replanting establishment, and environmental factors, such as temperature, to stimulate flower initiation. Exposure to greater temperatures during summer dormancy encourages flowering in the bulbs [13]. Warm storage treatments during dormancy are well-established in the floriculture industry, and in addition to their ability to accelerate or delay flowering, they increase quality attributes and yield percentages. Warm storage (between 25 °C and 30 °C) resulted in later flowering or vernalisation in several *Ornithogalum* bulb species, leading to higher percentage yields and earlier flowering [24]. Similarly, when subjected to consistently higher storage temperatures immediately after leaf wilting, saffron (*Crocus sativus* L.) bulbs flowered earlier [34]. Storage temperatures up to 25 °C promoted healthy vegetative development, inflorescence maturity and flower bud opening in *Watsonia* corms [35]. Storage significantly impacted the flower diameter and number of days before flowering in *Hippeastrum* [36] and accelerated flowering in *Eucomis* species [37]. Other Amaryllidaceae genera that flourish in hot, dry summer areas and respond to warm storage techniques include *Narcissus* species, *Nerine flexuosa*, and *N. sarniensis* [38]. According to [24], standardised forcing protocols make an analysis and comparison extremely challenging; therefore, clarifying the processes that modify these factors in floricultural crops is essential to achieve sustainable production and increased product quality.

There is a pressing need to advance specialized cultivation technologies and adaptations to overcome obstacles in furthering our understanding of indigenous and underutilised plant species [10,16,39]. Progress cannot advance without thorough scientific investigation, and the awareness that ornamental geophytes' intrinsic genetic variation and composition are increasingly valued [16,40,41], and research is needed to encourage growers to cultivate and reintroduce these species to the industry [42,43]. Furthermore, geophytes are a broad group of plants that impact agricultural production and as a result, each novel insight into a species's behavioural characteristics is crucial to its development [31,44,45]. The floriculture industry may implement strategies in contrast to the conventional mainstream that explore and establish alternative approaches to encourage producers to cultivate and reintroduce more sustainable and resilient indigenous and underutilised species for the SCF market, particularly in hot, dry climates with limited water availability. Although extending *A. belladonna's* flowering time would improve the seasonal period of successful SCF production, the competent number of flower stems grown could facilitate and regulate the expansion. These developments would enable greater financial returns through increased market penetration and the ability to manage a larger, more efficient single-harvest crop under sustainable growing methods at the same or lower cost per unit value within the growing season [33]. The most effective approach for inducing optimal flowering during *A. belladonna's* dormancy has not been clearly defined, and there is a scarcity of published scientific research assessing the precise duration of simulating warm temperatures and subsequent planting for these purposes. Therefore, to disseminate a better understanding of the species complex LCA and facilitate the efficiency and expansion of flower production as a sustainable niche crop for the SCF sector of the floricultural network, *Amaryllis belladonna* bulbs were evaluated to establish the most effective warm storage duration for the optimal flowering production, flowering time, visual quality, and foliage characteristics after planting in the greenhouse.

2. Materials and Methods

2.1. Experimental Location

A 10-month study was conducted from mid-November 2021 to the end of September 2022 in the research greenhouse facility of the Department of Horticultural Sciences at the Cape Peninsula University of Technology in Bellville, Cape Town, South Africa, 33°55′45″ S, 18°38′31″ E. The ventilated greenhouse includes a thermostatically regulated system and a transparent polycarbonate roof sheet (Envirowatch, Envirowatch Solutions, Estcourt, South Africa). Evaporative cooling walls, extractor fans, and heaters kept air temperatures in the greenhouse between 21 °C and 26 °C during the day and 14 °C and 18 °C at night. The relative humidity (RH) averaged 60%. Under natural light conditions, the daily average photosynthetic photon flux density (PPFD) was 420 $\mu mol/m^2/s$, with the intensity peaking at 1020 $\mu mol/m^2/s$. The photoperiod corresponded to the prevailing conditions between late spring and early winter (9–12 h).

2.2. Plant Material and Preparation

Eighty-four dormant, flower-sized *A. belladonna* bulbs (30–33 cm circumference, corresponding to a diameter of 9.5–11 cm) were obtained from Assegaaibosch Farm on the Agulhas Plain in the Western Cape, South Africa, in mid-November 2021 (mid–late spring in the southern hemisphere) at the commencement of their dormancy period. Two weeks prior, the contractile bulb roots were undercut deep beneath the soil to accelerate the progression rate of late leaf senescence. After two weeks, the bulbs were uprooted while preserving as many roots as possible and minimising damage, utilising traditional cultural practices and those particular to the Amaryllidaceae species, as [46] outlined. The bulbs were rinsed to remove any unwanted surface soil debris, stripped of senescent leaves, and sorted to ensure sample homogeneity. Selected bulbs were later dipped in Sporekill™ (ICA International Chemicals (Pty) Ltd., Stellenbosch, South Africa), a biocidal solution with didecyldimethylammonium chloride as the active ingredient, at a dilution rate of 0.1% for 5 min, removed, air dried, and placed in warm storage.

2.3. Experimental Design and Treatment Set-Up

Experimental treatments included bulbs randomly assigned to one of six storage regimes with intervals of 0 (no storage, control), 4, 6, 8, 10, or 12 weeks (Table 1). Storage-treated bulbs were conditioned in breathable containers at a constant temperature of 23 ± 1 °C and 60% relative humidity in a ventilated and darkened room in mid-November 2021. The continuous warm storage temperature was derivative of the relative mean ambient temperature of the bulb's endemic phenological region during the height of summer. After the prescribed storage intervals, the bulbs were planted individually into standard round plastic pots (20 cm diameter and 3 L volume) in mid-November (0-control), mid-December (4 weeks), end-December (6 weeks), mid-January (8 weeks), end-January (10 weeks), and mid-February (12 weeks) and placed in the greenhouse for further development, differentiation, and flowering (Table 1). The pots were filled with a growing substrate consisting of sieved compost, fine river sand, and pre-rinsed silica sand (Consol®, grade 6/17) at a ratio of 1:1:1 ($v/v/v$), ensuring that the neck of each bulb was visible on the surface [46]. A weekly soil drench with tap water (of equal quantities) was applied manually to maintain moisture levels in all planted bulb treatments. Experimental pots were placed on the greenhouse floor in a complete randomised block design (CRBD) with 14 sample replicates per storage interval (n = 14).

Table 1. Warm bulb storage duration of a continuous 23 ± 1 °C at 60% RH in a ventilated, darkened room and the subsequent planting time after storage in the greenhouse.

S/N	Code	Storage Duration Description	Subsequent Planting Time after Storage
1	D1	0-week bulb storage (c) (0 days)	(mid-November 2021)
2	D2	4-week bulb storage (28 days)	(mid-December 2021)
3	D3	6-week bulb storage (42 days)	(end-December 2021)
4	D4	8-week bulb storage (56 days)	(mid-January 2022)
5	D5	10-week bulb storage (70 days)	(end-January 2022)
6	D6	12-week bulb storage (94 days)	(mid-February 2022)

D = storage duration; (c) = control.

2.4. Data Collection

2.4.1. Determination of Inflorescence Morphological Development

Morphological data were recorded as markers of inflorescence growth and development (that remained attached to the potted bulb) using a standard soft cloth metric tape measure (Empisal EMT-001, Builders Warehouse, Boksburg, South Africa), a stainless-steel ruler (Sealy, Leroy Merlin, Boksburg, South Africa) with readability of 1500 × 12 mm and 450 × 25 mm, respectively, and a steel vernier calliper (Grip GV9370, Leroy Merlin, South Africa) with a readability of 0.02 mm.

The following parameters were used to assess inflorescence characteristics: percentage of bulbs that produced a flowering inflorescence, length and diameter of the stem (scape) (mean relative value determined both at the widest point and when rotated horizontally through 90°), number of florets per inflorescence, and length and diameter of a single floret (the first floret to develop). The diameter of the inflorescence crown, fullness ratio, and potted longevity were evaluated to assess the marketability of the potted bulb inflorescence. The diameter of the inflorescence crown was determined using the mean relative value of the distance measured from one side of the circumferential edge through the centre of the umbel arrangement to the outermost edge and at a horizontal rotation of 90° (Figure 2). The fullness ratio was calculated as the ratio of florets to crown diameter, and inflorescence longevity was characterised as the time interval between the opening of the first and the wilting of the last floret on a potted inflorescence scape.

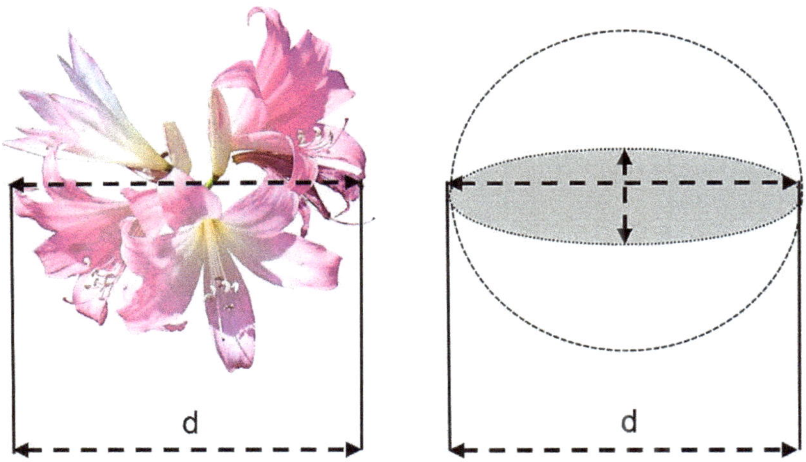

Figure 2. Determination of the relative average inflorescence crown diameter (d) of *Amaryllis belladonna*; measured from one side of the circumferential edge through the centre of the umbel arrangement to the other side and at a 90° horizontal rotation. (photo and diagram: C. Wilmot).

2.4.2. Determination of Inflorescence Flowering Time Course

Observations made up to five times per week were evaluated to determine different transition events during the flowering phase. Days to flowering were observed as the date of appearance of the visible flower buds, the opening of the first floret (anthesis), the opening of 50% of florets per inflorescence, and complete floral senescence (the wilting of the last floret on the inflorescence stem) and recorded as the number of days after planting (DAP) in the greenhouse.

2.4.3. Determination of Leaf Morphological Growth

During the resumptive vegetative leaf phase, morphological data were collected using the precision measuring devices described above to assess leaf growth and development. The number of leaves produced by each bulb, leaf length, leaf width, and leaf area were recorded. The number of leaves was determined manually, and leaf length was determined as the distance between the base of the longest leaf (where it first emerged from the bulb) and its apex, while leaf width was measured at the widest point of the leaf. The Montgomery equation (ME) described by [47,48] was used to calculate leaf area.

$$A_{leaf} \alpha L_{leaf} \times W_{leaf}$$

where A = area; L = length; W = width; and α = Montgomery parameter.

2.5. Statistical Analysis

Morphological data were calculated and analysed using statistical data analysis software (Minitab 17, Minitab LLC, Pennsylvania State University, USA). Data were subjected to one-way analysis of variance (ANOVA) for the factor bulb storage duration (6 levels) and presented as means with standard errors (S.E.s). Fisher's least significant difference (LSD) was used to further separate the means at a significance level of $p \leq 0.05$. Means with a different letter(s) differed significantly at the 95% confidence level.

3. Results

3.1. Effect of Warm Bulb Storage Period on Inflorescence Morphological Development

3.1.1. Percentage Flowering Yield

This study showed that bulb storage treatments significantly affected the number of bulbs that flowered in the greenhouse in a storage-dependent manner ($p \leq 0.05$). As shown in Figure 3, between 14.3% and 64.3% of the *A. belladonna* bulbs produced an emergent inflorescence and flowered, while the remaining bulbs persisted in a vegetative state. The practical flowering potential of the bulbs increased dramatically after 12 weeks of storage and planting in mid-February 2022, with the maximum proportion of emerging flower buds (64.3%). However, the variance was statistically marginal compared with the control without storage and immediate planting (50%). The percentage of flowering bulbs declined within the 4- and 10-week storage range, with the shorter storage durations and subsequent planting showing the most significant reduction (14.3%).

3.1.2. Inflorescence Stem Length and Stem Diameter

Inflorescence stem length was not affected by the storage duration ($p > 0.05$); however, there were significant visual differences (37.0–56.7 cm) between treatments (Figure 1B). In contrast, the stem diameter responded differently, and the bulb storage had a significantly positive effect ($p \leq 0.05$) on the inflorescence stem thickness (Table 2). Compared to the no-storage control (8.9 mm), the stem diameter was thicker in all storage-treated bulbs (9.6–12.7 mm) (Table 2). The bulbs stored for 8 weeks had the longest (56.7 cm) and thickest (12.7 mm) inflorescence stems.

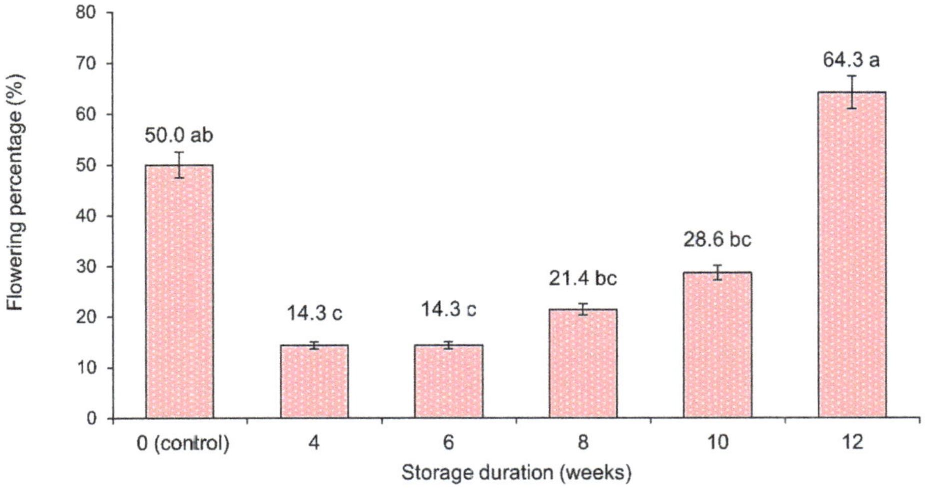

Figure 3. The effect of warm bulb storage duration and subsequent planting on flowering percentage, as defined by successful potted inflorescence maturation of *Amaryllis belladonna* bulbs. Bars represented by mean values followed by a different letter(s) are significantly different at $p \leq 0.05$, according to Fisher's least significant difference (L.S.D).

Table 2. Effects of warm bulb storage period over 12 weeks on inflorescence characteristics of stem length, stem diameter, number of florets, floret length, floret diameter, crown diameter, and fullness ratio of *Amaryllis belladonna* bulbs under greenhouse conditions.

Bulb Storage (Weeks)	**Inflorescence Characteristics**						
	Inflorescence Stem Length (cm)	**Inflorescence Stem Diameter (mm)**	**Number of Florets (n)**	**Floret Length (cm)**	**Floret Diameter (cm)**	**Inflorescence Crown Diameter (cm)**	**Inflorescence Fullness Ratio**
0 (control)	45.1 ± 2.89 b	8.9 ± 0.39 d	9.6 ± 0.90 b	12.0 ± 0.36 a	9.8 ± 0.21 a	22.7 ± 0.64 a	0.4 ± 0.04 b
4	39.2 ± 6.00 b	9.6 ± 0.66 cd	8.0 ± 4.00 b	11.4 ± 0.75 a	9.2 ± 0.55 a	21.6 ± 1.35 a	0.4 ± 0.21 b
6	37.0 ± 6.55 b	10.5 ± 0.09 bcd	12.5 ± 4.50 ab	11.8 ± 0.85 a	9.5 ± 0.70 a	22.4 ± 1.46 a	0.6 ± 0.24 ab
8	56.7 ± 6.08 a	12.7 ± 0.02 a	17.0 ± 5.78 a	11.5 ± 0.10 a	9.6 ± 0.07 a	21.9 ± 0.17 a	0.8 ± 0.03 a
10	43.2 ± 5.72 b	11.7 ± 0.75 ab	16.5 ± 1.50 a	10.8 ± 0.53 a	8.9 ± 0.42 a	21.0 ± 0.87 a	0.8 ± 0.09 a
12	44.0 ± 1.23 b	10.2 ± 0.36 c	10.2 ± 0.97 b	11.6 ± 0.39 a	9.5 ± 0.36 a	22.0 ± 0.72 a	0.5 ± 0.05 b
			One-way ANOVA F-statistic				
Bulb storage	2.15 ns	7.16 *	5.31 *	0.74 ns	0.67 ns	0.50 ns	4.51 *

Mean values ± standard error (S.E.) in the same column with a different letter(s) are significantly different at $p \leq 0.05$ (*) based on Fisher's least significant difference (L.S.D); ns = not significant.

3.1.3. Number of Florets

The analysis showed that the storage treatments had a significant ($p < 0.05$) effect on the number of florets produced by each inflorescence, as shown in Table 2. Bulbs stored for 8 weeks had the most florets (17.0); however, there were no statistically significant differences between the 6- and 10-week treatments. Although statistically comparable to the control and 12-week treatments, the lowest number of florets (8.0) was observed in inflorescences after 4 weeks of storage.

3.1.4. Floret Length and Diameter

The storage period did not significantly affect treatment comparisons of floret length ($p \leq 0.05$). The floret diameter showed a similar tendency. Nevertheless, the control treatment (12.0 and 9.8 cm) had the highest and the 10-week storage (10.8 and 8.9 cm) the lowest values for the floret length and diameter characteristics for all treatments, as indicated in Table 2.

3.1.5. Inflorescence Crown Diameter

At the 95% confidence level, the storage interval had no discernible effect on the spherical-ovate crown diameter of the inflorescence arrangement. Table 2 shows that although there was no statistically significant difference between treatments, the control (22.7 cm) had the largest crown diameter, about 2 cm wider than the smallest diameter in the 10-week storage period (21.0 cm).

3.1.6. Inflorescence Fullness Ratio

The visual quality of the inflorescence was assessed by comparing the fullness ratio. Storage duration strongly influenced this characteristic reception, as shown in Table 2 ($p \leq 0.05$). The highest ratios (0.8:1) were observed in inflorescences stored for 8 and 10 weeks; however, the crowns were not significantly fuller than those stored for 6 weeks (Table 2). In addition, although not statistically different and less compact, a ratio of 0.4:1 was observed in the control and at 4 and 12 weeks of storage.

3.1.7. Inflorescence Longevity

The data presented in Figure 4 show a significant effect of storage treatment ($p \leq 0.05$) and subsequent planting on the potted inflorescence longevity of *A. belladonna* scapes. Apart from the shortest storage duration of 4 weeks (10 days), the bulb scapes in the 10-week storage had the longest flowering interval (17.5 days) but did not differ significantly from the control or any other treatments (14.6–16.7 days).

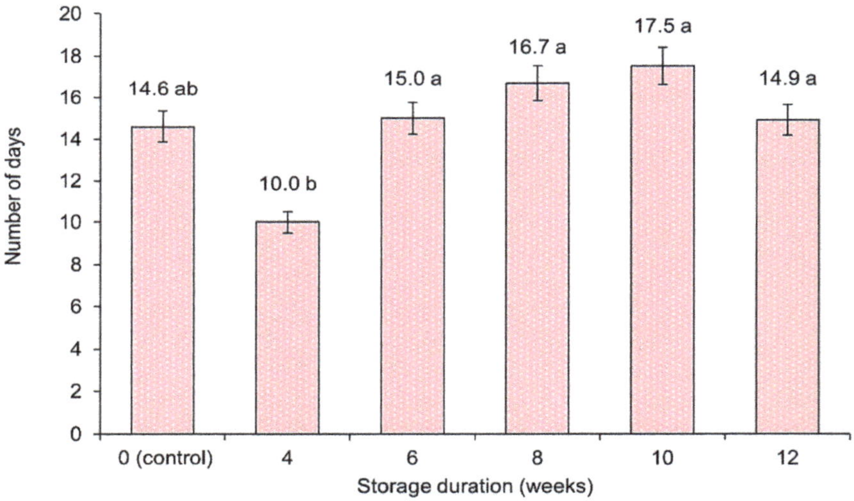

Figure 4. The effect of warm bulb storage duration on *Amaryllis belladonna* bulbs inflorescence potted longevity. Bars represented by mean values followed by a different letter(s) are significantly different at $p \leq 0.05$, according to Fisher's least significant difference (L.S.D).

3.2. Effect of Warm Bulb Storage Period on Flowering Time Course of Visible Flower Buds, Anthesis, Opening of 50% Florets, and Complete Floral Senescence

The results in Figure 5 show a significant difference ($p \leq 0.05$) in the occurrence of visible flower buds in response to storage treatments and subsequent planting. Compared to the immediately planted control, the bulbs stored for 10 and 12 weeks, and the last to be planted were the first to show flower buds 9 days after planting. In addition, the flower buds of the 10-week storage bulbs emerged two weeks earlier than those of the 12-week storage, although they were statistically similar. Compared to the control, the onset of flower bud emergence was delayed by an average of 7 days for shorter storage periods of 4 and 6 weeks.

The timing of the first floret opening (anthesis) occurred between 17.9 and 101.0 days after storage intervals and subsequent planting in the greenhouse. It was significantly reduced compared to days after planting in the control bulbs ($p \leq 0.05$) (Figure 5). The first anthesis was observed three days earlier in the 8 and 10-week treatments than in the control bulbs. In addition, the opening of the first floret was delayed by about 5 days in the 12-week storage and 10 days in the 4- and 6-week storage compared to the control. Although the developmental course of anthesis differed by a few days between treatments, it occurred 7.4–13.0 days after flower bud emergence.

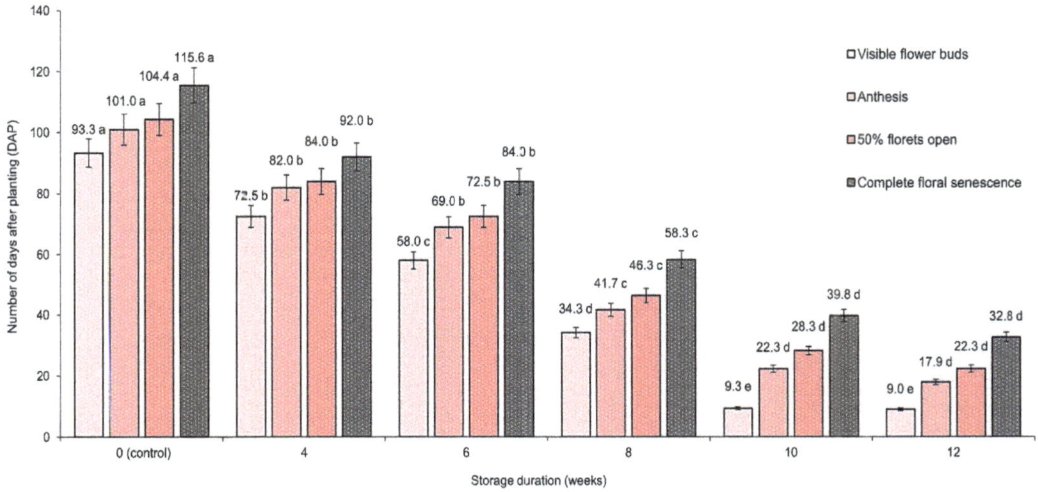

Figure 5. The effect of warm bulb storage duration on the number of days until the emergence of visible flower buds, anthesis, the opening of 50% florets, and complete floral senescence of the potted inflorescence of *Amaryllis belladonna* after planting. Bars represented by mean values followed by a different letter(s) are significantly different at $p \leq 0.05$, according to Fisher's least significant difference (L.S.D).

The influence on the opening of 50% florets on an inflorescence showed similar trends to the effect of the storage treatments on anthesis compared to the control. Figure 5 shows the flowering event between 22.3 and 104.4 days after storage intervals and subsequent planting. However, this trend was observed between 2 and 6 days after the first anthesis. The maximum number of days (6 days) was shorter than the minimum duration (7.4 days) observed for anthesis in all storage treatments, indicating that this period was shorter than the interval between the appearance of the visible flower bud and the opening of the first floret.

Complete floral senescence was observed in all potted inflorescence florets, showing a turgidity loss followed by complete wilting. The storage interval and subsequent DAP significantly ($p \leq 0.05$) affected this occurrence in the greenhouse and was observed between 32.8 and 115.6 DAP in all storage treatments (Figure 5). Compared to the control and the 8- and 10-week treatments, the bulbs in the 4-, 6-, and 12-week treatments were the last to show flower senescence.

3.3. *Effect of Warm Bulb Storage Period on Leaf Morphological Growth*

3.3.1. Number of Leaves

As shown in Table 3, there was no significant difference ($p > 0.05$) in the number of leaves after the storage treatments and subsequent planting time during the vegetative growth phase of the bulbs. In contrast to the control and 4- and 10-week treatments, the

12-week stored bulbs had the most significant number (13.1) and the least variability. In addition, 11 or more leaf sets were identified in the bulbs under all storage conditions.

Table 3. Effects of warm bulb storage period on leaf characteristics of the number of leaves, leaf length, width, and area of *Amaryllis belladonna*.

Bulb Storage (Weeks)	Leaf Characteristics			
	Number of Leaves (n)	Leaf Length (cm)	Leaf Width (cm)	Leaf Area (cm^2)
0 (control)	11.9 ± 0.43 ab	47.1 ± 2.21 a	3.1 ± 0.12 b	148.6 ± 10.40 b
4	11.9 ± 0.28 ab	49.4 ± 3.02 a	3.4 ± 0.13 ab	170.0 ± 13.30 ab
6	11.7 ± 0.34 b	47.1 ± 2.11 a	3.4 ± 0.12 ab	160.90 ± 12.0 ab
8	11.7 ± 0.66 b	50.6 ± 2.28 a	3.5 ± 0.11 a	178.6 ± 11.20 a
10	12.2 ± 0.43 ab	45.9 ± 2.08 a	3.4 ± 0.06 ab	154.22 ± 7.69 ab
12	13.0 ± 0.42 a	47.1 ± 1.41 a	3.5 ± 0.09 a	165.14 ± 8.17 ab
	One-way ANOVA F-statistic			
Bulb storage	1.26 ns	0.62 ns	1.59 ns	1.03 ns

Mean values ± standard error (S.E.) in the same column with a different letter(s) are significantly different at $p \leq 0.05$ based on Fisher's least significant difference (L.S.D); ns = not significant.

3.3.2. Leaf Length, Leaf Width, and Leaf Area

The leaf morphology assessment of bulbs of *A. belladonna*, as measured by leaf length, width, and area, was not significantly affected by the number of weeks of extended warm storage ($p > 0.05$), as shown in Table 3. Despite similar results in all treatments, leaf expansion ranged from 45.9 to 50.6 cm. Notably, leaf blade expansion was slightly broader in all storage treatments than in the control (3.1 cm). Although 8-week storage presented the highest total leaf area (178.6 cm^2), the differences were comparable and less pronounced in all storage conditions.

4. Discussion

This study found that a warm bulb storage period and the subsequent planting significantly impacted flowering precocity, flowering time, and visual characteristics of *A. belladonna* scapes in the greenhouse. Irrespective of the storage treatment and subsequent planting, flower abortion (blindness) was observed, as not all bulbs successfully flowered. This finding is consistent with the physiological anomaly of inflorescence abortion noted in the species [12,49] and seen in other bulbous species of *Hippeastrum*, *Iris*, *Lachenalia*, and *Nerine*, where an after-storage planting resulted in the cessation of flower development [31,50,51]. Aspects of cultivation and climatic conditions may have influenced this flowering anomaly before and during the bulb harvest in the previous season(s). Moreover, root disturbances may have caused flower abortion, as seen in many Amaryllidaceae species [46]. According to [35,52], contingent on the cultivation and environmental growth circumstances, which are not always contemporaneous, the shoot apical meristem (SAM) may take a different trajectory to reach a physiological stage. Furthermore, these authors deduce that the timing and form of this transition vary depending on the species. Authors [44,52,53] reached similar conclusions.

In further elucidating this flowering oddity, the most intriguing finding was that the collective proportion of the 0 and 12-week storage treatments accounted for 59.3% (more than half) of the overall maturation performance yield. It can be inferred that in this study, the longer continuous warm bulb storage conditions for the completion of the rest period were met, and flower emergence was prompted by a particular cumulative temperature range and duration for the last stages of floral differentiation before planting [26,37,38]. However, the shorter storage periods and the timing of the subsequent greenhouse temperature fluctuations because of an earlier planting time after storage may have affected the receptive signalling pathways and transitional apical meristematic activity, resulting in noticeable aberrations of stunted spike emergence and inflorescence abortion [26]. Further insight into these findings suggests that there appeared to be a difference in bulb rest immediately after leaf senescence and harvesting when bulbs were sensitive to temperature changes and later when not. According to [31,54], when dormancy is established, species are at different stages of internal bulb development, and the degree of dormancy at harvest may be related to how receptive the bulbs are to temperature fluctuations. Similarly, *Nerine sarniensis* [38] and *Lachenialia* species [51] reported findings of varied dormancy degrees, storage, and association with flowering performance. This responsiveness may explain why the control (no-storage), with an immediate planting, had a better flowering capacity and regulation than those subjected to shorter storage treatments and earlier subsequent greenhouse temperature differentials because they were only exposed to greenhouse temperature variations from the onset and not both.

The capacity for a bulb to flower depends on an optimal species-specific temperature range and duration for growth and reproduction, and deviating from this can cause unmanageable stress, lowering the floral meristem development rate or abortion, and thus degrading the inflorescence's productivity and quality [25,55]. This finding validates the research that the capacity for a bulb to flower depends on the key storage intervals, and straying from these threshold parameters increases the risk of inflorescence abortion [50]. According to this study's findings, the timing, length, and variety of subsequent temperature regimes affected the flowering response, aligning with those of [52,56]. The timing of floral induction after leaf development is typically governed by vegetative growth and senescence under the influence of temperature during a phase of limited vegetative growth [31]. Ref. [49] estimated that the bulbs of *A. belladonna* initiate a single inflorescence each year during summer dormancy, about one month before the previously formed inflorescence appears. As a result, morphogenesis during dormancy is incomplete as different developmental stages of the imminent and subsequent seasons' inflorescence are initiated. Temperature fluctuations can adversely or favourably impact both phases, with the results only presenting in the following seasons [38]. As a result, factors that impact the dormant period essentially define the competitive character of the commencement of reproductive activities in the species.

Flowering periodicity, productivity, and market quality do not depend exclusively on thermal stimuli; additional criteria are necessary to promote anthesis signalling pathways. These criteria include age, bulb size, and weight, all affecting the flowering capability, with critical parameters differing between taxa, species, and cultivars [23,31,36]. The dynamic metabolic processes associated with shoot apical meristematic transition activity during bulb dormancy necessitate energy, water transfer and delivery. These resources can only be derived from subsurface organ sources, and temperature influences their mobilisation and distribution [57,58]. Therefore, as a perennial hysteranthous taxon, *A. belladonna* relies heavily on the vegetative growth seasons where the emergence of flowers and leaves are succinctly divided to accumulate and maintain adequate carbohydrate reserves in larger underground storage organs for flowering and fruiting [59,60]. Although this study utilised bulbs with a circumference of 30-33 cm, which did not attain 100% flowering, the minimum flower-size bulb for *A. belladonna* is approximately 26 cm in circumference [12,49]. This disparity is reinforced by the fact that, although having a sufficient bulb size, the accumulation, supply, and distribution of resources under the influence of temperature are numerous and may dictate the rate at which growth, development, dormancy, and flowering occur [26,61].

A significant finding from this study was the considerable variation in the number of scape florets after warm bulb storage and planting; however, it had little effect on the morphology since the floret diameter and length were unaffected. These findings contradict that of [51], who found that delayed planting altered the diameter of solitary florets, not the length. In another study, a later planting date in the same species of *Lachenalia* enhanced the number of florets per inflorescence [62]. The quantity of flowers on an ornamental plant greatly reflects its aesthetic value and impacts its qualitative characteristics, according to [63]. Other key factors influencing decorative quality include crown diameter [64] and fullness ratio [65]. Given the disparities in floret numbers, they were likely already formed in the bulb before storage treatments, and the discrepancies may be due to several other factors, such as the genetic disposition of floriferous clones [66] and carbohydrate assimilation [56,61].

Interestingly, despite the variation in floret numbers across treatments, none of the bulbs showed evidence of floret bud atrophy. Observations from this study suggest that the differentiation of inflorescence emergence is more susceptible to exogenous temperature changes than the differentiation of florets opening after planting. Further results from this study found that flowering scapes of *A. belladonna* may retain their aesthetic appeal for at least 10 days, if not longer. *A. belladonna* floret's lifespan is about 2.5 days, opening in coordinated succession for an overall display in the vase [18]. The authors added that this is comparable to the longevity of florets attached to the bulbs, making them attractive to growers, sellers, and buyers as specimens for cut flowers and potted plants. The study also found that the inflorescence scape diameter was thicker in storage treatments compared to the control but minimal compared to commercial quality standards. Furthermore, the minimum stem length was 37 cm, although insignificant, making it a suitable and positive attribute for the cut flower market.

Except for the 10-week treatment, this study found that staggered bulb planting by extended storage treatments did not significantly accelerate the flowering morphogenesis of *A. belladonna*. Instead, they dramatically reduced the period following planting in a storage-dependent manner. This can be ascribed to the fact that the immediately planted bulbs of the control required 93 days (almost ten times as long) to reach this developmental stage, even though flower buds emerged simultaneously (calendar date) with the later greenhouse plantings in the 8- and 12-week treatments. As a result, compared to the immediately planted control, the 12-week-treated bulbs required at least one month to complete the flowering cycle after planting. In contrast, the immediately planted control took nearly four times as long (115.6 days) to achieve this last seasonal greenhouse development stage. This finding may be explained by the fact that long-term bulb storage reduced the time of floral events after planting due to the advanced internal morphogenesis of elongation and the altered carbohydrate content of the inflorescence during storage [57,58]. Similar results were obtained in *Lachenalia* species [62], *Nerine sarniensis* [38], and *Ornithogalum dubium* [67].

Conversely, shorter storage intervals and subsequent planting would have postponed or halted the final stages of flower development, resulting in delayed inflorescence bud emergence. In addition, the proportion of flowering bulbs would have been much lower if inflorescence abortion and lodging had occurred due to arrested development. Interestingly, both results were observed in this study; however, this shorter timeframe contradicts the findings that a 6-week storage period resulted in a quicker sprouting of Asiatic lily cv. "Royal Trinity" [68] and accelerated growth and flowering of *Ornithogalum thyrsoides* hybrids [24]. In this study, the total duration of the flowering cycle, from the appearance of visible buds to the complete senescence of florets, was about 7.5 weeks, consistent with numerous authors' conclusions about the species short flowering season [3,46].

The hysteranthous leaf emergence occurred after flowering and correlated with the onset and decline in temperature as the autumnal season approached. However, the precise physiological processes that initiate the onset are unclear. This study found that warm bulb storage did not affect the foliage quality parameters of the leaf number, length, width, and area. It is proposed that the leaf set had already been established before the bulb harvest and would not have changed significantly during the vegetative growth phase. However, [31] explains that the rate of expansion and production may have been impacted. Visual observations of emerging leaves initially made in the control and shorter storage treatments of 4 and 6 weeks support this finding. In addition, leaf emergence was slightly delayed in bulbs that flowered during the seasonal study period compared with those that remained vegetative in all treatments. According to [69,70], this delay is caused by mobilising compounds and ions from wilting tepals to other organs through the degradation of macromolecules when flowers perish.

This study unveils that the innocuous observation and proclivity of the emerging *Amaryllis belladonna* flower bud, seemingly appearing out of nowhere and rapidly extending to create a prescient curiosity of the imminent and lasting floral display and a conscious awareness of the seasonal shift that heralds the end of summer cannot be undervalued. This study highlights that focusing on market-driven initiatives, as opposed to the product-based strategies of the past, is the solution to SCF's success [4,6,71]. Focusing on the attractive qualities of unique native flora and underutilised species like *A. belladonna*, the potential of the SCF may be invigorated and conserved [4,10]. Along with [4,16], this study demonstrates the high potential of *A. belladonna* as a cut flower and potted plant, promoting its candidacy for local and international SCF markets as a seasonal and sustainable niche product, emblematic of South Africa's unique floral heritage.

5. Conclusions

This study found that warm bulb storage after lifting significantly affected the flowering production, flowering time, and flower quality attributes of *A. belladonna* but not the foliage growth. Aside from favouring a higher flowering capacity and shorter intervals under greenhouse cultivation after planting, extended warm bulb storage did not typically advance the flowering time. The control and shorter storage periods exhibited an unsustainable and uneven pattern. The findings of this study advocate that a cumulative temperature range during bulb dormancy is crucial for supporting inflorescence maturity's energy demands and floret buds' opening. If not adequately maintained during bulb storage and cultivation, *A. belladonna*'s flowering ability is compromised. Ideally, bulbs should be stored at an elevated temperature for 8–10 weeks after the harvest to achieve the highest floret-quality attributes and longevity. However, for optimal anthesis in the shortest interval, sustainable and economical greenhouse, and specialty cut flower production, 12-week warm bulb storage is recommended for this seasonal single-harvest species. Additionally, we recommend investigating a broader range of storage temperatures, durations, and initial lifting dates during dormancy to identify patterns and develop a more precise protocol for improving flowering competency and quality attributes, potentially extending the season.

Author Contributions: Conceptualization, C.M.W., M.O.J. and C.P.L.; Data curation, M.O.J.; Formal analysis, C.M.W. and M.O.J.; Funding acquisition, C.P.L.; Investigation, C.M.W.; Methodology, M.O.J. and C.P.L.; Project administration, C.P.L.; Resources, C.M.W. and C.P.L.; Software, C.M.W.; Supervision, M.O.J. and C.P.L.; Validation, M.O.J.; Writing—original draft, C.M.W.; Writing—review and editing, C.M.W., M.O.J. and C.P.L. All authors have read and agreed to the published version of the manuscript.

Funding: The Research Directorate Centre for Postgraduate Studies—CPUT Postgraduate Bursary Fund, Consolidated Research Fund (CRF-RJ17), and the Department of Horticultural Sciences of the Cape Peninsula University of Technology are gratefully acknowledged for funding and supporting this study.

Institutional Review Board Statement: This study was approved by the Office of the Chairperson, Research Ethics Committee, Faculty of Applied Sciences, Cape Peninsula University of Technology. Reference no: 205054196/06/2020.

Data Availability Statement: Data associated with the current study are presented in the manuscript.

Acknowledgments: The authors would like to thank the Assegaaibosch Farm on the Agulhas Plain for their generosity and donation of plant material for this study.

Conflicts of Interest: The authors state they have no conflict of interest. The funders were not involved in the study design, data collection, analysis, interpretation, article preparation, or decision to publish the results.

References

1. Sharma, P. Floriculture-World Wide Production, Trade, Consumption Pattern, Market Opportunities and Challenges. Available online: https://medium.com/@preetisharma_51610/floriculture-world-wide-production-trade-consumption-pattern-market-opportunities-and-challenges-e797d87c87a7 (accessed on 6 August 2023).
2. Anumala, N.V.; Kumar, R. Floriculture Sector in India: Current Status and Export Potential. *J. Hortic. Sci. Biotechnol.* **2021**, *96*, 673–680. [CrossRef]
3. Barnhoorn, C. *The Bulb Book: A South African Gardener's Guide*; Sunbird Publishers: Cape Town, South Africa, 2013; ISBN 1920289763.
4. Darras, A.I. Overview of the Dynamic Role of Specialty Cut Flowers in the International Cut Flower Market. *Horticulturae* **2021**, *7*, 51. [CrossRef]
5. Darras, A.I. Implementation of Sustainable Practices to Ornamental Plant Cultivation Worldwide: A Critical Review. *Agronomy* **2020**, *10*, 1570. [CrossRef]
6. Thörning, R.; Ahlklo, Å.K.; Spendrup, S. The Slow Flower Movement—Exploring Alternative Sustainable Cut-Flower Production in a Swedish Context. *Heliyon* **2022**, *8*, e11086. [CrossRef] [PubMed]
7. Ingram, D.L.; Hall, C.R.; Knight, J. Understanding Carbon Footprint in Production and Use of Landscape Plants. *Horttechnology* **2019**, *29*, 6–10. [CrossRef]
8. Lee, J.; Miller, W.B. Preplant Storage and Greenhouse Temperature Influence Flowering of Ornithogalum. *Hortscience* **2015**, *50*, 1784–1790. [CrossRef]
9. Miller, W.B. Commercial Flower Production Methodology. *Encycl. Appl. Plant Sci.* **2017**, *3*, 203–208. [CrossRef]
10. De Pascale, S.; Romano, D. Potential Use of Wild Plants in Floriculture. *Acta Hortic.* **2019**, *1240*, 87–98. [CrossRef]
11. Adams, T.D. Amaryllis Belladonna | PlantZAfrica. Available online: https://pza.sanbi.org/amaryllis-belladonna (accessed on 3 March 2023).
12. Theron, K.I.; De Hertogh, A.A. Amaryllidaceae: Geophytic Growth, Development, and Flowering. *Hortic. Rev. Am. Soc. Hortic. Sci.* **2001**, *25*, 1–70.
13. Manning, J.; Goldblatt, P.; Snijman, D. *The Color Encyclopedia of Cape Bulbs*; Timber Press: Cambridge, UK, 2002; ISBN 0881925.
14. Duncan, G.D.; Jeppe, B.; Voigt, L. *Field Guide to the Amaryllis Family of Southern Africa and Surrounding Territories*; Galley Press: Nelspruit, South Africa, 2020; ISBN 9780620885911.
15. Campos-Rocha, A.; Meerow, A.W.; Lopes, E.F.M.; Semir, J.; Mayer, J.L.S.; Dutilh, J.H.A. Eithea Lagopaivae, a New Critically Endangered Species in the Previously Monotypic Genus Eithea Ravenna (Amaryllidaceae). *PhytoKeys* **2017**, *85*, 45–58. [CrossRef]
16. Reinten, E.Y.; Coetzee, J.H.; Van Wyk, B.E. The Potential of South African Indigenous Plants for the International Cut Flower Trade. *South Afr. J. Bot.* **2011**, *77*, 934–946. [CrossRef]
17. Wilmot, C.M.; Laubscher, C.P. Amaryllis Belladonna: A Potential Urban Landscape Wonder. *Acta Hortic.* **2019**, *1237*, 287–293. [CrossRef]
18. Gul, F.; Tahir, I.; Shahri, W. Flower Senescence and Some Postharvest Considerations of Amaryllis Belladonna Cut Scapes. *Plant Physiol. Rep.* **2020**, *25*, 315–324. [CrossRef]
19. Duncan, G.D. Amaryllis Magic. *Veld. Flora* **2004**, *90*, 142–147.
20. Corbesier, L.; Coupland, G. The Quest for Florigen: A Review of Recent Progress. *J. Exp. Bot.* **2006**, *57*, 3395–3403. [CrossRef]
21. Horvath, D. Common Mechanisms Regulate Flowering and Dormancy. *Plant Sci.* **2009**, *177*, 523–531. [CrossRef]
22. Denay, G.; Chahtane, H.; Tichtinsky, G.; Parcy, F. A Flower Is Born: An Update on Arabidopsis Floral Meristem Formation. *Curr. Opin. Plant Biol.* **2017**, *35*, 15–22. [CrossRef] [PubMed]
23. De Hertogh, A.; Le Nard, M. *The Physiology of Flower Bulbs*; Elsevier: Amsterdam, The Netherlands, 1993; ISBN 9780444874986.
24. Roh, M.S.; Hong, D.K. Inflorescence Development and Flowering of Ornithogalum Thyrsoides Hybrid as Affected by Temperature Manipulation during Bulb Storage. *Sci. Hortic.* **2007**, *113*, 60–69. [CrossRef]
25. Khodorova, N.V.; Boitel-Conti, M. The Role of Temperature in the Growth and Flowering of Geophytes. *Plants* **2013**, *2*, 699. [CrossRef]
26. Capovilla, G.; Schmid, M.; Posé, D. Control of Flowering by Ambient Temperature. *J. Exp. Bot.* **2015**, *66*, 59–69. [CrossRef]
27. Howard, C.C.; Folk, R.A.; Beaulieu, J.M.; Cellinese, N. The Monocotyledonous Underground: Global Climatic and Phylogenetic Patterns of Geophyte Diversity. *Am. J. Bot.* **2019**, *106*, 850–863. [CrossRef] [PubMed]
28. Wang, Z.; Li, X.; Xu, J.; Yang, Z.; Zhang, Y. Effects of Ambient Temperature on Flower Initiation and Flowering in Saffron (*Crocus sativus*, L.). *Sci. Hortic.* **2021**, *279*, 109859. [CrossRef]
29. Le Nard, M.; De Hertogh, A.A. Bulb Growth and Development and Flowering. In *The Physiology of Flower Bulbs*; Le Nard, M., De Hertogh, A.A., Eds.; Elsevier: Amsterdam, The Netherlands, 1993; pp. 29–44.
30. Miller, W.B. Flower Bulbs Worldwide: Perspectives on the Production Chain and Research. *Acta Hortic.* **2017**, *1171*, 1–8. [CrossRef]
31. Rees, A.R. The Physiology of Ornamental Bulbous Plants. *Bot. Rev.* **1966**, *32*, 1–23. [CrossRef]
32. Kamenetsky-Goldstein, R. Geophyte Cultivation in Changing Climate: Environmental Effects on Development and Production. *Acta Hortic.* **2019**, *1237*, 277–285. [CrossRef]
33. Wilmot, C.M.; Laubscher, C.P. Flowering Initiation in Amaryllis Belladonna. *Acta Hortic.* **2019**, *1237*, 137–144. [CrossRef]
34. Molina, R.V.; Valero, M.; Navarro, Y.; Guardiola, J.L.; García-Luis, A. Temperature Effects on Flower Formation in Saffron (*Crocus sativus* L.). *Sci. Hortic.* **2005**, *103*, 361–379. [CrossRef]

35. Thompson, D.I.; Mtshali, N.P.; Ascough, G.D.; Erwin, J.E.; Van Staden, J. Flowering Control in Watsonia: Effects of Corm Size, Temperature, Photoperiod and Irradiance. *Sci. Hortic.* **2011**, *129*, 493–502. [CrossRef]
36. Inkham, C.; Piriyapongpitak, P.; Ruamrungsri, S. Storage and Growth Temperatures Affect Growth, Flower Quality, and Bulb Quality of Hippeastrum. *Hortic. Env. Biotechnol.* **2019**, *60*, 357–362. [CrossRef]
37. Carlson, A.S.; Dole, J.M. Determining Optimal Bulb Storage and Production Methods for Successful Forcing of Cut Pineapple Lily. *Horttechnology* **2015**, *25*, 608–616. [CrossRef]
38. Warrington, I.J.; Brooking, I.R.; Fulton, T.A. Lifting Time and Bulb Storage Temperature Influence *Nerine sarniensis* Flowering Time and Flower Quality. *N. Z. J. Crop. Hortic. Sci.* **2011**, *39*, 107–117. [CrossRef]
39. Viljoen, C.C.; Jimoh, M.O.; Laubscher, C.P. Studies of Vegetative Growth, Inflorescence Development and Eco-Dormancy Formation of Abscission Layers in Streptocarpus Formosus (Gesneriaceae). *Horticulturae* **2021**, *7*, 120. [CrossRef]
40. Kamenetsky, R.; Miller, W.B. The Global Trade in Ornamental Geophytes. *Chron. Hortic.* **2010**, *50*, 27–30.
41. Marasek-Ciolakowska, A.; Sochacki, D.; Marciniak, P. Breeding Aspects of Selected Ornamental Bulbous Crops. *Agronomy* **2021**, *11*, 1709. [CrossRef]
42. Loyola, C.E.; Dole, J.M.; Dunning, R. North American Specialty Cut Flower Production and Postharvest Survey. *Horttechnology* **2019**, *29*, 338–359. [CrossRef]
43. Salachna, P.; Mikiciuk, M.; Zawadzińska, A.; Piechocki, R.; Ptak, P.; Mikiciuk, G.; Pietrak, A.; Łopusiewicz, Ł. Changes in Growth and Physiological Parameters of ×Amarine Following an Exogenous Application of Gibberellic Acid and Methyl Jasmonate. *Agronomy* **2020**, *10*, 980. [CrossRef]
44. Kapczyńska, A. Effect of Bulb Size on Growth, Flowering and Bulb Formation in Lachenalia Cultivars. *Hortic. Sci.* **2014**, *41*, 89–94. [CrossRef]
45. Slezák, K.A.; Mazur, J.; Jezdinský, A.; Kapczyńska, A. Bulb Size Interacts with Lifting Term in Determining the Quality of *Narcissus Poeticus* L. Propagation Material. *Agronomy* **2020**, *10*, 975. [CrossRef]
46. Duncan, G.D. *Grow Bulbs*; Kirstenbosch Gardening Series; South African National Biodiversity Institute: Cape Town, South Africa, 2010; ISBN 9781919684567.
47. Yu, X.; Shi, P.; Schrader, J.; Niklas, K.J. Nondestructive Estimation of Leaf Area for 15 Species of Vines with Different Leaf Shapes. *Am. J. Bot.* **2020**, *107*, 1481–1490. [CrossRef]
48. Shi, P.J.; Li, Y.R.; Niinemets, Ü.; Olson, E.; Schrader, J. Influence of Leaf Shape on the Scaling of Leaf Surface Area and Length in Bamboo Plants. *Trees Struct. Funct.* **2021**, *35*, 709–715. [CrossRef]
49. Hartsema, A.M.; Leupen, F.F. Orgaanvorming En periodiciteit van Amaryllis Belladonna. *Meded. Landbouwhoogesch. Wagening.* **1942**, *46*, 1–30.
50. Van Kilsdonk, M.G.; Nicolay, K.; Franssen, J.M.; Kollöffel, C. Bud Abortion in Tulip Bulbs Studied by Magnetic Resonance Imaging. *J. Exp. Bot.* **2002**, *53*, 1603–1611. [CrossRef] [PubMed]
51. Kapczyńska, A. Effect of Planting Time on Flowering of Four Lachenalia Cultivars. *Acta Hortic.* **2012**, *937*, 575–579. [CrossRef]
52. Du Toit, E.S.; Robbertse, P.J.; Niederwieser, J.G. Effects of Growth and Storage Temperature on *Lachenalia* Cv. Ronina Bulb Morphology. *Sci. Hortic.* **2002**, *94*, 117–123. [CrossRef]
53. Roh, M.S. Flowering and Inflorescence Development of Lachenalia Aloides "Pearsonii" as Influenced by Bulb Storage and Forcing Temperature. *Sci. Hortic.* **2005**, *104*, 305–323. [CrossRef]
54. Rees, A.R. *Ornamental Bulbs, Corms and Tubers*; CAB International: Wallingford, UK, 1992; Volume 1.
55. Dole, J.M.; Wilkins, H.F. *Floriculture: Principles and Species*, 2nd ed.; Prentice Hall: Upper Saddle River, NJ, USA, 2005; ISBN 0130462500.
56. Kim, J.; Oh, W. Cold Storage Period of Bulbs Influences Their Sugar Contents and Post-Storage Growth and Flowering in Lilium Oriental Hybrids. *Agriculture* **2021**, *11*, 1080. [CrossRef]
57. Wendell, M.; Ripel, L.; Lee, Y.K.; Rognli, O.A.; Torre, S.; Olsen, J.E. Thermoperiodic Control of Floral Induction Involves Modulation of the Diurnal Flowering Locus T Expression Pattern. *Plant Cell Physiol.* **2017**, *58*, 466–477. [CrossRef]
58. Marković, M.; Momčilov, M.T.; Uzelac, B.; Jevremović, S.; Subotić, A. Bulb Dormancy in Vitro—*Fritillaria meleagris*: Initiation, Release and Physiological Parameters. *Plants* **2021**, *10*, 902. [CrossRef]
59. Dafni, A.; Cohen, D.; Noy-Mier, I. Life-Cycle Variation in Geophytes. *Ann. Mo. Bot. Gard.* **1981**, *68*, 652. [CrossRef]
60. Howard, C.C.; Cellinese, N. Tunicate Bulb Size Variation in Monocots Explained by Temperature and Phenology. *Ecol. Evol.* **2020**, *10*, 2299–2309. [CrossRef]
61. Miller, W.B. A Review of Carbohydrate Metabolism in Geophytes. *Acta Hortic.* **1992**, *325*, 239–246. [CrossRef]
62. Kapczyńska, A.; Kidawska, A. "Namakwa" Lachenalia's Response to Flurprimidol and Different Planting Dates. *Folia Hortic.* **2016**, *28*, 173–179. [CrossRef]
63. Chen, Q.X.C.; Guo, Y.; Warner, R.M. Identification of Quantitative Trait Loci for Component Traits of Flowering Capacity across Temperature in Petunia. *G3 Genes Genomes Genet.* **2019**, *9*, 3601–3610. [CrossRef] [PubMed]
64. Zhao, Y.; Huo, B.; Lin, S.; Zhang, S.; Mao, C.; Jiang, J.; Chen, S.; Fang, W.; Guan, Z.; Liao, Y.; et al. Germplasm Innovation and Establishment of Comprehensive Evaluation System for Hedgerow Garden Chrysanthemum. *Agronomy* **2022**, *12*, 1736. [CrossRef]
65. Kapczyńska, A.; Stodolak, B. The Morphological and Physiological Response of Lachenalia to Supplemental Irradiation. *Hortic. Env. Biotechnol.* **2019**, *60*, 455–465. [CrossRef]

66. Chandel, A.; Thakur, M.; Singh, G.; Dogra, R.; Bajad, A.; Soni, V.; Bhargava, B. Flower Regulation in Floriculture: An Agronomic Concept and Commercial Use. *J. Plant Growth Regul.* **2023**, *42*, 2136–2161. [CrossRef]
67. Luria, G.; Watad, A.A.; Cohen-Zhedek, Y.; Borochov, A. Growth and Flowering of Ornithogalum Dubium. *Acta Hortic.* **2002**, *570*, 113–119. [CrossRef]
68. Malik, K.M.; Shiekh, M.Q.; Nazki, I.T.; Mir, S.A. Influence of Storage Temperature, Storage Duration and Disbudding on Bulb Production in *Asiatic lilium* Cv. 'Royal Trinity'. *Biosci. Biotechnol. Res. Asia* **2017**, *14*, 577–585. [CrossRef]
69. Van Doorn, W.G.; Woltering, E.J. Physiology and Molecular Biology of Petal Senescence. *J. Exp. Bot.* **2008**, *59*, 453–480. [CrossRef]
70. Zhang, Y.; Zhong, D.; Liu, Z.; Gao, J. Study on the Physiological, Cellular, and Morphological Aspects of the Postharvest Development of Cut Lily Flowers. *Hortic. Plant J.* **2021**, *7*, 149–158. [CrossRef]
71. Reinten, E.Y.; Van Wyk, B.E. Floriculture Industry Benefits from Southern African Floral Biodiversity. *Acta Hortic.* **2018**, *1201*, 659–663. [CrossRef]

horticulturae

MDPI

Article

Use of Light Spectra for Efficient Production of PLBs in Temperate Terrestrial Orchids

Hossein Naderi Boldaji, Shirin Dianati Daylami * and Kourosh Vahdati

Department of Horticulture, College of Aburaihan, University of Tehran, Tehran 3391653755, Iran; hossein.naderi87@ut.ac.ir (H.N.B.); kvahdati@ut.ac.ir (K.V.)

* Correspondence: dianati@ut.ac.ir; Tel.: +98-9126359782

Abstract: Wild orchids, especially the terrestrial temperate ones are endangered species due to challenges in their natural habitats. Therefore, there is an urgent need to introduce efficient propagation methods to overcome the natural reproduction problems of these orchids. In this study, the effects of different light spectrums, explant types, wounding, and combinations of different plant growth regulators (PGRs) on direct somatic embryogenesis (DSE) of two species of these endangered orchids listed in the conservation category, were studied. The highest percentages of DSE formation and embryo germination were observed in *Dactylorhiza umberosa* protocorm explants exposed to white light (400–730 nm) and in *Epipactis veratifolia* protocorm explants exposed to a combination of red and far-red spectra (R: FR = 70:30). This occurred while red (610–700) alone and in combination with far-red (710–730 nm) spectrum induced embryogenesis more than the blue spectrum and dark condition in *E. veratifolia*. Thidiazuron (TDZ, 3 mg L^{-1}), produced the highest percentage of protocorm-like bodies (PLBs) on protocorm explants in both orchids. Kinetin (Kin, 2 mg L^{-1}) and Benzyladenine (BA 3 mg L^{-1}) had the most effect on the survival and growth of PLBs, respectively, in *D. umberosa* and *E. veratifolia*. Species did not show similar embryogenesis responses under light spectrums. In a medium containing 3 mg L^{-1} TDZ, white light and R-FR spectra produced the most PLBs on wounded protocorm explants of *D. umberosa* and *E. veratifolia* respectively. The developmental stage of apical meristem of PLBs in both species was more advanced under R-B spectra, compared to others.

Keywords: *Epipactis veratifolia*; *Dactylorhiza umberosa*; direct somatic embryogenesis; plant growth regulator; protocorm-like body; wounding

Citation: Naderi Boldaji, H.; Dianati Daylami, S.; Vahdati, K. Use of Light Spectra for Efficient Production of PLBs in Temperate Terrestrial Orchids. *Horticulturae* **2023**, *9*, 1007. https://doi.org/10.3390/horticulturae9091007

Academic Editor: Sumei Chen

Received: 9 July 2023
Revised: 24 August 2023
Accepted: 4 September 2023
Published: 7 September 2023

1. Introduction

Orchids are high-price ornamental crops that have attracted the attention of their production as pot plants and cut flowers [1]. Of the wild orchids, native orchids such as saleps have traditionally been harvested from their natural habitats for their use in the medicinal and food industries. The tubers of some terrestrial orchids like *Orchis mascula*, *Dactylorhiza umberosa*, and some other species are called salep [2]. Salep also refers to the powder of dried tubers that are used in the production of ice cream, drinks, medicines, confectionery, and hot drinks [3]. The scarcity of resources for salep has led to the uncontrolled harvesting of these plants from their natural habitats [4], making them an endangered species. *Epipactis veratrifolia* is one of the wild orchids used in traditional medicine and has a high potential for breeding as an ornamental plant [5]. Low seed germination, slow vegetative growth, and lack of proper coexistence have sometimes made their propagation by conventional methods impossible [6,7]. To achieve successful regeneration, in vitro techniques are now efficiently practiced and used for sexual and non-sexual propagation of orchids, especially for endangered orchid species [8,9]. There are various methods for mass in vitro culture of orchids such as culture of seeds [10], shoot tip and axillary bud culture [11], and using protocorm-like bodies (PLBs) [12], and flower buds [13]. Direct and indirect somatic embryogenesis (DSE), (ISE) and protocols have been offered for the induction of protocorm-like bodies (PLBs) in orchids [5–7]. These micropropagation methods have been used for somatic embryogenesis (SE) in many orchids such as *Cymbidium* [14], *Oncidium* [15], *Phalaenopsis* [1], and *Xenikophyton smeeanum* [16], however, there is not a wealth of information for in vitro micropropagation of temperate terrestrial orchids [17]. Explant type, genotype, medium composition, plant growth regulators (PGRs), and light regime can be mentioned as factors influencing somatic embryogenesis (SE) in Oncidium orchids [18]. Light is a determinant factor for the tissue culture of plants. Different light spectra have been used to study their effects on plant growth and organogenesis [19–21]. Light-emitting diodes (LEDs) have emerged as a new light source for in vitro culture. Different kinds of artificial light can play a key role in successful in vitro plant production, besides other factors such as gas exchange in the culture vessel, temperature, and composition of the culture medium [22]. The effects of the light spectrum have been investigated on the in vitro growth of many plant species such as *Lilium* 'Pesaro' [23], *Dianthus caryophyllus* [24], and *Zantedeschia jucunda* [25]. The emission of light in LED lighting systems allows the selection of spectra quality and provides the opportunity for the regulation of photosynthetic and photomorphogenic reactions required for an in vitro culture of plants [26]. Significant improvements have been achieved in increasing the fresh and dry weight of shoots and proliferation rate by changing the photoperiod regime from 16 h to a 4 h photoperiod, thereby allowing explants to do a better exchange of CO_2 [27,28]. Furthermore, the light has been introduced as one of the important inducers for the generation of SE [29]. Effects of various spectra of LED lights have been reported in some orchids such as *Oncidium* [18], *Phalaenopsis* [30], and *Cymbidium* [31], as well as other plant species such as *China Rose* [32], and *Agave tequilana* [33]. The induction of SE and callus was obtained under different light conditions in different plant species. For instance, red, red + far-red lights in *Phalaenopsis*, red light in *Cymbidium*, red and white light in *Agave tequilana,* and red light in *Rosa chinensis* [30–33]. PGRs are among the most important factors that influence SE. Auxins and cytokinins can be mentioned specifically as the most used ones [34]. Applying PGRs can effectively improve DSE or ISE [24]. Positive effects of PGRs on embryogenesis have also been reported in plant species such as *Phalaenopsis amabilis* [34], *Anthurium* [35], *Lilium ledebourii* [36], and *Cyclamen* [37]. It was reported that wounding of the explants can lead to faster SE induction in soybeans [38], while there is little information on the effects of wounding on embryogenesis in plants such as orchids [39]. Most studies have been limited to tropical orchids, and only a few studies have been conducted on SE and regeneration in temperate terrestrial orchids. To the best of our knowledge, there are no studies on the effects of light spectra and wounding on somatic embryogenesis and regeneration in terrestrial orchids. In the present study, we have successfully established an in vitro propagation protocol for *D. umberosa* and *E. veratifolia* by proliferation through PLBs to facilitate the conservation, cultivation, and introduction of these plants into future flower markets.

2. Materials and Methods

2.1. Plant Materials and Explant Preparation

In this study, two native Iranian orchids were utilized; namely the rhizomatous species, *E. veratrifolia* and the tuberous species *D. umberosa*. Capsules containing mature brown seeds of these plants were collected from their natural habitats in the Alborz Mountains. Capsules were sterilized for 20 min in 20% sodium hypochlorite (NaOCl), followed by 3 rinses with sterilized distilled water; then about 100 tiny, dusty seeds (There are more than a thousand seeds in each capsule) were cultured on modified FAST (MFAST) as a basal medium in each Petri dish [40]. Four types of explants were used. Protocorm was the first kind of explant. At the end of the germination process of orchid seeds, embryos form small spherical tuber-like bodies referred to as protocorms and were used in two forms: wounded and non-wounded in both species (Figure 1f). After seedling growth, three other types of explants were prepared, which included leaf segment (in both species Figure 1d), single node (in *E. veratrifolia*-Figure 1e), and crown of the plantlet (in *D. umberosa*). The explants were cultured on a basal medium, containing different concentrations of PGRs (including TDZ, NAA, BA, and Kin) according to Table 1.

Figure 1. The seed germination stage of Epipactis veratifolia; kinds of explants and seedlings (**a**) cultivated seeds, (**b**) swelling of the embryos (arrow) in the first stage of germination, (**c**) rupture of seed testa by swelled embryos, (**d**) leaf explant (basal part of leaf), (**e**) stem node explant, (**f**) intact and wounded protocorm explants; the end part without shoot apical meristem (arrow) used as explants, (**g**) in vitro rooted seedlings; their leaf and stem nodes (N) used as explants, (**h**) a protocorm resulting from the development of a Dactylorhiza umberosa seed embryo at the stage of converting apical meristem (AP) to leaf primotdia (LP), with many hairy rhizoids (R) at the bottom end.

Table 1. Effects of PGRs and explant types on the mean formation of somatic embryos (PLBs) and their germination rate, in *Epipactis veratifolia and Dactylorhiza umberosa.* (×) in column Shows the kinds of explant. Within each column, different letters indicate significant differences at $p < 0.05$ using ANOVA and Duncan's multiple range tests.

Plant	Concentration of PGRs (mg/L)				Explant Type				PLBs Formation (%)	PLBs Growth (%)
	TDZ	NAA	BA	Kin	Protocorm	Leaf	Node	Crown		
tiEpipactis verafolia	0	0	0	0	×	-	-	-	0	0
	3	0	0	0	×	-	-	-	100 [a]	100 [a]
	0	0	1	0	-	×	-	-	0	0
	0	0	1	0	-	-	×	-	0	0
	0	0	2	0	-	×	-	-	0	0
	0	0	2	0	-	-	×	-	0	0
	0	0	3	0	-	×	-	-	0	0
	0	0	3	0	-	-	×	-	0	0
	0	0.5	0	0	-	×	-	-	0	0
	0	0.5	0	0	-	-	×	-	25 [b]	25 [d]
	0	0.5	1	0	-	×	-	-	0	0
	0	0.5	1	0	-	-	×	-	25 [b]	75 [b]
	0	0.5	2	0	-	×	-	-	0	0

Table 1. *Cont.*

Plant	Concentration of PGRs (mg/L)				Explant Type				PLBs Formation (%)	PLBs Growth (%)
	TDZ	NAA	BA	Kin	Protocorm	Leaf	Node	Crown		
	0	0.5	2	0	-	-	×	-	0	0
	0	0.5	3	0	-	×	-	-	0	0
	0	0.5	3	0	-	-	×	-	0	0
	0	1	0	0	-	×	-	-	0	0
	0	1	0	0	-	-	×	-	38 ^{a}	20 ^{d}
	0	1	1	0	-	×	-	-	0	0
	0	1	1	0	-	-	×	-	25 ^{b}	50 ^{c}
	0	1	2	0	-	×	-	-	0	0
	0	1	2	0	-	-	×	-	18 ^{c}	75 ^{b}
	0	1	3	0	-	×	-	-	0	0
	0	1	3	0	-	-	×	-	0	0
	0	0	1	0	-	×	-	-	0	0
	0	0	1	0	-	-	×	-	0	0
	0	0	2	0	-	×	-	-	0	0
	0	0	2	0	-	-	×	-	25 ^{b}	50 ^{c}
	0	0	3	0	-	×	-	-	0	0
	0	0	3	0	-	-	×	-	13 ^{c}	100 ^{a}
Dactylorhiza umberosa	0	0	0	0	×	-	-	-	0	0
	3	0	0	0	×	-	-	-	100 ^{a}	90 ^{b}
	0	0	0	1	×	-	-	-	0	0
	0	0	0	1.5	×	-	-	-	25 ^{c}	75 ^{b}
	0	0	0	2	×	-	-	-	50 ^{b}	100 ^{a}
	2	0	2	0	-	-	-	×	0	0
	1	0.5	0	0	-	-	-	×	0	0
	0	0.5	2	0	-	-	-	×	0	0

2.2. Media and Culture Conditions

Cultivated seeds (Figure 1) were grown using a hormone-free modified FAST medium (MFAST) as the basal medium containing both macro and micro elements (Merck, Hesse-Darmstadt, Germany). The medium was supplemented with myo-inositol 100 mg L^{-1}, nicotinic acid 0.5 mg L^{-1}, pyridoxine HCl 0.5 mg L^{-1}, thiamine HCl 0.1 mg L^{-1}, glycine 2 mg L^{-1}, sucrose 3 g L^{-1}, peptone 2% and agar 4.8 g L^{-1}. [40]. The pH was adjusted to 5.5 ± 0.1 and the solution was autoclaved at 121 °C for 20 min. The same medium was used for continuing growth of explants for a period of 6 to 8 weeks. To study the effects of four kinds of growth regulators (PGRs), including thidiazuron (Sigma Aldrich, UAS) (TDZ 0, 2 and 3 mg L^{-1}), N6-benzyleadenine (BA 0, 1, 2 and 3 mg L^{-1}), ∞-naphthaleneacetic acid (NAA 0, 0.5 and 1 mg L^{-1}), kinetin (KIN 0, 1, 2 and 3 mg L^{-1}) and four kinds of explants (including five explants of each type of crown, node, leaf segment, and protocorm) were cultured in separate Petri dishes (five micro-samples in each of three replicates) and placed in a growth room under white LED lights (Figure 1). Then the interaction between TDZ (0 and 3 mg L^{-1}), wounding, and light spectra was investigated on SE of protocorm explants (wounded or un-wounded) in both species. The light treatments including white light (W, as the control range at 400–700 nm), blue (B) range at 460 nm, red (R) range at 660 nm, green (G) range at 530 nm, combination of red and blue (R:B = 70:30), also a combination of red and far-red (R:FR = 70:30) were provided using LED lamps at light intensity of 80 μmol m^{-2} s^{-1} as well as darkness condition, as the control. The cultures were placed in a growth room with a temperature of 22 ± 2 °C, a light period of 16 h and, a relative humidity of 70%. Wavelengths were measured using a Sekonic C7000 spectrometer (Sekonic Corp., Tokyo, Japan) within the range of 300–800 nm. After four weeks, induction of DSE (globular type) was evaluated using a microscope. Following SE induction, the cultures were transferred to the hormone-free MFAST medium and placed under the light with the same environmental conditions as previously described.

2.3. Germination and Acclimatization

Embryos produced via DSE were counted as germinated (Table 1) and transferred to a hormone-free basal medium for plant growth under the environmental conditions described above. For acclimatization, the plantlets (developed from DSE) were transferred to plastic pots filled with a sterile coco-peat: perlite mixture (3:1), and placed in an adapted chamber at a temperature of 28 ± 2 °C.

2.4. Statistical Analysis

Experiments were arranged as factorial in a completely randomized design. The data were subjected to analysis of variance (ANOVA) and means were compared using Duncan's multiple range tests at $p < 0.05$ probability level using the SAS 9.3 software.

3. Results

3.1. Effects of PGRs, Explant Type, and Wounding on DSE

Embryo formation was observed three weeks after placing the explants in the growth medium. The results of the statistical analysis showed a significant difference between the interaction of treatments (growth regulators, the type of explant, and the light spectrum). The percentage of DSE formation, PLBs germination, and final plantlets formation, especially on protocorm explants, in both species, had significant differences ($p < 0.05$) under various light spectra (Figures 2–8). The type of PGR caused a significant effect on embryo formation in both species (Table 1, Figure 4). Between the PGR treatments, *E. veratifolia* showed the highest embryogenesis frequency (100%) when grown on a medium supplemented with 3 mg L^{-1} TDZ, using protocorms and followed by 1 mg L^{-1} NAA (38%) in node explants (Table 1). The medium supplemented with 3 mg L^{-1} BA had the lowest SE rate at 13%. Table 1 shows an increase in DSE production with increasing NAA concentration (from 0 to 1 mg L^{-1}) in combination with BA. NAA treatment (0.5 mg L^{-1}) alone and in combination with 1 mg L^{-1} BA, caused 25% PLB formation on the node explants in *Epipactis veratifolia*, but in combination with BA, the survival of these somatic embryos (PLBs growth %) was 50% more (Table 1). The positive effect of BA on the survival of somatic embryos is also observed in combination with 1 mg L^{-1} of NAA (Table 1).

Different concentrations of the BA alone or in combination with NAA did not affect the leaf explants of *Epipactis veratifolia* (Table 1). Three weeks after placing the explants in the growth medium, somatic embryos formed on protocorm explants of *D. umberosa*, under all light spectra. Embryo formation and germination rate, percentage of DSE, embryo germination, and plantlet production were significantly ($p < 0.05$) influenced by the type of explant, the combination of PGRs, light spectrums, and wounding (Table 1, Figures 4–6 and 8). The kind of PGRs utilized had a considerable effect on the DSE percentage (Table 1). Only protocorm explants that developed globular embryos displayed somatic embryo initiation in *D. umberosa* (Table 1). The DSE was induced on the protocorm explants after three to four weeks. The maximum rate (100%) of embryo formation (DSE) was observed in a medium containing 3 mg L^{-1} TDZ (Table 1). In contrast, no embryos were observed on crown explants of *D. umberosa* in all PGR treatments. Kin in 1.5 and 2 mg L^{-1} concentrations, also caused 25% and 50% direct embryogenesis in protocorm explants of *D. umberosa*, respectively. All concentrations of PGRs (TDZ, BA, and NAA) did not affect the crown explant. SE was observed on both protocorm (100%) and stem node (38%) explants of *E. veratifolia* (Table 1). R-FR spectra caused most DSE on non-wounded protocorm explants of *E. veratifolia* (Figures 2 and 5a). In *D. umberosa* only one form of the explant (protocorm) produced a number of somatic embryos (Figure 3) with a high frequency of embryo germination (Table 1). Therefore, the protocorm explant was selected to proceed with the experiment. The most somatic embryo of *D. umberosa* was obtained under white light and wounding, which had a positive effect on the DSE of this species (Figure 5b).

Figure 2. PLBs formation of Epipactis veratifolia after direct somatic embryogenesis under different light spectra; on protocorm explants ((**a–g**). in medium contain 3 mg L^{-1} TDZ) and node explant ((**h**), in medium contain 0.5 mg L^{-1} NAA); (**a**) Cluster of PLBs at different developmental stages, including rudimentary PLBs (RP), PLBs with shoot apical meristem (SM), PLBs with first leaf primordia (LP) and PLBs with first leaf (FL) under blue spectrum. Each red circle represents a PLB which its stage or morphology is explained by arrow, (**b**) PLBs with huge diameter under green spectrum, (**c**) large PLBs with two leaf primordia (2LP) under red + blue spectra, (**d**) deformed PLBs with branched first leaf primordia under red spectrum, (**e**) many tiny PLBs with one leaf primordia without chlorophyll under red + far spectra, (**f**) PLBs with three leaf primordia (3LP) without chlorophyll under dark conditions, (**g**) many PLBs with three leaf primordia (3LP) under white light, (**h**) direct somatic embryo (one PLB) formation on node explant under dark condition by using 0.5 mg L^{-1} NAA, (**i**) growth of somatic plantlet on hormone-free medium.

Figure 3. Direct somatic embryogenesis (formation of PLBs) on protocorm explants of Dactylorhiza umberosa under different light conditions including; (**a**) Cluster of PLBs at different stages of development stages with many rhizoids (R) under blue spectrum (in media containing 1.5 mg L^{-1} Kin), (**b**) under white light (in media containing 2 mg L^{-1} Kin), (**c**) red + far-red spectra (in media containing 3 mg L^{-1} TDZ), (**d**) dark condition (in media containing 1.5 mg L^{-1} Kin), (**e**) red + blue spectra (in media containing 3 mg L^{-1} TDZ), (**f**) green spectra (in media containing 3 mg L^{-1} TDZ), (**g**) red spectra (in media contain 3 mg L^{-1} TDZ), (**h**) germination of somatic embryos (PLBs), (**i**) growth of somatic plantlets.

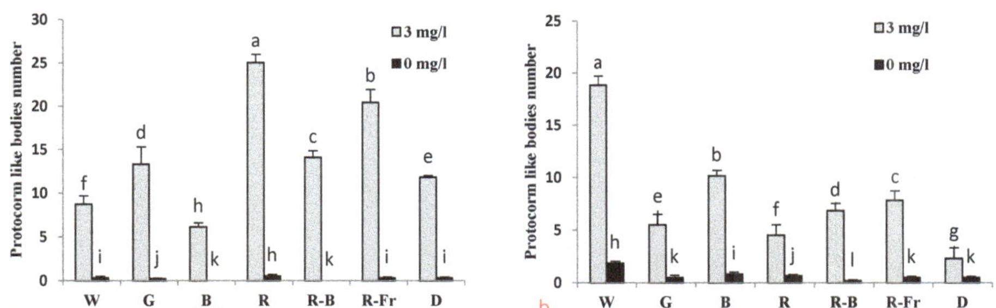

Figure 4. Interaction effects of TDZ concentrations (0 and 3 mg L^{-1}) and different light spectra on DSE of protocorm explants of Epipactis Veratifolia (**a**) and Dactylorhiza umberosa (**b**). The horizontal axis shows the light treatment (W = white, G = green, R = red, R-B = red + blue, RFR = red + far-red, and D = dark). The columns show somatic embryogenesis (the mean number of embryos (PLBs) per treatment). Values are the mean of three replicates and bars represent the standard errors. Data were recorded two months after embryo formation. Different letters indicate significant differences at $p < 0.05$ using Duncan's multiple range test.

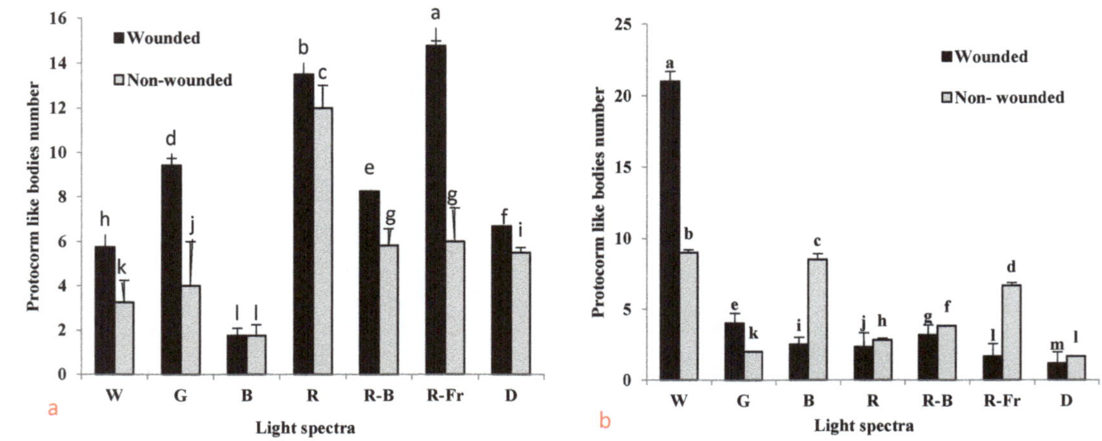

Figure 5. Interaction between different light spectra and wounding on embryogenesis (PLBs formation on protocorm explants) in Epipactis veratifolia (**a**) and Dactylorhiza umberosa (**b**) on media containing 3 mg L^{-1} TDZ. The horizontal axis shows the light treatment (W = white, G = green, R = red, R-B = red + blue, RFR = red + far-red, and D = dark). The vertical columns show the means of somatic embryogenesis with wounding/unwounding explants. Values are the means of three replicates and bars represent the standard errors. Data were recorded two months after the formation of embryos Different letters indicate significant differences at $p < 0.05$ using Duncan's multiple range test.

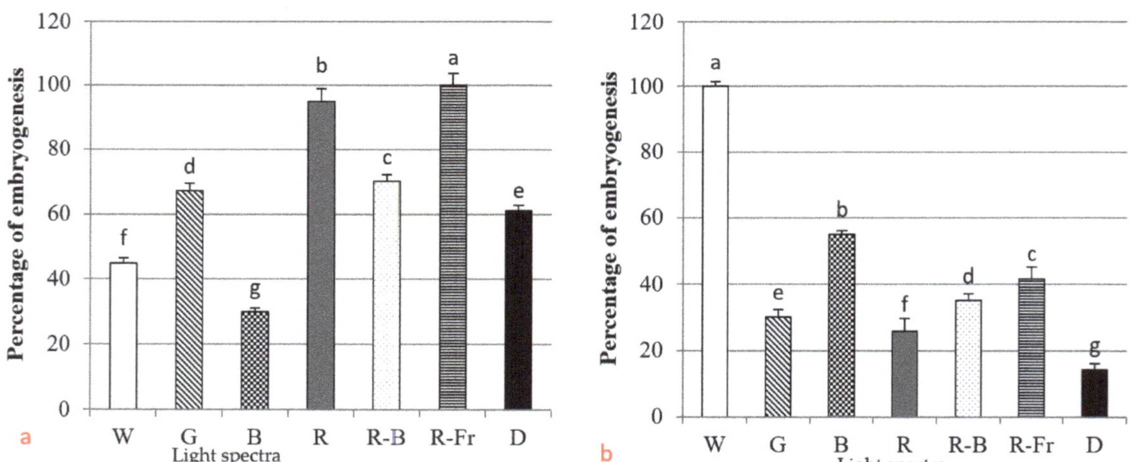

Figure 6. Influence of different light spectra on the induction of somatic embryos% on protocorm wounded explants from *Epipactis veratifolia* (**a**) and *Dactylorhiza umberosa* (**b**) in media containing TDZ. The horizontal axis shows the light treatment (W = white, G = green, R = red, R-B = red + blue, RFR = red + far-red, and D = dark). The vertical columns show the means of somatic embryogenesis under the wounding/unwounding and PGR-free/TDZ conditions. Values are the means of three replicates and bars represent the standard errors. Different letters indicate significant differences at $p < 0.05$ using Duncan's multiple range test.

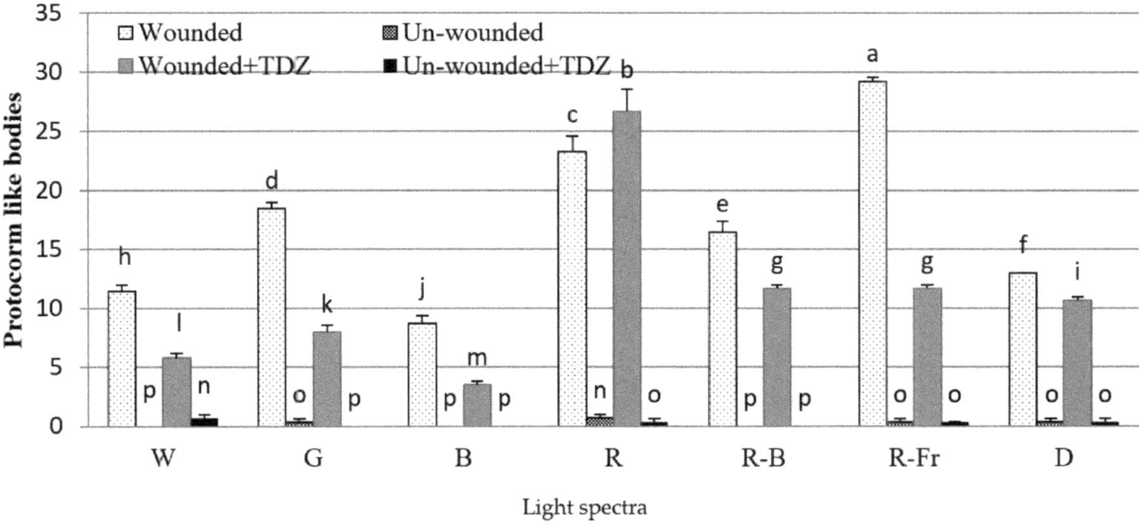

Figure 7. Interaction of different light spectra, TDZ application, and wounding on the induction of somatic embryos (PLBs) in Epipactis Veratifolia. The horizontal axis shows the light treatment (W = white, G = green, R = red, R-B = red + blue, RFR = red + far-red, and D = dark). The vertical columns show the means of somatic embryogenesis under the wounding/unwounding and PGR-free/TDZ conditions. Values are the means of three replicates and bars represent the standard errors. Different letters indicate significant differences at $p < 0.05$ using Duncan's multiple range test.

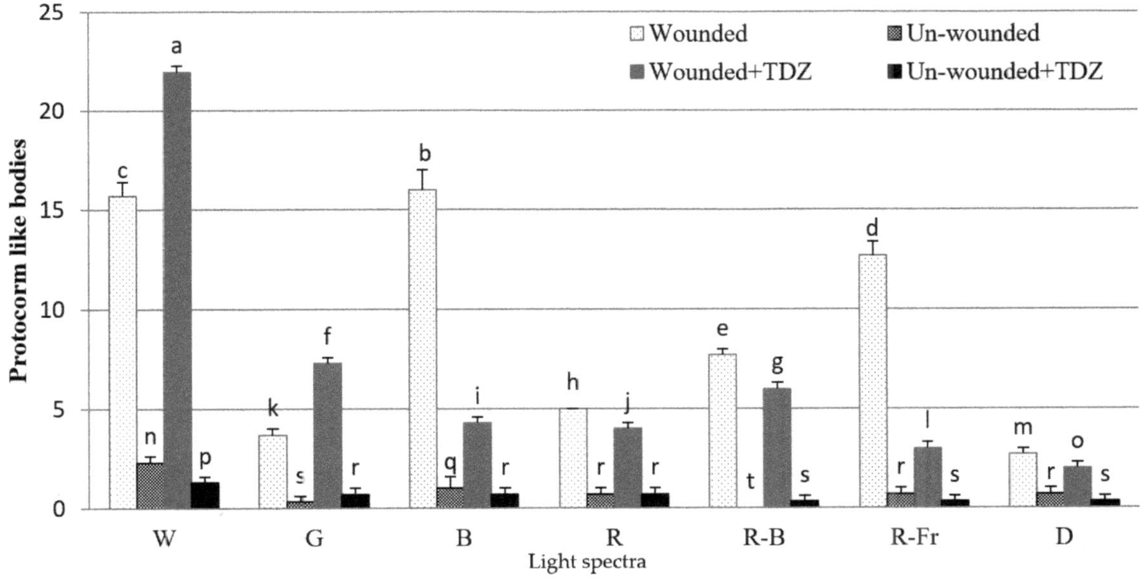

Figure 8. Interaction of different light spectra, TDZ application, and wounding on the induction of somatic embryos (PLBs) in Dactylorhiza umberosa. The horizontal axis shows the light treatment (W = white, G = green, R = red, R-B = red + blue, RFR = red + far-red, and D = dark). The vertical columns show the means of somatic embryogenesis under the wounding/unwounding and PGR-free/TDZ conditions. Values are the means of three replicates and bars represent the standard errors. Different letters indicate significant differences at $p < 0.05$ using Duncan's multiple range test.

3.2. Induction of SE under Different Light Spectra

Different light spectrums had a notable impact on DSE in both orchid types. Red light led to a higher percentage (100%) of DSE and a number of PLBs (25 PLB) in *E. veratifolia* were observed under red light (Figure 4). The percentages of DSE were also affected by the light spectrums ($p < 0.05$). The greatest DSE percentages were observed under R: FR (100%), R (95%), RB (70%), and G (67%) spectra, with the lowest DSE seen under B (30%), W (45%), and darkness (61%), respectively (Figure 6a). FR (100%), R (95%), RB (70%), and G (67%) spectra, with the lowest DSE seen under B (30%), W (45%), and darkness (61%), respectively (Figure 6a). In *D. umberosa*, W light induced the highest DSE (100%), followed by B (55%), R: FR (41.7%), RB (35%), G (30%), R (25.8%), and D (14.2%) (Figure 6b). The size of SEs was significantly larger when explants of *E. veratifolia* were exposed to B, R-B, and G spectra. Both white light and RB spectra led to greater development of PLBs with three-leaf primordia (Figure 2). Bigger PLBs of *D. umberosa* were observed under the G spectrum and dark conditions. RB spectra resulted in the development of more PLBs with three-leaf primordia in this species (Figure 3). Explants of *E. veratifolia* turned brown and subsequently exhibited necrosis when exposed to RB, W, and R lights. As a result, the survival rate of plantlets decreased (Figure 9). Furthermore, the light spectrum had an effect on embryo germination in both orchids. The embryos generated under R-Fr and W lights resulted in the highest percentage of embryo germination and plantlet development in *E. veratifolia* (100%) and *D. umberosa* (100%), respectively (not shown).

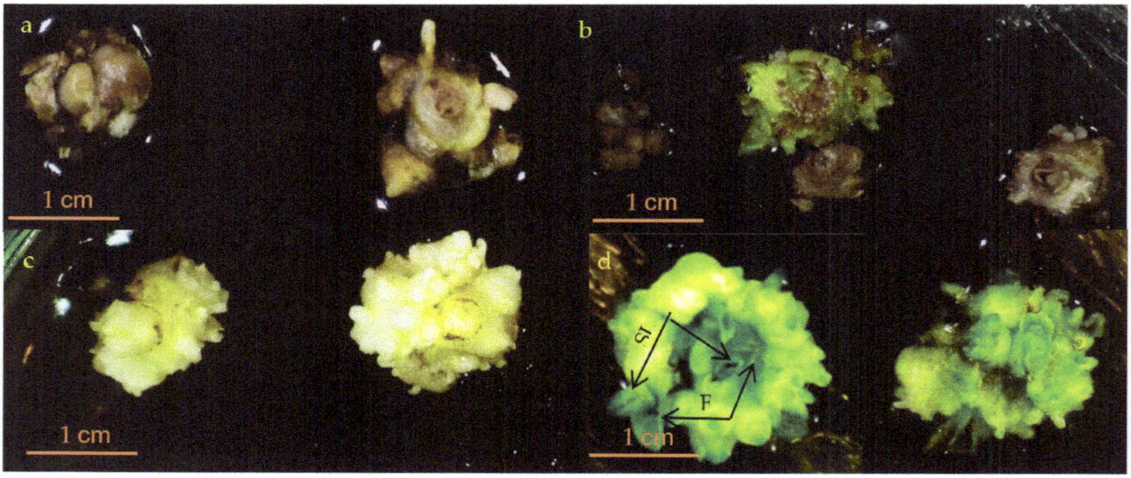

Figure 9. Growth and survival of PLBs after direct somatic embryogenesis in Epipactis veratifolia. (**a**) Production of the cluster with a number of large PLBs which became brown and necrotic under a blue and white LED light spectra. (**b**) Red + blue and green light spectra produced the cluster with a few numbers of PLBs that also turned brown and necrotic. (**c**) Red light produced the highest embryogenesis and PLBs proliferation rate. (**d**) Significantly larger PLBs with a higher developmental stage were observed in the cluster when explants were exposed to blue and red-blue light. In most of the PLBs, the first leaflet (FL) and in some others the second leaflet (SL) were visible under these spectra (arrow).

4. Discussion

The effect of different light spectra on the direct embryogenesis of temperate terrestrial native orchids and their combination with plant growth regulators, explant type, and wounding has not been previously investigated. Some researchers have studied the embryogenesis of native orchids, for example, an in vitro somatic embryogenesis method and regeneration for *Anoectochilus elatus* Lindley, an endangered orchid, by studying some PGRs and vitamins [7]. Similar methods have been developed for some of the other endemic terrestrial orchids such as *Anoectochilus elatus* and *Caladenia latifolia* [41]. A number of studies conducted in this field on orchids have used different types of explants for DSE; for example, stem nodes [5,6,42], protocorm [41–43], apical buds, crowns, and leaves [5]. In the current study, we focused on introducing an efficient method for DSE in *D. umberosa* and *E. veratifolia* using a novel combination of light spectra, PGRs, and wounding. Leaf explants of *E. veratifolia* did not produce any SE when placed in different media with different PGRs, but protocorm and nodal explants efficiently induced SE (Table 1 and Figure 2a–h). Wounding had effects on the genes encoding the induction of cell-wall proteins [44]. In the current research, we found that in *D. umberosa*, wounded protocorm explants efficiently induced DSE compared to non-wounded ones especially under white light and G spectrum (Figure 5b). This phenomenon is observed in *E. veratifolia* under all light conditions (Figure 5a). This finding is consistent with previous findings in soybean [38] and in tomato shoot regeneration by wounding of cotyledonary explants [45]. The results suggest that intact explants are more efficient for embryo formation in *D. umberosa* under B, R, RB, RFR spectra, and dark conditions (Figure 5b). It has been previously reported that the wounding of explants causes browning, which increases the phenolic compounds and eventually leads to the death of explants [46]. It can be said that wounding is an effective treatment to induce embryogenesis and the light spectrum may affect its efficiency.

PGRs are known to stimulate cell division and play an important role in the induction of SE [47]. The majority of previous studies have required auxins to induce SE in various plant species, while some reports have shown that cytokinins promote the formation of embryogenic cells with a role similar to that of auxins [48,49], for example, in orchids [5,15,50]. We used three concentrations of Kin (1, 1.5, and 2 mg L^{-1}), which had not previously been used for DSE in terrestrial orchids. Two concentrations of Kin (1.5 and 2 mg L^{-1}) resulted in positive effects on the DSE of *D. umberosa* (Table 1 and Figure 3). In other studies, the growth regulator kinetin has been used to induce protocorm formation in *Dactylorhiza majalis* [51,52]. The effect of these kinds of PGRs on embryogenesis has been reported on black iris [53], *Phalaenopsis* [54], and *Rhynchostylis retusa* [55]. Effective PGR treatments were 3 mg L^{-1} TDZ on protocorm and 1 mg L^{-1} NAA in nodal explant of *E. veratifolia* and 3 mg L^{-1} TDZ and 2 mg L^{-1} Kin on protocorm explant of *D. umberosa* (Table 1). The high concentrations of Kin and NAA (3 and 1 mg L^{-1}, respectively) had no positive effect on SE. It was reported that the exogenous addition of a high concentration of BA and Kin significantly inhibited SE formation in orchard grass [56]. Our results showed that TDZ had the highest efficiency for DSE than BA and NAA in *E. veratifolia*. This was consistent with the previous results in *Dendrobium* [50,57] and *Cymbidium* [33]. In general, we found that the application of TDZ and NAA on wounded explant was more effective for DSE in *E. veratifolia* and TDZ, Kin on wounded explant induced DSE in *D. umberosa*. There are only a few reports that mention the role of light spectra on SE induction in native orchid species. In the present experiment, the highest percentage of DSE and embryo germination rate in *E. veratifolia* was observed in protocorm explants exposed to R-Fr (100%) and then R (95%) spectra (Figure 6), while B spectra, W light, and D condition had less effect on embryo formation on explants (Figure 6a). This finding is in agreement with the results of previous studies on the initiation and development of SE by R and R-Fr spectra in *Araujia sericifera* [29], *Oncidium* [18], and *Rosa chinensis* [34]. In *E. veratifolia*, R and R-Fr spectra produced the highest rate of embryogenesis and proliferation of embryos (PLBs); although, the highest percentage of embryo proliferation was obtained under R-Fr light (Figures 5a and 7). The positive effects of the red light spectrum have been reported on the formation of 100% protocorms in orchids [58]. Consistently, R light has been reported to promote the induction and proliferation of SE in *Oncidium* [18]. Red light has been shown to affect plant reproduction through phytochrome, which in its active form increases the endogenous hormonal balance, increasing the amount of cytokinin in the tissues and counteracting the action of auxin [59]. In our study, in addition to R and R-Fr, G, R-B, B, W light and D conditions were used on explants to induce embryogenesis in *E. veratifolia*, but most of the explants produced a small number of embryos that also turned brown and necrotic under these conditions during three to four weeks (Figure 9a,b). This shows that G, R-B, B spectra, W light, and D conditions delayed the germination of SE in *E. veratifolia*. In the current study, the maturation of embryos as well as the germination of embryos was significantly different in different light treatments. In some of the previous studies, it has been reported that callus formation was high in the blue spectrum, which proves that rapid cell division occurred, although the organized center of cell division required for primordia formation was reduced and growth was delayed [60]. Interestingly, we found that G light increased the formation and germination of SE in protocorm explants of *E. veratifolia* (Figures 4a and 6a). The effect of the G spectrum was significant on the percentage of explants that had embryogenesis and on the number of embryos (Figures 5a and 6a). A study on carrot embryogenesis showed that G light caused the highest SE compared to the other light spectra [61]. The highest PLBs formation, root formation, and shoot formation rate in *Cymbidium insigne* was reported under the G spectrum [62]. The G light may have active photoreceptors [63], but unlike red and blue light, green light photoreceptors have not been discovered yet [64]. Embryos germinated when sub-cultured in the hormone-free medium. In both orchids, the somatic embryos had advanced developmental stages under the B and R-B spectrum, which rapidly increased the size of the embryos (Figure 9d). This result shows the positive effects of the B spectrum on the development of PLBs and it is in agreement with the previous report on the promotive effects of the B spectrum on embryo size in carnation [24].

The results of this study show that the highest percentage of DSE (100%) was observed in the explants exposed to W light, while the dark condition had the lowest effect on embryo induction rate in *D. umberosa* (Figure 6b). Researchers have also reported that SE induction and promotion are reduced in darkness in olive [65], soybean [66], and *Coffea Arabica* [67]. The highest percentage of embryogenesis in *D. umberosa* was produced under the W light and B spectrum (Figure 6b). The B spectrum had an efficient effect on DSE and ISE in carnation [24]. The highest number of the torpedo, globular, and heart embryos was observed under the B spectrum; in comparison to the effect of other light treatments that had no embryo formation [24]. The B light can affect plant auxin content (specifically indoleacetic acid-IAA). Therefore, it can affect PGR and morphotomorphogenesis [68]. Similar to the previous species, in *D. umberosa*, the apical meristem of PLBs had the most development under R-B spectra and produced two to three-leaf primordia (Figure 2e). This issue shows the positive and promotive effects of the B spectrum. The result showed that when the G, R spectra and D conditions were used to induce embryogenesis in *D. umberosa*, most of the explants did not produce embryos or a high number of embryos died (Figure 9). The PLB with larger diameters were observed under the G spectrum and D conditions (Figure 3d,f), but these conditions were not suitable for their survival. In *D. umberosa*, the embryos germinated when they were transferred and cultured on the PGR-free basal medium, and the germination level of embryos in different light spectra was significantly different. After exposure to white light, the B spectrum increased the induction and germination of SE from the protocorm explant (Figures 6b and 8). The B spectrum is important for photomorphogenesis [68]. Consistent with this, it has been reported in *Oncidium* that the B spectrum increases differentiation and enzyme activities in embryos [18]. Previous studies have already mentioned the relationship between SE production and photo-equilibrium and the increase in SE formation by increasing photo-equilibrium. There is a higher photo-equilibrium under W light compared to the B spectrum [69]. Our results indicate that LEDs are a forward light source that increases DSE induction. Formation with high efficiencies of plant regeneration in both terrestrial orchids studied. We recommend the use of white light to produce the most PLBs on wounded protocorm explants of *D. umberosa* (Figure 8) and R-Fr light for producing the highest number of PLBs on wounded protocorm explants of *E. veratifolia* (Figure 7), both in a culture medium containing 3 mg L^{-1} TDZ.

5. Conclusions

Some native species like *D. umberosa* have been exposed to extinction because of uncontrolled harvest from their natural habitats. Mass propagation of *E. veratrifolia* and *D. umberosa* is difficult and sometimes impossible by using conventional in vitro culture methods because of low seed germination, lack of suitable symbionts, slow vegetative growth, and absence of efficient methods for asexual reproduction. In the present study, for the first time, a successful method and an efficient in vitro propagation protocol were established for these species by using light spectra through DSE. This protocol will have a significant impact on commercial micropropagation and genetic resource conservation of these native orchids. Wounding on the protocorm explants of *D. umberosa* with the use of a medium containing 3 mg L^{-1} TDZ caused a 94.1% increase in the number of PLBs under white light. In *E. veratrifolia*, the use of wounded protocorm explants in the medium containing 3 mg L^{-1} TDZ under the R spectrum caused an increase in production by 98.8%.

Author Contributions: Conceptualization, H.N.B. and S.D.D.; Data curation, H.N.B.; Formal analysis, H.N.B. and S.D.D.; Investigation, H.N.B. and S.D.D.; Methodology, S.D.D.; Project administration, S.D.D.; Resources, S.D.D.; Software, H.N.B. and S.D.D.; Supervision, S.D.D.; Validation, S.D.D.; Writing—original draft, H.N.B.; Writing—review and editing, H.N.B., S.D.D. and K.V. All authors have read and agreed to the published version of the manuscript.

Funding: This research received no external funding.

Data Availability Statement: Not applicable.

Acknowledgments: The authors wish to thank the Orchid Breeding and Propagation Laboratory of the Horticultural Department, University of Tehran for the opportunity to carry out their research.

Conflicts of Interest: No potential conflict of interest were reported in this study.

References

1. Gow, W.P.; Chen, J.T.; Chang, W.C. Influence of growth regulators on direct embryo formation from leaf explants of *Phalaenopsis* orchids. *Acta Physiol. Plant.* **2008**, *30*, 500–507. [CrossRef]
2. Ghorbani, A.; Barbara, G.; Farzaneh, N.; Hugo, B. Wild orchid tuber collection in Iran: A wake-up call for conservation. *Biodivers. Conserv.* **2014**, *23*, 2749–2760. [CrossRef]
3. Starin, D. Salepi extinction, salepi survival: How a change in ingredients could help safeguard orchids. *Am. Orch. Soc. Bull.* **2012**, *81*, 490–494.
4. Sezik, E. Destroying of Ophrys species to obtain salep in Turkey. *J. Eur. Orch.* **2006**, *38*, 290–295.
5. Moradi, S.; Dianati Daylami, S.; Arab, M.; Vahdati, K. Direct somatic embryogenesis in *Epipactis veratrifolia*, a temperate terrestrial orchid. *J. Hortic. Sci. Biotechnol.* **2017**, *92*, 88–97. [CrossRef]
6. Jalal, J.S. Endemic orchids of peninsular India: A review. *J. Threat. Taxa* **2012**, *15*, 3415–3425. [CrossRef]
7. Sherif, N.; Ahamed, J.H.; Franklin, B.; Senthil, K.; Rao, M.V. Somatic embryogenesis, acclimatization and genetic homogeneity assessment of regenerated plantlets of *Anoectochilus elatus* Lindl an endangered terrestrial jewel orchid. *Plant Cell Tissue Organ Cult.* **2018**, *132*, 303–316. [CrossRef]
8. Kumar, S.; Narender, S.; Manisha, M. Micropropagation of *Simmondsia chinensis* (Link) Schneider through enhanced axillary branching from nodal segments. *J. Plant Biol.* **2009**, *36*, 75–81.
9. Dogan, Y.; Suleyman, B.; Gungor, A.; Hasan, H.M. the use of wild edible plants in western and central Anatolia (Turkey). *Econ. Bot.* **2004**, *58*, 684–690. [CrossRef]
10. Stewart, S.L.; Kane, M.E. Asymbiotic seed germination and in vitro seedling development of *Habenaria macroceratitis* (Orchidaceae) a rare Florida terrestrial orchid. *Plant Cell Tissue Organ Cult.* **2006**, *86*, 147–158. [CrossRef]
11. Pant, B.; Thapa, D. *In vitro* mass propagation of an epiphytic orchid, *Dendrobium primulinum* lindl through shoot tip culture. *Afr. J. Biotechnol.* **2012**, *11*, 9970–9974 [CrossRef]
12. Tanaka, M.; Hasegawa, A.; Goi, M. Studies on the clonal propagation of monopodial orchids by tissue culture I. Formation of protocorm-like bodies from leaf tissues in Phalaenopsis and Vanda. *J. Jap. Soc. Hort. Sci.* **1975**, *44*, 47–58.
13. Kosir, P.; Skof, S.; Luthar, Z. Direct shoot regeneration from nodes of *Phalaenopsis* orchids. *Acta Agric. Slov.* **2004**, *83*, 233–242.
14. Mahendran, G.; Bai, V.N. Direct somatic embryogenesis and plant regeneration from seed derived protocorms of *Cymbidium bicolor* Lindl. *Sci. Hortic.* **2012**, *135*, 40–44. [CrossRef]
15. Chen, J.T.; Chang, W.C. Effects of auxins and cytokinins on direct somatic embryogenesis on leaf explants of *Oncidium* 'Gower Ramsey'. *J. Plant Growth Regul* **2001**, *34*, 229–232. [CrossRef]

16. Mulgund, G.S.; Nataraja, K.; Malabadi, R.B.; Kumar, S.V. TDZ induced *in vitro* propagation of an epiphytic orchid *Xenikophyton smeeanum*. *Plant Biol.* **2011**, *1*, 235–241.
17. Arditti, J. *Micropropagation of Orchids*; John Wiley & Sons: Hoboken, NJ, USA, 2009.
18. Mengxi, L.; Zhigang, X.; Yang, Y.; Yijie, F. Effects of different spectral lights on *Oncidium* PLBs induction, proliferation, and plant regeneration. *Plant Cell Tissue Organ Cult.* **2011**, *106*, 1–10. [CrossRef]
19. Nguyen, Q.T.; Kozai, T.; Niu, G.; Van Nguyen, U. Photosynthetic characteristics of coffee (*Coffea arabusta*) plantlets *in vitro* in response to different CO_2 concentrations and light intensities. *Plant Cell Tissue Organ Cult.* **1998**, *55*, 133–139. [CrossRef]
20. Nhut, D.T.; Takamura, T.; Watanabe, H. Responses of strawberry plantlets cultured in vitro under superbright red and blue lightemitting diodes (LEDs). *Plant Cell Tissue Organ Cult.* **2003**, *73*, 43–52. [CrossRef]
21. Cybularz-Urban, T.; Hanus-Fajerska, E.; Swiderski, A. Effect of light wavelength on in vitro organogenesis of a Cattleya hybrid. *Acta Biol. Crac. Ser. Bot.* **2007**, *49*, 113–118.
22. Cavallaro, V.; Pellegrino, A.; Muleo, R.; Forgione, I. Light and plant growth regulators on in vitro proliferation. *Plants* **2022**, *11*, 844. [CrossRef] [PubMed]
23. Lian, M.L.; Murthy, H.N.; Paek, K.Y. Effects of light emitting diodes (LEDs) on the in vitro induction and growth of bulblets of *Lilium oriental* hybrid 'Pesaro'. *Sci. Hortic.* **2002**, *94*, 365–370. [CrossRef]
24. Aalifar, M.; Arab, M.; Aliniaeifard, S.; Dianati, S.; Mehrjerdi, M.Z.; Limpens, E.; Serek, M. Embryogenesis efficiency and genetic stability of *Dianthus caryophyllus* embryos in response to different light spectra and plant growth regulators. *Plant Cell Tissue Organ Cult.* **2019**, *139*, 479–492. [CrossRef]
25. Jao, R.C.; Chien-Chou, L.; Wei, F.; Sen-Fuh, C. Effects of red light on the growth of *Zantedeschia* plantlets in vitro and tuber formation using light-emitting diodes. *HortScience* **2005**, *40*, 436–438. [CrossRef]
26. Dutta Gupta, S.; Jatothu, B. Fundamentals and applications of light-emitting diodes (LEDs) in in vitro plant growth and morphogenesis. *Plant Biotechnol. Rep.* **2013**, *7*, 211–220. [CrossRef]
27. Biswal, A.K.; Pattanayak, G.K.; Pandey, S.S.; Leelavathi, S.; Reddy, V.S.; Govindjee; Tripathy, B.C. Light intensity-dependent modulation of chlorophyll b biosynthesis and photosynthesis by overexpression of chlorophyllide a oxygenase in Tobacco. *Plant Physiol.* **2012**, *159*, 433–449. [CrossRef]
28. Zhou, M.; Guan, Q.; Wei, Y.; Zhang, Z. Effects of Sucrose Concentration and light intensity on growth and photosynthesis of ginger plantlets in vitro. *Chin. J. Appl. Environ. Biol.* **2008**, *14*, 356–361.
29. Torne, J.M.; Moysset, L.; Santos, M.; Simón, E. Effects of light quality on somatic embryogenesis in *Araujia sericifera*. *Physiol. Plant.* **2001**, *111*, 405–411. [CrossRef] [PubMed]
30. Park, S.Y.; Yeung, E.C.; Paek, K.Y. Endoreduplication in *Phalaenopsis* is affected by light quality from light-emitting diodes during somatic embryogenesis. *Plant Biotechnol. Rep.* **2010**, *4*, 303–309. [CrossRef]
31. Le, V.T.; Tanaka, M. Effects of red and blue light-emitting diodes on callus induction, callus proliferation, and protocorm-like body formation from callus in *Cymbidium* orchid. *Environ. Control. Biol.* **2004**, *42*, 57–64.
32. Chen, J.R.; Wu, L.; Hu, B.W.; Yi, X.; Liu, R.; Deng, Z.N.; Xiong, X.Y. The influence of plant growth regulators and light quality on somatic embryogenesis in *China rose*. *J. Plant Growth Regul.* **2014**, *33*, 295–304. [CrossRef]
33. Rodríguez-Sahagún, A.; Acevedo-Hernández, G.; Rodríguez-Domínguez, J.M.; Rodríguez-Garay, B.; Cervantes-Martínez, J.; Castellanos-Hernández, O.A. Effect of light quality and culture medium on somatic embryogenesis of *Agave tequilana* Weber, Azul. *Plant Cell Tissue Organ Cult.* **2011**, *104*, 271–275. [CrossRef]
34. Mose, W.; Indrianto, A.; Purwantoro, A.; Semiarti, E. The influence of thidiazuron on direct somatic embryo formation from various types of explant in *Phalaenopsis amabilis* (L.) blume orchid. *Hayati J. Biosci.* **2017**, *24*, 201–205. [CrossRef]
35. Bhattacharya, C.; Dam, A.; Karmakar, J.; Bandyopadhyay, T.K. Direct somatic embryogenesis and genetic homogeneity assessment of regenerated plants of *Anthurium andraeanum* Linden cv. *Fantasia*. *Vitr. Cell. Dev. Biol. Plant* **2016**, *52*, 512–519. [CrossRef]
36. Bakhshaie, M.; Babalar, M.; Mirmasoumi, M.; Khalighi, A. Somatic embryogenesis and plant regeneration of *Lilium ledebourii* (Baker) Boiss an endangered species. *Plant Cell Tissue Organ Cult.* **2010**, *102*, 229–235. [CrossRef]
37. Prange, A.N.S.; Bartsch, M.; Serek, M.; Winkelmann, T. Regeneration of different Cyclamen species via somatic embryogenesis from callus, suspension cultures and protoplasts. *Sci. Hortic.* **2010**, *125*, 442–450. [CrossRef]
38. Santarem, E.R.; Pelissier, B.; Finer, J.J. Effect of explant orientation, pH, solidifying agent and wounding on initiation of soybean somatic embryos. *Vitr. Cell. Dev. Biol. Plant* **1997**, *33*, 13–19. [CrossRef]
39. Rojas-Herrera, R.; Loyola-Vargas, V. Induction of a class acidic chitinase in foliar explants of (*Coffea arabica* L.) during somatic embryogenesis and wounding. *Plant Sci.* **2002**, *163*, 705–711. [CrossRef]
40. Dianati, S.; Kafi, M.; Mirmaoumi, M.; Mozaffarian, V.; Salami, A.R. Introduce the suitable medium for asymbiotic germination of *Epipactis veratrifolia*. *J. Crops Improv.* **2014**, *15*, 161–177. [CrossRef]
41. Bustam, B.M.; Dixon, K.; Bunn, E. Proliferation and harvesting of secondary protocorms as a novel means for improving propagation of terrestrial orchids. *Aust. J. Bot.* **2015**, *62*, 614–621. [CrossRef]
42. Gangaprasad, A.N.; Decruse, W.S.; Seeni, S.; Menon, S. Micropropagation and restoration of the endangered Malabar daffodil orchid Ipsea malabarica. *Lindleyana* **1999**, *14*, 38–46.
43. Romero, C.; Cuba-Díaz, M.; Silva, R. In vitro culture of *Chloraea gavilu* Lindl., an endemic terrestrial orchid from Chile. *Plant Biosyst.* **2018**, *152*, 612–620. [CrossRef]

44. Showalter, A.M.; Bell, J.N.; Cramer, C.L.; Bailey, J.A.; Varner, J.E.; Lamb, C.J. Accumulation of hydroxyproline-rich glycoprotein mRNAs in response to fungal elicitor and infection. *Proc. Natl. Acad. Sci. USA* **1985**, *82*, 6551–6555. [CrossRef] [PubMed]
45. Bhatia, P.; Ashwath, N.; Midmore, D.J. Effects of genotype, explant orientation, and wounding on shoot regeneration in tomato. *Vitr. Cell. Dev. Biol. Plant.* **2005**, *41*, 457–464. [CrossRef]
46. Chamandoosti, F. The relationship between plant growth regualtors for organogenesis and phenolic compound in cotton. *Asian J. Dev. Biol.* **2010**, *2*, 16–22.
47. Vahdati, K.; Jariteh, M.; Niknam, V.; Mirmasoumi, M.; Ebrahimzadeh, H. Somatic embryogenesis and embryo maturation in Persian walnut. *Acta Hortic.* **2006**, *705*, 199–205. [CrossRef]
48. Thomas, C.; Jiménez, V.M. Mode of action of plant hormones and plant growth regulators during induction of somatic embryogenesis: Molecular aspects. *Somat. Embryog.* **2005**, *23*, 157–175. [CrossRef]
49. Singh, C.; Raj, S.R.; Jaiswal, P.; Patil, V.; Punwar, B.; Chavda, J.; Subhash, N. Effect of plant growth regulators on *in vitro* plant regeneration of sandalwood *Santalum album* L. via organogenesis. *Agrofor. Syst.* **2016**, *90*, 281–288. [CrossRef]
50. Gow, W.P.; Jen-Tsung, C.; Wei-Chin, C. Enhancement of direct somatic embryogenesis and plantlet growth from leaf explants of *Phalaenopsis* by adjusting culture period and explant length. *Acta Physiol. Plant.* **2010**, *32*, 621–627. [CrossRef]
51. Ponert, J.; Vosolsobě, S.; Kmecová, K.; Lipavská, H. European orchid cultivation–from seed to mature plant. *Eur. J. Environ. Sci.* **2011**, *26*, 5–15. [CrossRef]
52. Chen, L.R.; Chen, J.T.; Chang, W.C. Efficient production of protocorm-like bodies and plant regeneration from flower stalk explants of the *sympodial* orchid Epidendrum radicans. *Vitr. Cell. Dev. Biol. Plant.* **2002**, *38*, 441–445. [CrossRef]
53. Shibli, R.A.; Ajlouni, M. Somatic embryogenesis in the endemic black iris. *Plant Cell Tissue Organ Cult.* **2000**, *61*, 15–21. [CrossRef]
54. Kuo, H.L.; Chen, J.T.; Chang, W.C. Efficient plant regeneration through direct somatic embryogenesis from leaf explants of *Phalaenopsis* 'Little Steve'. *Vitr. Cell. Dev. Biol. Plant.* **2005**, *41*, 453–461. [CrossRef]
55. Bhattacharjee, B.; Shahinul Islam, S.M. Assessment of antibacterial and antifungal activities of the extracts of *Rhynchostylis retusa* Blume-A Medicinal Orchid. *WJPPS* **2015**, *2*, 74–87.
56. Wenck, A.R.; Conger, B.V.; Trigiano, R.N.; SAMS, C.E. Inhibition of somatic-embryogenesis in orchardgrass by endogenous cytokinins. *Plant Physiol.* **1988**, *88*, 990–992. [CrossRef] [PubMed]
57. Chung, H.H.; Chen, J.T.; Chang, W.C. Plant regeneration through direct somatic embryogenesis from leaf explants of *Dendrobium*. *Biol. Plant.* **2007**, *2*, 346–350. [CrossRef]
58. Kamal, M.M.; Shimasaki, K.; Akter, N. Effect of light emutting diode (LED) lamps and N-Acetylglucosamine (NAG) on organogenesis in protocorm-like bodies (PLBs) of a *Cymbidium* hybrid cultured in vitro. *Plant Tissue Cult. Biotechnol.* **2014**, *24*, 273–297. [CrossRef]
59. Finlayson, S.A.; Krishnareddy, S.R.; Kebrom, T.H.; Casal, J.J. Phytochrome regulation of branching in Arabidopsis. *Plant Physiol.* **2010**, *152*, 1914–1921. [CrossRef]
60. Lotfi, M.; Mars, M.; Werbrouck, S. Optimizing pear micropropagation and rooting with light emitting diodes and trans-cinnamic acid. *Plant Growth Regul.* **2019**, *88*, 173–180. [CrossRef]
61. Michler, C.H.; Lineberger, R.D. Effects of light on somatic embryo development and abscisic levels in carrot suspension cultures. *Plant Cell Tissue Organ Cult.* **1987**, *11*, 189–207. [CrossRef]
62. Golovatskaya, I.F.; Karnachuk, R.A. Role of green light in physiological activity of plants. *Russ. J. Plant Physiol.* **2015**, *62*, 727–740. [CrossRef]
63. Shinomura, T.; Nagatani, A.; Hanzawa, H.; Kubota, M.; Watanabe, M.; Furuya, M. Action spectra for phytochrome A-and B-specific photoinduction of seed germination in Arabidopsis thaliana. *Proc. Natl. Acad. Sci. USA* **1996**, *93*, 8129–8133. [CrossRef] [PubMed]
64. Nahar, S.J.; Haque, S.M.; Kazuhiko, S. Application of chondroitin sulfate on organogenesis of two *Cymbidium* spp. under different sources of lights. *Not. Sci. Biol.* **2016**, *8*, 156–160. [CrossRef]
65. Lavee, S.; Glenda, M. The effect of growth-regulating substances and light on olive callus growth in vitro. *J. Exp. Bot.* **1969**, *20*, 604–614. [CrossRef]
66. Lazzeri, P.; David, F.H.; Glenn, B.C. Soybean somatic embryogenesis: Effects of hormones and culture manipulations. *Plant Cell Tissue Organ Cult.* **1987**, *10*, 197–208. [CrossRef]
67. Gatica, A.; Andrés, M.; Griselda, A.; Ana, M.E. direct somatic embryogenesis in *Coffea arabica* L. Caturra and Catuaí: Effect of tricontanol, light condition, and medium consistency. *Agron. Costarric.* **2008**, *5*, 203–212.
68. Schuerger, C.; Christopher, S.; Brown, C. Anatomical features of pepper plants *Capsicum annuum* grown under red light-emitting diodes supplemented with blue or far-red light. *Ann. Bot.* **1997**, *79*, 273–282. [CrossRef]
69. D'Onofrio, C.; Morini, S.; Bellocchi, G. Effect of light quality on somatic embryogenesis of quince leaves. *Plant Cell Tissue Organ Cult.* **1998**, *53*, 91–98. [CrossRef]

horticulturae

MDPI

Article

Prediction Model for Breeding Hardy Geraniums

Mehdrad Akbarzadeh [1,2,3,*], Paul Quataert [1], Johan Van Huylenbroeck [1], Stefaan P. O. Werbrouck [2,*] and Emmy Dhooghe [1,2]

1 Plant Sciences Unit, Flanders Research Institute for Agriculture, Fisheries and Food (ILVO), Caritasstraat 39, 9090 Melle, Belgium; paul.quataert@ilvo.vlaanderen.be (P.Q.); johan.vanhuylenbroeck@ilvo.vlaanderen.be (J.V.H.); emmy.dhooghe@ugent.be (E.D.)

2 Department of Plants and Crops, Faculty of Bioscience Engineering, Ghent University, Coupure Links 653, 9000 Ghent, Belgium

3 Department of Horticultural Science, Faculty of Agriculture, Shahid Chamran University, Ahvaz 613, Iran

* Correspondence: mehrdad.akbarzadeh@ugent.be (M.A.); stefaan.werbrouck@ugent.be (S.P.O.W.); Tel.: +32-493121249 (M.A.)

Abstract: Key message. The success of interspecific hybridization in hardy geranium (*Geranium* sp.) can be predicted by considering the genetic distance between parental genotypes and using the logistic regression model developed in this study. Hardy geranium is a popular ornamental plant known for its architecture, hardiness, prolonged flowering, and diverse colors in leaves and flowers. In ornamental breeding, the pursuit of novel trait combinations is never-ending. Even in *Geranium*, certain combinations of valuable traits have not yet been achieved. Interspecific hybridization can increase diversity; however, success remains low due to pre- and postzygotic barriers. Crossing success can be predicted by response criteria such as pollen tube growth (tube_length), seed development (seed_dev), and seed setting (seed_set). Within a collection of 42 *Geranium* genotypes and during two consecutive breeding seasons (years), we evaluated tube_length, seed_dev, and seed_set for 150, 1155, and 349 crosses, respectively. These crosses varied in four parental differences (variables): chromosome number (Chrom), DNA/chromosome (DNA), style length (Style), and genetic distance expressed as the Jaccard distance (cJaccard = 1 − Jaccard). Using logistic regression models has confirmed that most often, the success rate decreased with increasing parental distance. The most consistent association was seen in seed_dev in combination with cJaccard. The model was used to predict the number of crosses necessary to have 10 successful crossing products by taking into account the uncertainty in the model. These findings provide valuable guidance for future planning of interspecific breeding experiments in *Geranium*. By incorporating the genetic distance between parental genotypes, breeders can enhance the efficiency and success of hybridization efforts.

Keywords: Geraniaceae; incongruity; interspecific hybridization; Jaccard; pollen tube; pre- and postzygotic barriers

Citation: Akbarzadeh, M.; Quataert, P.; Van Huylenbroeck, J.; Werbrouck, S.P.O.; Dhooghe, E. Prediction Model for Breeding Hardy Geraniums. *Horticulturae* **2023**, *9*, 617. https://doi.org/10.3390/horticulturae9060617

Academic Editor: Rossano Massai

Received: 16 April 2023
Revised: 17 May 2023
Accepted: 22 May 2023
Published: 24 May 2023

1. Introduction

Geranium, the largest genus of the Geraniaceae, is divided into three subgenera: *Erodioidea*, *Robertium*, and *Geranium* [1]. Hardy geraniums are well-suited to temperate climates and thrive in well-watered soils. These plants can grow in a wide range of environments, as long as the climate is suitable [2]. Successful intersubgeneric crosses have been carried out between species of *Geranium* and *Erodioidea*, although the most common combinations occur within the subgenus *Geranium* [2]. Under natural conditions, crosses between members of *Robertium* and the other two subgenera are not successful, confirming the occurrence of incongruity and fertilization barriers in hardy geranium. Hardy geraniums with commercially appealing traits typically showcase characteristics such as large flowers, an extended flowering period, unique flower or leaf colors, distinctive morphologies, and unusual plant architecture (e.g., compactness).

Creating new plant combinations, especially in ornamentals, remains a crucial approach to increasing diversity within a plant assortment [3]. Hardy geraniums show an enormous variation, with a minority of cultivars combining particularly attractive traits. Interspecific hybridization is the most obvious way to create new cultivars, by introducing new and desirable traits into existing cultivars. This method has been used in other herbaceous ornamental breeding programs, including *Asclepias* sp. [4], *Pavonia* [5], ornamental pepper [6], and many others.

Successful hybridization requires parent plants with viable pollen and receptive stigmas as well as essential congruity. Crossing barriers have either a genetic or a morphological background [7]. These barriers can be classified as different flowering times of the species, varietal differences, self or class incompatibility, pollen failure, low rate of pollen tube growth, fertilization failure, developmental arrest after the initial cell divisions after fertilization, or the production of nonviable seeds [7].

The majority of pollen grains capable of long-distance transport exhibit tolerance to desiccation and/or have the ability to form long pollen tubes [8]. In the absence of incongruence, the pollen on the stigma undergoes five successive events: adhesion, hydration, germination, penetration of the pollen tube, and growth through the transmitting tissue [9]. Prezygotic barriers occur from the start; in some crosses, pollen is even shed at the stigma surface. This is most common in species with dry stigmas [10,11]. Studies in *Arabidopsis* have confirmed that as the genetic distance between parent species increases, the attachment of pollen decreases [12]. Incongruity can also be observed as an atypical behavior of the pollen tube, such as growth in the wrong direction, callose plugging, or growth that stops before entering the ovules. Longer or more intense callose plugs in apples and tobacco, respectively, are indications of incongruity [13]. *Geranium* has a dry stigma [10], with five carpels and ten ovules producing five seeds under optimal conditions [14]. In addition, a study of pollen tube growth in *Geranium caespitosum* showed that after 2 h, fewer self- than cross-pollinated tubes entered the ovules, but after 24 h the difference between the number of tubes from self- and cross-pollination entering the ovules had decreased [15]. In *Geranium maculatum*, 30 min was sufficient for pollen tubes to reach the ovules with a growth rate of 0.133 mm/min [14]. However, achieving optimal conditions is not always possible. In such cases, embryo rescue has been well-documented to increase breeding success in interspecific hybridizations [16–21].

A highly significant positive correlation was observed between transgression frequency and genetic distance in eudicot plants [22], making it important for breeders to find the effect of parental differences on hybrid production rate. Some studies highlight that knowledge of genetic distance is a good predictor of success in breeding programs [23]. However, in *Sarcococca*, genetic distance, ploidy level, and genome size did not represent a true hybridization barrier [24], while in *Helleborus*, a general relationship between genetic distance and hybrid offspring was established [25]. Shortening the breeding cycle is a desirable goal in breeding programs and the use of breeding programs and modeling can greatly help breeders in parental selection, predicting crossing performance, and selecting strategies. The objective of this study was to develop a model that can predict crossbreeding success in hardy geraniums through interspecific hybridization. This novel model is based on analyzing various parental differences including chromosome numbers (Chrom), DNA/chromosome (DNA, pg/chromosome), style length (Style), and genetic distances (cJaccard). By observing pre- and postzygotic barriers during two crossing seasons of *Geranium*, we constructed a logistic regression model. To the best of our knowledge, there is no existing research that directly compares our unique combination of factors. This study offers a fresh and comprehensive approach to predicting cross-success in hardy geraniums, providing valuable insights for breeding programs aiming to enhance efficiency and achieve successful hybridization outcomes.

2. Materials and Methods

2.1. Plant Material and Growing Conditions

A *Geranium* collection comprising 42 genotypes was used for this study. The collection represented the three subgenera of *Geranium* (*Geranium*, *Erodioidea*, and *Robertium*) and included 18 species, 18 interspecific hybrids, and 6 genotypes with unknown backgrounds (Table 1).

Table 1. Different genotypes of *Geranium* sp. used in this study.

Codes	Genotypes	Subgenera	Parents
G01	*G.* 'Anne Thomson'	*Geranium*	Parent **1:** *G. procurrens*, Parent **2:** *G. psilostemon*
G02	*G.* 'Azure rush'	*Geranium*	Parent **1:** *G. wallichianum*, Parent **2:** *G.* 'Rozanne'
G03	*G.* 'Bob's Blunder'	*Geranium*	unknown
G04	*G.* 'Brookside'	*Geranium*	Parent **1:** *G. pretense*, Parent **2:** *G. clarkei*
G05	*G.* 'Catherine Deneuve'	*Geranium*	Parent **1:** *G. psilostemon*, Parent **2:** *G.* × *oxonianum* or *G.procurrens*
G06	*G.* 'Chantilly'	*Geranium*	Parent **1:** *G. gracile*, Parent **2:** *G. renardii*
G07	*G.* 'Dragon Heart'	*Geranium*	Parent **1:** *G. psilostemon*, Parent **2:** *G. procurrens*
G09	*G. pratense* 'Galactic'	*Geranium*	*G. pratense*
G10	*G. cinereum* 'Jolly Jewel Red'	*Erodioidea*	*G. cinereum*
G13	*G.* × *riversleaianum* 'Mavis Simpson'	*Geranium*	Parent **1:** *G. endressii*, Parent **2:** *G. traversii*
G14	*G.* 'Orion'	*Geranium*	Parent **1:** *G.* 'Brookside', Parent **2:** *G. himalayense*
G16	*G.* 'Rozanne'	*Geranium*	Parent **1:** *G. wallichianum*, Parent **2:** *G. himalayense*
G17	*G.* 'Salome'	*Geranium*	Parent **1:** *G. lambertii* × *G. procurrens*, Parent **2:** *G. sanguineum*
G18	*G.* 'Sanne'	*Geranium*	Parent **1:** *G. sessiliflorum*, Parent **2:** *G.* × *oxonianum*
G19	*G. nodosum* 'Silverwood'	*Geranium*	*G. nodosum*
G21	*G. wallichianum* 'Sylvia's Surprise'	*Geranium*	*G. wallichianum*
G22	*G.* 'Tanya Rendall'	*Geranium*	*G.* × *antipodeum*
G24	*G.* 'Tiny Monster'	*Geranium*	Parent **1:** *G. sanguineum*, Parent **2:** *G. psilostemon*
G27	*G.* × *cantabrigiense* 'Biokovo'	*Robertium*	Parent **1:** *G. maccrorhizum*, Parent **2:** *G. dalmaticum*
G30	*G. cinereum* 'Laurence Flatman'	*Erodioidea*	*G. cinereum*
G35	*G. endressii*	*Geranium*	*G. endressii*
G37	*G. endressii* 'Trevor Bath'	*Geranium*	*G. endressii*
G38	*G. himalayense* 'Baby Blue'	*Geranium*	*G. himalayense*
G39	*G. himalayense* 'Derrick Cook'	*Geranium*	*G. himalayense*
G42	*G. macrorrhizum* 'Czakor'	*Robertium*	*G. maccrorhizum*
G44	*G. macrorrhizum* 'White Ness'	*Robertium*	*G. maccrorhizum*
G45	*G. maculatum* 'Album'	*Geranium*	*G. maculatum*
G46	*G. maculatum* 'Elizabeth Ann'	*Geranium*	*G. maculatum*
G49	*G.* × *oxonianum* 'Katherine Adele'	*Geranium*	Parent **1:** *G. versicolor*, Parent **2:** *G. endressi*
G50	*G.* × *oxonianum* 'Southcombe Double'	*Geranium*	Parent **1:** *G. versicolor*, Parent **2:** *G. endressi*
G54	*G. phaeum* 'Angelina'	*Erodioidea*	*G. phaeum*
G57	*G. pratense* 'Algera Double'	*Geranium*	*G. pratense*
G61	*G. pratense* 'Purple Ghost'	*Geranium*	*G. pratense*
G62	*G. psilostemon*	*Geranium*	*G. psilostemon*

Table 1. *Cont.*

Codes	Genotypes	Subgenera	Parents
G64	*G. renardii*	*Geranium*	*G. renardii*
G69	*G. sanguineum* 'Album'	*Geranium*	*G. sanguineum*
G71	*G. sylvaticum* 'Album'	*Geranium*	*G. sylvaticum*
G73	*G. versicolor*	*Geranium*	*G. versicolor*
G75	*G.* 'Bloomtime'	*Geranium*	Parent **1:** *G. wallichianum*, Parent **2:** *G. himalayense*
G76	*G. wallichianum* 'Havana Blue'	*Geranium*	*G. wallichianum*
G77	*G. wlassovianum*	*Geranium*	*G. wlassovianum*
G80	*G.* 'Blushing Turtle'	*Geranium*	*G. sanguineum*

All plants were planted in duplicate in an open field at ILVO, Melle, Belgium (50°59′31.6″ N, 3°47′07.3″ E); in addition, at least two plants of each genotype were kept in the greenhouse. Plants in the greenhouse were grown in 2 L pots (peat substrate Saniflor NPK 12:14:24, EC 45 mS·m^{-1}), except *G. cinereum*, which was potted in a stone mixture (Kift). Plants were grown under natural conditions (greenhouse ventilation set point was 10 °C).

2.2. Parental Difference

To test pre- and postzygotic barriers, 1654 crosses were carried out, of which 41, 96, and 236 cross combinations were performed for pollen tube length (tube_length, 2020 and 2021, Table S2), seed setting (seed_set, 2020, Table S3), and seed development (seed_dev, 2021, Table S4), respectively. Four different measures were used to characterize the parental difference between two parents (P1 and P2) in each cross: Chrom, DNA, Style, and cJaccard (cJaccard = 1 − Jaccard, which is the complement of Jaccard similarity and represents Jaccard distance) (Tables S2–S4). Chrom expresses the parental difference between the chromosome number in a cross [26]. DNA is the difference in genome size/chromosome number (pg/chromosome) [26]. Style is the difference in style length (mm), which was measured using a ruler when the style was fully developed. The Jaccard similarity coefficient is calculated using statistical software based on AFLP markers (0 and 1), and the formula [27] is:

$$J = \frac{a}{a + b + c}$$

where J is the Jaccard similarity coefficient, a represents the total number of attributes where parent P1 and P2 both have a value of 1, b represents the total number of attributes where the attribute of P1 is 0 and the attribute of P2 is 1, and c represents the total number of attributes where the attribute of P1 is 1 and the attribute of P2 is 0.

By using cJaccard, all four indices represent a distance, thereby ensuring that the regression coefficients of the logistic regression have the same interpretation: a negative slope implies a lower success rate for an increasing parental distance.

2.3. Crossing Success Criteria

The success rate of these crosses was assessed based on the following crossing success response criteria:

(1) Pollen tube growth (tube_length) using aniline blue staining tests (see further);
(2) Seed development (seed_dev) is defined as the percentage of crossed flowers with at least one swollen ovule with white, yellow, or green testa or mature seed with brown testa, 7–35 days after pollination (for 2021);
(3) Seed set (seed_set) as the percentage of crossed flowers with at least one mature seed with brown testa, 25–35 days after pollination (for 2020).

The difference between seed_set and seed_dev is in the harvesting stage, in which seed_set means all seeds remained on the plant until full maturation, while for seed_dev, seeds were harvested immature or mature and rescued in tissue culture.

For the aniline blue staining 48 h after pollination, the pollinated pistils were harvested and placed in FAA (formaldehyde:acetic acid:ethanol (70%) 1:1:18) for 24 h. After a washing step with water, the pistils were transferred to NaOH (6 M) for 16 h. Then, the pistils were transferred into 0.033 M K_3PO_4 + 0.1% aniline blue (Acros Organics, Geel, Belgium) and kept in this solution for 3 h (in the dark), after which they were squashed and the pollen tubes were examined using fluorescence microscopy (Leica DMIRB, Wetzlar, Germany).

2.4. Statistical Analysis

To predict the success of hybridization in hardy geraniums, the following univariate logistic regression model was fitted:

$$log\frac{\pi}{1-\pi} = \beta_0 + \beta_1 pd$$

In the formula, π represents crossing success, *pd* represents the parental difference, and β_0 and β_1 represent the slopes of the regression line in the logit scale. For each response criterion, four parental difference measures (variables) were examined for their quality as predictors: Chrom, DNA, Style, and cJaccard.

A quasi-binomial error distribution was considered [28], i.e., an extension of the binomial distribution so that proportions (such as tube_length) can be modeled. Additionally, the quasi-model can cope with overdispersion, ensuring that standard errors (and therefore confidence limits) are correctly estimated [29].

If the probability of success is π, the average number of crosses required to achieve 10 successes is $10/\pi$. To take into account the uncertainty of the model, we also calculated this number for the lower and upper limits of the confidence intervals to predict the value of π. Statistical analysis was conducted using R [30] in conjunction with RStudio (version 2022.07.2). The regression models were fitted with the glm function from R-base followed by the emmeans function from the emmeans package [31] to calculate the predicted values and 95% confidence intervals. For drawing the regression models along with confidence bands, the function geom_smooth from the ggplot2 package [32] was used, with method = "glm" and family = "quasibinomial".

3. Results

3.1. Pre-Zygotic and Post-Zygotic Barriers

Prezygotic barriers were evaluated using aniline blue staining for some cross combinations (Table S2). This resulted in different scenarios. Sometimes, germinated pollen stopped early on the stigma, as depicted for the cross between *G. himalayense* 'Baby Blue' (G38) x *G. sanguineum* 'Album' (G69) (Figure 1a). The cross *G. nodosum* 'Silverwood' (G19) x *G. phaeum* 'Angelina' (G54) also clearly showed a high degree of prezygotic incongruity due to curled pollen tubes, disoriented pollen tube growth, and short pollen tubes (Figure 1b). Figure 1c shows the cross *G.* 'Blushing Turtle' (G80) x *G.* 'Brookside' (G04) in which the pollen tubes stopped before entering the ovules. Often, the pollen tubes contained many callose plugs, but they could grow throughout the style (Figure 1d, *G.* 'Brookside' (G04) x *G. sanguineum* 'Album' (G69)). Based on pollen tube growth, the parental difference between cJaccard was the variable that best expressed the success or failure of pollen tube growth (Tables S2–S4).

During two consecutive crossing seasons, more than 1504 crosses (crosses for tube_length excluded) were performed for harvesting seeds. In some combinations, seed development was observed by swelling of the carpels (ovules) and embryo formation, but not all combinations could produce mature seeds. This indicates the existence of postzygotic barriers. Their impact was assessed by comparing the seed production in the first crossing season (2020), where seeds were harvested at maturity, with the second season's success (2021), where immature seeds and embryos were saved (Table 2). In 2020, it was found that

5.2% of all combinations were successful in producing mature seeds, and 2.08% of them were able to obtain a healthy seedling, while in 2021, when in vitro rescue of seeds and embryos was included, successful combinations improved to 15.25%, and the probability of a successful combination with healthy seedlings increased to 6.35%. Although different crosses were performed in 2020 and 2021, the overall mean parental genetic distance (expressed as cJaccard) was similar for all crosses (Table 2). These data confirmed that some postzygotic barriers can be overcome by embryo rescue treatments. Our results showed that waiting for maturation reduced seed production from 3.03% to 0.85%, and healthy seedling development from 0.57% to 0.17%, suggesting that efficiency would increase with the rescue of immature seeds and embryos (Table 2).

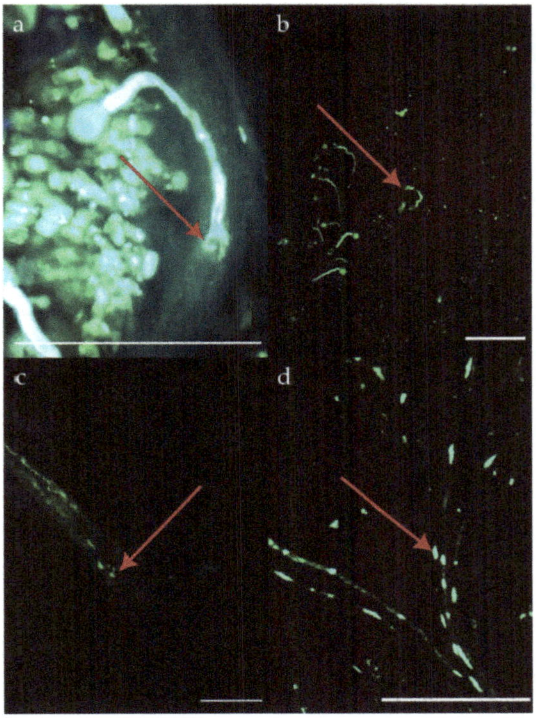

Figure 1. Details of pollen tube growth visualized by aniline blue staining after 48 h in different crossing combinations: (**a**) *G. himalayense* 'Baby Blue' (G38) x *G. sanguineum* 'Album' (G69) (arrow shows that the pollen tube stopped), (**b**) *G. nodosum* 'Silverwood' (G19) x *G. phaeum* 'Angelina' (G54) (arrow shows that the pollen tube grew in the wrong direction), (**c**) *G.* 'Blushing Turtle' (G80) x *G.* 'Brookside' (G04) (arrow shows that the pollen tube stopped before it entered the oocyte), and (**d**) *G.* 'Brookside' (G04) x *G. sanguineum* 'Album' (G69) (arrow shows the pollen tube with callose plug) (bar scale = 0.5 mm).

Table 2. Success rate of different crosses and combinations of the first and second breeding seasons.

Growing Season	2020	2021
Number of all crosses	349	1155
Number of combinations	96	236
Number of harvested seeds	15	175
Number of combinations that produced seeds	5	36
Number of combinations that produced healthy seedlings	2	15
Number of healthy seedlings	3	33
Average of cJaccard of crosses	0.81	0.78

3.2. Prediction of Crossing Success

To describe and predict the crossing success for future breeding programs of hardy geraniums, univariate logistic regression models were examined using tube_length, seed_dev and seed_set as the response criteria (Figures 2 and 3). The explanatory variables expressing parental difference were Chrom, DNA, Style, and cJaccard. Figure 2 displays the regression models, while Figure 3 shows the estimated slopes along with their 95% confidence intervals. Parameter estimates (intercept and slope), 95% confidence intervals, t-values, and corresponding *p*-values for all fitted models can be found in Tables S5–S7. For most combinations, the slope of the parental difference is negative (decreasing regression lines). With increasing parental difference, the success rate, as assessed by tube_length, seed_set, and seed_dev, decreases. In addition, some of the 95% confidence limits for the slope estimates do not cover zero, implying that we can reject the null hypothesis of no slope at a significance level of 5%. More specifically, for cJaccard, the slope is consistently negative, while for Style, Chrom, and DNA, the slopes were not. Conversely, when considering the response variables, seed_dev consistently exhibited a negative relationship with all explanatory variables, making it the most suitable predictor of breeding success in *Geranium*.

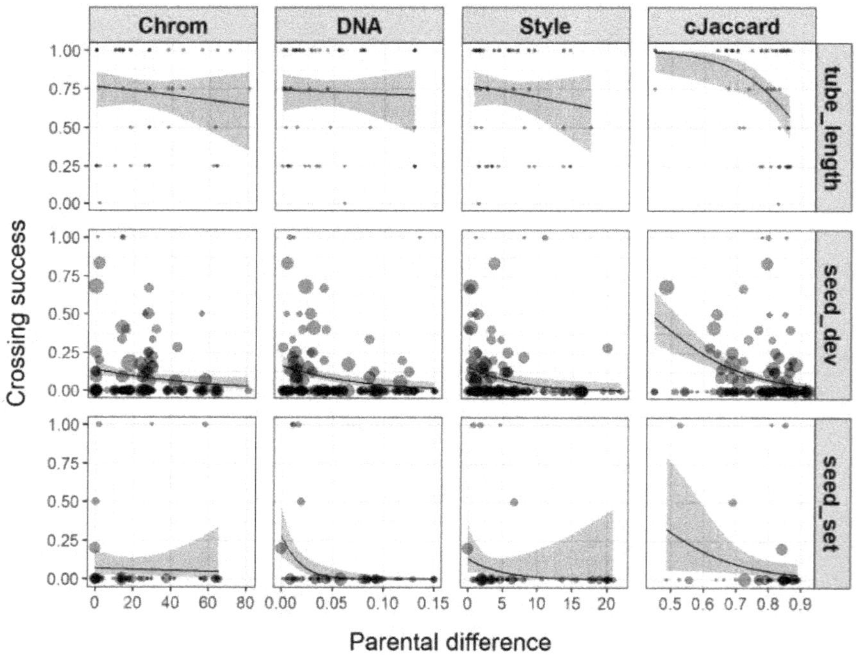

Figure 2. Univariate logistic regression model with response criteria tube_length, seed_dev, and seed_set with Chrom, DNA, Style, and cJaccard as explanatory variables. The size of the points is proportional to the number of observations with the same coordinates. The interval represents the 95% confidence limit. The graphs were created using the function geom_smooth from the R package ggplot with the options method = "glm" and family = "quasibinomial" [30].

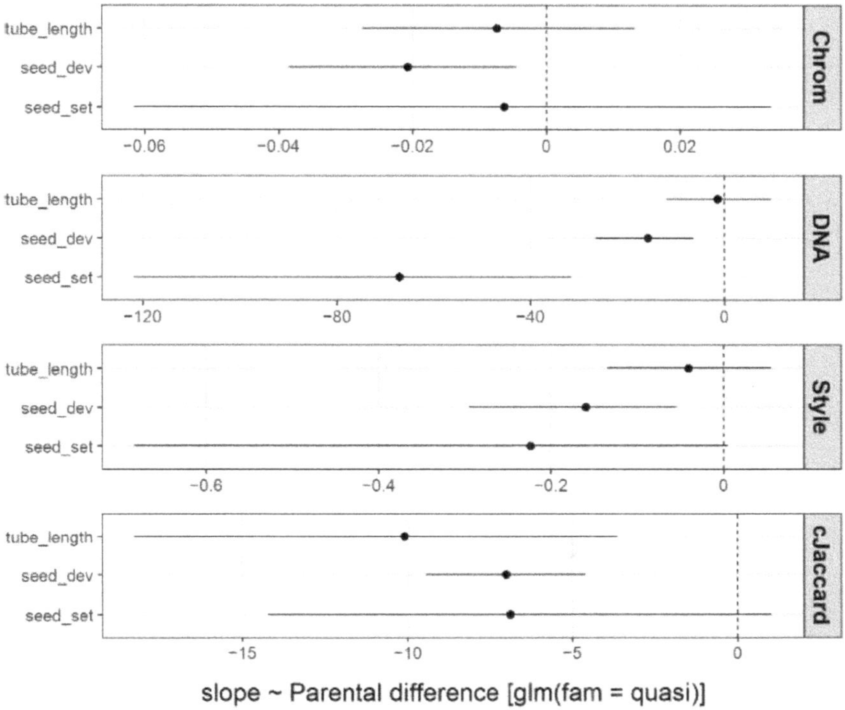

Figure 3. The slope parameters and their 95% confidence interval of the logistic regression model. Negative values whose confidence interval does not pass 0 on the X-axis are significantly negatively correlated with the variables Chrom, DNA, Style, and cJaccard. The confidence intervals were calculated with the emmeans function of the R package emmeans [30].

3.3. *Prediction of Required Cross Number to Have 10 Crossing Products*

The combination of Jaccard distance (cJaccard) and seed development (seed_dev) as the response variable yielded the best model. According to Table S7, the model for the probability π_{sd} of achieving at least one success for seed development (seed_dev) is as follows:

$$log\frac{\pi_{sd}}{1-\pi_{sd}} = 3.07 - 7.00\ cJaccard$$

We utilized the predicted probabilities to evaluate the number of crosses necessary to achieve 10 successful crossing products on average, which is represented by the equation $10/\pi_{sd}$. Figure 4 provides a graphical representation of this equation for both the fitted value and the lower and upper limits of the predictions. Although the confidence limits for π_{sd} decrease for decreasing Jaccard distance, the uncertainty around the predicted number of crosses increases due to the reduced probability of success at large parental distances. The table displayed within the graph presents the results for some specific values of the Jaccard distance. When the cJaccard of the parents is 0.5, the success rate of crossing is 39%. This implies that harvesting 10 developed seeds requires 25 (18–38) crosses. On the other hand, if the cJaccard value increases to 0.8, the success rate will drop to 7%, and to achieve 10 successful crosses, 136 (99–197) crosses are necessary (Figure 4).

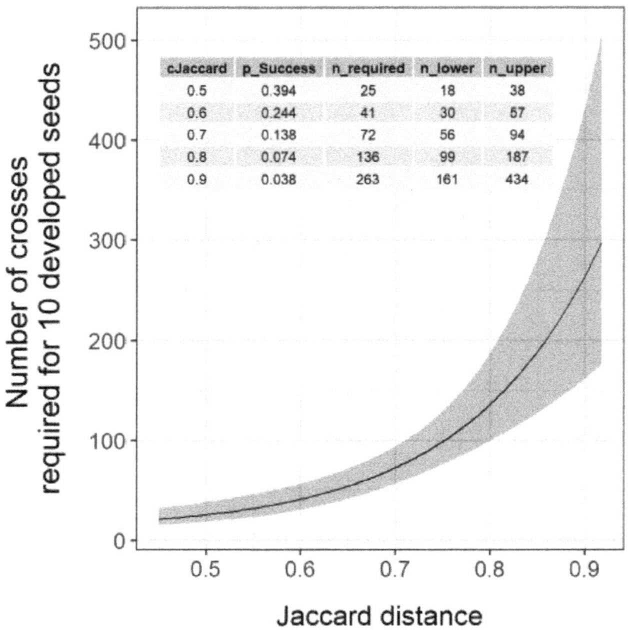

Figure 4. Number of crosses required to obtain 10 flowers with at least one developed seed as predicted by the model based on seed_dev and the Jaccard distance (cJaccard) as a prediction parameter. The estimated coefficients of the model are according to Table S7.

4. Discussion

Prezygotic barriers such as incorrect pollen growth directions, twisted and/or short pollen tubes, and high density of callose plugs observed in *Geranium* are often observed in interspecific or intergeneric crosses in ornamentals [33,34]. When pollen germinates, the accumulation of callose can indicate the incompatibility of pollen grains and tubes, which may also appear in the papillae of the stigma after rejection [35]. The number of callose plugs in *Geranium* is not a real impediment to fertilization since many callose plugs are present in the compatible parent combinations with a low cJaccard, but in plants with a larger size of plugs, the pollen tube may stop earlier. Callose also plays a crucial role in protecting the elongating pollen tube from tensile and compressive stresses and prevents reflux of pollen tube contents, thus maintaining turgor pressure and tube integrity [36,37]. Qin et al. confirmed that in *Arabidopsis thaliana*, pollen tubes without callose plugs were shorter than those with plugs [38]. Further, in *Hibiscus moscheutos*, the number of callose plugs can also be used as an indicator of pollen tube growth rate [39].

Our observations confirm that crossing barriers in *Geranium* primarily resulted from pollen failure, aberrant pollen tube growth, failed fertilization, and arrested growth before the formation of viable seeds. In some combinations, pollen tubes fail to adhere to the stigma, which is common in plants with dry stigmas. In other crosses, swollen pistils are observed without embryo formation, indicating successful pollen tube growth through the style tissue but unsuccessful fertilization. Eventually, in some cases, embryos were formed but never reached the mature stage. Techniques such as reciprocal crosses, mixed or mentor pollination, style manipulation, or stigma treatment can help overcome prezygotic barriers [34,40,41]. Postzygotic barriers often result from genetic discrepancy and chromosome degeneration during cell division of the zygote [42]. Lack of endosperm development, abortion of the embryo, albinism, or lack of vigor are the main observations in postzygotic abortion [3]. In the interspecific hybridization of *Lilium*, embryonic development is influenced by endosperm development, and the lack of embryo development in the endosperm can be resolved by early embryo rescue [42]. *Geranium* seeds typically have little endosperm [43], and during the mature stage, both the nucellus and endosperm have disappeared [44]. Embryo rescue has been successfully applied to *Pelargonium* sp., a close relative of *Geranium* [45,46].

Our findings strongly support the concept that crossing success diminishes as parental distance increases. In our study, the best measure of parental distance was the Jaccard distance, the complement of the Jaccard similarity. With all variables for parental differences of crossing success, the slope with cJaccard is negative. Notably, the strongest relationship was found with seed_dev, making it the most reliable predictor of crossing success in *Geranium*. This is probably due in part to the high number of replicates and good data coverage. Previous research also supports the importance of cJaccard in determining cross compatibility [26], and based on hybrids described in the literature, we hypothesized that good cross compatibility was correlated with a cJaccard of 0.5 and that the maximum value of cJaccard for hybridization is 0.87. According to our logistic regression model utilizing our hybridization data, cJaccard values of 0.5 and 0.87 correspond to a seed development crossing success of 39% or 4.6%, respectively. These rates are considered indicative of good and poor success, respectively. To our knowledge, this is the first instance where ornamental breeders can quantitatively estimate the number of crosses required to attain a desired success rate based on specific variables of parental differences. This newfound knowledge has the potential to significantly enhance the efficiency of hybrid production in hardy geraniums. By accurately predicting crossing success rates, breeders can strategically plan their breeding programs, optimize resource allocation, and ultimately achieve higher success rates in geranium hybridization.

5. Conclusions

In conclusion, our study demonstrates that the success of interspecific hybridization in *Geranium* can be predicted by considering the genetic distance between parental genotypes and utilizing the logistic regression models developed in this study. We have also discovered that harvesting immature seeds and saving embryos can increase the success rate of hybridization. The logistic regression model we have developed provides breeders with the ability to estimate the number of required crosses to achieve a specific success rate based on parental differences, enabling more realistic and efficient breeding programs in *Geranium*. The increasing success rate of interspecific hybridization through the use of predictive models and advanced breeding techniques can have significant implications for the development of new cultivars.

Supplementary Materials: The following supporting information can be downloaded at: https://www.mdpi.com/article/10.3390/horticulturae9060617/s1, Table S1: Different genotypes of *Geranium* sp. used in this study; Table S2: Cross combinations that were carried out for pollen tube growth evaluation with their respective parental difference variables; Table S3: Cross combinations that were carried out for seed_set evaluation with their respective parental difference variables; Table S4: Cross combinations that were carried out for seed_dev evaluation with their respective parental difference variables; Table S5: Intercept and slope of the logistic regression model in the logit scale for pollen tube growth; Table S6: Intercept and slope of the logistic regression model in the logit scale for seed_set; Table S7: Intercept and slope of the logistic regression model in the logit scale for seed_dev.

Author Contributions: Conceptualization, M.A. and E.D.; methodology, M.A., P.Q. and E.D.; data curation, M.A.; writing—original draft preparation, M.A.; writing—review and editing, M.A., P.Q., J.V.H., S.P.O.W. and E.D.; visualization, M.A. and P.Q.; supervision, E.D., J.V.H. and S.P.O.W. All authors have read and agreed to the published version of the manuscript.

Funding: The project received funding from the BOF number 20033086_30351392.

Data Availability Statement: The data presented in this study are available on request from the corresponding author. The data are not publicly available because this research was in collaboration with a private company.

Acknowledgments: We thank Roger Dobbelaere and Rina Vanhaecke for their help and management of the plants, Miriam Levenson for the English language editing, and Josephine Kinoo for her help in making crosses.

Conflicts of Interest: The authors declare no conflict of interest.

References

1. Aedo, C. Taxonomic Revision of *Geranium* Sect. *Brasiliensia* (Geraniaceae). *Syst. Bot.* **2001**, *26*, 205–215. [CrossRef]
2. Yeo, P.F. *Hardy Geraniums*; Batsford: London, UK, 2001; ISBN 0713485000.
3. Van Huylenbroeck, J.; Eeckhaut, T.; Leus, L.; Van Laere, K.; Dhooghe, E. Bridging the Gap: Tools for Interspecific and Intergeneric Hybridization in Ornamentals. In Proceedings of the XXVI International Eucarpia Symposium Section Ornamentals: Editing Novelty 1283, Erfurt, Germany, 1–4 September 2019; pp. 161–168. [CrossRef]
4. Lewis, M.; Chappell, M.; Thomas, P.A.; Maynard, R.C.; Greyvenstein, O. Development and Verification of an Interspecific Hybridization Protocol for Asclepias. *HortScience* **2021**, *56*, 831–837. [CrossRef]
5. Yue, Y.; Ruter, J.M. Pavonia 3rufula (Malvaceae): An Interspecific Hybrid between *Pavonia lasiopetala* and *Pavonia missionum*. *HortScience* **2021**, *56*, 732–735. [CrossRef]
6. Do Nascimento, N.F.F.; Do Rêgo, E.R.; Nascimento, M.F.; Bruckner, C.H.; Finger, F.L.; Do Rêgo, M.M. Evaluation of Production and Quality Traits in Interspecific Hybrids of Ornamental Pepper. *Hortic. Bras.* **2019**, *37*, 315–323. [CrossRef]
7. Blakeslee, A.F. Removing Some of the Barriers to Crossability in Plants. *Proc. Am. Philos. Soc.* **1945**, *89*, 561–574.
8. Dresselhaus, T.; Franklin-Tong, N. Male-Female Crosstalk during Pollen Germination, Tube Growth and Guidance, and Double Fertilization. *Mol. Plant* **2013**, *6*, 1018–1036. [CrossRef] [PubMed]
9. Hiscock, S.J.; Allen, A.M. Diverse Cell Signalling Pathways Regulate Pollen-stigma Interactions: The Search for Consensus. *New Phytol.* **2008**, *179*, 286–317. [CrossRef]
10. Heslop-Harrison, Y.; Shivanna, K.R. The Receptive Surface of the Angiosperm Stigma. *Ann. Bot.* **1977**, *41*, 1233–1258. [CrossRef]
11. Heslop-Harrison, Y. Control Gates and Micro-Ecology: The Pollen-Stigma Interaction in Perspective. *Ann. Bot.* **2000**, *85* (Suppl. S1), 5–13. [CrossRef]
12. Zinkl, G.M.; Zwiebel, B.I.; Grier, D.G.; Preuss, D. Pollen-Stigma Adhesion in *Arabidopsis*: A Species-Specific Interaction Mediated by Lipophilic Molecules in the Pollen Exine. *Development* **1999**, *126*, 5431–5440. [CrossRef]
13. Tupý, J. Callose Formation in Pollen Tubes and Incompatibility. *Biol. Plant. Prague* **1959**, *1*, 192–198. [CrossRef]
14. Mulcahy, D.L. Models of Pollen Tube Competition in *Geranium Maculatum*. In *Pollination Biology*; Elsevier Science & Technology Books: Amsterdam, The Netherlands, 1983; pp. 151–161. ISBN 0125839804.
15. Hessing, M.B. Differential Pollen Tube Success in *Geranium caespitosum*. *Bot. Gaz.* **1989**, *150*, 404–410. [CrossRef]
16. Sahoo, J.P.; Mohapatra, P.P.; Acharya, L.K.; Jena, C. Insights into the Embryo Rescue-a Modern in-Vitro Crop Improvement Approach in Horticulture. *Plant Cell Biotechnol. Mol. Biol.* **2021**, *22*, 20–33.
17. Shen, X.; Gmitter, F.G.; Grosser, J.W. Immature Embryo Rescue and Culture. In *Plant Embryo Culture: Methods and Protocols*; Thorpe, T.A., Yeung, E.C., Eds.; Humana Press: Totowa, NJ, USA, 2011; pp. 75–92. ISBN 978-1-61737-988-8. [CrossRef]
18. Lafon-Placette, C.; Johannessen, I.M.; Hornslien, K.S.; Ali, M.F.; Bjerkan, K.N.; Bramsiepe, J.; Glöckle, B.M.; Rebernig, C.A.; Brysting, A.K.; Grini, P.E. Endosperm-Based Hybridization Barriers Explain the Pattern of Gene Flow between *Arabidopsis lyrata* and *Arabidopsis arenosa* in Central Europe. *Proc. Natl. Acad. Sci. USA* **2017**, *114*, E1027–E1035. [CrossRef] [PubMed]
19. Giancaspro, A.; Mazzeo, A.; Carlomagno, A.; Gadaleta, A.; Somma, S.; Ferrara, G. Optimization of an In Vitro Embryo Rescue Protocol for Breeding Seedless Table Grapes (*Vitis Vinifera* L.) in Italy. *Horticulturae* **2022**, *8*, 121. [CrossRef]
20. Bhatia, R.; Dey, S.S.; Sharma, K.; Singh, S.; Kumar, S.; Pramanik, A.; Parkash, C.; Kumar, R. Back-Cross Introgression of 'Tour' Cytoplasm from *Brassica napus* through in Vitro Embryo Rescue Reveals Partial Restoration of Sterility in *B. oleracea*. *Sci. Hortic.* **2021**, *282*, 110014. [CrossRef]
21. Stewart, J.M. In Vitro Fertilization and Embryo Rescue. *Environ. Exp. Bot.* **1981**, *21*, 301–315. [CrossRef]
22. Stelkens, R.; Seehausen, O. Genetic Distance between Species Predicts Novel Trait Expression in Their Hybrids. *Evolution* **2009**, *63*, 884–897. [CrossRef]
23. Granados Mendoza, C.; Wanke, S.; Goetghebeur, P.; Samain, M.S. Facilitating Wide Hybridization in *Hydrangea* s. l. Cultivars: A Phylogenetic and Marker-Assisted Breeding Approach. *Mol. Breed.* **2013**, *32*, 233–239. [CrossRef]
24. Denaeghel, H.; Van Laere, K.; Leus, L.; Van Huylenbroeck, J.; Van Labeke, M.-C. Interspecific Hybridization in Sarcococca Supported by Analysis of Ploidy Level, Genome Size and Genetic Relationships. *Euphytica* **2017**, *213*, 149. [CrossRef]
25. Dhooghe, E.; Sparke, J.; Oenings, P.; Van Paemel, T.; Van Labeke, M.-C.; Winkelmann, T. *Helleborus*. In *Ornamental Crops*; Van Huylenbroeck, J., Ed.; Springer International Publishing: Cham, Switzerland, 2018; pp. 439–452. ISBN 978-3-319-90698-0. [CrossRef]
26. Akbarzadeh, M.; Van Laere, K.; Leus, L.; De Riek, J.; Van Huylenbroeck, J.; Werbrouck, S.P.O.; Dhooghe, E. Can Knowledge of Genetic Distances, Genome Sizes and Chromosome Numbers Support Breeding Programs in Hardy Geraniums? *Genes* **2021**, *12*, 730. [CrossRef] [PubMed]
27. Chung, N.C.; Miasojedow, B.; Startek, M.; Gambin, A. Jaccard/Tanimoto Similarity Test and Estimation Methods for Biological Presence-Absence Data. *BMC Bioinform.* **2019**, *20*, 644. [CrossRef] [PubMed]

28. McCullagh, P.; Nelder, J.A. *Generalized Linear Models*; Chapman and Hall: London, UK, 1989. [CrossRef]
29. Venables, W.N.; Ripley, B.D. Random and mixed effects. In *Modern Applied Statistics with S*; Venables, W.N., Ripley, B.D., Eds.; Springer: New York, NY, USA, 2002; pp. 271–300. ISBN 978-0-387-21706-2.
30. R Core Team. *R: A Language and Environment for Statistical Computing*; R Foundation for Statistical Computing: Vienna, Austria, 2013; Available online: https://www.R-project.org/ (accessed on 15 April 2023).
31. Lenth, R.; Lenth, M.R. Package 'Emmeans'. *Statistician* **2017**, *34*, 216–221.
32. Wickham, H. Data analysis. In *ggplot2: Elegant Graphics for Data Analysis*; Wickham, H., Ed.; Springer International Publishing: Cham, Switzerland, 2016; pp. 189–201. ISBN 978-3-319-24277-4. [CrossRef]
33. Parton, E. Flower Biology and Crossing Barriers in Bromeliaceae. Ph.D. Thesis, Katholieke Universiteit, Leuven, Belgium, 2001.
34. Dhooghe, E.; Reheul, D.; Van Labeke, M.-C. Overcoming Pre-Fertilization Barriers in Intertribal Crosses between *Anemone coronaria* L. and *Ranunculus asiaticus* L. *Horticulturae* **2021**, *7*, 529. [CrossRef]
35. Dumas, C.; Knox, R.B. Callose and Determination of Pistil Viability and Incompatibility. *Theor. Appl. Genet.* **1983**, *67*, 1–10. [CrossRef]
36. Parre, E.; Geitmann, A. More than a Leak Sealant. The Mechanical Properties of Callose in Pollen Tubes. *Plant Physiol.* **2005**, *137*, 274–286. [CrossRef]
37. Li, Y.-Q.; Moscatelli, A.; Cai, G.; Cresti, M. Functional Interactions among Cytoskeleton, Membranes, and Cell Wall in the Pollen Tube of Flowering Plants. *Int. Rev. Cytol.* **1997**, *176*, 133–199. [CrossRef]
38. Qin, P.; Ting, D.; Shieh, A.; McCormick, S. Callose Plug Deposition Patterns Vary in Pollen Tubes of *Arabidopsis thaliana* ecotypes and Tomato Species. *BMC Plant Biol.* **2012**, *12*, 178. [CrossRef]
39. Snow, A.A.; Spira, T.P. Differential Pollen-tube Growth Rates and Nonrandom Fertilization in *Hibiscus moscheutos* (Malvaceae). *Am. J. Bot.* **1991**, *78*, 1419–1426. [CrossRef]
40. Van Tuyl, J.M.; Lim, K.-B. Interspecific Hybridisation and Polyploidisation as Tools in Ornamental Plant Breeding. In Proceedings of the XXI International Eucarpia Symposium on Classical versus Molecular Breeding of Ornamentals—Part I 612, München, Germany, 25–29 August 2003; pp. 13–22. [CrossRef]
41. Van Tuyl, J.M.; De Jeu, M.J. Methods for overcoming interspecific crossing barriers. In *Pollen Biotechnology for Crop Production and Improvement*; Cambridge University Press: New York, NY, USA, 1997; pp. 273–292.
42. Van Tuyl, J.M.; Maas, I.W.G.M.; Lim, K.-B. Introgression in Interspecific Hybrids of Lily. In Proceedings of the VIII International Symposium on Flowerbulbs 570, Cape Town, South Africa, 28 August 2000; pp. 213–218. [CrossRef]
43. Xu, Z.; Deng, M. Geraniaceae. In *Identification and Control of Common Weeds: Volume 2*; Springer: Dordrecht, The Netherlands, 2017; pp. 629–637. [CrossRef]
44. Boesewinkel, F.D.; Been, W. Development of Ovule and Testa of *Geranium pratense* L. and Some Other Representatives of the Geraniaceae. *Acta Bot. Neerl.* **1979**, *28*, 335–348. [CrossRef]
45. Plaschil, S.; Budahn, H.; Klocke, E.; Wiedemann, M.; Olbricht, K. Spontaneous Polyploidisation of Interspecific and Intersectional *Pelargonium* Hybrids during Embryo Rescue. *J. Appl. Bot. Food Qual.* **2021**, *94*, 206–212. [CrossRef]
46. Denis-Peixoto, L.; Cadic, A.; Renou, J.-P. Interspecific Crosses between *Pelargonium* × *Hortorum* and *P. quinquelobatum* Using Embryo Rescue and Molecular Characterization of Hybrids by an Endogenous Chs Probe. *Plant Breed.* **1997**, *116*, 177–180. [CrossRef]

horticulturae

MDPI

Article

Effect of Seaweed-Based Biostimulants on Growth and Development of *Hydrangea paniculata* under Continuous or Periodic Drought Stress

Paulien De Clercq [1,2,*], Els Pauwels [2], Seppe Top [3], Kathy Steppe [3] and Marie-Christine Van Labeke [1,*]

1 Department of Plants and Crops, Faculty of Bioscience Engineering, Ghent University, 9000 Ghent, Belgium
2 Ornamental Plant Research (PCS, Proefcentrum voor Sierteelt), 9070 Destelbergen, Belgium
3 Laboratory of Plant Ecology, Department of Plants and Crops, Faculty of Bioscience Engineering, Ghent University, 9000 Ghent, Belgium
* Correspondence: paulien.declercq@ugent.be (P.D.C.); mariechristine.vanlabeke@ugent.be (M.-C.V.L.)

Abstract: To adapt to climate change and water scarcity during dry, hot summers, more sustainable, or even deficit, irrigation is required in the ornamental sector, as it uses large amounts of water to sustain high-value crop production. Biostimulants, especially seaweed extracts, could offer a sustainable solution against drought stress as they are known to increase plant tolerance to abiotic stress. The effect of four seaweed extracts based on *Ascophyllum nodosum, Soliera chordalis, Ecklonia maxima,* and *Saccharina latissima* and one microbial biostimulant were tested on container-grown *Hydrangea paniculata* under drought stress conditions for two years. During the first trial year, in 2019, overall irrigation was reduced by 20%. In 2021, plants were subjected to repeated drying and wetting cycles. In general, less irrigation, and thus a lower substrate moisture content, reduced stomatal conductance, biomass production, and root development, but increased plant compactness. The biostimulants showed minor effects, but these were not observed in both experiments. Treatment with the *A. nodosum* extract resulted in longer branches and more biomass under deficit irrigation but tended to accelerate flowering when repeated drying and wetting cycles were applied. The *E. maxima* extract negatively affected the branching of *Hydrangea* under repeated drying and wetting cycles.

Keywords: woody ornamentals; biostimulants; seaweed extracts; drought stress; pigments; stress metabolites; reflectance; stomata; dendrometer (LVDT)

Citation: De Clercq, P.; Pauwels, E.; Top, S.; Steppe, K.; Van Labeke, M.-C. Effect of Seaweed-Based Biostimulants on Growth and Development of *Hydrangea paniculata* under Continuous or Periodic Drought Stress. *Horticulturae* **2023**, *9*, 509. https://doi.org/10.3390/horticulturae9040509

Academic Editor: Aušra Brazaitytė

Received: 15 March 2023
Revised: 14 April 2023
Accepted: 18 April 2023
Published: 19 April 2023

1. Introduction

Ornamental horticulture is a small, but economically important, sector within agriculture in Belgium. The sector had a production value of 511 million euros in 2020, demonstrating its economic importance. To obtain high-quality plants, the hardy nursery sector uses large amounts of water [1,2] but, due to climate change, growers are increasingly facing periods of prolonged drought and heat waves, from which legal restrictions on water use and water shortages can arise. In the future, growers will be forced to use less water in their growing system, such that plants could suffer from drought stress [3].

Plant biostimulants are defined in the European Regulation (EC) No 2019/1009 for fertilizing products as follows: 'A plant biostimulant shall be an EU fertilizing product the function of which is to stimulate plant nutrition processes independently of the product's nutrient content with the sole aim of improving one or more of the following characteristics of the plant or the plant rhizosphere: (a) nutrient use efficiency, (b) tolerance to abiotic stress, (c) quality traits, or (d) availability of confined nutrients in the soil or rhizosphere' [4]. Biostimulants consist of a variety of ingredients and formulations and therefore can be classified into different groups, e.g., humic and fulvic acids, protein hydrolysates and other N-containing compounds, seaweed extracts and botanicals, chitosan and other polymers, inorganic compounds, beneficial fungi, and beneficial bacteria [5–7].

Since antiquity, seaweeds have been used in agriculture as a source of organic matter and fertilizer, but their biostimulant effects have only been discovered recently [6,8]. Seaweed extracts contain polysaccharides, e.g., laminarin, alginates, carrageenans, micro- and macronutrients, sterols, N-containing compounds such as betaines, and hormones as potentially bio-active components and can act on soils and plants. They can affect the physical, chemical, and biological properties of the soil, e.g., improvement of moisture-holding capacity and soil aeration, contribution to the fixation and exchange of cations, and promotion of beneficial soil microbes [6,8,9]. They may also affect root architecture by improving lateral root formation and increasing the total root volume thus facilitating the efficient uptake of nutrients and water [9,10]. Seaweed extracts influence photosynthesis through a reduced degradation of chlorophyll, possibly caused by betaines [11]. Hormones present in seaweed extracts, e.g., auxins, abscisic acid, gibberellins, and other classes of hormone-like compounds, are considered to be the major causes of biostimulant activity on crops. The hormonal effects may affect seed germination, plant establishment, and further growth and development [8,9]. Wally et al. (2013) found evidence that the hormonal effects of the brown seaweed *A. nodosum* are, to a lesser extent, related to the hormonal content of the seaweed extracts themselves, but are mainly linked with the up- and down-regulation of hormone biosynthetic genes in the plant tissues [12]. Furthermore, seaweed concentrates trigger early flowering and fruit set in several crop plants probably by initiating robust plant growth [9]. Finally, seaweed extracts have also been shown to alleviate a variety of abiotic stresses including drought, salinity, and nutrient stresses [13]. Many abiotic stress factors manifest as osmotic stress and cause secondary effects, such as oxidative stress, which will lead to an accumulation of reactive oxygen species (ROS). These are known to damage DNA, lipids, carbohydrates, and proteins and cause aberrant cell signaling [9,14]. The mode of action of seaweed extracts in alleviating abiotic stress is not well understood, but the presence of bioactive molecules in the extracts, such as betaines [15] and cytokinins [16], may play a role. Seaweed extracts also increase endogenous concentrations of stress-related molecules in treated plants, such as cytokinins, proline, antioxidants, and antioxidant enzymes [13].

In the current study, biostimulants based on different seaweed species were selected as they are well known to increase the drought tolerance of plants. The effects of three commercial biostimulants based on the seaweeds *A. nodosum*, *E. maxima*, and *S. chordalis*, and one experimental biostimulant based on *S. latissima*, are studied on *Hydrangea paniculata* grown under (a) deficit irrigation or (b) repeated drying and wetting cycles. A microbial biostimulant was also included as they are known to protect plants from adverse environmental conditions. We used specific plant monitoring tools, physiological plant parameters, and ornamental value to find the best biostimulant that allows reduced irrigation without a loss of the ornamental quality of *Hydrangea paniculata*.

2. Materials and Methods

2.1. Plant Material and Growing Conditions

This research was conducted in greenhouses at the Ornamental Plant Research station (Proefcentrum voor Sierteelt PCS, Destelbergen, Belgium), Destelbergen, Belgium (51°3′ N; 3°48′ E) during the growing seasons of 2019 and 2021.

In spring 2019, rooted plug plants of *Hydrangea paniculata* 'Phantom' were transplanted in 1.5 dm^3 containers filled with a commercial peat-based substrate (Agaris, Belgium), supplemented with a controlled-release fertilizer (3 kg m^{-3} Osmocote® Exact 5/6 M 15-9-12 + 2 MgO + 1 kg m^{-3} media trace elements). In spring 2021, rooted stem cuttings of *Hydrangea paniculata* 'Little Alf' were transplanted in 3 dm^3 containers with a commercial peat/coconut substrate (Agaris, Belgium), containing a PG-mix fertilizer for the first three months of growth (0.4 kg m^{-3}, NPK 14-16-18 + trace elements; Agaris, Belgium). The experiments were set up in a greenhouse (average air temperature of 18–20 °C; average relative humidity of 71–82%) to avoid interference with rainfall. Prior to the experimental setup, plants were uniformly irrigated according to good horticultural practices. The young plants acclimated 4–12 weeks, and were then randomly assigned to an irrigation-biostimulant treatment (control or drought, with or without biostimulants).

2.2. Experiment 1—Deficit Irrigation (2019)

In 2019, a total of five treatments were studied. Two irrigation treatments started four weeks after transplanting (13 June 2019). The control treatment, based on growers' advice, received 3 L m^{-2} overhead irrigation (standard irrigation), while the treatment with a deficit irrigation (80%) received 2.4 L m^{-2} (deficit irrigation). A reduction of 20% was chosen, as 10% reduction, tested in a preliminary screening, had minor effects on plant growth and stress level, and no biostimulant effects were observed under these growing conditions. Irrigation frequency was controlled by radiation sum when a threshold value of 20 MJ m^{-2} was exceeded.

Only the plants grown under deficit irrigation were treated with a foliar spray of biostimulants (three biostimulant treatments). Three commercial seaweed extracts, one based on the seaweed *Ascophyllum nodosum* (Phylgreen; Tradecorp), one on *Soliera chordalis* (SeaMelPure; Olmix), and another based on *Ecklonia maxima* (Kelpak; Kelp Products International), were tested, in comparison with a non-treated deficit control (DS Control) (Table 1). The doses and frequencies were specified by the manufacturer (Table 1).

Table 1. Application doses and frequencies of the tested biostimulants.

Trial Year	Biostimulant	Application Dose	Application Frequency
2019	*Ascophyllum nodosum* extract (Phylgreen, Tradecorp)	1.5 L ha^{-1}	Every 15 days
	Soliera chordalis extract (SeaMelPure, Olmix)	2 L ha^{-1}	One application two weeks after planting
	Ecklonia maxima extract (Kelpak, Kelp Products International)	2.5 L ha^{-1}	First application: 7–10 days after planting; repeated at 14–21 days intervals up to four applications
2021	*Ascophyllum nodosum* extract combined with product based on plant-based amino acids (Phylgreen + Delfan Plus V, Tradecorp)	Phylgreen: 0.5 mL L^{-1} Delfan Plus V: 2 mL L^{-1}	Phylgreen: Every 15 days Delfan Plus V: during stress
	Ecklonia maxima extract (Kelpak, Kelp Products International)	2.5 L ha^{-1}	First application: 7–10 days after planting; repeated at 14–21 days intervals up to four applications
	Saccharina latissima extract (experimental product)	3 mL L^{-1}	Every two weeks
	Previsan S (Agriton)	30 mL L^{-1}	Every two weeks

All five treatments were repeated four times in a randomized block design. An experimental unit consisted of nine measuring plants surrounded by eleven border plants distributed in seven trays. In total, 400 plants were present. This experiment was completed at the end of October 2019.

2.3. Experiment 2—Repeated Drying and Wetting Cycles (2021)

In 2021, plants were irrigated by a drip irrigation system, one dripper per plant. The supply was set at 250 mL per dripper, corresponding to an irrigation of 0.88 L m^{-2}. The irrigation frequency was controlled by radiation sum set at 8 MJ m^{-2}, so plants were watered one to three times a day during the summer season. During the summer months, three drying cycles were applied by turning off the irrigation, followed by a recovery period compared with a continuously well-irrigated control so that two irrigation treatments were present. The first drying cycle in June started 12 weeks after transplanting (23 June–1 July) and ended before the presence of wilting symptoms, due to high temperatures. During the second cycle in July (16–22 July) and the third and final cycle in August (19–25 August), the plants were kept under water deprivation until they showed wilting symptoms. Measurements were only performed during the first and last drying cycle. The trial ended at the end of the growing season in September.

In 2021, two commercial seaweeds applied in 2019 were tested again, namely the *E. maxima* extract and the *A. nodosum* extract, the latter in combination with the application of a biostimulant based on hydrolyzed proteins (Delfan Plus V; Tradecorp). In addition, an experimental seaweed extract based on *Saccharina latissima* (North Sea Farmers) and a commercial biostimulant based on micro-organisms (Previsan S; Agriton) were included (Table 1). Biostimulants (four treatments) were tested on plants grown under repeated drying and wetting cycles compared with a non-treated control (DS Control). No biostimulants were applied to the continuously well-irrigated plants (No stress Control). A total of six treatments were present, which were repeated four times in a randomized block design. An experimental unit consisted of ten measuring plants and twelve border plants. A total of 288 plants were present.

2.4. Substrate Moisture Content Measurements

Substrate characteristics (volumetric moisture content, electrical conductivity, and temperature) were determined using a WET sensor (Delta-T Devices Ltd., Cambridge, UK). During the deficit irrigation trial, measurements were performed every two weeks since continuous drought stress was expected, resulting in twelve measurements. During the drying–wetting cycle treatment, measurements were performed before the irrigation stop, (almost) every day of the drought period (no irrigation), and after drought recovery of three–four days. A total of eight (Exp. 2, cycle 1) or sixteen (Exp. 2, cycle 3) determinations of volumetric moisture content were performed.

2.5. Plant Physiological Responses

Chlorophyll and flavonoid levels in the leaves were determined non-destructively using a DUALEX® (Force A, Orsay, France). Hyperspectral reflectance spectra were determined at leaf level with a PolyPen RP410 (Photon Systems Instruments, Drásov, Czech Republic). Based on the hyperspectral data, selected indices were calculated (Table 2). The red edge inflection point (REIP) was calculated by determining the maximum value of the first deviation of the hyperspectral curve in the red region. A total of four measurements, on the two youngest fully developed leaves of two measurement plants were taken per experimental unit (n = 16). Twice as many measurements were performed in the second experiment in 2021.

Table 2. Reflectance indices calculated by the PolyPen RP410 based on hyperspectral data.

Reflectance Index	Formula	Reference
NDVI (Normalized Difference Vegetation Index)	$NDVI = \frac{R_{NIR} - R_{RED}}{R_{NIR} + R_{RED}}$	Ref. [17]
Lic1 (Lichtenthaler Index 1)	$Lic1 = \frac{R_{790} - E_{680}}{R_{790} + R_{680}}$	Ref. [18]
Ctr2 (Carter Index 2)	$Ctr2 = \frac{R_{695}}{R_{760}}$	Ref. [19]
ARI1 (Anthocyanin Reflectance Index 1)	$ARI1 = \frac{1}{R_{550}} - \frac{1}{R_{700}}$	Ref. [20]

Chlorophyll fluorescence was measured using a MINI-PAM II (Walz, Effeltrich, Germany). After 20–30 min of dark adaptation, the initial fluorescence (F_0) was determined, followed by a saturating flash (>4000 µmol m^{-2} s^{-1}; 8 s) to determine the maximum fluorescence level (F_M). The maximum quantum efficiency of PSII (F_V/F_M) was calculated as the ratio of the difference between F_M and F_0 over F_M [21].

Stomatal conductance was measured five hours after sunrise, around midday (between 11:00 a.m.–01:00 p.m.), using a porometer (Delta-T Devices Ltd., Cambridge, UK). A total of eight (Exp. 1) and sixteen (Exp. 2) replicates per treatment were obtained by measuring one leaf, the youngest fully developed, per plant and two plants per experimental unit each time.

During the last drying cycle of experiment two, continuous measurements of stem diameter variation using linear variable displacement transducers (LVDT; DF series, Solartron Metrology Ltd., Steyning Way, UK) were performed on fifteen plants to monitor one to three repetitions in all six treatments. The data obtained from the LVDTs result in variations in stem diameter thickness (mm) after calibration. Calibration was undertaken beforehand giving a linear regression with $R^2 > 0.998$ for each sensor. The continuous stem diameter variation data were used to calculate daily stem diameter growth (the difference between stem thickness at midnight between two consecutive days) and stem shrinkage (the difference between the thickest and smallest stem diameter during that day). These calculations were performed every day during the observed period. Three sensors per treatment were installed but due to the movement of the stems during growth, erroneous displacement of the sensor head could occur, resulting in fewer replicates.

2.6. Morphological Parameters

At the end of both experiments, the number of branches was counted, and the length of the longest branch was measured. The plants were then harvested by cutting the stems just above the substrate to determine fresh and dry weight. The latter was performed by heating the above-ground biomass at 70–90 °C for at least 48 h. Finally, substrates were removed from their containers to examine the visible root distribution on a relative scale from 1 to 5 (Exp 1): 1—almost no visible roots, 2—limited visible roots at the bottom, 3—well-developed roots, but not all around the pot, 4—good root development all around the pot, and 5—excellent rooting (Figure S1 in Supplementary Materials). For the second experiment, the containers were double in volume and roots were in general less developed, so a slightly different scale was used: 1—almost no visible roots, 2—limited visible roots at the side, 3—limited root development at both the side and the bottom, 4—root development all around the pot and limited at the bottom, 5—good root development all around the pot and at the bottom (Figure S2). In experiment 1, six plants were harvested from each experimental unit ($n = 24$). In experiment 2, fourteen plants were harvested from each experimental unit, and thus 56 plants per treatment. Because of the late pinching to stimulate branching in the first experiment, inflorescences could not be assessed. At the end of the second trial, the number of inflorescences was counted, and the development of each inflorescence was divided into five categories: 1—closed bud, 2—first elongation of the inflorescence with flower clusters still together, 3—second elongation of the inflorescence with the extension of the green flower clusters, 4—flowers open but still green, and 5—whitening of the flower (Figure S3).

2.7. Statistical Analysis

Statistical analysis was performed using Rstudio (R version 4.0.2) [22], completed with packages for specific statistical tests and making graphs [23–32]. First, data were checked on the presence of outliers. If the data complied with normality and homoscedasticity, results were subjected to a two-way analysis of variance with treatment and block as main effects (ANOVA). As no block effects were observed, the main effects were further analyzed by a post-hoc Tukey HSD test ($p \leq 0.05$). For the comparison of well-watered and drought stress without biostimulants, a Student's t-test was performed. Non-parametrical data were analyzed by a Scheirer–Ray–Hare test [33], a non-parametrical alternative for a two-way ANOVA and extension of the Kruskal–Wallis test, followed by a post hoc Dunn's test with 'Benjamini-Hochberg' correction ($p \leq 0.05$) in case of comparison of multiple treatments. Two treatments were compared using a Mann–Whitney U test. All results were expressed as means ± Standard Error (SE).

3. Results

3.1. Effect of Biostimulants under Deficit Irrigation in 2019

3.1.1. Effect of Deficit Irrigation

Deficit irrigation had a significant effect on most of the soil- and plant-related parameters determined during the first experiment (Table 3). The volumetric water content of the substrate, measured twice weekly to evaluate the effect of reduced irrigation, was on average 27 vol% under the standard irrigation, but was significantly lower under deficit irrigation. When less water was available in the substrate, a significantly lower electrical conductivity was observed. The Dualex and PolyPen RP410 were used to indirectly determine the effect of reduced irrigation on pigment and secondary metabolite contents. Chlorophyll seemed to be concentrated in the leaves of plants under deficit irrigation, as a significantly higher chlorophyll index was measured. Furthermore, a slight but significant increase in REIP (Red Edge Inflection Point) compared with standard irrigation was noted. The maximum quantum efficiency (F_V/F_M) was measured to evaluate the effect of reduced irrigation on photosystem 2. No significant effects were observed. To investigate the effect of reduced irrigation on stomatal conductance, measurements with the porometer were performed. Due to the lower substrate moisture content, a significantly lower stomatal conductance was measured. Table 3 shows that a reduced irrigation supply also had significant effects on the morphological parameters linked to the plant growth and quality of *Hydrangea*. The plants under deficit irrigation showed a higher branching degree but the branch length was reduced by 50%. Consequently, fresh and dry weights were reduced and these plants showed a less developed root system.

Table 3. Effect of deficit irrigation (trial 2019) on the substrate- and plant-related parameters of *Hydrangea* compared with a standard irrigation treatment. The average of each parameter over the trial is presented ± SE. Different letters (a and b) per parameter indicate a significant difference at $p \leq 0.05$.

Parameter	Standard Irrigation (100%)			Deficit Irrigation (80%) (DS Control)			Statistics
	Substrate-related parameters						
Volumetric moisture content [vol%]	26.6	±	1.1 a	20.9	±	1.2 b	$p = 0.0003$ [1]
EC [3] [mS.m^{-1}]	221.3	±	7.1 a	178.6	±	6.0 b	$p = 0.0005$ [1]
	Plant-related parameters						
Chlorophyll index [-]	22.25	±	0.50 b	25.00	±	0.49 a	$p < 0.0001$ [1]
Flavonol index [-]	0.675	±	0.019 a	0.712	±	0.026 a	$p = 0.2394$ [1]
REIP [4] [nm]	702.8	±	0.9 b	704.4	±	0.8 a	$p = 0.0464$ [2]
NDVI [5] [-]	0.505	±	0.005 a	0.506	±	0.004 a	$p = 0.8599$ [1]
Ctr2 [6] [-]	0.438	±	0.006 a	0.432	±	0.004 a	$p = 0.9867$ [2]
Lic1 [7] [-]	0.573	±	0.005 a	0.576	±	0.004 a	$p = 0.7523$ [2]
ARI1 [8] [-]	0.296	±	0.012 a	0.251	±	0.014 b	$p = 0.0136$ [1]
F_V/F_M [-]	0.794	±	0.003 a	0.784	±	0.008 a	$p = 0.0685$ [1]
Stomatal conductance [mmol m^{-2} s^{-1}]	239.8	±	20.2 a	132.1	±	11.8 b	$p = 0.0011$ [2]
Branch length [cm]	63.1	±	1.5 a	31.5	±	0.9 b	$p < 0.0001$ [2]
Number of branches [-]	9.8	±	0.6 b	11.6	±	0.5 a	$p = 0.0167$ [1]
Fresh weight [g]	89.6	±	4.3 a	51.2	±	2.5 b	$p < 0.0001$ [1]
Dry weight [g]	29.2	±	1.3 a	14.9	±	0.7 b	$p < 0.0001$ [2]
Water content [%] [(FW-DW)/FW × 100]	67.3	±	0.3 b	70.7	±	0.3 a	$p < 0.0001$ [2]
Root development score [-]	3.42	±	0.10 a	2.63	±	0.2 b	$p = 0.0002$ [2]

[1] treatment effect by a two-way analysis of variance (ANOVA). [2] treatment effect by a Scheirer–Ray–Hare test. [3] Electrical conductivity. [4] Red Edge Inflection Point. [5] Normalized Difference Vegetation Index. [6] Carter Index 2. [7] Lichtenthaler Index 1. [8] Anthocyanin Reflectance Index 1.

3.1.2. Effect of Biostimulants

The seaweed-based biostimulants were given as foliar applications, meaning that no influence on the volumetric substrate moisture content was expected and observed. The biostimulants had a limited influence on the chlorophyll and secondary metabolite contents, non-destructively determined with the Dualex and the PolyPen RP410 (Table S1). After four weeks of deficit irrigation, treatment with *E. maxima* significantly increased the chlorophyll index compared with the other biostimulant treatments but was not significantly different from the DS Control (non-treated deficit irrigation) (+10.7%). This initial positive effect disappeared during the growing season. One-time differences were also recorded for other spectral indices determined with the PolyPen RP410. After four weeks of deficit irrigation, a significantly different ARI1 index (Anthocyanin Reflectance Index 1) was observed between treatment with *S. chordalis* and *E. maxima*, where the last treatment had a 27.5% higher index. At the end of the trial, in September, differences in the Ctr2 index (Carter Index 2) and the REIP were observed between treatment with *A. nodosum* and treatment with *E. maxima*. Differences in the Lic1 index (Lichtenthaler Index 1) were observed between treatment with *A. nodosum* (+12.4%) and *S. chordalis*. Although these effects were not different from the DS Control. Stomatal closure is an adaptation response to drought stress. The commercial biostimulant *A. nodosum* tended to increase on average the stomatal conductance (147.4 mmol m^{-2} s^{-1} $\pm$ 12.0, +11.6%) compared with the DS Control (132.1 mmol m^{-2} s^{-1} $\pm$ 11.9), but this effect was not significant. The other biostimulants resulted on average in a slightly lower stomatal conductance, but again the differences were not significant (Table S2).

Figure 1 shows that the tested biostimulants did affect the morphological and plant quality parameters at the end of the trial. The *A. nodosum* treatment significantly increased the branch length by +27.9% but not the number of branches compared with the DS Control. The fresh and dry weights, as well as the root development, tended to increase for the plants treated with *A. nodosum* compared with the DS Control, though these effects were not significant. The water content was significantly lower for the plants treated with *A. nodosum* in comparison with the DS Control (−4.2%). The other tested biostimulants did not affect any of these parameters compared with the DS Control.

3.2. *Effect of Biostimulants under Repeated Drying and Wetting Cycles in 2021*

In the second trial, irrigation was turned off three times during a hot period in June, July, and August, resulting in a substantial decrease in the volumetric moisture content of the substrate from $\pm$60 vol% to 25 vol% or lower during the following days, daily measured with the WET-sensor. When visible wilting started in the youngest leaves, irrigation was turned on again so plants could recover (Figure 2). Biostimulants were sprayed on the leaves, so no effects of the biostimulants on the volumetric moisture content of the substrate were expected.

Few differences between irrigation and biostimulant treatments were detected in pigment and secondary metabolite contents during both measured drying cycles (Table S3). Water shortage significantly increased the chlorophyll index in the leaves compared with the no-stress treatment. The largest difference in chlorophyll index, determined with the Dualex, between the No stress Control treatment (24.07 $\pm$ 0.42) and the DS Control (29.84 $\pm$ 0.50; +24%) was observed at the end of the drying period. Furthermore, the stressed plants treated with biostimulants showed a significantly increased chlorophyll index compared with the No stress Control treatment; the effect of *A. nodosum* was less. Before the start of the first drying cycle, *E. maxima* resulted in a significant decrease in the flavonol index by 13.3% compared with the DS Control and a 15% decrease compared with the treatment with the *S. latissima* extract. During the drying period and after recovery, no differences in indices between treatments were observed. During the third drying cycle, there were also no effects of drought stress nor biostimulant treatment on the flavonol index, NDVI, Ctr2, Lic1, and REIP.

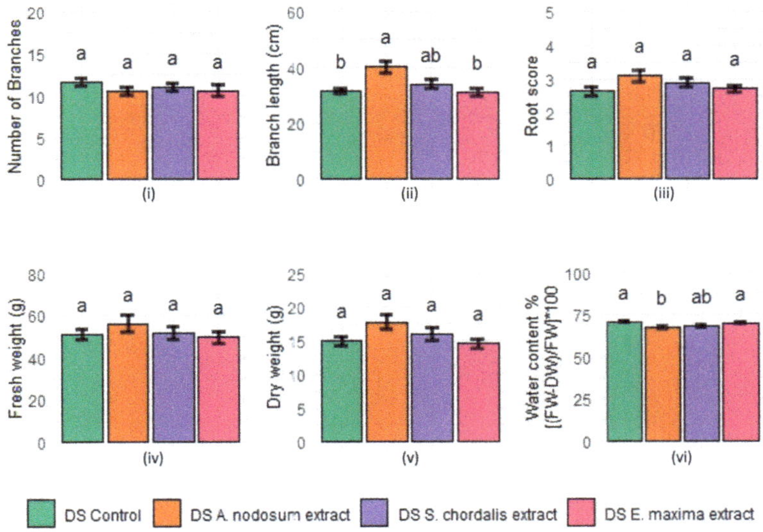

Figure 1. Effect of seaweed-based biostimulants on morphological parameters of *Hydrangea paniculata* under deficit irrigation (DS) (DS Control = non-treated deficit irrigation). (**i**) number of branches, (**ii**) length of the longest branch, (**iii**) root development score, (**iv**) fresh and (**v**) dry weight, (**vi**) plant water content. Fresh weight and water content were analyzed by a two-way ANOVA and a Tukey HSD test, the other parameters by a Scheirer–Ray–Hare test followed by a Dunn's test. Different letters (a and b) per parameter indicate a significant difference at $p \leq 0.05$ (Mean ± SE, n = 24).

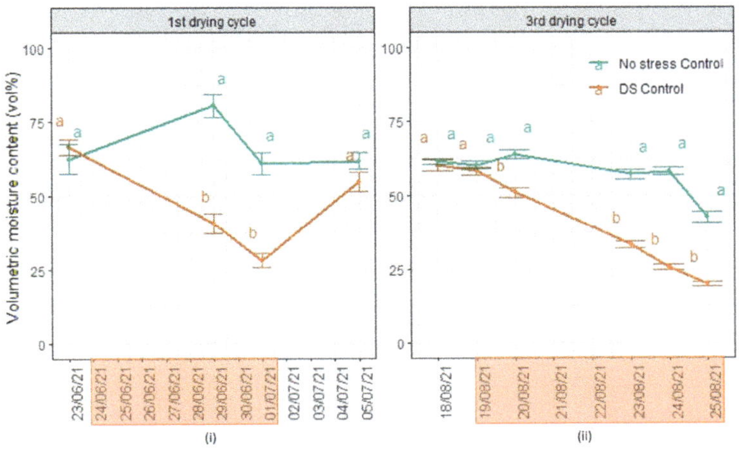

Figure 2. Effect of the drying cycles (DS Control) on the volumetric moisture content of the substrate of *Hydrangea* compared with optimal irrigation conditions (No stress Control). (**i**) first cycle: 23 June 2021–1 July 2021, (**ii**) third cycle: 19–25 August 2021. Highlighted dates indicate stress period. Treatments at 29 June 21, 1 July 2021, 5 July 2021, 19 August 2021, and 20 August 2021 were compared by a two-way ANOVA. Results on the other measurement days were analyzed by a Scheirer–Ray–Hare test. Different letters (a and b) between the treatments per measurement day indicate a significant difference at $p \leq 0.05$ (Mean ± SE, n_cycle1 = 8, n_cycle3 = 16).

The effect of the drying cycles and the decreasing volumetric moisture content of the substrate was also reflected in the stomatal conductance of *Hydrangea* leaves measured with the porometer. The results in Figure 3 show that, in both measured periods, there was a significant reduction in stomatal conductance during the drying cycle compared with the continuously well-irrigated treatment, starting from a substrate moisture content below 25 vol%. Furthermore, the influence of the biostimulants on the stomatal closure was investigated. Before the first irrigation stop (23 June 2021), the plants treated with Previsan S already had a significantly lower stomatal conductance compared with the DS Control and the plants treated with *E. maxima*. This difference disappeared during the drying period. Treatment with *E. maxima* tended to increase stomatal conductance during drought but this effect was not significant and also not present during the third drying cycle. Plants treated with the *S. latissima* extract had a significantly lower stomatal conductance compared with the No stress Control treatment. During the third drying cycle, no significant effects of the biostimulants on the decreasing stomatal conductance were observed.

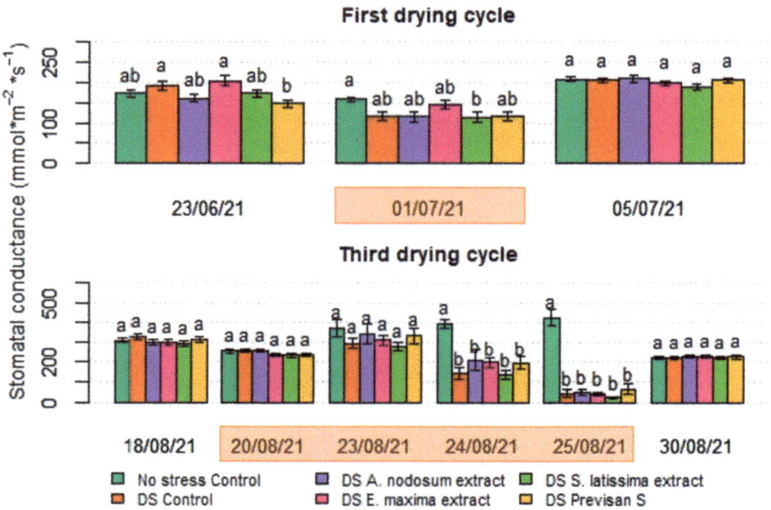

Figure 3. Effect of drying cycles without (DS Control) and with biostimulant treatments on the stomatal conductance of *Hydrangea*, compared with optimal irrigation (No stress Control). Highlighted dates indicate stress period. Data were analyzed by a two-way ANOVA followed by a Tukey HSD test, except for data on 23 June 2021, 24 August 2021, and 25 May 2021. These results were analyzed by a Scheirer–Ray–Hare test followed by a Dunn's test. Different letters (a and b) per measurement day indicate significant differences at $p \leq 0.05$ (Mean ± SE, n = 16).

The diel variations in the stem thickness of the control plants and the biostimulant-treated plants were measured continuously with LVDT (linear variable displacement transducer) sensors during the third drying cycle (Figure 4). It was mainly stem diameter shrinkage that showed the effects of drought stress on the plants. The No stress Control had a uniform and expected shrinkage over the consecutive days as these plants did not suffer from water shortage. From 23 August onwards, the stem shrinkage of plants without irrigation started to increase, which is one day before stomata started to close. At the time the stomata were almost closed at 18 vol% moisture content in the substrate (25 August 2021), the stem shrinkage was the largest. Plants treated with the biostimulant *A. nodosum* appeared to reduce stem shrinkage, whereas treatment with Previsan S increased it. No significant differences were calculated, as the number of repetitions for some treatments was too limited. These effects on stem thickness, induced by *A. nodosum* and Previsan S, were not observed on stomatal conductivity measurements.

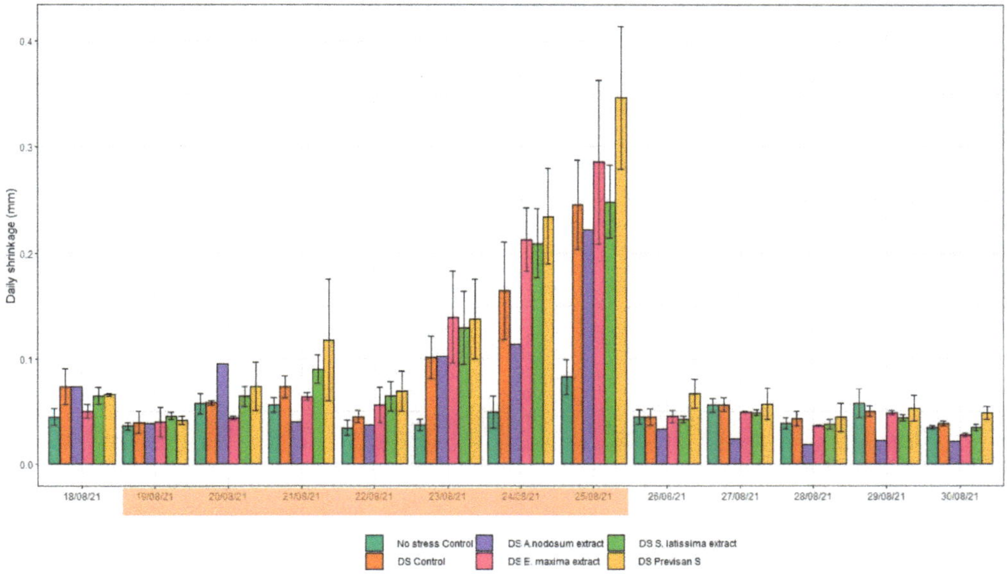

Figure 4. Effects of repeated drying cycles without (DS Control) and with biostimulant treatments on the daily stem shrinkage of *Hydrangea* before, during, and after the third drying cycle compared with optimal irrigation (No stress Control). Highlighted dates indicate stress period. (Mean ± SD (Standard Deviation)); *n_No stress Control* = 3, *n_DS Control* = 3, *n_DS A. nodosum extract* = 1, *n_DS E. maxima extract* = 2, *n_DS Previsan S* = 2).

At the end of the trial, the effects of repeated drying and wetting cycles and the biostimulant treatments on plant quality traits were evaluated. Figure 5 shows that the drying cycles with or without the application of biostimulants decreased the number of branches, but this was only significant for *E. maxima*. The repeated drying cycles with or without biostimulants significantly reduced branch length by 8.1% on average, though this was less pronounced for treatment with *E. maxima* (−5.5%). Water shortage negatively influenced root development compared with the No stress Control treatment. Here, the *A. nodosum* and *E. maxima* treatments improved the root development under stress conditions, up to the same root score as the No stress Control treatment. The lowest root score was obtained for the Previsan S treatment. The fresh and dry weight of the above-ground biomass decreased considerably due to the drying cycles by 24.7% and 24.2%, on average. Biostimulants had no additional effect.

At the end of the trial, the number of inflorescences was also counted. Each inflorescence was divided into five inflorescence development stages. *Hydrangea paniculata* has terminal inflorescences, so the number of branches had a strong influence on their total number. Again, treatment with *E. maxima* resulted in significantly fewer inflorescences, as this treatment also resulted in fewer branches (Figure 5). The drying cycles did not accelerate flowering; the DS Control had even more inflorescences in the first development stage compared with the No stress Control. Treatment with biostimulants affected the development of the inflorescences. *A. nodosum*, *S. latissima* extract, and Previsan S showed a more advanced development with fewer inflorescences in the first developmental stage (score 1) compared with the DS Control. This effect was most pronounced for *A. nodosum* with significantly more fully developed inflorescences (score 5) compared with all other treatments (Figure 6).

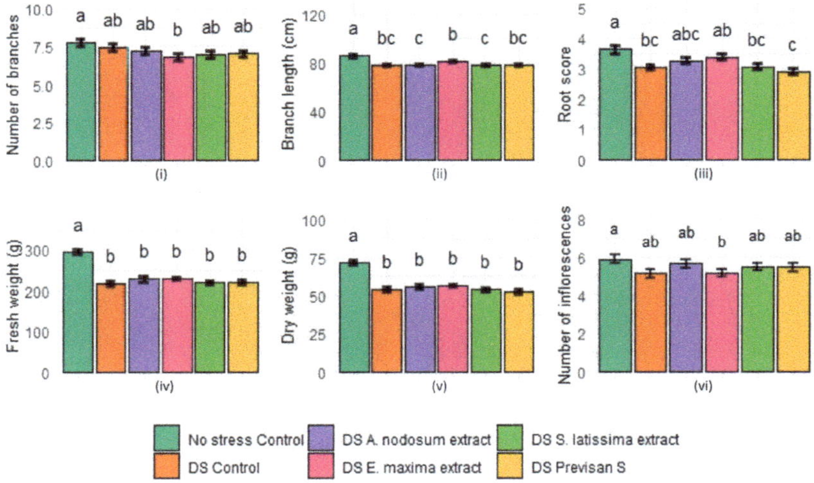

Figure 5. Effects of repeated drying cycles without (DS Control) and with biostimulants on plant quality traits compared with optimal irrigation (No stress Control): (**i**) number of branches, (**ii**) length of the longest branch, (**iii**) root development score, (**iv**) fresh and (**v**) dry weight, and (**vi**) total number of inflorescences. Results of branch length and dry weight were analyzed by a two-way ANOVA followed by a Tukey HSD test. Other results were analyzed by a Scheirer–Ray–Hare test combined with a Dunn's test. Different letters (a, b and c) per parameter indicate a significant difference at $p \leq 0.05$ (Mean $\pm$ SE; n = 56).

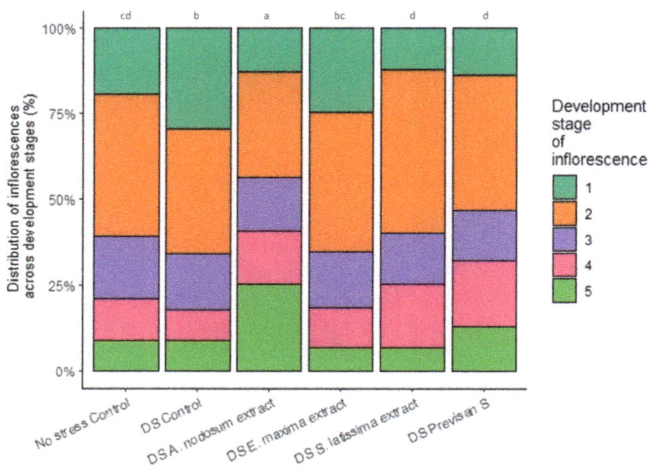

Figure 6. Effect of the drying cycles without (DS Control) and with biostimulants treatments on flower development of *Hydrangea paniculata* compared with optimal irrigation (No stress Control). Development of inflorescences was assessed using a 1–5 scoring system (1—closed bud; 2—first elongation of the inflorescence with flower clusters still together; 3—second elongation of the inflorescence with the extension of the green flower clusters; 4—flowers open but still green; 5—whitening of the flower). Differences in treatments on flower development were evaluated by a Scheirer–Ray–Hare test, followed by a post hoc Dunn's test. Different letters (a, b, c, and d) indicate a significant difference (n = 56).

4. Discussion

4.1 Effect of Deficit Irrigation and Repeated Drying Cycles on Hydrangea

Both deficit irrigation and repeated drying cycles significantly reduced the volumetric moisture content of the substrate compared with the well-watered (no stress) controls. This was accompanied by a significantly higher chlorophyll index measured in the plants under deficit irrigation compared with the standard irrigation. Furthermore, the REIP (red edge inflection point) shifted to higher wavelengths indicating a higher chlorophyll content [34]. The chlorophyll seemed to be more concentrated due to a lower leaf water content. Marenco et al. (2009) and Martínez and Guiamet (2004) also found a negative correlation between the leaf water content and the chlorophyll content measured with the SPAD-meter in different Amazonian tree species, wheat, and maize [35,36]. The production of secondary metabolites, especially phenylpropanoids such as flavonoids and flavonols, is induced by various biotic and abiotic environmental stresses [37]. Furthermore, oxidative and drought stress induced the increased the production of these secondary metabolites, including flavonols and anthocyanins, to mitigate the effects of stress with their strong radical scavenging activity [38,39]. In this research, no increased flavonol or anthocyanin content, determined by the optical sensors, was observed under reduced irrigation.

Moreover, no effect of reduced irrigation on maximal photochemical yields was observed. The photosynthetic system appears to remain intact above certain drought stress levels. Several studies reviewed by Flexas et al. (2004) show that the F_V/F_M-ratio remained constant as long as the stomatal conductance remained above 50 mmol m^{-2} s^{-1}, the general threshold for severe drought stress in C3 plants. However, the F_V/F_M-ratio abruptly decreased at lower stomatal conductance, indicating a down-regulation of the entire photosynthetic metabolism at this stress level [40,41]. In the research of Liu et al. (2010) on different woody ornamental species, the same pattern was observed during repeated cycles of drying and rewetting [42].

This reduced irrigation went along with a lower stomatal conductance. From the moment the volumetric moisture content fell below 25 vol%, plants responded to the water deficit by closing their stomata. Turner (1991) also observed stomatal response only from the moment a certain threshold of soil water content was exceeded [43]. The moment the moisture content was about 20 vol%, stomata were almost closed compared with the well-irrigated treatment. In *Hydrangea paniculata* and *Petunia x hybrida*, grown in another type of substrate, stomatal conductance was still between 300–500 mmol m^{-2} s^{-1} at a moisture content of 20–35 vol% [44,45]. The stomata of *Petunia x hybrida* almost closed when the plant experienced severe drought (moisture content around 10 vol%) [45]. Flexas and Medrano (2002) [40,46] defined four different phases of drought based on the daily maximum stomatal conductance (g_s) of different C3 crops (mild drought stress: $g_s > 150$ mmol m^{-2} s^{-1}; moderate drought: 150 mmol m^{-2} s^{-1} > g_s > 100 mmol m^{-2} s^{-1}; severe drought: 100 mmol m^{-2} s^{-1} > g_s > 50 mmol m^{-2} s^{-1}; very severe drought: $g_s < 50$ mmol m^{-2} s^{-1}). According to their definition, *Hydrangea* plants under deficit irrigation (experiment one) already experienced moderate drought stress while the standard irrigation was under mild drought stress because an average volumetric substrate moisture content of 26 vol% is rather low [45]. This low substrate moisture content, despite a standard irrigation scheme, can be explained by the overhead irrigation leading to an umbrella effect due to the plants' foliage preventing the irrigation water to reach the substrate. This can be prevented by drip irrigation, used in the second experiment. Here, in the last drying cycle (August 2021), the plants were subjected to severe drought stress. In general, stomatal conductance measured at midday during the first drying cycle in 2021 was half that measured during the last cycle. This first cycle was additionally characterized by a higher light intensity and vapor pressure deficit which explains the lower values [43,47,48].

Stem diameter variations show diel dynamics. Soon after dawn, there is a time delay between the water lost from the plant via leaf transpiration and the water uptake by the roots, causing plants to use water stored in their internal stem reserves, resulting in stem diameter shrinkage. Shortly after noon, the sap flow reaches its daily maximum, and the stem shrinks rapidly. In the afternoon, the sap flow and stem shrinkage both decrease. During the night, when sap flow is the lowest, internal water storage pools are replenished and the stem will swell [49]. When plants are depleted of water and there is not enough water available in the soil/substrate to respond to the evaporative demand of the atmosphere, the maximum daily stem shrinkage increases [49,50]. This pattern is also observed in *Hydrangea*.

Both deficit irrigation and repeated drying cycles reduced the biomass production and branch length of the hydrangeas in this study. Furthermore, in a study by Cameron et al. (2006), the vegetative growth of different woody ornamental species, e.g., *Forsythia*, *Cotinus* was reduced under deficit irrigation while the effect on *Hydrangea macrophylla* at the end of the season was limited. Reduced growth can be a favorable effect for ornamental plants because the combination of shorter shoot lengths/shorter internodes improves compactness and reduces the need for mid-season pruning to become a compact, well-branched plant [51]. In our experiment with deficit irrigation, a significantly higher branching rate was observed in the stressed plants. This was not the case when the hydrangeas were grown under repeated drying and rewetting cycles. Cameron et al. (2006, 2008) also observed no effect of deficit irrigation on the number of shoots and number of formative primary shoots [51,52]. Induced flowering can be a response in many plant species to stressors such as drought, poor nutrition, and light quality. This response is biologically advantageous especially in plants that produce fertile seeds [53]. For *Hydrangea*, no significant effect of the repeated drying and rewetting cycles on the number of inflorescences was observed, nor on inflorescence development with a higher percentage of inflorescences in a less advanced stage compared with well-irrigated plants. Also found in *Forsythia*, *Cotinus*, and *Hydrangea macrophylla*, the flower number per node was unaffected by the deficit irrigation [51].

4.2. Effects of Biostimulants on Hydrangea Grown under Deficit Irrigation or Repeated Drying Cycles

The betaines in seaweed extracts enhance the leaf chlorophyll content [15], which could be due to a reduction in chlorophyll degradation [11]. In grapevine, the negative effect of drought stress on the chlorophyll content was lower in plants treated with the different types of biostimulants compared with the untreated control plants grown under the same stress conditions [54]. In our study, chlorophyll did not appear to be degraded by the presence of drought. On the contrary, deficit irrigation increased the chlorophyll index and REIP (Red Edge Inflection Point) compared with the well-watered No stress Controls. The tested biostimulants did not provide any added value to the deficit irrigation. Seaweed extracts have been reported to increase important bioactive molecular concentrations such as phenolics, flavonoids, and anthocyanins in several crops, such as vegetables and grapevine grown under both optimal and stressed conditions [54–57]. Moreover, in *Calibrachoa* under optimal conditions, increases in phenolic and flavonol contents were found following treatment with a seaweed extract [58]. In this study on *Hydrangea paniculata* under drought, no increase in flavonoid or anthocyanin content, determined by optical sensors, was observed by any of the biostimulants. F_V/F_M was not affected by the deficit irrigation, and biostimulants did not affect this parameter. This was also the case for the treatment of spinach under drought with a seaweed-based biostimulant [59].

Stomatal closure is regulated, among other hormones and mechanisms, by the accumulation of abscisic acid (ABA), which is induced under drought stress [60]. In research on grapevine under drought stress, vines treated with different types of biostimulants, especially a seaweed-based product, accumulated higher levels of ABA compared with the untreated controls to reduce water loss and increase plant drought tolerance [54]. Biostimulants can also act by postponing drought stress, as in the study of Campobenedetto et al. (2021) where a seaweed-based biostimulant reduced the ABA concentration in tomato compared with the untreated control grown under the same mild drought conditions [61]. In other vegetables, such as spinach and broccoli, the application of seaweed-based biostimulants significantly increased stomatal conductance [59,62]. In contrast, the biostimulants in our trials had no significant effect on the stomatal conductance of *Hydrangea paniculata* grown under drought conditions. The *A. nodosum* extract under deficit irrigation slightly increased the stomatal conductance, but this was not linked to a better water-use efficiency as plants had a lower water content at the end of the trial. Treatment with *E. maxima* also showed some effect, but in both cases, they were not significant and not repetitive.

In this research on *Hydrangea paniculata* under water shortage, *A. nodosum* slightly reduced stem shrinkage during drought compared with the control (DS Control), indicating some alleviation of the imposed stress. Previsan S, on the other hand, increased the shrinkage compared with the stressed control (DS control). Top et al. (2023) tested similar seaweed-based products on tomato (*Solanum lycopersicum*) plants under deficit irrigation. For this crop, *A. nodosum* showed no beneficial effect compared with the stressed control. However, another tested *Ascophyllum nodosum* extract (Asco-N2) reduced stem shrinkage in drought-treated tomato plants, similar to *A. nodosum* in *Hydrangea*, and resulted in a similar performance and water uptake as untreated, well-watered control tomato plants [63]. These results might indicate that *A. nodosum*-derived biostimulants can mitigate drought stress to some extent.

The beneficial effects of seaweed extracts on shoot growth and yield were reported in several studies on different crops [9,13], including several ornamentals such as *Calibrachoa*, [58], rose [64], and *Pelargonium* [65]. The effect was mostly dependent on the dose and application method. The tested seaweed extracts in our study had limited effects on the morphological growth parameters and plant quality of hydrangea. *Hydrangea paniculata*, grown under deficit irrigation and treated with *A. nodosum*, had significantly longer branches and more dry weight, but this biostimulant had a rather negative effect on plant growth under repeated drying and wetting cycles. Seaweed extracts induce early flowering in several crops [9]. The positive effects of a seaweed extract on the flowering and fruit set numbers of eggplant were observed in field conditions [66]. Furthermore, in research on container-grown roses, the application of a seaweed extract increased the flowering [64]. The *A. nodosum* product in our research accelerated flowering, as (significantly) higher numbers of fully developed inflorescences were counted at the end of the growing season. The experimental product based on *S. latissima* and Previsan S also gave similar results. Plants treated with *E. maxima* had significantly fewer branches and thus significantly fewer inflorescences in total. This lower number of branches could be the effect of the biostimulant or of the pruning in June.

5. Conclusions

Reduced irrigation resulted in more compact hydrangeas, but this effect is more pronounced under continuous deficit irrigation than under repeated drying and wetting cycles. Although plants were more compact, they also produced less biomass as stomata closed under dry conditions and thus reduced gas exchanges, although photosystem 2 remained intact.

The tested biostimulants in this study had only limited effects on the morphological parameters of *Hydrangea paniculata*, depending on the applied drought stress treatment (deficit irrigation or repeated drying cycles). The *Ascophyllum nodosum* extract positively influenced plant growth under deficit irrigation, and flowering under repeated drying cycles. Flowering could not be assessed under deficit irrigation. This specific biostimulant also slightly reduced stem shrinkage under drought, which might indicate better plant-water relations. The *Ecklonia maxima* extract negatively influenced branching and flowering under repeated drying and wetting cycles, but not under deficit irrigation. From this study, it can be concluded that it is difficult to observe repeated effects under field conditions.

In general, research on biostimulants is complex, as effects seem to depend on many factors such as plant, cultivar, application dose, and method, but also growing conditions which fluctuate widely in the field. This latter aspect increases the challenge of determining the perfect combination between biostimulant, dose, application timing, crop, and growing conditions.

Supplementary Materials: The following supporting information can be downloaded at: https://www.mdpi.com/article/10.3390/horticulturae9040509/s1, Table S1: Effect of biostimulant treatments on pigment reflectance indices of *Hydrangea paniculata* grown under deficit irrigation compared with the untreated control (DS Control). Different letters per parameter per measurement day indicate a significant difference at $p \leq 0.05$ (Mean $\pm$ SE; $n = 16$), Table S2: Effect of biostimulant treatments on stomatal conductance of *Hydrangea paniculata* grown under deficit irrigation compared with the untreated control (DS Control). Different letters per parameter per measurement day indicate a significant difference at $p \leq 0.05$ (Mean $\pm$ SE; $n = 16$), Table S3: Effect of repeated drying and rewetting cycles and biostimulant treatments on pigment reflectance indices of *Hydrangea paniculata* (experiment 2) compared with an untreated control (DS Control) and an untreated control under optimal irrigation (No stress Control). Different letters per parameter per measurement day indicate a significant difference at $p \leq 0.05$ (Mean $\pm$ SE; $n = 16$), Figure S1: Examination of the root development of *Hydrangea paniculata* on a relative scale from 1 to 5 after experiment one: 1—almost no visible roots, 2—limited visible roots at the bottom, 3—well-developed roots, but not all around the pot, 4—good root development all around the pot and 5—excellent rooting, Figure S2: Examination of the root development of *Hydrangea paniculata* on a relative scale from 1 to 5 after experiment 2: 1—almost no visible roots, 2—limited visible roots at the side, 3—limited root development at both the side and the bottom, 4—root development all around the pot and limited at the bottom, 5—good root development all around the pot and at the bottom. Plants were grown in bigger containers, Figure S3: Evaluation of development of inflorescences of *Hydrangea paniculata* on a relative scale from 1 to 5 after experiment 2: 1—closed bud, 2—first elongation of the inflorescence with flower clusters still together, 3—second elongation of the inflorescence with the extension of the green flower clusters, 4—flowers open but still green, 5—whitening of the flower.

Author Contributions: Conceptualization, E.P., K.S. and M.-C.V.L.; methodology, P.D.C., S.T., E.P., K.S. and M.-C.V.L.; software, K.S. and S.T.; formal analysis, P.D.C. and S.T.; investigation, P.D.C. and S.T.; data curation, P.D.C. and S.T.; writing—original draft preparation, P.D.C.; writing—review and editing, P.D.C., S.T., E.P., K.S. and M.-C.V.L.; visualization, P.D.C.; supervision, M.-C.V.L.; project administration, E.P., M.-C.V.L. and K.S.; funding acquisition, E.P., M.-C.V.L. and K.S. All authors have read and agreed to the published version of the manuscript.

Funding: This research was funded by Interreg 2 Seas Mers Zeeën, grant number 2S03-029.

Data Availability Statement: Data will be available without any reservations by the authors to qualified researchers.

Acknowledgments: The author would like to thank Simon Van Kerkhove, Hanne Denaeghel, and Ellen Dams for organizing the trials, their on-site coordination, insights, and ideas about results and also, the entire Bio4safe consortium for the nice collaboration.

Conflicts of Interest: The authors declare no conflict of interest.

References

1. Departement Landbouw en Visserij. *Landbouwrapport 2018: Uitdagingen Voor de Vlaamse Land- En Tuinbouw*; Departement Landbouw en Visserij: Brussel, Belgium, 2018.
2. Departement Landbouw en Visserij. *Landbouwrapport 2020*; Departement Landbouw en Visserij: Brussel, Belgium, 2020.
3. Cook, B.I.; Mankin, J.S.; Anchukaitis, K.J. Climate Change and Drought: From Past to Future. *Curr. Clim. Chang. Rep.* **2018**, *4*, 164–179. [CrossRef]
4. European Union Regulation (EC). *No 765/2008 of the European Parliament and of the Council (4) Lays down Rules on the Accredi*; European Union Regulation (EC): Brussel, Belgium, 2019.
5. du Jardin, P. *The Science of Plant Biostimulants—A Bibliographic Analysis*; Ad hoc Study Report to the European Commission; European Union: Brussel, Belgium, 2012.
6. du Jardin, P. Plant Biostimulants: Definition, Concept, Main Categories and Regulation. *Sci. Hortic.* **2015**, *196*, 3–14. [CrossRef]
7. Yakhin, O.I.; Lubyanov, A.A.; Yakhin, I.A.; Brown, P.H. Biostimulants in Plant Science: A Global Perspective. *Front. Plant Sci.* **2017**, *7*, 2049. [CrossRef]
8. Craigie, J.S. Seaweed Extract Stimuli in Plant Science and Agriculture. *J. Appl. Phycol.* **2011**, *23*, 371–393. [CrossRef]

9. Khan, W.; Rayirath, U.P.; Subramanian, S.; Jithesh, M.N.; Rayorath, P.; Hodges, D.M.; Critchley, A.T.; Craigie, J.S.; Norrie, J.; Prithiviraj, B. Seaweed Extracts as Biostimulants of Plant Growth and Development. *J. Plant Growth Regul.* **2009**, *28*, 386–399. [CrossRef]
10. Battacharyya, D.; Babgohari, M.Z.; Rathor, P.; Prithiviraj, B. Seaweed Extracts as Biostimulants in Horticulture. *Sci. Hortic.* **2015**, *196*, 39–48. [CrossRef]
11. Whapham, C.A.; Blunden, G.; Jenkins, T.; Hankins, S.D. Significance of Betaines in the Increased Chlorophyll Content of Plants Treated with Seaweed Extract. *J. Appl. Phycol.* **1993**, *5*, 231. [CrossRef]
12. Wally, O.S.D.; Critchley, A.T.; Hiltz, D.; Craigie, J.S.; Han, X.; Zaharia, L.I.; Abrams, S.R.; Prithiviraj, B. Regulation of Phytohormone Biosynthesis and Accumulation in *Arabidopsis* Following Treatment with Commercial Extract from the Marine Macroalga *Ascophyllum Nodosum*. *J. Plant Growth Regul.* **2013**, *32*, 324–339. [CrossRef]
13. Calvo, P.; Nelson, L.; Kloepper, J.W. Agricultural Uses of Plant Biostimulants. *Plant Soil* **2014**, *383*, 3–41. [CrossRef]
14. Mittler R Oxidative Stress, Antioxidants and Stress Tolerance. *Trends Plant Sci.* **2002**, *7*, 405–410. [CrossRef]
15. Blunden, G.; Jenkins, T.; Liu, Y.-W. Enhanced Leaf Chlorophyll Levels in Plants Treated with Seaweed Extract. *J. Appl. Phycol.* **1996**, *8*, 535–543. [CrossRef]
16. Zhang, X.; Ervin, E.H. Cytokinin-Containing Seaweed and Humic Acid Extracts Associated with Creeping Bentgrass Leaf Cytokinins and Drought Resistance. *Crop Sci.* **2004**, *44*, 1737–1745. [CrossRef]
17. Rouse, J.W.; Haas, R.H.; Schell, J.A.; Deering, D.W.; Harlan, J.C. *Monitoring the Vernal Advancement and Retrogradation (Greenwave Effect) of Natural Vegetation*; Remote Sensing Center, Texas A&M University: College Station, TX, USA, 1974.
18. Lichtenthaler, H.K. Vegetation Stress: An Introduction to the Stress Concept in Plants. *J. Plant Physiol.* **1996**, *148*, 4–14. [CrossRef]
19. Carter, G.A. Ratios of Leaf Reflectances in Narrow Wavebands as Indicators of Plant Stress. *Int. J. Remote Sens.* **1994**, *15*, 697–703. [CrossRef]
20. Gitelson, A.A.; Merzlyak, M.N.; Chivkunova, O.B. Optical Properties and Nondestructive Estimation of Anthocyanin Content in Plant Leaves. *Photochem. Photobiol.* **2001**, *74*, 38–45. [CrossRef]
21. Maxwell, K.; Johnson, G.N. Chlorophyll Fluorescence-a Practical Guide. *J. Exp. Bot.* **2000**, *51*, 659–668. [CrossRef] [PubMed]
22. R Core Team. *R: A Language and Environment for Statistical Computing*; R Foundation for Statistical Computing: Vienna, Austria, 2021.
23. Auguie, B.; Antonov, A. *GridExtra: Miscellaneous Functions for "Grid" Graphics.* R Package Version 2.3. 2017. Available online: https://cran.r-project.org/web/packages/gridExtra/index.html (accessed on 23 January 2023).
24. Neuwirth, E. *RColorBrewer: ColorBrewer Palettes.* R Package Sersion 1.1-3. 2022. Available online: https://cran.r-project.org/web/packages/RColorBrewer/RColorBrewer.pdf (accessed on 23 January 2023).
25. Wickham, H.; François, R.; Henry, L.; Müller, K. *Dplyr: A Grammar of Data Manipulation.* R-Package Version 1.1.1. 2022. Available online: https://cran.r-project.org/web/packages/dplyr/dplyr.pdf (accessed on 23 January 2023).
26. Wickham, H. *Ggplot2: Elegant Graphics for Data Analysis*; Springer: New York, NY, USA, 2016; ISBN 978-3-319-24277-4.
27. Mangiafico, S. *Rcompanion: Functions to Support Extension Education Program Evaluation.* R-Package Version 2.4.26. 2022. Available online: https://cran.r-project.org/web/packages/rcompanion/index.html (accessed on 23 January 2023).
28. Ogle, D.H.; Doll, J.C.; Wheeler, P.; Dinno, A. *FSA: Fisheries Stock Analysis.* R Package Version 0.9.4. 2022. Available online: https://CRAN.R-project.org/package=FSA (accessed on 23 January 2023).
29. Fox, J.; Weisberg, S. *An {R} Companion to Applied Regression.* R Package Version 3.1-2. 2019. Available online: https://cran.r-project.org/web/packages/car/index.html (accessed on 23 January 2023).
30. de Mendiburu, F. *Agricolae: Statistical Procedures for Agricultural Research.* R Package Version 1.3-5. 2021. Available online: https://cran.r-project.org/web/packages/agricolae/agricolae.pdf (accessed on 23 January 2023).
31. Wickham, H.; Bryan, J. *Readxl: Read Excel Files. R Package.* R Package Version 1.4.2. 2022. Available online: https://cran.r-project.org/web/packages/readxl/index.html (accessed on 23 January 2023).
32. Wickham, H. *Forcats: Tools for Working with Categorical Variables (Factors).* R Package Version 1.0.0. 2021. Available online: https://cran.r-project.org/web/packages/forcats/index.html (accessed on 23 January 2023).
33. Sokal, R.R.; Rohlf, F.J. *Biometry: The Principles and Practice of Statistics in Biological Research*, 3rd ed.; W. H. Freeman: New York, NY, USA, 1995.
34. Gitelson, A.A.; Merzlyak, M.N.; Lichtenthaler, H.K. Detection of Red Edge Position and Chlorophyll Content by Reflectance Measurements near 700 Nm. *J. Plant Physiol.* **1996**, *148*, 501–508. [CrossRef]
35. Marenco, R.A.; Antezana-Vera, S.A.; Nascimento, H.C.S. Relationship between Specific Leaf Area, Leaf Thickness, Leaf Water Content and SPAD-502 Readings in Six Amazonian Tree Species. *Photosynthetica* **2009**, *47*, 184–190. [CrossRef]
36. Martínez, D.E.; Guiamet, J.J. Distortion of the SPAD 502 Chlorophyll Meter Readings by Changes in Irradiance and Leaf Water Status. *Agronomie* **2004**, *24*, 41–46. [CrossRef]
37. Dixon, R.A.; Paiva, N.L. Stress-Induced Phenylpropanoid Metabolism. *Plant Cell* **1995**, *7*, 1085. [CrossRef] [PubMed]
38. Nakabayashi, R.; Mori, T.; Saito, K. Alternation of Flavonoid Accumulation under Drought Stress in *Arabidopsis thaliana*. *Plant Signal. Behav.* **2014**, *9*, e29518. [CrossRef] [PubMed]

39. Nakabayashi, R.; Yonekura-Sakakibara, K.; Urano, K.; Suzuki, M.; Yamada, Y.; Nishizawa, T.; Matsuda, F.; Kojima, M.; Sakakibara, H.; Shinozaki, K.; et al. Enhancement of Oxidative and Drought Tolerance in *Arabidopsis* by Overaccumulation of Antioxidant Flavonoids. *Plant J.* **2014**, *77*, 367–379. [CrossRef] [PubMed]
40. Aroca, R. *Plant Responses to Drought Stress: From Morphological to Molecular Features*; Springer: Berlin/Heidelberg, Germany, 2013; ISBN 9783642326530.
41. Flexas, J.; Bota, J.; Loreto, F.; Cornic, G.; Sharkey, T.D. Diffusive and Metabolic Limitations to Photosynthesis under Drought and Salinity in C3 Plants. *Plant Biol.* **2004**, *6*, 269–279. [CrossRef]
42. Liu, C.C.; Liu, Y.G.; Guo, K.; Zheng, Y.R.; Li, G.Q.; Yu, L.F.; Yang, R. Influence of Drought Intensity on the Response of Six Woody Karst Species Subjected to Successive Cycles of Drought and Rewatering. *Physiol. Plant.* **2010**, *139*, 39–54. [CrossRef] [PubMed]
43. Turner, N.C. Measurement and Influence of Environmental and Plant Factors on Stomatal Conductance in the Field. *Agric. For. Meteorol.* **1991**, *54*, 137–154. [CrossRef]
44. Basiri Jahromi, N.; Fulcher, A.; Walker, F.; Altland, J. Photosynthesis, Growth, and Water Use of *Hydrangea Paniculata* 'Silver Dollar' Using a Physiological-Based or a Substrate Physical Properties-Based Irrigation Schedule and a Biochar Substrate Amendment. *Irrig. Sci.* **2020**, *38*, 263–274. [CrossRef]
45. Kim, J. Water Relations of Bedding Plants: Water Use, Drought Physiology, and Gene Expression. Ph.D. Thesis, University of Georgia, Athens, Greece, 2011.
46. Flexas, J.; Medrano, H. Drought-Inhibition of Photosynthesis in C3 Plants: Stomatal and Non-Stomatal Limitations Revisited. *Ann. Bot.* **2002**, *89*, 183–189. [CrossRef]
47. Barradas, V.L.; Jones, H.G. Responses of CO_2 Assimilation to Changes in Irradiance: Laboratory and Field Data and a Model for Beans (*Phaseolus vulgaris* L.). *J. Exp. Bot.* **1996**, *47*, 639–645. [CrossRef]
48. Turner, N.C.; Schulze, E.-D.; Gollan, T. The Responses of Stomata and Leaf Gas Exchange to Vapour Pressure Deficits and Soil Water Content. *Oecologia* **1984**, *63*, 338–342. [CrossRef]
49. Steppe, K.; Sterck, F.; Deslauriers, A. Diel Growth Dynamics in Tree Stems: Linking Anatomy and Ecophysiology. *Trends Plant Sci.* **2015**, *20*, 335–343. [CrossRef]
50. de Swaef, T.; de Schepper, V.; Vandegehuchte, M.W.; Steppe, K. Stem Diameter Variations as a Versatile Research Tool in Ecophysiology. *Tree Physiol.* **2015**, *35*, 1047–1061. [CrossRef] [PubMed]
51. Cameron, R.W.F.; Harrison-Murray, R.S.; Atkinson, C.J.; Judd, H.L. Regulated Deficit Irrigation—A Means to Control Growth in Woody Ornamentals. *J. Hortic. Sci. Biotechnol.* **2006**, *81*, 435–443. [CrossRef]
52. Cameron, R.; Harrison-Murray, R.; Fordham, M.; Wilkinson, S.; Davies, W.; Atkinson, C.; Else, M. Regulated Irrigation of Woody Ornamentals to Improve Plant Quality and Precondition against Drought Stress. *Ann. Appl. Biol.* **2008**, *153*, 49–61. [CrossRef]
53. Wada, K.C.; Takeno, K. Stress-Induced Flowering. *Plant Signal. Behav.* **2010**, *5*, 944–947. [CrossRef]
54. Irani, H.; ValizadehKaji, B.; Naeini, M.R. Biostimulant-Induced Drought Tolerance in Grapevine Is Associated with Physiological and Biochemical Changes. *Chem. Biol. Technol. Agric.* **2021**, *8*, 1–13. [CrossRef]
55. Fan, D.; Hodges, D.M.; Critchley, A.T.; Prithiviraj, B. A Commercial Extract of Brown Macroalga (*Ascophyllum Nodosum*) Affects Yield and the Nutritional Quality of Spinach In Vitro. *Commun. Soil Sci. Plant Anal.* **2013**, *44*, 1873–1884. [CrossRef]
56. Lola-Luz, T.; Hennequart, F.; Gaffney, M. Effect on Yield, Total Phenolic, Total Flavonoid and Total Isothiocyanate Content of Two Broccoli Cultivars (*Brassica oleraceae* Var Italica) Following the Application of a Commercal Brown Seaweed Extract. *Agric. Food Sci.* **2014**, *23*, 28–37. [CrossRef]
57. Lola-Luz, T.; Hennequart, F.; Gaffney, M. Enhancement of Phenolic and Flavonoid Compounds in Cabbage (Brassica Oleraceae) Following Application of Commercial Seaweed Extracts of the Brown Seaweed (Ascophyllum Nodosum). *Agric. Food Sci.* **2013**, *22*, 288–295. [CrossRef]
58. Elansary, H.O.; Norrie, J.; Ali, H.M.; Salem, M.Z.M.; Mahmoud, E.A.; Yessoufou, K. Enhancement of *Calibrachoa* Growth, Secondary Metabolites and Bioactivity Using Seaweed Extracts. *BMC Complement. Altern. Med.* **2016**, *16*, 341. [CrossRef]
59. Xu, C.; Leskovar, D.I. Effects of *A. Nodosum* Seaweed Extracts on Spinach Growth, Physiology and Nutrition Value under Drought Stress. *Sci. Hortic.* **2015**, *183*, 39–47. [CrossRef]
60. Jones, R.J.; Mansfield, T.A. Suppression of Stomatal Opening in Leaves Treated with Abscisic Acid. *J. Exp. Bot.* **1970**, *21*, 714–719. [CrossRef]
61. Campobenedetto, C.; Agliassa, C.; Mannino, G.; Vigliante, I.; Contartese, V.; Secchi, F.; Bertea, C.M. A Biostimulant Based on Seaweed (*Ascophyllum Nodosum* and *Laminaria Digitata*) and Yeast Extracts Mitigateswater Stress Effects on Tomato (*Solanum Lycopersicum l.*). *Agriculture* **2021**, *11*, 557. [CrossRef]
62. Kałuzewicz, A.; Krzesiński, W.; Spizewski, T.; Zaworska, A. Effect of Biostimulants on Several Physiological Characteristics and Chlorophyll Content in Broccoli under Drought Stress and Re-Watering. *Not. Bot. Horti Agrobot. Cluj Napoca* **2017**, *45*, 197–202. [CrossRef]
63. Top, S.; Vandoorne, B.; Pauwels, E.; Perneel, M.; Van Labeke, M.-C.; Steppe, K. Plant Sensors Untangle the Water-Use and Growth Effects of Selected Seaweed-Derived Biostimulants on Drought-Stressed Tomato Plants (*Solanum lycopersicum*). *J. Plant Growth Regul.* **2023**. [CrossRef]
64. Sumangala, K.; Srikrishnah, S.; Sutharsan, S. Roses Growth and Flowering Responding to Concentration and Frequency of Seaweed (*Sargassum crassifolium* L.) Liquid Extract Application. *Curr. Agric. Res. J.* **2019**, *7*, 236–244. [CrossRef]

65. Krajnc, A.U.; Ivanuš, A.; Kristl. J.; Šušek, A. Seaweed Extract Elicits the Metabolic Responses in Leaves and Enhances Growth of *Pelargonium* Cuttings. *Eur. J. Hortic. Sci.* **2012**, *77*, 170–181.
66. Pohl, A.; Grabowska, A.; Kalisz, A.; Ekara, A.S. Biostimulant Application Enhances Fruit Setting in Eggplant—An Insight into the Biology of Flowering. *Agronomy* **2019**, *9*, 482. [CrossRef]

horticulturae

MDPI

Article

Exploring Genetic Diversity in an *Ilex crenata* Breeding Germplasm

Esther Geukens [1], Annelies Haegeman [1], Jef Van Meulder [2], Katrijn Van Laere [1], Erik Smolders [3], Tom Ruttink [1,4] and Leen Leus [1,*]

[1] Plant Sciences Unit, Flanders Research Institute for Agricultural, Fisheries and Food Research (ILVO), Caritasstraat 39, 9090 Melle, Belgium; esther.geukens@ilvo.vlaanderen.be (E.G.); annelies.haegeman@ilvo.vlaanderen.be (A.H.); katrijn.vanlaere@ilvo.vlaanderen.be (K.V.L.); tom.ruttink@ilvo.vlaanderen.be or tom.ruttink@ugent.be (T.R.)

[2] Arboretum Provincial Domain Bokrijk, Bokrijk, 3600 Genk, Belgium

[3] Division of Soil and Water Management, KU Leuven, Kasteelpark Arenberg 20, 3001 Leuven, Belgium; erik.smolders@kuleuven.be

[4] Department of Plant Biotechnology and Bioinformatics, Ghent University, Technologiepark 71, 9052 Ghent, Belgium

* Correspondence: leen.leus@ilvo.vlaanderen.be

Abstract: Knowledge of genetic identity, genetic relationships, ploidy level, and chromosome numbers can enhance the efficiency of ornamental plant breeding programs. In the present study, genome sizes, chromosome numbers, and genetic fingerprints were determined for a collection of 94 *Ilex* accessions, including 69 *I. crenata*. The genome size of the entire collection ranged from 1.50 ± 0.03 to 8.01 ± 0.18. Within the species of *I. crenata*, genome sizes varied (mean ± sd) between 1.98 ± 0.08 and 2.30 ± 0.04, with three outliers: 3.06 ± 0.04, 4.04 ± 0.09, and 4.19 ± 0.08. The chromosome counting results showed 2n = 40 for *I. crenata* accessions and confirmed the outliers as one triploid and two tetraploids. A high intra-specific genetic diversity in *Ilex crenata* was found, after genetic fingerprinting using genotyping-by-sequencing (GBS). The species *I. crenata* was separated into three clades, which coincided with intraspecific differences in genome sizes (mean ± sd) of 2.09 ± 0.006, 2.07 ± 0.05, and 2.19 ± 0.06 pg/2C per clade as mean values for the diploids. Applying a principal coordinate analysis (PCoA) to the genetic fingerprinting data of all species in the collection revealed a wide genetic variation, which has not yet been commercially exploited. These findings could form the basis for selectively breeding parents, in order to create more genetic diversity via intra- and interspecific crosses.

Keywords: holly; genotyping-by-sequencing; 2C-value; chromosome number; genome size

Citation: Geukens, E.; Haegeman, A.; Van Meulder, J.; Van Laere, K.; Smolders, E.; Ruttink, T.; Leus, L. Exploring Genetic Diversity in an *Ilex crenata* Breeding Germplasm. *Horticulturae* **2023**, *9*, 485. https://doi.org/10.3390/horticulturae9040485

Academic Editor: Carlo Nicoletto

Received: 13 March 2023
Revised: 7 April 2023
Accepted: 11 April 2023
Published: 13 April 2023

1. Introduction

The genus *Ilex* belongs to the monogeneric family Aquifoleaceae and is a large dioecious genus with 571 accepted species [1]. The genus can be found in regions of mesic growing conditions around the world. While most *Ilex* species originate from East Asia and South America, a multitude of species can also be found in Southeast Asia and North America. Only four species occur in Europe, one occurs in Africa, and another species occurs in Northern Australia. A few species can be found in Hawaii, the Caribbean, the Canary Islands, the Azores, Madeira, New Caledonia, and Fiji [2]. Previous phylogenetic studies have used fossil records [2–4] and nuclear and plastid markers to unravel the evolution of the *Ilex* genus, revealing a complex evolutionary history. Moreover, the taxonomy of the genus is not yet fully resolved. There are multiple heterotypic and homotypic synonyms within the genus, which occasionally cause confusion. For example, *I. maximowicziana* var. *kanehirae* (Yamamoto) Yamazaki and *I. crenata* var. *mutchagara* (Makino) Hara are two names given to the same plant [5]. The same applies to *I. leucoclada* (Maxim.) Makino and *I. integra* var. *leucoclada* Maxim. [1]. Recently, the increasing interest in the genus and decreasing cost of DNA sequencing have led to reference genomes for *I. latifolia* [6], *I. polyneura* [4], and *I. asprella* [7]. In some parts of the world, *Ilex* species such as *I. opaca*, *I. vomitoria*, and *I. paraguariensis* are used for brewing teas, while *I. asprella* is used as a medicinal plant [8]. *I. aquifolia* (common holly, English holly, European holly, or Christmas holly) is well known as an ornamental garden plant, with typical wintergreen branches and red berries used as a Christmas decoration. Other hollies used for their ornamental properties include *I. polyneura* and *I. crenata* (box-leaved holly or Japanese holly). *I. crenata* resembles boxwood (*Buxus* L.) and is also used for hedges and topiary [9].

I. crenata originates from Japan, Korea, China, the Sakhalin Islands, the Kuril Islands, Taiwan, the Philippines, and the Himalayas. In Japan, the species naturally grows from sea level up to an altitude of 1000 m, with local populations exhibiting wide morphological variation [9]. The species was introduced to the garden of the Russian Czar Alexander II in 1864 by Russian botanist Carl Maximowicz. Starting from that introduction, distribution spread throughout Europe. In 1898, *I. crenata* was introduced to the United States. This evergreen, small-leaved shrub gained popularity in the 1930s as an ornamental plant. Since then, *I. crenata* has been introduced multiple times to other parts of the world, and the plant has been extensively used for breeding [5], leading to the release of numerous new cultivars. These cultivars were mainly lucky findings, selected based on phenotypic variation and subsequently vegetatively propagated [9,10]. With more targeted breeding, value could be added, by overcoming abiotic and biotic stress challenges, such as high soil pH and infection with black root rot caused by the soil-dwelling fungus *Berkeleyomyces basicola*. Knowledge about the genetic relations and variation in existing species and cultivars could guide breeders selecting breeding parents, thus paving the way for novel or improved, and thus commercially marketable, varieties. Moreover, knowledge of the ploidy level and chromosome number of genotypes can help to predict cross-compatibility. The genome sizes of *I. cornuta*, *I. mucronata*, and *I. verticillata* have previously been reported in the range of 1.31 pg/2C [11] to 4.13 pg/2C [12]. The Chromosome Count Database describes 41 *Ilex* species ranging from 34 = 2n for *I. crenata* and *I. leucoclada* [13,14] to 120 = 2n for *I. pedunculosa* [15,16]. For both *I. crenata* and *I. leucoclada*, Galle reports the chromosome number to be 40 = 2n [9]. Darlington and Wylie mention a basic number of x = 9 or 10 [17].

Some efforts have been made to clarify the phylogeny of the different species within the genus, but knowledge about the underlying genetic variation within the species *I. crenata* is still limited.

The present study included an investigation into the genetic diversity within the species *I. crenata* and a selection of other hardy *Ilex* species. The genetic identity of the plants in the collection was also examined in light of their taxonomy. Naming mistakes can occur for many reasons, including ex situ conservation over the years, selling plants under an incorrect name for economic reasons, erroneous determinations, etc. We assembled a collection of 94 accessions, comprising 69 accessions labeled as *I. crenata* and 28 accessions from 17 other species. The genomic constitution was characterized at three levels: (1) genome sizes were determined using flow cytometry, (2) the chromosome number and ploidy level were investigated through fluorescence microscopy, and (3) genotyping-by-sequencing (GBS) [18] was performed as a reduced-representation genome sequencing technique, which allowed for high-resolution genetic fingerprinting using thousands of genetic markers spread across the genome.

2. Materials and Methods

2.1. Collection of Plant Material for the Ilex Collection

A germplasm collection was compiled of 94 *Ilex* accessions, belonging to 18 species and hybrids including *I.* x *attenuata*, *I. aquifolium*, *I.* x *aquipernyi*, *I. crenata*, *I. glabra*, *I. integra*, *I. leucoclada*, *I. macrocarpa*, *I.* x *makinoi*, *I. maximowicziana*, *I.* x *meserveae*, *I. mutchagara*, *I. opaca*, *I. rugosa*, *I. serrata*, *I. sugerokii*, *I. vomitoria*, and *I. yunnanensis*. The collection also contains a total of 69 accessions labeled as *I. crenata* (Table 1). Plant material from *I. crenata*

'Fastigiata' was sampled multiple times from independent sources. Thirteen accessions were purchased as potted plants from Esveld Plant Garden (The Netherlands), 78 genotypes were collected as cuttings from the arboreta of Bokrijk (Genk, Belgium) and Het Leen (Eeklo, Belgium), and 3 genotypes were obtained as potted plants from the company Plant Select. The plants were maintained as potted plants in a container field at ILVO (51°00′ N, 3°78′ E, Melle, Belgium) (Table 1).

Table 1. Overview of the *Ilex* collection. The table shows all the accessions and for every genotype the accession number used at ILVO; the species and cultivar name; the collection of origin, along with the accession number from the arboretum; the mean genome size; and the chromosome number. Genetically identical accessions are indicated with the same superscript letter (a–d), accessions for which a naming problem was discovered are indicated with an asterisk.

ILVO Accession Number	Species and Cultivar Name	Collection of Origin	Arboretum Accession Number	Mean Genome Size (pg/2C ± sd)	Chromosome Number
IL001	*Ilex* x *aquipernyi*	Esveld		1.80 ± 0.09	40
IL002	*Ilex crenata* 'Caroline Upright'	Esveld		2.11 ± 0.06	
IL003	*Ilex crenata* 'Convexa'	Esveld		2.13 ± 0.05	
IL004	*Ilex crenata* 'Dwarf Pagoda'	Esveld		2.30 ± 0.05	
IL005 [a]	*Ilex crenata* 'Fastigiata'	Esveld		2.16 ± 0.02	
IL006	*Ilex crenata* 'Glorie Dwarf'	Esveld		1.99 ± 0.07	
IL007	*Ilex crenata* 'Green Hedge'	Esveld		2.10 ± 0.06	
IL008	*Ilex crenata* 'Green Lustre'	Esveld		2.02 ± 0.14	
IL009	*Ilex crenata* 'Microphylla'	Esveld		2.16 ± 0.05	
IL010	*Ilex crenata* 'Stokes'	Esveld		2.10 ± 0.04	
IL011	*Ilex crenata* 'Twiggy'	Esveld		2.13 ± 0.08	
IL012	*Ilex glabra*	Esveld		1.52 ± 0.05	40
IL013 [b,*]	*Ilex maximowicziana* var. *kanehirae*	Esveld		2.14 ± 0.09	40
IL014 [c]	*Ilex crenata* 'Dark Green'	Plant Select		2.18 ± 0.05	
IL015 [b]	*Ilex crenata* 'Blondie'	Plant Select		2.30 ± 0.04	
IL016	*Ilex crenata* 'Braddock Heights'	Het Leen	00004670	1.98 ± 0.08	40
IL017	*Ilex crenata* 'Cherokee'	Het Leen	00004020	4.19 ± 0.08	80
IL018	*Ilex crenata* 'Convexa Gold'	Het Leen	00004672	2.05 ± 0.06	
IL019	*Ilex crenata* 'Ellipta Gold'	Het Leen	00004017	2.25 ± 0.04	
IL020	*Ilex crenata* 'Erecta'	Het Leen	00004021	2.12 ± 0.09	
IL021 [a]	*Ilex crenata* 'Fastigiata'	Het Leen	00001571	2.09 ± 0.04	40
IL022 [a]	*Ilex crenata* 'Fastigiata'	Het Leen	00001904	2.05 ± 0.04	
IL023	*Ilex crenata* 'Golden Gem'	Het Leen	00001905	2.08 ± 0.08	40
IL024 [d]	*Ilex crenata* 'Golden Helleri'	Het Leen	00004014	2.09 ± 0.05	
IL025	*Ilex crenata* 'Green Dragon'	Het Leen	00004013	2.13 ± 0.07	
IL026	*Ilex crenata* 'Green Island'	Het Leen	00004090	2.12 ± 0.06	
IL027	*Ilex crenata* 'Hetzii'	Het Leen	00004022	1.99 ± 0.05	40
IL028	*Ilex crenata* 'Ivory Tower'	Het Leen	00004673	2.08 ± 0.05	
IL029 [d]	*Ilex crenata* 'Lancaster Yellow'	Het Leen	00004024	2.03 ± 0.07	
IL030	*Ilex crenata* 'Luteovariegata'	Het Leen	00005561	2.09 ± 0.05	
IL031	*Ilex crenata* 'Mariesii'	Het Leen	00004010	2.22 ± 0.05	
IL032	*Ilex crenata* 'Nakada'	Het Leen	00004016	2.23 ± 0.03	
IL033	*Ilex crenata* 'Picas'	Het Leen	00004091	2.09 ± 0.06	
IL034	*Ilex crenata* 'Piccolo'	Het Leen	00004004	2.13 ± 0.06	
IL035	*Ilex crenata* 'Pride's Tiny'	Het Leen	00004023	1.98 ± 0.06	40
IL036	*Ilex crenata* 'Red Lion'	Het Leen	00004018	2.22 ± 0.03	
IL037	*Ilex crenata* 'Robert Culpepper'	Het Leen	00004011	2.04 ± 0.03	
IL038	*Ilex crenata* 'Rotundifolia'	Het Leen	00004671	2.03 ± 0.05	
IL039	*Ilex crenata* 'Snowflakes'	Het Leen	00004669	2.25 ± 0.06	
IL040	*Ilex crenata* 'Tee Dee'	Het Leen	00004025	2.26 ± 0.05	
IL042	*Ilex crenata* f. watanabeata	Het Leen	00005253	2.04 ± 0.04	
IL043	*Ilex crenata* 'Wiesmoor Silber'	Het Leen	00004012	2.09 ± 0.07	
IL044	*Ilex crenata* 'Yunnanensis'	Het Leen	00004019	2.18 ± 0.09	
IL045	*Ilex* x *aquipernyi*	Het Leen	00005002	1.78 ± 0.04	
IL046	*Ilex* x *meserveae* 'Blue Girl'	Het Leen	00001829	1.93 ± 0.02	40
IL047 [b,*]	*Ilex mutchagara* Makinoi	Het Leen	00001901	2.17 ± 0.06	
IL048 [b,*]	*Ilex yunnanensis*	Het Leen	00004449	2.20 ± 0.03	
IL049	*Ilex aquifolium* 'Angustifolia'	Het Leen	00001859	1.85 ± 0.04	
IL050	*Ilex aquifolium* 'Crispa'	Het Leen	00001873	1.90 ± 0.07	
IL051 [c]	*Ilex* x *makinoi* Hara	Het Leen	00004998	2.00 ± 0.06	40
IL052	*Ilex* x *attenuata* 'Foster'	Het Leen	00004994	1.89 ± 0.05	
IL053	*Ilex opaca* 'Longwood Gardens'	Het Leen	00004993		
IL055	*Ilex crenata* 'Black Beauty'	Bokrijk	19991318	3.06 ± 0.04	60
IL056	*Ilex crenata* 'Border Gem'	Bokrijk	19991317	2.07 ± 0.09	
IL059	*Ilex crenata* 'Cape Fear'	Bokrijk	19991320	2.09 ± 0.05	
IL060	*Ilex crenata* 'Chesapeake'	Bokrijk	19900267	4.04 ± 0.09	80
IL061	*Ilex crenata* 'Cole's Hardy'	Bokrijk	1983ACQI175	2.14 ± 0.08	
IL062 [a]	*Ilex crenata* 'Fastigiata'	Bokrijk	19900529	2.16 ± 0.08	
IL063	*Ilex crenata* 'Goldstaub'	Bokrijk	19940813	2.16 ± 0.08	
IL065	*Ilex crenata* 'Hatfieldii'	Bokrijk	19950491	2.06 ± 0.10	

Table 1. *Cont.*

ILVO Accession Number	Species and Cultivar Name	Collection of Origin	Arboretum Accession Number	Mean Genome Size (pg/2C ± sd)	Chromosome Number
IL066	*Ilex crenata* 'Helleri'	Bokrijk	19940347	2.25 ± 0.05	
IL067	*Ilex crenata* 'Highlander'	Bokrijk	19950296	2.12 ± 0.07	
IL069	*Ilex crenata* 'Horizontalis'	Bokrijk	19910107	2.18 ± 0.08	
IL070	*Ilex crenata* 'Ivory Hall'	Bokrijk	19991306	2.09 ± 0.08	40
IL071	*Ilex crenata* 'Kingsville Dwarf'	Bokrijk	19940329	2.29 ± 0.08	
IL072	*Ilex crenata* 'Midas Touch'	Bokrijk	19900222	2.13 ± 0.07	
IL073	*Ilex crenata* 'Monmouth'	Bokrijk	19950499	2.18 ± 0.06	
IL075	*Ilex crenata* 'Oleafera'	Bokrijk	19980688	2.04 ± 0.06	
IL076	*Ilex crenata* var. paludosa	Bokrijk	19991354	2.05 ± 0.10	40
IL077	*Ilex crenata* 'Piedmont Pyramidal'	Bokrijk	19900271	2.16 ± 0.03	
IL078	*Ilex crenata* 'Schwoebel's Compact'	Bokrijk	19950345	2.15 ± 0.03	
IL079	*Ilex crenata* 'Sentinel'	Bokrijk	1988ACQI162	2.01 ± 0.09	
IL082 [a]	*Ilex crenata* 'Sky Pencil'	Bokrijk	20040562	2.15 ± 0.07	
IL084	*Ilex crenata* 'Tennyson'	Bokrijk	19900395	2.28 ± 0.06	
IL086	*Ilex crenata* 'Valeria Rankin'	Bokrijk	19900274	2.08 ± 0.11	
IL087	*Ilex crenata* 'Variegata'	Bokrijk	19980670	2.22 ± 0.04	
IL088	*Ilex crenata* 'Fructu Lutea'	Bokrijk	19890583	2.01 ± 0.07	
IL094	*Ilex vomitoria* 'Folsom Weeping'	Bokrijk	20200779	1.57 ± 0.10	
IL095	*Ilex vomitoria* 'Hoskins shadow'	Bokrijk	20110310	1.58 ± 0.06	
IL096	*Ilex sugerokii*	Bokrijk	19980060	1.78 ± 0.07	
IL097	*Ilex sugerokii* var. *longipedunculata*	Bokrijk	19900505	1.75 ± 0.02	
IL098	*Ilex rugosa*	Bokrijk	20070853	1.97 ± 0.04	40
IL099	*Ilex leucoclada*	Bokrijk	20020979	2.00 ± 0.05	40
IL100	*Ilex opaca*	Bokrijk	19940225	1.94 ± 0.02	
IL101	*Ilex serrata* 'Rakusogu'	Bokrijk	0000ACQI10	1.86 ± 0.07	
IL102	*Ilex integra*	Bokrijk	20120839	1.81 ± 0.04	
IL103	*Ilex macrocarpa*	Bokrijk	19960445		
IL104	*Ilex yunnanensis* var. *gentilis*	Bokrijk	00004515	8.01 ± 0.18	160
IL105	*Ilex crenata* 'Jeran' (Bokrijk select)	Bokrijk	20140097	2.15 ± 0.07	
IL106	*Ilex crenata* 'Convexa Variegata'	Bokrijk	19900459	2.17 ± 0.05	
IL107	*Ilex crenata* 'Nummularia'	Bokrijk	19991312	2.23 ± 0.08	40
IL108	*Ilex vomitoria* 'Will Fleming'	Bokrijk	20080892	1.50 ± 0.03	
IL121	*Ilex leucoclada* seedling	Bokrijk			
IL122	*Ilex crenata* 'Luxus'	Plant Select		2.15 ± 0.08	

2.2. Genome Size of the Collected Ilex Accessions

The genome size of 91 accessions in the collection was measured using flow cytometry (Table 1), according to the methods described by Van Oost et al. 2021 [19]. Four accessions did not develop rooted cuttings and could not be used. The genome size is expressed as pg/2C. Terminology used is according to Greilhuber et al. [20]. All samples were analyzed using a Quantum P flow cytometer and CyPAD software (Quantum Analysis, Münster, Germany). For all genotypes, except IL104, the internal standard used was *Zea mays*, with a known genome size of 5.43 pg/2C [21]. *Pisum sativum* (9.09 pg/2C [22]) was used as an internal standard for IL104, because of its deviant genome size. Young leaf material of the sample and standard was mixed, co-chopped, and used for sample preparation with a CyStain PI kit (Sysmex, Münster, Germany), according to the manufacturer's protocol with minor modifications: using 0.4 mL extraction buffer, and 1.2 mL staining buffer, to which 1% (*w/w*) PVP10 was added. The stained samples were incubated in the dark at 4 °C for at least 20 min. At least three replicates were analyzed for every genotype, preferably from three plants of this genotype and on three different days. The sample genome size was derived by calculating the ratio of the peak positions of the sample and internal standard in the histograms of both the FL2 and FL3 detectors. Mean values and standard deviations of the sample genome size were calculated as the mean of a minimum of six values, specifically the peak position ratio of both detectors for at least three replicates.

2.3. Chromosome Number of the Collected Ilex Accessions

Based on the flow cytometry results, a subset of 17 *I. crenata* genotypes (Table 1), covering the full range of genome sizes, was selected to count the chromosomes. The chromosomes of the following species were also counted: *I.* x *aquipernyi* (IL001), *I. glabra* (IL012), *I. maximowicziana* var. *kanehirae* (IL013), *I.* x *meserveae* (IL046), *I.* x *makinoi* (IL051), *I. rugosa* (IL098), and *I. leucoclada* (IL099) (Table 1). Young root tips of newly rooted cuttings were harvested and incubated in 0.1% colchicine for 3 h at room temperature, to arrest mitosis at the metaphase stage. Subsequently, the root tips were fixed in 3:1 ethanol:acetic acid for 1 h at room temperature. Cell suspensions were made by digesting the root tips in 0.6% enzyme solution (0.6% cellulase, 0.6% pectolyase, and 0.6% cytohelicase in 0.1 M citrate buffer (0.1 M sodium citrate tribasic dehydrate, and 0.1 M citric acid)) at 37 °C for 45 min. Chromosome slides were prepared using the SteamDrop protocol [23], using 1:1 ethanol:acetic acid solution for both fixation steps. Subsequently, the chromosome slides were stained with 1% DAPI (100 g/mL) diluted in Vectashield mounting medium and analyzed with a fluorescence microscope (AxioImager M2, Carl Zeiss MicroImaging, Belgium) at 1000× magnification. Images were captured using an Axiocam MRm camera and ZEN software (Carl Zeiss MicroImaging, Belgium). Chromosomes were counted on at least five images of well-spread metaphase chromosomes per genotype using DRAWID v0.26 [24].

2.4. GBS Fingerprinting of the ILEX Collection

2.4.1. Library Preparation and Sequencing

DNA was extracted from all genotypes of the collection using a modified CTAB protocol [25]. Preparation of genotyping-by-sequencing (GBS) libraries and sequencing was performed by LGC Genomics (Berlin, Germany), using the following steps: (1) quality control (agarose gel check); (2) library preparation, including indexing and quality control using a double digest with *Pst*I and *Mse*I; and (3) PE-150 sequencing using an Illumina NextSeq 500/550 instrument.

2.4.2. GBS Data Analysis

All data analysis steps described below are also available, together with accompanying scripts, on https://gitlab.com/ilvo/GBS_Ilex (accessed on 25 May 2021) and on Zenodo with doi 10.5281/zenodo.7669149 under an MIT license.

Data preprocessing. The obtained GBS reads were processed using GBprocesS v3.0.3 [26]. This software makes use of Cutadapt [27] and PEAR [28], and comprises the following steps: (1) trimming of adapters, barcodes, restriction site remnants from 5′- and 3′-ends, and discarding reads that are shorter than 30 bp using Cutadapt; (2) merging of forward and reverse reads with a minimum overlap of 10 bp and a minimal final size of 40 bp with PEAR [28]; (3) filtering of reads that have a mean Phred quality score lower than 25 or two consecutive bases with a Phred quality score below 20 and discarding reads containing Ns; and (4) filtering of reads with intact internal restriction enzyme recognition sites (due to partial restriction digest). The raw PE-150 read data, as well as the merged preprocessed read data, were submitted to NCBI's Sequence Read Archive (SRA) under the BioProject number PRJNA895194.

Loci identification. Data were analyzed reference-free, using GIbPSs software [29]. Unless mentioned otherwise, standard settings were used and the recommended workflow of GIbPSs was followed. More detailed information can be found in the accompanying software repository, as mentioned above. Briefly, loci were identified in all merged reads using *indloc*, *poploc*, and *indpoploc*, and errors were corrected using *indloc*. Loci of lengths between 32 and 300 bp were selected using *data selector*. Loci containing indels were detected using *indelchecker* and discarded. Loci that were extremely deeply sequenced were identified using *depth analyzer* and discarded. The minimum value for the median depth percentage and the maximum value for the median scaled depth were set to 1 and 0.15, respectively. Additionally, split loci were removed from the dataset. Finally, a GIbPSs genotype call table was exported, containing the locus genetic information for all samples.

Determination of common loci and allele similarity. All samples were compared to each other using a custom python script available in the script repository [30], which uses the GIbPSs genotype call table as input. This script calculates (1) the number of loci per sample; (2) the number of common loci between every pair of samples and; (3) the allele similarity of the common loci of every pair of samples. The allele similarity is defined as the percentage of common loci that share at least one allele. An allele similarity of 100% means that all common loci have at least one shared allele (=a locus sequence with a combination of SNPs) (see Supplementary Materials). The results are represented in a false-colored similarity matrix, displaying the total number of loci for every genotype on the diagonal, the number of loci in common between two genotypes in the upper half of the matrix, and the allele similarity in the lower half of the matrix.

Concatenated alignment and phylogeny. Using a custom perl script available in the script repository, all SNP-containing loci from the GIbPSs genotype call table were concatenated and saved as an alignment file. In addition, a text file that kept track of the original locus (and positions in that locus) in the concatenated alignment was saved. Next, again using a custom perl script, positions where more than 20% of the samples contained missing data were filtered out. Invariant positions were removed using the 'ascbias.py' script (available at https://github.com/btmartin721/raxml_ascbias) (accessed on 25 May 2021). The resulting filtered concatenated alignment was used for (1) principal coordinate analysis (PCoA); and (2) phylogenetic tree construction. The PCoA was performed by reading the multiple alignments in R v4.2.2 in R Studio Server v2022.07.2 build576, using the Biostrings package. Next, the pairwise distance between all individuals was calculated using the Hamming distance (Decipher package), while ignoring gap–letter matches and performing the PCoA using the 'cmdscale' function (stats package). The R scripts are available in the script repository. This analysis was performed for the entire collection of 94 individuals and two technical replicates, as well as for the 69 individuals in the *I. crenata* group only and one technical replicate. A phylogenetic tree was constructed containing the collection of all 94 individuals using the maximum likelihood method with RAXML v8.2.12, including a GTRCAT model without rate heterogeneity (-V), with ascertainment bias correction (–asc-corr=lewis), and including 100 bootstraps. *I. vomitoria* 'Hoskins Shadow' (IL095) was used as an outgroup. The tree was then visualized using iTOL v6 [31].

3. Results and Discussion

This study explored the genetic diversity within the genus of *Ilex*, more specifically within the species *I. crenata*. We assembled a large collection of 94 *Ilex* accessions from arboreta and commercial growers, 69 of which were labeled *I. crenata,* as well as three others with other names that were genetically identical to *I. crenata*. For the accessions in this collection, we delineated the genetic identity, investigated the genetic relationships, and studied the genetic diversity between the accessions of the collection.

3.1. Genetic Identity

GBS was performed on all 94 accessions of the *Ilex* collection. The number of preprocessed merged reads per sample (see Materials and Methods above) ranged from 1.8 M to 8.4 M, with a median of 3.6 M. Reference-free locus delineation was performed using GIbPSs and resulted in a total of 218,024 loci for all samples combined. Rarefaction analysis of a few selected samples revealed that the number of identified loci using GIbPSs reached a plateau at ca. 3–4 M reads, suggesting that most of the samples had been sequenced deep enough to reach locus saturation (data not shown). The number of identified loci per sample ranged from 9513 to 53,403, with a mean of 18,079. These loci had a mean length of 184 bp (range 40 to 270 bp), a mean read depth of 57, and a median read depth of 20. In total, 70% of the loci had a mean read depth per sample of at least 15. The GBS loci were compared between samples in a pairwise manner to estimate the number of common loci and the fraction of loci with shared alleles (for definitions of 'common loci' and 'allele similarity', see Materials and Methods and [30]). The heat map (Figure S1) shows clear clusters, indicating the genetic substructure in the *Ilex* collection. The lowest number of common loci between species was 2880 for *I. crenata* 'Luxus' (IL122) and *I. macrocarpa* (IL103); the highest number was 11,249 for *I. crenata* 'Pride's Tiny' (IL035) and *I. sugerokii* var. *longipedunculata* (IL097). Within the species *I. crenata*, the lowest number of common loci between two accessions was 6340 for *I. crenata* 'Luxus' (IL122) and *I. crenata* 'Nummularia' (IL107); the highest was 15,219 for *I. crenata* 'Braddock Heights' (IL016) and *I. crenata* 'Pride's Tiny'

(IL035). The lowest allele similarity between species was 8.85% between *I. macrocaropa* (IL103) and *I. vomitoria* 'Will Fleming' (IL108); the highest allele similarity between species was 60.25% for *I. opaca* (IL100) and *I. integra* (IL102).

Next, we analyzed the allele similarity of replicates, to set a threshold for identifying genetically identical accessions across the *Ilex* collection. For two accessions (IL014 and IL051), we used two technical replicates to analyze the reproducibility of GBS fingerprinting. The total number of loci of *I. crenata* 'Dark Green' (IL014) replicates was similar (11,312 and 11,945), with 10,335 loci in common. Likewise, technical replicates of *I.* x *makinoi* 'Hara' (IL051) had 16,716 and 19,756 loci, with 15,195 loci in common. The allele similarity between these technical replicates was 99.88% and 99.77% for IL014 and IL051, respectively. Furthermore, we collected *I. crenata* 'Fastigiata', a common cultivar, from four sources as 'biological' replicates (IL005, IL021, IL022 and, IL062). As expected, this group of Fastigiata accessions displayed high allele similarities, ranging from 98.13% to 99.68%. Taken together these results showed that the allele similarity method for pairwise comparison based on GBS genetic fingerprints was especially accurate. The number of common loci was less reliable for identifying similar genotypes, which was probably due to datasets that were not completely saturated. Consequently, we used a 98% threshold of minimal allele similarity to consider GBS genetic fingerprints as identical (=the same cultivar) and screened the *Ilex* collection for genetically identical accessions. For instance, *I. crenata* 'Sky Pencil' (IL082), morphologically identical to the Fastigiata genotypes with an upright phenotype can also be considered genetically identical to the Fastigiata genotypes (allele similarity ranging from 98.19% with IL022 to 99.50% with IL005). Furthermore, the allele similarity of *I. crenata* 'Golden Helleri' (IL024) and *I. crenata* 'Lancaster Yellow' (IL029) was 99.89%, thus surpassing this threshold. Dudley and Eisenbeiss (1992) previously reported that a mutation of 'Helleri' was commercialized under the name 'Lancaster Yellow' [5], thus confirming that these accession names are genetically identical. Comparison of the allele similarities of other genotypes revealed that *I. yunnanensis* (IL048) is highly related to *I. mutchagara* Makino (IL047), *I. crenata* 'Blondie' (IL015), and *I. maximowicziana* var. *kanehirae* (IL013), with allele similarity values ranging from 98.83% to 99.76% (Figure S1). There are two reasons for these high allele similarities. The first is that *I. yunnanensis* (IL048) is actually an *I. crenata* genotype. Visual assessment of the original plants in the arboretum revealed that the mother plant of *I. yunnanensis* (IL048) looked like an *I. crenata* and not like the other *I. yunnanensis* in the collection, leading to the conclusion that the mother plant, and subsequently the *I. yunnanensis* (IL048) accession in our collection, was initially incorrectly labeled as *I. yunnanensis*. The other reason for the observed high allele similarities is that *I. maximowicziana* var. *kanehirae* (Yamam.) T. Yamaz. and *I. mutchagara* Makino are both synonyms of one another and are also synonyms of *I. crenata* var. *mutchagara* (Makino) Ohwi, as previously reported by Dudley and Eisenbeiss (1992) [5]. More recently, many more heterotypic synonyms have been added to the list of synonyms, including *I. crenata* var. *kanehirae* Yamam., *I. crenata* var. *scoriatum* Yamam., and *I. maximowicziana* Loes. [1]. Our data show that both *I. maximowicziana* var. *kanehirae* (IL013) and *I. mutchagara* Makino (IL047) are not more genetically different from *I. crenata* than the different *I. crenata* genotypes are from one another (more than 98% allele similarity) (Figure S1). Moreover, they are genetically almost identical to the *I. crenata* cultivar 'Blondie', with allele similarities only slightly lower than that of the technical replicates. In conclusion, our data confirm that *I. maximowicziana* var. *kanehirae* (Yamam.) T. Yamaz. and *I. mutchagara* Makino are synonyms and further show that they are also synonyms of *I. crenata* Thunb.

3.2. Genetic Relationships among Ilex Species

Pairwise comparison of allele similarities across the collection showed blocks of samples that are more closely related to each other. These blocks were consistent with the clades of the phylogenetic tree (Figure 1) and the clusters of the PCoA (Figure 2). Four clusters were distinguished using PCoA analysis of the GBS data of the *Ilex* collection. Cluster 1, containing *I. crenata* accessions, was separated on the PC1 axis. Cluster 2, containing

I. vomitoria accessions, was separated on the PC2 axis. Cluster 3, containing *I. opaca* accessions, was separated from Cluster 4, containing multiple other species. The phylogenetic tree generally agreed with this genetic structure. The species *I. vomitoria* was used as an outgroup, as it became clear during preliminary analyses that this species was phylogenetically less related to the other species than the other species were related to one another. Besides *I. vomitoria*, *I. opaca* was also less related to the other species than the remaining species were related to each other. The phylogenetic analysis generally confirmed the species boundaries, as the different species were clearly grouped in separate clusters in the phylogenetic tree, with high bootstrap support (100% for species).

In previously published phylogenetic trees based on a limited number of nuclear markers (Cuénoud et al. [2] and Manen et al. [3]), *I. sugerokii* and *I. yunnanensis* were closely related. However, in phylogenetic trees based on plastid markers, *I. sugerokii* is more closely related to *I. crenata* [2,3]. In our study, *I. sugerokii* and *I. yunnanensis* are more closely related. In the studies of Cuénoud et al. [2] and Manen et al. [3], *I. opaca* and *I.* x *attenuata* are placed in the same group in the phylogenetic trees resulting from both nuclear and plastid markers. In addition, *I. rugosa, I. aquifolium, I. leucoclada, I.* x *makinoi*, and *I. integra* were placed in the same group. Those results are consistent with our findings. Furthermore, our phylogenetic tree (Figure 1) was consistent with the traditional classifications of Galle [9], but also revealed some deviations. First, similar to Galle's classification, *I. integra, I. leucoclada*, and *I. aquifolium* were also closely related in our data. According to Galle, they belong together in subgenus Aquifolium, section Aquifolium, series Aquifoliodes. Second, *I.* x *makinoi, I.* x *meserveae*, and *I. rugosa* were closely related in our study. Galle [9] places them under subgenus Aquifolium, section Aquifolium, series Hookerianae. Third, in our study, *I. vomitoria* was not closely related to any other species in our collection (the highest allele similarity is 17.45% with *I.* x *meserveae*). This species is also classified separately in Galle's classification, under subgenus Aquifolium, series Vomitoriae, along with one other species, *I. fuertisiana* (not in our collection). Our GBS data also revealed some discrepancies with previous classifications. Galle [9] places *I.* x *attenuata, I. glabra, I. opaca, I. sugerokii*, and *I. yunnanensis* together in the subgenus Prinos, section Paltoria, series Cassinoides. However, in the phylogenetic tree shown above, *I.* x *attenuata* and *I. opaca* were placed separately from the others. Furthermore, *I. serrata* and *I. macrocarpa*, both belonging to different sections, were placed close to *I. glabra, I. yunnanensis*, and *I. sugerokii*.

The genome size of *I. aquifolium* (included in our collection) is previously reported to be 2.30 pg/2C [32] as measured by Feulgen densitometry, an older and less accurate technique. The results for *I. aquifolium* in our study were lower, with 1.85 ± 0.04 pg/2C for the cultivar 'Angustifolia' and 1.90 ± 0.07 for the cultivar 'Crispa'. The genome sizes of other accessions in our collection have not been previously reported. We determined the chromosome numbers of several species in our collection: *I. aquifolium, I.* x *attenuata, I. crenata, I. glabra, I. integra, I. opaca, I. vomitoria, I. rugosa, I. serrata*, and *I. leucoclada* [9,16]. Fluorescence microscopy images of stained nuclei are shown in Figure 3. The genome sizes and chromosome numbers for all analyzed accessions in the collection are presented in Table 1. For most species, our data confirmed the results of previous studies. All diploid species studied in this research had 2n = 40 chromosomes. In addition, we observed 2n = 40 for *I. crenata* and *I. leucoclada* (confirmed by Galle, 1997 for *I. crenata* [9]), which does not agree with Sugiura [13] and Fedorov [14], who reported a chromosome number of 2n = 34 for both species (found in the Chromosome Count Database (CCDB)) [16]. The largest reported chromosome number in the Chromosome Count Database is that of *I. pedunculosa* Miq. (2n = 120). In our collection, a larger chromosome number of 2n = 160 was observed for *I. yunnanensis* var. *gentilis* (IL104). This species also had the largest genome size in our collection, at 8.01 ± 0.18 pg/2C.

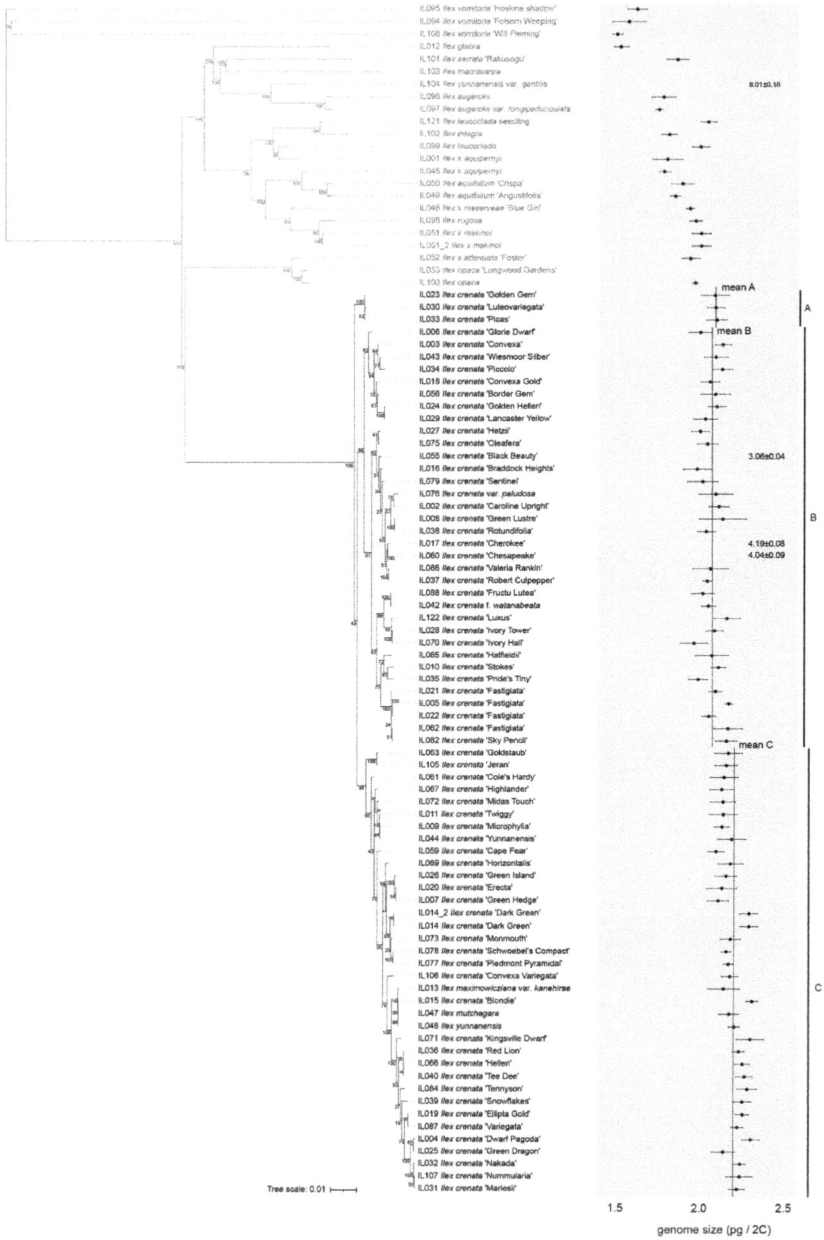

Figure 1. Phylogenetic tree of the *Ilex* collection along with the genome size (pg/2C-value). The clades containing non-*Ilex crenata* spp. are depicted in grey and the clades containing *I. crenata* accessions in black. The three clades within *I. crenata* are indicated on the right (Clade A, Clade B, and Clade C). For every genotype a dot represents the mean genome size (n = 6) and the horizontal bar the standard deviation, as measured using flow cytometry. Vertical lines labeled mean A, mean B, and mean C indicate the mean genome sizes per clade (polyploids not included) of 2.09 ± 0.006; 2.07 ± 0.05, and 2.19 ± 0.06 pg/2C ± sd for clade A, B, and C, respectively. The genome size of genotypes with a genome size exceeding the scale are depicted on the right.

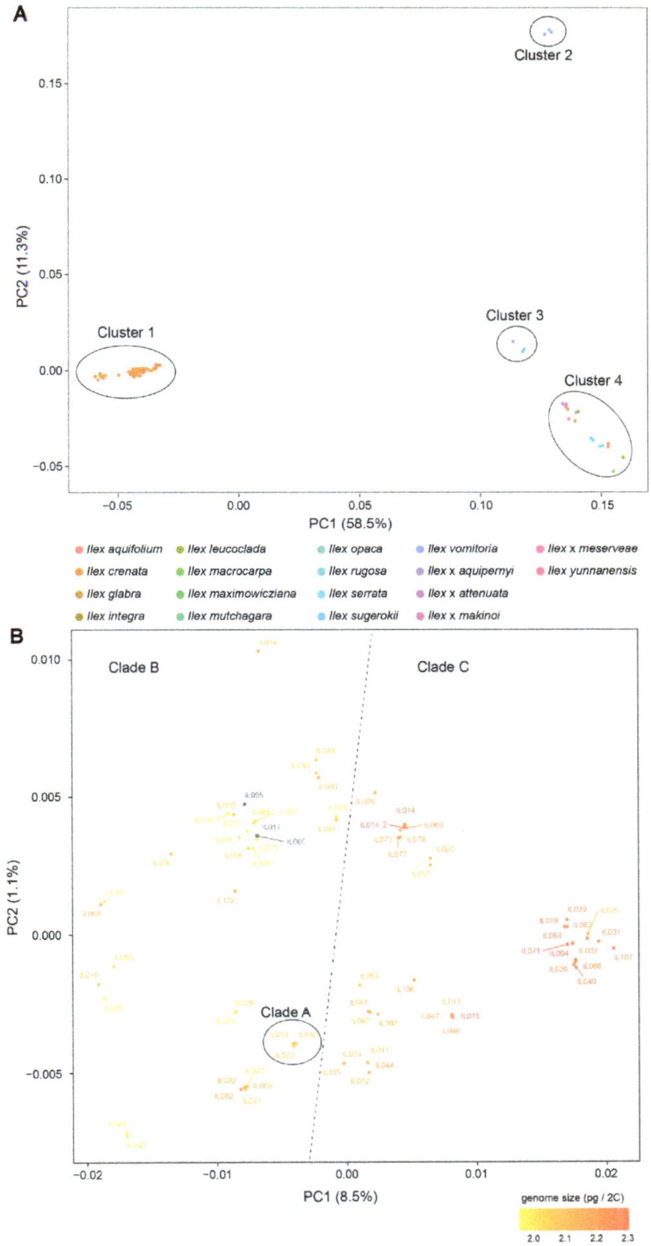

Figure 2. PCoA of the 94 accessions of the *Ilex* collection (**A**) and the subset of 72 *I. crenata* accessions (**B**). (**A**) Samples are colored by species and grouped in four clusters. (**B**) Samples are false-colored according to their genome size (range 1.98–2.30 pg/2C), while the polyploids are pictured in gray. Accessions of Clade A are grouped using a circle. The dotted line separates accessions of Clade B and Clade C.

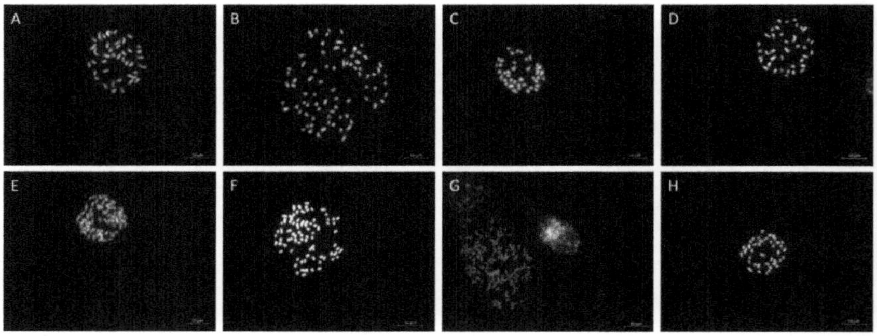

Figure 3. Fluorescence microscopy images of stained nuclei reveal chromosome numbers of *Ilex* spp. (**A**) *I. crenata* 'Braddock Heights', 2n = 40; (**B**) *I. crenata* 'Cherokee', 2n = 80; (**C**) *I. crenata* 'Fastigiata', 2n = 40; (**D**) *I.* x *makinoi* 'Hara', 2n = 40; (**E**) *I. crenata* 'Black Beauty', 2n = 60; (**F**) *I. crenata* 'Chesapeake', 2n = 80; (**G**) *I. yunnanensis* var. *gentilis*, 2n = 160; and (**H**) *I. leucoclada*, 2n = 40.

3.3. Genetic Relationships of Ilex Crenata Cultivars

When studying the phylogenetic tree (Figure 1), combined with the pairwise allele similarities of the different *I. crenata* accessions (Supplementary Materials), a segregation into three groups emerged within the species of *I. crenata,* which was also observed based on the number of common loci and based on the allele similarity. Clade A consists of *I. crenata* 'Golden Gem' (IL023), *I. crenata* 'Luteovariegata' (IL030), and *I. crenata* 'Picas' (IL033) (bootstrap value of 100%). Clade B contains 34 *I. crenata* genotypes (bootstrap value of 95%), and Clade C contains 36 *I. crenata* genotypes. Clade B has three highly supported subgroups (bootstrap values of 83%, 93% and 99%), while Clade C has some highly supported subgroups but also some genotypes (such as *I. crenata* 'Monmouth' (IL073) and *I. crenata* 'Tee Dee' (IL040)) with less clear phylogenetic positions. The PCoA of *I. crenata* (Figure 2B) roughly confirmed this structure, with clear separation of clusters A and B from C. Contrary to clusters B and C, there was only very little variation within Clade A. In Figure 1, the color gradient shows the distribution of the genome size of *I. crenata*. For all analyzed samples, the mean genome size and standard deviation are shown in Table 1. The genome size of the *I. crenata* genotypes varied from 1.98 pg/2C to 2.30 pg/2C (all diploids 2n = 2x, data shown in Table 1), with three outliers: *I. crenata* 'Black Beauty' (3.06 ± 0.04 pg/2C, triploid: 2n = 3x = 60), *I. crenata* 'Chesapeake' (4.04 ± 0.09 pg/2C, tetraploid: 2n = 4x = 80), and *I. crenata* 'Cherokee' (4.19 ± 0.08 pg/2C tetraploid: 2n = 4x = 80). The observed genome size variation in *I. crenata* (1.16-fold difference) was larger than generally expected within a species. Interestingly, the genome sizes were more concise when calculated in accordance with the observed genetic clade structure. The phylogenetic clades had a significant difference in genome size, as depicted in the right panel of Figure 1. Clade A had a mean genome size of 2.09 pg/2C ± 0.006, Clade B had a mean genome size of 2.07 pg/2C ± 0.05, and Clade C contains the *I. crenata* samples with the largest genomes (mean of 2.19 pg/2C ± 0.06). All three clades showed little variation within them and have low standard deviations, which is the normal variation expected within a species. There was, however, a significant difference in the mean genome size of Clade B and Clade C (p-value = 7.44×10^{-13}). As Clade A only comprises three genotypes, statistical differences were not calculated for this clade.

Intraspecific genome size variation should be interpreted with caution, as illustrated by Noirot et al. (2000) [33] and Greilhuber (2005) [34]. Noirot et al. showed that the use of an internal standard can reduce, but not eliminate, measurement errors [35]. As *Ilex* is a dioecious plant and sex chromosomes can differ in size, a possible link between genome size and sex was investigated but not found, indicating that the observed variations in genome size and grouping in clades were not due to the sex chromosomes. A more plausible explanation is that the genome size differences are linked to the evolutionary or breeding history of the plants. However, information about the geographic origin, breeding parents, and year of discovery of the different cultivars in this study is scarce and did not lead to unraveling the origin of the observed clades. In the species *I. polyneura*, the emergence of two groups and some in between genotypes was described by Yao et al. using RADseq population genetic analyses [4]. A link between genome size and evolution has been proposed in maize [36,37]. Bilinsk et al. observed small differences in genome size in maize and described a negative correlation between genome size and cell production rate, as well as a negative correlation between cell production rate and flowering time. Their research suggested that a mechanistic relationship between genome size, cell production, and developmental rate may lead to differences in optimal flowering times across altitudes, thus affecting genome size [36].

3.4. Genetic Diversity via Crossing and Hybridization

In addition to identifying very closely related genotypes, GBS-derived calculations of pairwise common loci and allele similarities also make it possible to identify hybrids, while only considering a phylogenetic tree often does not [30]. In this study, we were able to identify IL051 *I. x makinoi* as the result of a spontaneous hybridization event between *I. leucoclada* (IL099) and *I. rugosa* (IL098), and *I. x meserveae* 'Blue Girl' (IL046) as the result of a manual cross between *I. rugosa* (IL098) and *I. aquifolium* (IL050) (Figure 4). Furthermore, the table with common loci and allele similarities in the supplemental materials (Figure S1) shows the known hybrid origin of *I. x aquipernyi* (IL045 and IL001) as being the result of a manual cross of *I. aquifolium* and *I. pernyi* (not in dataset) as parents for *I. x aquipernyi* (IL045). This accession has many loci in common with *I. aquifolium* (IL049 and IL050). For *I. x aquipernyi* (IL001), however, the number of common loci with *I. aquifolium* (IL049 and IL050) is not as high as for *I. x aquipernyi* (IL045). It is possible that *I. x aquipernyi* (IL001) has a higher similarity to its other parent, *I. pernyi* (not included in the data).

	IL050 *Ilex aquifolium* 'Crispa'	IL046 *Ilex x meserveae* 'Blue Girl'	IL098 *Ilex rugosa*	IL051 *Ilex x makinoi*	IL099 *Ilex leucoclada*
IL050	100	85.34	34.08	36.30	35.22
IL046	85.34	100	82.79	73.35	41.57
IL098	34.08	82.79	100	82.86	41.59
IL051	36.30	73.35	82.86	100	67.90
IL099	35.22	41.57	41.59	67.90	100

Figure 4. Parental relations of hybrids *I.* x *meserveae* (IL046) (*I. aquifolium* x *I. rugosa*) and *I.* x *makinoi* (IL051) (*I. leucoclada* x *I. rugosa*). The high allele similarity of GBS fingerprints reflects the relationship of *I.* x *meserveae* (IL046) to *I. aquifolium* 'Crispa' (IL050) (85.34%) and *I. rugosa* (IL098) (82.79%), whereas *I. aquifolium* 'Crispa' and *I. rugosa* have an allele similarity of only 34.08%. Similarly, the relation of *I.* x *makinoi* (IL051) to *I. leucoclada* (IL099) and *I. rugosa* (IL098) can be derived from their high allele similarities of 82.86% and 67.90%, respectively, whereas *I. leucoclada* and *I. rugosa* have a much lower allele similarity of 41.59%.

I. crenata 'Pride's Tiny' (IL035) appears to have been derived from a spontaneous or manual hybridization event involving *I. sugerokii* var. *longipedunculata* (IL097) in its ancestral lineage. The allele similarities within *I. crenata* reflect the clades of the phylogenetic tree. Note the position of *I. crenata* 'Dark Green' (IL014), *I. crenata* 'Monmouth' (IL073), *I. crenata* 'Horizontalis' (IL069), *I. crenata* 'Schwoebel's Compact' (IL078), and *I. crenata* 'Piedmont

Pyramidal' (IL077). They cluster with genotypes in Clade C of the phylogenetic tree, but they also had a high allele similarity with *I. crenata* 'Convexa' (IL003), *I. crenata* 'Convexa Gold' (IL018), and *I. crenata* 'Wiesmoor Silber' (IL043). The same pattern is observed for *I. crenata* 'Twiggy' (IL011), *I. crenata* 'Midas Touch' (IL072), *I. crenata* 'Microphylla' (IL009), and *I. crenata* 'Yunnanensis' (IL044), all of which had high allele similarities with *I. crenata f. watanabeana* Mak. (IL042) and *I. crenata* 'Fructu Lutea' (IL088). Furthermore, *I. crenata* 'Green Island' (IL026) is part of Clade C, but has a high allele similarity with *I. crenata* 'Golden Helleri' (IL024) and *I. crenata* 'Lancaster Yellow' (IL029). These 'in between' genotypes could be the result of past intraspecific hybridization events between genotypes of Clade B and Clade C.

The variety of commercial *I. crenata* cultivars is rather limited, comprised mostly of a small selection of cultivars that are multiplied. The PCoA analyses revealed the breadth of currently unexploited genetic diversity. There were no genotypes in our collection that are positioned in the spaces between clusters, indicating a great potential for crosses between clusters. Interspecific hybridization in *Ilex* is usually more complicated than intraspecific crossing, and chromosome numbers can be used as a basis for estimating the rate of success [9]. Nevertheless, hybridization has led to many commercially important cultivars of different *Ilex* species in the past [9], such as *I. x meserveae* cultivars. Our data even confirm the possibility of inter-species hybridization. To obtain cultivars that are more robust to challenges such as high soil pH and black root rot, the knowledge from this paper needs to be combined with knowledge about phenotypic traits such as tolerance and resistance. In the future, both intra- and interspecific hybridization represent promising approaches to creating novel genetic diversity.

4. Conclusions

Despite the popularity of *I. crenata* as an ornamental shrub, a handful of cultivars are commercially dominant. The aim of this study was to investigate the genetic diversity within the species *I. crenata*, and between *I. crenata* and a few other hardy *Ilex* species. Genome size measurements, chromosome numbers, and genetic fingerprinting were used to unravel the genetic identity, genetic relationships, and genetic diversity of 72 *I. crenata* accessions and 22 accessions of other species. Our data showed that *I. crenata* has a high intra-specific genetic diversity. This was reflected in the differences in genome size of diploid *I. crenata* genotypes related to three phylogenetic clades identified through GBS genetic fingerprinting. Furthermore, the genetic fingerprinting data showed that large parts of this genetic variation might be interesting for commercial use. Interspecific hybridization or intraspecific crossings are a possible strategy to further enlarge the genetic diversity in the commercial assortment of *I. crenata*. This research is valuable for breeding programs, as it allows deliberately selecting for genetically diverse parental material, creating novel genetic diversity, and screening for novel morphological traits.

Supplementary Materials: The following supporting information can be downloaded at: https://www.mdpi.com/article/10.3390/horticulturae9040485/s1, Figure S1: Common loci and allele similarities of the complete collection.

Author Contributions: Conceptualization, E.G., L.L., A.H. and T.R.; methodology, E.G., L.L., A.H., T.R. and K.V.L.; formal analysis, E.G. and A.H.; investigation, E.G.; resources, J.V.M.; data curation, E.G. and A.H.; writing—original draft preparation, E.G.; writing—review and editing, L.L., A.H., T.R. and K.V.L.; visualization, A.H. and E.G.; supervision, L.L. and E.S.; funding acquisition, E.G. All authors have read and agreed to the published version of the manuscript.

Funding: This research is part of a Baekeland mandate (HBC.2020.2847) awarded by Flanders Innovation and Entrepreneurship (VLAIO).

Data Availability Statement: The raw PE-150 read data, as well as the merged preprocessed read data were submitted to NCBI's Sequence Read Archive (SRA) under BioProject number PRJNA895194.

Acknowledgments: The authors would like to sincerely thank the arboreta of Bokrijk (Genk, Belgium) and Het Leen (Eeklo, Belgium) for providing plant material to start our collection. Special thanks to Reine Audenaert for DNA extractions, and Magali Losschaert and Laura Schelfhout for chromosome counting. We would also like to thank Roger Dobbelaere and Rina Vanhaecke for their excellent management of the plants in the ILVO collection, and Miriam Levenson for English-language editing.

Conflicts of Interest: The authors declare no conflict of interest.

References

1. POWO. Plants of the World Online. Facilitated by the Royal Botanic Gardens, Kew. 2022. Available online: https://powo.science.kew.org/ (accessed on 27 February 2023).
2. Cuénoud, P.; Del Pero Martinez, M.A.; Loizeau, P.-A.; Spichiger, R.; Andrews, S.; Manen, J.-F. Molecular phylogeny and biogeography of the genus *Ilex* L. (Aquifoliaceae). *Ann. Bot.* **2000**, *85*, 111–122. [CrossRef]
3. Manen, J.-F.; Barriera, G.; Loizeau, P.-A.; Naciri, Y. The history of extant *Ilex* species (Aquifoliaceae): Evidence of hybridisation within a Miocene radiation. *Mol. Phylogenetics Evol.* **2010**, *57*, 961–977. [CrossRef]
4. Yao, X.; Lu, Z.; Song, Y.; Hu, X.; Corlett, R.T. A chromosome-scale genome assembly for the holly (Ilex polyneura) provides insights into genomic adaptations to elevation in Southwest China. *Hortic. Res.* **2022**, *9*, uhab049. [CrossRef]
5. Dudley, T.R.; Eisenbeiss, G.R. *International Checklist of Cultivated Ilex: Part 2, Ilex crenata Thunberg ex. J.A. Murray*; U.S. Department of Agriculture: Washington, DC, USA; U.S. National Arboretum Contribution: Washington, DC, USA, 1992; Volume 6, p. 91.
6. Xu, K.-W.; Wei, X.-F.; Lin, C.-X.; Zhang, M.; Zhang, Q.; Zhou, P.; Fang, Y.-M.; Xue, J.-Y.; Duan, Y.-F. The chromosome-level holly (*Ilex latifolia*) genome reveals key enzymes in triterpenoid saponin biosynthesis and fruit color change. *Front. Plant Sci.* **2022**, *13*, 982323. [CrossRef] [PubMed]
7. Kong, B.L.-H.; Nong, W.; Wong, K.-H.; Law, S.T.-S.; So, W.-L.; Chan, J.J.-S.; Zhang, J.; Lau, T.-W.D.; Hui, J.H.-L.; Shaw, P.-C. Chromosomal level genome of *Ilex asprella* and insight into antiviral triterpenoid pathway. *Genomics* **2022**, *114*, 110366. [CrossRef] [PubMed]
8. Yao, X.; Zhang, F.; Corlett, R.T. Utilization of the Hollies (*Ilex* L. spp.): A Review. *Forests* **2022**, *13*, 94. [CrossRef]
9. Galle, F.C. *Hollies: The genus Ilex*; Timber Press, Inc.: Portland, OR, USA, 1997; p. 573.
10. Molnar, T.; Capik, J.M. The Rutgers University woody ornamentals breeding program: Past, present, and future. *Acta Hortic.* **2013**, *990*, 11. [CrossRef]
11. Zhang, L.; Cao, B.; Bai, C. New reports of nuclear DNA content for 66 traditional Chinese medicinal plant taxa in China. *Caryologia* **2013**, *66*, 375–383. [CrossRef]
12. Bai, C.; Alverson, W.S.; Follansbee, A.; Waller, D.M. New reports of nuclear DNA content for 407 U.S. plant species. *Ann. Bot.* **2012**, *110*, 1623–1629. [CrossRef]
13. Sugiura, T. A list of chromosome numbers in angiospermous plants. *Shokubutsugaku Zasshi* **1931**, *45*, 353–355. [CrossRef]
14. Fedorov, A. *Chromosome Numbers of Flowering Plants*; Reprint 1974; Academy of Sciences of the USSR: Moscow, Russia; Komarov Botanical Institute: Leningrad, Russia, 1969.
15. Goldblatt, P. New or noteworthy chromosome records in the angiosperms. *Ann. Mo. Bot. Gard.* **1976**, *63*, 889–895. [CrossRef]
16. Rice, A.; Glick, L.; Abadi, S.; Einhorn, M.; Kopelman, N.M.; Salman-Minkov, A.; Mayzel, J.; Chay, O.; Mayrose, I. The Chromosome Counts Database (CCDB)—A community resource of plant chromosome numbers. *New Phytol.* **2014**, *206*, 7. [CrossRef]
17. Darlington, C.D.; Wylie, A.P. *Chromosome Atlas of Flowering Plants*; Allen & Unwin: Crows Nest, Australia, 1955.
18. Poland, J.A.; Rife, T.W. Genotyping-by-sequencing for plant breeding and genetics. *Plant Genome* **2012**, *5*, 92–102. [CrossRef]
19. Van Oost, E.; Leus, L.; De Rybel, B.; Van Laere, K. Determination of genetic distance, genome size and chromosome numbers to support breeding in ornamental Lavandula species. *Agronomy* **2021**, *11*, 2173. [CrossRef]
20. Greilhuber, J.; Dolezel, J.; Lysák, M.A.; Bennet, M.D. The origin, evolution and proposed stabilization of the terms 'genome size' and 'C-value' to describe nuclear DNA contents. *Ann. Bot.* **2005**, *95*, 255–260. [CrossRef] [PubMed]
21. Doležel, J.; Greilhuber, J.; Lucretti, S.; Meister, A.; Lysák, M.A.; Nardi, L.; Obermayer, R. Plant genome size estimation by flow cytometry: Inter-laboratory comparison. *Ann. Bot.* **1998**, *82*, 17–26. [CrossRef]
22. Lysak, M.A.; Dolezel, J. Estimation of nuclear DNA content in *Sesleria* (Poaceae). *Caryologia* **1998**, *51*, 123–132. [CrossRef]
23. Kirov, I.; Divashuk, M.; Van Laere, K.; Soloviev, A.; Khrustaleva, L. An easy "SteamDrop" method for high quality plant chromosome preparation. *Mol. Cytogenet.* **2014**, *7*, 21. [CrossRef] [PubMed]
24. Kirov, I.; Khrustaleva, L.; Van Laere, K.; Meeus, S.; Romanov, D.; Soloviev, A.; Fesenko, I. DRAWID: User-friendly java software for chromosome measurements and idiogram drawing. *Comp. Cytogenet.* **2017**, *11*, 10. [CrossRef]
25. Doyle, J.J.; Doyle, J.L. A rapid DNA isolation procedure for small quantities of fresh leaf tissue. *Phytochem. Bull.* **1987**, *19*, 11–15.
26. Schaumont, D. GBprocesS: Genotyping-by-Sequencing Data Processing Toolkit. Available online: https://gitlab.com/ilvo/GBprocesS. (accessed on 27 May 2021).
27. Martin, M. Cutadapt removes adapter sequences from high-throughput sequencing reads. *EMBnet. J.* **2011**, *17*, 10–12. [CrossRef]
28. Zhang, J.; Kobert, K.; Flouri, T.; Stamatakis, A. PEAR: A fast and accurate Illumina Paired-End reAd mergeR. *Bioinformatics* **2013**, *30*, 614–620. [CrossRef] [PubMed]

29. Hapke, A.; Thiele, D. GIbPSs: A toolkit for fast and accurate analyses ofgenotyping-by-sequencing data without a reference genome. *Mol. Ecol. Resour.* **2016**, *16*, 11. [CrossRef] [PubMed]
30. Van Poucke, K.; Haegeman, A.; Goedefroit, T.; Focquet, F.; Leus, L.; Jung, M.H.; Nave, C.; Redondo, M.A.; Husson, C.; Kostov, K.; et al. Unravelling hybridization in Phytophthora using phylogenomics and genome size estimation. *IMA Fungus* **2021**, *12*, 16. [CrossRef]
31. Letunic, I.; Bork, P. Interactive Tree of Life (iTOL) v5: An online tool for phylogenetic tree display and annotation. *Nucleic Acids Res.* **2021**, *49*, W293–W296. [CrossRef]
32. Mowforth, M.A. *Variation in Nuclear DNA Amounts in Flowering Plants: An Ecological Analysis*; University of Sheffield: Sheffield, UK, 1986.
33. Noirot, M.; Barre, P.; Louarn, J.; Duperray, C.; Hamon, S. Nucleus-cytosol interactions—A source of stoichiometric error in flow cytometric estimation of nuclear DNA content in plants. *Ann. Bot.* **2000**, *86*, 309–316. [CrossRef]
34. Greilhuber, J. Intraspecific variation in genome size in angiosperms: Identifying its existence. *Ann. Bot.* **2005**, *95*, 91–98. [CrossRef]
35. Noirot, M.; Poncet, V.; Barre, P.; Hamon, P.; Hamon, S.; De Kochko, A. Genome size variations in diploid African *Coffea* species. *Ann. Bot.* **2003**, *92*, 709–714. [CrossRef]
36. Bilinski, P.; Albert, P.S.; Berg, J.J.; Birchler, J.A.; Grote, M.N.; Lorant, A.; Quezada, J.; Swarts, K.; Yang, J.; Ross-Ibarra, J. Parallel altitudinal clines reveal trends in adaptive evolution of genome size in *Zea mays*. *PLoS Genet.* **2018**, *14*, e1007162. [CrossRef]
37. Díez, C.M.; Gaut, B.S.; Meca, E.; Scheinvar, E.; Montez-Hernandez, S.; Eguiarte, L.E.; Tenaillon, M.I. Genome size variation in wild and cultivated maize along altitudinal gradients. *New Phytol.* **2013**, *199*, 264–276. [CrossRef]

horticulturae

Article

Comparative Analysis of Morphological, Physiological, Anatomic and Biochemical Responses in Relatively Sensitive *Zinnia elegans* 'Zinnita Scarlet' and Relatively Tolerant *Zinnia marylandica* 'Double Zahara Fire Improved' under Saline Conditions

Sara Yasemin [1,2,*] and Nezihe Koksal [3]

1 Department of Horticulture, Faculty of Agriculture, Siirt University, Kezer, 56100 Siirt, Turkey
2 Flanders Research Institute for Agriculture, Fisheries and Food, 9090 Melle, Belgium
3 Department of Horticulture, Faculty of Agriculture, Cukurova University, Balcalı, 01330 Adana, Turkey
* Correspondence: sara.yasemin@siirt.edu.tr

Abstract: Salinity is one of the major abiotic stresses in plants. The aim of the present study was to determine the effects of salinity on relatively sensitive *Zinnia elegans* Jacq. 'Zinnita Scarlet' and relatively tolerant *Zinnia marylandica* D.M. Spooner et al. 'Double Zahara Fire Improved' through a comparative analysis of morphological, physiological, anatomic, and biochemical traits. Plants were irrigated at five levels of salt concentrations (0 [control], 50, 100, 150, 200 mM NaCl) for three weeks at one-day intervals in pots under greenhouse conditions. The effects of salinity stress on plant growth parameters, ion leakage, the loss of turgidity, minimum fluorescence (F_O'), plant nutrient elements, leaf anatomic parameters, stoma response to the application of light and abscisic acid perfusion, proline content, chlorophyll a, b and total chlorophyll, and carotenoid content were investigated. Differences in the stages and levels of plant response in the relatively sensitive and relatively tolerant cultivar were determined. Proline accumulation appeared to be higher in Double Zahara Fire Improved (D.Za.F.I.) than Zinnita Scarlet (Zi.S.) in the low concentration of salinity. After the application of abscsic acid perfusion to intact leaf surfaces, the stomata of the relatively tolerant cultivar D.Za.F.I. closed earlier (7 min) than Zi.S. (29 min). Ion leakage (32.3%) and Na accumulation (0.9%) in the aerial parts increased dramatically for Zi.S in the 50 mM NaCl treatment. Moreover, values of plant growth parameters, minimum fluorescence (F_O'), photosynthetic pigments, and plant nutrient elements all showed a greater decreasing percentage in Zi.S. compared to D.Za.F.I. Stomatal densities for both the abaxial and adaxial surfaces of the leaf decreased in parallel with the increase in salt stress. Palisade parenchyma cell height and leaf thickness values decreased in Zi.S. as salinity increased. In D.Za.F.I., leaf thickness increased by up to 100 mM NaCl while the height of palisade parenchyma cells decreased under high salt stress conditions (100 mM and above). Recommendations for future research include molecular-level evaluations and the study of how to increase salt tolerance in these potentially valuable ornamental cultivars.

Keywords: abscisic acid; ion leakage; photosynthetic pigments; plant nutrients; proline; stoma

Citation: Yasemin, S.; Koksal, N. Comparative Analysis of Morphological, Physiological, Anatomic and Biochemical Responses in Relatively Sensitive *Zinnia elegans* 'Zinnita Scarlet' and Relatively Tolerant *Zinnia marylandica* 'Double Zahara Fire Improved' under Saline Conditions. *Horticulturae* **2023**, *9*, 247. https://doi.org/10.3390/horticulturae9020247

Academic Editor: Aisheng Xiong

Received: 6 January 2023
Revised: 5 February 2023
Accepted: 7 February 2023
Published: 11 February 2023

1. Introduction

Abiotic stress in plants can affect growth, quality, and yield. The salinity of soil and water, already an important problem in arid and semiarid regions, is increasing due to climate change, irrigation, and fertilization. Salinity caused by $\geq$4 dS m^{-1} EC (electrical conductivity) negatively affects all plant growth stages, from seed germination through phenological stages and productivity [1]. Under saline conditions, plants cannot take water from the soil and can accumulate toxic levels of Na^+ and Cl^- ions. This osmotic and ionic stress leads to reductions in cell and tissue elongation, and nutritional imbalance and oxidative stress occur [2,3]. The negative effects of these physiological and biochemical processes cause plant growth to slow. Ratios of Na/K and Na/Ca are important for the homeostasis of plants [4], but calcium and potassium uptake are reduced by sodium uptake. High concentrations of chloride can cause chlorophyll to degrade, which leads to a reduction in photosynthesis [5]. In combination, these effects significantly reduce plant productivity. Plants respond to salinity in two phases: (1) stomata closure and reduced growth and (2) a cytotoxic stage leading to death [6]. Many anatomical characteristics that help the plant cope with salinity is found in salt-tolerant plants [7]. Anatomical modifications of leaf tissues defined as markers of adaptation to abiotic stress include the thickness of the mesophyll and epidermis and the size and density of the stomata [8]. Plants exposed to salinity generally show traits of succulence, thick cuticles, hairs on stems and leaves, salt glands, and sunken stomata. Some reports indicate that salt stress reduces the number of stomata, while others have determined that stomata and epidermis cell numbers increase with the degree of salinity [7,9–11]. Waqas et al. [12] reported that morphological traits, such as stomata density and aperture on the upper and lower sides of leaves, greatly influence the response of the plant to changes in salinity. Abscisic acid (ABA) is a hormone that helps plants respond to environmental stressors such as salinity and drought by regulating stomatal closure and gene expression. Under environmental stress, ABA plays a significant role in regulating the water status of plants by controlling stomatal movement. ABA does this by manipulating the ion fluxes of the guard cells of the stomata, which, in turn, controls the transpiration water loss of the plant and helps it conserve water [13]. In this case, the determination of ABA perception capability with stomata behavior might be important for the detection of the plant's tolerance.

The determination of the salt stress tolerance of ornamental plants is crucial for their selection and use in salty areas, as well as for the use of alternative (saline) water sources. Bedding plants are an important group of ornamental plants that are widely used in landscaping. *Zinnia* (Asteraceae family) is grown as an annual bedding plant used throughout the world during the summer season. It can also be used as a cut flower and potted plant. *Zinnia* has a wide diversity of vegetative characteristics, flower colors, and flower morphology. These features make Zinnia a popular choice for use in landscaping, gardening, and as a cut flower [14].

Studies examining the tolerance of *Zinnia* cultivars against salinity are limited to a few cultivars. Markovic et al. [15] found that *Z. elegans* (MagellanTM Scarlet) was sensitive under saline conditions. They revealed that this cultivar should not be planted in saline areas due to negative effects on plant growth and flower parameters. In previous work [16], we reported on antioxidant defense mechanisms of *Zinnia marylandica* 'Double Zahara Fire Improved' and *Zinnia elegans* 'Zinnita Scarlet' cultivars under saline conditions. These two cultivars can tolerate salinity up to 150 mM NaCl in terms of antioxidant defense, SOD, and CAT enzyme activities, which increased significantly with 150 mM NaCl in both Zinnia species but decreased with 200 mM NaCl. The highest GR enzyme activity was found in 200 mM salinity at *Z. marylandica* 'Double Zahara Fire Improved'. In another work, 20 Zinnia cultivars were screened for germination under saline conditions [17]. While the majority of Zinnia cultivars showed relatively high sensitivity to salt stress at the germination stage, Dreamland Ivory and Dreamland Coral were more tolerant. Macherla and McAvoy [18] determined that *Zinnia elegans* 'Dreamland' could be irrigated with saline water up to 0.5 g L^{-1} NaCl (an EC of 1 dS m^{-1}) in a 5-week production cycle without negative effects on plant growth. Manivannan et al. [19] suggested that salt stress led to important decreases in plant growth, biomass, photosynthetic parameters, and pigments and increased the electrolyte leakage potential (ELP), lipid peroxidation, and hydrogen peroxide level. Escalona et al. [20] found that salinity negatively affected plant growth but not flowering in *Zinnia elegans*. *Z. elegans* cv. 'Magellan' was shown to be relatively tolerant to salinity [21]. Niu et al. [22] reported that *Z. marylandica* 'Zahara Yellow', 'Zahara White', 'Zahara Scarlet', 'Zahara Rose Starlight', 'Zahara Fire', and 'Zahara Coral Rose' and *Z. maritima* 'Solcito' cultivars were sensitive to salinity (investigated saline concentrations were 1.4 dS m^{-1} (nutrient solution, control), 3.0, 4.2, 6.0, and 8.2 dS m^{-1}

EC). Villarino and Mattson [23] reported that Z. angustifolia 'Star Gold' was sensitive to salinity. Zivdar et al. [24] found that salinity reduced the germination parameters of *Zinnia*, in contrast to Carter and Grieve [25], who showed that marketable flowers could be produced up to 10 dS m^{-1} salinity when using *Z. elegans* cv. 'Benary's Giant Salmon Rose' and 'Benary's Giant Golden Yellow'. Based on these studies of *Zinnia* and the reports that plants react differently to salinity at different developmental stages, we aimed to determine the differences in the effects of salt stress on two cultivars, *Z. elegans* 'Zi.S.' and *Z. marylandica* 'D.Za.F.I', which are considered to be relatively salt-sensitive and relatively salt-tolerant [26], respectively. The study compared the morphological, biochemical, anatomical, and physiological changes in the two cultivars in response to irrigation with saline water.

2. Materials and Methods

2.1. Plant Material and Experimental Conditions

This study was carried out in the greenhouse of the Department of Horticulture, Cukurova University in Adana, Turkey, during the summer of 2019 (19 May 2019–6 July 2019). In this study, seeds of *Zinnia elegans* Jacq. 'Zinnita Scarlet' (Z.S.) and *Zinnia marylandica* D.M. Spooner et al. 'Double Zahara Fire Improved' (D.Za.F.I.) cultivars obtained from a local seed distributor (Tasaco Farm, Antalya, Turkey) were used as plant material. The average temperature was 32.9/19.7 °C; the average humidity was 54%.

2.2. Experimental Design and Treatments

In May 2019, the seeds of Zinnia cultivars were germinated in trays of peat-filled cells, each with a diameter of 3 cm and a height of 4.5 cm. Municipal water was used for irrigation. Once the seedlings had developed four leaves, they were transferred to 2-L plastic pots containing a mixture of peat and perlite in a 2:1 ratio. After a 5-day adaptation period, irrigation with NaCl solutions of different concentrations (0 mM [control], 50 mM, 100 mM, 150 mM, and 200 mM) was initiated and continued for three weeks on a daily basis. Irrigation solutions were prepared using municipal water. The control group and the salt treatment group were provided with macro and micronutrient solutions. The experiment was conducted in a completely randomized design, with four replicates of five plants each, for a total of 20 plants. The salinity treatments were terminated as soon as the first visible symptoms of the damage, such as necrosis on the leaves or differences in plant height, were observed.

2.3. Plant Growth Parameters

At the end of the study, the harvested plants were gently washed first with tap water and then with distilled water to remove any remaining peat and perlite. Shoot length, stem diameter, branch number, branch length, leaf width, and leaf length were determined. After plants were divided into roots and shoots (leaves and stems), the fresh weights of plants were recorded. The plants were then dried in an oven, and their dry weight was determined.

2.4. Physiological Parameters

2.4.1. Ion Leakage

In order to determine ion leakage of plants, leaf discs (1 cm diameter) were taken from young, fully expanded, and same-type leaves (second and third leaves from the apex). The leaf discs were washed in distilled water and gently blotted dry. Leaf discs (n = 3) were placed in each test tube, and the tubes were shaken for 4 h. The ion leakage in each sample was determined with the EC meter (EC600 model, Extech Instruments) and accepted as the first measurement (EC1). Leaf discs in the same solution were autoclaved; ion leakage at room temperature was determined, and the final measurement (EC2) was accepted. Ion leakage was calculated using the following Equation (1) [27].

$$\text{Ion leakage (\%)} = (\text{EC1}/\text{EC2}) \times 100 \quad (1)$$

2.4.2. Loss of Turgidity

Fully expanded and young leaves (second and third leaves from the apex) were used to determine the loss of turgidity under salt stress. First, the fresh weight (FW) of the leaf discs (1 cm) was recorded, and the turgor weight (TW) was determined after the leaf discs were soaked in distilled water for 4 h. The leaves were dried at 70 °C for 24 h, after which the dry weight (DW) was determined [28].

For the calculation of turgor loss, the following Equation (2) was used:

$$\text{Loss of Turgidity (\%)} = [(\text{TW} - \text{FW})/\text{TW}] \times 100 \quad (2)$$

FW: Fresh Weight, TW: Turgor Weight

2.4.3. Minimum Fluorescence

The OJIP curve is a graph that shows how Chl fluorescence changes over time when measured on dark-adapted samples. It has four distinct stages: "O", "J", "I", and "P". "O" represents the minimal fluorescence (F_O') caused by energy loss in antenna pigments before it reaches the reaction centers. "J" represents the fluorescence at 2 ms, "I" represents the fluorescence at 30 ms, and "P" represents the highest fluorescence. The OJIP curve is a useful tool for understanding the process of photosynthesis [29]. In this study, we measured the minimum fluorescence (F_O') using a FluorPen FP100 fluorimeter (FluorPen FP100, Photon System Instruments Ltd., Drasov, Czech Republic). At the end of the experiment, readings were made on three leaves from each plant [30].

2.4.4. Ion Concentration Analysis

The dried sample of the roots and shoots was used to determine plant nutrient concentrations. The dry materials were ground and digested using the dry digestion method. Sodium (Na^+), calcium (Ca^{2+}), magnesium (Mg^{2+}), potassium (K^+), phosphorus (P), Copper (Cu^{2+}), manganese (Mn^{2+}), iron (Fe^{2+}), and zinc (Zn^{2+}) concentrations were determined by inductively coupled plasma-atomic emission spectrometry (ICP-AES) [31]. After determining the ion concentrations, the Na^+/K^+ and Na^+/Ca^{2+} ratios were calculated. Furthermore, Cl^- concentration was determined with a scientific chloride analyzer—Sherwood [32].

2.5. *Anatomical Parameters*

2.5.1. Preparation for Stomatal Examination

Leaf samples were taken early in the morning for stoma counting and measurements. The stomatal density and size were measured from replicas of the abaxial and adaxial epidermis of the leaf. In these samples, a piece of nail polish (transparent) was applied to both sides of the lamina. The molds of the leaves were removed by pressing the leaves with tape. The image of the removed molds was taken with a light microscope (Motic, BA210 Trinocular, Xiamen, Hong Kong, China). The Motic Images Plus 2.0 analyzer program was used to determine the number, length, and width of the stomata in the photographed preparations. Stoma width and length on the upper and lower surface of the leaf were measured and determined as µm in the preparation photographs. In the preparations photographed under the microscope, the stomata on the upper and lower surface of the leaf were determined by counting the number of stomata per mm^2.

2.5.2. Investigation of Stoma Behaviors against Light Application

In order to verify the behavior of the stomata against ABA perfusion, the responses of the closed stomata in plant leaves to light were investigated. In order to examine the motility of the stoma, the leaf of the intact plant was placed on a vertical microscope table (Leica DM1000 LED, Leica Microsystems GmbH, Wetzlar, Germany) and images were recorded automatically every 60 s with the camera (Leica ICC500W) which equipped by microscope. In order to test the ability of stomata to open under light and to observe the differences resulting from the application under light, the white light intensity was obtained from the microscope lamp with approximately 300 μ mol m^{-2} s^{-1}. The upper surface of the plant leaves was fixed on an acrylic glass block using double-sided adhesive tape. Stomatal openings were determined from the images taken using the ImageJ program. Then, the measurements taken were graphed. Stoma clearance measurements were carried out on control plants of both cultivars between 08:00 and 16:00 h [33].

2.5.3. Investigation of Stoma Behaviors against ABA Perfusion

In order to observe the instantaneous effect of ABA application on stoma behaviors, intact *Zinnia* plants were fixed on the microscope table as described above, and control (pure water) and (±) ABA (50 μM) applications were carried out on the leaf surface using the perfusion system as described by [33] and [34]. Aqueous-immersion lenses were used to examine the stomata and apply the ABA solution. During the studies, the working solution (5 mM KCl, 5 mM potassium citrate (pH 5.0), 0.1 mM $CaCl_2$, and 0.1 mM $MgCl_2$] was perfused to allow 0.3 mL of the solution to flow between the leaf and the lens at a continuous flow of 1.5 mL [33,34]. The closing times of the stomata were observed after perfusing ABA from the leaf surface (intact) on the plant in the D.Za.F.I and Zi.S cultivars. After perfusing a solvent of the ABA solution, as a control, the stomata were not closed. This process was carried out to support the ABA. Measurements were carried out at least 8 times on control plants of both cultivars. Stomatal behaviors were investigated using the Image J program.

2.5.4. Preparation for Leaf Cross-Section Examinations

The young and fully expanded leaves (third leaf from the apex) were collected at the end of the study. The leaves were cut from the central part of the middle leaflet, near the widest point of each leaf [35]. Paraffin embedding and microtome sectioning were performed in order to obtain leaf cross sections [36]. The samples were transferred to FPA-70 fixation liquid. All the leaf samples were dehydrated using ethanol and tertiary butyl alcohol series (70, 85, 95, 100%) for 4–5 h in each solution, then embedded in paraffin, sectioned longitudinally (10 μm) with a rotary microtome (Leica RM2245), stained with 0.125% hematoxyline, and mounted in Entellan [37]. Serial sections of 10 μm thickness were made from the leaf samples with a microtome (Leica RM2245; Leica Microsystems GmbH, Wetzlar, Germany). Leaf thickness, upper and lower epidermis length, the length of the palisade parenchyma, and the width of sponge parenchyma from five predetermined points of five randomly chosen photographs of each leaf lamina cross-section were determined. The measurements were made using image processing software (Motic Images Plus 2.0 analyzer program).

2.6. Biochemical Parameters

2.6.1. Proline Content

Plant samples (250 mg) were crushed and ground in a mortar in 5 mL 3% sulfosalicylic acid solution. The ground samples were centrifuged at 3000 rpm for 4 min. After centrifugation, 1 mL of the supernatant part of the samples was taken, and the reaction mixture was prepared by adding acid ninhydrin (1 mL) and glacial acetic acid (1 mL). This prepared mixture reacted in a water bath at 100 °C for 1 h, and the reaction was terminated in an ice bath. The reaction mixture was extracted with 2 mL of toluene while in an ice bath. The spectrophotometer readings were performed at 520 nm irradiance [38].

2.6.2. Photosynthetic Pigment and Total Carotenoid Content

The total chlorophyll, chlorophyll a, chlorophyll b, and carotenoid contents were determined according to the Arnon [39] and Lichtenthaler and Wellburn [40] methods. In brief, 150 mg of fresh leaf samples taken from equal points from the plants were crushed and ground in 15 mL of an 80% acetone solution. After the samples were filtered through filter paper, they were read in a spectrophotometer (Shimadzu, UV 1800, Japan) at 470, 645, and 663 nm wavelengths. The obtained data were calculated according to Arnon [39] and Lichtenthaler and Wellburn [40] using the following Equations (3)–(6)

$$\text{Total chlorophyll} = [(20.2 \times \text{Abs}_{645}) + (8.02 \times \text{Abs}_{663})] \times (-\text{Acetone (mL)})/\text{Leaf (mg)} \quad (3)$$

$$\text{Chlorophyll-a} = [(12.7 \times \text{Abs}_{663}) - (2.6 \times \text{Abs}_{645})] \times (\text{Acetone (mL)})/\text{Leaf (mg)} \quad (4)$$

$$\text{Chlorophyll-b} = [(22.9 \times \text{A}_{645}) - (4.68 \times \text{A}_{663})] \times (\text{Acetone (mL)})/\text{Leaf (mg)} \quad (5)$$

$$\text{Total Carotenoid} = [(1000 \times \text{A}_{470}) - (3.27 \times \text{Chl-a}) - (104 \times \text{Chl-b})]/229 \quad (6)$$

2.7. Statistical Analysis

Stress treatment was carried out in a completely randomized experimental design with two factors (salinity and cultivar). Each treatment had four replicates of five plants. Data were subjected to ANOVA, and the means were separated using the LSD (least significant differences) multiple range test at $p < 0.05$. All the statistical analyses were performed using the JMP8.1 Software package.

3. Results

3.1. Morphological Parameters

The effects of salt stress on the plant growth parameters and physiological, anatomical, and biochemical properties of two *Zinnia* cultivars grown under different salt concentrations (0, 50, 100, 150, and 200 mM) were investigated. The effects of salinity and cultivar on plant growth parameters were important statistically, but we focused on salt and cultivar interaction results to see the different response stages and levels in the study. So, separately, the effects of cultivar and salt on all parameters are given in Tables S1–S10. The effects of the salinity and cultivar interaction were not important in terms of plant growth parameters except for branch length, stem diameter, and root collar diameter (Table 1). Branch length was higher in 0 mM NaCl—D.Za.F.I (9.92 cm) and 50 mM NaCl—D.Za.F.I (9.72 cm) and 0 mM NaCl—Zi.S (9.48 cm). The percentage of the decrease in branch length from the control to 200 mM NaCl was higher in Zi.S (52.6%) than in D.Za.F.I (36.5%). The percentage of the decrease from the control to 200 mM NaCl of root fresh (71.7%), and dry (72.9%) weights, shoot fresh (65.3%) and dry (57.4%) weights were also higher in Zi.S. The appearance of D.Za.F.I and Zi.S under the application of saline irrigation is shown in Figure 1.

Figure 1. The appearence of D.Za.F.I and Zi.S under salt stress.

Table 1. Effects of salinity and cultivar interaction on plant growth parameters of D.Za.F.I and Zi.S.

Plant	NaCl (mM)	Shoot Lenght (cm)	Branch Number (Unit)	Branch Length (cm)	Stem Diameter (mm)	Root Collar Diameter (mm)	Leaf Width (mm)	Leaf Length (mm)	Root Fresh Weight (g)	Root Dry Weight (g)	Shoot Fresh Weight (g)	Shoot Dry Weight (g)
D.Za.F.I	0	15.6 ± 1.4	14.9 ± 1.6	9.92 + 1.4 a	5.22 ± 0.2 a	6.16 ± 0.2 a	30.4 + 1.7	99.6 ± 6.3	6.66 ± 0.7	0.48 ± 0.06	30.7 ± 0.7	3.51 ± 1.0
	50	12.7 ± 1.1	14.8 ± 0.9	9.72 ± 0.8 a	4.97 ± 0.5 ab	5.77 ± 0.3 ab	28.2 ± 3.2	92.5 ± 6.0	6.51 ± 0.6	0.39 ± 0.02	22.1 ± 1.5	2.04 ± 0.2
	100	11.4 ± 0.8	13.1 ± 1.1	8.73 ± 0.4 b	5.01 ± 0.2 ab	5.71 ± 0.1 bc	25.7 ± 3.0	84.5 ± 6.9	6.11 ± 0.5	0.38 ± 0.02	19.4 ± 3.1	2.14 ± 0.0
	150	10.3 ± 0.5	11.9 ± 1.3	6.82 ± 0.7 d	4.44 ± 0.2 c	5.33 ± 0.1 cd	24.8 ± 2.5	80.0 ± 6.1	5.39 ± 0.5	0.31 ± 0.04	14.9 ± 1.1	1.57 ± 0.1
	200	9.9 ± 0.6	11.8 ± 0.9	6.30 ± 0.7 d	4.64 ± 0.1 bc	5.26 ± 0.3 d	22.3 ± 2.2	78.2 ± 3.8	4.78 ± 0.8	0.28 ± 0.05	13.3 ± 2.6	1.49 ± 0.2
Zi.S	0	14.9 ± 2.2	8.6 ± 2.2	9.48 ± 1.2 a	5.21 ± 0.4 a	5.63 ± 0.4 bcd	41.4 ± 5.1	74.7 ± 9.8	7.90 ± 3.6	0.59 ± 0.24	24.8 ± 3.2	2.70 ± 0.4
	50	13.4 ± 1.2	8.4 ± 1.3	7.53 ± 1.3 c	4.82 ± 0.5 abc	4.71 ± 0.1 e	39.2 ± 4.1	67.1 ± 7.1	5.62 ± 2.3	0.36 ± 0.11	20.0 ± 2.0	2.07 ± 0.3
	100	12.8 ± 3.1	7.0 ± 1.1	6.46 ± 1.0 d	4.82 ± 0.2 abc	4.49 ± 0.2 e	36.0 ± 5.8	63.4 ± 7.4	3.88 ± 1.6	0.30 ± 0.10	12.4 ± 4.3	1.59 ± 0.3
	150	10.2 ± 1.3	7.0 ± 1.2	5.06 ± 0.9 e	4.74 ± 0.1 bc	4.43 ± 0.1 e	33.8 ± 3.6	60.9 ± 6.7	3.99 ± 0.5	0.28 ± 0.05	12.0 ± 0.7	1.55 ± 0.1
	200	10.1 + 1.0	6.7 ± 1.3	4.49 ± 0.6 c	3.64 ± 0.1 d	3.39 ± 0.1 f	33.3 ± 5.3	58.1 ± 6.6	2.23 ± 1.0	0.16 ± 0.07	8.6 ± 1.4	1.15 ± 0.1
LSD		—NS	—NS	0.708 **	0.445 **	0.404 **	—NS	—NS	—NS	—NS	—NS	—NS

** $p < 0.01$, —NS: nonsignificant. The differences between the averages were indicated by separate letters.

3.2. Anatomical Parameters

3.2.1. Leaf Stomatal Parameters on Abaxial and Adaxial Epiderma

The results of the study revealed that there were statistically significant differences in stomatal parameters between the two cultivars under varying salinity levels, as seen in Table 2. The abaxial stomata width was higher in 0 mM NaCl—D.Za.F.I cultivar (27 µm). The abaxial stomata width decreased slightly in D.Za.F.I as the salt stress increased. A slight increase was observed at 50 mM (20.2 µm) and 100 mM (22.8 µm) of NaCl in the Zi.S cultivar. The longest abaxial stomata length was found in 0 mM NaCl—D.Za.F.I cultivar (46.3 µm); in this cultivar, abaxial stomata length remained stable under saline conditions. In contrast, the abaxial stomata length of Zi.S decreased as salinity increased. The highest stomatal density was under 0 mM (227 units) and 50 mM (238 units) NaCl for Zi.S, with dramatic decreases observed up to 100 mM NaCl (117 units). The abaxial stomatal density of D.Za.F.I was stable from 0 mM NaCl to 150 mM NaCl, but it dramatically decreased at the 200 mM NaCl (88 units) level. Adaxial stomata width and length were greater at 100 mM (width: 28.5 µm—length: 52.6 µm) and 150 mM (width: 28.0 µm—length: 51.0 µm) NaCl in the D.Za.F.I cultivar. Adaxial stomatal width and length increased at 150 and 200 mM NaCl in both cultivars. Adaxial stomatal density was higher in 0 (85 unit) and 50 (71 unit) mM NaCl in Zi.S cultivar. However, when compared with the control group, the decreasing percentage in adaxial stomatal density was found at 27% in 150 mM NaCl and 25% in 200 mM NaCl treatments in D.Za.F.I plants. The adaxial stomata density decreasing percentage was found at 14%, 13%, and 29% under 100, 150, and 200 mM NaCl treatments, respectively, in the Zi.S. cultivar. Stoma width, length, and density on the abaxial and adaxial epiderma of cultivars are shown in Figure 2.

Table 2. Effects of salinity and cultivar interaction on stomatal parameters of D.Za.F.I and Zi.S.

Cultivar	NaCl (mM)	Abaxial Stomatal Parameters			Adaxial Stomatal Parameters		
		Width (µm)	Length (µm)	Density (unit)	Width (µm)	Length (µm)	Density (Unit)
D.Za.F.I	0	27.0 ± 3.0 a	46.3 ± 4.4 a	147 ± 32 c	25.4 ± 2.4 b	46.7 ± 3.4 b	85 ± 24 d
	50	24.2 ± 2.0 bc	44.4 ± 3.9 ab	135 ± 30 c	22.3 ± 1.8 c	46.5 ± 2.8 b	71 ± 11 de
	100	24.1 ± 1.7 bc	43.1 ± 4.6 b	141 ± 17 c	28.5 ± 3.9 a	52.6 ± 3.4 a	66 ± 10 e
	150	24.3 ± 2.0 bc	43.3 ± 5.6 b	139 ± 18 c	28.0 ± 3.2 a	51.0 ± 3.5 a	62 ± 15 e
	200	24.9 ± 2.4 b	44.6 ± 5.3 ab	88 ± 16 d	25.3 ± 3.7 b	48.6 ± 3.9 b	64 ± 14 e
Zi.S	0	19.7 ± 2.2 d	32.1 ± 2.9 d	227 ± 34 a	17.9 ± 1.6 d	30.4 ± 2.1 ef	175 ± 19 a
	50	20.2 ± 2.1 d	32.4 ± 3.7 d	238 ± 36 a	20.5 ± 1.7 c	32.5 ± 2.2 de	173 ± 25 a
	100	22.8 ± 2.8 c	36.4 ± 2.4 c	177 ± 9 b	18.6 ± 2.1 d	33.5 ± 2.9 cd	150 ± 16 b
	150	19.8 ± 1.9 d	30.3 ± 1.9 de	179 ± 20 b	21.2 ± 3.0 c	35.5 ± 5.2 c	152 ± 28 b
	200	19.5 ± 2.7 d	28.3 ± 3.5 e	182 ± 25 b	17.9 ± 2.0 d	29.4 ± 2.0 f	124 ± 17 c
LSD		1.612 ***	2.802 ***	23.811 ***	1.314 ***	2.290 **	17.590 *

* $p < 0.05$, ** $p < 0.01$, *** $p < 0.001$. The differences between the averages were indicated by separate letters.

In the study, the opening stomatal aperture behavior of the control (0 mM) group plants was investigated by shining light on the closed stomata on the abaxial surface of the leaves. As a result of microscopic examinations, the opening duration of the stomatal aperture of D.Za.F.I and Zi.S. were found to be close to each other under applications of light. The maximum opening of the stomatal aperture was obtained after 36 min in both cultivars (Figure 3a, Supplementary data Video S1). After ABA perfusion, the stomatas of D.Za.F.I (7 min) closed more quickly than Zi.S (29 min) (Figure 3b, Supplementary data Videos S2 and S3). In Figure 4, sample views in two cases with the stomata open and closed in the control group of D. Za.F.I and Zi.S cultivars are presented.

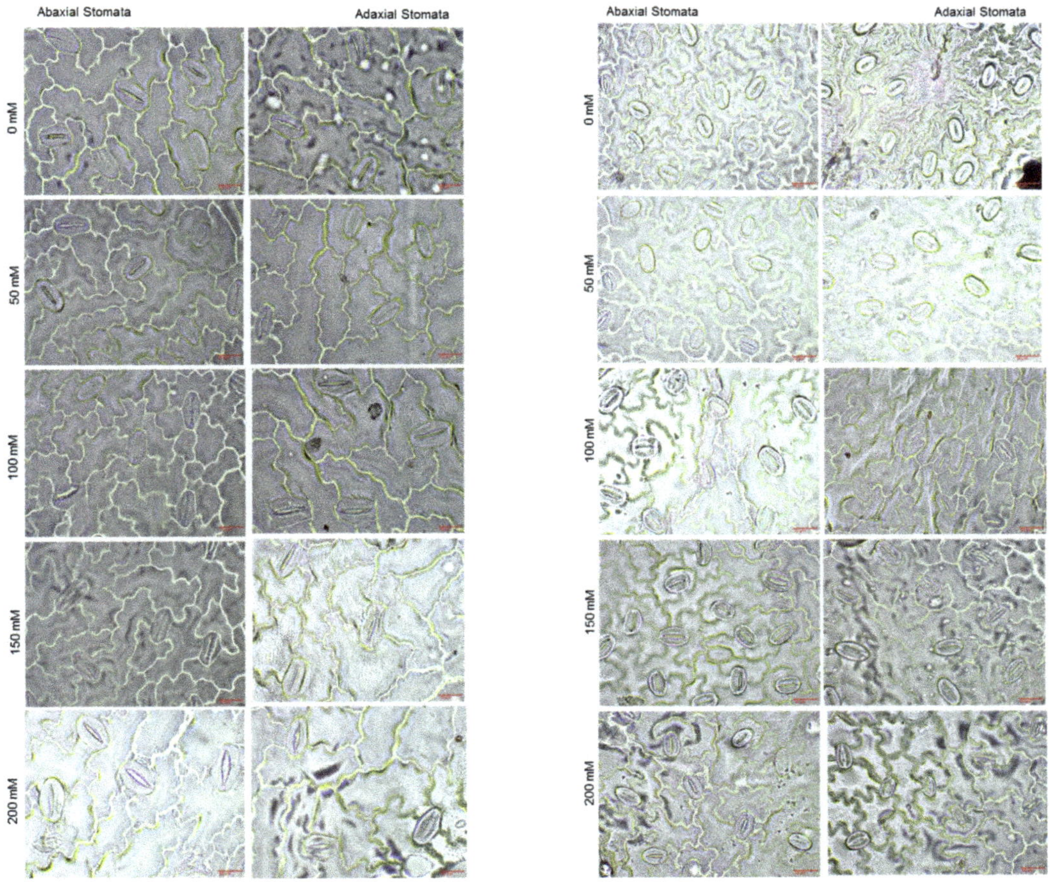

Figure 2. The effects of salt stress on stoma width, length, and density on abaxial and adaxial epiderma of D., Za., F., I, and Zi.S. Scale bar: 30 μm.3.2.2. Stoma Behavior against Light Application and ABA Perfusion.

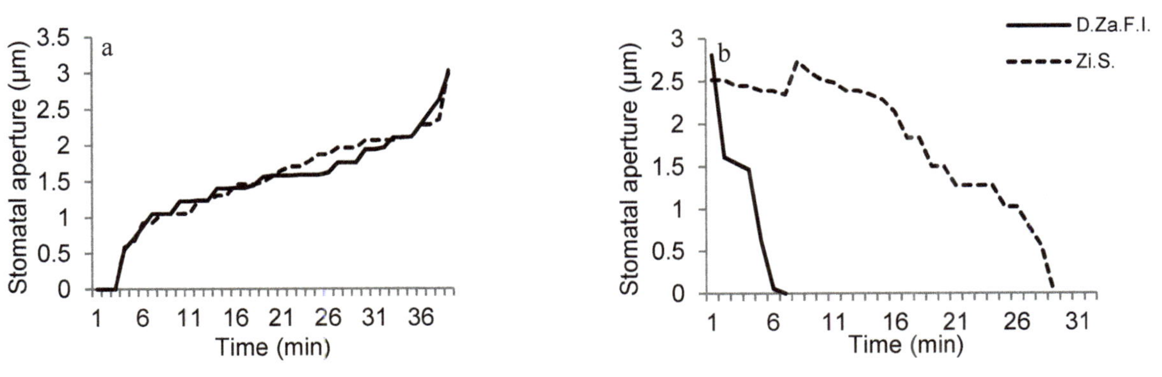

Figure 3. The change in stomatal aperture of D.Za.F.I and Zi.S under light application (**a**) and ABAperfusion (**b**).

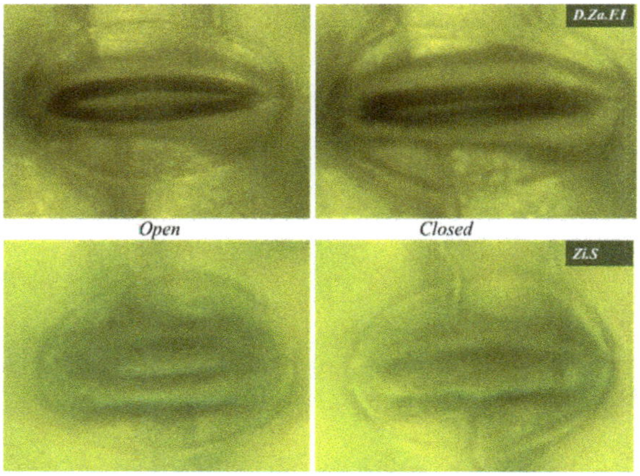

Figure 4. The view of open and closed stomata of intact control (0 mM NaCl) plants of D.Za.F.I and Zi.S.

3.2.2. Leaf Cross-Sections

The study found that the salinity levels had a significant effect on leaf anatomical parameters, with the exception of the length of the lower surface epidermal cells in both cultivars (Table 3). The highest leaf thickness (LT) value was found at 0 mM NaCl (359.7 µm)—Zi.S. cultivar, and the lowest values were at 150 mM (246.2 µm) and 200 mM (247 µm) NaCl—D. Za.F.I cultivar. When the control (0 mM NaCl) and 200 mM NaCl treatments were compared, leaf thickness decreased by a total of 9% in D.Za.F.I and 28% in Zi.S. The length of the palisade layer (LPL) was lower in 150 mM (54 µm) and 200 mM NaCl (54.4 µm)—D.Za.F.I. The width of spongy parenchyma was higher in the 100 mM NaCl—Zi.S cultivar (23.9 µm). The lowest values were obtained from 100 mM, 150 mM, and 200 mM NaCl—D.Za.F.I cultivar. The length of adaxial epidermal cells was higher in 100 mM NaCl—D.Za.F.I cultivar (19.8 µm). The aspects of leaf cross-sections of D.Za.F.I and Zi.S are presented in Figure 5. As salinity increased, glandular hairs (trichomes) were observed on the leaves of the *Z. marylandica* Double Zahara Fire Improved cultivar (data not presented) (Figure 6).

Table 3. Effects of salinity and cultivar interaction on stomatal parameters of D.Za.F.I and Zi.S.

Cultivar	NaCl (mM)	LT (µm)	LAbEC (µm)	LPL (µm)	WSPC (µm)	LAdEC (µm)
D.Za.F.I	0	272.8 ± 21 c	25.5 ± 5	73.0 ± 5 b	17.4 ± 4 de	15.6 ± 3 cd
	50	270.5 ± 12 c	24.8 ± 6	73.0 ± 8 b	15.7 ± 2 ef	15.1 ± 3 d
	100	292.1 ± 23 b	33.5 ± 8	57.8 ± 7 d	14.2 ± 4 f	19.8 ± 6 a
	150	246.2 ± 12 d	28.1 ± 5	54.0 ± 5 e	14.4 ± 2 f	18.1 ± 3 abc
	200	247.0 ± 26 d	27.6 ± 6	54.4 ± 8 de	14.5 ± 3 f	18.3 ± 5 ab
Zi.S	0	359.7 ± 22 a	20.5 ± 3	84.1 ± 7 a	22.0 ± 3 ab	11.3 ± 6 e
	50	303.1 ± 10 b	23.9 ± 4	72.2 ± 4 b	20.0 ± 3 bc	18.2 ± 7 ab
	100	303.4 ± 36 b	33.1 ± 5	82.2 ± 4 a	23.9 ± 5 a	17.2 ± 5 bcd
	150	302.5 ± 57 b	28.5 ± 5	80.5 ± 7 a	19.2 ± 3 cd	15.7 ± 8 bcd
	200	255.6 ± 33 cd	25.0 ± 4	67.5 ± 4 c	17.6 ± 4 de	15.2 ± 7 d
LSD		17.996 ***	—NS	3.826 ***	2.180 ***	2.611 **

** $p < 0.01$, *** $p < 0.001$, —NS: nonsignificant, LT: leaf thickness, LAbEC: length of abaxial epidermal cells, LPL: length of palisade layer, WSPC: width of spongy parenchyma cells, LAdEC: length of adaxial epidermal cells. The differences between the averages were indicated by separate letters.

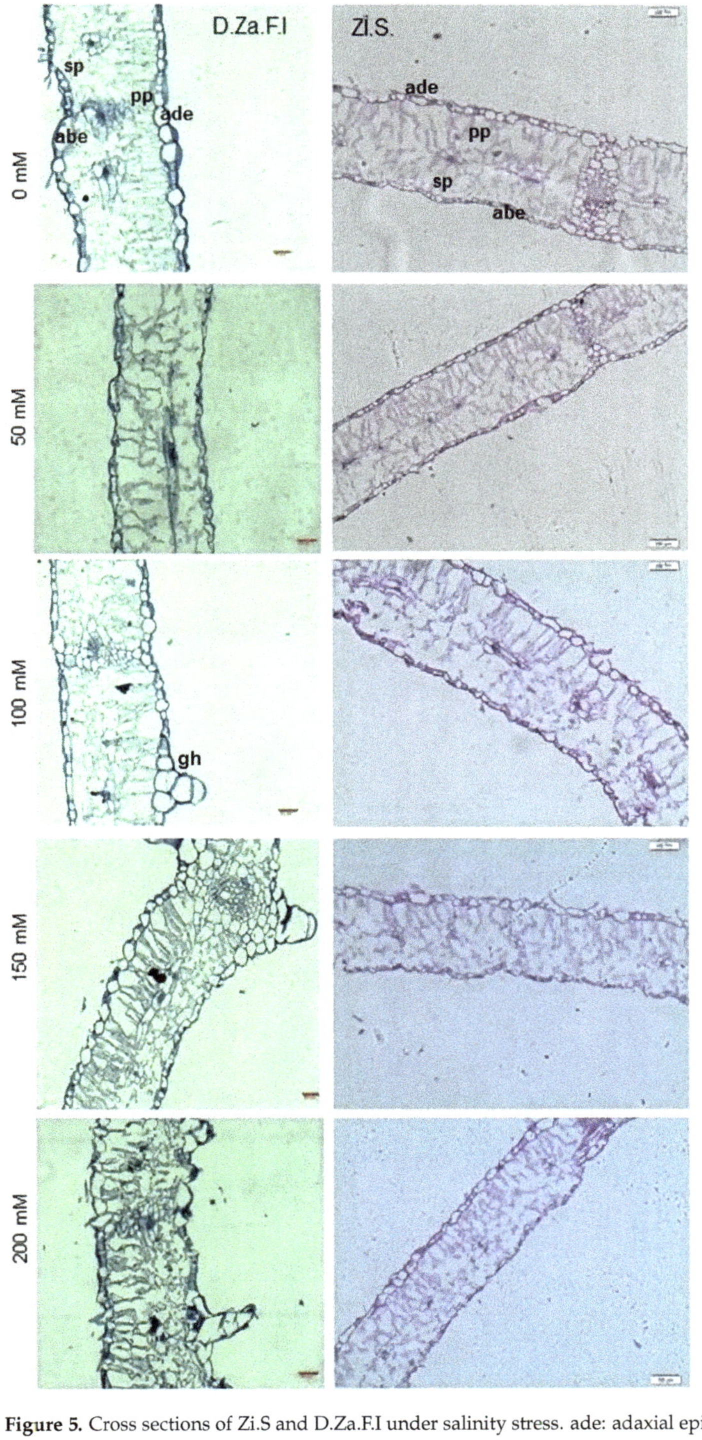

Figure 5. Cross sections of Zi.S and D.Za.F.I under salinity stress. ade: adaxial epiderma, abe: abaxial epiderma pp: palisade parenchyma, sp: spongy parenchyma, gh: glandular hair (trichome) (scale bar for Zi.S.: 50 µm, scale bar for D.Za.F.I: 100 µm).

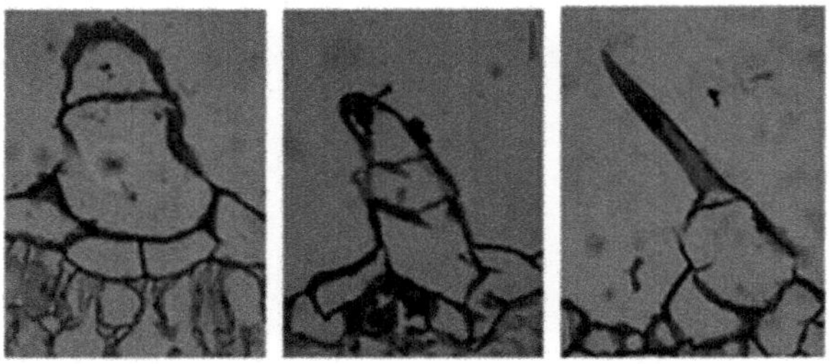

Figure 6. Glandular hairs on D.Za.F.I plants under saline conditions, scale bar: 100 µm.

3.3. Physiological and Biochemical Parameters

Ion leakage, proline content, F_O', and photosynthetic pigments were all affected by salt stress and cultivar interactions, as shown by the statistical analysis (Figure 7). Ion leakage had the lowest value in the control group of D.Za.F.I (25.8%), and the control group of Zi.S (27.9%) and 50 mM NaCl of D.Za.F.I (28.1%) followed. The ion leakage increased drastically in 100 mM NaCl of Zi.S cultivar (50.9%); the increasing rate from the control to 100 mM NaCl was 82%. The ion leakage increase was in 150 mM NaCl for D.Za.F.I cultivar (41.7%) (Figure 7A). The loss of turgidity increased in both cultivars, but interaction effects did not differ between the cultivars (Figure 7B). While proline content increased in the 50 mM NaCl level in D.Za.F.I (7.14 mg/g FW), it was 3.74 mg/g FW in the same concentration of Zi.S. The proline increased in 100 mM NaCl in Zi.S (12.3). A greater proline content was found in the D.Za.F.I cultivar (20.7 mg/g FW) compared to Zi.S (14.7 mg/g FW) under 200 mM NaCl. (Figure 7C). The minimum fluorescence value (F_O') had important differences in both cultivars under different salt concentrations. While F_O' was stable and slightly increased in 100 mM NaCl in D.Za.F.I, it decreased as the salinity increased in Zi.S (Figure 7D). It was found that the content of photosynthetic pigments in both cultivars was affected by salinity, and this effect was statistically significant. Photosynthetic pigments were stable and slightly increased by 150 mM NaCl in D.Za.F.I. As the salinity increased, photosynthetic pigment contents decreased in Zi.S (Figure 7E–H).

Based on the data presented in Table 4, it was determined that there were no statistically significant differences in the root content of P, K, Ca, Mg, Fe, Cu, Mn, and Zn among the Zinnia cultivars that were evaluated when subjected to salt stress conditions.

Table 4. Effects of salinity and cultivar interaction on plant nutrient elements in roots of D.Za.F.I and Zi.S.

Plant	NaCl (mM)	P	K	Ca	Mg	Fe	Cu	Mn	Zn
D.Za.F.I.	0	0.59 ± 0.10	1.86 ± 0.6	0.25 ± 0.02	1.36 ± 0.2	278.2 ± 72	40.1 ± 2.5	24.4 ± 5	117.0 ± 19
	50	0.61 ± 0.12	2.16 ± 1.0	0.29 ± 0.09	1.12 ± 0.2	270.2 ± 76	41.5 ± 1.5	24.4 ± 2	166.2 ± 40
	100	0.70 ± 0.05	0.97 ± 0.9	0.34 ± 0.06	1.05 ± 0.4	252.7 ± 6	41.1 ± 1.2	28.8 ± 11	150.8 ± 65
	150	0.79 ± 0.01	1.42 ± 0.4	0.32 ± 0.05	0.88 ± 0.2	165.8 ± 18	38.2 ± 2.9	23.0 ± 6	132.3 ± 14
	200	0.61 ± 0.11	0.75 ± 0.4	0.31 ± 0.06	0.92 ± 0.3	215.4 ± 42	42.3 ± 0.5	28.5 ± 12	163.7 ± 54
Zi.S.	0	0.55 ± 0.15	1.51 ± 0.5	0.34 ± 0.08	1.12 ± 0.3	237.8 ± 90	44.7 ± 7	63.7 ± 23	86.9 ± 15
	50	0.55 ± 0.02	1.45 ± 0.9	0.42 ± 0.09	1.30 ± 0.1	277.3 ± 63	56.9 ± 24	68.2 ± 32	86.8 ± 13
	100	0.49 ± 0.15	0.62 ± 0.0	0.53 ± 0.07	0.83 ± 0.2	260.6 ± 110	45.5 ± 8	57.2 ± 28	94.1 ± 24
	150	0.52 ± 0.25	1.28 ± 0.9	0.50 ± 0.14	0.77 ± 0.3	255.2 ± 74	48.4 ± 7	57.9 ± 24	95.4 ± 6
	200	0.39 ± 0.19	0.85 ± 0.1	0.70 ± 0.30	0.45 ± 0.3	191.6 ± 52	59.9 ± 13	43.8 ± 25	84.6 ± 11
LSD		—NS	—NS	—NS	—NS	—NS	—NS	—NS	—NS

—NS: nonsignificant.

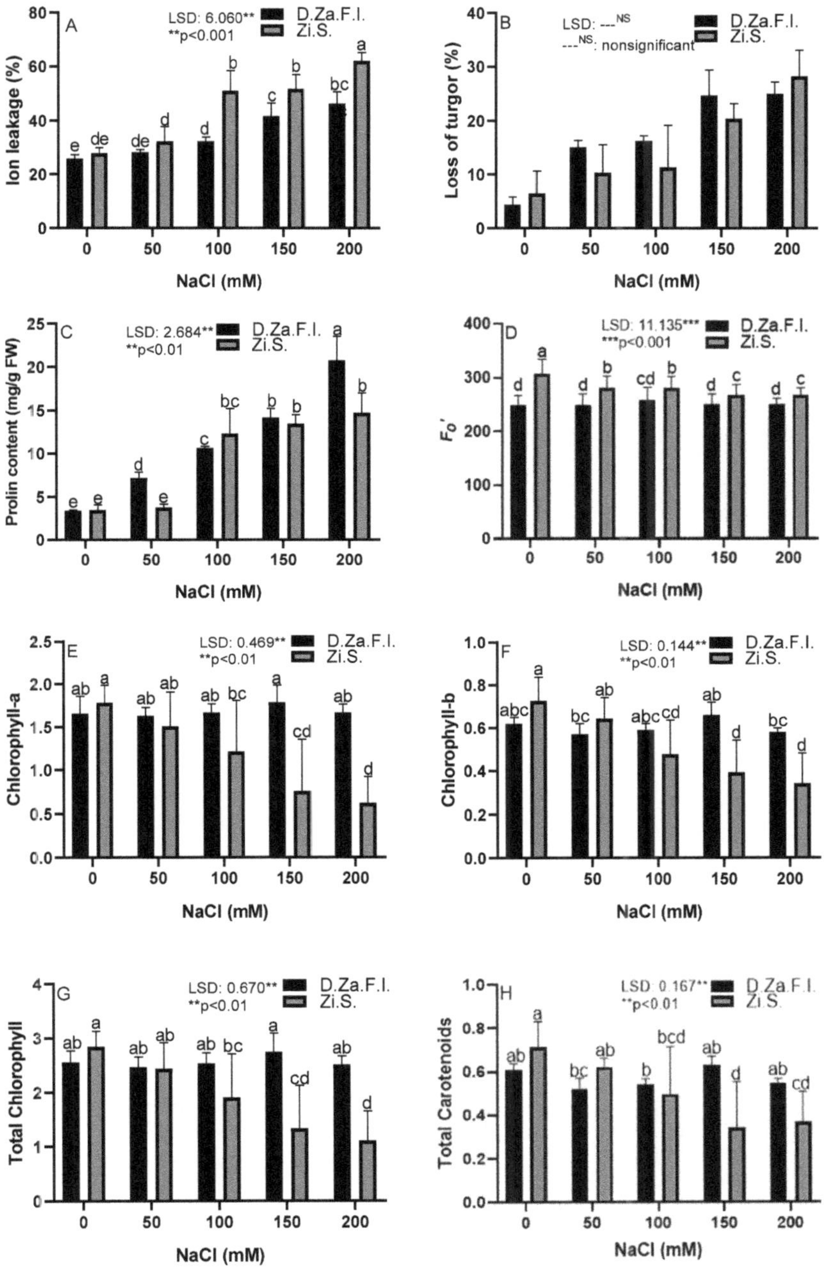

Figure 7. The change in ion leakage (**A**), turgidity (**B**), proline content (**C**), minimum fluorescence (F_O') (**D**) Chlorophyll-a (**E**), Chlorophyll-b (**F**), total chlorophyll (**G**), and total carotenoids (**H**) of D.Za.F.I and Zi.S under different levels of salinity. —^{NS}: nonsignifcant, ** $p < 0.01$ *** $p < 0.001$. The differences between the averages were indicated by separate letters.

Under saline stress conditions, the concentrations of P, K, Ca, Mg, Fe, Cu, Mn, and Zn in the aerial parts of both cultivars exhibited similar trends of decline or enhancement, with no statistically significant variation between the cultivars. (Table 5). However, the cultivar and salinity interaction effect were important for the N content. The highest N content was in 200 mM NaCl—D.Za.F.I (4.29), 50 mM NaCl—Zi.S (4.48), and 100 mM NaCl—Zi.S (4.46) combinations. The N content in the D.Za.F.I increased with increasing salinity. However, it dramatically decreased to 150 mM (1.87) and 200 mM NaCl (0.30) in the Zi.S.

Table 5. Effects of salinity and cultivar interaction on plant nutrient elements in aerial parts of D.Za.F.I and Zi.S.

Plant	NaCl (mM)	N	P	K	Ca	Mg	Fe	Cu	Mn	Zn
D.Za.F.I.	0	2.68 ± 0.37 bc	0.77 ± 0.03	5.30 ± 0.9	0.57 ± 0.03	1.54 ± 0.03	203.9 ± 31	46.1 ± 5	204.3 ± 15	174.2 ± 15
	50	2.88 ± 0.60 bc	0.79 ± 0.01	6.19 ± 0.4	1.03 ± 0.30	1.82 ± 0.11	188.1 ± 27	43.1 ± 1	206.1 ± 41	180.9 ± 18
	100	3.33 ± 0.78 ab	0.73 ± 0.05	6.12 ± 0.1	0.86 ± 0.13	1.74 ± 0.15	162.9 ± 14	42.6 ± 6	226.9 ± 17	168.2 ± 27
	150	3.94 ± 0.24 ab	0.76 ± 0.03	5.84 ± 0.8	1.05 ± 0.11	1.87 ± 0.15	149.9 ± 11	39.9 ± 2	239.2 ± 11	179.6 ± 27
	200	4.29 ± 0.64 a	0.71 ± 0.03	5.84 ± 0.7	1.11 ± 0.20	1.91 ± 0.36	232.6 ± 10	42.4 ± 2	273.8 ± 10	157.1 ± 23
Zi.S.	0	3.84 ± 0.36 ab	1.01 ± 0.07	4.10 ± 1.4	0.46 ± 0.07	1.91 ± 0.17	144.0 ± 19	38.7 ± 3	223.3 ± 36	200.4 ± 65
	50	4.48 ± 0.79 a	1.14 ± 0.09	4.96 ± 0.6	0.55 ± 0.12	1.91 ± 0.39	143.3 ± 6	38.7 ± 4	209.7 ± 49	142.6 ± 60
	100	4.46 ± 0.07 a	1.21 ± 0.11	4.43 ± 0.6	0.67 ± 0.06	1.99 ± 0.20	151.1 ± 18	44.2 ± 5	298.4 ± 50	165.4 ± 69
	150	1.87 ± 1.99 c	1.05 ± 0.16	4.58 ± 0.5	0.79 ± 0.08	1.94 ± 0.10	144.3 ± 6	38.9 ± 3	272.5 ± 18	225.1 ± 12
	200	0.30 ± 0.29 d	1.09 ± 0.17	4.97 ± 1.9	0.95 ± 0.19	1.87 ± 0.37	138.9 ± 8	38.7 ± 2	241.6 ± 23	201.8 ± 54
LSD		1.361 ***	—^{NS}	—^{NS}	—^{NS}	—^{NS}	—^{NS}	—^{NS}	—^{NS}	—^{NS}

*** $p < 0.001$, —^{NS}: non-significant. The differences between the averages were indicated by separate letters.

The effects of root and aerial part Na, Cl content, Na/K, and Na/Ca ratio under salinity are shown in Figure 8. The root Na content of both cultivars increased as salinity increased, but the interaction was non-significant (Figure 8A). While the content of Na in the aerial parts of D.Za.F.I. increased at 100 mM NaCl, Na in the aerial parts of Zi.S. increased to 50 mM NaCl and increased dramatically at 100 mM (Figure 8B). The Na and Cl content in the aerial parts of the sensitive Zi.S cultivar was higher than that in the tolerant D.Za.F.I cultivar. Increased Na ions were found in both cultivars under salinity conditions, with a higher increase percentage observed in the sensitive cultivar Zi.S. Compared to the control (0 mM NaCl) treatment, the Na+ aerial content (0.23%) in D.Za.F.I increased by 78% at 50 mM NaCl, 171% at 100 mM, 892% at 150 mM, and 1135% at 200 mM NaCl. In Zi.S, the Na+ content was increased by 263% at 50 mM, 880% at 100 mM, 1238% at 150 mM, and 1355% at 200 mM NaCl compared to the control (0.25%). Although the effects of interaction salinity and cultivar on the ratio of Na/K (Figure 8C) and Na/Ca (Figure 3E) in the root were non-significant, ratios increased with salinity. The change in the Na/K and Na/Ca ratios was significant in aerial parts (Figure 8D,F). Both ratios increased with salinity, but the increases in Zi.S were higher than in D.Za.F.I. The content of Cl in the root and shoot increased with salinity in both cultivars, but interaction effects were non-significant (Figure 8G,H). The Cl accumulation in the root and aerial parts increased in both D.Za.F.I and Zi.S cultivars with increasing NaCl concentrations. In D.Za.F.I, the increase was 136% at 50 mM, 200% at 100 mM, 209% at 150 mM, and 264% at 200 mM NaCl compared to the control (1.1%). In Zi.S, the increase was 120% at 50 mM, 160% at 100 mM, 280% at 150 mM, and 300% at 200 mM NaCl, compared to the control (1.0%).

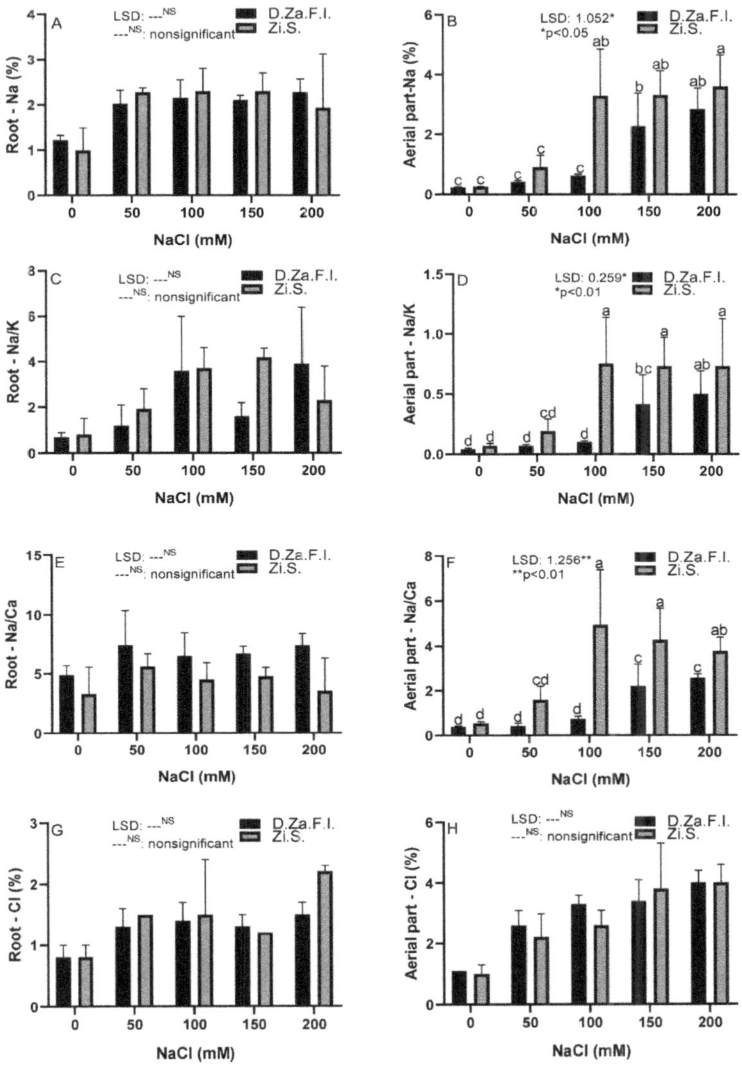

Figure 8. The change in root-Na (**A**), aerial part-Na (**B**) root Na/K (**C**) aerial part-Na/K (**D**), root Na/Ca (**E**), aerial part-Na/Ca (**F**), root Cl (**G**), aerial part Cl (**H**) content of D.Za.F.I, and Zi.S under different levels of salinity. The differences between the averages were indicated by separate letters.

4. Discussion

Salt (NaCl) taken up by plants does not directly affect plant growth; it first affects the turgor state, photosynthesis, and some enzyme activities [41]. In addition, Munns [41] reported that growth retardation begins with a decrease in soil water potential and then continues with specific effects such as salt damage which primarily affects old leaves due to the excessive accumulation of salt ions in the cell wall or cytoplasm. Salt ions accumulating in old leaves accelerate cell death and prevent the transport of carbohydrates and growth hormones to growth tissues. This excessive accumulation of salt ions slows plant growth as a result of the decrease in the rate of photosynthesis and the formation of growth-inhibiting metabolites. In this study, shoot length, branch number, branch length, stem diameter, root collar diameter, leaf width, leaf length, root fresh weight, and root dry weight values decreased in general as the salt stress level increased. Many previous studies on ornamental plants show that plant growth parameters decreased with salt stress [21,22,42–48].

In relatively sensitive Zi.S leaf thickness, the length of palisade parenchyma cells and the width of sponge parenchyma cells decreased as salt stress increased. The length of lower and upper epidermis cells increased up to 100 mM NaCl and then decreased as salt stress increased. The leaf thickness of the D.Za.F.I cultivar, which is relatively tolerant to salt stress, reached a peak value at 100 mM and decreased as NaCl concentrations increased. The length of the palisade parenchyma cells and the width of the sponge parenchyma cells decreased as salt concentrations increased. The lengths of the lower and upper epidermis cells increased with the increasing salt stress. Navarro et al. [49] reported that in *Arbutus unedo*, no change was observed in the palisade parenchyma of the first layer in the leaves exposed to salt stress compared to the control group, while the size of the palisade parenchyma cells in the second layer increased significantly in parallel with the increase in stress. They also found a significant decrease in the intercellular spaces of the sponge parenchyma. Fernandez-Garcia et al. [50] reported that increased leaf thickness could be observed in *Lawsonia inermis* L. plants under highly saline conditions. Acosta-Motos et al. [51,52] investigated leaf anatomy in *Myrtus communis* and *Eugenia myrtifolia* plants under salt-stress conditions. While they observed no anatomical changes in the palisade parenchyma in the *Myrtus communis* plant, they found a decrease in the cells of the sponge parenchyma and an increase in the intercellular spaces. However, the size of palisade parenchyma cells increased significantly. Likewise, Gomez-Bellot et al. [53] reported an increase in leaf thickness along with an increase in the palisade parenchyma cells in *Vibirnum tinus* plants. Hameed et al. [54] revealed that 200 mM NaCl increased succulence in *Imperata cylindrica* (L.) Raeuschel plants. It is known that the increase in leaf thickness and in the palisade parenchyma as salt stress increases helps to facilitate the diffusion of carbon dioxide (CO_2) and its progress through the layer, helping the chloroplasts to reach higher rates of CO_2 in the palisade parenchyma. Such anatomical changes are particularly important for the maintenance and advancement of the photosynthetic performance of plants under stress situations, which tend to reduce the stomatal opening but also help the plants cope with salt stress [51,52]. The study by Li et al. [55] emphasizes the crucial role of leaf anatomy in regulating the balance between water and CO_2 diffusion during drought conditions. The research suggests that by understanding the relationship between leaf anatomy and drought stress, it may be possible to develop more drought-tolerant crops in the future. Plants have the ability to alleviate the negative effects of salt stress by adjusting the density and size of their stomata. This is thought to be an adaptive mechanism that allows plants to respond to changes in environmental factors such as temperature and water availability [12,55,56]. Waqas et al. [12] suggested that the increase in stomatal density in quinoa plants under salinity stress may be a result of the shrinkage of pavement cells, which provides more surface area for CO_2 assimilation. This allows the plant to maximize its CO_2 uptake, continue its photosynthesis process, and improve its water usage efficiency in response to salinity. Stomatal density is a trait that is dependent on the species, the duration, and the intensity of salinity. An increase in stomatal density leads to an increase in the ion requirements of charge balancing and maintaining osmotic potential at the plasma membrane of guard cells. This is thought to contribute to salinity tolerance. We found that salinity had a negative impact on stomatal density in this study. The stomatal aperture was observed to decrease more rapidly in the cultivar and was relatively tolerant to salinity (D.Za.F.I) compared to the cultivar that was relatively sensitive to salinity (Zi.S).

Leaf water potential and changes in osmotic potential depend on the osmotic potential of the root environment and the amount of stress the plant is experiencing [57]. The plant's stress level can thus be measured by using some parameters that reveal the plant's water content under salt stress conditions. The results of the present study indicate that turgor loss in the leaves of the two *Zinnia* cultivars was low when irrigated with 100 mM NaCl. As salinity increased to 150 and 200 mM, the loss of turgidity increased. Salt stress did not affect the relative water content of *Pelargonium* [58] and calla lily [59]. In rose, Carvalho et al. [44] reported that the relative water content decreased as salt stress increased. The maintenance of relative water content and turgidity under salt stress conditions is associated with an increase in Ca and Mg accumulation in leaf tissues as well as Na accumulation in plants [58]. Salt stress in plants leads to damage to the cellular membrane. In the present study, increased an ion leakage was observed as salt stress levels increased in both salt-sensitive and salt-tolerant *Zinnia* cultivars. Cell damage rates were found to be higher in the relatively sensitive Zi.S than the relatively tolerant D.Za.F.I. Trivellini et al. [60] reported that ion leakage increased under 200 mM salt stress conditions in *H. rosa-chinensis*.

Salt stress can negatively impact the process of photosynthesis in plants. In the short-term, high salt concentrations can cause the plant's stomata to close, reducing the rate of photosynthesis. This can lead to a halt in plant growth within a few hours of exposure [61]. In the long term, salt accumulation in young leaves can lead to a decrease in chlorophyll and carotenoid levels, which are essential for the process of photosynthesis [22,62–67]. Many studies have shown that photosynthesis, specifically the PSII (Photosystem—II) process, is negatively impacted by salt stress [50,68–70]. The present study revealed that the minimum fluorescence (F_O') decreased in the relatively sensitive Zi.S as salt stress increased while it was preserved in relatively tolerant D.Za.F.I under the same conditions. The chlorophyll-a and total chlorophyll content of cv. D.Za.F.I, which is relatively tolerant to salt stress, did not change as the salt stress increased; a slight decrease in chlorophyll-b content was observed. In contrast, the sensitive cv. Zi.S showed a significant decrease in chlorophyll-a, chlorophyll-b, and total chlorophyll contents under salt stress. Chlorophyll content decreased in salt-sensitive plants [65,71]. Mukarram et al. [72] reported that chlorophyll fluorescence, chlorophyll content, and plant growth were minimized under high salt concentrations (240 mM) in lemongrass (*Cymbopogon flexuosus*). Several reports show that the total chlorophyll content decreases with salt stress: Vernieri et al. [73] in *Acacia cultriformis, Callistemon citrinus, Carissa edulis microphylla, Gaura lindheimeri, Jasminum sambac, Westringia fruticosa*; Eom et al. [74] in *Alchemilla mollis, Nepeta faassenii, Phlox subulata, Solidago cutleri, Thymus praecox*; Lee and van Iersel [75] in *Chrysanthemum morifolium*; Bahadoran and Salehi [76] in *Polyanthes tuberosa*; and Cantabella et al. [77] in *Stevia rebaudiana*. We further found that the total carotenoid content of the plants of both cultivars decreased as salt stress increased. When the total carotenoid content of the plant leaves in the highest salt stress condition (200 mM NaCl) was compared with the content of the control group, and the decrease percentage (9%) in the D.Za.F.I was quite low; it was found to be quite high in the Zi.S (48%). Carotenoids are a crucial class of biochemicals, such as antioxidants, which protect membrane lipids against oxidative stress induced by environmental stressors such as salt stress, thus promoting plant health and survival [78,79]. Furthermore, carotenoids can interconvert and thus contribute to increasing tolerance under stress conditions. [80,81].

Another common plant response to salt stress is an increase in intracellular osmotic regulators. Among organic osmolytes, proline is one of the most important and effective substances. In addition to its role as an osmoprotectant, proline helps plants cope with a variety of environmental stresses, as it has antioxidant properties and acts as a molecular chaperone to protect the structure of biological macromolecules during water loss from the cell [82,83]. In general, proline accumulation in salt-tolerant plants increases after exposure to salt stress. In this study, we found that the amount of proline increased in two *Zinnia* cultivars that were relatively sensitive and tolerant, respectively. However, in the salt-tolerant D.Za.F.I cultivar, proline content increased as soon as the plant was exposed to salt (50 mM NaCl). Similarly, a 20% increase in the leaf proline content was determined in *Eugenia myrtifolia* L., which was tolerant to 8 dS m^{-1} NaCl. [51]. In addition, Bizhani et al. [19] in *Zinnia elegans* 'Magellan' cultivar, Bres et al. [58] in *Pelargonium*, and Li et al. [84] in *Crysahthemum* reported a large increase in proline content. Kumar et al. [45] reported a low level of increase in the proline content during stress in oleander (*Nerium oleander*). Garcia-Capparos et al. [85] observed a peak in proline content at the 60 mM

NaCl level in the root tissues and in the control treatments in the leaves of the *Lavandula multifida* L. plant after the irrigation at different salt concentrations (0, 10, 30, 60, 100, and 200 mM NaCl). Mukarram et al. [72] revealed an upward trend in the proline concentration with increasing salt levels. They found that the concentration of proline increased about 2.2 times at a salt level of 240 mM NaCl compared to the control group.

Salt stress can affect the nutritional balance of a plant through a complex network of interactions, including restriction during the uptake and/or transport of nutrients from root to shoot [3]. In general, in ornamental plants grown under saline conditions, a decrease in the concentration of N, P, K, and Ca in leaf tissues was observed, while the Na and Cl concentration increased due to the antagonistic interactions associated with Na and Cl. Plant behavior regarding nitrogen uptake can differ widely under conditions of salt stress. While nitrogen uptake under salt stress may decrease, usually due to antagonism between NH_4 and Na or [86] Cl and NO_3 [87,88], the N content may also increase as N-containing amino acids such as proline increase in response to salt stress [66]. Different trends have also been observed in the P uptake of plants under salt stress conditions. Salt stress can reduce P availability due to the antagonism between Cl and H_2PO_4 [57]. While some researchers have detected a decrease in phosphorus content due to the competition between the mentioned Cl and H_2PO_4 [89], others have reported an increase in P due to the energy (ATP) required to transport the ions, which is more than necessary [90]. In our study, the nitrogen content of the salt-sensitive Zi.S increased up to 100 mM NaCl, while a sharp decrease was observed at higher concentrations. In the tolerant cultivar D.Za.F.I, N uptake increased as salt stress increased. Salt stress treatments did not affect the phosphorus content in both cultivars. Simon et al. [91] reported that salt stress led to decreased N and P content in *Chamaerops humilis* and *Washingtonia robusta* plants, and Navarro et al. [92] observed similar changes in *Dianthus caryophyllus*. Garcia Capparos et al. [93] stated that there was no consistent change in N and P concentrations in the root and leaf tissues of some ornamental plant species (*Aloe vera* L. Burm, *Kalanchoe blossfeldiana* Poelln and *Gazania splendens* Lem sp.) under salt stress. Jampeetong and Brix [94] determined that the N content of *S. natans* increased in leaf tissues but decreased in root tissues at 50, 100, and 150 mmol L^{-1} NaCl concentrations.

The increase in toxic elements such as Na and Cl in the leaves of plants exposed to salt stress conditions caused visual damage such as tip and marginal blights, which negatively affected the decorative value of the ornamental plant [95]. The typical symptom of sodium (Na) accumulation was leaf blight, which occurred first on the oldest leaves and along the leaf margins. As the stress level increased, the leaf dried further towards the leaf center until all the tissue died. However, symptoms due to Cl toxicity typically begin at the leaf tip of older leaves and progress toward the stem as the stress level increases [96]. Although sodium (Na) is the main ion that causes toxicity related to high salinity, some plants are particularly sensitive to Cl. In our study, the Na and Cl content of the aerial parts of the sensitive Zi.S cultivar were found to be higher than the tolerant D.Za.F.I cultivar. Although increased sodium (Na) ions were detected in both cultivars as the salt stress increased, the percentage of the Na-increase was found to be higher in the sensitive cultivar Zi.S. The study found that the root Na content of D.Za.F.I was 1.5 times higher than its aerial part Na content, while the root and aerial part Na content of Zi.S was almost equal. The root Na content of both cultivars was found to be similar, but the aerial part Na content of Zi.S was 1.6 times higher than D.Za.F.I, indicating that the relatively sensitive cultivar transferred Na+ to the shoots at a higher rate compared to the relatively tolerant cultivar, which retained more Na+ in the root zone. As the salinity level increased, the concentration of Cl in both cultivars also increased. According to Cassaniti et al. [95], an increase in the Cl concentration in mature plants of Leptospermum scoparium led to a decrease in growth. Picchioni and Graham [97] also found that the increasing Cl concentration in seedlings of Crataegus opaca caused a decrease in growth. Controlling the salt concentration in the upper part of the plant by limiting the entry of salt ions from the roots and preventing their transport to the shoots is an important mechanism that ensures the survival and growth of plants growing in salty conditions [3,98]. The retention of Na and/or Cl ions in root or leaf tissues is also important for salt stress tolerance [99,100]. In our study, in the relatively tolerant cultivar D.Za.F.I, this feature was manifested in the presence of higher Na ions in the roots than in the leaves. Similar results were expressed by Cassaniti et al. [95], with a higher ion concentration in the roots than in the leaves of *Viburnum lucidum*.

The physical and chemical similarities between K and Na allow Na to compete with K for binding sites on plasma membranes of root cells, leading to a reduced K uptake and decreased K availability for the plant [101,102]. The decrease in calcium intake is due to the replacement of Ca with Na in the cell membrane and the antagonistic interaction between Ca and Na ions, which impairs membrane integrity and selectivity and affects membrane properties [103]. In the present study, the aerial part K and Ca content of the tolerant D.Za.F.I was found to be higher than in the Zi.S, and the Ca content increased in both cultivars with salt stress. However, the Na/K and Na/Ca ratios in the aerial part were higher in the Zi.S than the D.Za.F.I. The concentration of K^+ decreased in *Celosia argentea* [104], *Limonium sinuatum* and *L. perezii* [105] under saline conditions. In addition, Carter and Grieve [106] in *Antirrhinum majus*, Navarro et al. [107] in *Arbutus unedo*, Simon et al. [91] in *Washingtonia robusta*, Navarro et al. [85] in *Dianthus caryophyllus*, Grieve et al. [108] in *Matthiola incana*, and Niu et al. [109] in *Rosa hybrida* found that the content of K and Ca decreased, while the content of Na and Cl increased. The ability of plant genotypes to maintain high levels of the K/Na ratio in their tissues is a key mechanism contributing to the expression of salt stress tolerance [110–113]. The conservation of Ca and K content in the plant under saline conditions helps maintain turgor status and cell membrane integrity [114]. Acosta-Motos et al. [51] reported an increase in Ca in different parts of *Eugenia* plants under salt stress. An increase in Ca concentrations in response to salt stress conditions has also been reported in other plant species such as *Vicia faba* L. and *Myrtus communis* L. [115,116]. Koksal et al. [117] emphasized that with the increase in the salt stress level in the *Hyacinthus orientalis* L. plant, Na intake increased significantly, the K content decreased, and the NaCl of 75 mM and above caused a sharp decrease in K/Na and Ca/Na ratios. In addition, Koksal et al. [118] stated that depending on the increase in salt stress, Ca, Mg, and Na concentrations increased while K decreased in both the roots and shoots of *Tagetes erecta*. In addition, in the same study, the roots and shoots of K/Na and Ca/Na ratios were found to be lower than the control at all salt levels. The researchers emphasized that the determination of these rates is important in terms of revealing the plant's tolerance level.

Excessive salinity also reduced the Mg absorption of plants [119]. In our study, the Mg content in the root and aerial part of D.Za.F.I cultivar did not change under salt stress conditions the root-Mg content in Zi.S cultivar decreased, while the Mg content in the aerial parts did not change. An increase in manganese content was found in the shoot tissues of the D.Za.F.I cultivar under stress conditions. At increasing salt concentrations, Rout and Shaw [120], who studied *Hydrilla erticillate* Esteves and Suzuki [119], who studied *Typha domingensis* and Jampeetong and Brix [94], who studied *S. natans*, all reported decreases in Mg content. Niu et al. [22] showed that the change in Mg content in *Zinnia marylandica* cultivars (Zahara Coral Rose, Zahara Fire, Zahara Rose Starlight, Zahara Scarlet, Zahara Yellow, and Zahara White) was minimal when compared to the changes in the amount of Na and Cl ions. Manganese is a very important trace element for plants and acts as an activator for different enzymes, which are involved in many biological events, such as oxidation, reduction, decarboxylation, and hydrolytic reactions in plant systems [121]. In our study, the effect of salt stress on the intake of microelements, the interaction effects of salt, and the cultivar were not important in roots and aerial parts tissues.

5. Conclusions

The study investigated the differences in salt tolerance between two Zinnia cultivars, Zinnita Scarlet (relatively sensitive) and Double Zahara Fire Improved (relatively tolerant). The results showed that the sensitive cultivar had high Na content, high ion leakage, slow stomatal closure, reduced photosynthetic pigments, and decreased stomatal number under salt stress, while the tolerant cultivar showed quicker stomatal closure, early proline synthesis, maintained photosynthetic pigments, and low ion leakage (in 50 and 100 mM NaCl). Further studies can focus on understanding the differences from a molecular perspective and enhancing salt tolerance in Zinnia.

Supplementary Materials: The following supporting information can be downloaded at: https://www.mdpi.com/article/10.3390/horticulturae9020247/s1, Tables S1–S10: Cultivar and salinity effects, separately, on all plant parameters; Video S1: *Zinnia marylandica* D.Za.F.I opening stomata Video S2: *Zinnia marylandica* ABA perfusion Video S3: *Zinnia elegans* Zi.S. ABA perfusion.

Author Contributions: Conceptualization, S.Y. and N.K.; methodology, S.Y. and N.K.; investigation S.Y.; data analysis, S.Y.; writing—original draft preparation, S.Y.; writing—review and editing, S.Y and N.K.; visualization, S.Y. and N.K; supervision, N.K. All authors have read and agreed to the published version of the manuscript.

Funding: This research received no external funding.

Data Availability Statement: The data presented in this study are available on request from the corresponding author.

Acknowledgments: Data presented in the study were obtained from the thesis of Sara Yasemin (corresponding author). Special thanks to the Çukurova University Scientific Research Projects Coordinating Office (Project No: FBA-2019-11481) for their support of this work.

Conflicts of Interest: The authors declare no conflict of interest.

References

1. García-Caparrós, P.; Lao, M.T. The effects of salt stress on ornamental plants and integrative cultivation practices. *Sci. Hortic* **2018**, *240*, 430–439. [CrossRef]
2. Hasegawa, P.M.; Bressan, R.A.; Zhu, J.-K.; Bohnert, H.J. Plant cellular and molecular responses to high salinity. *Annu. Rev. Plant Physiol. Plant Mol. Biol.* **2000**, *51*, 463–499. [CrossRef] [PubMed]
3. Munns, R.; Tester, M. Mechanisms of salinity tolerance. *Annu. Rev. Plant Biol.* **2008**, *59*, 651–681. [CrossRef] [PubMed]
4. Carillo, P.; Cirillo, C.; De Micco, V.; Arena, C.; De Pascale, S.; Rouphael, Y. Morpho-anatomical, physiological and biochemical adaptive responses to saline water of *Bougainvillea spectabilis* Willd. trained to different canopy shapes. *Agric. Water Manag.* **2018** *212*, 12–22. [CrossRef]
5. Tavakkoli, E.; Fatehi, F.; Coventry, S.; Rengasamy, P.; McDonald, G.K. Additive effects of Na+ and Cl− ions on barley growth under salinity stress. *J. Exp. Bot.* **2011**, *62*, 2189–2203. [CrossRef] [PubMed]
6. Hewage, K.A.H.; Yang, J.; Wang, D.; Hao, G.; Yang, G.; Zhu, J. Chemical Manipulation of Abscisic Acid Signaling: A New Approach to Abiotic and Biotic Stress Management in Agriculture. *Adv. Sci.* **2020**, *7*, 2001265. [CrossRef] [PubMed]
7. Akcin, T.A.; Akcin, A.; Yalcın, E. Anatomical changes induced by salinity stress in *Salicornia freitagii* (Amaranthaceae). *Braz. J. Bot* **2017**, *40*, 1013–1018. [CrossRef]
8. Souza, P.U.; Lima, L.K.S.; Soares, T.L.; de Jesus, O.N.; Filho, M.A.C.; Girardi, E.A. Biometric, physiological and anatomical responses of Passiflora spp. to controlled water deficit. *Sci. Hortic.* **2018**, *229*, 77–90. [CrossRef]
9. Çavuşoğlu, K.; Kılıç, S.; Kabar, K. Some morphological and anatomical observations during alleviation of salinity (NaCI) stress on seed germination and seedling growth of barley by polyamines. *Acta Physiol. Plant.* **2007**, *29*, 551–557. [CrossRef]
10. Boughalleb, F.; Denden, M.; Ben Tiba, B. Anatomical changes induced by increasing NaCl salinity in three fodder shrubs, *Nitraria retusa, Atriplex halimus* and *Medicago arborea. Acta Physiol. Plant.* **2009**, *31*, 947–960. [CrossRef]
11. Kılıc, S.; Cavusoglu, K.; Kabar, K. Effects of 24-epibrassinolide on salinity stress induced inhibition of seed germination, seedling growth and leaf anatomy of barley. *Suleyman Demirel Univ. Fac. Art. Sci. J. Sci.* **2007**, *2*, 41–52.
12. Waqas, M.; Yaning, C.; Iqbal, H.; Shareef, M.; Rehman, H.U.; Bilal, H.M. Synergistic consequences of salinity and potassium deficiency in quinoa: Linking with stomatal patterning, ionic relations and oxidative metabolism. *Plant Physiol. Biochem.* **2020**, *159*, 17–27. [CrossRef] [PubMed]
13. Ali, S.; Hayat, K.; Iqbal, A.; Xie, L. Implications of Abscisic Acid in the Drought Stress Tolerance of Plants. *Agronomy* **2020**, *10*, 1323. [CrossRef]
14. Stimart, D.; Boyle, T. Zinnia. In *Flower Breeding and Genetics*; Anderson, N.O., Ed.; Springer: Dordrecht, The Netherland, 2007, pp. 337–357. [CrossRef]
15. Marković, M.; Šoštarić, J.; Kojić, A.; Popović, B.; Bubalo, A.; Bošnjak, D.; Stanisavljević, A. Zinnia (*Zinnia elegans* L.) and Periwinkle (*Catharanthus roseus* (L.) G. Don) Responses to Salinity Stress. *Water* **2022**, *14*, 1066. [CrossRef]

16. Yasemín, S.; Güzel Değer, A.; Çevík, S.; Koksal, N. Benchmarking of the effects of salinity on antioxidant enzymes ac-tivities, lipid peroxidation and H2O2 levels in the leaves of two Zinnia species. *Kahramanmaraş Sütçü İmam Üniversitesi Tarım Doğa Derg.* **2021**, *24*, 31–39. [CrossRef]
17. Yasemin, S.; Değer, A.G.; Köksal, N. The effects of salt stress in Zinnia (Zinnia sp.) cultivars during seed germination and at the early stages of seedling growth. *Turk. J. Agric. Res.* **2020**, *7*, 253–265.
18. Macherla, K.; McAvoy, R.J. The Effect of Salinity on the Growth and Nutrient Status of Zinnia Grown Under Short- and Long-cycle Subirrigation Management. *Hortscience* **2017**, *52*, 770–773. [CrossRef]
19. Manivannan, A.; Soundararajan, P.; Arum, L.S.; Ko, C.H.; Muneer, S.; Jeong, B.R. Silicon-mediated enhancement of physiological and biochemical characteristics of Zinnia elegans 'Dreamland Yellow' grown under salinity stress. *Hortic. Environ. Biotechnol.* **2015**, *56*, 721–731. [CrossRef]
20. Escalona, A.; Salas-Sanjuán, M.C.; Santos, C.D.; Guzmán, M. The effect of water salinity on growth and ionic con-centration and relation in plant tissues in *Zinnia elegans* and *Tagetes erecta* for use in urban landscasping. *ITEA* **2014**, *110*, 325–334.
21. Bizhani, S.; Jowkar, A.; Abdolmaleki, M. Growth and antioxidant response of *Zinnia elegans* under salt stress conditions. *Tech. J. Eng. Appl. Sci.* **2013**, *3*, 1285–1292.
22. Niu, G.; Wang, M.; Rodriguez, D.; Zhang, D. Response of Zinnia Plants to Saline Water Irrigation. *Hortscience* **2012**, *47*, 793–797. [CrossRef]
23. Villarino, G.H.; Mattson, N.S. Assessing Tolerance to Sodium Chloride Salinity in Fourteen Floriculture Species. *Horttechnology* **2011**, *21*, 539–545. [CrossRef]
24. Zivdar, S.; Khaleghi, E.; Dehkordi, F.S. Effect of salinity and temperature on seed germination indices of *Zinnia elegans* L. *J. Appl. Hortic.* **2011**, *13*, 48–51. [CrossRef]
25. Carter, C.T.; Grieve, C.M. Growth and Nutrition of Two Cultivars of Zinnia elegans Under Saline Conditions. *Hortscience* **2010**, *45*, 1058–1063. [CrossRef]
26. Yasemin, S. The Changes on Morphological Anatomic, Physiological and Biochemical Features of Zinnia (*Zinnia* sp.) Species under Salt Stress. Ph.D. Thesis, Cukurova University, Sarıçam/Adana, Turkey, 2020; 248p.
27. Arora, R.; Pitchay, D.S.; Bearce, B.C. Water-stress-induced heat tolerance in geranium leaf tissues: A possible linkage through stress proteins? *Physiol. Plant.* **1998**, *103*, 24–34. [CrossRef]
28. Kesici, M.; Gulen, H.; Ergin, S.; Turhan, E.; Ipek, A.; Koksal, N. Heat-stress Tolerance of Some Strawberry (*Fragaria* × *ananassa*) Cultivars. *Not. Bot. Horti Agrobot. Cluj-Napoca* **2013**, *41*, 244. [CrossRef]
29. Padhi, B.; Chauhan, G.; Kandoi, D.; Stirbet, A.; Tripathy, B.; Govindjee, G. A comparison of chlorophyll fluorescence transient measurements, using Handy PEA and FluorPen fluorometers. *Photosynthetica* **2021**, *59*, 399–408. [CrossRef]
30. Çimen, B. Rootstock influences on photosynthetic performance of young 'Interdonato' trees grown in calcareous soil. *Akad. Ziraat Derg.* **2019**, *8*, 185–194. [CrossRef]
31. Torun, A.A.; Erdem, N.; Kacar, Y.; Serçe, S. Screening of Wild Strawberry Genotypes against Iron Deficiency under Greenhouse Conditions. *Not. Bot. Horti Agrobot.* **2013**, *41*, 560–566. [CrossRef]
32. Incesu, M.; Cimen, B.; Yesiloglu, T.; Yilmaz, B. Growth and Photosynthetic Response of Two Persimmon Rootstocks (*Diospyros kaki* and *D. virginiana*) under Different Salinity Levels. *Not. Bot. Horti Agrobot. Cluj-Napoca* **2014**, *42*, 386–391. [CrossRef]
33. Guzel Deger, A.; Scherzer, S.; Nuhkat, M.; Kedzierska, J.; Kollist, H.; Brosche, M.; Unyayar, S.; Boudsocq, M.; Hedrich, R.; Roelfsema, M.R.G. Guard cell SLAC 1-type anion channels mediate flagellin-induced stomatal closure. *New Phytol.* **2015**, *208*, 162–173. [CrossRef]
34. Roelfsema, M.R.G.; Levchenko, V.; Hedrich, R. ABA depolarizes guard cells in intact plants, through a transient activation of R- and S-type anion channels. *Plant J.* **2004**, *37*, 578–588. [CrossRef] [PubMed]
35. Åström, H.; Metsovuori, E.; Saarinen, T.; Lundell, R.; Hänninen, H. Morphological characteristics and photosynthetic capacity of *Fragaria vesca* L. winter and summer leaves. *Flora-Morphol. Distrib. Funct. Ecol. Plants* **2015**, *215*, 33–39. [CrossRef]
36. Karabıyık, Ş.; Gundesli, M.A.; Eti, S.; Kafkas, S.; Güney, M.; Zarıfıkhosroshahı, M.; Kafkas, N. Detection of bud abscission of pistachio via histological analysis. *Acta Hortic.* **2018**, *1219*, 193–198. [CrossRef]
37. Johansen, D.A. *Plant Microtechnique*, 1st ed.; McGraw-Hill Publishing Company, Ltd.: London, UK, 1940; pp. viii+523.
38. Pérez-Clemente, R.M.; Montoliu, A.; Zandalinas, S.I.; de Ollas, C.; Gómez-Cadenas, A. Carrizo citrange Plants Do Not Require the Presence of Roots to Modulate the Response to Osmotic Stress. *Sci. World J.* **2012**, *2012*, 795396. [CrossRef]
39. Arnon, D.I. Copper enzymes in isolated chloroplasts. Polyphenoloxidase in *Beta vulgaris*. *Plant Physiol.* **1949**, *24*, 1–15. [CrossRef]
40. Lichtenthaler, H.K.; Wellburn, A.R. Determinations of total carotenoids and chlorophylls a and b of leaf extracts in different solvents. *Biochem. Soc. Transac.* **1983**, *11*, 591–592. [CrossRef]
41. Munns, R. Comparative physiology of salt and water stress. *Plant Cell Environ.* **2002**, *25*, 239–250. [CrossRef] [PubMed]
42. Sun, Y.; Niu, G.; Perez, C.; Pemberton, H.B.; Altland, J. Responses of Marigold Cultivars to Saline Water Irrigation. *Horttechnology* **2018**, *28*, 166–171. [CrossRef]
43. Oliveira, F.I.F.; de Medeiros, W.J.F.; de Lacerda, C.F.; Neves, A.L.R.; Oliveira, D.R. Saline water irrigation managements on growth of ornamental plants. *Rev. Bras. Eng. Agrícola Ambient.* **2017**, *21*, 739–745. [CrossRef]
44. Carvalho, D.R.; Vasconcelos, M.W.; Lee, S.; Vreugdenhil, D.; Heuvelink, E.; Carvalho, S.M. Moderate salinity improves stomatal functioning in rose plants grown at high relative air humidity. *Environ. Exp. Bot.* **2017**, *143*, 1–9. [CrossRef]

45. Kumar, D.; Al Hassan, M.; Vicente, O.; Agrawal, V.; Boscaiu, M. Mechanisms of Response to Salt Stress in Oleander (*Nerium oleander* L.). *Bull. UASVM Hortic.* **2016**, *73*, 249. [CrossRef]
46. Salachna, P.; Piechocki, R. Effects of sodium chloride on growth and mineral nutrition of purpletop vervain. *J. Ecol. Eng.* **2016**, *17*, 148–152. [CrossRef]
47. Li, X.; Wan, S.; Kang, Y.; Chen, X.; Chu, L. Chinese rose (*Rosa chinensis*) growth and ion accumulation under irrigation with waters of different salt contents. *Agric. Water Manag.* **2015**, *163*, 180–189. [CrossRef]
48. Zapryanova, N.; Atanassova, B. Effects of Salt Stress on Growth and Flowering of Ornamental Annual Species. *Biotechnol. Biotechnol. Equip.* **2009**, *23*, 177–179. [CrossRef]
49. Navarro, A.; Bañon, S.; Olmos, E.; Sánchez-Blanco, M. Effects of sodium chloride on water potential components, hydraulic conductivity, gas exchange and leaf ultrastructure of Arbutus unedo plants. *Plant Sci.* **2007**, *172*, 473–480. [CrossRef]
50. Fernández-García, N.; Olmos, E.; Bardisi, E.; la Garma, J.G.-D.; López-Berenguer, C.; Rubio-Asensio, J.S. Intrinsic water use efficiency controls the adaptation to high salinity in a semi-arid adapted plant, henna (*Lawsonia inermis* L.). *J. Plant Physiol.* **2014**, *171*, 64–75. [CrossRef]
51. Acosta-Motos, J.-R.; Diaz-Vivancos, P.; Álvarez, S.; Fernández-García, N.; Sanchez-Blanco, M.J.; Hernández, J.A. Physiological and biochemical mechanisms of the ornamental *Eugenia myrtifolia* L. plants for coping with NaCl stress and recovery. *Planta* **2015**, *242*, 829–846. [CrossRef]
52. Acosta-Motos, J.R.; Diaz-Vivancos, P.; Álvarez, S.; Fernández-García, N.; Sánchez-Blanco, M.J.; Hernández, J.A. NaCl-induced physiological and biochemical adaptative mechanisms in the ornamental *Myrtus communis* L. plants. *J. Plant Physiol.* **2015**, *183*, 41–51. [CrossRef] [PubMed]
53. Gómez-Bellot, M.J.; Nortes, P.A.; Ortuño, M.F.; Romero-Trigueros, C.; Fernández-García, N.; Sánchez-Blanco, M.J. Influence of arbuscular mycorrizal fungi and treated wastewater on water relations and leaf structure alterations of *Viburnum tinus* L. plants during both saline and recovery periods. *J. Plant Physiol.* **2015**, *188*, 96–105. [CrossRef] [PubMed]
54. Hameed, M.; Ashraf, M.; Naz, N. Anatomical adaptations to salinity in cogon grass [*Imperata cylindrica* (L.) Raeuschel] from the Salt Range, Pakistan. *Plant Soil* **2009**, *322*, 229–238. [CrossRef]
55. Li, S.; Hamani, A.K.M.; Zhang, Y.; Liang, Y.; Gao, Y.; Duan, A. Coordination of leaf hydraulic, anatomical, and economical traits in tomato seedlings acclimation to long-term drought. *BMC Plant Biol.* **2021**, *21*, 536. [CrossRef]
56. Becker, V.I.; Goessling, J.W.; Duarte, B.; Caçador, I.; Liu, F.; Rosenqvist, E.; Jacobsen, S.-E. Combined effects of soil salinity and high temperature on photosynthesis and growth of quinoa plants (*Chenopodium quinoa*). *Funct. Plant Biol.* **2017**, *44*, 665–678. [CrossRef] [PubMed]
57. Parihar, P.; Singh, S.; Singh, R.; Singh, V.P.; Prasad, S.M. Effect of salinity stress on plants and its tolerance strategies: A review. *Environ. Sci. Pollut. Res.* **2014**, *22*, 4056–4075. [CrossRef]
58. Breś, W.; Bandurska, H.; Kupska, A.; Niedziela, J.; Frąszczak, B. Responses of pelargonium (*Pelargonium* × *hortorum* L.H. Bailey) to long-term salinity stress induced by treatment with different NaCl doses. *Acta Physiol. Plant.* **2015**, *38*, 26. [CrossRef]
59. Veatch-Blohm, M.E.; Malinowski, M.; Keefer, D. Leaf water status, osmotic adjustment and carbon assimilation in colored calla lilies in response to saline irrigation. *Sci. Hortic.* **2012**, *144*, 65–73. [CrossRef]
60. Trivellini, A.; Gordillo, B.; Rodríguez-Pulido, F.J.; Borghesi, E.; Ferrante, A.; Vernieri, P.; Quijada-Morín, N.; González-Miret, M.L.; Heredia, F.J. Effect of Salt Stress in the Regulation of Anthocyanins and Color of *Hibiscus* Flowers by Digital Image Analysis. *J. Agric. Food Chem.* **2014**, *62*, 6966–6974. [CrossRef]
61. Parida, A.K.; Das, A.B. Salt tolerance and salinity effects on plants: A review. *Ecotoxicol. Environ. Saf.* **2005**, *60*, 324–349. [CrossRef]
62. Acosta-Motos, J.R.; Ortuño, M.F.; Bernal-Vicente, A.; Diaz-Vivancos, P.; Sanchez-Blanco, M.J.; Hernandez, J.A. Plant responses to salt stress: Adaptive mechanisms. *Agronomy* **2017**, *7*, 18. [CrossRef]
63. Acosta-Motos, J.R.; Hernández, J.A.; Álvarez, S.; Barba-Espín, G.; Sánchez-Blanco, M.J. Long-term resistance mechanisms and irrigation critical threshold showed by Eugenia myrtifolia plants in response to saline reclaimed water and relief capacity. *Plant Physiol. Biochem.* **2017**, *111*, 244–256. [CrossRef]
64. Duarte, B.; Santos, D.; Marques, J.C.; Caçador, I. Ecophysiological adaptations of two halophytes to salt stress: Photosynthesis, PS II photochemistry and anti-oxidant feedback—Implications for resilience in climate change. *Plant Physiol. Biochem.* **2013**, *67*, 178–188. [CrossRef]
65. Stepien, P.; Johnson, G.N. Contrasting Responses of Photosynthesis to Salt Stress in the Glycophyte Arabidopsis and the Halophyte Thellungiella: Role of the Plastid Terminal Oxidase as an Alternative Electron Sink. *Plant Physiol.* **2008**, *149*, 1154–1165. [CrossRef]
66. Parida, A.; Das, A.B.; Das, P. NaCl stress causes changes in photosynthetic pigments, proteins, and other metabolic components in the leaves of a true mangrove, *Bruguiera parviflora*, in hydroponic cultures. *J. Plant Biol.* **2002**, *45*, 28–36. [CrossRef]
67. Wu, S.; Sun, Y.; Niu, G.; Pantoja, G.L.G.; Rocha, A.C. Responses of Six Lamiaceae Landscape Species to Saline Water Irrigation. *J. Environ. Hortic.* **2016**, *34*, 30–35. [CrossRef]
68. Miralles, J.; Franco, J.; Sánchez-Blanco, M.; Bañón, S. Effects of pot-in-pot production system on water consumption, stem diameter variations and photochemical efficiency of spindle tree irrigated with saline water. *Agric. Water Manag.* **2016**, *170*, 167–175. [CrossRef]
69. Cai, X.; Niu, G.; Starman, T.; Hall, C. Response of six garden roses (*Rosa* × *hybrida* L.) to salt stress. *Sci. Hortic.* **2014**, *168*, 27–32. [CrossRef]

70. Bañón, S.; Miralles, J.; Ochoa, J.; Franco, J.; Sánchez-Blanco, M. Effects of diluted and undiluted treated wastewater on the growth, physiological aspects and visual quality of potted lantana and polygala plants. *Sci. Hortic.* **2011**, *129*, 869–876. [CrossRef]
71. Ashraf, M.; Harris, P.J.C. Photosynthesis under stressful environments: An overview. *Photosynthetica* **2013**, *51*, 163–190. [CrossRef]
72. Mukarram, M.; Khan, M.M.A.; Zehra, A.; Petrik, P.; Kurjak, D. Suffer or Survive: Decoding Salt-Sensitivity of Lemongrass and Its Implication on Essential Oil Productivity. *Front. Plant Sci.* **2022**, *13*, 903954. [CrossRef]
73. Vernieri, P.; Trivellini, A.; Malorgio, F.; Ferrante, A.; Serra, G. Effect of salt spray on six ornamental species. *Acta Hortic.* **2010**, *881*, 463–468. [CrossRef]
74. Eom, S.H.; Setter, T.; DiTommaso, A.; Weston, L. Differential Growth Response to Salt Stress Among Selected Ornamentals. *J. Plant Nutr.* **2007**, *30*, 1109–1126. [CrossRef]
75. Lee, M.K.; Van Iersel, M.W. Sodium Chloride Effects on Growth, Morphology, and Physiology of Chrysanthemum (*Chrysanthemum* × *morifolium*). *Hortscience* **2008**, *43*, 1888–1891. [CrossRef]
76. Bahadoran, M.; Salehi, H. Growth and flowering of two tuberose (*Polianthes tuberosa* L.) cultivars under deficit irrigation by saline water. *J. Agric. Sci. Technol.* **2015**, *17*, 415–426.
77. Cantabella, D.; Piqueras, A.; Acosta-Motos, J.R.; Bernal-Vicente, A.; Hernández, J.A.; Díaz-Vivancos, P. Salt-tolerance mechanisms induced in *Stevia rebaudiana* Bertoni: Effects on mineral nutrition, antioxidative metabolism and steviol glycoside content. *Plant Physiol. Biochem.* **2017**, *115*, 484–496. [CrossRef] [PubMed]
78. Edge, R.; McGarvey, D.; Truscott, T. The carotenoids as anti-oxidants—A review. *J. Photochem. Photobiol. B Biol.* **1997**, *41*, 189–200. [CrossRef]
79. Falk, J.; Munné-Bosch, S. Tocochromanol functions in plants: Antioxidation and beyond. *J. Exp. Bot.* **2010**, *61*, 1549–1566. [CrossRef] [PubMed]
80. Zhang, Q.-Y.; Wang, L.-Y.; Kong, F.-Y.; Deng, Y.-S.; Li, B.; Meng, Q.-W. Constitutive accumulation of zeaxanthin in tomato alleviates salt stress-induced photoinhibition and photooxidation. *Physiol. Plant.* **2012**, *146*, 363–373. [CrossRef] [PubMed]
81. Huang, C.; He, W.; Guo, J.; Chang, X.; Su, P.; Zhang, L. Increased sensitivity to salt stress in an ascorbate-deficient Arabidopsis mutant. *J. Exp. Bot.* **2005**, *56*, 3041–3049. [CrossRef]
82. Tang, X.; Mu, X.; Shao, H.; Wang, H.; Brestic, M. Global plant-responding mechanisms to salt stress: Physiological and molecular levels and implications in biotechnology. *Crit. Rev. Biotechnol.* **2014**, *35*, 425–437. [CrossRef]
83. Ashraf, M.; Foolad, M.R. Roles of glycine betaine and proline in improving plant abiotic stress resistance. *Environ. Exp. Bot.* **2007**, *59*, 206–216. [CrossRef]
84. Li, W.; Meng, R.; Liu, Y.; Chen, S.; Jiang, J.; Wang, L.; Zhao, S.; Wang, Z.; Fang, W.; Chen, F.; et al. Heterografted chrysanthemums enhance salt stress tolerance by integrating reactive oxygen species, soluble sugar, and proline. *Hortic. Res.* **2022**, *9*, uhac073. [CrossRef]
85. García-Caparrós, P.; Llanderal, A.; Pestana, M.; Correia, P.J.; Lao, M.T. *Lavandula multifida* response to salinity: Growth, nutrient uptake, and physiological changes. *J. Plant Nutr. Soil Sci.* **2016**, *180*, 96–104. [CrossRef]
86. Naidoo, G. Effects of salinity and nitrogen on growth and water relations in the mangrove, *Avicennia marina* (forsk.) vierh. *New Phytol.* **1987**, *107*, 317–325. [CrossRef]
87. Abdelgadir, E.M.; Oka, M.; Fujiyama, H. *Characteristics of Nitrate of Vegetables and Ornamentals*; Savvas, D., Passam, H., Eds.; Embryo Publications: Athens, Greece, 2005; pp. 211–261.
88. Munns, R.; Gilliham, M. Salinity tolerance of crops—What is the cost? *New Phytol.* **2015**, *208*, 668–673. [CrossRef]
89. Kaya, C.; Kirnak, H.; Higgs, D. Enhancement of growth and normal growth parameters by foliar application of potassium and phosphorus in tomato cultivars grown at high (NaCl) salinity. *J. Plant Nutr.* **2001**, *24*, 357–367. [CrossRef]
90. Mengel, K.; Kirkby, E.A. *Principles of Plant Nutrition*; Kluwer Academic Publishers: Dordrecht, The Netherlands, 2001; p. 849.
91. Simó, M.; Nieves-Cordones, M.; Nieves, M. Differences in growth and ornamental parameters between young *Chamaerops humilis* L. and *Washingtonia robusta* H. Wendl palm trees in response to salinity. *J. Hortic. Sci. Biotechnol.* **2010**, *85*, 7–11. [CrossRef]
92. Navarro, A.; Elia, A.; Conversa, G.; Campi, P.; Mastrorilli, M. Potted mycorrhizal carnation plants and saline stress: Growth, quality and nutritional plant responses. *Sci. Hortic.* **2012**, *140*, 131–139. [CrossRef]
93. García-Caparrós, P.; Llanderal, A.; Pestana, M.; Correia, P.J.; Lao, M.T. Tolerance mechanisms of three potted ornamental plants grown under moderate salinity. *Sci. Hortic.* **2016**, *201*, 84–91. [CrossRef]
94. Jampeetong, A.; Brix, H. Effects of NaCl salinity on growth, morphology, photosynthesis and proline accumulation of *Salvinia natans*. *Aquat. Bot.* **2009**, *91*, 181–186. [CrossRef]
95. Cassaniti, C.; Leonardi, C.; Flowers, T.J. The effects of sodium chloride on ornamental shrubs. *Sci. Hortic.* **2009**, *122*, 586–593. [CrossRef]
96. Cassaniti, C.; Romano, D.; Hop, M.E.C.M.; Flowers, T.J. Growing floricultural crops with brackish water. *Environ. Exp. Bot.* **2013**, *92*, 165–175. [CrossRef]
97. Picchioni, G.; Graham, C. Salinity, growth, and ion uptake selectivity of container-grown Crataegus opaca. *Sci. Hortic.* **2001**, *90*, 151–166. [CrossRef]
98. Colmer, T.D.; Munns, R.; Flowers, T.J. Improving salt tolerance of wheat and barley: Future prospects. *Aust. J. Exp. Agric.* **2005**, *45*, 1425–1443. [CrossRef]
99. Boursier, P.; Läuchli, A. Growth Responses and Mineral Nutrient Relations of Salt-Stressed Sorghum. *Crop. Sci.* **1990**, *30*, 1226–1233. [CrossRef]

100. Alfocea, F.P.; Balibrea, M.E.; Alarcón, J.J.; Bolarín, M.C. Composition of Xylem and Phloem Exudates in Relation to the Salt tolerance of Domestic and Wild Tomato Species. *J. Plant Physiol.* **2000**, *156*, 367–374. [CrossRef]
101. Adams, E.; Shin, R. Transport, signaling, and homeostasis of potassium and sodium in plants. *J. Integr. Plant Biol.* **2014**, *56*, 231–249. [CrossRef]
102. Benito, B.; Haro, R.; Amtmann, A.; Cuin, T.A.; Dreyer, I. The twins K+ and Na+ in plants. *J. Plant Physiol.* **2014**, *171*, 723–731. [CrossRef]
103. Kopittke, P.M. Interactions between Ca, Mg, Na and K: Alleviation of toxicity in saline solutions. *Plant Soil* **2011**, *352*, 353–362. [CrossRef]
104. Carter, C.T.; Grieve, C.M.; Poss, J.A.; Suarez, D.L. Production and ion uptake of Celosia argentea irrigated with saline wastewaters. *Sci. Hortic.* **2005**, *106*, 381–394. [CrossRef]
105. Grieve, C.M.; Poss, J.; Grattan, S.; Shouse, P.; Lieth, J.; Zeng, L. Productivity and Mineral Nutrition of Limonium Species Irrigated with Saline Wastewaters. *Hortscience* **2005**, *40*, 654–658. [CrossRef]
106. Carter, C.T.; Grieve, C.M. Mineral Nutrition, Growth, and Germination of *Antirrhinum majus* L. (Snapdragon) when Produced Under Increasingly Saline Conditions. *Hortscience* **2008**, *43*, 710–718. [CrossRef]
107. Navarro, A.; Bañón, S.; Conejero, W.; Sánchez-Blanco, M. Ornamental characters, ion accumulation and water status in Arbutus unedo seedlings irrigated with saline water and subsequent relief and transplanting. *Environ. Exp. Bot.* **2008**, *62*, 364–370. [CrossRef]
108. Grieve, C.; Poss, J.; Amrhein, C. Response of Matthiola incana to Irrigation with Saline Wastewaters. *Hortscience* **2006**, *41*, 119–123. [CrossRef]
109. Niu, G.; Starman, T.; Byrne, D. Responses of Growth and Mineral Nutrition of Garden Roses to Saline Water Irrigation. *Hortscience* **2013**, *48*, 756–761. [CrossRef]
110. Maathuis, F.J.M. K+Nutrition and Na+Toxicity: The Basis of Cellular K+/Na+Ratios. *Ann. Bot.* **1999**, *84*, 123–133. [CrossRef]
111. Mansour, M.; Salama, K.; Al-Mutawa, M. Transport proteins and salt tolerance in plants. *Plant Sci.* **2003**, *164*, 891–900. [CrossRef]
112. Zeng, L.; Poss, J.A.; Wilson, C.; Draz, A.-S.E.; Gregorio, G.B.; Grieve, C.M. Evaluation of salt tolerance in rice genotypes by physiological characters. *Euphytica* **2003**, *129*, 281–292. [CrossRef]
113. Colmer, T.D.; Flowers, T.; Munns, R. Use of wild relatives to improve salt tolerance in wheat. *J. Exp. Bot.* **2006**, *57*, 1059–1078. [CrossRef] [PubMed]
114. Bernstein, L. Effects of Salinity and Sodicity on Plant Growth. *Annu. Rev. Phytopathol.* **1975**, *13*, 295–312. [CrossRef]
115. Gadallah, M. Effects of Proline and Glycinebetaine on Vicia Faba Responses to Salt Stress. *Biol. Plant.* **1999**, *42*, 249–257. [CrossRef]
116. Acosta-Motos, J.R.; Álvarez, S.; Hernández, J.A.; Sánchezblanco, M.J. Irrigation of *Myrtus communis* plants with reclaimed water: Morphological and physiological responses to different levels of salinity. *J. Hortic. Sci. Biotechnol.* **2014**, *89*, 487–494. [CrossRef]
117. Koksal, N.; Kulahlioglu, I.; Ertargin, E.; Alkan-Torun, A. Relationship between salinity stress and ion uptake of hyacinth (*Hyacinthus orientalis*). *Turk. J. Agric. Nat. Sci.* **2014**, 578–583.
118. Koksal, N.; Alkan-Torun, A.; Kulahlioglu, I.; Ertargin, E.; Karalar, E. Ion uptake of marigold under saline growth conditions. *Springerplus* **2016**, *5*, 139. [CrossRef] [PubMed]
119. Esteves, B.S.; Suzuki, M.S. *Typha domingensis* Pers. subject to interactions among water level and fire event in a tropical lagoon. *Acta Limnol. Bras.* **2008**, *20*, 227–234.
120. Rout, N.P.; Shaw, B.P. Salt Tolerance in Aquatic Macrophytes: Ionic Relation and Interaction. *Biol. Plant.* **2001**, *44*, 95–99. [CrossRef]
121. Marschner, H. Mineral nutrition of higher plants. *J. Ecol.* **1995**, *76*, 1250.

horticulturae

MDPI

Article

Flowering Time and Physiological Reaction of *Dendrobium nobile* Lindl in Response to TDZ Application

Shuxian Ren [1,†], Menglu Hu [1,†], Qian Wu [1], Lin Wang [1], Huaishan Gu [1], Ziyue Chen [1], Zhu Ming [1,2] and Zongyan Li [1,2,*]

1 Landscape Architecture and Horticulture Faculty, Southwest Forestry University, Kunming 650224, China
2 Landscape Engineering Institute of National Forestry Bureau in Southwest Area, Southwest Forestry University, Kunming 650224, China
* Correspondence: lizyan@swfu.edu.cn
† These authors contributed equally to this work.

Abstract: The objective of this work was to analyze the effect of Thidiazuron (TDZ) treatment on floral initiation, flowering time, ornamental characteristics and physiological metabolism of potted *Dendrobium nobile*. Three TDZ concentrations (200, 500 and 1000 mg L^{-1}) were applied as solution to water the root zone of the plants. Control plants (plants watered with water) showed a good vegetative development but no floral branches. TDZ greatly influenced the flowering process. For all the tested TDZ concentrations, the first flower bud occurred at 55–60 days after the last irrigation (DAI), the highest TDZ concentration showing the major delay in its occurrence. The initial flowering (30% of flowered plants) began 47 days after the first flower bud initiation with no statistical differences among the treatments. Plants treated with TDZ 500–1000 mg L^{-1} showed the longest period of flowering (about 32 days) and the single flowers delayed the withering of about 2–3 days compared to the lowest TDZ treatment (200 mg L^{-1}). The number of flowers, floral branches and flowering percentage were distinctly influenced by the TDZ concentration. The highest percentage of flowering (40%) was scored when plants were watered with a TDZ solution at 500 mg L^{-1} and this was a performant treatment providing the best morphological flower features for the ornamental value of this plant. Among the physiological factors affecting the flowering, this study showed that TDZ increased the relative membrane permeability which facilitated the transport of macromolecular flower-forming substances into and out of the membrane. Therefore, the membrane permeability change could be an indicator of shifts in physiologically active substances during the flowering transition process in *Dendrobium nobile* plants.

Keywords: *Dendrobium nobile*; TDZ; flower bud formation; early flowering; physiological metabolism

Citation: Ren, S.; Hu, M.; Wu, Q.; Wang, L.; Gu, H.; Chen, Z.; Ming, Z.; Li, Z. Flowering Time and Physiological Reaction of *Dendrobium nobile* Lindl in Response to TDZ Application. *Horticulturae* **2023**, *9*, 129. https://doi.org/10.3390/horticulturae9020129

Academic Editors: Margherita Beruto, Emmy Dhooghe and Bruce Dunn

Received: 19 October 2022
Revised: 11 November 2022
Accepted: 11 November 2022
Published: 18 January 2023

1. Introduction

The regulation of flowering time is an important issue to guarantee the year-round production of ornamental crops and to enable the flower synchronization of cross-parents for successful fertilization in a crossbreeding protocol. In addition, flower initiation is one of the critical life-history phases for ornamental plants. Generally, the environmental response and endogenous pathways stimulate flower bud formation to ensure successful reproduction [1,2].

Endogenous pathways, including the genes of flowering initiation, hormones, carbohydrate levels, miRNA and floral competence, form an integrated network to regulate the flowering time [3–7]. Plant growth regulators (PGR) have been proven to act as exogenous hormones and therefore they are widely used to interfere with floral initiation in ornamental crops [8]. Thidiazuron (TDZ), which is one of the phenylurea derivatives considered as the most active cytokinin-like substance, acts as a multifunctional PGR in regulating floral induction, explant regeneration and dormancy breaking, as well as the leaf postharvest senescence process [9–11].

TDZ successfully induced the in vitro florigens of soybean, *Pennisetum glaucum*, and contributed to early flowering by shortening the required time at the vegetative growth stage [10,12]. Normally, TDZ addition to Murashige and Skoog Medium (MS media) can stimulate explant in vitro regeneration, resulting in adventitious shoots, callus production and protocorm-like body (PLB) proliferation [9,13–15].

As a preservative ingredient, 100 μM TDZ applied as a foliar spray at the preharvest stage or added in post-harvest solution significantly prolonged the vase life of *Dianthus caryophyllus* [16], *Gladiolus grandiflora* [11] and *Calendula officinalis* [17] by enhancing solution uptake, superoxide dismutase (SOD) activity and membrane stability index (MSI). Additionally, TDZ could inhibit carotenoid degradation, flower abscission and leaf senescence [18,19].

Genus *Dendrobium*, consisting of approximately 1000 species around the world, is mainly distributed in tropical and subtropical areas. China has more than 100 wild species of *Dendrobium*, with the southwest region being particularly rich; many species have high ornamental value [20]. *D. nobile* is one of the appreciated Chinese medicinal plants but it is used in breeding programs to obtain hybrids of *D. nobile* 'Spring Dream'.

Previous studies indicated that *Dendrobium* flowering could be influenced by TDZ directly through in vivo application or indirectly by influencing the flowering of the ex vitro plantlets when applied during the in vitro culture. Wen et al. (2013) proved that TDZ could promote the flowering of *D. nobile* by affecting the transcriptions of some transcription factors and signal genes [21]. Supplementing TDZ to the MS medium successfully induced the early flowering in ex vitro plantlets of *D. capra* and *D. nobile* [22,23]. The tissue-cultured plantlets of *D. officinale* produced a high percentage of inflorescences (83.2%) and normal flowers (73.6%) provided that the in vitro culture was carried out for 9 weeks on MS medium supplemented with TDZ 0.1 mg L^{-1} [24]. Some studies indicated that TDZ can further increase the frequency of floral bud formation when used in combination with other PGRs [10,23]. It was reported that TDZ combined with auxins such as 1-naphthlcetic acid (NAA) and Paclobutrazol (PP_{333}) efficiently improved floral buds compared to its single application [22]. An optimal in vitro protocol to induce a high rate of flower buds and blossomed flowers in tissue-cultured plants of *D. officinale* indicated the use of a combination of TDZ with other PGRs (0.3 mg L^{-1} PP_{333} + 0.5 mg L^{-1} 6-Benzylaminopurine (6-BA) + 0.5 mg L^{-1} NAA + 0.06 mg L^{-1} TDZ) [25]. Foliar spraying and localized irrigation at root level are appropriate methods for the in vivo application. TDZ (30 mg L^{-1}) when applied as foliar spray and through root zone watering was beneficial in inducing floral bud formation but decreased the flower size and shortened the flowering duration in potted plants of *Dendrobium* 'Sunya Sunshine' [26]. Generally, the beneficial effect of TDZ on *Dendrobium* flowering is related to several factors such as the species, application method, the concentration of this growth regulator and if it is used alone or in combination with other PGRs.

The flowering switch in higher plants is profoundly affected by various pathways. Usually, the plant hormone status is a traditional way to check the effects of the growth regulator. Additionally, endogenous carbohydrate levels and some autonomous pathways can modulate the process of floral initiation. There is little information available for TDZ effects on floral induction and the physiological metabolism. Current evidence suggests that carbohydrate and chlorophyll metabolisms involve the molecular regulatory mechanism of floral initiation [24,27,28]. Generally, the transport of macromolecular flower-forming substances in and out of the membrane could lead to the membrane permeability change.

The objective of this research was to analyze the effect of three concentrations of TDZ applied via root irrigation on the floral morphogenesis and flowering time of *D. nobile*. In particular, the flowering features and the physiological metabolism in response to TDZ rates were studied.

2. Materials and Methods

2.1. Plant Material and Experimental Conditions

The work was carried out in the greenhouses of the Southwest Forestry University Department of Horticulture and Landscape Architecture, Kunming, Yunnan, China.

The experimental plant material was set with 2–3-year-old seedlings of *D. nobile* obtained through in vivo division. Attention was paid to select homogeneous seedlings (3–4 branches, 22–25 cm height) which were transplanted into pots (diameter: 18 cm) filled with a commercial potting mix, composed of 50% coconut fiber and 50% tree bark (NATURAL COCONUT FIBER, China). From May to September 2020, seedlings were cultivated in the greenhouse according to standard cultivation practices (T/HCHMA 0002-2021 Technical regulation of *Dendrobium nobile*) and maintained under a 65% polywoven shading net. Then, the plants were moved to the incubator at 25 °C/12 °C (day/night) temperature with a daily light intensity of 90 μ mol m^{-2} s^{-1}, and 80% relative air humidity (HR). After three months under these conditions, the experiment started.

The plant growth regulator TDZ (Shanghai EKEAR Biotechnology Limited, Shanghai, China) was applied at concentrations of 200, 500 and 1000 mg L^{-1}; plants were watered with 220 mL solution, volume which was considered to saturate the water content of the culture substrate. Plants watered with 220 mL water were used as the control. All samples were subjected to initial watering on 27 December 2020 when the soil water potential dropped at a threshold value equal to 40–50% of saturated water content. Plants were watered every 15 days and were completely irrigated three times during the whole process.

The experimental unit consisted of 8 uniform plants and each treatment had three replicates.

2.2. Flowering and Morphological Characterization of Flowers

Plants were monitored from December 2020 to April 2021 and the morphological observations on floral differentiation and formation were carried out after the last watering. The number of branches and the flowering characteristics were observed in each experimental unit and recorded as number of flower spikes, number of inflorescences per branch, number of flowers per inflorescence and total number of flowers. Early flower bud initiation was recorded and the flowering process was followed from the first flowering opening to the fading flowering.

When plants reached the full bloom, the flower morphology was assessed through the length of flower inflorescence, the length and width of the petal, middle sepal and lateral sepal and the lip height.

2.3. Physiological Characteristics Measurements

Three physiological parameters were examined at 13, 23, 33 and 102 days after the last irrigation (DAI), corresponding to the different stages of transition from vegetative growth to reproductive growth. More precisely, 13, 23, 33 and 102 DAI represented the time related to vegetative bud stage (stage I), transitional stage (stage II), inflorescence development stage (stage III) and bloom stage (stage IV), respectively.

The relative membrane permeability was assessed indirectly by detecting electrolyte leakage or conductivity of the leaf tissues according to the method described by Sairam, 1994 [27]. The leaf tissue (500 mg) was picked and placed in 20 mL double-distilled water (DDW) in two separate tubes. The first was incubated at 25 °C for 30 min and its conductivity was recorded as C1. The second tube was incubated at 100 °C for 20 min and its conductivity was designated as C2. The conductivity value was measured using an Elico DDS-307 Conductivity meter (INESA Scientific instrument Co., Ltd., Shanghai, China). Membrane stability index (MSI) was determined by a ratio of C1 to C2, expressed in percent.

The relative chlorophyll content (%) was determined using a chlorophyll meter (SPAD-502, Konica Minolta, Japan) by detecting the optical concentration difference at two wavelengths between 650 nm and 940 nm. SPAD value is widely used to stand for the relative amount of chlorophyll in vivo. The fully expanded leaf from the upper one-third of the canopy was considered for the measurement. Three samples were randomly chosen for measurements and 10 leaves were tested for each treatment. The chlorophyll relative content was measured with SPAD (SPAD-502, Konica Minolta, Japan) meter from 11.00 a.m. to 13.00 p.m. at the three above said times (13, 23, 33 DAI) and the additional time (102 DAI) corresponding to the reproductive bloom stage. The measurement of each clean leaf was repeated three times.

The soluble sugars were estimated by the anthrone method with glucose as the standard [28]. The leaf was sampled from the upper branch. Approximately, 10 g of tested and fresh leaves was milled to a fine powder and dried. Each sample was obtained in three portions, and put into graduated Pyrex tubes (150 mm × 25 mm); 5 mL distilled water was added and sealed with sealing film. Then, it was extracted in boiling water for 30 min (twice); the extract was filtered into a 25 mL flask and the tubes and residue were rinsed repeatedly to fix the volume on the scale. The anthrone reagent (5 mL) was pipetted into Pyrex tubes and frozen in ice water. A 1 mL solution was layered on the acid, cooled for 5 min and then thoroughly mixed. Afterwards, it was incubated in boiling water and then cooled in water for 5 min. Absorbance was measured at 620 nm with a UD751GD spectrophotometer (INESA Scientific instrument Co., Ltd., Shanghai, China) using blank as a reference, and the standard curve was plotted with standard solution.

2.4. Data Collection and Statistical Analysis

The data collection started after the last irrigation. An axillary bud was considered to be present when the bud reached 0.20 cm diameter. The time to reach initial flowering and full bloom was assumed to occur when the percentage of blooming plants reached 30% and 60%, respectively. Fading flowering was considered when 80% of the flowers withered. The bloom duration referred to the days from the initial flowering to the fading stage. The duration of a single flower was assessed by the average flowering period of 10 flowers from each treatment.

The influence of TDZ was assessed by evaluating 24 plants used as the population for the observation on the number of flowers/plant, number of flowers/inflorescence, number of inflorescences/branch, number of floral branches/plant and number of branches/plant. Flowering percentage for each treatment was calculated according to the formula: (Total number of floral branches/Total number of branches) × 100. A floral bud of *D. nobile* usually produces an inflorescence. The number of flowers per inflorescence was determined by the number of flowers emerging from each flower bud.

Measurements on flower morphological characteristics were performed on 10 flowers randomly selected from each treatment. The length of flower inflorescence, the length and width of the petal, middle sepal and lateral sepal and the height of the lip were examined with a vernier caliper.

The leaves for the physiological test were obtained from the upper branch [29]. Three representative plants were randomly chosen for measurements in each treatment and measurements for each plant were repeated three times repetitively from December 2020 to April 2021.

The significance of TDZ rates on each growth index and physiological parameter was assessed by one-way analysis of variance (*ANOVA*) using EXCEL (Microsoft Company, Albuquerque, NM, USA). In the case of significance tests of differences, data were analyzed using an *ANOVA* and Fisher's LSD test at a significance level of 5% ($p = 0.05$).

3. Results

3.1. The Influence of TDZ on Flowering

Plants treated with TDZ showed a consistent flowering increase (Table 1). When plants were watered only with water (control), no flowering was observed although plants showed a vigorous vegetative growth. TDZ increased the number of stems bearing flowers, the lower TDZ concentrations (200 and 500 mg L^{-1}) being more effective.

Table 1. Influence of TDZ treatments on vegetative growth and flowering of *Dendrobium nobile* plants watered at root level with TDZ solution. The percentage of flowering was calculated according to the formula: (Floral branches per plant/Branches per plant) × 100. Data were collected after the last irrigation. For each variable, values followed by the same small letter are not significantly different at $p = 0.05$.

TDZ Level (mg L^{-1})	Number					Percentage of Flowering (%)
	Flowers/Plant	Flowers/ Inflorescence	Inflorescences/ Branch	Floral Branches/Plant	Branches/Plant	
0	-	-	-	-	92 ± 0.81 a	0.00 d
200	68 ± 0.77 b	2.60 ± 0.44 a	1.60 ± 0.38 a	24 ± 0.52 b	90 ± 0.81 a	26.84 ± 6.21 b
500	97 ± 1.01 a	1.50 ± 0.33 a	2.70 ± 0.48 a	38 ± 0.62 a	95 ± 1.05 a	40.38 ± 7.41 a
1000	52 ± 0.74 c	1.60 ± 0.34 a	1.70 ± 0.37 a	17 ± 0.40 c	87 ± 0.87 a	19.66 ± 4.89 c
$LSD_{0.05}$	6.58 *	1.34	1.60	2.60 *	7.45	3.78 *

$LSD_{0.05}$ means the least squared difference at the $p \leq 0.05$. * means within the same column followed by different lowercase letters are significantly different at 0.05 level of single-factor *ANOVA*.

Nevertheless, the number of flowers per inflorescence and the number of inflorescences per branch showed no statistical significance for all the tested TDZ concentrations (Table 1).

Generally speaking, 55–60 days were necessary from the last irrigation with TDZ (DAI = days after the last irrigation) to notice the first appearance of the flower buds with significant differences among the TDZ treatments (Table 2). The time span from flower bud appearance to the onset of flowering was the same for all TDZ concentrations tested (47 d) as well as the time required to reach the full bloom was similar for all the TDZ treatments (52–54 d). The highest TDZ concentrations (500 mg L^{-1} and 1000 mg L^{-1}) significantly retarded the flower withering and prolonged the flowering period, compared to the low TDZ concentration (200 mg L^{-1}) (Table 2). ANOVA showed that TDZ treatment had a significantly positive impact on blooming period of single flowers too.

Table 2. Influence on flowering of *Dendrobium nobile* plants watered at root zone with TZD solutions. Initial flowering and full bloom were considered when 30% and 60% of flowered plants were observed, respectively. Fading flowering was when 80% of the flowers withered.

TDZ Level (mg L^{-1})	Days after the Last Irrigation (DAI)				Flowering Period (No. Days)	Blooming Period of Single Flower (No. Days)
	First Flower Bud Initiation	Initial Flowering	Full Bloom	Fading Flowering		
0	-	-	-	-	-	-
200	55 ± 0.32 c	102 ± 0.58 a	109 ± 0.60 a	128 ± 0.66 b	27 ± 0.32 b	20.30 ± 0.40 c
500	58 ± 0.39 b	105 ± 0.75 a	112 ± 0.66 a	136 ± 0.78 a	32 ± 0.40 a	21.70 ± 0.31 b
1000	60 ± 0.45 a	107 ± 0.79 a	112 ± 0.63 a	138 ± 0.82 a	32 ± 0.38 a	23.20 ± 0.34 a
$LSD_{0.05}$	1.46 *	4.77	3.64	5.29 *	1.27 *	1.15 *

$LSD_{0.05}$ means the least squared difference at the $p \leq 0.05$. * means within the same column followed by different lowercase letters are significantly different at 0.05 level of single-factor *ANOVA*.

3.2. The Influence of TDZ on Flower Morphology

Flower morphology of *D. nobile* was greatly influenced by TDZ treatments. Furthermore, the *ANOVA* analyses revealed that there were significant differences among the three TDZ tested concentrations. Treatment with 500 mg L^{-1} TDZ had a good inductive effect on the largest size of floral organs, including the length and height of the petal, sepal and lip. Specifically, petal length, middle sepal length and lateral sepal length reached 5.16 cm, 5.12 cm and 4.87 cm, respectively (Figure 1). This TDZ treatment was proven to be the best in enhancing the flower characteristics compared to the other TDZ concentrations (200 and 1000 mg L^{-1}).

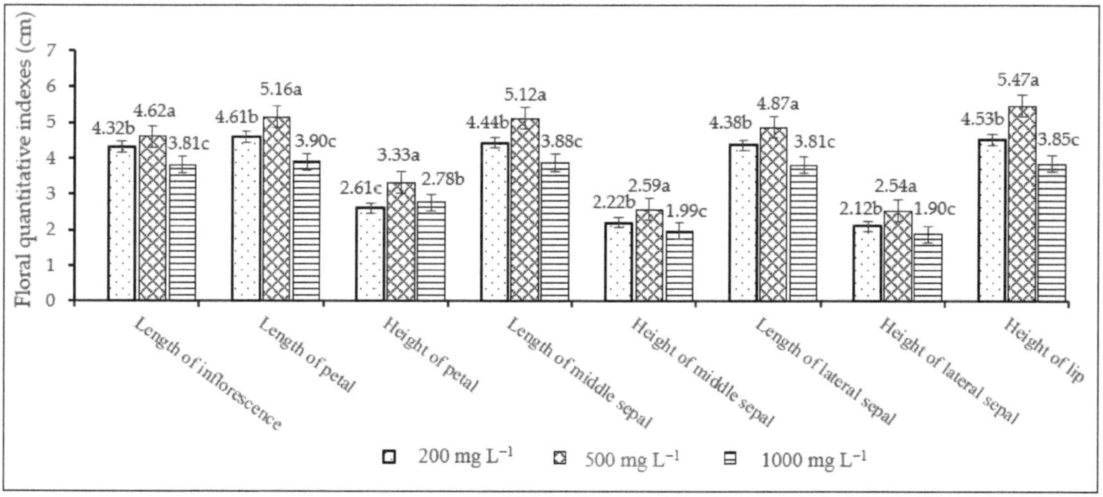

Figure 1. The effect of different TDZ concentrations on eight floral quantitative characteristics of *D. nobile*. $LSD_{0.05}$ means the least squared difference at the $p \leq 0.05$. For each characteristic, different lowercase letters are significantly different at 0.05 level of single-factor ANOVA.

3.3. *Effect of TDZ on Soluble Sugar Content during the Flowering Development*

According to Zhang et al. (2019), *Dendrobium nobile* plants sprayed with TDZ (30 mg L^{-1}) were able to switch to the flowering phase after one month of the treatment [26]. In our experiment, compared to the control (plants watered only with water), TDZ significantly affects the soluble sugar content of *Dendrobium* at all four developmental stages (Figure 2A). Generally, when TDZ was applied, the soluble sugar content was significantly raised from the vegetative bud stage to the inflorescence development stage. Plants treated with TDZ 200 mg L^{-1} showed an increase in sugar content over the three developmental stages according to a linear correlation ($y = 1.093x - 6.5557$ $R^2 = 1$); the highest sugar content value (56.18 ± 0.48 mg g^{-1}) was reached at the inflorescence development stage, and afterwards, a significant decrease was observed (25.11 ± 0.31 mg g^{-1}). When 500 mg L^{-1} TDZ was used, a slight increase in sugar content was scored between the inflorescence development stage (25.10 ± 0.52 mg g^{-1}) and the bloom stage (29.09 ± 0.97 mg g^{-1}). Plants treated with 1000 mg L^{-1} TDZ showed a significant decrease in sugar content from the inflorescence development stage (40.58 ± 0.67 mg g^{-1}) to the bloom stage (25.33 ± 0.58 mg g^{-1}).

At bloom stage, the sugar content was almost similar in the control plants and in plants treated with the three TDZ concentrations (Figure 2A).

These results suggest that the TZD treatment on *D. nobile* affected the soluble sugar content already at the lowest TDZ concentration tested (200 mg L^{-1}). A moderate rate (500 mg L^{-1} TDZ) did not significantly upgrade the soluble sugar at the transitional stage. Taking into consideration the floral characteristics scored when plants were treated with 500 mg L^{-1} TDZ, it could be argued that at this concentration, no important fluctuation of the soluble sugar content was observed in the critical period from vegetative bud until the inflorescence development stage.

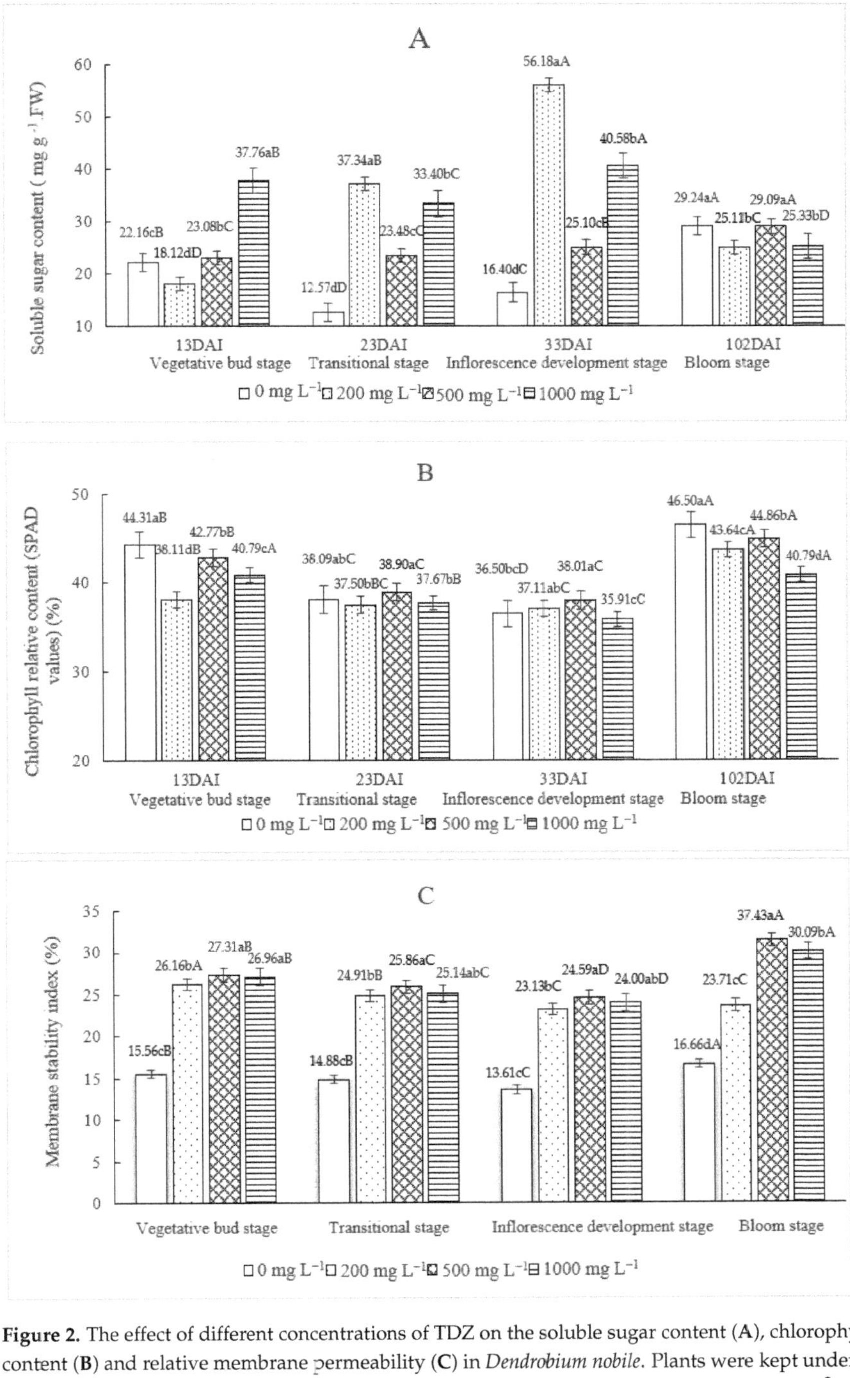

Figure 2. The effect of different concentrations of TDZ on the soluble sugar content (**A**), chlorophyll content (**B**) and relative membrane permeability (**C**) in *Dendrobium nobile*. Plants were kept under a controlled temperature in the incubator (temp: 25 °C/12 °C (day/night), light: 90 μ mol m^{-2} s^{-1}, relative air humidity (HR): 80%). For each stage, different lowercase letters indicated significantly different means (one-way *ANOVA*, $p \leq 0.05$). For each TDZ concentration, different capital letters indicated significantly different means (one-way ANOVA, $p \leq 0.05$).

3.4. Effect of TDZ on Chlorophyll Content during the Flower Development

The chlorophyll relative content (SPAD value) in the control plants showed a significant decrease until the inflorescence development stage; at the bloom stage, a significant increase was observed in both the control and TDZ treatments (Figure 2B). Generally speaking, and also in the TDZ-treated plants, a decrease in the SPAD values was scored during the three developmental stages followed by a significant increase at the bloom stage (Figure 2B). However, in the control plants, a more pronounced decrease could be observed until the inflorescence developmental stage (17.63% decrease in SPAD value in the control compared to 2.62%, 11.13% and 11.96% decrease in SPAD value when 200, 500 and 1000 mg L^{-1} TDZ was applied). At blooming stage, the leaf chlorophyll relative content in the control and the treated plants was not significantly different (Figure 2B).

3.5. The Change in Relative Membrane Permeability

Relative membrane permeability increased significantly under the three TDZ concentrations and it was significantly higher than the control at all the developmental stages (Figure 2C). Generally, the MSI% slightly decreased over the three stages until the bloom stage; at the bloom stage, a significant increase in MSI% was scored both for the control and the three TDZ applications; in the case of TDZ 200 mg L^{-1}, the MSI% was stable between the inflorescence development stage and the bloom stage. Looking at the bloom stage, it is interesting to highlight that the MSI% significantly increased from the control until the 500 mg L^{-1} TDZ treatment. At this TDZ concentration, the highest values were scored and a consequent significant decrease was then observed at 1000 mg L^{-1} TDZ.

4. Discussion

Flowering is a critical event which implies the transition from the vegetative to the reproductive phase. This process is influenced by several environmental and endogenous signals [30]. Among them, the effects of phytohormones on *orchid* flowering have been studied [31]. TDZ applied as the only growth regulator during the in vitro propagation of *D. capra* and *D. nobile* caused a low rate of flower bud formation in the ex vitro plantlets [10,22]. Some researchers considered that a combination of various hormones was more effective to induce the flower transition in *Dendrobium*. An inflorescence induction of 100% was observed when microshoots of *D. wangliangii* were cultured on 1/2 MS medium with 2 mg L^{-1} TDZ and 0.5 mg L^{-1} NAA [32]. Besides NAA, paclobutrazol (PP_{333}) was found to be beneficial for a high frequency of floral induction. Wang et al. (2014) proved that a combination of 0.05–0.1 mg L^{-1} TDZ and 1.0 mg L^{-1} NAA could produce more than 50% floral bud induction and 84% normal flowers in *D. officinale* [23]. Similarly, they could successfully induce the early flowering if the plantlets of *D. capra* were in vitro cultured in MS medium containing TDZ and NAA over an 11-month period [22]. High TDZ rates with a warm temperature (exceeding 25 °C) usually led to flower deformation.

When the in vivo irrigation technique was applied, the TDZ effect could be influenced mostly by the watering method and cultivation environment, such as substrates, air temperature and humidity. Sphagna moss and coconut shell could achieve a relatively better water holding capacity than pine bark. *D.* 'Sunya Sunshine' treated with 30–120 mg L^{-1} TDZ in a combination of foliar spaying and root irrigation exhibited no significant difference in floral induction rate, which was more than 80% in all treatments. Compared with the control group, the flower number per inflorescence and flower size decreased and the flowering duration was shortened [26]. Nevertheless, the floral bud formation percentage in nodes of *D. wardianum* was 84.3% after in vivo spraying twice with 200 mg L^{-1} Paclobutrazol (PP_{333}) and 200 mg L^{-1} TDZ mixed solution in vivo [33]. It was presumed that the species greatly determined the TDZ concentration response. Undoubtably, an appropriate rate of TDZ would enhance the *Dendrobium* flower traits and the early flowering. Our research found that a moderate rate of TDZ (500 mg L^{-1}) could induce a high percentage of flower formation, retard the flower withering and prolong the flowering period. Moreover, TDZ significantly improved the number of flowers, number of floral branches and the size of floral organs.

Our study showed that TDZ greatly improve the blooming rate of *D. nobile* but, compared to previous results (Zhang et al., 2019), the percentage of flowering (about 20–40% depending on TDZ concentration) was poor [26]. Several reasons could account for that. Firstly, the substrate greatly affected the TDZ effect because the pine bark reduced TDZ adsorption. Secondly, foliar spraying combined with root irrigation, as used in the previous studies, could better enhance the TDZ uptake and utilization with an effect on flowering percentage. The results suggested that an integrated network constructed by different ways of application and the use of several growth regulators could effectively regulate the flowering transition of *Dendrobium*.

However, some studies revealed the adverse effect of TDZ application on abnormal bud appearance [34,35]. No abnormal flowers of *D. nobile* were observed under our experimental conditions. Endogenous regulation involves many crucial chemical signals to modulate the repression and activation of florigen gene expression, such as *FLC* (FLOWERING LOCUS C), *FLD* (FLOWERING LOCUS D) and *FT* (FLOWERING LOCUS T) [36–38]. The previous studies revealed that TDZ upregulated flowering genes' expression [39,40]. In orchids, a low temperature (10 °C) was beneficial for the vernalization of *D. nobile* by upregulating the expression of DnFT and, DnVRN1 expression but decreasing DnMFT (*Dendrobium nobile* FLOWERING LOCUS T) and DnVRN1 (*Dendrobium nobile* VERNALIZATION 1) expression [40,41].

Zheng et al. 2017 reported that *D. spring* could not finish the floral differentiation at 26/21 °C (day/night). A relative low night temperature (≤17 °C) was conductive to its floral initiation [42]. Tian et al. 2007 found that its floral differentiation rate could reach 64.74% at 10 °C for 30 d [43]. A continuous temperature of 12 °C at night could enhance the flowering induction and early flowering of *D. nobile* in our experiment, although it produced a relatively low flower formation rate. Further research should be carried out to discuss whether a lower temperature could improve a high rate of floral differentiation in *Dendrobium*.

Carbohydrates are believed to play a crucial role in the regulation of flowering. *TPS1* (TREHALOSE-6-PHOSPHATE SYNTHASE 1) could inhibit its floral initiation regardless of favorably inductive environmental conditions [31]. Seo et al. (2011) proved that the INDETERMINATE DOMAIN transcription factor AtIDD8 regulated photoperiodic flowering by modulating the sugar transport and metabolism of *Arabidopsis thaliana* [44].

In our experiment, the soluble sugar content (SSC) was correlated with the occurrence of flower buds, whereas there was no distinct fluctuation at the transition stage. The 500 mg L^{-1} TDZ treatment sustained a stable level of soluble sugar content. We could speculate that during the reproductive development, a sugar signal regulation occurred mainly at the inflorescence differentiation and blooming phase under this average TDZ concentration. An intensive study should be performed to obtain a sequential and dynamic examination of its content based on the precise scheduling of the whole transition phase.

A previous study indicated that chloroplasts worked as essential sensors to cause nuclear transcriptional changes at the developmental transition stage. In *Arabidopsis*, PTM, a PHD transcription factor modulating chloroplast retrograde signaling, could mediate the transcriptional repression of *FLC* through the recruitment of *FVE* (a component of the histone deacetylase complex) [45]. In our research, TDZ irrigation decreased in the chlorophyll content in *D. nobile* from the vegetative phase to the inflorescence development stage (13 DAI, 23 DAI and 33 DAI). However, a more pronounced decrease was shown in the control plants. These results seemed to be irrelevant to the flower transition. We speculated that the chloroplast regulatory network might be involved in multiple responses including carbohydrate synthesis and flowering gene expression. Further research is required regarding the chloroplast signaling regulatory mechanism to determine the tradeoff between vegetative growth and reproductive growth.

TDZ used as a preservative solution could improve the membrane integrity of cut flowers [11,17]. In contrast to the control group, TDZ increased the cell membrane permeability and had a positive rate effect. TDZ facilitated the transport of macromolecular flower-forming substances into and out of the membrane. Therefore, the membrane permeability change could be an indicator of shifts in physiologically active substances during the flowering transition process.

5. Conclusions

Investigations addressed to promote flowering in *Dendrobium* are well appreciated due to the long vegetative period scored in these plants. In our experiments, we found that the TDZ applied to water the root zone of *D. nobile* plants was beneficial in affecting the flowering. In fact, the untreated plants were able to show a good vegetative growth but no flowering. The TDZ-treated plants flowered and the highest percentage of flowering (40%) was recorded when this growth regulator was applied at 500 mg L^{-1}. No significant differences were scored in the time frame necessary to reach the initial and full bloom stage when different TDZ concentrations were tested. On the contrary, when 500 mg L^{-1} TDZ was applied, the important flower features able to enhance the ornamental value, flower number and flowering period were improved. Among the different physiological parameters studied to elucidate the positive effect of TDZ on *D. nobile* flowering, the relative membrane permeability seems to account for a more efficient transport of nutrients and macromolecules during the flowering process and this could explain the best performing flowering.

Author Contributions: Investigation, Q.W. and Z.C.; methodology, L.W.; data curation, H.G. and Z.C.; draft preparation, S.R.; writing—editing, M.H. and S.R.; writing—review and supervision, Z.L.; resources, Z.M. All authors have read and agreed to the published version of the manuscript.

Funding: The project is funded by the National Key Research and Development Project of the Ministry of Science and Technology (2019YFD10010003). This research belongs to the sub-project of efficient breeding technology and variety creation of Orchid (2019YFD10010003). This item is also funded by Innovation and Entrepreneurship Training for College students of the Ministry of Education (20180554027) and Yunnan Postgraduate Adviser Team Construction Project (No. 2019-101).

Institutional Review Board Statement: Not applicable.

Informed Consent Statement: Not applicable.

Data Availability Statement: Not applicable.

Acknowledgments: The authors thank anonymous reviewers for comments on this manuscript.

Conflicts of Interest: The authors declare no conflict of interest. The funders had no role in the design of the study; in the collection, analyses, or interpretation of data; in the writing of the manuscript, or in the decision to publish the results.

References

1. Song, Y.H.; Ito, S.; Imaizumi, T. Flowering time regulation: Photoperiod-and temperature-sensing in leaves. *Trends Plant Sci.* **2013**, *18*, 575–583. [CrossRef] [PubMed]
2. Verhage, L.; Angenent, G.C.; Immink, R.G.H. Research on floral timing by ambient temperature comes into blossom. *Trends Plant Sci.* **2014**, *19*, 583–591. [CrossRef]
3. Amasino, R.M.; Michaels, S.D. The timing of flowering. *Plant Physiol.* **2010**, *154*, 516–520. [CrossRef] [PubMed]
4. Waheed, S.; Zeng, L. The critical role of miRNAs in regulation of flowering time and flower development. *Genes* **2020**, *11*, 319. [CrossRef] [PubMed]
5. Quiroz, S.; Yustis, J.C.; Chávez-Hernández, E.C.; Martínez, T.; Sanchez, M.D.; Garay-Arroyo, A.; Álvarez-Buylla, E.R.; García-Ponce, B. Beyond the genetic pathways, flowering regulation complexity in *Arabidopsis thaliana*. *Int. J. Mol. Sci.* **2021**, *22*, 5716. [CrossRef]
6. Higuchi, Y. Florigen and anti-florigen: Flowering regulation in horticultural crops. *Breed. Sci.* **2018**, *68*, 109–118. [CrossRef]
7. Kim, J.J. The miR156-SPL3 module regulates ambient temperature-responsive flowering via FT in *Arabidopsis thaliana*. *Plant Physiol.* **2012**, *159*, 461–478. [CrossRef]
8. Anawal, V.V.; Narayanaswamy, P.; Yallesh Kumar, H.S. Effects of plant growth regulators on induction of flowering in pomegranate (*Punica granatum* L.) cv. Bhagwa. *Int. J. Sci. Res.* **2015**, *4*, 7–9.

9. Chen, L.R.; Chen, J.T.; Chang, W.C. Efficient production of protocorm-like bodies and plant regeneration from flower stalk explants of the sympodial orchid *Epidendrum radicans*. *In Vitr. Cell. Dev. Biol. Plant* **2002**, *38*, 441–445. [CrossRef]
10. Kurumwanshi, R.S.; Jadhav, P.V.; Surbhaiyya, S.D.; Moharil, M.P.; Nandanwar, R.S.; Wandhare, M.R.; Manjaya, J.G. In vitro florigenesis: An efficient regeneration system avoiding time consuming vegetative phase in popular Indian soybean variety JS-335. *Indian J. Exp. Biol.* **2021**, *59*, 458–466.
11. Singh, M.; Tiwari, N. Thidiazuron outpaces 6-benzylamino purine and kinetin in delaying flower senescence in *Gladiolus grandiflora* by alleviating physiological and biochemical responses. *J. Appl. Biol. Biotechnol.* **2021**, *9*, 56–62.
12. Maity, P.J.; Kulkarni, V.M.; Vishnu, B. Thidiazuron-induced multiple shoot regeneration and in vitro flowering in *Pennisetum glaucum* (L.). *R. Br. Phytomorphol.* **2016**, *66*, 45–50.
13. Sujjaritthurakarn, P.; Kanchanapoom, K. Efficient direct protocorm-like bodies induction of dwarf *Dendrobium* using Thidiazuron. *Not. Sci. Biol.* **2011**, *3*, 88–92. [CrossRef]
14. Celikel, F.G.; Reid, M.S.; Jiang, C.Z. Water uptake and vase life of cut *Gardenia jasminoides* flowers. In Proceedings of the XXX International Horticultural Congress IHC 2018: International Symposium on Ornamental Horticulture and XI International 1263, Istanbul, Turkey, 12–16 August 2018; pp. 335–342.
15. Hajong, S.; Kumaria, S.; Tandon, P. Effect of plant growth regulators on regeneration potential of axenic nodal segments of *Dendrobium chrysanthum* Wall. ex Lindl. *J. Agric. Sci. Technol.* **2013**, *15*, 1425–1435.
16. Chamani, E.; Esmaeilpour, B. Thidiazuron effects on physiochemical characteristics of carnation during pre and postharvest periods. *J. Appl. Hort.* **2007**, *9*, 115–117. [CrossRef]
17. Lone, M.L.; Farooq, S.; Parveen, S.; Tahir, I. 6-Benzylamino purine outperforms Kinetin and Thidiazuron in ameliorating flower longevity in *Calendula officinalis* L. by orchestrating physiological and biochemical responses. *Ornam. Hortic.* **2021**, *27*, 183–195. [CrossRef]
18. Ferrante, A.; Vernieri, P.; Serra, G.; Tognoni, F. Changes in abscisic acid during leaf yellowing of cut stock flowers. *Plant Growth Regul.* **2004**, *43*, 127–134. [CrossRef]
19. Sankhla, N.; Mackay, W.A.; Davis, T.D. Reduction of flower abscission and leaf senescence in cut phlox inflorescences by thidiazuron. In Proceedings of the XXVI International Horticultural Congress: Issues and Advances in Postharvest Horticulture 628, Toronto, ON, Canada, 11–17 August 2002; pp. 837–841.
20. Chen, S.C.; Tsi, Z.H.; Luo, Y.B. Flora Reipublicae Popularis Sinicae Tomus 19. In *Dendrobium*; Science Press: Beijing, China, 1999; pp. 75–146.
21. Wen, Z.Z.; Zhang, E.; Liu, Y.G.; Liu, W. Primary screening of differentially expressed genes during TDZ induced floral initiation with SSH in *Dendrobium nobile*. *Acta Hortic. Sin.* **2013**, *40*, 1591–1599.
22. Lawrie, M.D.; Layina, Z.; Ningtias, D.R.; Alifianto, F.N.; Indrianto, A.; Purwantoro, A.; Semiarti, E. In vitro germination and flowering of *Dendrobium capra* JJ smith, an endemic orchid of java. *Hayati J. Biosci.* **2021**, *28*, 172. [CrossRef]
23. Wang, A.; Li, J.; Zeng, S.; Chen, Z. Preliminary report on the study of in vitro flowering and fruiting of *Dendrobium officinale*. *Guizhou Agric. Sci.* **2014**, *42*, 34–37.
24. Qian, X.; Wang, C.; Ouyang, T.; Tian, M. In vitro flowering and fruiting in culture of *Dendrobium officinate* kimura et migo. (Orchidaceae). *Pak. J. Bot.* **2014**, *46*, 1877–1882.
25. Cen, X.F.; Huang, C.H.; Wei, P.X. Effects of hormone factors on the in vitro culture flowering induction of *Dendrobium officinate* Kimura et Migo. *Agric. Sci. Technol. Hunan* **2010**, *11*, 75–79.
26. Zhang, D.; Liao, Y.; Lu, S.; Li, C.; Shen, Z.; Yang, G.; Yin, J. Effect of thidiazuron on morphological and flowering characteristics of *Dendrobium* 'Sunya Sunshine' potted plants. *N. Z. J. Crop Hortic. Sci.* **2019**, *47*, 170–181. [CrossRef]
27. Sairam, R.K. Effect of moisture-stress on physiological activities of two contrasting wheat genotypes. *Indian J. Exp. Biol.* **1994**, *32*, 594.
28. Yemm, E.W.; Willis, A.J. The estimation of carbohydrates in plant extracts by anthrone. *Biochem. J.* **1954**, *57*, 508. [CrossRef] [PubMed]
29. Alkio, M.; Schubert, A.; Diepenbrock, W.; Grimm, E. Effect of source-sink ratio on seed set and filling in sunflower (*Helianthus annuus* L.). *Plant Cell Environ.* **2003**, *26*, 1609–1619. [CrossRef]
30. Wahl, V.; Ponnu, J.; Schlereth, A.; Arrivault, S.; Langenecker, T.; Franke, A.; Feil, R.; Lunn, J.E.; Stitt, M.; Schmid, M. Regulation of flowering by trehalose-6-phosphate signaling in *Arabidopsis thaliana*. *Science* **2013**, *339*, 704–707. [CrossRef]
31. Jiang, Z.; Sun, L.; Wei, Q.; Ju, Y.; Zou, X.; Wan, X.; Liu, X.; Yin, Z. A new insight into flowering regulation: Molecular basis of flowering initiation in Magnolia × soulangeana 'Changchun'. *Genes* **2019**, *11*, 15. [CrossRef]
32. Zhao, D.; Hu, G.; Chen, Z.; Shi, Y.; Zheng, L.; Tang, A.; Long, C. Micropropagation and in vitro flowering of *Dendrobium wangliangii*: A critically endan- gered medicinal orchid. *J. Med. Plants Res.* **2013**, *7*, 2098–2110.
33. Wang, Z.H.; Zhu, G.F.; OUM, C.; Wang, B.Q. Induction of plant growth regulators on early in vivo flowering of nobile-type *Dendrobium*. *Guangdong Agric. Sci.* **2008**, *10*, 37–39.
34. Wu, H.H.; Chen, F.C. Effect of plant growth regulators on shoot multiplication from flower stalk nodal buds of *Phalaenopsis* and *Doritaenopsis*. *J. Taiwan Soc. Hortic. Sci.* **2008**, *54*, 151–159.
35. Wang, Z.H.; Wang, L.; Ye, Q.S. High frequency early flowering from in vitro seedlings of *Dendrobium nobile*. *Sci. Hortic.* **2009**, *122*, 328–331. [CrossRef]
36. Xiang, L.; Li, X.; Qin, D.; Guo, F.; Wu, C.; Miao, L.; Sun, C. Functional analysis of FLOWERING LOCUST orthologs from spring orchid (*Cymbidium goeringii* Rchb. f.) that regulates the vegetative to reproductive transition. *Plant Physiol. Biochem.* **2012**, *58*, 98–105. [CrossRef]

37. Wang, Y.; Liu, L.; Song, S.; Li, Y.; Shen, L.; Yu, H. DOFT and DOFTIP1 affect reproductive development in the orchid *Dendrobiun* Chao Praya Smile. *J. Exp. Bot.* **2017**, *68*, 5759–5772. [CrossRef]
38. Lee, P.L.; Chen, J.T. Plant regeneration via callus culture and subsequent in vitro flowering of *Dendrobium huoshanense*. *Acta Physiol. Plant.* **2014**, *36*, 2619–2625. [CrossRef]
39. Wen, Z.; Guo, W.; Li, J.; Lin, H.; He, C.; Liu, Y.; Zhang, Q.; Liu, W. Comparative transcriptomic analysis of vernalization-and cytokinin-induced floral transition in *Dendrobium nobile*. *Sci. Rep.* **2017**, *7*, 45748. [CrossRef]
40. Wang, H.M.; Tong, C.G.; Jang, S. Current progress in orchid flowering/flower development research. *Plant Signal. Behav.* **2017**, *12* e1322245. [CrossRef]
41. Wang, S.L.; An, H.R.; Tong, C.G.; Jang, S. Flowering and flowering genes: From model plants to orchids. *Hortic. Environ Biotechnol.* **2021**, *62*, 135–148. [CrossRef]
42. Zheng, B.Q.; Deng, X.M.; Li, K.; Miao, K.; Wang, Y. Effects of temperature treatment on flower bud differentiation and developmen of *Dendrobium*. *For. Res.* **2017**, *30*, 460–464.
43. Tian, D.Q.; Cao, Q.Y.; Ding, H.Q.; Yu, L.J. A preliminary study on flower forcing of *Dendrobium spring* at low temperature. *Zhejiang Agric. Sci.* **2007**, *1*, 38–40.
44. Seo, P.J.; Ryu, J.; Kang, S.K.; Park, C.M. Modulation of sugar metabolism by an INDETERMINATE DOMAIN transcription factor contributes to photoperiodic flowering in *Arabidopsis*. *Plant J.* **2011**, *65*, 418–429. [CrossRef] [PubMed]
45. Feng, P.; Guo, H.; Chi, W.; Chai, X.; Sun, X.; Xu, X.; Ma, J.; Rochaix, J.D.; Leister, D.; Wang, H.; et al. Chloroplast retrograde signa regulates flowering. *Proc. Natl. Acad. Sci. USA* **2016**, *113*, 10708–10713. [CrossRef] [PubMed]

horticulturae

MDPI

Article

The Composition of Anthocyanins and Carotenoids Influenced the Flower Color Heredity in Asiatic Hybrid Lilies

Jiewen Li [1], Jiawei Chen [1], Qian Zhang [1,2], Pengcheng Yu [1], Yanping Zhou [1,3] and Guixia Jia [1,*]

1 Beijing Key Laboratory of Ornamental Plants Germplasm Innovation & Molecular Breeding, National Engineering Research Center for Floriculture, Beijing Laboratory of Urban and Rural Ecological Environment, Key Laboratory of Genetics and Breeding in Forest Trees and Ornamental Plants of Education Ministry, School of Landscape Architecture, Beijing Forestry University, Beijing 100083, China

2 College of Forestry, Shanxi Agricultural University, Jinzhong 030801, China

3 College of Agriculture and Forestry, Longdong University, Qingyang 745000, China

* Correspondence: gxjia@bjfu.edu.cn

Abstract: Flower color, including color hues and spot patterns, is an extremely important trait in Asiatic hybrid lilies. Investigating flower color variation across different cross combinations may help us select breeding parents efficiently. Thus, three hybridizations with different compositions of anthocyanins and carotenoids were performed, and 65, 289 and 125 offspring were obtained, respectively. For the parents and progenies, flower color hues were quantified by the CIELAB system, and the total number of raised spots on tepals was counted. Then, a cluster analysis and a statistical analysis were used to explore the inheritance patterns of flower color. The results showed that an orange flower color with high levels of carotenoids was highly heritable and that the progenies were less segregated, showing orange, orange-red, and yellow flowers. Parents showing red and purple-red flowers with differing levels of carotenoids and anthocyanins produced offspring with extensive segregation, including pink-white, pink, orange-pink, orange-red, yellow, and orange flowers. Meanwhile, different contents of anthocyanins and carotenoids resulted in variable proportions in color groups. Additionally, for F1 generation, the number of raised spots was continuously separated, demonstrating distinct quantitative genetic characteristics. For parents with few or intermediate spots, the hybrid plants showed both prominent negative and positive heterosis. For parents with many spots that covered almost the full tepals, the hybrid seedlings all exhibited raised spots.

Keywords: Asiatic hybrid lilies; flower color; raised spots; CIELAB color system; anthocyanins; carotenoids

Citation: Li, J.; Chen, J.; Zhang, Q.; Yu, P.; Zhou, Y.; Jia, G. The Composition of Anthocyanins and Carotenoids Influenced the Flower Color Heredity in Asiatic Hybrid Lilies. *Horticulturae* **2022**, *8*, 1206. https://doi.org/10.3390/horticulturae8121206

Academic Editors: Margherita Beruto, Emmy Dhooghe and Bruce Dunn

Received: 21 November 2022
Accepted: 10 December 2022
Published: 16 December 2022

1. Introduction

Lilium is a major commercial floricultural crop around the world, due to its large, showy, fragrant, and multicolored flowers. The genus, which includes 110–115 species, is classified into eight sections, including *Sinomartagon*, *Martagon*, *Pseudolirium*, and others [1–3]. Interspecific and intersection hybridizations are the principal methods used for lily breeding [4–6]. Currently, more than 10,000 cultivars are registered with the Royal Horticulture Society, which are classified into different established hybrid groups, such as Asiatic, Oriental, and Longiflorum hybrids [7].

Asiatic hybrid lilies refer to the large hybrid population in the *Lilium* genus, which is derived from interspecific crosses of the *Sinomartagon* section [4]. These lilies are popular and are widely cultivated for landscaping due to their extensive variation in flower color [8], which is one of the most important characteristics determining the commercial value of lilies. Anthocyanins [9] and carotenoids [10–12] are the main pigments accumulated in their flowers. Generally, anthocyanins are responsible for pink colors, whereas carotenoids are primarily responsible for orange and yellow hues, and a combination of the two is responsible for red hues [9,10,13–15].

Some lily breeders have focused on the heredity of flower color hues for Asiatic hybrid cultivars to improve the aesthetics of their flowers. It was found that in orange × yellow and orange × orange crosses, the seedlings primarily bore orange flowers, and a few of them showed yellow flowers. For example, hybrids with orange-red as their dominant flower color were produced when the orange-red-flowered cultivars ('Compass', 'Gran Paradiso' and 'Alisa') were selected as parents [16]. The yellow-flowered Asiatic hybrid cultivar 'Dreamland' was crossed with the orange-flowered wild species *L. lancifolium* to produce F1 plants with flowers showing varying degrees of orange and a few individuals with yellow flowers [17]. Similarly, 'Kotnali', an orange-flowered cultivar, and 'Tiny Bee', a yellow-flowered cultivar, were crossed to produce hybrid progeny, which separated into plants with orange and yellow flowers [18]. Furthermore, the offspring from the following crosses, white × salmon-orange, orange-red × pink, and pink × orange, displayed parental codominance for the color hues. In the cross between 'Olga' (apricot-white) and 'Arabeska' (salmon-orange), the progenies showed flowers with intermediate hues of rosy-pink, apricot, and salmon-orange [16]. Progenies of the cross between the pink-flowered Asiatic hybrid cultivar 'Renoir' and the orange-flowered wild species *L. davidii* var. *willmottiae* presented flowers with different degrees of pink, orange-pink, and yellow [19]. However, due to the lack of hybrid populations and quantitative descriptions, the separation regularity of flower color controlled by different compositions and contents of anthocyanin and carotenoid remains unclear.

Currently, the quantification of color phenotypes is an important and objective method for defining flower color. The CIELAB (International Commission on Illumination L*a*b*) scale has been widely utilized in quantifying flower and fruit color phenotypes. The CIELAB color system is a color space that characterizes certain colors into geometric coordinates by employing the coordinate axes L^*, a^*, and b^*, which represent lightness, redness (red to green), and yellowness (yellow to blue), respectively [20,21]. As this color system can measure the target color rapidly and precisely using colorimetry, it is frequently used in the determination and classification of flower color hues in ornamental plants, such as *Magnolia biondii*, *Alstroemeria pallida*, *Narcissus*, *Rhododendron*, *Hemerocallis*, *Chrysanthemum*, *Zantedeschia hybrida*, *Gerbera hybrida*, rose, and *Lilium* [22–31].

In addition to color hues, spot patterns, including raised spots (also known as papillae), splatter-type spots, and brush marks, are also important factors affecting flower color [8,32–34]. Among them, raised spots are a unique pigmentation morphology in lilies [35]. The interior surfaces of tepals are raised to develop bumps where parenchymal and epidermal cell numbers increase and anthocyanin pigments accumulate [36]. Several studies have focused on the heredity of raised spots. In the Asiatic hybrid lily crosses with spotted cultivars ('Montreux', 'Renoir', and 'Orlito') × a non-spotted cultivar ('Connecticut King'), F1 plants with and without spots segregated at a 1:1 ratio, which indicated that the presence or absence of raised spots was regulated by a single locus [32,35]. Moreover, progenies of the cross between the 'White Fox' (no spots) and the 'Connecticut King' (no spots) showed that the segregation ratio of individuals with and without spots was 1:3, which suggested that a pair of alleles suppressed spot formation [37]. However, Zhao et al. reported that dominant genes promote spot formation in Asiatic hybrid lilies [38]. Additionally, the raised spot patterns on the perianths of Asiatic hybrid lilies are multifarious. They range in distribution from the basal part and marginal zone to half the tepals and almost the full tepals. Therefore, to clarify the intricate inheritance patterns of raised spots in lilies, more hybrid combinations and larger F1 populations are imperative.

In the present research, four Asian hybrid cultivars with different flower colors were selected as parents to design three cross combinations. The pink-white-flowered cultivar 'Easy Waltz' (no carotenoids, medium anthocyanins, few spots) was used as the maternal parent in three combinations. The orange-flowered cultivar 'Tresor' (high carotenoids, no anthocyanins, intermediate spots), the red-flowered cultivar 'Red Life' (high carotenoids, medium anthocyanins, many spots), and the purple-red-flowered cultivar 'Pearl Loraine' (low carotenoids, high anthocyanins, intermediate spots) were all used as paternal parents (pigment composition and content from Wang et al., 2021 [39]). The flower color (color hues and raised spots) of hybrid plants was determined. Then, according to the color phenotype, a cluster analysis and a statistical analysis were used to explore the heredity and variation of these decorative features for the three combinations. The objective of this study was to better understand the flower color changes from the different combinations of pigment composition and content in Asiatic hybrid lilies. The results could provide some insights into how to select parents to develop more lily cultivars with high aesthetic value.

2. Materials and Methods

2.1. Plant Materials

Four Asiatic hybrid cultivars with different flower colors were used as parents to obtain hybrid populations (Figure 1). Three cross combinations, 'Easy Waltz' × 'Tresor', 'Easy Waltz' × 'Red Life', and 'Easy Waltz' × 'Pearl Loraine', were performed, and 65, 289 and 125 seedlings were obtained, respectively (Table 1). All the cultivars were tetraploid (2n = 4x = 48). The hybrid seeds with embryos were stored in sand at 4 °C for a month. Then, they were sown and grown for three years until blossoming occurred. All these materials were grown in the germplasm resource nursery of Beijing Forestry University.

Figure 1. The parents of hybridizations. ♀: maternal parent; ♂: paternal parent.

Table 1. Pigment composition and content of parents and number of F1 hybrids in three cross combinations.

No.	Cross Combination	♀	Pigment Composition and Content (μg/g, DW)	♂	Pigment Composition and Content (μg/g, DW)	Hybrid Number
1	EW01	'Easy Waltz'	no carotenoid medium anthocyanin (94.36 ± 7.36)	'Tresor'	high carotenoid (141.38 ± 5.32) no anthocyanin	65
2	EW02	'Easy Waltz'	no carotenoid medium anthocyanin (94.36 ± 7.36)	'Red Life'	high carotenoid (120.78 ± 9.56) medium anthocyanin (91.33 ± 3.94)	289
3	EW03	'Easy Waltz'	no carotenoid medium anthocyanin (94.36 ± 7.36)	'Pearl Loraine'	low carotenoid (4.66 ± 0.16) high anthocyanin (144.43 ± 2.51)	125
Total						479

Note: pigment composition and content from Wang et al., 2021 [39]. DW, Dry weight.

2.2. Flower Color Measurement

The fresh tepal colors of all hybrids and parents were measured using a colorimeter (NF555, Nippon Denshoku Industries Co., Ltd., Tokyo, Japan) under a C/2° light source.

The inner tepals were placed on clean white paper, and the light source was aligned to the central part of the adaxial surface (Ad. 24), based on The International Union for the Protection of New Varieties of Plants [40]. The measurement was repeated three times. The average values of the color parameters, L^*, a^*, and b^*, of the CIELAB system were obtained. In this system, the parameter L^* value indicates lightness, ranging from 0 (black) to 100 (white). The parameter a^* value represents redness, with positive values for red colors and negative values for green. The parameter b^* value represents yellowness, with positive values for yellow colors and negative values for blue [20,41]. In addition, the total numbers of raised spots (abbreviated as SN) on the inner and outer tepals were counted in sequence. For the parents, the average of the spot numbers of the three individuals was calculated.

2.3. Statistical Analysis

Basic data management and statistics were performed using Microsoft Office Excel 2016. Maximum, minimum, and mean values, and the standard deviation of crossing parents and hybrids were calculated using the SPSS 25 software. The following formulae were used to determine the mid-parent value (*MPs*), index of heterosis (*Hi*), and coefficient of variation (*CV*):

$$MPs = \frac{P_1 + P_2}{2}$$

$$Hi = \frac{\overline{X}}{MPs} \times 100\%$$

$$CV = \frac{\sigma}{\overline{X}} \times 100\%$$

where P_1 and P_2 represent the values of the parents, $\overline{X}$ represents the mean value of the hybrids, and σ is the standard deviation. Then, cluster analysis (between-group linkage, squared Euclidean distance) with L^*, a^*, and b^* values of all hybrids was carried out using SPSS 25 software. Finally, box plots, three-dimensional coordinate scatterplots, and frequency distribution histograms were all created using the Origin 2021 software.

3. Results

3.1. Colorimetric Evaluation and Numerical Classification of the Hybrids

The cluster analysis with L^*, a^*, and b^* values for all 479 F1 hybrids divided them into 6 flower color groups (Euclidean distance = 7.5), including orange (236 individuals), pink (18 individuals), pink-white (27 individuals), yellow (43 individuals), orange-pink (134 individuals), and orange-red (21 individuals) (Figure 2). The flower color groups accounted for 49.27%, 3.76%, 5.63%, 8.98%, 27.97%, and 4.38% of the total hybrids, respectively.

Subsequently, to clearly describe and distinguish the flower color characteristics of the 6 groups, box plots were created (Figure 3). The color coordinates showed that the L^* values of hybrids ranged from 30.71 to 83.19. The L^* values of the pink-white and yellow flower color groups were higher than those of the other groups, while that of the orange-red flower color group was the lowest. The a^* values varied from −2.68 to 38.41; the orange-red flower color group had the highest a^* value, while the yellow and pink-white groups had the lowest. The b^* values ranged from 2.98 to 73.35, and that of the orange flower color group was the highest, with an order of orange > yellow > orange-red > orange-pink > pink > pink-white. Furthermore, although the distribution areas of the b^* values of the pink-white and pink flower color groups overlapped, their a^* value distributions were obviously different. Similarly, the orange-red and orange-pink flower color groups could be distinguished based on the a^* value regions, even though their b^* values overlapped.

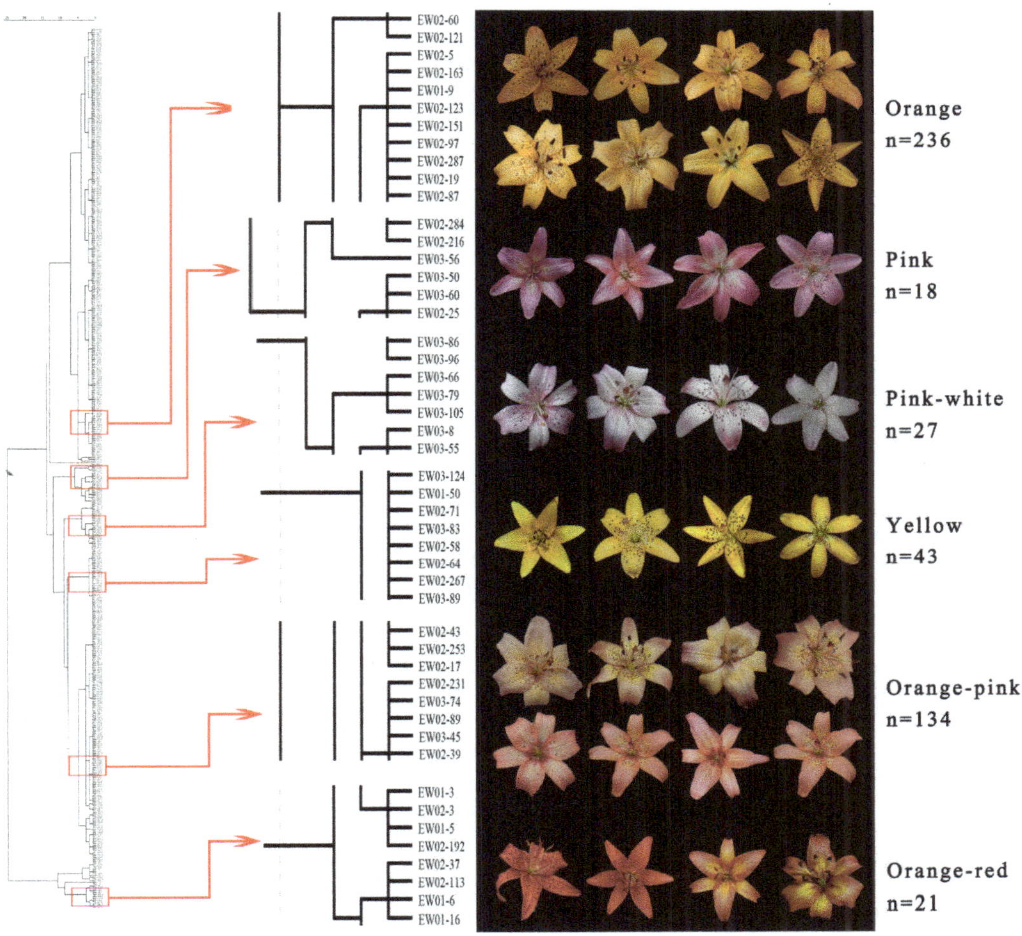

Figure 2. The cluster analysis with L^*, a^*, and b^* values for 479 F1 hybrids. The left part of this figure represents the cluster graph based on the flower color data (L^*, a^*, and b^* values) of each F1 individual; the middle part of this figure represents an enlargement of the partial branches of the cluster graph; the right part of this figure displays typical pictures of the 6 color groups (n represents the number of individuals in each group).

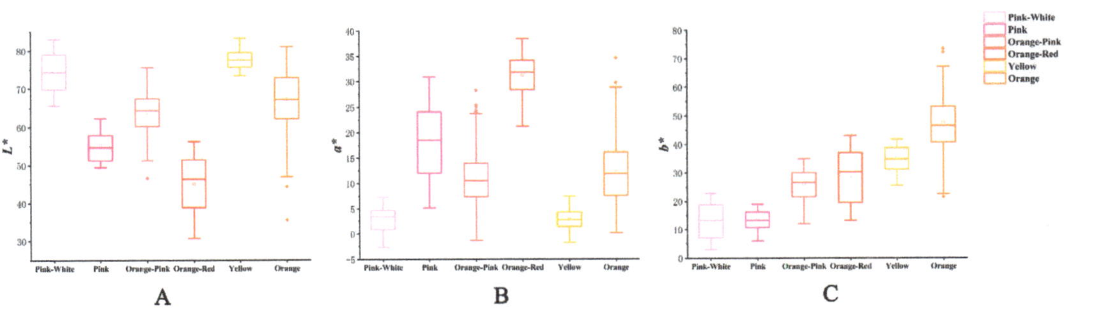

Figure 3. Box plots of different color groups for F1 hybrids with L^* (**A**), a^* (**B**), and b^* (**C**) values. *X*-coordinate: 6 color groups; *Y*-coordinate: color space value L^*, a^*, and b^*.

3.2. Variation in Flower Color of the Cross Combinations

To clarify the flower color diversity of F1 hybrids among the three combinations, we determined the sizes of the flower color groups for each cross (Table 2). The typical pictures of different color groups in each combination were shown (Figure 4). Meanwhile, the color space values, L^*, a^*, and b^*, of the progeny from each combination were assigned three-dimensional coordinates (Figure 5).

Table 2. The flower color segregation of F1 plants for three cross combinations.

Cross Combination	Pink-White	Pink	Orange-Pink	Orange-Red	Yellow	Orange
EW01	0	0	0	6	2	57
EW02	4	11	80	14	21	159
EW03	23	7	54	1	20	20

For the EW01 combination, there were three flower color groups (Figures 4A and 5A), including orange-red, yellow, and orange. The orange flower color group was the largest (Table 2). The offspring of EW02 were divided into 6 flower color groups (Figures 4B and 5B), which were orange, orange-pink, yellow, orange-red, pink and pink-white in order of proportion. Among them, the pink and pink-white groups had the lowest numbers of individuals, 11 and 4, respectively (Table 2). Progenies of EW03 were also divided into the same 6 flower color groups (Figures 4C and 5C). The orange-pink group accounted for the largest proportion, followed by the pink-white group, while the orange-red group accounted for the smallest proportion (Table 2).

Thus, compared with those of EW01, the progeny plants of EW02 and EW03 had more extensive and abundant variation in flower color. Comparison between EW02 and EW03 showed that the proportions of offspring in the six flower color groups were different.

To elucidate the inheritance of flower color, we investigated the variation and distribution of floral color measurement indices (L^*, a^*, and b^* values) for the parents and their offspring (Figure 6). Comparing the L^* values of the three combinations, we discovered that the averages of the L^* value for the offspring (66.45, 65.22, and 68.12, respectively) were closer to that of the maternal parents (68.25). In contrast, the a^* value distributions of the hybrids from the three hybridizations (−1.14~38.41, −1.15~38.41, and −2.68~28.77, respectively) fell within those of the parents, and the average a^* values (12.02, 13.09, and 8.1, respectively) were closer to that of the maternal parent (the low-value parent). The average b^* values for the three combinations (51.16, 37.6, and 25.66, respectively) were higher than those of the paternal parents (the high-value parent). More than half of the individuals in each cross combination (75.38%, 84.43%, and 69.60%) showed transgressive segregation of b^* values.

Figure 4. The typical pictures of different color groups in each combination. (**A–C**): F1 plants of EW01, EW02 and EW03; The F1 plants color groups are arranged from low to high according to the b^* value.

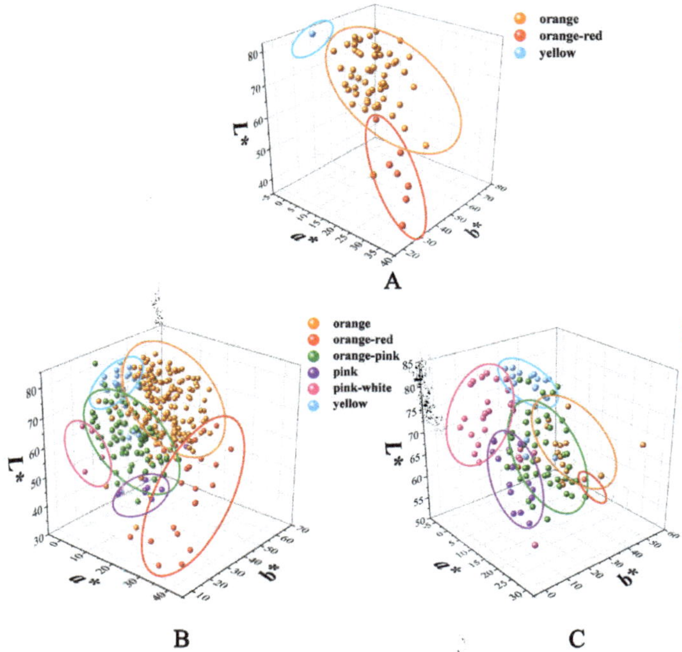

Figure 5. Three-dimensional (L^*, a^* and b^*) coordinates of F1 plants in each combination. (**A–C**): F1 plants of EW01, EW02, and EW03, respectively.

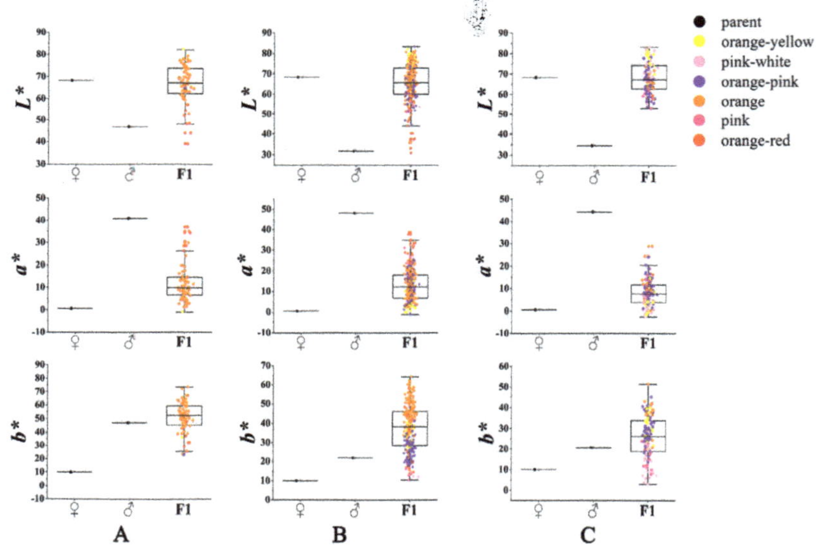

Figure 6. Box plots showing the variation and distribution of flower color data (L^*, a^*, and b^* value) of parents and hybrids in three combinations. (**A–C**): L^*, a^*, and b^* value distribution in combinations of EW01, EW02, and EW03; *X*-coordinate: parents (♀, maternal parent; ♂, paternal parent) and F1 plants; *Y*-coordinate: L^*, a^*, and b^* value, respectively.

3.3. Variation in Raised Spots on Tepals of Hybrids from Different Cross Combinations

Spot patterns constitute an important trait affecting flower color. In the present study, the tepal spot type for all parents and hybrids was raised spots. To understand the inheritance patterns of raised spots, we measured and evaluated the spot numbers of the parents and hybrids in each combination.

For the maternal parent, the average number of spots was 15, while that of the paternal parent was variable. In the EW01 and EW03 hybridizations, 'Tresor' and 'Pearl Loraine' had similar average numbers of spots, 147 and 164, respectively, which were distributed mainly in the basal area and marginal zone of the perianths (Figure 1) (Ad. 28) [34]. For F1 plants of these two combinations, the presence and absence of raised spots were separated at a ratio of 5:1 (54:11) and 4:1 (98:27), respectively (Table 3). However, in the EW02 cross, 'Red Life' had an average of 437 spots, which almost covered the entire perianth (Figure 1). The tepals of all EW02 hybrids exhibited raised spots (Table 3).

Table 3. The spot numbers of cross parents and different hybrid populations.

Cross Combination	The Average Spots Number of Parents				Hybrids		
	♀	SN	♂	SN	No. without Spots	No. with Spots	Proportion
EW01	'Easy Waltz'	15	'Tresor'	147	11	54	1:5
EW02	'Easy Waltz'	15	'Red Life'	437	0	289	——
EW03	'Easy Waltz'	15	'Pearl Loraine'	164	27	98	1:4

Note: SN, spots number.

Afterward, to investigate the effect of the parents' spot number on genetic segregation, the frequency distribution of spot numbers in the F1 generation were analyzed (Figure 7), and a heterosis analysis of the spot numbers in the three cross combinations was performed (Table 4). For all three combinations, the number of raised spots on tepals was continuously separated in the F1 generation (Figure 7), which showed clear quantitative genetic characteristics. The mean values of the offspring of all combinations were close to the mid-parent values. However, for the *CV* index, EW02 had the lowest value, followed by EW01 and then EW03 (Table 4). In EW02, spot number showed a normal distribution, and the numbers of almost all F1 plants were within the ranges of their parents' numbers (Figure 7B), i.e., 98.27% of progenies showed a distribution concentrated between those of their parents (Table 4). In EW01 and EW03, spot number showed an obviously skewed distribution (Figure 7A,C). For the indices of BLP and OHP, 20.00% and 24.62% of the progenies of EW01 and 29.60% and 17.60% of the progenies of EW03 exhibited transgressive segregation (Table 4).

Table 4. Heterosis analysis of spots number in three cross combinations.

Cross Combination	Parent			F1			*Hi*/%	Hybrid Ratio/%		
	♀	♂	*MPs*	$\overline{X} \pm \sigma$	Range	*CV*/%		BLP/%	BP/%	OHP/%
EW01	15.00	147.00	81.00	88.14 ± 84.04	0.00~339.00	95.35	108.81	20.00	55.38	24.62
EW02	15.00	437.00	226.00	219.38 ± 89.92	2.00~526.00	40.99	97.07	0.35	98.27	1.38
EW03	15.00	164.00	90.00	81.49 ± 82.98	0.00~341.00	101.83	90.54	29.60	52.80	17.60

Note: *MPs*, mid-parent value; $\overline{X}$, mean value; σ, standard deviation; range, from minimum to maximum; *CV*, Coefficient of Variation; *Hi*, index of heterosis; BLP, percentage of F1 individuals below low-value parent; OHP, percentage of F1 individuals over high-value parent; BP, percentage of F1 individuals distributed between parents.

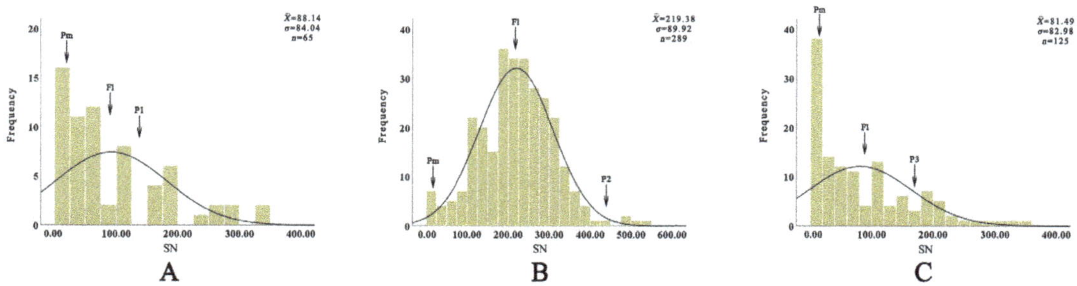

Figure 7. Frequency distribution diagrams of raised spots number (SN) for F1 plants. (**A–C**): spots number of F1 plants in combinations EW01, EW02, and EW03; X-coordinate: spots number (SN); Y-coordinate: frequency; F1 and $\overline{X}$: mean value of hybrids; σ: standard deviation; n: number of individuals; Pm: spots number of maternal parent; P1, P2, and P3: spots number of paternal parents in EW01, EW02, and EW03.

4. Discussion

Conventional crosses, including interspecific and intraspecific hybridization, are the main way to create new lily cultivars with distinctive flower colors. Thus, understanding the heredity of flower color can save lily breeders much time, trouble, and effort in selections among several thousand plants [16]. It is well known that the composition and content of anthocyanins and carotenoids mainly affect the flower color hues of Asiatic hybrid lilies [8]. Therefore, cultivars with four different flower colors and different types and concentrations of pigments were selected as parents. For the cross EW01 between parents with pink-white (no carotenoids, medium anthocyanins) and orange (high carotenoids, no anthocyanins) flowers, the primary flower color of F1 plants was orange. The similarity of this finding with those of previous studies [17,18] suggested that high carotenoid levels were highly heritable and that the progenies were less segregated in flower color. For the crosses EW02 and EW03, the simultaneous presence of both carotenoids and anthocyanins in the paternal parent's tepals produced offspring with extensive flower color segregation (Figure 4B,C). Meanwhile, different contents of anthocyanins and carotenoids resulted in variable levels of progeny isolation. As the carotenoid content decreased and the anthocyanin content increased in the parents, the proportions of the pink and pink-white flower color groups increased in the offspring, while that of the orange flower color group decreased (Table 2). Therefore, these results provide guidance on the selection of parents in cross-breeding practices, e.g., if our desired character is light-colored flowers, such as pink or pink-white flowers, varieties with low carotenoids and high anthocyanins should be selected as parents. When we wish to create deep-colored flowers, such as orange-red flowers, varieties with high levels of carotenoids and anthocyanins should be chosen for artificial crosses. Furthermore, if a cross between cultivars with high carotenoids is conducted, the majority of the first generation (F1) seedlings will predominantly produce flowers that are orange in hue, and very few will display yellow flowers.

Moreover, numerous studies have reported close relationships between color space values (L^*, a^*, and b^*) and the contents of anthocyanins and carotenoids. In *Chrysanthemum*, there was a significant negative correlation between L^* and a^* [27,42], which was consistent with the results presented in Figure 3A,B from our research. The pink-white and pink flower color groups had the highest L^* value and the lowest a^* value, whereas the orange-red group had the lowest L^* value and the highest a^* value. In addition, it was discovered that a^* and b^* were strongly positively correlated with total anthocyanins and total carotenoids, respectively [27,43]. In sweet-orange, there was a significant positive correlation between a^* and total carotenoids [44]. According to these findings, the variation and distribution of the flower color measurement indices, L^*, a^*, and b^*, of parents and offspring were investigated to determine the inheritance patterns of the pigments (Figure 6). Among the three combinations, the average b^* value in the hybrids was higher than that in the high-value parent (Figure 6, b^*), suggesting that carotenoids exhibited transgressive segregation. The offspring of EW01 had the largest b^* value (Figure 6, b^*), followed by those of EW02 and then EW03, which may explain why the last two hybridizations produced offspring with more flower colors. The average a^* value of the F1 plants was closer to that of the low-value parent than to that of the high-value parent (Figure 6, a^*), indicating that the effect of anthocyanins on progeny flower color separation was weaker than that of carotenoids. Furthermore, the discovery that L^* was negatively correlated with total anthocyanins and total carotenoids [44,45] may also explain why the average L^* value of F1 plants was closer to that of the maternal parent (Figure 6, L^*). Although our research provides fundamental information on the inheritance of pigments, how the metabolic pathways of anthocyanins and carotenoids interact to regulate flower color requires in-depth investigation in future studies.

In addition to color hues, raised spots on tepals are a significant component influencing flower color [8,32]. Investigating the inheritance patterns of raised spots in lilies is an important issue. Several studies have reported that the presence or absence of raised spots is regulated by a single locus [32,35]; however, whether dominant or recessive alleles regulate spot formation remains unclear [37,38]. In the present research, the crosses 'Easy Waltz' (with few spots, n = 15) × 'Tresor' (with intermediate spots, n = 147), and 'Easy Waltz' (with few spots, n = 15) × 'Pearl Loraine' (with intermediate spots, n = 164) yielded F1 plants with and without raised spots separated at a ratio of 5:1 (54:11) and 4:1 (98:27), respectively. For the cross 'Easy Waltz' (with few spots, n = 15) × 'Red Life' (with many spots, n = 437), the tepals of all F1 hybrids exhibited raised spots. It follows that dominant genes promote spot formation in Asiatic hybrid lilies. The discrepancy with previous results [37] may be due to the complex chromosomal ploidy and genetic background of lilies. Additionally, the number of the F1 plant spots segregated continuously, showing quantitative genetic characteristics (Figure 7) consistent with the findings of previous studies [32,35]. These results have implications for breeding, i.e., if varieties with few/intermediate spots are selected as parents, the progeny will be spotless or will have more spots than the parents. If our goal is to select new cultivars with numerous spots covering the full tepals, multi-spotted varieties are a better choice as parents.

5. Conclusions

In conclusion, anthocyanins and carotenoids are responsible for regulating the complex flower color of Asiatic hybrid lilies. It is crucial to clarify how parental pigment composition and content influence the separation of flower colors in offspring. Our research revealed that a high carotenoid content was highly heritable and that cultivars with this trait generally generated offspring with orange flowers. When the parents have differing levels of carotenoids and anthocyanins, the progenies exhibit more extensive color segregation and variable proportions of color groups. Additionally, the number of raised spots was a quantitative trait regulated by multiple genes, and dominant alleles promoted spot formation. These patterns of inheritance can help breeders predict the outcomes of particular crosses. Meanwhile, they will contribute to the breeding of novel lily cultivars with high aesthetic value.

Author Contributions: Conceptualization, G.J.; Methodology, G.J., J.L. and J.C.; Data curation, J.L. and J.C.; Investigation, J.L. and J.C.; Formal analysis, P.Y. and Y.Z.; Writing—original draft preparation, J.L., Q.Z. and P.Y.; Writing—review and editing, J.L., Y.Z. and G.J.; Supervision, G.J.; Project administration, G.J.; Funding acquisition, G.J. All authors have read and agreed to the published version of the manuscript.

Funding: This research was supported by the National Key R&D Program of China (Grant No.2019YFD1000400) and the Key R&D Program of Hebei Province (Grant No. 21326317D).

Institutional Review Board Statement: Not applicable.

Informed Consent Statement: Not applicable.

Data Availability Statement: The data supporting the results are already mentioned in the main text.

Conflicts of Interest: The authors declare no conflict of interest.

References

1. Comber, H.F. A new classification of the genus *Lilium*. In *Royal Horticultural Society Lily Year Book*; Chittenden, F.J., Ed.; The Royal Horticultural Society: London, England, 1949; pp. 85–105.
2. Smyth, D.R.; Kongsuwan, K.; Wisudharomn, S. A survey of C-band patterns in chromosomes of *Lilium* (Liliaceae). *Plant Syst. Evol.* **1989**, *163*, 53–69. [CrossRef]
3. Du, Y.P.; He, H.B.; Wang, Z.X.; Wei, C.; Li, S.; Jia, G.X. Investigation and evaluation of the genus *Lilium* resources native to China. *Genet Resour. Crop. Ev.* **2014**, *61*, 395–412. [CrossRef]
4. Leslie, A.C. *The International Lily Register*, 3rd ed.; The Royal Horticultural Society: London, England, 1982.
5. Lim, K.B.; van Tuyl, J.M. Lily: *Lilium* hybrids. In *Flower Breeding and Genetics: Issues, Challenges and Opportunities for the 21st Century*; Anderson, N.O., Ed.; Springer: Berlin, Germany, 2006; pp. 513–532.
6. Barba-Gonzalez, R.; Van Silfhout, A.A.; Visser, R.G.F.; Ramanna, M.S.; Van Tuyl, J.M. Progenies of allotriploids of Oriental × Asiatic lilies (*Lilium*) examined by GISH analysis. *Euphytica* **2006**, *151*, 243–250. [CrossRef]
7. Van Tuyl, J.M.; Arens, P.; Shahin, A.; Marasek-Ciołakowska, A.; Barba-Gonzalez, R.; Kim, H.T.; Lim, K.-B. *Lilium* . In *Ornamental Crops, Handbook of Plant Breeding*; Van Huylenbroeck, J., Ed.; Springer: Berlin, Germany, 2018; pp. 481–512. [CrossRef]
8. Yamagishi, M. How genes paint lily flowers: Regulation of colouration and pigmentation patterning. *Sci. Hortic.* **2013**, *163*, 27–36. [CrossRef]
9. Nørbæk, R.; Kondo, T. Anthocyanins from flowers of *Lilium* (Liliaceae). *Phytochemistry* **1999**, *50*, 1181–1184. [CrossRef]
10. Yamagishi, M.; Kishimoto, S.; Nakayama, M. Carotenoid composition and changes in expression of carotenoid biosynthetic genes in tepals of Asiatic hybrid lily. *Plant Breed.* **2010**, *129*, 100–107. [CrossRef]
11. Jeknić, Z.; Morré, J.T.; Jeknić, S.; Jevremović, S.; Subotić, A.; Chen, T.H.H. Cloning and functional characterization of a gene for capsanthin-capsorubin synthase from tiger lily (*Lilium lancifolium* Thunb. 'Splendens'). *Plant Cell Physiol.* **2012**, *53*, 1899–1912. [CrossRef]
12. Wang, X.; Yamagishi, M. Mechanisms suppressing carotenoid accumulation in flowers differ depending on the hybrid groups of lilies (*Lilium* spp.). *Sci. Hortic.* **2019**, *243*, 159–168. [CrossRef]
13. Deli, J.; Molnár, P.; Matus, Z.; Tóth, G.; Steck, A.; Pfander, H. Isolation and characterization of 3,5,6-trihydroxy-carotenoids from petals of *Lilium tigrinum*. *Chromatographia* **1998**, *48*, 27–31. [CrossRef]
14. Lai, Y.S.; Shimoyamada, Y.; Nakayama, M.; Yamagishi, M. Pigment accumulation and transcription of *LhMYB12* and anthocyanin biosynthesis genes during flower development in the Asiatic hybrid lily (*Lilium* spp.). *Plant Sci.* **2012**, *193*, 136–147. [CrossRef]
15. Yamagishi, M. White with partially pink flower color in *Lilium cernuum* var. *album* is caused by transcriptional regulation of anthocyanin biosynthesis genes. *Sci. Hortic.* **2020**, *260*, 108880. [CrossRef]
16. Balode, A. Color analysis of flowers in *Lilium* sp. breeding. *Acta Hortic.* **2007**, *755*, 213–218. [CrossRef]
17. Hwang, Y.J.; Kang, S.Y.; Rhee, H.Y.; Lim, K.B. The characteristics of progenies derived from *Lilium lancifolium* and Asiatic hybrid 'Dreamland'. *Korean J. Breed. Sci.* **2009**, *41*, 451–455.
18. Jo, Y.K.; Ramzan, F.; Son, B.G.; Kim, H.Y.; Lim, K.B. Crossing of Asiatic hybrids for breeding of new Lily cultivars. *Korean J. Breed. Sci.* **2018**, *50*, 1–12. [CrossRef]
19. Wang, H.; Kong, Y.; Dou, X.Y.; Lang, L.X.; Bai, J.R. Genetic variation of phenotypic traits in F1 generation between Asiatic hybrid lily 'Renoir' and *Lilium davidii* var. *willmottiae*. *Mol. Plant Breed.* **2021**, *19*, 6111–6119. [CrossRef]
20. Hanbury, A.; Serra, J. Mathematical morphology in the CIELAB space. *Image. Anal. Stereol.* **2002**, *21*, 201–206. [CrossRef]
21. Afonso, T.; Moresco, R.; Uarrota, V.G.; Navarro, B.B.; Nunes, E.D.C.; Maraschin, M.; Rocha, M. UV-vis and CIELAB based chemometric characterization of Manihot esculenta carotenoid contents. *J. Integr. Bioinform.* **2017**, *14*, 20170056. [CrossRef]
22. Wang, N.H.; Dai, M.Y.; Zheng, G.; Chang, P.J.; Xuan, L.J.; Liu, Z.G.; Wang, Y.L.; Cheng, S.Y.; Wang, Z.W.; Wang, H.L. Flavonoid components and gene expression analysis reveal flower pigmentation difference between *Magnolia biondii* and its variety *M. biondii* var. *purpurascens*. *Trees-Struct. Funct.* **2021**, *36*, 583–591. [CrossRef]
23. Donoso, A.; Rivas, C.; Zamorano, A.; Pena, A.; Handford, M.; Aros, D. Understanding *Alstroemeria pallida* flower colour: Links between phenotype, anthocyanins and gene expression. *Plants* **2021**, *10*, 55. [CrossRef]
24. Li, X.; Lu, M.; Tang, D.Q.; Shi, Y.M. Composition of carotenoids and flavonoids in *Narcissus* cultivars and their relationship with flower color. *PLoS ONE* **2015**, *10*, e0142074. [CrossRef]
25. Liu, L.; Zhang, L.Y.; Wang, S.L.; Niu, X.Y. Analysis of anthocyanins and flavonols in petals of 10 *Rhododendron* species from the Sygera Mountains in Southeast Tibet. *Plant Physiol. Bioch.* **2016**, *104*, 250–256. [CrossRef]
26. Cui, H.L.; Zhang, Y.A.; Shi, X.L.; Gong, F.F.; Xiong, X.; Kang, X.P.; Xing, G.M.; Li, S. The numerical classification and grading standards of daylily (*Hemerocallis*) flower color. *PLoS ONE* **2019**, *14*, e0216460. [CrossRef]
27. Lu, C.F.; Li, Y.F.; Wang, J.Y.; Qu, J.P.; Chen, Y.; Chen, X.Y.; Huang, H.; Dai, S.L. Flower color classification and correlation between color space values with pigments in potted multiflora chrysanthemum. *Sci. Hortic.* **2021**, *283*, 110082. [CrossRef]
28. Lei, T.; Song, Y.; Jin, X.H.; Su, T.Y.; Pu, Y.W. Effects of pigment constituents and their distribution on spathe coloration of *Zantedeschia hybrida*. *Hortscience* **2017**, *52*, 1840–1848. [CrossRef]

29. Zhou, Y.W.; Yin, M.; Abbas, F.; Sun, Y.; Gao, T.; Yan, F.L.; Li, X.Y.; Yu, Y.Y.; Yue, Y.C.; Yu, R.C. Classification and association analysis of Gerbera (*Gerbera hybrida*) flower color traits. *Front. Plant Sci.* **2022**, *12*, 779288. [CrossRef]
30. Wang, H.; Fan, Y.W.; Yang, Y.; Zhang, H.; Li, M.F.; Sun, P.; Zhang, X.Z.; Xue, Z.; Jin, W.M. Classification of rose petal colors based on optical spectrum and pigment content analyses. *Hortic. Environ. Biote.* **2022**. [CrossRef]
31. Wang, J.; Xu, L.F.; Wang, L.; Qi, X.Y.; Song, M.; Cao, Y.W.; He, G.R.; Tang, Y.C.; Yang, P.P.; Ming, J. The numerical classification of flower color phenotype in lily. *Acta Hortic. Sin.* **2021**, *49*, 571–580. [CrossRef]
32. Abe, H.; Nakano, M.; Nakatsuka, A.; Nakayama, M.; Koshioka, M.; Yamagishi, M. Genetic analysis of floral anthocyanin pigmentation traits in Asiatic hybrid lily using molecular linkage maps. *Theor. Appl. Genet.* **2002**, *105*, 1175–1182. [CrossRef]
33. Yamagishi, M. MYB19LONG is involved in brushmark pattern development in Asiatic hybrid lily (*Lilium* spp.) flowers. *Sci. Hortic.* **2020**, *272*, 109570. [CrossRef]
34. He, X.F.; Xu, S.F.; Leng, P.S.; Wang, W.H. Transcriptome sequencing reveals genes involved in petal spot formation of Asiatic hybrid lily cultivar 'Easy Dance'. *Int. J. Agric. Biol.* **2018**, *20*, 939–944. [CrossRef]
35. Yamagishi, M.; Akagi, K. Morphology and heredity of tepal spots in Asiatic and Oriental hybrid lilies (*Lilium* spp.). *Euphytica* **2013**, *194*, 325–334. [CrossRef]
36. Yamagishi, M. Isolation and identification of *MYB* transcription factors (*MYB19Long* and *MYB19Short*) involved in raised spot anthocyanin pigmentation in lilies (*Lilium* spp.). *J. Plant Physiol.* **2020**, *250*, 153164. [CrossRef]
37. Shahin, A.; Arens, P.; Van Heusden, A.W.; Van Der Linden, G.; Van Kaauwen, M.; Khan, N.; Schouten, H.J.; Van De Weg, W.E.; Visser, R.G.F.; Van Tuyl, J.M. Genetic mapping in *Lilium*: Mapping of major genes and quantitative trait loci for several ornamental traits and disease resistances: Genetic mapping in *Lilium*. *Plant Breed.* **2011**, *130*, 372–382. [CrossRef]
38. Zhao, X.Y.; Wang, S.D.; Wang, W.H.; Zhang, K.; Liu, C.Y. *Tingyuan Baihe Shiyong Jishu*, 1st ed.; China Agriculture Press: Beijing, China, 2016; pp. 69–78.
39. Wang, Q.; Lv, T.; Lv, Y.M. The impact of tepal structure and pigment composition on the flower colour of lily. *Acta Hortic. Sin.* **2021**, *48*, 1873–1884. [CrossRef]
40. UPOV. Guidelines for the conduct of tests for distinctness, uniformity and stability, *Lilium*, TG/59/7. In Proceedings of the International Union for the Protection of New Varieties of Plants, Geneva, Switzerland, 24 March 2010; Available online: https://www.upov.int/portal/index.html.en (accessed on 10 October 2022).
41. Schmitzer, V.; Veberic, R.; Osterc, G.; Stampar, F. Color and phenolic content changes during flower development in groundcover rose. *J. Am. Soc. Hortic. Sci.* **2010**, *135*, 195–202. [CrossRef]
42. Akbari, R.; Hatamzadeh, A.; Sariri, R.; Bakhshi, D. Relationship of flower color parameters and metal ions of petal tissue in fully opened flowers of gerbera. *J. Plant Stud.* **2013**, *2*, 89–96. [CrossRef]
43. Itle, R.A.; Kabelka, E.A. Correlation between L*a*b* color space values and carotenoid content in pumpkins and squash (*Cucurbita* spp.). *Hortscience* **2009**, *44*, 633–637. [CrossRef]
44. Abdelaali, S.B.; Rodrigo, M.J.; Saddoud, O.; Zacarías, L.; Hajlaoui, M.R.; Mars, M. Carotenoids and colour diversity of traditional and emerging Tunisian orange cultivars (*Citrus sinensis* (L.) Osbeck). *Sci. Hortic.* **2018**, *227*, 296–304. [CrossRef]
45. Xue, L.; Wang, Z.G.; Zhang, W.; Li, Y.X.; Wang, J.; Lei, J. Flower pigment inheritance and anthocyanin characterization of hybrids from pink-flowered and white-flowered strawberry. *Sci. Hortic.* **2016**, *200*, 143–150. [CrossRef]

horticulturae

MDPI

Article

Breeding and Selection of Nursery Plants Assisted by High-Throughput Field Phenotyping Using UAV Imagery: Case Studies with Sweet Box (*Sarcococca*) and Garden Rose (*Rosa*)

Irene Borra-Serrano *, Katrijn Van Laere, Peter Lootens and Leen Leus

Plant Sciences Unit, Flanders Research Institute for Agriculture, Fisheries and Food (ILVO), Caritasstraat 39, 9090 Melle, Belgium

* Correspondence: irene.borra-serrano@ilvo.vlaanderen.be

Abstract: Breeding and selection of nursery plants require evaluation of a wide variety of traits. Characteristics that are visually scored in the field include aesthetic beauty as well as tolerance to (a)biotic stresses, among others. This work proposes methodologies based on vegetation indices and canopy height data derived from visual imagery captured using an RGB (red, green, blue) camera embedded in a drone to rank and select genotypes. This approach relies on quantitative evaluation standards that exclude breeder bias and speed up data collection. A proof of concept for nursery plants was developed in two woody ornamentals: sweet box (*Sarcococca* Lindl.) and garden rose (*Rosa* L.). This work aimed to compare methodologies and to propose how drones (unmanned aerial vehicles, UAV) and high throughput field phenotyping (HTFP) can be used in a commercial plant selection program. Data was gathered in 2019 in three well-established breeding trials, two for sweet box and one for rose. Characteristics discussed include plant architecture traits such as plant height and shape in the sweet box and rose as well as floribundity, continuous flowering and disease resistance in roses. Correlations were calculated between on-ground measurements and UAV-derived methods, obtaining significant results. The advantages and disadvantages of the methodology and the approach for each of the traits are discussed.

Keywords: UAV; drone; image analysis; HTFP; nursery plant; ornamental; plant breeding; plant architecture; flowering; disease resistance

Citation: Borra-Serrano, I.; Van Laere, K.; Lootens, P.; Leus, L. Breeding and Selection of Nursery Plants Assisted by High-Throughput Field Phenotyping Using UAV Imagery: Case Studies with Sweet Box (*Sarcococca*) and Garden Rose (*Rosa*). *Horticulturae* **2022**, *8*, 1186. https://doi.org/10.3390/horticulturae8121186

Academic Editor: Youping Sun

Received: 25 October 2022
Accepted: 9 December 2022
Published: 13 December 2022

1. Introduction

Interest in and the use of drones (unmanned aerial vehicles, UAV) have dramatically increased over the last decade. Advances have been made in the development of image capture, open-source spatial data analysis tools, and (affordable) sensors, creating many opportunities for more in-depth non-destructive high throughput field phenotyping (HTFP). These advances can accelerate breeding and selection [1,2]. Particularly when using UAV, images can be collected at a higher spatial and temporal resolution than with other remote sensing platforms. Drones and other proximal sensing installations equipped with sensors are widely used in agronomic crops to monitor crop growth and development [2]. More specific applications for agricultural crops include biotic and abiotic stress detection [3–6], biomass estimation and agricultural traits [7–10], ground cover estimation and growth monitoring [11,12], grain yield prediction [13] and evaluation of other yield components such as spike density and number of kernels [14,15], among others.

In horticulture (including cultivation of woody ornamentals and nursery plants), crop monitoring, pesticide treatments, crop inventory, sales and marketing, phenotyping, and selection of traits in breeding programs would benefit from increased automation [16]. However, the use of UAVs in horticulture is not largely exploited, mostly due to the small size of production areas and the diversity of crops and cultivation systems [16]. In addition, for nursery plants, market shares of individual crops are sometimes too limited to justify the use and costs of these technologies. Currently, the use of drones is limited to economically important fruit trees and orchards [17] to study water stress (in vines, olive, citrus, almond, apricot, peach, lemon, and orange orchards), to monitor diseases to support spray management decisions (in olive orchards and vineyards), to monitor the crop and weeds (in vineyards, apple, orange, and olive orchards) ([16] and references herein), to evaluate tree architecture and flowering (in almond [8]), to estimate height and site-specific crop management (of grapevine [7]), and to evaluate the impact of different pruning strategies on canopy area, tree height and crown volume (plant architecture) over time (in olive production or breeding [18–20]).

For woody ornamentals, to the best of our knowledge, small UAVs have only been used to identify water-stressed and non-stressed containerized ornamental shrubs of 6 different species [21], to count plants [22], and remotely sensed images were used to assess the effect of irrigation salinity on *Hibiscus rosa-sinensis* L. and *Mandevilla splendens* (Hook. f.) Woodson [23]. Furthermore, only a few lab-scale phenotyping studies in breeding programs of woody plants have been done using image capture and analysis in roses [24], *Rhododendron* L. [25,26], and *Escallonia* Mutis ex L.f. [27,28]. The use of drones in woody ornamental breeding programs is not reported, although it could improve and, in some cases, even speed up the multi-year breeding and selection process. For many phenotypic parameters, quantitative and objective evaluation standards can be improved by using image analysis [29], which excludes visual selection and breeder bias. The major cost areas of phenotyping are plant handling and labor; total costs per pot or microplot are similar in robotized platforms or field experiments with drones and hand-held or robotic ground vehicles [30]. The cost of vehicles carrying sensors is only 5% to 26% of the total cost, and data analysis represents 10% to 20% of the total costs in cases where pipelines are already available. Although there is an initial high cost for pipeline development compared to labor costs of manual operations [30], this will result in usefully integrating the enormous volume and velocity of imaging and remote-sensing data generated by automated phenotyping into breeding programs.

In this study, we aim to discuss the use of UAV-based image analysis on relevant traits for woody ornamental nursery plants to support selection in breeding programs. Case studies were developed on *Sarcococca* spp. (sweet box or Christmas box) and *Rosa* hybrids (garden roses) represent two important segments for nursery plants, evergreen and flowering shrubs. *Sarcococca* is a genus of 11 species in the family Buxaceae that contains evergreen, winter-flowering shrubs with strongly scented flowers. Species are cultivated to be used as a groundcover or are planted as low hedges. *Sarcococca* breeding mainly focuses on growth habits, i.e., compact growth, round shape, uniformity, and leaf greenness [31]. For garden roses, besides selection for positive and novel traits such as attractive flower color and flower shape, disease resistance (e.g., resistance to black spot, *Diplocarpon rosae* Wolf, a cause of leaf drop), uniform plant architecture and flowering characteristics like floribundity (the abundant production of flowers on each branch), and continuous blooming are the most important traits in breeding [32]. In ongoing breeding programs in *Sarcococca* and garden roses, a UAV equipped with an RGB (Red, Green, Blue) camera and subsequent image analysis was evaluated to determine plant density and compactness in *Sarcococca*. It was also evaluated its value in assisting in the selection of garden roses for plant architecture, flowering, and disease resistance. The correlation between on-ground measurements and data obtained from UAV image analysis was examined. Based on these experiences and observations, best practices, and recommendations for setting up UAV experiments for woody ornamentals are presented and discussed below.

2. Materials and Methods

2.1. Plant Material and Field Trials

For *Sarcococca*, different species were used as parent plants in the breeding program: *confusa* Lindl. (S1, S12), *saligna* (D. Don) Mull.-Arg. (S4, S10), *wallichii* Stapf (S5), *ruscifolia* Stapf var. *chinensis* (S6), *ruscifolia* 'Dragon Gate' (S7), *hookeriana* Baill. var. *digyna* (S08), *hookeriana* var. *humilis* (S9), *orientalis* Wu (S3, S11), *humilis* Stapf (S16) [31]. Seeds from interspecific crosses and open pollinations were harvested and sown (summer of 2014 and 2015) as described in [31]. Year-old seedlings were planted on the field (spring 2016 or 2017 under natural conditions (Melle, Belgium, lat. 50.59 N, long. 3.47 E) at a planting distance of 30 cm. Morphological examination (see below) was performed in 2019 on individual genotypes that are two (Experiment 1, SarcoE1) or three (Experiment 2, SarcoE2) years established on the field (Figure 1). Field trials are composed of 552 plants in SarcoE1 and 583 plants in SarcoE2 (Table S1).

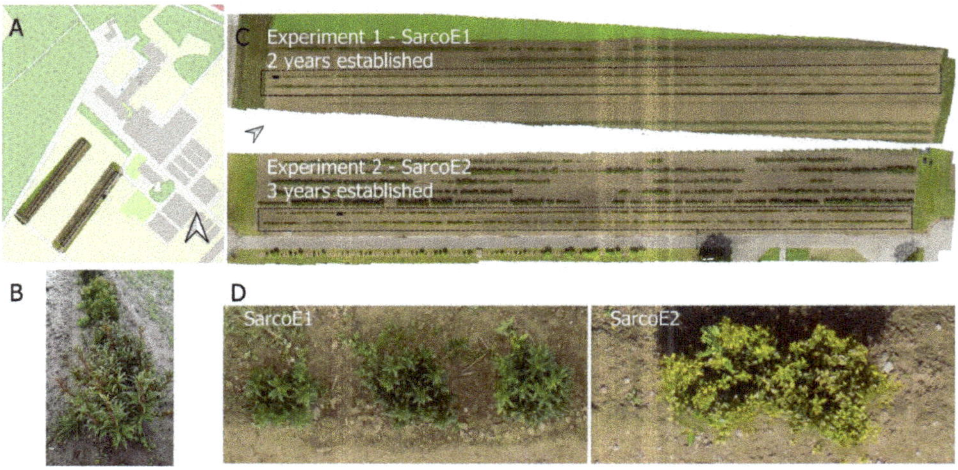

Figure 1. (**A**) Location of the two *Sarcococca* trials; (**B**) *Sarcococca* plants in the field; (**C**) SarcoE1 and SarcoE2 indicated with black dashed lines; and (**D**) close-up of the established plants (from the location marked in C with a black rectangle).

Cultivation of the rose genotypes was started in the winter of 2015–2016 by planting the rootstock *Rosa canina* L. 'Pfänder' in seven rows on a selection field (Melle, Belgium lat 50.59 N, long. 3.46 E). In July 2016, the rootstock was budded with 87 different floribunda and shrub rose genotypes obtained from cross hybridizations created in the years 2007 to 2014. Seedlings obtained from cross hybridizations were clonally multiplied by winter grafting in the first year after sowing and T-budding in July of all following years. During the winter prior to the UAV flights and on-ground evaluation, the plants were pruned to approximately 20 cm. Plants of the same genotype (between 1 and 51 plants/genotype) succeed each other in a row. Seventy-five rose genotypes with more than seven plants (in total, 1646 plants) were included in the observations. The three-year-old plants were evaluated per genotype with UAV imagery and on-ground observations in 2019. An overview of the trial is shown in Figure 2.

For both plant species, *Sarcococca* and *Rosa*, morphological characteristics were evaluated as described in Table 1. In the rose field, the plants were evaluated per genotype, whereby one genotype was represented by different cloned plants, while for *Sarcococca*, each plant was evaluated as an individual genotype.

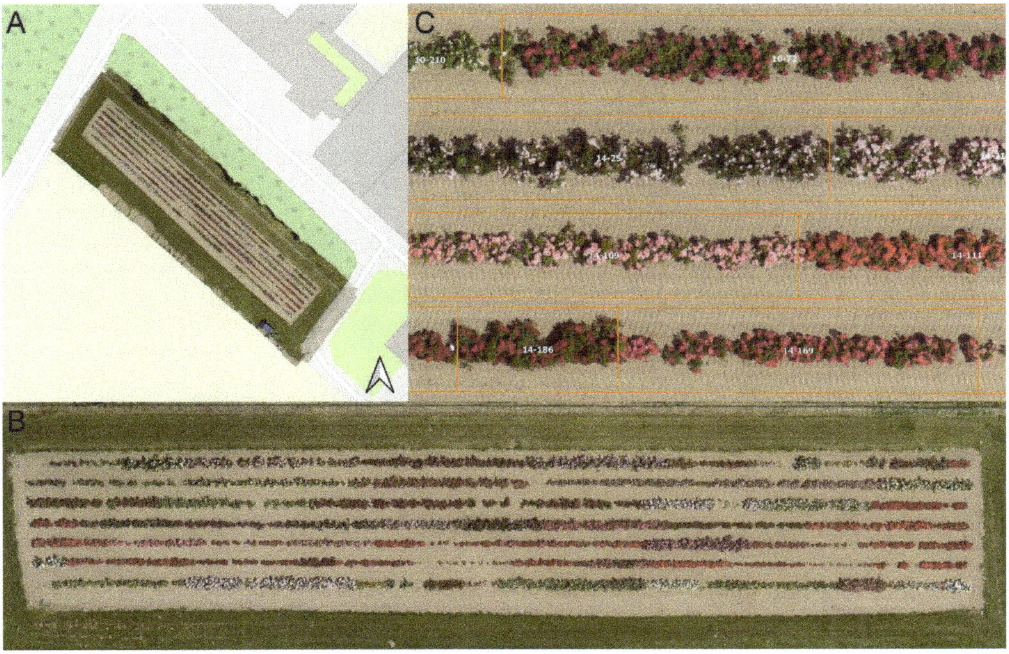

Figure 2. (**A**) Location of the rose trial; (**B**) Orthophoto of the entire trial based on a UAV flight done on 1 July 2019; (**C**) detail of a part of the trial with polygons defining the different genotypes.

Table 1. Overview of the morphological characteristics evaluated for *Sarcococca* and *Rosa*.

Case Study	Measured Plant Traits		
	Plant Architecture	**Flowering**	**Disease Resistance**
Sarcoccoca	Plant height Plant shape		
Rosa	Plant height Plant uniformity	Floribundity Continuous flowering	Black spot resistance

2.2. On-Ground Morphological Measurements

For *Sarcococca*, an evaluation was made for plant architecture based on plant height and plant shape. For this, the maximum plant height (GR_Hmax = height of the longest branch, measured from the soil to the top of the branch), average plant height (GR_Hmean = height of the plant measured from the soil to the top of the plant excluding protruding branches) and plant width in two perpendicular directions (in the direction of the line and 90°, GR_W1 and GR_W2) were manually measured on the individual plants using a ruler. In June 2019, 94 genotypes from SarcoE1 and 583 from SarcoE2, and in September 2019, only 94 genotypes from SarcoE1 were measured. Manual on-ground plant area (GR_Area) was calculated by multiplying the two plant widths (Table S2).

In the rose trial, on-ground measurements for plant architecture were based on two parameters: plant height and plant uniformity (Table S2). On 5 September, canopy height and height uniformity were assessed for the 75 individual genotypes of the rose breeding trial. Canopy height (GR_Hmax) was measured using a ruler from the soil to the top of the plant. Plant height uniformity (GR_U) was scored in the week of 5 September using a range between 0–3, whereby uniformity was lowest with a score of 0 and highest in plants with a score of 3. Floribundity was quantified by the visual evaluation of flower abundance at a single time point. Floribundity at consecutive time points during summer is used to assess continuous flowering. On-ground scoring of floribundity was done by estimating the total number of flowers showing color or the amount of color on the plant as a percentage in 10% steps. Another scoring evaluated the flowering stages whereby a total score of 100% is divided over 4 stages: (1) no buds, (2) buds, (3) open flowers, and (4) faded flowers. This makes it possible to estimate repeated flowering as, ideally, the plants develop new buds as the old buds fade. Data for floribundity were collected during week 32 (5 August). Data for flowering stages in week 28 (8 July), week 32 (5 August), and week 36 (2 September). The assessment of black spot infestation on the rose genotypes was scored using two parameters. Typically, the disease appears during the rainy period in autumn, starting at the bottom of the plant and traveling upwards, with black leaf spots first developing, followed by the leaf drop. Therefore, the first score is related to the detection of black spots on the leaves (leaf spot, GR_BSS). The leaf spot score ranges from 0 to 3, with 0 indicating no black spots present on the leaves and 3 indicating a severe infestation. The second score is based on the resulting effect of the black spot disease being early leaf drop (GR_BSD) and showing only bare stems too early in the season. The leaf drop score ranges from 0 to 3, with 0 indicating no leaf drop and 3 indicating severe leaf drop. Plants infected by rust earlier in the season were excluded from the analysis. Thus only 66 genotypes were scored. Black spot was scored on 20 August (early disease development) and 15 October (full disease pressure).

2.3. *UAV Flights and Image-Based Morphological Measurements*

For both case studies, a UAV DJI Matrice 600 Pro (DJI, Shenzhen, China) platform equipped with an RGB (visible spectral range) camera (model α6000, Sony Corporation, Tokyo, Japan) with 6000 × 4000 pixels and a 35 mm lens was used to capture the images at 20 m flight height with 80% overlap (forward-lap and side-lap) at a speed of 1.8 m/s. The settings of the RGB camera (shutter speed, aperture, and ISO) were adjusted in the field before the flight, and the sensor was triggered automatically based on flight planning. Correction for white balance and exposure was applied to the images using Lightroom v6.5. (Adobe Systems Incorporated, San Jose, CA, USA) using a gray reference card (18% reference gray, Novoflex Präzisionstechnik GmbH, Memmingen, Germany) placed in the field during the flight campaign prior to processing with Agisoft Photoscan v1.2.6 Professional Edition (Agisoft LLC, St. Petersburg, Russia) to build the orthophoto and digital elevation model (DEM) with the coordinates of nine ground control points (GCP). In addition, for *Sarcococca*, a digital terrain model (DTM) with the same spatial resolution as the DEM was built based on point classification (in Agisoft Photoscan) to calculate the Canopy Height Model (CHM) for each date by subtracting the DTM from the DEM. For the rose trial, the DTM was interpolated based on the nine GCPs using the TIN interpolation tool available in QGIS. The data extraction for both cases and interpolation were done in QGIS 3.18.2 (QGIS.org, 2021. QGIS Geographic Information System. QGIS Association. http://www.qgis.org, accessed on 20 April 2021).

For *Sarcococca*, one UAV flight was executed per field, over SarcoE1 on 5 September and over SarcoE2 on 29 May 2019. The spatial resolution for the orthophoto and the DEM (both flights) was 0.2 cm and 1.5 cm, respectively. From the CHM, the mean and Q90 height values (UAV_Hmean and UAV_HQ90) of all individual plants were extracted. Other parameters derived from the orthophoto were the (top) plant area (UAV_Area) and the surrounding box dimensions and area (UAV_B1, UAV_B2, and UAV_BoxArea).

For roses, the UAV flights were performed on 5 September 2019 for height estimation and plant uniformity. In the CHM, a threshold of 25 cm was set to eliminate weeds and remove the soil within the defined polygons. Each polygon included the plants of one genotype. Only polygons with at least 0.5 m^2 of plants or >7 plants were kept for the analysis. This results in 71 polygons/genotypes being evaluated. For every polygon, the 90th quantile of the height (UAV_HQ90) was calculated, allowing assessment of the top of the canopy, excluding real extreme height values. The coefficient of variation (UAV_Hcv) was calculated to assess plant height uniformity.

A relevant trait In roses is the number/percentage of flowers; abundant and continuous flower coverage is preferred. Therefore, three flights were considered to estimate the flower coverage over time (1 July, 6 August, and 5 September). An evaluation was done to select the most appropriate index (between 12 common RGB indices) or spectral band to separate colors. The highest f-values were obtained for the Rband and the Woebbecke index (WBI = G − B/R − G [33]). A clear separation between green (leaves) and the other colors (flowers) was observed (Figure S1). The overlap between the soil and flower colors is not a problem, as the pixels representing soil were not considered (a threshold based on the CHM was applied to remove them first). The pixel classification was done using a threshold to separate the pixels corresponding to green leaves and flowers. The number of pixels for the entire plant (leaves and flowers included) and the ones classified as flowers was extracted (Figure S2). The proportion between the two sets of pixels was calculated to obtain the percentage. A comparison between flowering based on on-ground scores and UAV data is presented in Section 3.2 below for a set of the genotypes.

On the same date of on-ground scoring of black spot infestation (15 October 2019), a UAV flight was carried out and compared with the UAV flight of 20 August to evaluate the progress of black spot development. These dates are used for evaluation of black spot occurrence as the disease typically develops most symptoms at the end of the season resulting in an early leaf drop. Based on the RGB orthophoto, the Excess Green vegetation index was calculated (UAV_ExG = 2 * G − R− B/R + G + B [33]). The median value of ExG was determined for every genotype (only considering pixels corresponding to vegetation based on the threshold applied to the canopy height model). For every genotype, the canopy cover (UAV_CC) of 20 August was used as a mask to assess leaf loss on 15 October. Then the UAV_ExG values were compared for both dates within the mask, using the first date as a reference.

2.4. Comparison between On-Ground and Image-Based Trait Measurements and Application as Selection Criteria

The ideal *Sarcococca* plant is compact and has a spherical shape. Plant evaluation was therefore based on compactness, height uniformity, and plant shape. To validate the UAV image-based trait measurements and evaluate their performance in plant selection, the UAV-based measurements were correlated to the on-ground measurements. For compactness (or circular shape), the ratio between the perpendicular widths, manually measured in the field (GR_Wratio) and derived from the image-derived surrounding box (UAV_Bratio), was calculated. For height uniformity, the ratio between the maximum and mean height manually measured in the field (GR_Hratio) was used and compared to the height derived from the CHM (UAV_Hratio). Another way to evaluate height uniformity is by the coefficient of variation (UAV_Hcv), which was possible to derive from the detailed UAV imagery (average 250 pixels per plant for the CHM). Variability among half-sib families (originating from the same mother plant) was evaluated by data distributions (box plots) in both on-ground and UAV measurements to determine the value of the mother plant for the future breeding program. In SarcoE2, after 3 years of field evaluation, a final selection of 12 plants was made in collaboration with *Sarcococca* growers. This selection, which was based on field inspection and intuition, was evaluated in this study based on defined selection criteria, which combines all parameters measured and is compared to a similar selection based on UAV data.

For roses, the ideal plant has a uniform canopy height, shows a high floribundity throughout the season (continuous flowering), and is resistant to diseases. Plant height is not a selection criterion but helps to assign roses for specific use (e.g., smaller roses can be used as a potted patio plant). Correlations between on-ground and UAV canopy height were calculated as well between uniformity scores and UAV_Hcv. The flower percentage derived from UAV imagery was compared to the visual scoring during three different weeks and continuity during summer. Resistance to black spot was compared by correlating on-ground scores with UAV_ExG. Leaf loss was calculated based on the UAV_CC of an earlier date. As a final selection was carried out, these results mainly focused on genotypes with good uniformity, high floribundity, continuous flowering, and resistance to black spot.

2.5. Statistical Analysis and Representation of the Data

Statistical analysis was carried out in R v3.6.3 using rStudio v1.2.5033 (rStudio: Integrated Development Environment for R, rStudio Inc., Boston, MA, USA). All phenotypic variables' names and definitions can be found in the supplementary information (Table S2). For calculating correlations based on continuous data, a Pearson's correlation was used; if discrete data (i.e., scores) were included, then the Spearman method was used.

3. Results

3.1. Phenotyping Capturing Plant Architecture

For *Sarcococca*, manual measurements and UAV-derived parameters were performed to capture plant architecture. Correlations between all measurements determined on *Sarcococca* on SarcoE1 (94 individual plants) are shown in Figure 3A. In the case of SarcoE2 (583 plants), it was difficult to separate plants with connected canopies; therefore, results on SarcoE2 are not shown. A significant correlation of 0.63 was obtained between the on-ground and UAV height measurements and 0.69 for plant area (SarcoE1 plants). In Figure 3B, linear relations between on-ground and UAV measurements specific for canopy height and plant area are shown. The general trend showed lower values derived from the UAV data, as the points were not on the 1-1 line. This is partially explained by the difference between how measurements were done manually in the field and those derived from the images.

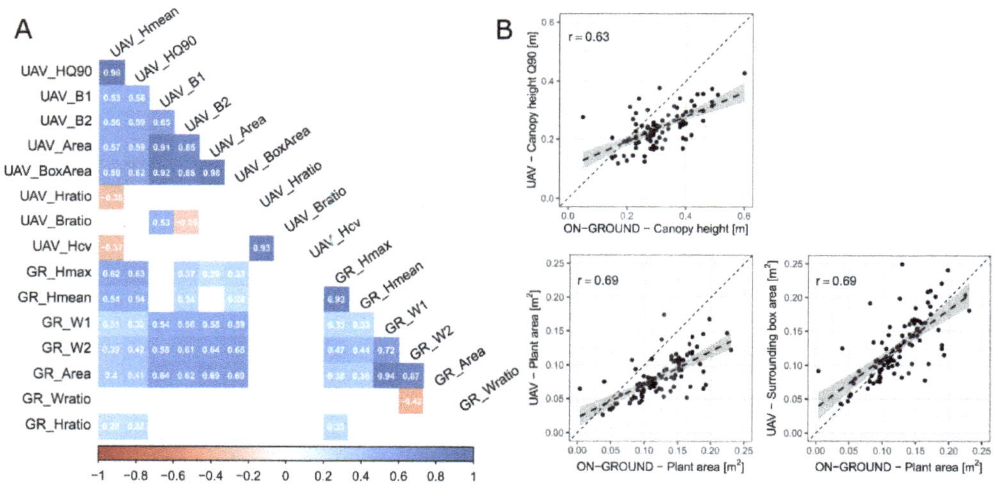

Figure 3. (**A**) Correlogram and (**B**) linear relations between on-ground measurements and UAV-derived plant architectural (plant height and width) parameters in *Sarcococca*, SarcoE1 (n = 94 plants). For the abbreviations, see Table S2.

Assessment of plant architecture for the rose genotypes was done using ruler-based measurements (canopy height, GR_Hmax) and visual scores (plant uniformity, GR_U), as well as with the data extracted from the images taken by the UAV. UAV_HQ90 correlates very well ($r = 0.73$ and $p < 0.01$) with the ruler-based measurements (Figure 4A). The height based on UAV_HQ90 is similar for less tall plants and lower for taller plants compared to the on-ground measurements. Related to the uniformity, we found a significant negative correlation between the uniformity score and the UAV_Hcv ($r = -0.38$; Figure 4B) for the observations on 5 September. This is to be expected as plants that received a lower score is less uniform.

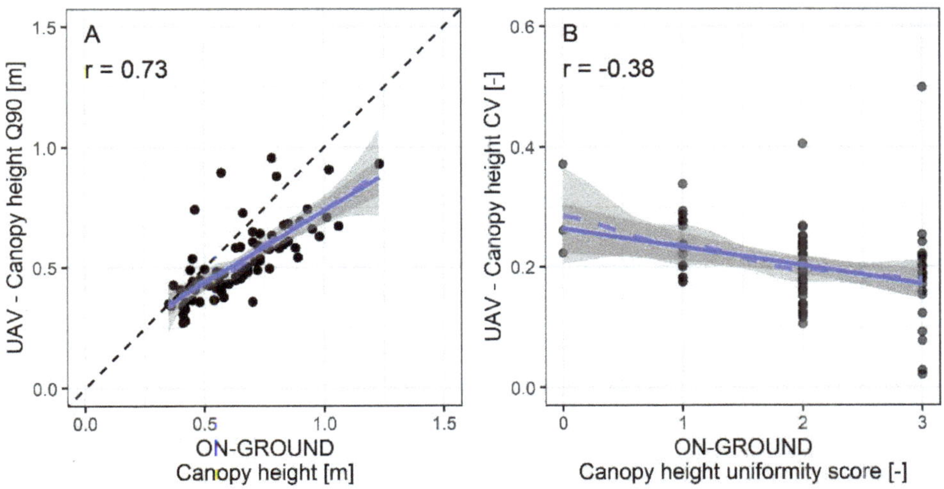

Figure 4. (**A**) Correlations between the manually measured canopy height (GR_Hmax) and the drone-derived canopy height (UAV_HQ90) in the rose trial, and (**B**) between the visual score of the canopy height uniformity (GR_U) and the drone-derived canopy height coefficient of variation (UAV_Hcv), respectively, for 5 September 2019.

3.2. Phenotyping: Capture of Flowering

When flowering is scored visually in the field, several parameters are considered in the final score, which has to be parsed to match data derived from UAV imagery (e.g., the relation between the number of flowers showing color and the size of the flower, different stages of flowering). The approach we took distinguished between flowers and other plant parts. A pixel classification in two classes (leaf or flower) was carried out based on the Rband threshold to differentiate colors. Then the percentage of flowers was calculated as the ratio between the pixels corresponding to the entire bush or canopy cover (UAV_CC) and the ones classified as a flower. Based on the percentage of flowers (at a single time point or multiple to catch the evolution during the flowering period), a comparison could be made between genotypes, and the ones with repeated flowering could be selected by evaluating different dates. A comparison between the on-ground scores and the percentage of flowers derived from the UAV imagery is presented for a representative selection of genotypes (Figure 5). In the graph based on UAV data, the three time points are clearly distinguishable, and the trend of the flowering can be derived. More blooms in week 27, a subsequent decrease with almost no presence of flowers in week 32, and a mild second blooming event in week 36 were observed. This trend is hardly observable using the on-ground scores where the flowering stages were used. According to the flowering stages across all genotypes, in week 28, 17% of the genotypes had buds, 30% had open flowers, and 51% were fading; in week 32, only 14% had open buds, and 66.9% were fading; and in week 36, 19% had open buds, and 56% had faded flowers. The correlation between the on-ground scores for color and open buds in week 32 is 0.73.

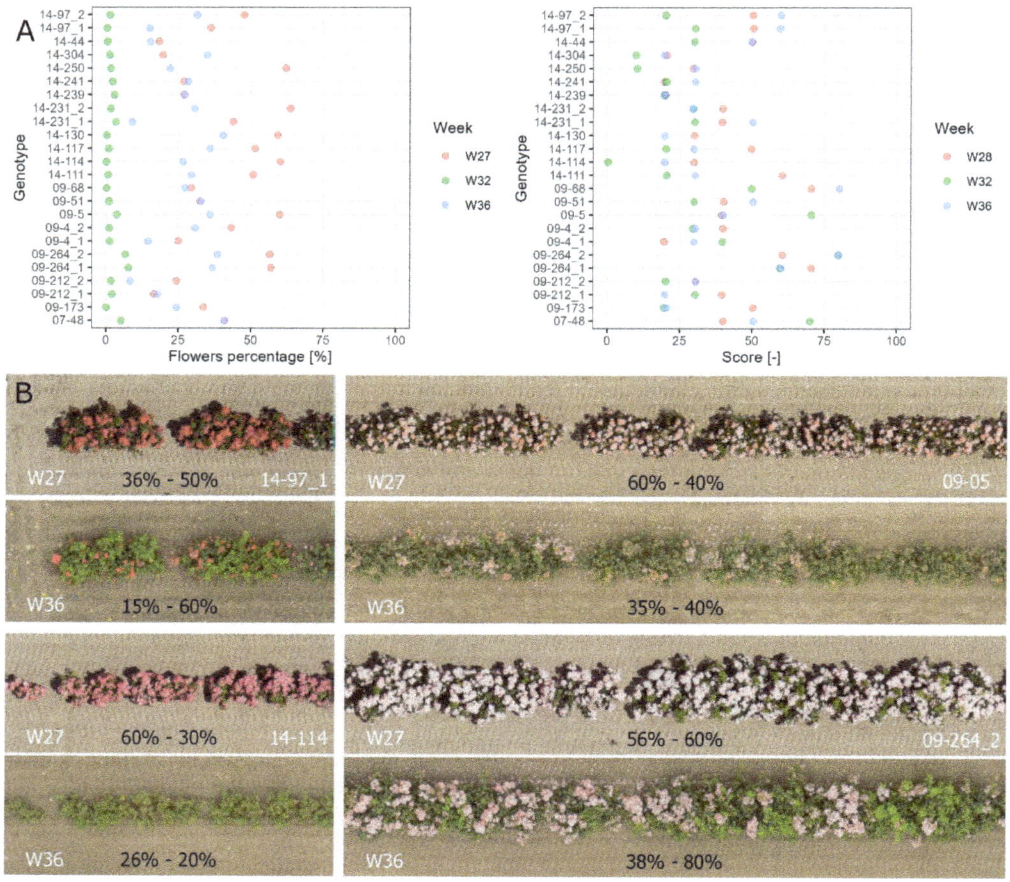

Figure 5. (**A**) Evolution of flower percentage and scores per week for a set of genotypes and (**B**) imagery of four genotypes in the first and last week with percentages derived from the UAV imagery and scores.

3.3. Assessment of Black Spot Occurrence

Genotypes are scored on-ground for leaf spots and leaf drops to determine the severity of black spot disease. The correlation between the two on-ground measured scores on 15 October was 0.79 ($p < 0.01$), showing a significant link as expected. As an alternative for the scoring, we tested the correlation between both on-ground determined scores and the UAV median ExG for the same date. This resulted in a negative correlation of −0.74 and −0.76 (both $p < 0.01$) with the leaf spot and the leaf drop scores, respectively. This indicated that the canopy was less green and showed increased disease symptoms. This demonstrates the clear link between the variables and the ability of UAV data to quantify disease severity between the genotypes in the breeding trial. Figure 6 compares the ExG values of the reference date versus the black spot assessment date in autumn. The figure clearly shows that interesting genotypes (green dots above the line) had an increase in canopy instead of leaf drop.

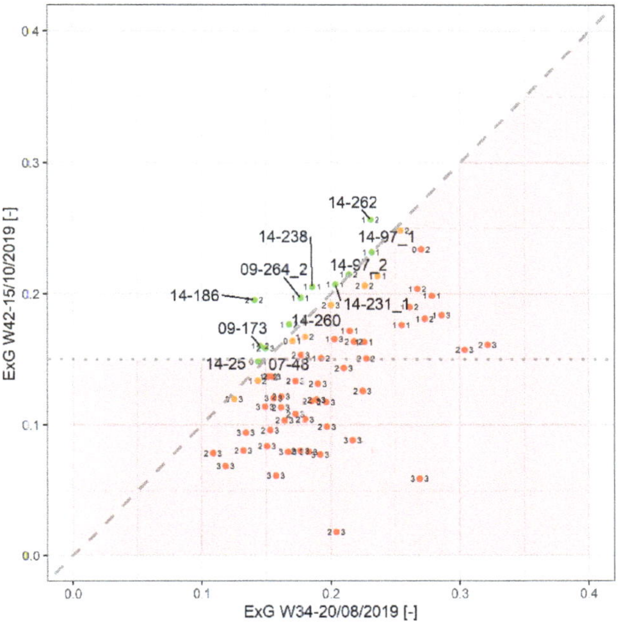

Figure 6. Evaluation of black spot disease in rose genotypes. All dots on and above the 1:1 line represent genotypes where the ExG value in the middle of October is equal to or higher than at the end of August (=reference) (green). Genotypes with an ExG reduction of ≤10% are shown in orange, or >10% reduction is indicated in red. For comparison, the scores of leaf spot and leaf drop per genotype (5 October 2019) are shown at the right or left of the symbol, respectively. The transparent red area indicates genotypes that can be omitted as they did not perform well for this trait (1:1 line and, e.g., a minimum ExG threshold for 5 October 2019 set by the breeder).

3.4. Application of UAV-Derived Parameters as Selection Criteria in Breeding Programs

Variability in families and populations is an important feature that allows for improved selections: the more segregation, the more progress can be made toward the creation of new cultivars. Figures 7 and S3 give an overview of the variability per mother plant and trial for open-pollinated half-sib populations of *Sarcococca*. Most mother plants used in SarcoE1 and SarcoE2 trials produce offspring showing a large variation in canopy height and plant area.

As a final step, plants were selected based on a visual inspection of the selection criteria. From the 3-year-old seedlings in SarcoE2, 12 seedlings were selected. A closer look at these selected plants compared to the rest of the seedlings from SarcoE2 is shown in Figure 8B. The plants selected by the breeders are mainly from S10 and S8 mother plants, taller (higher plants) and bigger (larger areas) than the median of all species. Aiming to select for compact seedlings with a round shape, "ideal" plant selection criteria can be defined as the following ranges: UAV_Hratio 1–1.5 (uniformity), UAV_Bratio 0.85–1.15 (compactness), UAV_Area 0.1–0.2, UAV_HQ90 0.3–0.45 and UAV_Hcv 0.2–0.5 (see blue rectangles in Figure 8B). A closer look at the resulting seedlings when combining all "ideal" selection criteria is presented in Figure S4. Three out of the 12 seedlings (nrs. 68, 175, 223) would be selected based on either UAV or on-ground field inspection.

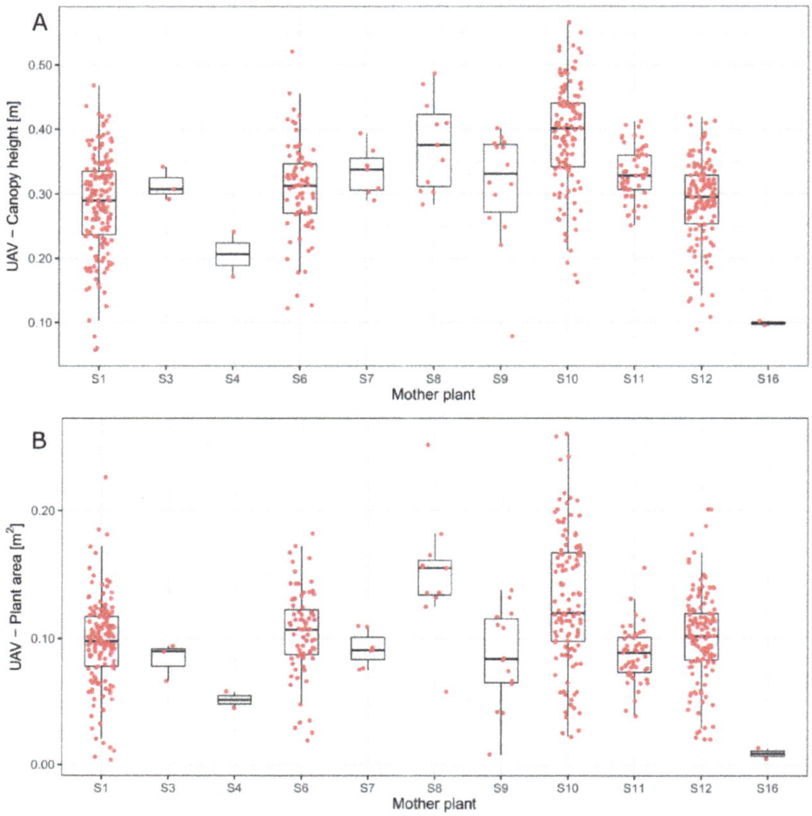

Figure 7. Box plots per mother plant for all *Sarcococca* plants in SarcoE2 for (**A**) canopy height and (**B**) plant area. Every dot corresponds to an individual plant.

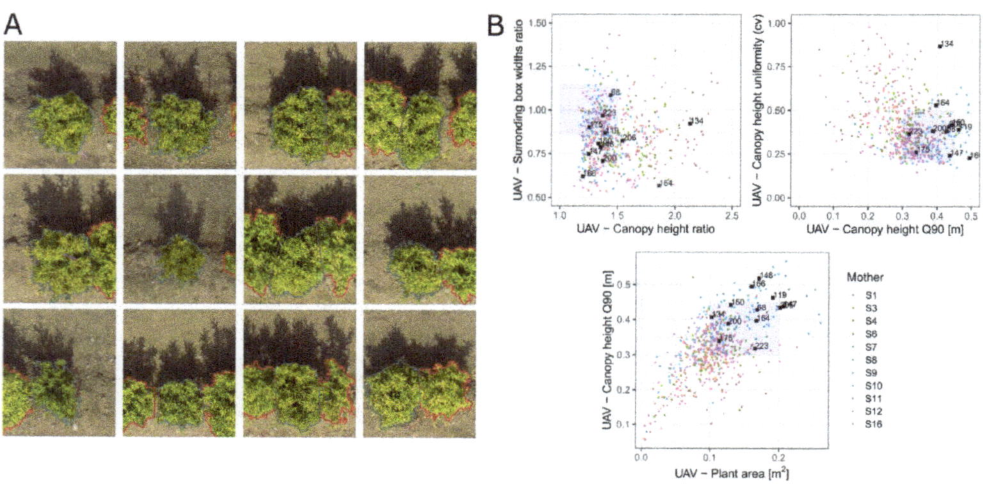

Figure 8. (**A**) *Sarcococca* plants selected (blue border) in the field identified by a unique ID (SarcoE2) and (**B**) comparison with the rest of the plants. The blue rectangle corresponds to the defined "ideal" selection criteria.

For roses, an example of how the UAV data can be used to select plant growth habits based on canopy height and canopy uniformity can be seen in Figure 9.

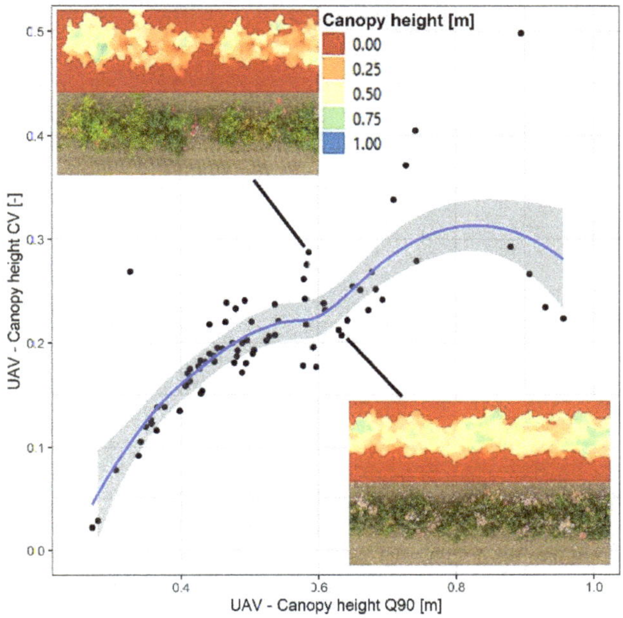

Figure 9. Variations present in the rose trial related to canopy height (UAV_HQ90) and uniformity (UAV_Hcv) with a top view image and a height image based on the CHM for two genotypes.

A final selection of roses was made, and four roses were commercialized: 'Mel Bee' (09-264), 'Bee Gold' (09-212), 'Bee Proud' (09-05), and 'Tsarina' (09-173). 'Mel Bee' (09-264) was among the best-scoring genotypes for floribundity in on-ground and UAV evaluation showing high scores later in the season (Figure 5). For disease resistance, this genotype is also among the genotypes with increased ExG value (Figure 6). 'Bee Proud' (09-05) also is among the genotypes with high floribundity in the on-ground scores and UAV data (Figure 5). 'Tsarina' (09-173) and 'Bee Gold' (09-212) show average scores for floribundity. 'Tsarina' scores very well for disease resistance and was selected for its healthy dark foliage in combination with attractive flower color.

4. Discussion

The multitude of UAV phenotyping applications either currently under development or already applied in agricultural crops are paving the way for the introduction of UAV and image analysis in woody ornamental and nursery plant breeding programs. For woody ornamentals, breeding towards compactness, improved or novel leaf and flower morphology, abundant flowering, and improved resistance to diseases, drought, cold, and other abiotic stresses is important. Knowledge from many applications in agricultural crops, e.g., the capture of (a)biotic stress symptoms, biomass estimation, growth monitoring, spike density, etc., can be translated for use in woody ornamentals (reviewed in [34] and references herein). Major advantages of UAV imaging include the (high) resolution of the images (to mm), chosen depending on the feature being evaluated (e.g., a lower resolution is needed to infer plant area as compared to flower detection), the objectivity of UAVs compared to multiple experts involved in on-site scoring measurements, the possibility of automation, the available sensors with different spectral ranges and the fast and large-scale image capture.

4.1. Experimental Setup and UAV Image Capture according to Trait(s) under Evaluation

The diversity in crop types and species of nursery plants is enormous. Each demands a dedicated approach when it comes to the selection of specific characteristics. In our study, an RGB camera and the calculation of the canopy height model and a small number of vegetation indices were successful for the evaluation of traits related to plant architecture, biotic stress, and flowering. However, selection towards improved aesthetic characteristics, which is very important for ornamentals, depends on the type of plant considered: trees, shrubs, bushes, perennials, or cut flowers [29]. It is not always easy to capture "beauty" in numbers. The development of crop-specific tools is required, and a good experimental setup is essential to implement the use of UAV imagery for high-throughput phenotyping in this field. This involves, among other things, the timing of flights in accordance with the breeding cycle of the particular crop, the number of flights, sensor and flight settings linked to the specific trait of interest, and the field setup. Selection using UAV imagery could be used on single genotypes of the crop, as done in the case of *Sarcococca*, or in a later phase after clonal propagation of the plant material, as in the case of the garden rose. Instead of the evaluation of individual genotypes, the evaluation of families or populations can also be valuable, as done in *Sarcococca*. Information on the mother plants could be obtained by comparing different half-sib families. According to the objectives, one should also choose the appropriate unit of plant phenotyping in the field to maximize the efficiency of selection and the corresponding measurable genetic gain. Today the most common unit used is the multi-plant unit, i.e., a densely grown field plot. Another approach could be the selection of individual plants grown in the absence of competitive interactions to maximize phenotypic expression and the corresponding phenotypic variance [35]. The field setup should be adapted to the parameters examined using image analysis. In the case of *Sarcococca*, the architecture of the individual plants needed to be evaluated, which imposed a plant density that made it possible to distinguish individual plants for proper image analysis. In our study on *Sarcococca*, the SarcoE1 trial contained smaller and younger plants compared to the SarcoE2 trial. A planting distance of 30 cm for the SarcoE1 plants enabled the good separation of individual canopies. In contrast, for the SarcoE2, the plants were already too large to be separated on the UAV images. For roses, we used a selection field with the roses planted in rows according to standard practice in this crop. However, it could be useful to work with small square plots filled with plants, for example, to examine ground coverage via image analysis. In all cases, weeds must be removed before flights are carried out to facilitate image processing. Remote sensing technology could enable simultaneous selection at different locations, such as in cut rose breeding, where selection is performed remotely, and rose genotypes are sent to the most appropriate regions (e.g., South America or Africa) to be tested under specific conditions of altitude and climate [32].

The number of flights needed and the timing of the flights both depend on the trait under evaluation and require thorough consideration, including cost efficiency in phenotyping, as discussed by Reynolds et al. [30]. For plant architecture, as assessed in *Sarcococca* and rose, a single flight during the growing season could suffice. For dynamic traits like recurrent flowering or disease development, the assessment at different time points is essential [8]. In the example of black spot disease in rose, two time points were compared using the first time point as a reference. Symptoms of black spot mainly occur in autumn, and plant height uniformity shows more variation later in the season after the formation of new branches. In rose selection, flights for evaluation of the aforementioned traits are therefore best planned for the second half of the growing season (early autumn). Nursery plants are typically evaluated over several years in the same field [28]. Attention to the comparison of data obtained by image analysis over different years is required, as this can help to adjust plant evaluation in relation to year effects, e.g., differences in the type of weather during the season in outdoor selection. For UAV flights, a strict protocol (see description in the Section 2) needs to be followed, including stable light conditions during the flight to avoid unnecessary variability in the data. Recommendations about optimal flight parameters and settings to capture and generate accurate data have been assessed in several publications [36–39]. In our rose evaluation, the images from the flight carried out in W25 had to be discarded because of bad quality, while this week would have been ideal for evaluating flowering. The use of a multispectral camera allows for radiometric correction and can avoid this problem [10]. In addition to well-considered setups of the experiment and the field trial, the choice of parameter for the best estimation of the trait is also essential. For the evaluation of plant height uniformity, the coefficient of variation was found to be a better parameter than the standard deviation, as the former is normalized for the height of the plant and is, therefore, more comparable for plants with different plant heights, such as in roses.

4.2. UAV Imagery-Based Measurements Compared to On-Ground Measurements

In general, in our study, there was a good correlation between on-ground measurements and measurements based on UAV images. For example, leaf drop caused by black spot was assessed in roses and used to compare rose genotypes under selection. A high correlation of 0.79 with on-ground scores was found, which could support the implementation of this UAV-based methodology to quantify disease severity in a more standardized manner compared to on-ground evaluation. For some parameters, UAV imagery showed an underestimation or lower correlation. In *Sarcococca*, plant heights derived from UAV imagery are lower than from manual measurements in the field, which could be explained by the difficulty of modeling/detecting ("narrow") individual branches (present in a high number in bigger plants) and the use of percentile Q90 to avoid possible outliers/errors from the DEM construction against the measurement of the tallest branch top of the individual plant. For the plant area in *Sarcococca*, differences between on-ground and UAV data can be explained by the simplification to calculate the area by multiplying the two directions plant width measurements (therefore, the correlation with the UAV box area is higher (rectangle shape)). However, the area estimated by the UAV is more accurate as it is possible to capture the border and the shape of the plants. For roses, a high correlation between the field measurements and UAV-derived heights was achieved. In roses also, the Q90 value of the CHM was used as this value gave the best correlation to the on-ground measurements, excluding possible outliers.

Discrepancies can arise between the selection by visual inspection and the selection by UAV imagery: the "ideal" criteria cannot be totally mimicked as these criteria are also influenced by (subjective) expert experience. For *Sarcococca*, one of the breeding goals was to select more compact varieties. But analysis of the selected plants chosen by the experts revealed that they had intuitively selected somewhat taller plants with bigger plant areas. Only 3 out of 12 manually selected plants would have been selected using UAV-based selection criteria for height and plant area. In the case of the evaluation of floribundity in roses, differences can arise if the experts are more interested in plant vitality or visual quality, which differs from counting all flowers regardless of their stage of development or only the fully open flowers [29]. In this work, UAV imagery evaluated the amount of color on the plant (number of pixels classified as vegetation with a different color than green). In contrast, visual inspection considered the number of individual flowers. For example, 'Bee Gold' (09-212), one of the selected and commercialized genotypes, is a mini or patio rose with a low canopy height. The single flowers in this genotype are small and yellow. Within this type of rose, it is more difficult to select good genotypes that combine floribundity with disease resistance. Therefore, it can be difficult to impose the criteria when imagery is used to evaluate flowering: e.g., buds not visible, different views (lateral vs. top). Specific analytical tools for specific subtypes of plants could be developed, e.g., floribunda roses (branched shrubs with large flower sprays) need another approach compared to hybrid tea roses, which have upright stems with a single large flower. A few large flowers could be seen in the images of the top and upper sides of the bush, and at the same time, a large number of flowers and buds could be seen from the side in the on-ground observation, thus leading to differences in the evaluation. An improved and better approach to capture the number of flowers per branch (as the 2D orthophoto does not give enough information could be the use of 3D point clouds instead.

4.3. Possible Future Applications of UAV Imagery in Breeding Programs of (Woody) Ornamentals

Despite the remaining challenges associated with UAV imagery, many applications can already be valuable and useful for open-field selection in woody plant breeding programs. Breeding in woody ornamentals requires a cost-efficient system. Typically, in breeding woody plants, a long and costly observation period is required to accurately evaluate the performance of plant genotypes [32]. A more efficient and less labor-intensive selection, e.g. the use of UAV images, could improve effective knowledge acquisition of the genotypes and reduce costs over the long term. For morphological traits, such as fruit and berry development, plant shape and architecture, and flower characteristics in the entire bush more detailed 3D reconstructions can be useful. 3D point-based deep learning models were used in synthetic rose bushes for the phenotyping of plant architecture and morphology [40]. This would be ideal for training models to relate architecture and flowering to aesthetic values. Furthermore, the use of multispectral and thermal sensors could allow the evaluation of (a)biotic stress symptoms in different crops. In container fields, water gift is essential, and in [41], the water stress of woody ornamentals grown on a container field was assessed using sensor technology. The effect of salinity has been observed by multispectral image analysis and the use of a normalized difference vegetation index (NDVI) calculated from the Near Infrared Reflectance (NIR) [23]. The detection of diseased plants due to biotic stress by use of remote sensing data [42] or by means of UAV [43] is a hot topic in plant cultivation and breeding, e.g., monitoring of powdery mildew in apple orchards [44] and monitoring of rust on coffee plants [45]. Although detection of disease emergence during the period of latency seems to be the current holy grail [42], in nursery plant breeding, this is less important, as the evaluation of the resistance and comparison between genotypes under selection is only carried out after some years on the field.

5. Conclusions

Using UAV-based measurements of plant characteristics for selection purposes has many advantages for breeding woody plants. Characteristics such as plant architecture, flowering, and disease resistance may be assessed more objectively and faster using UAV imaging as compared to expert on-ground observations. We have shown some of the advantages in case studies on an evergreen shrub, *Sarcococca*, a flowering garden plant, *Rosa*. Significant correlations (r > 0.65) were obtained between UAV imaging and on-ground measurements. Therefore, UAV-based phenotyping is valuable for traits such as plant height and shape, especially for large breeding trials. In addition, more detailed data on the trial, plots, and individual plants are obtained, which allows us to better examine variability and uniformity in plant architecture and flowering traits. Besides objective morphological assessment, disease evaluation using UAV data assists in the selection of resistant genotypes. In conclusion, with a well-considered design of the field trial and experiment, and with well-chosen parameters for explaining the traits based on different sensor types, many more useful applications can be developed for breeding and cultivation monitoring of woody ornamental plants to support selection.

Supplementary Materials: The following supporting information can be downloaded at: https://www.mdpi.com/article/10.3390/horticulturae8121186/s1, Table S1: Overview of Sarcococca genotypes under field evaluation; Table S2: Overview of the phenotypic parameters for Sarcococca and rose experiments; Figure S1: Evaluation of spectral bands and vegetation indices to separate colors. Separation of the colors by the index with the highest F-value (WBI); Figure S2: Flight 1 July, examples of genotypes with different flower colors with the classification of pixels as flower marked in yellow and bush area delimited in white in roses; Figure S3: Box plots per mother plant of all Sarcococca plants in SarcoE1 for (A) canopy height and (B) plant area. Every dot corresponds to a plant; Figure S4 UAV plants selected according to "ideal" plant selection criteria (central part of the image) in Sarcococca.

Author Contributions: Conceptualization, I.B.-S., K.V.L., L.L. and P.L.; methodology, K.V.L., L.L., I.B.-S. and P.L.; validation, I.B.-S., K.V.L. and L.L.; formal analysis, I.B.-S. and P.L.; investigation, I.B.-S., K.V.L., L.L. and P.L.; resources, K.V.L. and L.L.; data curation, I.B.-S.; writing—original draft preparation, I.B.-S., K.V.L. and L.L.; writing—review and editing, I.B.-S., K.V.L., L.L. and P.L.; visualization, I.B.-S. and P.L.; supervision, I.B.-S.; project administration, K.V.L. and L.L.; funding acquisition, K.V.L. and L.L. All authors have read and agreed to the published version of the manuscript.

Funding: This research received no external funding.

Data Availability Statement: Not applicable.

Acknowledgments: We thank the UAV pilot, the field workers, and the thesis students for their data-gathering efforts and Miriam Levenson for English language corrections.

Conflicts of Interest: The authors declare no conflict of interest.

References

1. Araus, J.L.; Kefauver, S.C.; Zaman-Allah, M.; Olsen, M.S.; Cairns, J.E. Translating High-Throughput Phenotyping into Genetic Gain. *Trends Plant Sci.* **2018**, *23*, 451–466. [CrossRef]
2. Xie, C.; Yang, C. A Review on Plant High-Throughput Phenotyping Traits Using UAV-Based Sensors. *Comput. Electron. Agric.* **2020**, *178*, 105731. [CrossRef]
3. Elazab, A.; Ordóñez, R.A.; Savin, R.; Slafer, G.A.; Araus, J.L. Detecting Interactive Effects of N Fertilization and Heat Stress on Maize Productivity by Remote Sensing Techniques. *Eur. J. Agron.* **2016**, *73*, 11–24. [CrossRef]
4. di Gennaro, S.F.; Battiston, E.; di Marco, S.; Facini, O.; Matese, A.; Nocentini, M.; Palliotti, A.; Mugnai, L. Unmanned Aerial Vehicle (UAV)-Based Remote Sensing to Monitor Grapevine Leaf Stripe Disease within a Vineyard Affected by Esca Complex. *Phytopathol. Mediterr.* **2016**, *55*, 262–275. [CrossRef]
5. Kerkech, M.; Hafiane, A.; Canals, R. Deep Leaning Approach with Colorimetric Spaces and Vegetation Indices for Vine Diseases Detection in UAV Images. *Comput. Electron. Agric.* **2018**, *155*, 237–243. [CrossRef]
6. De Swaef, T.; Maes, W.H.; Aper, J.; Baert, J.; Cougnon, M.; Reheul, D.; Steppe, K.; Roldán-Ruiz, I.; Lootens, P. Applying RGB- and Thermal-Based Vegetation Indices from UAVs for High-Throughput Field Phenotyping of Drought Tolerance in Forage Grasses. *Remote Sens.* **2021**, *13*, 147. [CrossRef]
7. de Castro, A.I.; Jiménez-Brenes, F.M.; Torres-Sánchez, J.; Peña, J.M.; Borra-Serrano, I.; López-Granados, F. 3-D Characterization of Vineyards Using a Novel UAV Imagery-Based OBIA Procedure for Precision Viticulture Applications. *Remote Sens.* **2018**, *10*, 584. [CrossRef]
8. López-Granados, F.; Torres-Sánchez, J.; Jiménez-Brenes, F.M.; Arquero, O.; Lovera, M.; De Castro, A.I. An Efficient RGB-UAV-Based Platform for Field Almond Tree Phenotyping: 3-D Architecture and Flowering Traits. *Plant Methods* **2019**, *15*, 160. [CrossRef]
9. Borra-Serrano, I.; De Swaef, T.; Muylle, H.; Nuyttens, D.; Vangeyte, J.; Mertens, K.; Saeys, W.; Somers, B.; Roldán-Ruiz, I.; Lootens, P. Canopy Height Measurements and Non-destructive Biomass Estimation of *Lolium Perenne* Swards Using UAV Imagery. *Grass Forage Sci.* **2019**, *74*, 356–369. [CrossRef]
10. Pranga, J.; Borra-Serrano, I.; Aper, J.; De Swaef, T.; Ghesquiere, A.; Quataert, P.; Roldán-Ruiz, I.; Janssens, I.A.; Ruysschaert, G.; Lootens, P. Improving Accuracy of Herbage Yield Predictions in Perennial Ryegrass with Uav-Based Structural and Spectral Data Fusion and Machine Learning. *Remote Sens.* **2021**, *13*, 3459. [CrossRef]
11. Borra-Serrano, I.; De Swaef, T.; Aper, J.; Ghesquiere, A.; Mertens, K.; Nuyttens, D.; Saeys, W.; Somers, B.; Vangeyte, J.; Roldán-Ruiz, I.; et al. Towards an Objective Evaluation of Persistency of Lolium Perenne Swards Using UAV Imagery. *Euphytica* **2018**, *214*, 142. [CrossRef]
12. Di Gennaro, S.F.; Rizza, F.; Badeck, F.W.; Berton, A.; Delbono, S.; Gioli, B.; Toscano, P.; Zaldei, A.; Matese, A. UAV-Based High-Throughput Phenotyping to Discriminate Barley Vigour with Visible and near-Infrared Vegetation Indices. *Int. J. Remote Sens.* **2018**, *39*, 5330–5344. [CrossRef]
13. Kefauver, S.C.; Vicente, R.; Vergara-Díaz, O.; Fernandez-Gallego, J.A.; Kerfal, S.; Lopez, A.; Melichar, J.P.E.; Serret Molins, M.D.; Araus, J.L. Comparative UAV and Field Phenotyping to Assess Yield and Nitrogen Use Efficiency in Hybrid and Conventional Barley. *Front. Plant Sci.* **2017**, *8*, 1733. [CrossRef]
14. Jin, X.; Liu, S.; Baret, F.; Hemerlé, M.; Comar, A. Estimates of Plant Density of Wheat Crops at Emergence from Very Low Altitude UAV Imagery. *Remote Sens. Environ.* **2017**, *198*, 105–114. [CrossRef]
15. Fernandez-Gallego, J.A.; Lootens, P.; Borra-Serrano, I.; Derycke, V.; Haesaert, G.; Roldán-Ruiz, I.; Araus, J.L.; Kefauver, S.C. Automatic Wheat Ear Counting Using Machine Learning Based on RGB UAV Imagery. *Plant J.* **2020**, *103*, 1603–1613. [CrossRef] [PubMed]
16. Robbins, J.A. Small Unmanned Aircraft Systems (SUAS): An Emerging Technology for Horticulture. *Hortic. Rev. (Am. Soc. Hortic. Sci.)* **2018**, *45*, 33–71. [CrossRef]
17. Zhang, C.; Valente, J.; Kooistra, L.; Guo, L.; Wang, W. Orchard Management with Small Unmanned Aerial Vehicles: A Survey of Sensing and Analysis Approaches. *Precis. Agric.* **2021**, *22*, 2007–2052. [CrossRef]

18. Jiménez-Brenes, F.M.; López-Granados, F.; Castro, A.I.; Torres-Sánchez, J.; Serrano, N.; Peña, J.M. Quantifying Pruning Impacts on Olive Tree Architecture and Annual Canopy Growth by Using UAV-Based 3D Modelling. *Plant Methods* **2017**, *13*, 55. [CrossRef] [PubMed]
19. Díaz-Varela, R.; de la Rosa, R.; León, L.; Zarco-Tejada, P. High-Resolution Airborne UAV Imagery to Assess Olive Tree Crown Parameters Using 3D Photo Reconstruction: Application in Breeding Trials. *Remote Sens.* **2015**, *7*, 4213–4232. [CrossRef]
20. De Castro, A.I.; Rallo, P.; Suárez, M.P.; Torres-sánchez, J.; Casanova, L.; Jiménez-brenes, F.M.; Morales-sillero, A.; Jiménez, M.R.; López-granados, F. High-Throughput System for the Early Quantification of Major Architectural Traits in Olive Breeding Trials Using UAV Images and OBIA Techniques. *Front. Plant Sci.* **2019**, *10*, 1472. [CrossRef]
21. Freeman, D.; Gupta, S.; Hudson Smith, D.; Maja, J.M.; Robbins, J.; Owen, J.S.; Peña, J.M.; de Castro, A.I. Watson on the Farm: Using Cloud-Based Artificial Intelligence to Identify Early Indicators of Water Stress. *Remote Sens.* **2019**, *11*, 2645. [CrossRef]
22. Bayraktar, E.; Basarkan, M.E.; Celebi, N. A Low-Cost UAV Framework towards Ornamental Plant Detection and Counting in the Wild. *ISPRS J. Photogramm. Remote Sens.* **2020**, *167*, 1–11. [CrossRef]
23. Yu, X.; Her, Y.; Chang, A.; Song, J.-H.; Campoverde, E.V.; Schaffer, B. Assessing the Effects of Irrigation Water Salinity on Two Ornamental Crops by Remote Spectral Imaging. *Agronomy* **2021**, *11*, 375. [CrossRef]
24. Li-Marchetti, C.; Le Bras, C.; Chastellier, A.; Relion, D.; Morel, P.; Sakr, S.; Oyant, L.H.-S.; Crespel, L. 3D Phenotyping and QTL Analysis of a Complex Character: Rose Bush Architecture. *Tree Genet. Genomes* **2017**, *13*, 112. [CrossRef]
25. Susko, A.Q.; Rinehart, T.A.; Bradeen, J.M.; Hokanson, S.C. An Evaluation of Two Seedling Phenotyping Protocols to Assess PH Adaptability in Deciduous Azalea (Rhododendron Sect. Pentanthera G. Don). *HortScience* **2018**, *53*, 268–274. [CrossRef]
26. Wang, S.; Leus, L.; Lootens, P.; Van Huylenbroeck, J.; Van Labeke, M.-C. Germination Kinetics and Chlorophyll Fluorescence Imaging Allow for Early Detection of Alkalinity Stress in Rhododendron Species. *Horticulturae* **2022**, *8*, 823. [CrossRef]
27. Denaeghel, H.E.R.; Van Laere, K.; Leus, L.; Lootens, P.; Van Huylenbroeck, J.; Van Labeke, M.C. The Variable Effect of Polyploidization on the Phenotype in Escallonia. *Front. Plant Sci.* **2018**, *9*, 354. [CrossRef] [PubMed]
28. Van Laere, K.; Hokanson, S.C.; Contreras, R.; Van Huylenbroeck, J. *Woody Ornamentals of the Temperate Zone*; Springer: Cham, Switzerland, 2018; pp. 803–887.
29. Santagostini, P.; Demotes-Mainard, S.; Huché-Thélier, L.; Leduc, N.; Bertheloot, J.; Guérin, V.; Bourbeillon, J.; Sakr, S.; Boumaza, R. Assessment of the Visual Quality of Ornamental Plants: Comparison of Three Methodologies in the Case of the Rosebush. *Sci. Hortic. (Amst.)* **2014**, *168*, 17–26. [CrossRef]
30. Reynolds, D.; Baret, F.; Welcker, C.; Bostrom, A.; Ball, J.; Cellini, F.; Lorence, A.; Chawade, A.; Khafif, M.; Noshita, K.; et al. What Is Cost-Efficient Phenotyping? Optimizing Costs for Different Scenarios. *Plant Sci.* **2019**, *282*, 14–22. [CrossRef]
31. Denaeghel, H.; Van Laere, K.; Leus, L.; Van Huylenbroeck, J.; Van Labeke, M.C. Interspecific Hybridization in Sarcococca Supported by Analysis of Ploidy Level, Genome Size and Genetic Relationships. *Euphytica* **2017**, *213*, 149. [CrossRef]
32. Leus, L.; Van Laere, K.; De Riek, J.; Van Huylenbroeck, J. *Ornamental Crops*; Handbook of Plant Breeding; Van Huylenbroeck, J., Ed.; Springer: Cham, Switzerland, 2018; Volume 11, pp. 719–767. [CrossRef]
33. Woebbecke, D.M.; Meyer, G.E.; Von Bargen, K.; Mortensen, D.A. Color Indices for Weed Identification under Various Soil, Residue, and Lighting Conditions. *Trans. ASAE* **1995**, *38*, 259–269. [CrossRef]
34. Borra-Serrano, I.; Leus, L.; Wang, S.; Van Huylenbroeck, J.; Lootens, P.; Van Laere, K. A View from above: The Use of Drones and Image-Based Phenotyping in Breeding and Production of Woody Ornamentals. *Acta Hortic.* **2021**, *1331*, 1–9. [CrossRef]
35. Fasoula, D.A.; Ioannides, I.M.; Omirou, M Phenotyping and Plant Breeding: Overcoming the Barriers. *Front. Plant Sci.* **2020**, *10*, 1713. [CrossRef] [PubMed]
36. Mesas-Carrascosa, F.J.; Torres-Sánchez, J.; Clavero-Rumbao, I.; García-Ferrer, A.; Peña, J.M.; Borra-Serrano, I.; López-Granados, F. Assessing Optimal Flight Parameters for Generating Accurate Multispectral Orthomosaicks by Uav to Support Site-Specific Crop Management. *Remote Sens.* **2015**, *7*, 12793–12814. [CrossRef]
37. Torres-Sánchez, J.; López-Granados, F.; Borra-Serrano, I.; Manuel Peña, J. Assessing UAV-Collected Image Overlap Influence on Computation Time and Digital Surface Model Accuracy in Olive Orchards. *Precis. Agric.* **2018**, *19*, 115–133. [CrossRef]
38. Jang, G.J.; Kim, J.; Yu, J.K.; Kim, H.J.; Kim, Y.; Kim, D.W.; Kim, K.H.; Lee, C.W.; Chung, Y.S. Review: Cost-Effective Unmanned Aerial Vehicle (UAV) Platform for Field Plant Breeding Application. *Remote Sens.* **2020**, *12*, 998. [CrossRef]
39. Guo, W.; Carroll, M.E.; Singh, A.; Swetnam, T.L.; Merchant, N.; Sarkar, S.; Singh, A.K.; Ganapathysubramanian, B. UAS-Based Plant Phenotyping for Research and Breeding Applications. *Plant Phenomics* **2021**, *2021*, 9840192. [CrossRef]
40. Turgut, K.; Dutagaci, H.; Galopin, G.; Rousseau, D. Segmentation of Structural Parts of Rosebush Plants with 3D Point-Based Deep Learning Methods. *Plant Methods* **2022**, *18*, 20. [CrossRef]
41. De Castro, A.I.; Maja, J.M.; Owen, J.; Robbins, J.; Peña, J.M. Experimental Approach to Detect Water Stress in Ornamental Plants Using SUAS-Imagery. *Auton. Air Ground Sens. Syst. Agric. Optim. Phenotyping III* **2018**, *10664*, 178–188. [CrossRef]
42. Oerke, E.-C. Remote Sensing of Diseases. *Annu. Rev. Phytopathol.* **2020**, *58*, 225–252. [CrossRef]
43. Neupane, K.; Baysal-Gurel, F. Automatic Identification and Monitoring of Plant Diseases Using Unmanned Aerial Vehicles: A Review. *Remote Sens.* **2021**, *13*, 3841. [CrossRef]
44. Chandel, A.K.; Khot, L.R.; Sallato, B. Apple Powdery Mildew Infestation Detection and Mapping Using High-Resolution Visible and Multispectral Aerial Imaging Technique. *Sci. Hortic. (Amst.)* **2021**, *287*, 110228. [CrossRef]
45. Marin, D.B.; Ferraz, G.A.e.S.; Santana, L.S.; Barbosa, B.D.S.; Barata, R.A.P.; Osco, L.P.; Ramos, A.P.M.; Guimarães, P.H.S. Detecting Coffee Leaf Rust with UAV-Based Vegetation Indices and Decision Tree Machine Learning Models. *Comput. Electron. Agric.* **2021**, *190*, 106476. [CrossRef]

horticulturae

MDPI

Article

Air Layering Improves Rooting in Tree Peony Cultivars from the Jiangnan Group

Ying Zhang, Shui-Yan Yu * and Yong-Hong Hu *

Shanghai Key Laboratory of Plant Functional Genomics and Resources, Shanghai Chenshan Botanical Garden, Shanghai 201602, China
* Correspondence: yushuiyan1982@163.com (S.-Y.Y.); huyonghong@csnbgsh.cn (Y.-H.H.)

Abstract: Tree peony (*Paeonia suffruticosa* Andr.), a unique traditional flower in China, is famous for its ornamental value, medical use, and edible oil production. Traditional propagation methods, such as sowing, dividing, and grafting do not allow the large-scale production of selected peony varieties. Therefore, the objective of our study is to evaluate an air-layering technique on the rooting success of three tree peony cultivars ('Baoqing Hong', 'Quehao', and 'Xishi'). The experiments were established through consideration of the influence of the time of the year the rooting was performed (mid-May, mid-June, or mid-July) and the growth regulators (1-naphthaleneacetic acid-NAA and indole-3-butyric acid-IBA) applied at different concentrations (1000 mg/L, 1500 mg/L, 2000 mg/L). The results showed that the rooting rate was the highest when the air-layering time occurred in mid-June, and the rooting rate of 'Quehao' was found to be the most significant, reaching 100%. The rooting percentages of 82.86% and 77.14% were obtained for 'Baoqing Hong' and 'Xishi', respectively. The growth regulators affected the rooting performance of the three cultivars differently. The rooting parameters of 'Quehao' were negatively correlated with the concentration of NAA but positively correlated with IBA, whereas the 'Baoqing Hong' and 'Xishi' cultivars showed no dose dependence for the supplied growth regulators. Root number, root tip number, and maximum root length in 'Quehao' were higher than those of the other two cultivars. The conclusion of our study is that the air-layering technique is a suitable method for achieving satisfactory propagation of selected tree peony cultivars.

Keywords: propagation; tree peony; air layering; rooting parameters; growth regulators

Citation: Zhang, Y.; Yu, S.-Y.; Hu, Y.-H. Air Layering Improves Rooting in Tree Peony Cultivars from the Jiangnan Group. *Horticulturae* **2022**, *8*, 941. https://doi.org/10.3390/horticulturae8100941

Academic Editors: Margherita Beruto, Emmy Dhooghe and Bruce Dunn

Received: 6 September 2022
Accepted: 12 October 2022
Published: 14 October 2022

1. Introduction

Tree peony (*Paeonia suffruticosa* Andr.) is a traditional flower that is unique to China. It is a rare ornamental flower as well as a medicinal and food plant. Its roots have hypoglycemic, antibacterial, anti-inflammatory, anti-atherosclerosis, anti-arrhythmia, anticonvulsion, liver-protection, immunity-enhancing, and other pharmacological effects [1,2]. Tree peony seed oil contains a large amount of unsaturated fatty acids, such as α-linolenic acid (ω-3 family), linoleic acid, and oleic acid, which have anti-inflammatory, anti-tumor, lipid-lowering, and immunity-improvement effects [3,4]. In addition, tree peony stamens can be used to prepare tea, and the petals can be processed and refined into essential oils that are widely used in daily health care and various industries [5,6]. With the development of deep-processing products, the tree peony industry has gradually developed into a comprehensive and multi-product industry.

Increased market demand has promoted the rapid development of the tree peony seedling industry, but the capacity for domestic tree peony seedling production cannot meet market demands. This is because of the long life cycle of peony, which takes 4–6 years from sowing to flowering and fruiting [7]. This biological characteristic severely restricts peony industrialization. The traditional propagation methods for tree peony include sowing, dividing, and grafting, which have low propagation coefficients and long seedling periods, as well as inconsistent quality [8]. Although tissue culture could represent a valid method for vegetatively propagating tree peony, several drawbacks are still to be addressed with this method, including poor rooting, vitrification, a low multiplication rate, difficult acclimatization, and a low in vivo survival rate [9–18]. Large-scale seedling production is based on grafting and dividing methods, but several issues limit breeding and industrial development [19–21]. The main factors affecting the success of grafting include grafting time, rootstock, scion, and grafting method. The grafting time for tree peony is generally October–November, with graft survival being limited at other times. Even when plant grafting is carried out at the optimal time, diseases and insect pests may occur, and sudden low temperatures in early spring and other environmental factors can inhibit survival. The grafting survival rate of various peony varieties in the second year is only 50%–85% [22].

Tree peonies with high ornamental value are mostly seedless and cannot be sexually reproduced. The roots of *P. lactiflora* and *P. ostii* are typically selected as the rootstock. Due to the characteristics of the rootstock, the branches of the rootstock are often budded from the root, affecting the growth of the grafted seedlings [8]. The grafting affinity of some varieties is also weak, affecting the grafting success. Although some tree peony varieties can be sexually reproduced, the growth cycle is very long. It takes 5–6 years from sowing to flowering plants and production of seeds. Thus, the seed yield is low, and the plants cannot be quickly placed into the market [7]. The characteristics of the offspring of superior plants propagated through grafting are greatly affected by the rootstock, and the screened high-quality tree peony plants cannot be placed into the market quickly and efficiently, thereby hindering the development of the peony industry.

Air-layering propagation technology can retain the favorable traits of the mother plants as well as encourage early flowering and fruiting [23]. This method has been applied to the propagation of different plants with low propagation coefficients and has achieved good results, such as in *Phoebe bournei* [24], *Camellia japonica* [25], and *Saraca* 'Siji Flower' [26]. The progeny obtained by air layering have high purity and excellent traits, are not influenced by the rootstock, and exhibit early flowering and fruiting. The seedlings obtained by air layering grow better in the second year, at which point they can be placed on the market. Although the air-layering propagation technique is widely used, no relevant studies on this topic can be found for tree peony or Paeonia (data source: Web of Science). Therefore, it is a novel approach to adopt an air-layering technique for tree peony propagation. The purpose of this study is to explore whether the air-layering technique can be successfully implemented in tree peony to establish an effective propagation method that is different from the traditional propagation method.

2. Materials and Methods

2.1. Plant Material

Three varieties (Baoqing Hong', 'Quehao', and 'Xishi') belonging to the Jiangnan group were selected from plants previously grown at the Luoyang Shenzhou Peony Garden, and were subsequently transferred to the Shanghai Chenshan Botanical Garden (31°4′52″ N, 121°10′14″ E). The plants have been kept under the same environmental and cultivation conditions since 2014, and currently there are hundreds of living plants growing in the tree peony planting base of the Shanghai Chenshan Botanical Garden. The present experiment was conducted from May to November in 2020 with plants that were 8–10 years old.

2.2. Air-Layering Procedure

On the mother plant, an area about 1 cm away from the end node of two-year-old branches was selected, and the phloem was peeled off completely to expose the xylem. The ring-stripping incision was about 1 cm. A cotton swab was used to apply the rooting treatment by dipping into the incised surface. The center of a 20 × 20 cm square film was cut to the midpoint of one side and used as a funnel. The film funnel was filled with soaking medium, which consisted of 50% moss and 50% peat (Klasmann-Deilmann 413, pH 5.5–6.5). The moss was cut appropriately in advance, and the length was controlled within 1 cm. After filling the soaking medium, the top of the film was secured with a rust-proof black tie wire so that it was easy to naturally collect rain. If it did not rain for a long time, water was properly supplied to keep the medium moist. In late September or October, when the tree peonies were ready for transplanting, the branches below the film opening were removed and the rooted seedlings were transplanted into the flowerpot. The seedlings were then propagated using standard procedures [27].

2.3. Influence of the Time of the Year on Air Layering

The experiment was performed with the three already cited tree peony varieties and three periods for air layering were considered: mid-May in Spring, and mid-June and mid-July in Summer. The recorded climate conditions during our experiment were:

a. average temperature: 16–24 °C in May, 21–27 °C in June, and 25–32 °C in July
b. average precipitation: 112 mm in May, 169 mm in June, and 151 mm in July

For each period, seven individual mother plants replicates were considered, and 5 fresh branches of the current year were selected from each plant for the air layering. Then, 1000 mg/L of an equal ratio of 1-naphthaleneacetic acid (NAA) and indole-3-butyric acid (IBA) was applied to all ring strips.

2.4. Influence of Growth Regulators on Air Layering

The experiment was performed in mid-June 2020, and the three cultivars already mentioned were considered. In accordance with previous studies on air layering and with experience in tree peony cultivation, NAA and IBA were used as rooting growth regulators at concentration 1000 mg/L, 1500 mg/L, and 2000 mg/L [28–30]. A control treatment (no growth regulators applied) was considered too. For each cultivar, 3 mother plants and 5 branches for each mother plant were considered, resulting in a total of 105 treatments for each tested cultivar.

2.5. Evaluation of the Air-Layering Propagation in Tree Peony

In October (a suitable season for transplanting tree peony), 3–4 months after having applied air layering, the branches were cut below the membrane opening. After removing the tie wire, film, and the medium, the relevant growth parameters were measured.

Determination of air-layering efficiency: each air layer was classified according to the rooting efficiency, as follows: (i) air layers with more than 10 growing first-order roots (referred to as 1) and (ii) air layers with fewer than 10 growing first-order roots or with no roots (referred to as 0). According to this classification, for each plant, a 'rooting number' was derived, and then the air-layering efficiency was calculated as follows:

$$\text{Air} - \text{layering efficiency} = \left(\text{Total}\,'\text{rooting number}'\ /\ \text{Total air layers}\right) \times 100\% \quad (1)$$

In order to highlight the root architecture that influenced plant growth and stability, the following parameters were taken into account:

a. Root system quantity, which was evaluated through the number of taproots and root tips for each air layer. This was considered an indication of the rooting efficiency
b. Root system quality, which was determined by considering the primary root length. The taproots were determined one by one by measuring the length from the base to the top with a ruler, accurate to 1 mm. The sum of all primary root lengths of a single air layering represented the total length of the primary roots of the air layering, and the longest primary root length indicated the maximum primary root length of the air layering.

2.6. Data Processing

Excel 2017 (Microsoft Corp., Redmond, MA, USA), GraphPad Prism 7.0 (GraphPad Software, San Diego, CA, USA), and Adobe Illustrator CS6 (Adobe Mercury, San Jose, CA, USA) were used for experimental data processing and mapping. Standard deviation analysis was performed using STDEV in Excel 2017, and a variance analysis was performed using ANOVA analysis ($p < 0.05$) in GraphPad Prism 7.0.

3. Results

3.1. Influence of the Time of the Year on Air Layering

Figure 1 shows the air-layering procedure in tree peony roots and the growth of the air layers. A one-centimeter ring-stripping incision is shown in Figure 1A. After growth regulators were applied and a thin film was placed on the ring-stripped surface (Figure 1B), callus formation began on the ring-stripped branches in the upper air layer at about 15 d, rooting began at 25–30 d, and white roots filled the matrix at 60–70 d (Figure 1C,D). Figure 1E,F shows the air layer growth in pots one year from the air layering.

Figure 1. Steps in air layering of tree peony. (**A**) One-centimeter ring-stripping incision. (**B**) Matrix dressing. (**C**) The substrate is covered with new roots. (**D**) The new roots under air layering. (**E**,**F**) Air-layer growth in pots one year from the air layering.

The number of first-order roots of each layer was counted and calculated according to the formula already stated for air-layering efficiency. Table 1 highlights that the best results were reached when the air layering was carried out in mid-June, and this was true for all the three tested varieties, although a different performance could be scored (the air-layering efficiency was 82.86% ± 10.1, 100% ± 0% and 77.14% ± 8.9 for 'Baoqing Hong', 'Quehao' and 'Xishi' respectively). Generally speaking, the worst results were obtained when the air layering was performed in mid-July, except for the variety 'Baoqing Hong', for which the efficiency of the air layering was similar when the procedure was performed in May or July.

Table 1. Air-layering efficiency for the three tested tree peonies.

Cultivars	Time	Total Air Layers	Rooting Number	Air-Layering Efficiency (100%)
'Baoqing Hong'	mid-May	35	27	77.14 ± 9.1
'Baoqing Hong'	mid-June	35	29	82.86 ± 10.1 ^{a}
'Baoqing Hong'	mid-July	35	27	77.14 ± 9.3
Quehao	mid-May	35	33	94.29 ± 3.8 ^{b}
Quehao	mid-June	35	35	100 ± 0 ^{b}
Quehao	mid-July	35	30	85.71 ± 10.3 ^{a}
Xishi	mid-May	35	26	74.29 ± 9.8
Xishi	mid-June	35	27	77.14 ± 8.9
Xishi	mid-July	35	22	62.86 ± 12.1

Note: Air layering was applied over three different periods, and a solution of NAA + IBA (1000 mg/L each) was applied to stimulate the rooting. Data are mean average values ± standard error. ^{a} indicates significant differences, $0.01 < p < 0.05$; and ^{b} indicates highly significant differences, $0.01 < p < 0.05$.

3.2. Influence of Growth Regulator on Air Layering

As indicated in Figure 2A, the number of taproots in 'Baoqing Hong' under NAA treatment was twice as high as in the control (CK). Under IBA treatment, the number of taproots was slightly higher than that of the control group, though the difference was not significant. The number of root tips of 'Baoqing Hong' was lower than that of the control group under the two growth-regulator treatments (Figure 2A). Figure 2B indicates that the total length and maximum length of the primary roots of 'Baoqing Hong' under NAA treatment were similar to those of the control group. When IBA was applied, the total and maximum root length were slightly lower than those of the control group. Figure 2C showed that the numbers of primary roots and root tips of 'Quehao' with the growth regulator were significantly higher than those of the control (CK), which under the NAA treatment were slightly higher than those under the IBA treatment. As indicated in Figure 2D, the total length and maximum root length of the primary roots in 'Quehao' with the growth regulator were significantly higher than those of the control (CK), which under the IBA treatment were similar to those under the NAA treatment. As shown in Figure 2E,F, the number of main roots in the root system of 'Xishi' under the NAA and IBA treatments was significantly higher than that of the control group. The number of root tips under the NAA treatment was slightly higher than that of the control group, but significantly higher under IBA treatment. The total length and maximum length of the primary roots were slightly higher under NAA treatment than in the control (CK), but lower under IBA treatment. The results showed that NAA and IBA could affect the number and length of roots differently, although a varied response to the treatments was noted.

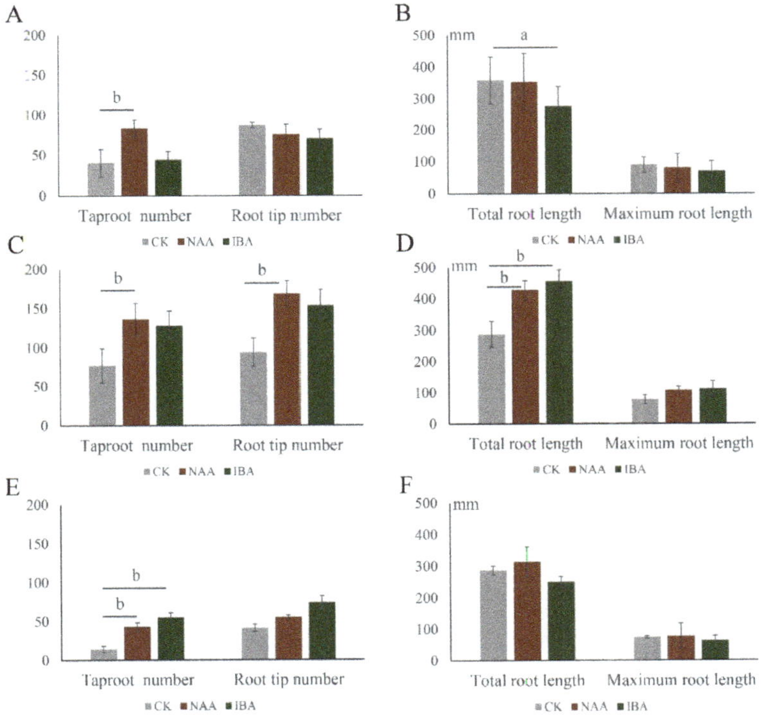

Figure 2. Rooting efficiency for air layering of tree peony when different growth regulators (IBA and NAA) were supplied at the concentration 1000 mg/L. The evaluation was carried out for three tree peony varieties: 'Baoqing Hong' (**A**,**B**), Quehao (**C**,**D**) and Xishi (**E**,**F**). A control treatment (no growth regulators applied = CK) was considered for each variety. [a] indicates significant differences, $0.01 < p < 0.05$; and [b] indicates highly significant differences, $0.01 < p < 0.05$.

3.3. *Influence of Different Concentrations of Growth Regulators on Air Layering*

In a subsequent experiment, three concentrations of the growth regulators NAA and IBA (1000 mg/L, 1500 mg/L and 2000 mg/L) were applied in the air layering carried out in mid-June. Table 2 shows that growth regulator choice and the concentration applied could greatly affect the root architecture; also in this case, we have to highlight that there was a genotype-dependent response.

Table 2. Air-layered root traits of the tree peony varieties under the different treatments.

Cultivars	Treat Agents	Concentration (mg/L)	Taproot Number	Root Tip Number	Total Root Length (mm)	Maximum Root Length (mm)
'Baoqing Hong'	CK	0	40 ± 6.2	101 ± 6.5	357.7 ± 11.9	88.6 ± 6.7
	NAA	1000	83 ± 10.7 [b]	75 ± 12.9	335.1 ± 31.1	79.2 ± 12.6
		1500	62 ± 5.1 [a]	84 ± 2.1	250.7 ± 26.6[a]	60.3 ± 4.1
		2000	64 ± 13.1 [a]	87 ± 4.3	282.4 ± 26.1 [a]	67.5 ± 5.5
	IBA	1000	44 ± 4.6	70 ± 2.3	274.8 ± 22.1 [a]	71.6 ± 4.5
		1500	114 ± 12 [a]	155 ± 10.5 [a]	324.5 ± 27.7	72.5 ± 6.1
		2000	45 ± 7.5	67 ± 6.1	340.4 ± 11.7	78.7 ± 7.6
'Quehao'	CK	0	77 ± 5.1	94 ± 18.2	364.7 ± 21.1	77.7 ± 5.6
	NAA	1000	136 ± 17.1 [a]	168 ± 8.9 [a]	426.8 ± 30.4 [a]	107.3 ± 12.6 [a]
		1500	110 ± 13.5 [a]	138 ± 3	396.4 ± 22.2	91.5 ± 9.6
		2000	69 ± 2.5	79 ± 7.5	312.9 ± 24.6	72.7 ± 3.5
	IBA	1000	128 ± 7.8 [a]	153 ± 28.2 [a]	455.9 ± 38.9 [a]	110.9 ± 8.3 [a]
		1500	204 ± 3.8 [b]	198 ± 5.2	471.3 ± 24.5	111.3 ± 3.4 [a]
		2000	222 ± 5.1 [b]	255 ± 23.8 [b]	564.8 ± 38.2 [b]	120.6 ± 18.9 [b]
'Xishi'	CK	0	21 ± 0.6	41 ± 3	286.3 ± 13.9	73.5 ± 4.1
	NAA	1000	43 ± 5.3 [b]	55 ± 2.5	312.2 ± 5.1 [a]	76.5 ± 3.1
		1500	37 ± 1.5	42 ± 2	167.4 ± 16.2 [a]	43.3 ± 2.2
		2000	31 ± 10.3	41 ± 1.7	225.6 ± 11.8	60.3 ± 3.7
	IBA	1000	55 ± 6.1 [b]	74 ± 3.6	248.5 ± 18.4	62.2 ± 7.1
		1500	43 ± 2.5 [b]	52 ± 5.1	263.9 ± 18.8	71.63 ± 8.6
		2000	55 ± 0.6 [b]	64 ± 4.9 [a]	278.5 ± 25.9	63.4 ± 4.6

Note: Data are the mean average values ± standard errors. [a] indicates significant differences, $0.01 < p < 0.05$; and [b] indicates highly significant differences, $0.01 < p < 0.05$. CK, blank control; NAA, 1-naphthaleneacetic acid; IBA, indole-3-butyric acid.

Generally speaking, the taproot number was enhanced when a growth regulator was applied. In the case of 'Baoqing Hong' variety, the highest number of taproots (114 ± 12) was obtained when the growth regulator IBA was applied at 1500 mg/L. Furthermore, with this IBA concentration the highest number of root tip was obtained (155 ± 10.5). Satisfactory results for all the root parameters were obtained in 'Quehao' variety at IBA 1500 or 2000 mg/L, the latter concentration being the best (taproots number: 222 ± 5.1; root tip number: 255 ± 23.8; total root length: 564.8 ± 38.2 mm and maximum root length 120 ± 18.9 mm). Interestingly, for this variety, we found that when the NAA concentration was increased, the root parameters decreased (particularly the number of taproots and root tips). On the contrary, decreasing the IBA concentration from 2000–1000 mg/L, the number of taproots was reduced by 40%. (Table 2). In the case of 'Xishi' variety, the lowest number of taproots were always obtained compared to the other varieties; the best results were obtained when IBA was provided at 1000 mg/L or 2000 mg/L (number of taproots: 55 ± 6.1 and 55 ± 0.6 respectively). For this variety, the root parameters showed an independence from the growth-regulator concentration applied (Table 2).

3.4. *Evaluation of the Air-Layering Propagation in the Various Tree Peony Cultivars*

The rooting characteristics of the three tree peony cultivars 'Baoqing Hong', Quehao', and 'Xishi' in response to the NAA and IBA treatments were compared at the same concentration of 1000 mg/L. The results in Figure 3 show that 'Quehao' had a significantly higher number of taproots and root tips, root lengths, and maximum root lengths than

'Baoqing Hong' and 'Xishi', exhibiting highly advantageous root growth under air layering (Figure 3A–D). Under NAA treatment, the root parameters of 'Baoqinghong', particularly the number of taproots and the total number of root tips, were significantly higher than those of 'Xishi'. However, there was no significant difference in the total length and maximum root length of the primary roots between the two cultivars. The root parameters (taproots number, root tip number, total root length and maximum root length) between 'Baoqing Hong' and 'Xishi' were similar under IBA treatment.

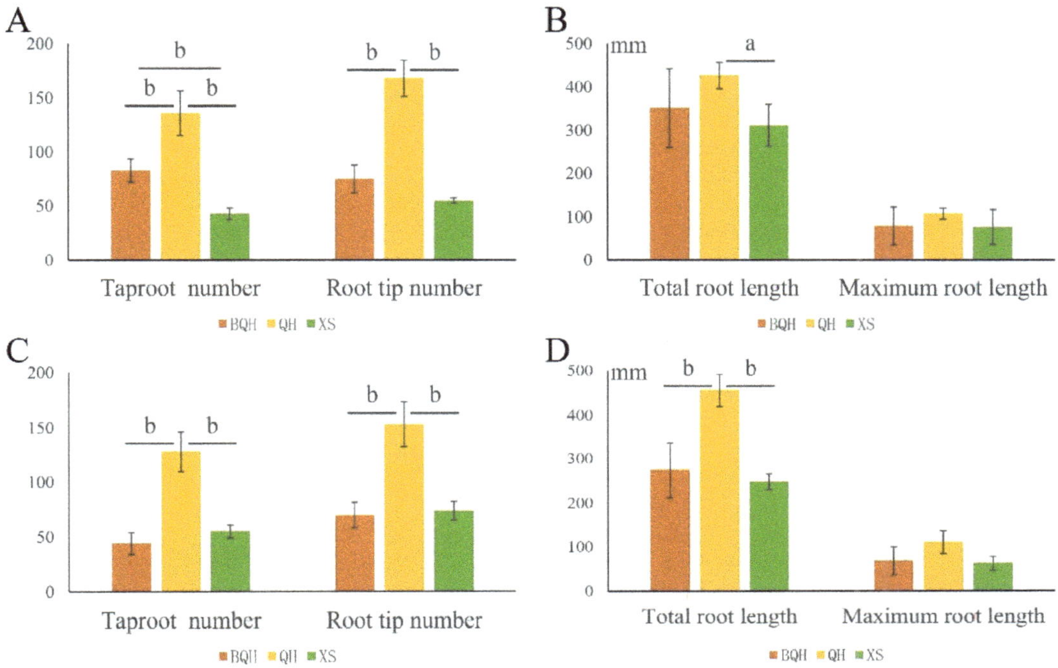

Figure 3. Comparison of the root characteristics among the three cultivars of tree peonies under air layering. (**A**) Comparison of the number of taproots and root tips of three cultivars treated with NAA at 1000 mg/L; (**B**) the total length and maximum root length of the primary roots of three cultivars treated with NAA at 1000 mg/L; and (**C**,**D**) comparison of the same root characteristics among the three cultivars treated with IBA at 1000 mg/L. BQH: Baoqing Hong; QH: Quehao; XS: Xishi. [a] indicates significant differences, $0.01 < p < 0.05$; and [b] indicates highly significant differences, $0.01 < p < 0.05$.

4. Discussion

Tree peonies prefer dry and cold growth conditions, while the humid and hot climate in Jiangnan area of China is not conducive to the growth of tree peony. Some varieties of tree peony have better growth adaptability, but the performance of different varieties differs greatly [7,31]. The three tree peony cultivars, 'Baoqing Hong', 'Quehao', and 'Xishi', belong to the traditional Jiangnan cultivar group, and they have good growth adaptability in the Jiangnan area [7,31]. To avoid poor experimental data due to the bad growth, these three cultivars were selected to undergo air-layering propagation experiments. In our experiment designed to evaluate the air-layering method for propagating the tested tree peony varieties, we found that the variety 'Quehao' performed better than the varieties 'Baoqing Hong' and 'Xishi'. The air-layering efficiency for this variety was good, and the root features were satisfactory, showing a dose dependence in response to the growth regulator treatment. This could be ascribed to the existence of root primordia in the cortexes of the branches of the tree peonies, the number of which varies by cultivar [7]. 'Quehao' belongs to the Huizi series of the Jiangnan cultivar group, which has a long cultivation history and better growth adaptability than 'Baoqing Hong' and 'Xishi' [7,31]. Considering the times of the air layering and flowering, it is speculated that the differences in rooting after air layering may have been related to the early and late flowering times among cultivar varieties.

The three air-layering times chosen were mid-May at the end of spring, mid-June in early summer, and mid-July in midsummer, during tree peony fruit development. The air-layering experiment results showed that air-layering time had a significant effect on the rooting rate among the three tree peony cultivars. Similar findings have also been reported in guava [32,33] and pomegranate cv. *Bhagwa* [34]. The rooting rates of the three tree peony cultivars were significantly higher when layered in early summer than in spring and midsummer. In early summer (mid-June), the seeds of tree peonies in the Jiangnan region are in the period of nutrient accumulation [35], and therefore, the plant needs to be supplied with a large quantity of nutrients for the seeds. It is possible that the phloem of the branches is more active during this time, and this would account for the rapid rooting on the callus. To corroborate this statement, it is reported in the literature that air layering in *Ilex verticillate* provided the best results in June, and this was ascribed to the fact that the fruits were forming and the branches were growing rapidly in that period [36]. Furthermore, in the Shanghai area, mid-June is the Meiyu season [37]. In our experiment, the rooting of the air layers started 25–30 days after the air layering, and this happened precisely in the Meiyu season. Humidity and temperature (22–25 °C) were very suitable for air-layered branch rooting and the promotion of root growth. This result was consistent with other studies, which found that air layering was beneficial during the wet season when there was high humidity [29,30]. It has been revealed that the rooting rate of layered shoots in the shea tree (*Vitellaria paradoxa*) under wet conditions is higher than that under dry conditions [29]. The air-layered rooting percentage in guava (*Psidium guajava*) is due to a particular combination of humidity, rainfall, and temperature [30], and similar findings have also been reported by Shrivastava in *Puncia granatum* [38], Sharma and Grewal in lithci [39], and Sarker and Ghose in guava [40].

The growth regulators NAA and IBA have a great influence on rooting characteristics [41,42]. The rooting features of 'Quehao' under air layering were negatively correlated with the concentration of NAA at 1000–2000 mg/L and positively correlated with the concentration of IBA at 1000–2000 mg/L. These correlations were not observed for the other tested varieties. It has been observed that IBA had a better rooting effect than NAA in air-layered rooting in 'Quehao'. According to our findings, the air-layered rooting characteristics had a significant difference between the treatment types and concentrations among the three cultivars. In air-layering studies of *Lasiococca comberi*, the rooting effect of IBA has been reported to be much higher than that of NAA and indole-3-acetic acid (IAA) [43]. In contrast, in a study of *Phoebe bournei* using air layering, the rooting percentage was significantly higher under the NAA treatment than that of IBA and IAA [24]. Therefore, NAA as a treatment agent has a less clear effect on layering rooting than the IBA treatment. The concentration of IBA used as the air-layering rooting growth regulator is obviously different among plant species. In the case of propagation by air layering in avocado (*Persea* sp.), the application of IBA increased the rooting percentage to a maximum of 74% at a concentration of 10,000 mg/L [44]. However, in the case of propagation by air layering in guava (*Psidium guajava*), callus formation was reduced with increasing concentrations of IBA (2000, 4000, and 6000 mg/L), and its formation was completely inhibited at a concentration of 6000 mg/L IBA [45]. A similar trend was obtained in uva camarona (*Macleania rupestris*) propagation by air layering, where a reduction in the dry and fresh weights of the roots and callus was the result of concentrations greater than 1500 mg/L IBA [28]. We could argue that in air layering, the choice of the growth regulators and the concentration should be evaluated based on the crop, the genotype, and the growth and environmental features.

There was a great difference in the rooting rates and rooting characteristics among varieties of tree peony cultivars. This implied that there is significant research space for establishing efficient air-layering propagation protocols for various tree peony cultivars. The time cycle for sexual reproduction of tree peonies is too long, and thus asexual reproduction should be promoted in the future, particularly in the breeding of novel tree peony cultivars as well as ancient tree peonies. The current experiments considered the effects of the air-layering times as well as the growth regulator types and concentrations, but did not consider the effects of the wrapping matrix shape, the ring stripping incision, the growing media and substrate from the rhizosphere of this species, or microorganisms. These issues require further experimental investigation.

5. Conclusions

This study clearly demonstrated that the application of the air-layering technique in tree peonies could achieve successful propagation, and this has not been previously reported. It was suggested that air-layering time had a significant effect on the rooting rates among the three tree peony cultivars. The maximum rooting percentage (100%) was present in the 'Quehao' cultivar at the air-layering time of early summer (mid-June). The rooting parameters of 'Quehao' were negatively correlated with the concentration of NAA (1000–2000 mg/L), but positively correlated with IBA (1000–2000 mg/L), whereas those of 'Baoqing Hong' and 'Xishi' showed no dose dependence for the growth regulator concentration. The parameters of the root number, root tip number, and maximum root length of 'Quehao' were greater than those of the other two cultivars. The propagation of tree peonies with air layering could be an effective method for consistently obtaining plant material, especially for the promotion of high-quality tree peonies. Furthermore, this also provides an avenue for the study of the rooting mechanisms and nutritional breeding technologies of tree peonies.

Author Contributions: Y.Z. conducted the main experiments, collected and analyzed the data, and prepared the manuscript. Y.-H.H. and S.-Y.Y. conceived and revised the paper. All authors have read and agreed to the published version of the manuscript.

Funding: This research was funded by the Shanghai Science and Technology program (21DZ1202000), Bureau of Facility Support and Budget Chinese Academy of Sciences (ZSZY-001-8), Special Fund for Scientific Research of Shanghai Landscaping & City Appearance Administrative Bureau (G222412).

Institutional Review Board Statement: Not applicable.

Informed Consent Statement: Not applicable.

Data Availability Statement: All datasets generated during this study are included in the article.

Acknowledgments: We thank Jian-Feng Lv for his guidance on tree peony cultivation.

Conflicts of Interest: The authors declare that they have no conflict of interest.

References

1. Ghayur, M.N.; Gilani, A.H.; Rasheed, H.; Khan, A.; Iqbal, Z.; Ismail, M.; Saeed, S.A.; Janssen, L.J. Cardiovascular and airway relaxant activities of peony root extract. *Canad. J. Physiol. Pharma.* **2008**, *86*, 793–803. [CrossRef] [PubMed]
2. Hu, S.L.; Shen, G.; Zhao, W.G.; Wang, F.; Jiang, X.D.; Huang, D.B. Paeonol, the main active principles of *Paeonia* Moutan, ameliorates alcoholic steatohepatitis in mice. *J. Ethnopha.* **2010**, *128*, 100–106. [CrossRef]
3. Shahidi, F.; Miraliakbari, H. Omega-3 fatty acids in health and disease: Part 2-health effects of omega-3 fatty acids in autoimmune diseases, mental health, and gene expression. *J. Medici. Food* **2005**, *8*, 133–148. [CrossRef]
4. Yu, S.; Du, S.; Yuan, J.; Hu, Y. Fatty acid profile in the seeds and seed tissues of *Paeonia* L. species as new oil plant resources. *Sci. Rep.* **2016**, *6*, 26944d. [CrossRef] [PubMed]
5. Zhu, H.X.; Yuan, T.; Li, D.T.; Song, G.F.; Liu, A.D. Advances in application research of *Paeonia suffruticosa*. *Shandong For. Sci. Technol.* **2004**, *150*, 49–52.
6. Wu, X.Y.; Lv, C.P.; Zhang, X.X.; Xue, J.Q. Development status and countermeasures of tree peony industry in China. *Xiandai Nongye Keji* **2017**, *12*, 156–157.
7. Li, J.J.; Zhang, X.F.; Zhao, X.Q. *Tree Peony of China*; China Encyclopedia Press: Beijing, China, 2011; ISBN 978-7-5000-8522-5.
8. Shen, G.X.; Zhang, X.Y. Propagation and cultivation techniques of tree peony. *Nor. Hortic.* **2012**, *8*, 63–66.
9. Han, J.G.; Qin, J.; Hu, Y.H.; Zhu, M.L. Research progress of tranditional propagation and in vitro regeneration of *Paeonia suffruticosa*. *Heilongjiang Agric. Sci.* **2014**, *4*, 77–80.
10. Li, Y.L.; Wu, D.Y.; Pan, S.L.; Xu, S.L.; Wei, Z.M.; Xu, Z.H.; Li, X.J. In vitro propagation of *Paeonia suffruticosa*. *Chin. Sci. Bull.* **1984**, *8*, 500–502.

11. Kong, X.S.; Zhang, M.X. Study on in vitro rapid propagation of tree peony. *North. Horticul.* **1998**, *3*, 87–89.
12. Beruto, M.; Lanteri, L.; Portogallo, C. Micropropagation of tree peony (*Paeonia suffruticosa*). *Plant Cell Tissue Organ Cult.* **2004**, *79*, 249–255. [CrossRef]
13. Meng, Q.X.; Luo, X.F.; Zhang, J.Q.; Kang, X.Y. 'Linghua Zhanlu'; peony bud scales of tissue culture. *Chin. Agricul. Sci. Bull.* **2011**, *27*, 102–107.
14. Wang, M.M.; Bu, X.P.; Zhang, Q.; Huo, J.R.; Chao, L.J.; Wang, H.F. Study on in vitro rapid propagation technology on *Paeonia ostii* var. lishizheni. *Mol. Plant Breeding* **2018**, *16*, 526–534.
15. Wen, S.S.; Cheng, F.Y.; Zhong, Y.; Wang, X.; Li, L.Z.; Huang, N.Z. Protocol for the micropropagation of tree peony (*Paeonia* × *lemoinei* 'High Noon'). *Plant Sci. J.* **2016**, *34*, 143–150.
16. Wen, S.S.; He, R.R.; Zheng, J.K.; Tian, R.N. Research advances in tissue culture of tree peony. *Sci. Silva. Sinic.* **2018**, *54*, 143–155.
17. Wen, S.S.; Chen, L.; Tian, R.N. Micropropagation of tree peony (*Paeonia sect. Moutan*): A review. *Plant Cell Tissue Organ Cult.* **2020**, *141*, 1–14. [CrossRef]
18. Wen, S.S.; Cheng, F.Y.; Zhong, Y. Optimization of the micropropagation protocol of *Paeonia* × *lemoinei* 'High Noon'. *Sci. Silva Sinic.* **2021**, *57*, 68–78.
19. Zeng, D.X.; Yin, W.L.; Wang, Y.H.; Zhao, X.Q.; Wang, H.F. Propagation with etiolated softwood cuttings of five dwarf cultivars of Chinese tree peony. *Acta Horticul. Sini.* **2005**, *32*, 725–728.
20. Cheng, F.Y. Advances in the breeding of tree peonies and a cultivar system for the cultivar group. *Interna. J. Plant Breeding* **2007**, *1*, 89–104.
21. Du, Y.M.; Cheng, F.Y.; Zhong, Y. Induction of direct somatic embryogenesis and shoot organogenesis and histological study in tree peony (*Paeonia sect. Moutan*). *Plant Cell Tissue Organ Cult.* **2020**, *141*, 557–570. [CrossRef]
22. An, C.M. Analysis of the influence of survival rate of peony grafting method. *J. Henan For. Sci. Technol.* **2017**, *37*, 6–8.
23. Liao, W.H.; Liu, J.; Shi, X.D.; Peng, J.G.; Pai, Z.K. Analysis of influencing factors of the rooting rate of air-layers in trees. *Sichuan Sci. Technol.* **2018**, *39*, 103–107.
24. Zheng, Y.P.; Yang, J.C.; Zou, W.T.; Wang, X.; Yu, N.; Liu, X.D.; Lai, X.E.; Li, R.S. Effects of rooting plant growth regulators on rooting of air-layers of *Phoebe bournei*. *Chin. J. Tropic. Crops* **2020**, *41*, 1803–1807.
25. Yang, Q.G.; Guo, C.; Liao, Y.M.; Zhong, H.J. Preliminary study on propagation technology of air-layers in *Camellia japonica*. *J. Green Sci. Technol.* **2015**, *1*, 61–62.
26. Weng, C.X.; Ning, Z.L. Study on air-layers of four seasons carefree flower. *Xiandai Yuanyi* **2018**, *11*, 39–40.
27. Zeng, D.X.; Yin, W.L.; Zhao, X.Q.; Wang, H.F. Propagation of Chinese tree peony (*Paeonia suffruticosa* Andr.). *J. Beijing For.Univ.* **2000**, *22*, 90–95.
28. Duran-Casas, S.; Veloza-Suan, C.; Magnitskiy, S.; Lancheros, H.O. Evaluation of uva camarona (*Macleania rupestris* Kunth A.C Smith) propagation with air layering. *Agron. Colomb.* **2013**, *31*, 18–26.
29. Yeboah, J.; Branoh Banful, B.; Boateng, P.; Amoah, F.; Maalekuu, B.; Lowor, S. Rooting Response of Air-Layered Shea (*Vitellaria paradoxa*) Trees to Media and Hormonal Application under Two Different Climatic Conditions. *Am. J. Plant Sci.* **2014**, *5*, 1212–1219. [CrossRef]
30. Naithani, D.C.; Nautiyal, A.R.; Rana, D.K.; Mewar, D. Effect of Time of Air Layering, IBA Concentrations, Growing Media and their Interaction on the Rooting Behaviour of Pant Prabhat Guava (*Psidium guajava* L.) under Sub-Tropical Condition of Garhwal Himalaya. *Int. J. Pure App. Biosci.* **2018**, *6*, 169–180. [CrossRef]
31. Hu, Y.H.; Han, J.G. *Tree Peony, Its Varieties. Cultivation and Use in East China*; Science Press: Beijing, China, 2018; ISBN 978-7-03-042066-4.
32. Kumar, S.; Sakthivel, T.; Chithiraichelvan, R.; Karunakaran, G. Effect of time, media and root inducing auxins on the rooting of guava (cv. *Allahabad Safeda*) air-layers under high rainfall zone of kodagu. *Ann. Agric. Res. New Ser.* **2007**, *28*, 10–14.
33. Rymbai, H.; Reddy, G.S. Effect of IBA, time of layering and rooting media on air-layers and plantlets survival under different growing nursery conditions in guava. *Indian J. Hort.* **2010**, *67*, 99–104.
34. Tayade, S.A.; Joshi, P.S.; Raut, H.S. Shete, M.B. Effect of time and air-layer per shoot on rooting and survival of air-layers in pomegranate cv. *Bhagwa. Inter. J. Minor Fru. Medi. Aroma. Plan.* **2017**, *3*, 20–24.
35. Yu, S.Y.; Zhang, X.; Huang, L.B.; Lyu, Y.P.; Zhang, Y.; Yao, Z.J.; Zhang, X.X.; Yuan, J.H.; Hu, Y.H. Transcriptomic analysis of α-linolenic acid content and biosynthesis in *Paeonia ostii* fruits and seeds. *BMC Genom.* **2021**, *22*, 297. [CrossRef] [PubMed]
36. Zhang, J.L.; Fan, J.; Jiang, L.Y.; Li, B.; Yu, Y.X.; Shen, B.C.; Wu, G.H. Study on air-layers of fruit branches in *Ilex verticillate*. *For. Sci. Technol.* **2016**, *6*, 63–66.
37. Xin, F.; Chen, B.M.; Sun, G.W. Low-frequency characteristics of heavy rain and its extended forecast during Meiyu season in Shanghai. *J. Meteo. Environ.* **2014**, *30*, 61–67.
38. Shrivastava, R.P. Propagation of pomegranate (*Punica granatum* L.) by air-layering. *Sci. Agri.* **1961**, *27*, 250–251.
39. Sharma, R.C.; Grewal, G.P.S. A note on propagation studies in litchi. *Haryana J. Dort. Ski.* **1989**, *18*, 74–76.
40. Sarker, A.; Ghosh, B. Air-layering in guava cv. L-49 as affected by plant growth regulators, wrappers and time of operation. *Environ. Ecol.* **2006**, *24*, 820–823.
41. Chen, T.T.; Chang, W.C. Effects of auxins and cytokinins on direct somatic embryogenesison leaf explants of Oncidium 'Gower Ramsey'. *Plant Growth Regul.* **2001**, *34*, 229–232. [CrossRef]

42. Ren, X.X.; Liu, Y.; Jeong, B.R. Enhanced somatic embryo induction of a tree peony, *Paeonia ostii* 'Fengdan', by a combination of 6-benzylaminopurine (BA) and 1-naphthylacetic acid (NAA). *Plants* **2020**, *9*, 3. [CrossRef]
43. Kamila, P.K.; Panda, P.C. Large-scale vegetative propagation of *Lasiococca comberi* by air-layers. *J. Tropic.For. Sci.* **2019**, *31*, 37–42.
44. Alves-de Oliveira, A.; Carlos-Koller, O.; Villegas-Monter, A. Propagación vegetativa de aguacate selección 153 (*Persea* sp.) por acodo en contenedor. *Rev. Chapingo Ser. Hort.* **1999**, *5*, 221–225.
45. Sánchez, A.; Suárez, E.; González, M.; Amaya, Y.; Colmenares, C.; Ortega, J. Efecto del ácido indolbutírico sobre el enraizamiento de acodos aéreos de guayabo (*Psidium guajava* L.) en el municipio Baralt, Venezuela. Evaluación preliminar. *Rev. Cient. UDO Agríc.* **2009**, *9*, 113–120.

horticulturae

MDPI

Article

Jasmonic Acid and Salicylic Acid Levels in Defense Response of Azalea (*Rhododendron simsii* Hybrid) to Broad Mite (*Polyphagotarsonemus latus*)

Leen Leus [1,*], Gil Luypaert [1], Emmy Dhooghe [1], Johan Witters [1], Els Pauwels [2], Christof Van Poucke [3], Els Van Pamel [3], Johan Van Huylenbroeck [1] and Joachim Audenaert [2]

1. Plant Sciences Unit, ILVO (Flanders Research Institute for Agriculture, Fisheries and Food), 9090 Melle, Belgium
2. PCS Ornamental Plant Research, 9070 Destelbergen, Belgium
3. Technology and Food Sciences Unit, ILVO (Flanders Research Institute for Agriculture, Fisheries and Food), 9090 Melle, Belgium

* Correspondence: leen.leus@ilvo.vlaanderen.be; Tel.: +32-92-72-2884

Abstract: The broad mite (*Polyphagotarsonemus latus* (Banks)) is an important pest in many crops, including azalea (*Rhododendron simsii* Planch. hybrid). Broad mites cause the malformation of shoot tips, leaves and flowers in azalea. It is known that the jasmonic acid (JA)- and salicylic acid (SA)-dependent signaling pathways are related to the presence of herbivorous mites. Here, we describe the levels of the two main plant defense-related hormones, SA and JA, in mite-infested plants. The plant hormones were analyzed using liquid chromatography combined with tandem mass spectrometry (LC-MS/MS). We studied both short-term hormonal responses under controlled conditions with artificial inoculation, and long-term responses under culture conditions with natural infestation. The long-term development of broad mite populations and hormone response were studied during two subsequent growing seasons on 3 and 18 different cultivars, respectively. During the experiments on 18 azalea cultivars under natural infestation, the presence of different species of tarsonemid mites was also examined. JA concentrations only showed variation in the early phase of infestation. Subsequently, the SA levels increased significantly for all the cultivars where broad mites were detected. Based on the observed timing of the defense responses, we suggest that the interaction of the JA and SA pathways as a defense response for pot azalea against *P. latus* involves a primary plant response through the JA pathway. In the presence of the mites, the production of SA increased in the plants in a later phase as the *P. latus* population grew. Our results also show that the hormone response depends on type of mite. Changes in hormone levels were found upon infestation with *P. latus*, but not in the presence of another frequently occurring tarsonemid mite, *Tarsonemus confusus* Ewing.

Keywords: defense suppression; LC-MS/MS; mite defense; mite resistance; plant defense; *Tarsonemus confusus*

Citation: Leus, L.; Luypaert, G.; Dhooghe, E.; Witters, J.; Pauwels, E.; Van Poucke, C.; Van Pamel, E.; Van Huylenbroeck, J.; Audenaert, J. Jasmonic Acid and Salicylic Acid Levels in Defense Response of Azalea (*Rhododendron simsii* Hybrid) to Broad Mite (*Polyphagotarsonemus latus*). *Horticulturae* **2022**, *8*, 840. https://doi.org/10.3390/horticulturae8090840

Academic Editor: Dariusz Piesik

Received: 19 July 2022
Accepted: 31 August 2022
Published: 13 September 2022

1. Introduction

Pot azalea, commercially sold as a flowering indoor pot plant, is the main ornamental plant produced in Belgium. In Europe, the main producers of azalea and *Rhododendron* are Belgium and Germany, with export values of EUR 36 and 7 million, respectively. The modern pot azalea traces its roots to at least four species, the most important being *Rhododendron simsii* Planch. The introduction of *R. simsii* in Belgium dates back to 1818 [1]. Other parental species are *Rhododendron indicum* (L.) Sweet, *R. mucronatum* G.Don and *R. scabrum* G.Don [2,3]. The production area in Belgium is around 230 ha, and almost all companies are located around Ghent [4,5]. Pot azalea is propagated by cuttings and its production cycle takes almost two years, combining greenhouse cultivation with outdoor-container field cultivation during summer.

The broad mite, *Polyphagotarsonemus latus* (Banks) (Acari: Tarsonemidae), is an important pest on many commercial crops and this mite has been reported in most parts of the world. The first record in Europe dates back to 1961 when the broad mite was found in Italy [6]. Since then, broad mites have been found in other European countries, and in Belgium, it was reported on pot azalea in the mid-1980s [7]. Restrictions on the use of broad-spectrum pesticides are the main reason for the significant increase in the pest status of broad mites in pot azalea in recent years. The broad mite causes esthetic damage; on azalea, it is responsible for the bronzing, browning and curling of apical leaves, and can cause leaves to become brittle, rigid and stunted with the edges curled downwards. Flowers also become malformed and discolored. Broad mites are very small (<200 μm) and can only be detected using a microscope. A small number of 20 mites or less can cause serious economic damage in crops such as pepper [8,9], cucumber [10], lime [9] and chili [9,11]. Also in pot azalea, it has been shown that low mite numbers can lead to high damage rates [12]. Even after treatment, symptoms may continue to develop [13] and damage initiated in a plant progresses even when mites are no longer present. In his review on broad mites, Gerson [13] suggested that the symptoms are caused by toxins. This conclusion was based on the persistence of symptoms for some weeks post-treatment. These putative toxins would explain the appearance of injuries very shortly after the initiation of a mite attack, and the later occurrence of symptoms at a distance from the site of mite feeding.

Differences in susceptibility to broad mites have been observed in pot azalea, along with the damage rate, and in relation to the population sizes of *P. latus* on the tested azalea cultivars [12]. Trichomes (plant hairs) are potentially involved in broad mite resistance in pot azalea [14]. Additionally, in *Capsicum* species, trichomes are known to act as a limiting factor for the development of broad mites [15]. In plants, trichomes can act as a physical barrier and chemical repellent to insect or mite attacks [16]. In pot azalea, four types of trichome are described. All pot azaleas have long non-glandular trichomes; in addition, there are three types of short trichomes. They are thorn-like trichomes, short non-sticky trichomes and short trichomes with glandular ends (plants with these trichomes also have sticky leaves). There are indications that the latter are related to broad mite resistance. However, this is not the only discriminative trait, as in certain susceptible cultivars, trichomes with glandular ends were observed [17]. Recently it was shown that wounding or exogenous jasmonic acid (JA) treatments are able to promote trichome development and enhance insect resistance [16].

For decades, researchers have already known that plants have a defense network dependent on phytohormones, which translates stresses of biotic or abiotic origin into plant defense responses. Jasmonic acid (JA) and salicylic acid (SA) are known to serve as the primary signals of the plant's immune response. SA and JA work antagonistically and use other hormones to fine-tune immune responses built on SA and JA [18].

SA is most recognized as primarily being induced in the mediation of attacks by biotrophs [19], whereas JA is involved in defense against necrotrophic pathogens [20] and herbivores [21]. As reviewed by Pieterse et al. [18], well-known trade-offs exist between SA-dependent resistance and JA-dependent defense in disease and pest resistance. However, neutral and synergistic interactions have also been described [18]. Most studies focus on plant pathogens, but in recent years, a large number of studies have also examined plant defense responses for herbivores. For example, in a study on tomato and the two-spotted spider mite, different types of interaction have been described. These different interaction types, including direct and indirect defenses, were reviewed by Blaazer et al. [22]. Their review shows that both JA and SA are related to the presence of herbivorous mites. In brief, (1) the accumulation of SA and JA can be triggered in tomato as a plant response at the start of mite infestation; (2) the levels of both JA and SA can decrease as a result of defense suppression by the mites; and (3) intraspecific variation in natural mite populations, especially in polyphagous mites, can have an effect on the plant–mite interaction. Not only is the herbivore–plant relationship involved, but so is the attraction of natural enemies by the emission of volatile compounds. Specifically, for *P. latus* in cucumber leaves, both the JA and SA-dependent pathways are activated, and possibly an oxidative stress response [23]. To examine the possible involvement of the JA signaling pathway in the defense response against *P. latus*, [23] evaluated feeding on wild-type (WT) tomato 'Castlemart' and on *def-1* a mutant impaired in JA biosynthesis. In the WT plants, broad mite feeding did not affect growth or leaf development, but in the *def-1* plants, 50–60% inhibition was observed. Mite populations developed only on *def-1* plants, as was shown by progeny counts. Furthermore *P. latus* prefers JA-defective mutants, and therefore, is able to actively discriminate between resistant and susceptible plants [23,24]. Therefore, it was concluded by [23] that the JA pathway is likely to play an important role in the resistance response against *P. latus* [23].

Different JA and SA biosynthesis marker genes in azalea have been studied [25]. To obtain insight into the molecular mechanisms behind the interaction between *P. latus* and the *R. simsii* hybrid, the expression levels of genes involved in the infestation of leaves with broad mites were analyzed. The focus was on JA and it was shown that transcript levels of marker genes for the JA biosynthesis pathway were significantly induced upon *P. latus* infestation in the *R. simsii* hybrid.

In this research, we aimed to study the relationship between the presence of broad mites and plant defense hormone levels in pot azalea (*R. simsii* hybrid). The hormonal responses were evaluated shortly after infestation, but the long-term responses were also analyzed. Another aim was to evaluate the variation in susceptibility to broad mites in different pot azalea cultivars. Additionally, we also wanted to observe other mites, such as *Tarsonemus confusus*, on pot azalea, for which it is uncertain if they should be considered harmful.

2. Materials and Methods

2.1. Plant Material and Sampling

2.1.1. Experiment 1—Defense Hormone Analysis after Broad Mite Infestation under Controlled Conditions in Two Cultivars of Pot Azalea Shortly after Infestation (Up to 11 Days)

In a controlled experiment and immediately after infestation with broad mites, the JA and SA levels in azalea plants were analyzed. Two cultivars with well-known resistance—'Nordlicht' (susceptible cultivar) and 'Elien' (resistant cultivar)—were used [12]. The levels of SA and JA were compared between infested and uninfested plants.

The plants were grown on ebb and flow tables. The plants were infested artificially by surrounding them with *Hedera helix* L. infested with broad mites, as described in Luypaert et al. [25]. The samples for LC-MS/MS analysis and for the quantification of the number of mites were taken as shoot tips at 4, 6, 8 and 11 days post-inoculation (dpi). Five independent biological replicates were taken at each time point (n = 5), with one shoot tip per replicate for LC-MS/MS analysis and another for mite counts. The samples for LC-MS/MS analysis were flash-frozen in liquid nitrogen and stored until analysis at −80 °C. The broad mites were extracted from the sampled shoots following the three-step detection method including (1) sampling of shoot tips, (2) isolation of the mites in 70% ethanol and (3) sieving and vacuum filtration to facilitate counting [26]. Subsequently, the mites were counted using a binocular microscope (Olympus SZX16, Antwerp, Belgium; max 115×).

2.1.2. Experiment 2—Defense Hormone Analysis under Natural Infestation in Three Cultivars of Pot Azalea during the Growing Season (Up to 32 Weeks)

In experiment 2, the natural occurrence of broad mites was monitored and correlated to the hormone levels. Three azalea cultivars were studied during a growing season of 32 weeks. The three cultivars used—'Nordlicht' (susceptible cultivar), 'Mevrouw Gerard Kint' (intermediate response) and 'Elien' (resistant cultivar)—differed in their susceptibility to *P. latus*: [12].

Each cultivar had one replicate with 100 plants and was placed on a separate greenhouse bed comparable to the setup in commercial cultivation. Luypaert et al. [12] tested a randomized cultivar setup and a setup grouped per cultivar. They concluded that randomization leads to a bias caused by the attractiveness of different azalea cultivars to broad mite; thus, in the present experiment, the plants were grouped according to cultivar. In experiment 2, no tarsonemid mites were artificially introduced into the crop. Instead, the natural distribution of *P. latus* was monitored closely.

The plants were sampled between 19 January (W3) and 13 September (W37). Every two weeks, samples were taken for the estimation of mite numbers. The dates and week (W) numbers of sampling were: 19 January (W3), 1 February (W5), 15 February (W7), 3 March (W9), 16 March (W11), 30 March (W13), 13 April (W15), 27 April (W17), 11 May (W19), 8 June (W23), 21 June (W25), 6 July (W27), 19 July (W29), 2 August (W31) and 31 August (W35) (there was no sampling for mite counts in W21 and W33). For the counting of mite numbers at every time point, three biological replicates were taken (n = 3) with three replicate shoot tips per replicate.

The shoot tips were sampled for stress hormone analysis every month and, once the presence of broad mites was observed, every two weeks. The dates and week numbers of sampling were: 15 February (W7), 16 March (W11), 13 April (W15), 11 May (W19), 9 June (W23), 22 June (W25), 5 July (W27), 19 July (W29), 2 August (W31), 16 August (W33), 30 August (W35) and 13 September (W37). For 'Mevrouw Gerard Kint', samples were taken separately on plants showing symptoms and plants without symptoms in W33 and W35, and for 'Nordlicht', separate samples were taken in W31, W33 and W35. Three independent biological replicates were taken per cultivar and per sampling date (n = 3) with three to five shoot tips per sample of the different plants. The samples were flash-frozen in liquid nitrogen and stored until analysis at −80 °C.

According to standard cultivation procedures, the plants were cut back in W5 (after sampling), W20 and W35 (after sampling). Sampling for mite counts stopped with the last cut in W35. One extra sample for defense hormone analysis was taken from the fresh shoots in W37.

Broad mites were extracted and counted as described in experiment 1. The morphological identification of mites was conducted using a light microscope (Leica DM2000, Diegem, Belgium; max 400×).

2.1.3. Experiment 3—Defense Hormone Analysis under Natural Infestation in 18 Cultivars of Pot Azalea and Occurrence of Mite Species during the Growing Season (30 Weeks)

In total, 18 commercial cultivars of pot azalea were used in the experiment. Eight of these cultivars were previously tested for susceptibility [12] (Table 1). For each replicate, 24 plants of one cultivar were placed in a greenhouse. To avoid dispersion of the mites between cultivars, sticky plates (Biobest Bug Scan IVOG system—yellow) were installed between the blocks of plants. No tarsonemid mites were introduced in the crop, but the natural distribution of the different species of tarsonemid was monitored closely.

To estimate mite numbers, the plants were sampled every two weeks between 20 October (W42) and 9 May (W19). On four dates—1 February (W5), 6 March (W10), 28 March (W13) and 27 April (W17)—shoot tips were sampled for JA and SA analysis. Sampling for defense hormones was conducted according to experiment 2: three independent biological samples were taken per cultivar and per sampling date with 3 to 5 shoot tips per sample of the different plants (n = 3). Tarsonemid mites were sampled (3 shoot tips per cultivar) and counted using a binocular and identified morphologically using a light microscope, as described in experiment 2. Damage was calculated per cultivar by counting the number of plants with damage.

In this experiment, we looked in more detail at the different mite species known for their occurrence on pot azalea. Besides *Polyphagotarsonemus latus*, identification keys were used for four *Tarsonemus* species: *Tarsonemus confusus*, *T. bilobatus* Suski, *T. floricolus* Canestrini & Fanzago, *T. lacustris* Schaarschmidt, species of the genus *Xenotarsonemus* Beer, and the species *Phytonemus pallidus* (Banks). For the identification of these mite species, all the collected specimens were mounted on slides for microscopic observation [27–29].

Table 1. *Rhododendron simsii* hybrids used in the different experiments; cultivars marked with * are also used in Luypaert et al. [12].

Cultivar Name	Parentage / Background	* Used in Luypaert et al. [12]
'Amélie'	sport from 'Thesla' ^{y}	
'Elien'	sport from 'Mistral' ^{x}	*
'Fluostern'	sport from 'Sachsenstern' ^{y}	
'Franziska R.'	sport from 'Michelle Marie' ^{x}	
'Huelsten'	sport from 'Helmutt Vogel' ^{x}	
'Inka'	sport from a 'Helmutt Vogel' sport ^{x}	
'Leonardo'	seedling with unknown parents ^{y}	
'Luntera'	sport from a 'Helmutt Vogel' sport ^{x}	
'Mevrouw Edmond Troch'	sport from a 'Helmutt Vogel' sport ^{x}	
'Mevrouw Gerard Kint' (picotee)	sport from 'Glaser Nummer 10' ^{x}	*
'Mevrouw Gerard Kint' (red)	sport from 'Mevrouw Gerard Kint' ^{y}	
'Michelle Marie'	seedling x 'Rosali' ^{x}	*
'Nordlicht'	sport from 'Helmutt Vogel' ^{x}	*
'Otto'	'Friedhelm Scherrer' x seedling ^{x}	*
'Renato'	derived from 'Glaser Nummer 10' ^{x}	
'Sachsenstern'	Sport from unknown seedling ^{x}	*
'Tamira'	Seedling with unknown parents ^{x}	*
'Thesla'	Seedling 'Sankt Valentin' x 'Mevrouw Gerard Kint' ^{x}	*

^{x} Parentage and background information based on Heursel et al. [2], ^{y} Parentage and background information obtained from growers.

2.2. JA and SA Analysis

In experiment 1, the extraction and analysis of phytohormones SA and JA were conducted according to Bosco et al. [30]. In brief, at −80 °C, frozen leaf samples were kept in liquid nitrogen and were ground using a mortar and pestle before 20 mg of homogenized sample material was extracted with 2 mL solvent mixture methanol:water:acetic acid (10:89:1) and spiked with $2H^{4}$-SA (Olchemim, Olomouc, Czech Republic) and $2H^{6}$-JA internal standards (Olchemim, Olomouc, Czech Republic). The samples were extracted at 4 °C on a shaker (210 rpm) for 16 h. Subsequently, the samples were vortexed and filtered using a 0.22 µm Millex-GV filter (Millipore, Billerica, MA, USA), and the extract was collected in glass tubes. Then, 1 mL of the extract was loaded onto Oasis MCX 1 cm^3 cartridges (Waters, Milford, MA, USA) (preconditioned with 1 mL pure methanol and 1 mL milliQ water). After loading the samples, 0.5 mL milliQ water was used to wash the cartridges. The flow-through was discarded and the columns were eluted under vacuum with 1 mL methanol. In a warm water bath (35 °C, nitrogen atmosphere), the eluates were evaporated to dryness. For resuspension of the eluates, 0.5 mL 75% methanol was added. The extract was again filtered through a 0.22 µm Millex-GV filter and transferred to a micro-vial prior to the injection of 10 µL into the LC-MS/MS. External calibration curves were used to calculate the final concentrations of the samples for JA (Olchemim, Czech Republic) and SA (Sigma-Aldrich, Bornem, Belgium).

In experiments 2 and 3, extraction was based on Sanchez-Bel et al. [31] using diethyl ether as solvent. Before analysis, the frozen samples were ground and lyophilized. All the steps were performed on ice. Per sample, 20 mg of lyophilized plant material was placed in a glass tube and spiked with internal standard $2H^{4}$-SA (Olchemim, Olomouc, Czech Republic) and $2H^{6}$-JA internal standard (Olchemim, Czech Republic) prepared in MeOH/water (75/25). Subsequently, 2.5 mL ultrapure water was added. The samples were shaken on ice for 30 min (150 rpm) and centrifuged (cooled; 5000× g) for 40 min. The supernatant was adjusted to pH 2.8 with 6% acetic acid. The samples were partitioned twice against diethyl ether. Subsequently, the total eluate was evaporated using nitrogen gas at 35 °C, and the residues were dissolved in 1 mL water/methanol 25:75 and put on a 0.22 μm PVDF filter. The filtrate was put in a micro-vial and stored at 4 °C prior to injection (10 μL) into the UHPLC−MS/MS system. In experiments 2 and 3, the level of JA present was calculated using a calibration curve (relative), but without the use of an internal standard.

The advantage of the protocol based on Sanchez-Bel et al. [31] in experiments 2 and 3, compared to Bosco et al. [30] used in experiment 1, is that the extraction is conducted on dry plant material. This results in lower variation in the measured values. Moreover, the protocol using dry plant material is more simple and makes no use of expensive solid-phase extraction columns that can also compromise the recovery of some compounds. Therefore, in experiment 1, the hormone levels are expressed as fresh weight (fw), and in experiments 2 and 3, the hormone levels are expressed as dry weight (dw).

For all the experiments, the analysis was run on an Acquity UHPLC system coupled with a Xevo TQ-S triple quadrupole MS detector (Waters, Milford, MA, USA), as described in Bosco et al. [30].

2.3. Statistical Data Analysis

The LC-MS/MS data were acquired using MassLynx, version 4.1 (Waters, Milford, MA, USA), and data processing of the acquired LC-MS/MS data was performed using Targetlynx (Waters, Milford, MA, USA). The types of statistical test used are mentioned with the different experiments in the Section 3 and in tables and graphs.

3. Results

3.1. Experiment 1— Defense Hormone Analysis after Broad Mite Infestation under Controlled Conditions in Two Cultivars of Pot Azalea Shortly after Infestation (Up to 11 Days)

In experiment 1, the phytohormone level was determined in *R. simsii* 'Elien' and 'Nordlicht' at four time points upon artificial infestation. On 'Nordlicht', the first mites were observed at 6 dpi, while on 'Elien', the first observation was at 8 dpi. At 11 dpi, the final day of observations, the number of mites on 'Nordlicht' increased to 12.4 ± 4.4, while on 'Elien' at the same time point, 1.4 ± 1.2 mites were counted. Figure 1 shows the levels of JA over the course of 11 days. The infestation of *R. simsii* 'Nordlicht' had no effect on the JA level (Figure 1). In contrast, 'Elien' had a significantly higher level of JA when comparing the *P. latus*-infested plants to the control plants (mock) at 6 dpi.

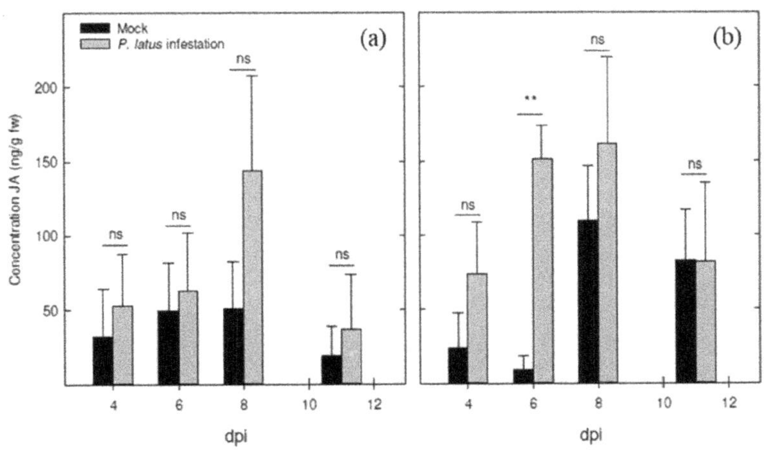

Figure 1. Levels of jasmonic acid (JA) in experiment 1 analyzed 4, 6, 8 and 11 days post-infestation (dpi) in the pot azalea cultivars 'Nordlicht' (**a**) and 'Elien' (**b**). Treatments are mock or *Polyphagotarsonemus latus*-infested leaflets and presented as the average (+SE) (n = 5). Significant differences compared to the respective mock are marked: ns = not significant; ** $p < 0.01$ (Mann–Whitney U).

The SA levels are expressed as the relative response: area analyte/area internal standard. As sample preparation required ground fresh leaves, the hormone levels are expressed as fresh weight. The levels of free SA in 'Nordlicht' were increased by 1.6, 10.4, 6.3 and 18.7 at 4, 6, 8 and 11 dpi, respectively (Table 2). Statistical differences were only observed at 6 and 11 dpi (Mann–Whitney U, $p < 0.05$). In contrast, in 'Elien' the level of free SA remained relatively stable. Across all the time points after *P. latus* infestation the increase in the level of free SA was 1 to 1.6 times baseline; these differences were not statistically significant (Mann–Whitney U, $p < 0.05$).

Table 2. The relative difference in salicylic acid content in the pot azaleas 'Nordlicht' and 'Elien'. Leaflets are analyzed 4, 6, 8 and 11 days post-infestation (dpi) with *Polyphagotarsonemus latus* or mock treatment (experiment 1). Relative differences are calculated as the ratio of area analyte/internal standard of infested leaves and mock treatment ($n = 5$) (Mann–Whitney U, p-values are given).

Genotype	Time Point (dpi)	Treatment	Area Analyte/Area Internal Standard	Relative Difference of SA Content	Level of Significance (p-Value)
'Nordlicht'	4	Mock	0.11	1.6	0.09
		P. latus infestation	0.16		
	6	Mock	0.08	10.4	0.01
		P. latus infestation	0.83		
	8	Mock	0.10	6.3	0.09
		P. latus infestation	0.60		
	11	Mock	0.12	18.7	0.01
		P. latus infestation	2.17		
'Elien'	4	Mock	0.10	1.48	0.19
		P. latus infestation	0.15		
	6	Mock	0.16	1.21	0.69
		P. latus infestation	0.19		
	8	Mock	0.15	0.99	1.00
		P. latus infestation	0.14		
	11	Mock	0.13	1.62	0.31
		P. latus infestation	0.22		

3.2. Experiment 2—Defense Hormone Analysis under Natural Infestation in Three Cultivars of Pot Azalea during the Growing Season (Up to 32 Weeks)

The plants were examined for the presence of mites from January onward, but it took until W19 for the first *P. latus* mite to be observed on 'Mevrouw Gerard Kint'. A large number of 28 mites was found on 'Nordlicht' in W25 (Figure 2). In W29, large numbers of mites, 36.3 (SD $\pm$ 48.5), were found on 'Mevrouw Gerard Kint'. On 'Elien', during the whole experiment, only a limited number of two mites was counted (W29). No mites were extracted on the other dates. When symptoms of mite infection were observed in a cultivar, this was always locally at first, and subsequently spread to a larger area.

A significant increase in SA was found in 'Mevrouw Gerard Kint' and 'Nordlicht' after observations of mite infestations. In 'Elien', the levels of SA were very low during the whole growing period (Figure 2). The differences between the plants of 'Nordlicht' with and without visual symptoms were not significant in W31. In W33 and W35, the differences between symptomatic and asymptomatic plants were significant for both 'Mevrouw Gerard Kint' and 'Nordlicht' (t-test; $p < 0.05$). The levels of JA were low in April but increased in all three cultivars from W19 onward. From W35, a decrease in JA levels was observed.

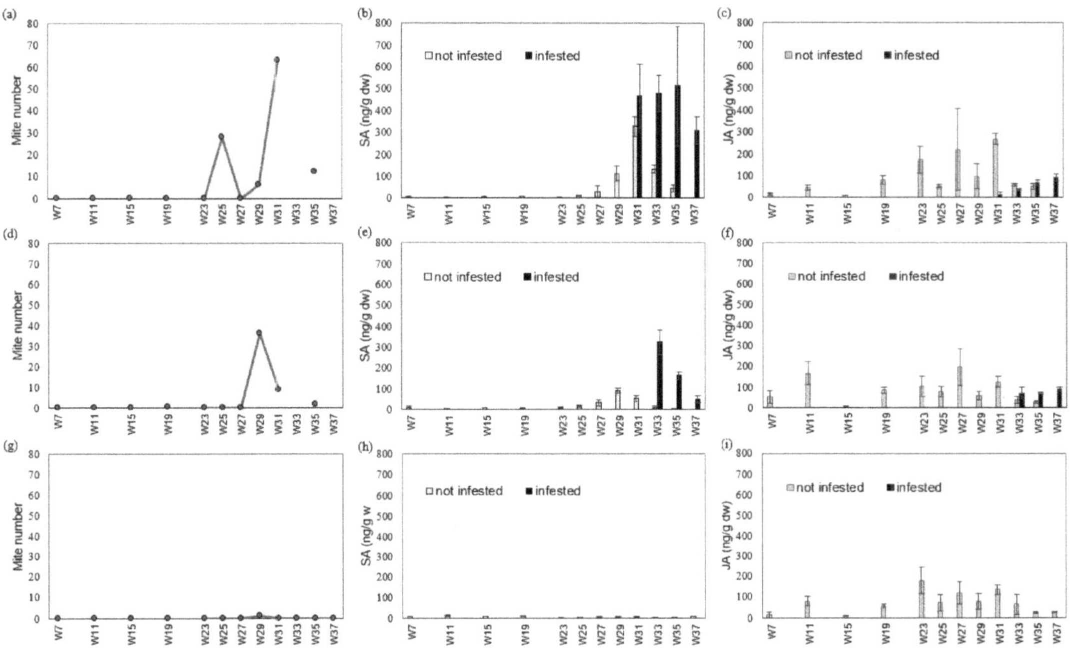

Figure 2. Experiment 2: Mean numbers of the mite *Polyphagotarsonemus latus* (n = 3) counted on different sampling dates (12 time points) for 'Nordlicht' (**a**), 'Mevrouw Gerard Kint' (**d**) and 'Elien' (**g**), and respective levels of salicylic acid (SA) (**b**,**e**,**h**) and JA (**c**,**f**,**i**) (data are means ± SD) (n = 3). In 'Nordlicht' (weeks 31, 33 and 35) and 'Mevrouw Gerard Kint' (weeks 33 and 35) samples were taken separately for plants showing symptoms and plants without symptoms (n = 3).

3.3. Experiment 3—Defense Hormone Analysis under Natural Infestation in 18 Cultivars of Pot Azalea and Occurrence of Mite Species during the Growing Season (30 Weeks)

Different mite types were detected during the experiment. Only four mite species occurred; in order of prevalence (total numbers counted during the experiment are given): *P. latus* (counted mites: 3889), *T. confusus* (counted mites: 552), *Xenotarsonemus* spp. (counted mites: 34), *P. pallidus* (counted mites: 10) and *T. floricolus* (counted mites: 4). No specimens were found for *T. bilobatus* or *T. lacustris*.

No correlations (Pearson) were found between the mite numbers of *P. latus* and *T. confusus*. Nevertheless, we observed that the cultivars with no observations of *P. latus* also showed low numbers of *T. confusus*. This results in a moderate Spearman's rank correlation: r = 0.55 ($p < 0.01$). Plant damage was caused by the occurrence of *P. latus*, as shown by a high Pearson correlation between the total number of *P. latus* mites and the number of plants showing damage: r = 0.89 ($p < 0.0001$). The same correlation calculated between *T. confusus* and plants showing damage was low and not significant.

The first *P. latus* mites were found at the start of the experiment in W42 on 'Leonardo'. The first symptoms were only observed in W47 on 'Franziska R.'. This cultivar showed increasing damage afterwards. Later, symptoms were also seen on the cultivars 'Michelle Marie', 'Fluostern', 'Mevrouw Gerard Kint' (red), 'Otto', 'Leonardo' and 'Renato'.

No *P. latus* mites were observed during the experiment on the cultivars 'Elien', 'Inka', 'Huelsten' or 'Sachsenstern'. On 'Mevrouw Edmond Troch', only a single *P. latus* mite was counted on the last sampling date. While *P. latus* was not found on several cultivars, *T. confusus* was found on all the cultivars. *Xenotarsonemus* spp. were present on all the cultivars except 'Schachsenstern', 'Mevrouw Gerard Kint' (picotée), 'Luntera' and 'Otto'.

The highest total number was counted on 'Otto' (counted mites: 741), followed by 'Leonardo' (counted mites: 521) and 'Mevrouw Gerard Kint' (red) (counted mites: 451). All these cultivars showed infestation at an early stage of the experiment. On 'Michelle Marie', infestation was already present in December (W51), although the mite numbers did not increase considerable compared to the other cultivars. An overview of the occurrence of *P. latus* on the different cultivars is given in Figure 3.

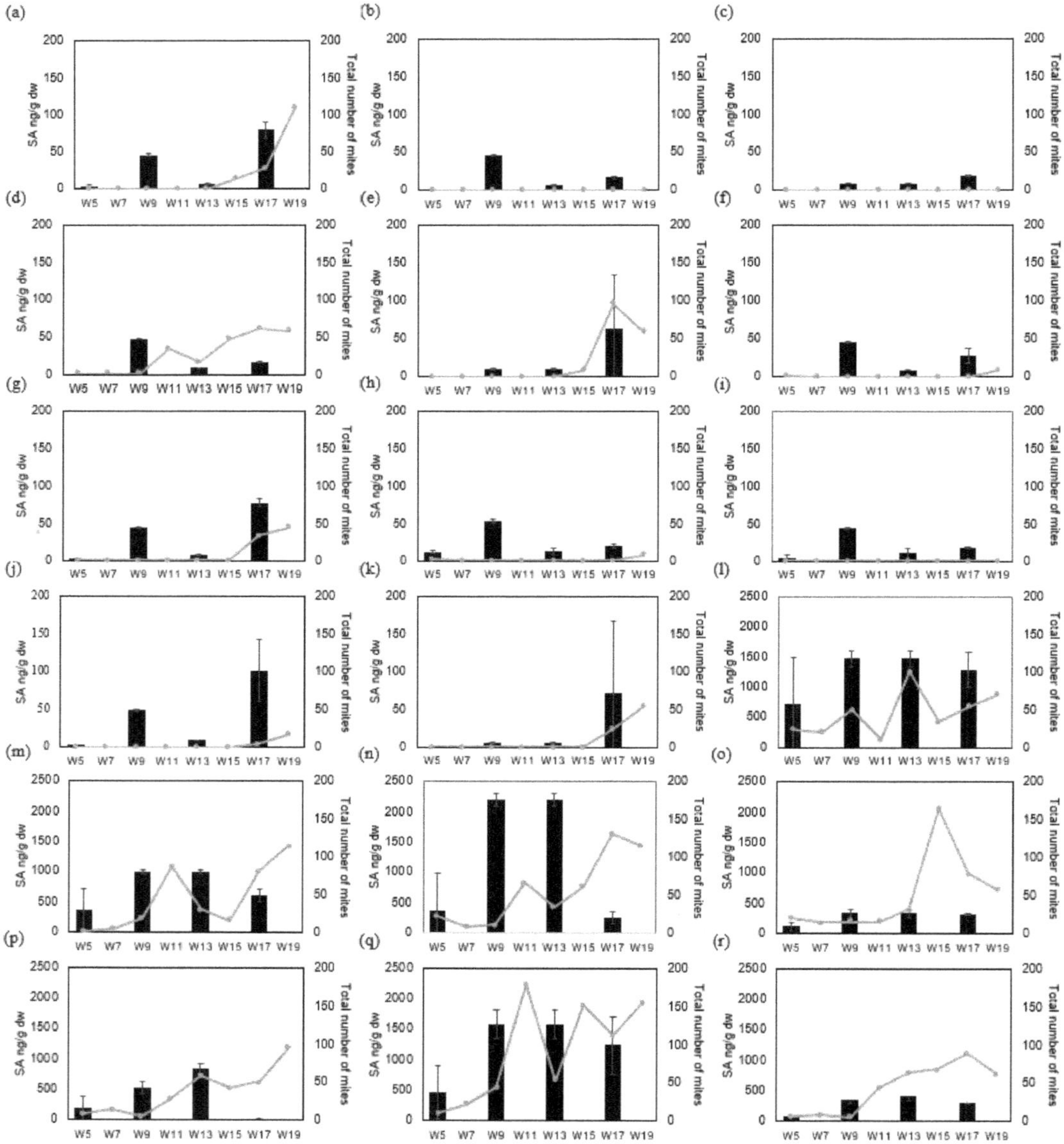

Figure 3. Experiment 3: mite numbers of *Polyphagotarsonemus latus* (total of 3 shoot tips) counted on different sampling dates (every two weeks, 15 time points) and levels of salicylic acid (SA) (+ SD) ($n = 3$) in 18 cultivars of pot azalea. With low SA levels (Y-axis up to 200 ng/g dw): 'Amélie' (**a**), 'Elien' (**b**), 'Huelsten' (**c**), 'Inka' (**d**), 'Luntera' (**e**), 'Mevrouw Edmond Troch' (**f**), 'Mevrouw Gerard Kint' (picotee) (**g**), 'Nordlicht' (**h**), 'Sachsenstern' (**i**), 'Tamira' (**j**) and 'Thesla' (**k**). With high SA levels (Y-axis up to 2500 ng/g dw): 'Fluostern' (**l**), 'Franziska R.' (**m**), 'Leonardo' (**n**), 'Mevrouw Gerard Kint' (red) (**o**), 'Michelle Marie' (**p**), 'Otto' (**q**) and 'Renato' (**r**).

In all the cultivars, the levels of SA increased in March (W9 to W13) compared to W5. A Pearson correlation was calculated between the maximum number of mites and the SA levels in W10. The correlation was high, with, respectively, r = 0.75 (p = 0.0002) and r = 0.84 (p < 0.0001) when the mite numbers of W9 and W11 were correlated with the SA levels on W10. Moreover, for all the cultivars, the correlation was very high, with r = 0.84 (p < 0.00001) when the total mite numbers across the entire experiment were correlated to the SA level found in W10.

For 11 cultivars, no *P. latus* mites were observed before W9. This was the case for: 'Elien', 'Huelsten', 'Inka', 'Sachsenstern', 'Troch', 'Nordlicht', 'Tamira', 'Mevrouw Gerard Kint ' (red), 'Thesla', 'Amélie', 'Luntera' and 'Mevrouw Gerard Kint' (picotee). All these cultivars had an SA level between 43.5 and 53.4 ng/g dw in W9. For cultivars with only a few mites, such as 'Michelle Marie' (counted mites: 5) and 'Renato' (counted mites: 6), in W9, the SA levels rose to 539.9 and 344.8 ng/g dw, respectively. The highest number of mites in W9 was counted on 'Otto' (counted mites: 44) and 'Fluostern' (counted mites: 50); these cultivars had SA levels of 1498.1 and 1475.1 ng/g dw, respectively (Figure 3).

Although in W13, the non-infected cultivars showed lower levels of SA, the correlation between the SA levels and mite numbers for all cultivars at this time point was still very strong (r = 0.85; p < 0.0001). Two cultivars showed symptoms only on part of the plant. Therefore, the samples taken from symptomatic vs. asymptomatic plants were analyzed separately. In both cultivars, huge differences in the SA level between symptomatic and asymptomatic plants were observed. For 'Michelle Marie', the plants with symptoms had an SA level of 852.0 ng/g dw compared to 15.1 ng/g dw in plants without symptoms. For 'Leonardo', symptomatic plants had an SA level of 2204.1 and asymptomatic plants 254.3 ng/g dw. Only these two cultivars, 'Michelle Marie' and 'Leonardo', had relevant levels of JA when comparing plants with and without symptoms (Figure 4). In both cultivars, the levels of JA in plants showing symptoms were around 10 and 20 times higher in 'Michelle Marie' and 'Leonardo', respectively, in comparison to plants showing no symptoms.

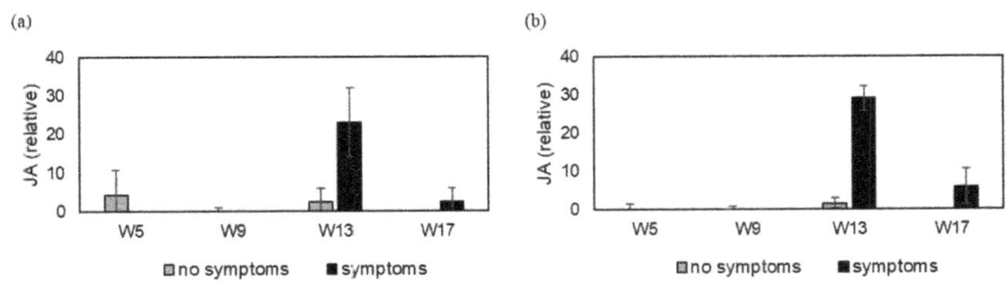

Figure 4. Jasmonic acid (JA) levels on the pot azalea cultivars 'Michelle Marie' (**a**) and 'Leonardo' (**b**). Plants showing symptoms and plants without symptoms of infestation by *Polyphagotarsonemus latus* were sampled separately. Data are means ± SD (n = 3).

4. Discussion

The study of plant–arthropod interactions lags behind the study of plant–pathogen interactions. However, in recent years, plant–herbivore interactions have received increasing attention. For mites, plant reactions to the presence of the two-spotted spider mite (*Tetranychus urticae* Koch) have been studied most, as reviewed by Agut et al. [32] and Blazer et al. [22]. In the present study, we have focused on the relationship between the two main plant defense hormones, SA and JA, in response to broad mite presence in pot azalea. Our results are in line with the general accepted idea that the JA pathway is involved in defense against herbivores [21]. It is known that endogenous bioactive plant defense hormones such as JA are present in low levels in the plant, but accumulate upon herbivore feeding or wounding [33]. The results of experiment 1 under controlled conditions and of experiment 3 under natural infestation during the growing season showed that an increase in JA is only found at specific time points and early after infestation. In experiment 2, no significant changes in JA levels were observed. As suggested by Luypaert et al. [17], the timing of sampling is essential while measuring hormonal changes. Additionally, infestation pressure at the time of sampling plays a role in the observed hormone levels. The present study confirms the findings of Luypaert et al. [25], that *P. latus* infestation activates JA-dependent pathways most strongly at earlier time points (5–12 dpi) in *R. simsii* hybrids. In preliminary laboratory tests using floating leaf discs, it was shown that after the artificial application of 50 or 100 µM MeJA, the population growth of *P. latus* on *R. simsii* 'Nordlicht' significantly decreased [17]. Luypaert et al. [25] isolated different JA biosynthesis marker genes in azalea to obtain insight into the molecular mechanisms behind the interaction between *P. latus* and the *R. simsii* hybrid. Transcript levels of marker genes for the JA biosynthesis pathway were significantly induced upon *P. latus* infestation in the *R. simsii* hybrid [25]. This strengthens our hypothesis of an early-induced plant defense based on the JA response upon *P. latus* infestation in pot azalea.

The high increase in SA levels upon broad mite infestation is in contrast with the expected plant defense response. The SA pathway is often associated with defense against biotrophic pathogens [19]. Arena et al. [34] found, in *Arabidopsis thaliana* (L.) Heynh., a simultaneous increase in the levels of SA and JA when the plants were infested with false spider mites of the genus *Brevipalpus* Donnadieu. They also evaluated mite performance on *Arabidopsis* mutants impaired in their SA or JA response and showed a function for the SA-mediated pathway in improved mite reproduction. Therefore, it is suggested that *Brevipalpus* mites manipulate the plant defensive response via the SA-mediated pathway for their own benefit [34].

In our study, significant increases in SA level were strongly related to the presence of broad mites. Levels remained low when no mites were present. In experiment 3 with 18 cultivars tested, an especially strong correlation was found between the rising number of mites in W9 (r = 0.75) and W11 (r = 0.84) and the SA level in W10. An increase of up to almost 100 times the SA level was measured in the susceptible cultivar 'Nordlicht' (experiment 2). These high SA levels where still observed during the following months of infestation. At the same moment, in 'Mevrouw Gerard Kint', a cultivar with moderate susceptibility, the increase in SA was lower, while in 'Elien', a resistant cultivar, no change in the SA level was observed.

In preliminary tests with a floating disc assay, the application of 50 or 100 µM SA had no significant effect on the broad mite population size [35]. Other preliminary data suggest that the application of SA can lead to increasing population sizes of broad mites, but this depends on the SA concentration applied. Additionally, the expression of SA marker genes isolated after *P. latus* infestation in the *R. simsii* hybrid was not clear [25]. Our SA analysis, however, could not confirm their conclusion that the SA biosynthesis pathway, led by either RsPAL or RsICS, was suppressed at later time points in response to *P. latus* infestation in pot azalea. If no local increase in transcript levels is observed, a systemic response of signaling molecules in plant defense for stronger and faster defense responses could be a hypothesis [36]. JA, for example, will not only accumulate at the sites of local wounding, but also in undamaged, more distal leaves. This increase can be found within two minutes [33]. For SA, more data are available in relation to pathogen-induced SA, which was also shown to act systemically in the plant and was found in the phloem [37].

In tomato, *P. latus* prefers plants deficient in the JA pathway compared to wild-type plants [38]. Treatment of the JA-deficient plants made them as unattractive as the wild-type plants. The wild-type plants infested by broad mites showed an increase in JA-related transcripts. In a study by Grinberg et al. [9] using Northern blot analysis, it was shown that both the JA and SA pathways are activated in response to broad mites on cucumber. In citrus, *P. latus* infestation activated both the JA- and the SA-dependent pathways in sour orange but no activation was observed in Cleopatra mandarin [39].

From our results, it is probable that in an early phase of infestation, the plant responds by activating the JA-dependent pathway; subsequently, the JA levels drop quickly and a switch is made to the production of SA. Because of the small timeframe, it is possible that the increase in JA was not observed in some of our experiments under natural infestation, as the exact moment of mite infestation is difficult to determine. In experiment 1, under controlled infestation conditions, it is shown that the JA level can change in a time period of only two days. Under culture conditions, as applied in experiments 2 and 3, not all cultivars and not all plants within one cultivar experience the same infestation pressure. Differences in JA content could be observed when symptomatic and asymptomatic plants were compared at the moment the mite population start to build up, as was the case for 'Michelle Marie' and 'Leonardo' (experiment 3). Because timing is crucial, further research would be needed to elucidate the switch from the JA- to the SA-dependent pathway.

The key question to be elucidated is that of whether *P. latus* is able to suppress plant defense by modulating the action of SA in response to JA. If so, a plausible answer might be that *P. latus* mites inject effectors, interfering with host immune responses. To date, no data for *P. latus* are available, but this phenomenon is known to occur in phytopathogens [40], aphids [41] and nematodes [42]; additionally, spider mites might secrete effectors via their saliva [43,44]. For example, strains of the spider mites *Tetranychus urticae* and *T. evansi* Baker & Pritchard can suppress the expression of SA and/or JA marker genes [21,44–46]. The genomic data of *P. latus*, in accordance with the whole spider mite genome, would provide evidence if this mite species is also able to encode putative salivary proteins [47]. Another alternative might be that *P. latus* is a vector for phytopathogens that have been implicated in the suppression of plant defenses. The hypothesis of virus transmission by *P. latus* feeding on plants is experimentally rejected [48,49], still leaving the possibility of transmitting bacteria or fungi.

For whitefly, *Bemisia tabaci (Gennadius)*,[2] under laboratory and semi-field conditions, examples are known of JA suppression in correlation with the induction of SA [50–52]. Negative crosstalk between the SA and JA defense pathways is often displayed. For example, the suppression of JA can be an effect of antagonistic crosstalk through strong SA induction [18]. However, suppression can also occur downstream of phytohormone accumulation and independently of JA-SA crosstalk [53].

The conclusion on the role of JA in the defense of pot azalea against broad mite is supported by the findings that the application of MeJA had a negative effect on the fitness of *P. latus* in the pot azalea cultivar 'Nordlicht' when compared to a control treatment [35]. This is in accordance with the reports of decreases in spider mite population growth after the application of MeJA [32,54–59]. In pot azalea, the negative effect of MeJA on the population of *P. latus* was related to the concentration applied: for 50 and 100 µM MeJA, the population growth decreased by a factor 3.2 and 2.3, respectively [35]. Choh et al. [60] reported a similar effect as the number of eggs produced by *T. urticae* on lima beans (*Phaseolus lunatus* L.) differs in relation to the JA concentration applied. Additionally, in the *T. urticae*-sensitive Cleopatra mandarin, MeJA treatment significantly reduced the number of eggs produced by the spider mite, whereas salicylic acid had no effect on the oviposition of *T. urticae* [32].

For the eight cultivars used in the study of Luypaert et al. [12] and in experiment 3, correlations between the experiments were found for mite numbers and for damage rates, but the correlations were not significant. Previous studies on pot azalea and broad mite interaction showed that leaf damage does not always correlate well with the number of mites found on the plant [12]. However, our results regarding susceptibility confirmed the study of Luypaert et al. [12], with 'Elien' as the most resistant and 'Otto' being the most susceptible cultivar. Similar results were found for cultivars with a shared genetic background. 'Huelsten', 'Inka', 'Lunterra', 'Mevrouw Edmond Troch' and 'Nordlicht' all are sports of 'Helmutt Vogel'. In these five cultivars, the levels of SA were relatively low upon infestation by broad mites. For 'Michelle Marie' and its sport 'Franziska R.', both mite numbers and SA levels were high, suggesting that these cultivars are susceptible to broad mites. Similar results are also found for 'Thesla' and its sport 'Amélie' in mite numbers and SA levels. Both cultivars showed infestation later in the growing season. 'Mevrouw Gerard Kint' (red) and (picotee) and 'Renato' share a common genetic background and show a relatively low increase in SA levels; nevertheless, higher numbers of mites can occur. Only on 'Sachsenstern' were the level of SA and mite number low, in contrast to the results obtained for its sport 'Fluostern'. 'Sachsenstern' probably avoided infestation. Additionally, 'Nordlicht' had lower mite numbers and a lower SA level compared to experiment 2. Additionally, in other crops, a differential response towards broad mite was found, e.g., in citrus, sour orange supported larger densities than Cleopatra mandarin but the levels of injury were similar [39], and in 36 genotypes of common bean (*Phaseolus vulgaris*), differences in *P. latus* infestation were found [61].

The present study showed that not all types of tarsonemid mites cause damage in pot azalea. In culture conditions, the herbivorous and damage-causing mite *P. latus* is most prevalent. Nevertheless, not all pot azalea cultivars are equally attractive to this mite. 'Elien', the most resistant cultivar towards *P. latus* mites, possibly exhibits an antixenotic effect based on the presence of glandular trichomes with a sticky end [17]. For the other cultivars, i.e., 'Huelsten', 'Inka' and 'Sachstenstern', the nature of their resistance to *P. latus* is less clear. Another frequently occurring mite, *T. confusus*, was present on all the azalea cultivars without causing any symptoms. Until now, it was not fully understood whether this mite species is herbivorous. This study could only relate the presence of *P. latus* to plant damage. *P. pallidus* could be harmful, but so far, this has not been observed and the number of mites found in our experiments was probably too low.

5. Conclusions

In conclusion, it is clear that the presence of broad mite on pot azalea increases both JA and SA levels. A primary plant response is through the JA pathway, but later, this changes to an increase in the levels of SA. Further efforts, e.g., experiments under controlled conditions, are needed to better understand whether the increase in SA is a result of crosstalk between JA and SA and whether it can help to elucidate the timing and factors influencing the JA and SA responses.

Another conclusion has importance for azalea cultivation. In 18 cultivars, a large variation in susceptibility to broad mite was found, with 'Otto' being the most susceptible and 'Elien' the most resistant cultivar. This variation in susceptibility to broad mite opens up perspectives for using more resistant genotypes in cultivation and suggests that more genotypes with improved resistance can be developed though breeding.

Author Contributions: Conceptualization, J.A., G.L., E.P., J.W., C.V.P., E.P. and L.L.; methodology, J.A., E.D., G.L., E.P., E.V.P. and C.V.P.; formal analysis, C.V.P., J.W. and L.L.; writing—original draft preparation, L.L., J.W. and J.A.; writing—review and editing, L.L., G.L., E.D., J.W., E.P., C.V.P., E.V.P., J.V.H. and J.A.; visualization, L.L.; supervision, J.V.H. and J.A.; project administration, J.A.; funding acquisition, J.A. All authors have read and agreed to the published version of the manuscript.

Funding: This research project UNIFORCE (Unification of IPM Forces to Control Mites in Berries, Soft Fruits and Woody Ornamentals) was funded by EU C-IPM and the Agentschap voor Innoveren en Ondernemen.

Institutional Review Board Statement: Not applicable.

Informed Consent Statement: Not applicable.

Data Availability Statement: The raw data were generated at ILVO and PCS. The derived data supporting the findings of this study are available from the corresponding author [L.L.] on request.

Acknowledgments: The authors wish to thank Veerle Buysens, Sophie Carbonelle, Kristien Janssens and Stijn Degroote for their dedicated technical assistance in the LC-MS/MS analysis, and Ruth Verhoeven and Bart Haleydt for the sampling, counting and identification of the mite populations in the greenhouse experiments. The authors also thank Miriam Levenson, the English-language editor at ILVO, for her helpful English corrections. Gil Luypaert's present affiliation is De Ceuster Meststoffen nv, 2280 Grobbendonk, Belgium, glu@dcm-info.com.

Conflicts of Interest: The authors declare no conflict of interest.

References

1. Scheerlinck, H.; Baumann, J.; De Somer, F.; Haerens, A.; Van Laere, A. *Tuinbouwencyclopedie I. De Azalea Indica L. De Sikkel*; Wereldbibliotheek: Amsterdam, The Netherlands, 1938; 527p.
2. Heursel, J.; Lootens, P.; Mertens, M.; Saverwijns, A. *Azalea's. Oorsprong, Veredeling en Cultivars*; Lannoo: Tielt, Belgium, 1999; p. 192.
3. Van Trier, H. *Gentse Azalea*; Stichting Kunstboek: Oostkamp, Belgium, 2012; 144p.
4. VLAM 2020. Belgische Export Van Azalea's en Rhododendrons (2011–2020)—Duitsland Sierteelt 2021—Belgische Sierteeltoppervlakte (2011–2020). Available online: https://www.vlaanderen.be/vlam/kennisbank?f%5B0%5D=facet_sector%3A58 (accessed on 25 February 2022).
5. VLAM 2021. Belgische Export Van Azalea's en Rhododendrons (2011–2020)—Duitsland Sierteelt 2021—Belgische Sierteeltoppervlakte (2011–2020). Available online: https://www.vlaanderen.be/vlam/kennisbank?f%5B0%5D=facet_sector%3A58 (accessed on 25 February 2022).
6. Navajas, M.; Migeon, A.; Estrada-Pena, A.; Mallieux, A.C.; Servigne, P.; Petanovic, R. Chapter 7.4: The Acari. Invasive terrestrial invertebrates in Europe. *BioRisk* **2010**, *4*, 149–192. [CrossRef]
7. Heungens, A. Soft-skinned mites in azalea culture and comparable control results on other host plants. *Verb. Voor De Belg. Sierteelt* **1986**, *30*, 257–269.
8. De Coss-Romero, M.; Peña, J.E. Relationship of broad mite (Acari: Tarsonemidae) to host phenology and injury levels in *Capsicum annuum*. *Fla. Entomol.* **1998**, *81*, 515–526. [CrossRef]
9. Grinberg, M.; Perl-Treves, R.; Palevsky, E.; Shomer, I.; Soroker, V. Interaction between cucumber plants and the broad mite, *Polyphagotarsonemus latus*: From damage to defense gene expression. *Entomol. Exp. Appl.* **2005**, *115*, 135–144. [CrossRef]
10. Peña, J.E. Relationships of broad mite (Acari: Tarsonemidae) density to lime damage. *J. Econ. Entomol.* **1990**, *83*, 2008–2015. [CrossRef]
11. Dhooria, J.C.; Bindra, O.S. *Polyphagotarsonemus latus* (Banks) a mite pest of chilli and potato in Punjab. *Acarol. Newsl.* **1977**, *4*, 7–9.
12. Luypaert, G.; Van Huylenbroeck, J.; De Riek, J.; De Clercq, P. Screening for broad mite susceptibility in *Rhododendron simsii* hybrids. *J. Plant Dis. Protect.* **2014**, *121*, 260–269. [CrossRef]
13. Gerson, U. Biology and control of the broad mite, *Polyphagotarsonemus latus* (Banks) (Acari: Tarsonemidae). *Exp. Appl. Acarol.* **1992**, *13*, 163–178. [CrossRef]
14. Van Huylenbroeck, J.; Calsyn, E.; De Keyser, E.; Luypaert, G. Breeding for biotic stress resistance in *Rhododendron simsii*. *Acta Hort.* **2015**, *1104*, 375–379. [CrossRef]
15. Nasrin, M.; Amin, M.R.; Miah, M.R.U.; Akanda, A.M.; Miah, M.G.; Kwon, O.; Suh, S.J. Occurrence and severity of mite *Polyphagotarsonemus latus* (Tarsonemidae) on chili plants: Analysis of pest-weather and host plant characteristics. *Entomol. Res.* **2020**, *51*, 273–281. [CrossRef]
16. Wang, T.; Jia, Q.; Wang, W.; Hussain, S.; Ahmed, S.; Adnan; Zhou, D.X.; Ni, Z.; Wang, S. GCN5 modulates trichome initiation in Arabidopsis by manipulating histone acetylation of core trichome initiation regulator genes. *Plant Cell Rep.* **2019**, *38*, 755–765. [CrossRef] [PubMed]
17. Luypaert, G.; Van Huylenbroeck, J.; De Riek, J.; Mechant, E.; Pauwels, E.; De Clercq, P. Opportunities to breed for broad mite resistance in *Rhododendron simsii* hybrids. In Proceedings of the XXV International EUCARPIA Symposium Section Ornamentals: Crossing Borders 1087, Melle, Belgium, 28 June 2015; pp. 479–484.
18. Pieterse, C.M.; Van Der Does, D.; Zamioudis, C.; Leon-Reyes, A.; Van Wees, S.C. Hormonal modulation of plant immunity. *Annu. Rev. Cell Dev. Biol.* **2012**, *28*, 489–521. [CrossRef] [PubMed]
19. Vlot, A.C.; D'Maris, A.D.; Klessig, D.F. Salicylic acid, a multifaced hormone to combat disease. *Annu. Rev. Phytopathol.* **2009**, *47*, 177–206. [CrossRef] [PubMed]
20. Glazebrook, J. Contrasting mechanisms of defense against biotrophic and necrotrophic pathogens. *Annu. Rev. Phytopathol.* **2005**, *43*, 205–227. [CrossRef]
21. Howe, G.A.; Jander, G. Plant immunity to insect herbivores. *Annu. Rev. Plant Biol.* **2008**, *59*, 41–66. [CrossRef]
22. Blaazer, C.J.H.; Villacis-Perez, E.A.; Chafi, R.; Van Leeuwen, T.; Kant, M.R.; Schimmel, B.C. Why do herbivorous mites suppress plant defenses? *Front. Plant Sci.* **2018**, *9*, 1057. [CrossRef]
23. Grinberg, M.; Alagarmalai, J.; Sharon, L.; Perl-Treves, R.; Soroker, V. The effect of jasmonic acid signalling pathway on tomato defense to broad mite behaviour. *Isr. J. Entomol.* **2007**, *37*, 389–390.
24. Alagarmalai, J.; Grinberg, M.; Reneh, S.; Sharon, L.; Perl-Treves, R.; Soroker, V. Host selection behavior of broad mite, *Polyphagotarsonemus latus* (Acari: Tarsonemidae). *IOBC-WPRS Bull.* **2007**, *30*, 1–7.
25. Luypaert, G.; Witters, J.; Van Huylenbroeck, J.; De Clercq, P.; De Riek, J.; De Keyser, E. Induced expression of selected plant defence related genes in pot azalea, *Rhododendron simsii* hybrid. *Euphytica* **2017**, *231*, 227. [CrossRef]
26. Mechant, E.; Luypaert, G.; Van Delsen, B.; Pauwels, E.; Witters, J.; Van Huylenbroeck, J.; Gobin, B. Development and validation of a three-step detection protocol for broad mites (*Polyphagotarsonemus latus*) in pot azalea (*Rhododendron simsii* hybrids). *Entomol. Exp. Appl.* **2015**, *156*, 99–104. [CrossRef]
27. Vacante, V. *The Handbook of Mites of Economic Plants: Identification, Bio-Ecology and Control*; CABI: Wallingford, UK, 2015.
28. Krantz, G.W.; Walter, D.E. *A Manual of Acarology*; Texas Tech University Press: Lubbock, TX, USA, 2009; 807p.

29. Zhang, Z.Q.; Bejakovich, D.; Martin, N.A. *Key to Tarsonemidae of New Zealand*; Final report to MAF Science Policy for Project FMA 102; Ministry of Business, Innovation and Employment: Wellington, New Zealand, 2000; pp. 1–35.
30. Bosco, R.; Daeseleire, E.; Van Pamel, E.; Scariot, V.; Leus, L. Development of an Ultrahigh-Performance Liquid Chromatography−Electrospray Ionization−Tandem Mass Spectrometry Method for the Simultaneous Determination of Salicylic Acid, Jasmonic Acid, and Abscisic Acid in Rose Leaves. *J. Agric. Food Chem.* **2014**, *62*, 6278–6284. [CrossRef] [PubMed]
31. Sánchez-Bel, P.; Sanmartín, N.; Pastor, V.; Mateu, D.; Cerezo, M.; Vidal-Albalat, A.; Pastor-Fernández, J.; Pozo, M.J.; Flors, V. Mycorrhizal tomato plants fine tunes the growth-defence balance upon N depleted root environments. *Plant Cell Environ.* **2018**, *41*, 406–420. [CrossRef] [PubMed]
32. Agut, B.; Pastor, V.; Jaques, J.A.; Flors, V. Can plant defence mechanisms provide new approaches for the sustainable control of the two-spotted spider mite *Tetranychus urticae*? *Int. J. Mol. Sci.* **2018**, *19*, 614. [CrossRef] [PubMed]
33. Wang, J.; Wu, D.; Wang, Y.; Xie, D. Jasmonate action in plant defense against insects. *J. Exp. Bot.* **2019**, *70*, 3391–3400. [CrossRef]
34. Arena, G.D.; Ramos-González, P.L.; Rogerio, L.A.; Ribeiro-Alves, M.; Casteel, C.L.; Freitas-Astúa, J.; Machado, M.A. Making a Better Home: Modulation of Plant Defensive Response by Brevipalpus Mites. *Front. Plant Sci.* **2018**, *9*, 1147. [CrossRef]
35. Luypaert, G. The Broad Mite, *Polyphagotarsonemus latus* (Acari: Tarsonemidae), and Its Interactions with Pot Azalea, *Rhododendron simsii* Hybrid. Ph.D. Thesis, Ghent University, Ghent, Belgium, 2015.
36. Heil, M.; Ton, J. Long-distance signalling in plant defence. *Trends Plant Sci.* **2008**, *13*, 264–272. [CrossRef]
37. Lim, G.-H.; Sine, M.B.; de Lorenzo, L.; Yu, K.; Cui, W.; Navarre, D.; Hunt, A.G.; Lee, J.Y.; Kachroo, A.; Kachroo, P. Plasmodesmata localizing proteins regulate transport and signaling during systematic acquired immunity in plants. *Cell Host Microbe* **2016**, *19*, 541–549. [CrossRef]
38. Grinberg-Yaari, M.; Alagarmalai, J.; Lewinsohn, E.; Perl-Treves, R.; Soroker, V. Role of jasmonic acid signaling in tomato defense against broad mite, *Polyphagotarsonemus latus* (Acari: Tarsonemidae). *Arthropod-Plant Interact.* **2015**, *9*, 361–372. [CrossRef]
39. Cabedo-López, M.; Cruz-Miralles, J.; Peris, D.; Ibáñez-Gual, M.V.; Flors, V.; Jaques, J.A. The response of citrus plants to the broad mite *Polyphagotarsonemus latus* (Banks) (Acari: Tarsonemidae). *Agric. For. Entomol.* **2021**, *23*, 411–419. [CrossRef]
40. De Jonge, R.; Bolton, M.D.; Thomma, B.P. How filamentous pathogens co-opt plants: The ins and outs of fungal effectors. *Curr. Opin. Plant Biol.* **2011**, *4*, 400–406. [CrossRef]
41. Rodriguez, P.A.; Bos, J.I.B. Toward understanding the role of aphid effectors in plant infestation. *Mol. Plant-Microbe Interact.* **2013**, *26*, 25–30. [CrossRef] [PubMed]
42. Haegeman, A.; Mantelin, S.; Jones, J.T.; Gheysen, G. Functional roles of effectors of plant-parasitic nematodes. *Gene* **2012**, *492*, 19–31. [CrossRef] [PubMed]
43. Alba, J.M.; Glas, J.J.; Schimmel, B.C.J.; Kant, M.R. Avoidance and suppression of plant defenses by herbivores and pathogens. *J. Plant Interact.* **2011**, *6*, 221–227. [CrossRef]
44. Alba, J.M.; Schimmel, B.C.; Glas, J.J.; Ataide, L.M.; Pappas, M.L.; Villarroel, C.A.; Schuurink, R.C.; Sabelis, M.W.; Kant, M.R. Spider mites suppress tomato defenses downstream of jasmonate and salicylate independently of hormonal crosstalk. *New Phytol.* **2015**, *205*, 828–840. [CrossRef] [PubMed]
45. Kant, M.R.; Sabelis, M.W.; Haring, M.A.; Schuurink, R.C. Intraspecific variation in a generalist herbivore accounts for differential induction and impact of host plant defences. *P. Roy. Soc. B Biol. Sci.* **2008**, *752*, 443–452. [CrossRef]
46. Sarmento, R.A.; Lemos, F.; Bleeker, P.M.; Schuurink, R.C.; Pallini, A.; Oliveira, M.G.A.; Lima, E.R.; Kant, M.; Sabelis, M.W.; Janssen, A. A herbivore that manipulates plant defence. *Ecol. Lett.* **2011**, *14*, 229–236. [CrossRef]
47. Grbić, M.; Van Leeuwen, T.; Clark, R.M.; Rombauts, S.; Rouzé, P.; Grbić, V.; Osborne, E.J.; Dermauw, W.; Thi Ngoc, P.C.; Ortego, F.; et al. The genome of *Tetranychus urticae* reveals herbivorous pest adaptations. *Nature* **2011**, *479*, 487–492. [CrossRef]
48. Higa, S.Y.; Namba, R. Vectors of the papaya mosaic virus in Hawaii. *Proc. Hawaii. Entomol. Soc.* **1971**, *21*, 93–96.
49. Aubert, B.; Lossois, P.; Marchal, J. Mise en evidence des dégâts causés par *Polyphagotarsonemus latus* (Banks) sur papayers à l'ile de la Réunion. *Fruits* **1981**, *36*, 9–24.
50. Zarate, S.I.; Kempema, L.A.; Walling, L.L. Silverleaf whitefly induces salicylic acid defenses and suppresses effectual jasmonic acid defenses. *Plant Physiol.* **2007**, *143*, 866–875. [CrossRef]
51. Zhang, P.J.; Li, W.D.; Huang, F.; Zhang, J.M.; Xu, F.C.; Lu, Y.B. Feeding by whiteflies suppresses downstream jasmonic acid signaling by eliciting salicylic acid signaling. *J. Chem. Ecol.* **2013**, *39*, 612–619. [CrossRef] [PubMed]
52. Zhang, P.-J.; He, Y.-C.; Zhao, C.; Ye, Z.-H.; Yu, X.-P. Jasmonic Acid-Dependent Defenses Play a Key Role in Defending Tomato Against *Bemisia tabaci* Nymphs, but Not Adults. *Front. Plant Sci.* **2018**, *20*, 1065. [CrossRef] [PubMed]
53. Schimmel, B.C.J.; Alba, J.M.; Wybouw, N.; Glas, J.J.; Meijer, T.T.; Schuurink, R.C.; Kant, M.R. Distinct Signatures of Host Defense Suppression by Plant-Feeding Mites. *Int. J. Mol. Sci.* **2018**, *19*, 3265. [CrossRef]
54. Omer, A.D.; Granett, J.; Kerban, R.; Villa, E.M. Chemically-induced resistance against multiple pests in cotton. *Int. J. Pest Man.* **2001**, *47*, 49–54. [CrossRef]
55. Li, C.; Williams, M.M.; Loh, Y.T.; Lee, G.I.: Howe, G.A. Resistance of cultivated tomato to cell content-feeding herbivores is regulated by the octadecanoid-signaling pathway. *Plant Physiol.* **2002**, *130*, 494–503. [CrossRef]
56. Warabieda, W.; Miszczak, A.; Olszak, R.W. The influence of methyl jasmonate (JA-Me) and βglucosidase on introduction of resistance mechanisms of strawberry against two-spotted spider mite (*Tetranychus urticae* Koch). *Commun. Agric. Appl. Biol. Sci.* **2005**, *70*, 829–836.

57. Warabieda, W.; Olszak, R.W. Aspects of the influence of exogenous methyl jasmonate on strawberry plants and on the population size of the two-spotted spider mite (*Tetranychus urticae* Koch). *IOBC-WPRS Bull.* **2012**, *83*, 7–13.
58. Rohwer, C.L.; Erwin, J.E. Spider mites (*Tetranychus utricae*) perform poorly on and disperse from plants exposed to methyl jasmonate. *Entomol. Exp. Et Appl.* **2010**, *137*, 143–152. [CrossRef]
59. Agut, B.; Gamir, J.; Jacas, J.A.; Hurtado, M.; Flors, V. Different metabolic and genetic responses in citrus may explain relative susceptibility to Tetranychus urticae. *Pest Manag. Sci.* **2014**, *70*, 1728–1741. [CrossRef]
60. Choh, Y.; Ozawa, R.; Takabayashi, J. Effects of exogenous jasmonic acid and benzo (1,2,3) thiadiazole7-carbothioic acid S-methyl ester (BTH), a functional analogue of salicylic acid, on the egg production of a herbivorous mite *Tetranychus urticae* (Acari: Tertanychidae). *Appl. Entomol. Zool.* **2004**, *39*, 311–314. [CrossRef]
61. Androcioli, H.G.; Hoshino, A.T.; Ventura, M.U.; Hata, F.T.; Brugnerotto, M.d.R.; Constantino, L.V.; Marques, F.d.A. Resistance of Common Bean Genotypes to the Broad Mite, *Polyphagotarsonemus latus* (Banks, 1904) (Acari: Tarsonemidae): Offspring Development and Biochemical Basis. *Insects* **2021**, *12*, 910. [CrossRef] [PubMed]

horticulturae

MDPI

Review

A Review on Flower Bulb Micropropagation: Challenges and Opportunities

Sara Yasemin [1,*] and Margherita Beruto [2]

[1] Department of Horticulture, Faculty of Agriculture, University of Siirt, Kezer, 56100 Siirt, Turkey
[2] International Society for Horticultural Science, Ornamental Plant Division, 18038 Sanremo, Italy; margheberuto@gmail.com
* Correspondence: sara.yasemin@siirt.edu.tr

Abstract: This comprehensive review scrutinizes tissue culture and micropropagation methodologies in geophytes, focusing on bulbous plants. The examination encompasses key stages, including somatic embryogenesis, bulb growth, dormancy breaking, and planting. Studies underscore the pivotal role of plant growth regulators (PGRs) in plant regeneration and bulb growth. Bioreactor systems for healthy plant regeneration, rooting methods, acclimatization strategies, and considerations for ex vitro survival are elucidated. The review also delves into somaclonal variation dynamics and acknowledges the burgeoning field of gene editing, particularly Clustered Regularly Interspaced Short Palindromic Repeats (CRISPR) studies, as a promising avenue for enhancing valuable compound content in geophytes. In addition to addressing challenges in flower bulb micropropagation, this review briefly highlights emerging opportunities, including the potential integration of artificial intelligence (AI) to optimize culture conditions, predict growth parameters, and enhance efficiency in bulb production. The conclusion emphasizes the necessity of a multifaceted approach integrating biochemistry, physiology, and molecular biology to address existing challenges and improve tissue culture protocols for diverse geophyte species. This review article also intends to highlight how tissue culture techniques could contribute to the development and valorization of flower bulbs in today's scenario of the ornamental industry.

Keywords: flower bulbs; organogenesis; somaclonal variations; somatic embryogenesis

Citation: Yasemin, S.; Beruto, M. A Review on Flower Bulb Micropropagation: Challenges and Opportunities. *Horticulturae* **2024**, *10*, 284. https://doi.org/10.3390/horticulturae10030284

Academic Editor: Haiying Liang

Received: 4 January 2024
Revised: 28 February 2024
Accepted: 13 March 2024
Published: 15 March 2024

1. Ornamental Geophyte History

For centuries, flower bulbs have captivated societies worldwide, drawing attention for their enticing fragrance, vibrant colors, and diverse blooming periods. These bulbs have held a significant place in various forms of art [1–3]. Greek philosophy dating back to 300 BC mentions the medicinal and consumable properties of flower bulbs like *Crocus*, *Colchicum*, and *Gladiolus*. Greek mythology also references bulbs such as *Lilium*, *Hyacinthus*, *Crocus*, *Iris*, and *Narcissus* [4,5]. Furthermore, Warren [6] reported that *Anemone coronaria*, *Ranunculus asiaticus*, and *Lilies* were shown in Knossos Palace in Crete, Greece, during the Bronze Age. Day [7] emphasized that the *Crocus* has been recognized in Minoan art as an important motif during the Aegean Bronze Age. Japan and China are known to have numerous wild *Lilium* species that are endemic to their regions. The United Kingdom is the origin of *Narcissus pseudonarcissus*, commonly known as the lenty lily, which serves as the ancestor for many cultivated daffodil varieties. The Persian-style gardens of the Taj Mahal, an iconic 17th-century monument, feature essential flower bulbs like *Anemone*, *Ranunculus*, *Fritillaria*, *Iris*, *Lilium*, *Pancratium*, and *Gloriosa* [8]. Flower bulbs such as Aztec lily, canna, dahlias, and tuberose are reported to have originated in the Americas [9]. In the 1180s, flower bulbs, particularly those with iris motifs, adorned the military uniforms of Italians from Florence-Tuscany and the French [10].

The worldwide spread of tulips has sparked numerous hypotheses. Originally found in the western regions of China, the Caucasus, Turkey, Iran, and Central Asia, tulips first appeared in Anatolia in 12th-century Konya motifs. The breeding of tulips began during the Ottoman Empire in the 16th century [11]. Tulip mania reached its peak between 1673 and 1736 during the Ottoman era, earning that period its name. The arrival of tulips in Europe has triggered several hypotheses. One suggests that Busbeck, an ambassador representing the Austro-Hungarian Empire in 1554, brought tulip bulbs from Istanbul to his friend Carolus Clusius in Austria. Another possibility involves a French physician named Belon, who traveled to the Near East in 1549 and visited Istanbul. Belon recorded in his memoirs that numerous foreigners journeyed to Istanbul by ship to acquire tulip bulbs, which he referred to as the "red lily". These accounts shed light on the early introduction of tulips to Europe and the growing fascination with these captivating flowers. In the 17th century, tulip mania took hold in the Netherlands. The tulip, originally imported from Ottoman lands in the 16th century, became a highly sought-after commodity in the 18th century, primarily sourced from the Netherlands [11–14].

As mentioned by Rees [15], the habitats of the ancestors of some flower bulbs were reported by Bailey [16], namely *Tulipa*: in the Mediterranean and across Asia to Japan; *Hyacinthus*: in Greece, Syria, and Asia Minor; *N. pseudonarcissus* L.: from Sweden to Spain and Romania; *Narcissus tazetta* L.: from the Canary Islands to Japan; *Narcissus poeticus* L.: from France to Greece; *Iris tingitana* Boiss. & Reut.: in Morocco; *Iris reticulata* Bieb.: in the Caucasus; *Iris xiphium* L.: in Spain and North Africa; *Freesia:* in South Africa; *Lilium longiflorum* Thunb.: in Japan; and *Lilium regale* Wils.: in West China.

Over time, flower bulbs have been recognized for their unique characteristics and have been cultivated with consideration for their potential uses and economic and social values. They have transitioned from being sold as individual bulbs to being traded in bulk quantities and have found commercial applications as cut flowers and potted plants. By the conclusion of the twentieth century, the Netherlands had emerged as the leading global player in the flower bulb trade, asserting its dominance in this industry [10].

2. World Ornamental Plants Sector: Situation of Flower Bulbs in the Sector

The global production of ornamental plants has witnessed a widespread increase, playing a crucial role in boosting the economies of more than 50 countries. According to Gabellini and Scaramuzzi [17], the total worldwide cut flower and potted plant production is about 35.5 billion, with a cultivation area of 745,000 ha. Asia–Pacific is the largest region, covering about 79% of the total surface area dedicated to ornamentals. The European Union occupies about 10% of the world flower production surface [18], but thanks to its high productivity per hectare, the ornamental industry represents about 40% of the worldwide production (value of more than 11 billion EUR). The biggest producers are the Netherlands (32%), France (12%), Germany (12%), Spain (12%), and Italy (11%). Outside the EU, Mexico, Colombia, and Ecuador are important producer countries in Central America. Thanks to favorable climatic conditions, foreign investments, and relatively cheap labor, Latin American countries have become an important export region for both the American and European markets. In South America, Brazilian floriculture has developed in the past five years with a cultivated surface of about 15,600 ha and a production value of 1.7 billion EUR; the internal market absorbs most of this production [19]. Countries such as Kenya, Tanzania, Ethiopia, Uganda, and Zambia in Africa, especially in the equatorial belt, have boosted the production and export of cut flowers. Advantages such as the economy being based on agriculture, suitable climatic conditions, and cheap labor in African countries have led to the development of ornamental plant production. Production areas are generally managed by professional companies and consist of large-scale nurseries owned by foreign investors. The global trade volume of ornamental plants is expected to surpass 60 billion USD (about 54.6 billion EUR) in 2027 at a CAGR of 8.8% (source: Flower and Ornamental Plant Global Market Report; https://www.thebusinessresearchcompany.com/report/flower-and-ornamental-plant-global-market-report, accessed on 25 February 2024), with a particular emphasis on markets with strong purchasing power, including the European Union, USA, and Japan. The Netherlands confirms its central and dominant role in the international trade of flowers and ornamental plants thanks to the combination of national production and re-exported products. Other important exporting countries in the world are Colombia, Kenya, Ecuador, and Israel. The EU is a net exporter of pot plants, conifers, and hardy perennial plants, bulbs, and corms, and a net importer of cut flowers and cut foliage. The most important European importing countries for flowers and potted plants are Germany, France, and the UK.

As the global economic situation improves and people want to spend more, the demand for high-quality products increases. This trend also applies to the global market for flowers and ornamental plants. Consequently, there is a growing demand for high-quality bouquets from different flower collections. Thanks to the huge diversity of the genera, flower bulbs can satisfy the need for new products; in addition, increasing research progress enables the establishment of a productive chain, enhancing the local flower industry. This can be of particular interest for countries that are aiming to diversify their production from the major popular floriculture products. An interesting example is provided by Turkey, which is very rich in natural resources of ornamental geophytes and quite recently started to develop a commercial production for flower bulbs with 51 ha dedicated (Turkish Statistical Institute, 2023).

Ornamental geophytes, also called flower bulbs, contribute significantly to the global ornamental industry and are used for commercial bulbs, cut flowers or potted flowering plant production, landscaping, and private gardening. In 2010, Benschop et al. [1] estimated the global value of the flower bulb industry at more than 1 billion USD, with perspectives of expansion all over the world. In 2022, the product category "Bulbs and Roots", including dormant and in growth bulbs, tubers, roots, corms, crowns, and rhizomes, registered a global value of export of over 2 billion USD (about 1.82 billion EUR), with a slight decrease with respect to the year 2021 counterbalanced by a 4% increase in the period 2018–20 (source: Trade statistics for international business development; https://www.trademap.org/, accessed on 17 October 2023). In 2022, the EU exported a total of over 100 million EUR worth of orchid, hyacinth, narcissi, and tulip bulbs in growth or in flower; the imports were 30.9 million EUR. Of all EU countries, the Netherlands was responsible for 81% of all exports of these bulbs, and other exporter countries were Lithuania (7%), Poland (5%), Denmark (2%) and Latvia (2%). The majority of these exported bulbs are addressed to the European area (Switzerland, UK, Norway, Russia, and Ukraine; Eurostat database, 2023 https://ec.europa.eu/eurostat/web/products-eurostat-news/-/edn-20230406-1, accessed on 17 October 2023). Due to the versatility of the marketable ornamental geophytes and the evolving consumption trend, in recent years the ratio for bulb forcing and landscaping uses has varied in the different countries [1]; moreover, as with any other product of the floriculture industry, the ornamental geophytes have also had to adapt their quality to the "big-box" selling system, requiring new products, standard quality, availability of product quantity, consistency in price, and regular supply. Flower bulbs can address these challenges thanks to their great diversity in morphology, growth, and physiological responses to environmental factors, which allow for scheduled production. Moreover, the globalization of the horticulture trade has led to new production centers located in Latin America, Africa, and Asia, expanding the production of high-quality flower bulbs once limited in countries with temperate climates. Another important aspect that can enhance the value of flower bulbs in the global economic situation is the increasing use of these plants in landscaping or gardens to satisfy the new customer-driven approach in the ornamental industry [10]. On this basis, the current megatrend of "neo-ecology" has also increased interest in natural products. Consequently, drugs from medicinal plants are more frequently included in clinical treatments, and many herbal extracts are authorized to be used in therapy [20]. Flower bulbs are well recognized for their utility as food or medicinal and aromatic plants [18,21–24]. The World Health Organization (WHO), aware of the fact that a poor scientific literature is addressed to evaluate medicinal herbs and that in most countries the market for herbal medicines is poorly regulated, published monographs where scientific information on the safety, efficacy, and quality control of widely used medicinal plants is provided [25–29]. Ornamental geophytes are considered in these monographs, and it can be envisaged that the number of flower bulbs used for the extraction of natural products will increase due to the worldwide floristic diversity of these plants and the various underground organs that can be used for the extraction of bioactive compounds, such as alkaloids. For example, it is reported that an increase in research activity is foreseen for Narcissus due to the increasing demand for galanthamine-based oral medicines and other alkaloids (the so-called Amaryllidaceae alkaloids) that promise to be of value to medicine [30].

Although ornamental geophytes consist of more than 800 different genera, the industry is still dominated by a few genera. Besides the genera that comprise most of the worldwide bulb production (*Tulipa, Lilium, Narcissus, Gladiolus, Hyacinthus, Crocus*, and *Iris*), there are other ornamental geophytes of economic importance in the cut flower industry (e.g., *Freesia, Altroemeria, Hippeatrum,* and *Zantedeschia*). *Anemone* and *Ranunculus* are two important genera for cut flower production under Mediterranean conditions.

In the following paragraphs, we want to review the most salient and recent findings for the micropropagation of flower bulb crops by providing examples of successful application of tissue culture techniques and outlining concluding remarks on the state of the art of flower bulb in vitro culture and future perspectives.

3. Methodology

The literature review on geophytes micropropagation followed the https://guides.library.uq.edu.au/research-techniques/literature-reviews#s-lg-box-16401198 (accessed on 4 February 2024) and Pautasso [31] rules guidelines to ensure a transparent and rigorous methodology. The review process began with a clearly defined research question focusing on tissue culture and micropropagation methodologies in ornamental geophytes, with a specific emphasis on bulbous plants. A comprehensive literature search was conducted across relevant databases, including Web of Science, Google Scholar, PubMed, ScienceDirect, and Scopus, using predetermined search terms.

The inclusion and exclusion criteria were established to filter studies based on their relevance and quality. Initial screening involved reviewing titles and abstracts, followed by a full-text assessment of potentially eligible articles. Data extraction included key information on somatic embryogenesis, bulb growth, dormancy breaking, planting, and other relevant aspects (Table 1). The quality of the included studies was assessed, and potential bias was considered during the analysis. The relevant literature was saved in the paper management system, Endnote Library. The flow diagram presented in Table 2 illustrates the study selection process, and the resulting evidence was synthesized and presented in accordance with Pautasso [31] guidelines. The gray literature not available in traditional channels and not represented in indexing sources addressed to commercial applications was not considered in this review article. The paper concluded with a discussion on the challenges and opportunities in geophyte micropropagation, emphasizing the necessity of a multidisciplinary approach integrating biochemistry, physiology, and molecular biology to advance tissue culture protocols for diverse geophyte species.

Table 1. The determined flow before starting to write the literature review.

Inputs	Outputs
Research Question	Micropropagation methodologies, challenges, and opportunities in geophytes, flower bulbs
Databases	Web of Science, Google Scholar, PubMed, ScienceDirect, Scopus
Sectioning	The basic flow (history, economy, flower bulb propagation and challenges, micropropagation, flower bulb micropropagation, stages, challenges, opportunities, conclusion) was determined for the literature review, and titles were added

Table 1. *Cont.*

Inputs	Outputs
Searching	For each section, keywords were determined
Screening literature	First, titles and abstracts; selecting the quality and relevance Second, full text assessment
Paper Management	Endnote

Table 2. The keywords used for relevant and quality literature search.

Sections	Keywords
History	Geophytes history Flower bulbs history
World Ornamental Plants Sector–Flower Bulbs situation in the sector	Flower bulbs production rate Flower bulbs marketing Flower bulbs sector Ornamental plants sector Ornamental industry Flower market Ornamental plants sector reports
Flower bulb propagation and challenges	Geophytes propagation Flower bulbs propagation Flower bulbs traditional propagation Storage organs Flower bulbs seed propagation Flower bulbs vegetative propagation Flower bulbs micropropagation
Micropropagation	Micropropagation Somatic embryogenesis Organogenesis Plant regeneration Factors affecting plant regeneration Explant choice for plant regeneration Culture medium for plant regeneration Environmental conditions for plant regeneration
Micropropagation of Flower Bulbs	Geophytes micropropagation Flower bulbs micropropagation Flower bulbs tissue culture Flower bulbs in vitro culture Flower bulbs micropropagation stages Flower bulbs plant material preparation Flower bulbs mother stock material Flower bulbs surface sterilization Flower bulbs disinfection Flower bulbs multiplication Flower bulbs dormancy breaking Flower bulbs acclimatization Flower bulbs planting Tissue culture and artificial intelligence
Somaclonal variation	Flower bulbs somaclonal variation Geophytes somaclonal variation
Conclusion	Flower bulbs genome editing Flower bulbs CRISPR Flower bulbs

4. Flower Bulbs: Traditional Propagation and Challenges

Flower bulbs are geophytes with storage organs that are of several kinds, morphologically and physiologically (Figure 1). These storage organs can allow the plant to survive unfavorable periods (whether too dry, too hot, or too cold) [15,32,33]. Thanks to the nutrient reserves within these modified structures, plant viability and further plant development are ensured despite the severe external conditions [34,35]. Prior to the onset of favorable environmental conditions, the differentiation of new buds occurs, allowing for shoot development through the ongoing physiological processes in the underground organs. [18,33,36]. Bulbs go through distinct developmental phases: juvenile vegetative, adult vegetative, and reproductive. The transition to the adult vegetative phase is necessary for flower initiation, which occurs after several years in the juvenile phase for certain species like *Tulipa* and *Narcissus*. The shift to the reproductive phase is triggered by high temperatures, leading to flower bud formation. Dormancy follows, requiring a prolonged period of cold for dormancy release and preparation for spring growth. This life cycle is common among bulbous plants, including *Tulipa*, *Crocus*, and *Hyacinthus* [15,33,37].

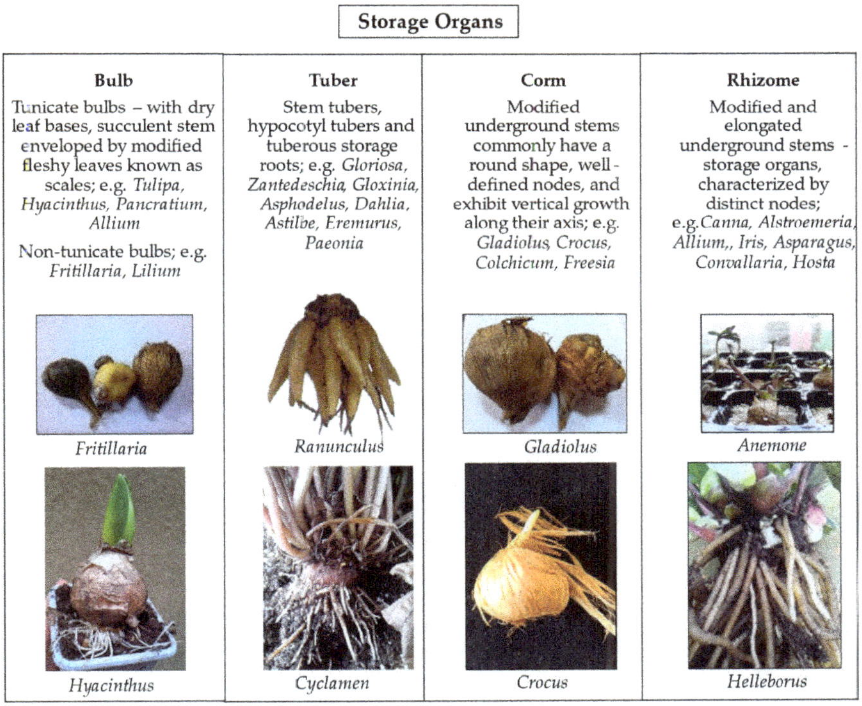

Figure 1. Types of storage organs in ornamental geophytes [32,33].

Geophytes can be propagated through generative (seed production) and vegetative methods [18]. The commercial production of flower bulbs through seed propagation is not widely preferred due to several reasons [36,38]. Firstly, seed production could result in offspring that are not identical to the parent plant, leading to increased diversity. This lack of uniformity makes this kind of propagation unsuitable for commercial production, except for specific seed-raised crops. Furthermore, the seed of some bulb plants has limited viability, and precise requirements for germination could be necessary to produce seedlings. Another difficulty in applying seed multiplication is the extended juvenile phase, lasting up to 5–7 years before flowering, which could be faced by certain species [38]. Nevertheless, seed production is highly valuable in breeding programs, allowing for the development of new cultivars. Seed propagation is particularly advantageous if a high number of seeds can be obtained, the juvenile phase is relatively short, and variability is not a concern compared to vegetative propagation [32]. Another advantage of seed propagation is that in almost all cases, plants produced are free of virus infection; vice versa, through vegetative propagation methods, there is the risk of spreading pathogenic microorganisms and compromising production. Examples of geophytes that can be commercially propagated through seed production include *Anemone, Allium, Begonia, Cyclamen, Freesia, Fritillaria*, and *Ranunculus* [18,33,36,38,39].

New techniques and strategies have been developed to improve the seed set, increase seed germination, and shorten the time required to flower [40]. Interesting findings were presented by Anderson [41] for reducing the generation time in *Lilium* and *Gladiolus* (less than 1 year from seeding to flowering). His laboratory developed a selection strategy to reach rapid generation cycling (RGC) in geophytes. A "toothpicking" technique by selecting germinated seedlings having different colors based on the week of germination was proposed to select early flowering hybrids of *Gladiolus*. In addition, comprehensive selection work was undertaken for all the phases of the life cycle of geophytes to correlate the features with the early flowering. The author's conclusions were that the geophytes breeding program can be shortened by selecting genotypes with good seed germination and seedlings with rapid leaf growth or with enhanced leaf unfolding rates, followed by further observations on the geophytic structures and contractile root generation, as well as the dormancy status and the subsequent stalk elongation and flowering.

In nature, the vegetative propagation of flower bulbs occurs through natural division. Natural division can be with daughter bulbs (*tulip, crocus*), offsets (*hyacinths, muscari*) bulblets (*lily*), bulbils (*lily*), cormels, rhizomes, branched rhizomes, and stolons (*allium oxalis*) [32,38,42]. The multiplication rates of the natural division are quite low [15,18,32]. For this reason, it has been envisaged that flower bulbs need alternative propagation methods. Traditional propagation methods include scaling, twin scaling, chipping, scooping, and scoring, as well as stem and leaf cuttings [42]. Scaling is the method where individual scales are removed from the bulb to produce bulblets. Twin scaling, a variation of scaling involves attaching two scales to a piece of the basal plate. Chipping, a mechanical technique similar to twin scaling, is employed to separate scales. Scooping involves removing the basal plate from the scales, while scoring entails cutting the basal plate. Additionally, stem and leaf cuttings can be used as propagation methods for certain bulbous plants. These techniques have been widely practiced by horticulturists and gardeners to propagate and expand their bulb collections, ensuring the continued availability and diversity of these plants [18,38,39]. It can also be considered a bulbous structure and propagation method. Tunicate bulbs can be multiplied using techniques such as offsets, scoring, scooping, coring, sectioning, and cottage. Non-tunicate bulbs, on the other hand, can be propagated through scaling after they have bloomed. Corms can be multiplied using cormels, which are small corms that develop between the old and new corms. It typically takes one to two years of growth for cormels to reach the size at which they can produce flowers. Tubers can be propagated by either planting the entire tuber or dividing it into sections, ensuring each section has at least one bud (eye). Rhizomes are divided into sections, each containing at least one shoot bud or active shoot. Some rhizomes have roots attached to their bottoms and can be individually planted in containers. Rhizomes can also be planted in nursery beds and used as a source of bare-root stock for planting or for cultural purposes such as basketry. Crown division is another method of propagation that differs slightly from dividing rhizomes [32,43]. However, these methods also present notable limitations. Firstly, generating a substantial number of plants within a practical time frame can be challenging. Secondly, the propagation process carries a significant risk of disease transmission. Tissue culture techniques offer effective solutions to conventional challenges in flower bulb propagation, and this is explained in Section 5.

5. Micropropagation

Micropropagation is a widely recognized method of vegetative propagation that allows for the production of numerous offspring plants. Podwyszyńska et al. [44] reported that the global annual production of ornamental plants through in vitro cultures has surged from 800 million to 2 billion in the last decade. According to a recent research report, the micropropagation market size is expected to rise by 2030, reaching a value of 2.8 billion USD (about 2.5 billion EUR; report "Micropropagation Market-Global Outlook and Forecast 2022–2027"-https://www.reportlinker.com/p06313600/?utm_source=GNW, accessed on 25 February 2024). Europe is one of the most prominent geographical segments in the global micropropagation market, with 140 commercial in vitro laboratories for which the production of ornamental plants is pivotal [44].

The micropropagation process involves multiplying plants under sterile conditions on a nutrient culture medium with a known composition and optimal artificial culture conditions [45]. Micropropagation represents an intricate plant multiplication technique involving substantial capital investment, significant operational costs, and a need for specialized expertise [46–48]. However, tissue culture methods offer various benefits, particularly in micropropagation, which is employed to multiply diverse plants, including genetically modified or conventionally bred ones. It proves useful for producing plantlets from seedless or challenging-to-reproduce plants economically and quickly, significantly reducing the time needed for abundant plantlet production [36].

5.1. In Vitro Regeneration Pathways

In vitro plant regeneration is a process where explants, through cell division and differentiation, give rise to the formation of organs and tissues [49]. This comprehensive process involves the regeneration of an entire plant from various sources, including adult tissues, organs, unorganized calli, or even a single cell [50,51].

Somatic embryogenesis is a remarkable process in which plant somatic cells undergo dedifferentiation, acquire embryonic cell characteristics, and subsequently, through embryonic development, form complete plants [52–54]. This process highlights the totipotency of plant cells, demonstrated by the formation of embryogenic calli [51,55]. It involves the development of structures resembling zygotic embryos from cells that are not associated with zygotes or the original tissue, closely mirroring the characteristic stages observed during zygotic embryogenesis [54]. Somatic embryos can differentiate through two pathways: direct somatic embryogenesis, where they differentiate directly from explants without the formation of a callus, and indirect somatic embryogenesis, where differentiation occurs after passing through a callus stage [56]. Somatic embryogenesis is favored for mass propagation due to its higher proliferation rate, ease of liquid culture medium utilization, efficient handling of numerous embryos, and greater adaptability for bioreactors [36,49].

Freshly formed structures like shoots, roots, or embryos can emerge on plant tissues without preexisting meristems. These recently developed organs are termed adventive or adventitious, and this initiation of a novel structure and organization is termed organogenesis [57,58]. Organogenesis involves the perception of phytohormones, differentiation of specialized cells to gain organogenic competence, cell division regulation, cell expansion, and the overall patterning of the organ [59–61]. The presence or absence of the callus stage results in two distinct developmental stages. The developmental stage that includes the callus phase is called indirect organogenesis. In indirect organogenesis, the likelihood of observing somaclonal variation is higher. The developmental stage that occurs without the callus phase is termed direct organogenesis [62,63]. The first step in the propagation of plant material using the organogenesis technique is the selection of the explant source from the donor plant. Usually, embryos, seedling parts, apical meristems (root or shoot tips), primordial organs, protoplasts, and young flower buds are used [58,61].

Many factors can affect the efficiency of the regeneration pathway, such as explant type, genotype, concentration, and type of PGRs, regeneration medium, other chemicals that indirectly affect plant growth, stress conditions, subculture, electrical excitation, or gene expression modification [51–53,64,65]. The shoot regeneration capacity depends on the establishment of a complex process involving the endogenous cytokinin metabolism of tissues and how the explants react to the exogenously supplied plant regulators in the culture medium [60]. Consequently, many factors should be considered when an explant is chosen to initiate an in vitro culture. Beside the medium composition and the culture conditions, the physiological and ontogenic age of the tissues, the season in which the explant is taken, the size and location of the explant, and the quality of the donor plant are key factors, which affect the explant viability and the degree of contamination in in vitro culture [51]. In addition, explant age and origin, as well as the period of the year in which the in vitro culture is initiated, affect the phenolic content of the tissues, which, in turn, can cause browning and subsequent death of tissues [66]. Figure 2 shows the aforesaid regeneration pathways.

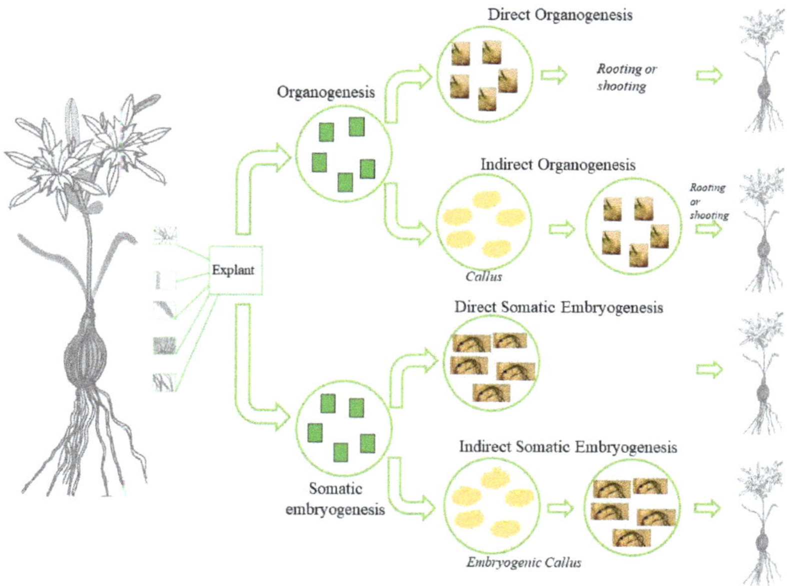

Figure 2. In vitro regeneration pathways.

5.2. Micropropagation of Flower Bulbs

Micropropagation of ornamental plants has been studied since the 1950s [67,68]. The impact of this technology on flower bulbs is also testified to by the literature cited in this review. Micropropagation involves axillary bud development through organogenesis, adventitious shoot formation, and somatic embryogenesis [69].

Plant biotechnology significantly contributes to the production of ornamental geophytes through various approaches. These methods focus on propagating specific genotypes, acquiring virus-free plant material, and supporting breeding and crop improvement programs [18,36,69,70]. Techniques such as callus culture, embryo rescue, in vitro pollination, somatic hybridization, induction of somaclonal variation, protoplast culture, synthetic seed production, in vitro ploidy manipulation, genetic transformation, gene mapping, and DNA fingerprinting play essential roles [22,69,71]. In vitro pollination and fertilization, embryo, ovary, ovule, anther, and pollen culture have been reported for some geophytes such as *Lilium*, *Cyclamen*, *Anemone*, *Ranunculus*, and *Amaryllis* [72–80]. Further scientific inquiry is warranted to evaluate the feasibility and potential applicability of protoplast culture and somatic cell hybridization techniques in the context of geophytes such as *Crocus cancellatus* as studied by Karamian and Ebrahimzadeh [81]. Recently, Koetle et al. [82]

reported about the procedures using *Agrobacterium* strains to transform into geophytes such as *Crocus, Allium, Agapanthus, Lilium, Tulip, Gladioulus, Hyacinthus,* and *Narcissus* [83–92].

Micropropagation systems were developed for major bulbous plants, but commercial propagation was achieved only for a few bulbous plants, such as *Lily* and *Zantadeschia* [36]. For other ornamental geophytes, more efforts are necessary to reach commercial applications. Nevertheless, it is well recognized that the availability of proper tissue culture protocols would enable the production of high-quality stock plant material that is virus-free and the advancement of breeding programs or the propagation of selected genotypes.

5.3. The Key Factors Affecting Micropropagation of Flower Bulbs

Establishing a successful micropropagation protocol is a complex task, influenced by various interacting factors, including plant materials, culture conditions, and culture media ingredients. Despite the development of micropropagation protocols for many flower bulb crops, challenges hinder commercial viability. Key limitations include the high cost of tissue-culture plantlets, inefficiencies in the initiation phase, low propagation rates for numerous genotypes, challenges during the acclimatization of ex vitro plantlets, and the occurrence of off-types.

5.3.1. Explant Choice

Thanks to its totipotent character, the potential for entire plant regeneration exists within every living plant cell. However, cells or tissues displaying active growth and robust physiological metabolism are commonly utilized as explants for the regeneration process in many studies [51]. A variety of tissue sources, including shoot tips (apical and axillary buds), bulb scales, leaves, stems, and different parts of the inflorescence, are utilized as explants in the initial stages of in vitro culture [38,93]. Below, in Section 5.4.2, some examples of the establishment of an in vitro culture for flower bulbs are provided.

As previously explained (Section 5.1), the selection and preparation of suitable explants play a crucial role in successful in vitro culture. Factors such as the regeneration capacity, physiological state, and hormonal balance of the donor plant, the origin, age, and size of the explants, their polarity, and pre-treatment methods are key considerations for the successful initiation of ornamental bulbous plants [94]. The induction of in vitro bulblets is primarily influenced by the temperatures and storage durations of the mother plant. Yasemin et al. [95] emphasized the significance of explant types regarding factors such as callus color, callus hardness, callus fragility, callus formation rate, embryogenic callus rate, and callus growth rate of *Pancratium maritimum*.

5.3.2. Culture Medium

The composition of the culture medium is a critical factor influencing the growth and morphogenesis of plant tissues in culture. The Murashige and Skoog (1962; MS) [96] formulation is widely used, initially designed to support the optimal growth of tobacco calli through extensive dose–response curve studies for essential minerals. Furthermore, N6 [97], Woody Plant Medium (WPM) [98], and B5 [99] are used as culture media in tissue culture [51]. Plant tissue and cell culture media are generally made up of some or all of the following components: macronutrients, micronutrients, sugar(s), vitamins, amino acids or other nitrogen supplements, other undefined supplements, solidifying agents or support systems, and growth regulators [100]. Undefined elements can be coconut milk, meat, malt, yeast extracts, potato extracts, juices, pulps, fruit extracts, plant/plant pars/seedling extracts. Osmotic balance, pH, and buffers of the culture medium also affect the target in the tissue culture [101]. Elicitors, such as chitosan, aminolevulinic acid (ALA), alginate (ALG), N-acetylglucosamine (NAG), salicylic acid (SA), hyaluronic acid (HA), silver nitrate ($AgNO_3$), jasmonic acid (JA), methyl jasmonate (MeJA), phloroglucinol (PG), and pectin, are extensively employed to induce secondary metabolites in plant tissue culture [71]. Methyl-Jasmonate (MeJA) and various polyamines also prove advantageous in tulip tissue culture by promoting efficient bulb formation during micropropagation [102]. Moreover, the medium's physico-chemical properties can alter the nutrient status of microshoots. The selection of the gelling agent is a crucial factor in the process of in vitro plant regeneration. The medium should be sufficiently firm to support explants, avoiding excessive rigidity that may hinder proper contact. The phenomenon of hyperhydricity, linked to agar, is also a recognized concern [103]. The experiments carried out over a ten-year period, highlighted that a relationship can be found between gel properties and the biological performance of in vitro *Ranunculus* shoots, and we correlate it with the different water and nutrient availability created in the different gelled media [104]. Moreover, our studies identified the concentration and diffusion through the gel of major phenolic impurities that can be found in the agar powder used to jellify the media. These findings proved that agar impurities can greatly affect the growth and multiplication of microshoots (Figure 3). Therefore, attention should be paid to choosing the agar brand and concentration and, even more importantly, creating a standardized method for the preparation of the gelled media.

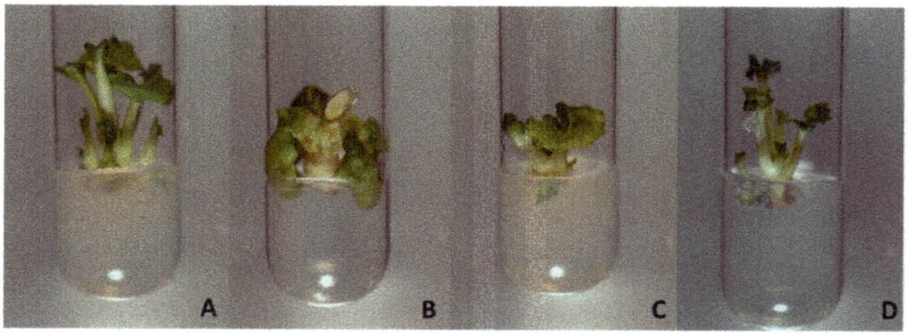

Figure 3. Biological performance of *Ranunculus* shoots cultured in vitro on basal MS medium gelled with different agar brands. (**A**) agar Oxoid (OX; cd L13(3)), 8 g/L; (**B**) agar Roth (RT; cd 4508), 8 g/L; (**C**) agar OX (8 g/L) supplemented with the impurities collected from dialysis of 8 g of RT agar powder; (**D**) washed agar RT, 18 g/L. It is possible to see that the gelling agent is responsible for quite different performance, with the growth on OX gels being satisfactory (**A**), while poor development and hyperhydricity were observed when RT gels were used (**B**). Removing the impurities from RT agar and increasing the agar concentration to reach a good gel firmness also allowed for good growth on RT gels (**D**). Vice versa, adding the RT impurities to OX gels decreased the quality of the microshoots (**C**).

The use of solidified media in micropropagation offers advantages, including easy visibility and recovery of small explants, maintained explant orientation, improved aeration without special measures, and orderly growth of shoots and roots. In contrast, liquid media can cause disoriented growth and difficulties in shoot separation. However, there are drawbacks to semi-solid media, such as the presence of the above-said inhibitory substances in agars, slow growth rates, and limited diffusion of toxic exudates. Poor oxygen diffusion to developing roots is a concern, even in aerated liquid media. Adherence of gel to roots can pose issues during plantlet transfer to soil, and cleaning containers for re-use becomes time-consuming with semi-solid media [105]. Different plants have varying nutritional requirements, affecting optimal growth and morphogenesis. Tissues from different plant parts may have unique needs for satisfactory growth [51].

5.3.3. Environmental Conditions

Temperature and light have key roles in micropropagation. Under in vitro conditions, younger tissues typically import sucrose and have limited photosynthesis. However, light (color and intensity) influences plant growth, metabolism, and organogenesis and has proved to be crucial for the in vitro regeneration of the storage organs of geophytes [38,70,71]. White, fluorescent light (350~750 nm) is conventionally used in in vitro culture but has drawbacks like high electricity consumption and uneven radiation. Monochromatic LEDs with specific wavelengths are now widely used for more energy-efficient and controlled in vitro plant propagation [71,106]. Phytochrome governs the plant's light response, influencing organogenesis induction and direction. Blue light and its receptor, cryptochrome, play a role in storage organ formation [107]. Examining light quality effects during adventitious organogenesis in bulbous plants, researchers utilized monochromatic fluorescent lamps. The results revealed that in in vitro cultures of *Hyacinthus orientalis*, blue light stimulates adventitious shoot development, while red light encourages the formation of adventitious bulbs [108]. Studies have explored the impact of light on plant growth and development parameters in various lily species and cultivars during in vitro organogenesis. [109–111]. According to Bach and Sochacki [38], the choice of light may interact with tissue cytokinin metabolism, potentially influencing in vitro tissue dormancy and subsequent development. While light is crucial for typical green shoot and plantlet growth, unorganized cell and tissue cultures can thrive without it. Surprisingly, darkness may enhance growth and morphogenesis, with brief dark periods observed to promote shoot morphogenesis in certain cases [105]. Embryo explants from *P. maritimum* were cultivated under both photoperiod and dark conditions. The results, based on unpublished data from Yasemin S., indicate that the dark condition was more successful in generating calli, and Figure 4 illustrates the outcomes after one month of incubation.

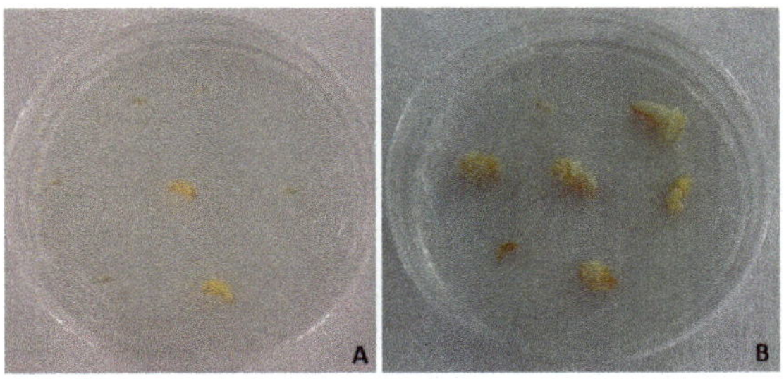

Figure 4. Embryo cultures of *P. maritimum* under 16/8 h photoperiod (**A**) and dark (**B**) conditions after one month of incubation.

Temperature influences both growth rates and the transition between vegetative and reproductive phases in plant development. In flower bulbs undergoing in vitro culture, temperature is a vital factor influencing regeneration, bulbing, and dormancy. For temperate ornamental geophytes, lower temperatures mimic winter, prompting dormancy. Under in vitro conditions, reduced temperatures induce storage organs to withstand unfavorable growth conditions [67]. Temperature is another key factor that is able to influence the regeneration, bulbing, and dormancy of flower bulbs during in vitro culture [112]. In vitro-regenerated bulblets necessitate dormancy breaking to facilitate ongoing development. A cold treatment was proven to be effective for bulblet initiation and for breaking dormancy in the regenerated bulblets [38].

5.4. *Stages of the Micropropagation of the Flower Bulbs*

Many scientific papers have been published on flower bulbs, highlighting the discrepancy between the potentiality of the micropropagation technique and its practical application. Next, we would like to analyze the most important findings related to the major steps in micropropagation of flower bulb crops, which are: preparation of mother plants, initiation, multiplication, bulbing/bulb growth, dormancy breaking, and planting. Dormancy breaking is a different step from the other herbaceous and woody plants [18,36].

5.4.1. Stage 0: Preparation of Mother Stock Plant Material

Happy plants make happy plantlets. Stage 0 in ornamental geophyte propagation represents the initial and crucial phase dedicated to ensuring the selection and cultivation of healthy plants [47]. This pivotal stage aims to pre-process the chosen specimens to prevent contamination, ultimately fostering the development of thriving plantlets [42]. Key pre-treatments, such as regulating humidity, irrigation methods, temperature, and light exposure, are implemented to curtail the potential source of contamination and maintain the genetic integrity of the desired traits. Additionally, the duration of storage impacts the formation of bulb-like structures in geophytes, further emphasizing the significance of meticulous handling during this phase [18,46,70,113–116].

To initiate the in vitro culture of tulip, Podwyszyńska and Sochacki [117] selected healthy, true-to-type, and virus-free plants during their flowering phase in spring. After having carried out ELISA tests to verify the virus-free status of their plants, they harvested the bulbs in June and stored them at 17–20 °C until October, followed by dry cooling at 5 °C. The selected healthy bulbs (the recommended bulb size is 10 cm in diameter) were then planted in pots with a fungicide (Captan)-soaked perlite substrate and forced in darkness at 9 °C for 5–14 days prior to initiating the in vitro culture. The effect of a bulb pre-treatment at 5 ± 2 °C for a period of 3–4 weeks was also positively referred by Muraseva and Novikova [118], who developed an efficient protocol for in vitro propagation from bulb scale explants of *Fritillaria ruthenica* Wikstr., an endangered and rare species. In other cases, natural conditions made it advisable to prepare the mother plants. Kumar et al. [119] successfully established an efficient plant regeneration system for *Lachenalia. viridiflora* via somatic embryogenesis by growing the mother stock plants in greenhouses under controlled conditions that closely resembled the natural habitat of this plant species, which is collected from a specific sub-population in St. Helena Bay, Western Cape, South Africa, credited to Prof. Graham Duncan. Around 10 mature bulbs of *L. viridiflora* were carefully selected and planted in terracotta pots with a standardized diameter of 200 mm. The potted plants were then nurtured under ambient temperature conditions and exposed to the natural photoperiod in the controlled environment of the University of KwaZulu-Natal Botanical Gardens' greenhouse. The researchers took special care to ensure regular watering and maintained a weed-free environment to promote optimal growth and development, mimicking the plants' natural conditions as closely as possible. Mirici et al. [120] studied in vitro bulblet regeneration from fresh bulb scale and immature embryos of the endangered geophyte *Sternbergia fischeriana*. Before explant sterilization, the collected bulbs were subjected to a drying process in the dark at room temperature for a duration of 6 weeks.

In conclusion, it could be argued that closed environments, such as glasshouses and tunnels, are high-quality places for plant cultivation with a reduced risk of contamination. Treating plants with fungicides and insecticides before obtaining plant material helps to decrease the risk of contamination. Various practices are available to enhance in vitro development, altering the physiological state of the parent plant from which explants will be taken. These practices include implementing long-day treatments, cultivating under red light, maintaining bulbous plants at low temperatures, and applying PGRs to the leaves [48]. By focusing on the careful selection and pre-processing of mother plants, stage 0 lays the foundation for a successful propagation process, leading to the cultivation of healthy ornamental geophytes [47].

5.4.2. Stage 1: Establishment of Aseptic Culture

In this stage, the selected explants are excised from the stock plant material and surface sterilized with certain chemicals before they are inoculated in the culture medium. Surface sterilization of plant explants for successful establishment in in vitro conditions, particularly in the context of plant tissue culture, is very important. The primary goal is to eliminate microbial contamination, both exogenous and endogenous, without causing harm to the plant tissues. The success of the sterilization process is crucial for the subsequent stages of tissue culture [121–124]. Various chemicals, such as antibiotics, fungicides, sodium hypochlorite, mercuric chloride, ethanol, hydrogen peroxide, calcium hypochlorite, and silver nitrate, are commonly used for surface sterilization. The concentration and duration of the disinfection process need to be carefully considered to achieve effective sterilization without compromising the viability and regeneration ability of the explants. The choice of disinfectants depends on the specific plant species and type of explant. The goal is to create aseptic conditions for plant tissue culture by addressing the challenges associated with microbial contamination and maintaining the health of the explants [122,125]. The use of seeds as an initial material in in vitro studies is advantageous because it decreases the risk of contamination. After germination of the seeds, in vitro seedling parts can be used as explants for some research [124,126,127]. Yasemin et al. [124] conducted the surface sterilization experiments for the seeds of *P. maritimum*. They used 70% ethanol and washed the seed with distilled water, followed by treatment with different concentrations of sodium hypochlorite (NaOCl 1.6% and 3.2%) at different durations (15, 20, and 25 min). High NaOCl and long durations had negative effects on the seeds viability, germination, and plant formation. Kumar et al. [119] described the successful surface sterilization process for *L. viridiflora* leaves. The leaves were treated on a laminar flow bench using 70% ethanol (v/v) for 60 s, followed by immersion in 2% sodium hypochlorite (NaOCl) for 10 min. To aid in the process, a few drops of Tween 20 were added as a surfactant. After sterilization, the leaves were rinsed three times with sterile distilled water.

Unfortunately, the contamination issue represents a substantial concern for the flower bulb tissue culture, particularly when underground plant tissues are used to initiate the culture. Although surface sterilization is successful, the plant tissues can be contaminated due to endogenous microorganisms in the bulbs, and this could also appear in further culture cycles [36]. Therefore, a thorough sterilization process is indispensable to eliminate contaminants from the geophytic storage organ explants. In the study conducted by Lagram et al. [128], a protocol was employed for the preparation of saffron daughter corms (*Crocus sativus* L.). The corms were subjected to a series of treatments, including a 30 min immersion in tap water, gentle brushing with Tween-20, a 1 min exposure to 80% ethanol, a 20 min treatment with freshly prepared mercuric chloride ($HgCl_2$) at a concentration of 0.2%, and, finally, rinsing four times for 4 min each with sterile distilled water. Sochacki and Orlikowska [129], in their studies aimed at improving the micropropagation protocol for two *Narcissus* cultivars, tested several pre-treatments of bulbs, including the use of fungicides and hot water treatment (44.4 °C for 3 h), followed by a disinfection procedure with $HgCl_2$ and chloramine T. By using these methods, they were successful in reducing the initial contamination to 14–17%, and they showed that different genotypes could differently react to the same treatment.

Seeds of *P. maritimum* were used as explants in some studies, and surface sterilization was successful [95,124,126,130]. However, surface sterilization of the flower bulbs is not quite easy. According to studies conducted by Yasemin S (unpublished data), *Gladioulus* corms and *Fritillaria* bulbs were washed under tap water, then treated with 70% ethanol and 2.5% NaOCl and washed by sterile distilled water (three times). Unfortunately, all the explants were contaminated due to the fact that the mother plant was already not healthy. This could show the importance of mother plant material selection. Some explant photos are shown in Figure 5.

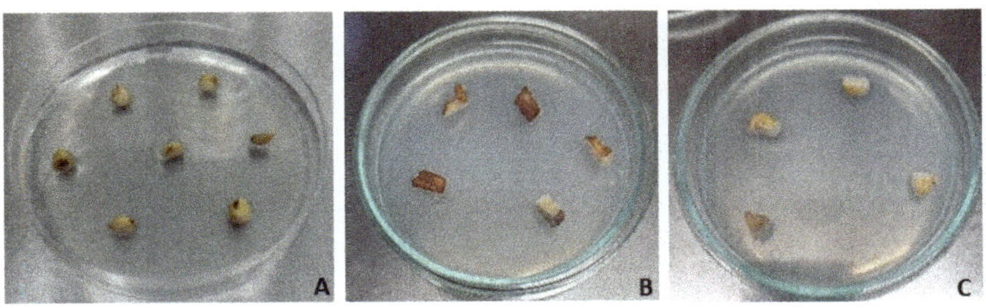

Figure 5. Different explant examples from in vitro propagation of flower bulbs: sea daffodil seeds (**A**), *Gladioulus* corm (**B**), and *Fritıllaria* bulbs (**C**).

Evaluating the negative impacts of sterilizing agents on plant tissues is crucial, with explant survival data serving as a key parameter. Rafiq et al. [131] and Rather et al. [132] indicate that combined chemicals were more effective for the achievement of disinfection, but the use of combined sterilants lead to a reduction in explant survival. Rafiq et al. [131] experimented with diverse sterilization approaches, incorporating NaOCl (1%) and $HgCl_2$ (0.1%) at varying durations (10 and 20 min), along with combinations involving 70% ethyl alcohol and carbendazim (200 ppm). The explants were bulb scale-base and bulb scale-tip from the oriental hybrid *Lilium* cv. Ravenna. The most effective surface sterilization method entailed treating healthy bulb scales with carbendazim (200 ppm) for 30 min, followed by 0.1% $HgCl_2$ for 10 min, and then 70% ethyl alcohol for 30 s. Farooq et al. [133] used bulb scales and young leaves to initiate the culture of *Lilium* LA hybrids 'Indian Summerset' and 'Nashville'. A pre-treatment of both explants has been made by washing the tissues in a water solution containing Tween-20 and the fungicide carbendezim. Afterwards, the surface sterilization of the explants was performed using carbendenzim, $HgCl_2$, and ethyl alcohol at different concentrations and durations. The highest surface sterilization success for bulb scales (Indian Summerset: 85.41%, Nashwille: 89.58%) was higher in the combinational sterilant. Chib et al. [84] also detailed an optimized two-step surface sterilization method for *C. sativus* L. This involved separately employing 0.1% $HgCl_2$ and 4% NaOCl, leading to effective asepsis and a noteworthy 86% survival rate of explants. The disinfection of corms was successfully achieved through this process. Application of antioxidants, such as ascorbic acid and citric acid, can effectively prevent browning caused by oxidation. Furthermore, some additives, such as polyvinylpyrrolidone (PVP), charcoal and fungicides, can also prevent the browning of explants and endogenous contamination risks. Appleton et al. [134] performed in vitro regeneration of *Hypoxis colchicifolia*. When establishing in vitro cultures, they have experimented with PVP, activated charcoal, ascorbic acid, citric acid to inhibit browning, and benomyl solutions as fungicides to inhibit contamination. They have solved the browning problem with PVP and partially solved the contamination problem with benomyl. However, these applications do not create the same response in every plant or plant part. This should be analyzed and optimized for each plant. Devi et al. [135] have used charcoal and ascorbic acid to reduce the phenolic exudates, but charcoal inhibited tissue growth, and ascorbic acid caused somatic embyo death in saffron (*C. sativus* L.).

In general, young tissues and organs have a higher regeneration capacity than older ones. Apical and axillary buds from tubers or bulbs can be used to initiate the in vitro culture. Furthermore, adventitious buds are generally induced by bulb scales or flower stems. Twin scales are useful explants, although the use of underground storage organs leads to serious contamination problems during the initiation and further multiplication phases. Flower stems have the advantage of having low endogenous contamination rates compared to other tissues [18,24,36]. Vegetative segments of plants often regenerate more easily in vitro than generative ones. Rafiq et al. [131] used basal and tip bulb scales to micropropagate the oriental hybrid *Lilium* cv. Ravenna. They also found that the explant survival rate of basal scales was higher than that of tip bulb scales. Lapiz-Culqui et al. [136] and Patil et al. [137] used the bulb scales to cultivate different *Lilium* cultivars. Youssef et al. [138] used the leaf as an explant in their study to increase *Lilium* bulb number and size. Ozel et al. [139] used a twin-scale bulb explant to micropropagate the endemic *Muscari muscarimi*. Twin scales were also used by Santos et al. [140] to propagate *Narcissus asturiensis* and Kukulczanka et al. [141] to propagate *Fritilaria melagris* through the same system. Kumar et al. [142] used the bulb scales of the critically endangered *Fritillaria roylei* for in vitro culture. Sevindik and Mendi [143], Taheri-Dehkordi et al. [144] used the corms as explants to propagate the *C. sativus* L. Furthermore, Slimani et al. [145] informed us that the vegetative apices, apical and axillary buds, meristematic zone, segments, leaves, ovaries, protoplasts, corms, and roots are useful to induce the somatic embryogenesis of *C. sativus* L.

Success in this initiation stage can be influenced by environmental factors encompassing nutrient composition, PGRs, light exposure, temperature, atmospheric composition, and various culture methodologies [18]. Devi et al. [135] reported that the dark conditions were effective for somatic embryo proliferation in saffron.

5.4.3. Stage 2: Multiplication

The process of multiplication, which is a continuation of in vitro initiation, constitutes one of the most critical stages in micropropagation. As mentioned in previous stages, leaves, inflorescences, and bulbous structures (such as bulbs, corms, tubers, etc.) are utilized for multiplication [53,69]. The selection of the medium to be used (solid, liquid, semi-solid, or media types MS, B5, WPM, etc.) is crucial for achieving successful responses, in addition to the use of different explants. The optimization of factors such as salt mixtures (essential macro-micro elements, carbon sources, vitamins), sugar derivatives, ratios, types, and concentrations of PGRs, light, temperature, and inductive agents is necessary [69,70]. The concentration of the PGRs (e.g., cytokinin) is critical, as it can simultaneously promote multiplication and development while causing adverse effects [146–149]. Each step implemented here will influence both multiplication and subsequent planting stages; hence, optimal requirements need to be determined. Unfortunately, due to the variability in results obtained for each species and even within varieties, specific optimization efforts are required for each plant. This situation poses one of the challenges of tissue culture. Additionally, high production costs add to these challenges [150].

To achieve effective results at this stage, understanding the plant's physiology and mimicking its natural requirements based on the choices made during this phase can be impactful. The selections made during this stage lead to different differentiations according to the meristematic zones that will be formed as a result of stimulations in explants (somatic tissues) [69]. In the regeneration stage, unipolar meristems give rise to shoots or roots, which is organogenesis. Bipolar meristems, on the other hand, lead to the development of a complete plant or microcorm, representing somatic embryogenesis [151].

PGRs can stimulate or inhibit the in vitro development of geophytes. Sochacki et al. [152] aimed at evaluating the influence of PGRs on in vitro shoot multiplication of *Tulipa* L. 'Heart of Warsaw', and they yielded the best performance (9.14 shoots/clump) when MS medium was supplemented with N6-(-isopentyl)adenine (2iP) 0.1 mg/L, 1-Naphthaleneacetic acid (NAA) 0.1 mg/L, and meta-topolin (mT) 5.0 mg/L. Additionally, this study showed that the type of carbohydrate added to the nutrient solution had a significant effect on the shoot proliferation of the tulip cultivar; the highest multiplication of the shoots (number of shoots for one starting clump) was achieved with sucrose (34.33 shoots/per clump) and glucose (35.88 shoots/per clump) as carbohydrates. Lagram et al. [128], used excised mother corm buds under different 6-benzyl aminopurine (BAP), 2,4-Dichlorophenoxy acetic acid (2.4-D) and NAA concentrations in MS medium. The highest bud sprouting (96.67%) and shoot growth (8.87 cm) were obtained from 1 mg/L BAP and 1 mg/L NAA. They also evaluated adventitious shoot formation in the explants. The highest adventitious shoot regeneration (80%) was obtained from 0.5 mg/L NAA and 2.75 mg/L BAP. Recently, micropropagation via organogenesis has been on the rise. Successful adventitious bud regenerations were obtained in *Lilium*, *Narcissus*, *Sternbergia*, *Hippeastrum*, *Firttilaria*, *Muscari*, *Tulip*, *Iris*, *Lachenelia*, and *Hyacinthus* [111,112,118,120,138,139,153–162]. As seen in Figure 6, organogenic structures were obtained from bulb scales in *P. maritimum*.

Figure 6. Organogenic structure, leaf, and bulb growth from *P. maritimum* callus derived from bulb scales.

Slimani et al. [145] explained the importance of somatic embryogenesis for *C. sativus* L. to propagate healthy corms. MS, LS, and B5 media were supplemented with the growth regulators BA, NAA, Kinetin (Kn), Thidiazuron (TDZ), and 2.4 D tested at different concentrations. Ebrahimzadeh et al. [163], obtained somatic embryos by using meristem explants of *C. sativus* L. cultured in LS culture medium, which included 2 μM benzyladenine (BA) and 2 μM NAA. Sheibani et al. [164] used corm explants in MS medium supplemented with TDZ (0, 0.1, 0.25, and 0.5 mg/L), and they obtained somatic embryos from 0.5 mg/L TDZ. Marković et al. [165] delved into the intricate world of somatic embryogenesis in *F. meleagris*, emphasizing the crucial role of PGRs in this process. Their study illuminated that the medium without PGRs exhibited exceptional efficiency throughout the experiment, showcasing a robust somatic embryogenic response. Moreover, the medium with lower concentrations of PGRs also proved to be conducive to somatic embryogenesis. The pinnacle of morphogenetic success was achieved in a BAP/2,4-D-containing medium with the lowest PGR concentrations. This insight into the optimal conditions for somatic embryogenesis opens avenues for enhanced propagation strategies, potentially revolutionizing the large-scale production of *F. meleagris*. Kocak et al. [166] investigated somatic embryogenesis potential in various explants (ovules, divided ovary parts, leaves, and petiole segments) from 15 different genotypes of the wild species *Cyclamen persicum* Mill. The explants were cultured on a medium with specific PGRs to induce embryogenic callus. The study found significant variations in embryogenic potential among explants and genotypes. While petiole explants produced the most callus, ovary explants were most efficient in forming somatic embryos. Recently, an efficient method for tulip regeneration via SE was developed [167,168]. Some reports on SE are also available by using leaf, petiole, ovary, anther, roots, and aseptic seedling tissues as explants in cyclamen [166,169–174]. Aseptic seedling tissues (cotyledons, petioles, tubers, and roots) have also been used as explants to initiate SE culture in many medicinal and endemic geophytes (e.g., *Crocus*, *Iris*, *Hypoxis*, *Colchicum*, etc.) [135,143,163,164,175–178]. Some embryo-like structures from *P. maritimum* bulb explants are shown in Figure 7A–C. Moreover, embryogenic callus and regeneration in *R. asiaticus* were shown in Figure 7D,E. Thalamus-derived calli in *R. asiaticus* L. have been initiated on MS medium containing the growth regulator 2,4-D and cytokinins (BA and kinetin), according to Beruto et al. [179].

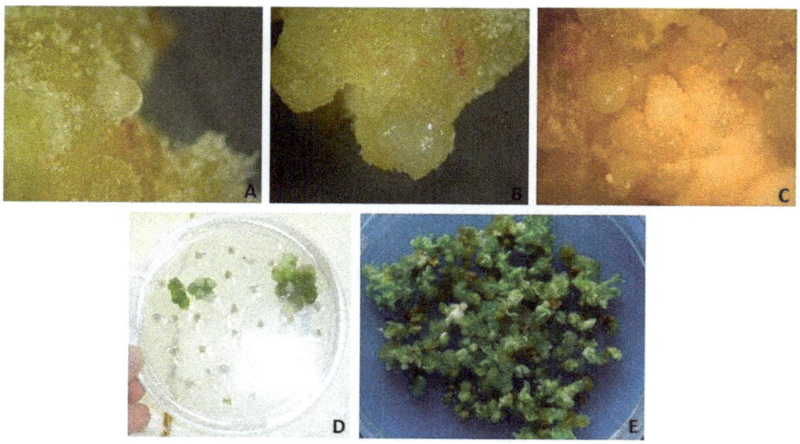

Figure 7. Embryo like structures (ELS) in the callus of *P. maritimum* (**A–C**) and the embryogenic callus and regeneration in *R. asiaticus* (**D**,**E**).

5.4.4. Stage 3: Bulb Growth

Following the multiplication phase, in non-bulbous plants, the optimization of media plays a crucial role in enhancing the successful rooting of developed shoots. However, this phenomenon is not uniformly applicable to geophytes, where the acquisition of storage organs proves to be challenging and time-intensive [152]. This phase in geophytes is directed towards the acquisition of shoots, primordia, and bulbs (Figure 8).

Figure 8. Bulbs formation from seeds and callus derived from bulb scales in *P. maritimum*.

In some plant species, the sequence involves the initial rooting of shoots followed by subsequent bulb formation, while in others, bulbs are generated directly [18]. The induction of bulb formation occurs through diverse mechanisms. Investigations into bulb formation have explored the influence of various sugar types and concentrations, PGRs, light quality, temperature, and distinct tissue culture conditions (solid, liquid, medium, and bioreactor utilization). Sochacki et al. [152] conducted an investigation on the bulb formation of previously multiplied *Tulipa* L. 'Heart of Warsaw' plantlets, examining various carbohydrate types, different phase media, and PGRs. The highest bulb formation (28.00 bulbs) and the most mature bulbs (14.50 bulbs) were achieved in a two-phase (liquid) system with 1 mg/L PBZ and glucose. Pałka et al. [111] utilized bulb scales of *Lilium candidum* as explants in in vitro MS media and assessed the impact of different light qualities (spectral compositions of red (100%), blue (100%), and red and blue light (RB ratio 7:3), RB light was mixed in equal proportions (50%) with green (RBG), yellow (RBY), UV (RBUV), and far-red (RBfR) light, white LED (Wled), fluorescent lamp light (Fl), and darkness (D)) on bulb formation without the use of PGRs. The study resulted in a bulb formation rate of 79–100% across all explants. The best results in terms of bulb number were obtained from Fl (16.3), RBY (13.33), and RBfR (13.33), B (12.69), RB (12.63), and Wled (14.36) treatments. The lowest rate (3.00) was observed in plants kept in darkness. The largest bulb diameter (5.41 mm) and the highest photosynthetic pigment content were obtained from the RBG treatment. Darkness and red light induced etiolation. Additionally, soluble sugars in bulbs were stimulated by darkness and blue light. Lagram et al. [128] used different sucrose, IBA, and NAA concentrations on corm production and root regeneration in $\frac{1}{2}$ MS. Furthermore, they considered the photoperiod effect (16 h/8 h and dark conditions) on the parameters. They obtained the best results in $\frac{1}{2}$ MS, 6% sucrose, 1 mg/L NAA, and dark conditions with 100% corm production, 7.9 g mini-corm weight, 93.8% root formation, and a root number of 14.9 per mini-corm. In the realm of bulb growth in *F. meleagris*, Marković et al. [165] unraveled key factors influencing the development of bulbs in vitro. Their investigation revealed that bulbing ability remained relatively high in the absence of PGRs and, for the most part, was unaffected by varying concentrations of the tested PGRs. Notably, the study identified that the lowest cytokinin concentration, in combination with low auxin, significantly enhanced bulb formation when the cultures were transferred to a higher temperature. This finding underscores the importance of temperature modulation and specific PGR combinations for optimizing bulb growth in *F. meleagris*. Understanding these dynamics can contribute to refining propagation methods and accelerating the scale-up of bulb production for this species. Azeri and Öztürk [180] investigated the most effective hormone treatment to induce and produce *Lilium monodelphum* M. Bieb, var. Armenum in tissue culture rapidly and efficiently. Optimal bud regeneration (11.67) was achieved in a medium containing 3.0 mg/L TDZ, 0.25 mg/L NAA, and 0.1 mg/L GA3. Additionally, the highest microbulb formation (15.83) was observed in a medium comprising 2.0 mg/L PAC, 0.2 mg/L NAA, and 0.1 mg/L GA3.

In recent years, the use of bioreactor systems has shown an increase in healthy plant regeneration. Studies on the regeneration of bulbs, corms, rhizomes, microtubers, shoots, and subsequent rooting, as well as somatic embryogenesis, have been on the rise [181]. In their published article, Murthy et al. [181] indicated the proliferation of *Lilium* hybrids and *Allium sativum* bulbs, *C. sativus* and *Alocasia amazonica* corms, microtubers of *Solanum tuberosum* varieties, and rhizomes of *Cymbidium sinense* using various bioreactor systems [110,182–194]. According to Murthy et al. [181] the selection of an appropriate bioreactor system is crucial due to various factors (design, principle, inoculation density, aeration, temperature, light intensity, etc.) that influence the regeneration of propagules. To minimize losses of small bulblets and prevent dormancy, direct transplantation of plantlets is often preferred.

Rooting can be improved through various methods, such as adding auxin and/or activated charcoal (AC) to the culture medium, adjusting the auxin-to-cytokinin ratio, and using half-strength salts and sucrose. In the study by Azeri and Öztürk [180], microbulbs were subjected to rooting media with indole 3-butyric acid (IBA 0.5 and 1.0 mg/L). Successful plant development, with root numbers of 6.4 and 5.9, respectively, was observed in media containing 0.5 and 1 mg/L IBA.

For successful acclimatization, hardening is essential. This process enhances tolerance to moisture stress and prevents hyperhydricity [18]. In Rafiq et al. [131] study on Oriental *Lilium* hybrid cv. Ravenna, varying combinations of IBA and NAA significantly influenced microshoot rooting behavior. IBA outperformed NAA, with the highest rooting (92.71%) and primary root characteristics observed in a medium supplemented with 1.50 mg/L IBA. Explants showed differences, with basal scale segments exhibiting the highest rooting (92.71%), percentage root number/shoot (10.02), and length of primary roots (2.17 cm) while tip scale segments had the lowest (77.55%, 9.08, 1.65 cm). Notably, primary hardening influenced plantlet survival, with the highest ex vitro survival (98.96%) in plantlets from IBA (1.5 mg/L)-fortified media, particularly in basal scale segments. Rooted plantlets were hardened in media containing perlite and vermiculite (1:1). This suggests that IBA-treated cultures resulted in superior ex vitro survival, possibly due to enhanced rooting characteristics such as root number and length.

Direct bulblet induction in certain genotypes can provide several advantages, including the elimination of in vitro rooting, prevention of hyperhydricity, avoidance of the need for hardening, increased survival rates, and a shortened bulb production period. Conditions conducive to bulb formation involve high sucrose concentrations, the application of plant growth retardants, exposure to low temperatures, and the utilization of aged shoots. It's worth noting that in vitro formed bulblets may experience spontaneous dormancy, as reported by Kim and De Hertogh [18]. In the research conducted by Chib et al. [84], they experimented with a high sucrose concentration combined with PGRs to enhance efficiency over a 90-day timeframe. The most favorable outcomes were observed when using a combination of MS medium, TDZ, IAA, activated charcoal, and 4% sucrose, resulting in a notable 68% efficiency. Lower sucrose concentrations led to sluggish growth, while concentrations exceeding 4% resulted in cell death, as evidenced by the blackening of the callus. Different sucrose, BA, and 2,4D concentrations in MS media affected the growing bulbs in in vitro conditions for *P. maritimum* [95].

5.4.5. Stage 4: Dormancy Breaking

Some geophytes produce new buds that enable vegetative propagation through their underground storage organs [195,196]. After an active growth and flowering period, senescence of aboveground tissues is followed by root senescence, leading the plant into a dormant phase with no visible organogenesis. To survive under unfavorable environmental conditions that are not conducive to their development and proliferation, these organs enter a state of dormancy. Seeds, apical and vegetative buds, floral buds, bulbs, corms, and tubers can enter dormancy. Most geophytes show dormancy as an integral part of their life cycle [197]. Dormancy is characterized by the inability to initiate growth from meristems under favorable conditions. In geophytes, dormancy can manifest in three different types: endodormancy (internal inability to grow), ecodormancy (environmental conditions), and paradormancy (apical dominance, hormonal status, metabolite-sugar levels) [196,198–200].

When evaluating the concepts of dormancy and true dormancy, it is important to note that true dormancy persists until dormancy is completely terminated, even if favorable environmental conditions are present. Unlike many other plants, geophytes do not exhibit true dormancy, as activities continue even during dormant phases. Most bulbous structures obtained through tissue culture display dormancy. During in vitro regeneration, bulbs and other storage organs such as tubers and corms typically undergo dormancy, similar to their behavior in natural conditions. The level of dormancy may vary depending on factors such as sucrose concentration, age of the bulb, and environmental conditions [18,116,201]. The growth and sprouting of geophytes in vitro are influenced by dormancy, bulb size, and maturity [202]. Therefore, an efficient in vitro protocol, including dormancy release, is crucial for the rapid, efficient, and valid commercial exploitation of all horticultural geophytes. To overcome this constraint, in some *Fritillaria* species, different temperature regimes were tested under in vitro conditions. Kizil and Khawar [158] found the bulblet diameter increased on MS medium with 50 mg/L sucrose after 30 days at 4 °C. Successful rooting of *Fritillaria* bulblets was achieved on MS medium with 0.5 mg/L NAA. Marković et al. [165] carried out bulb-scale culture with different PGR combinations for four weeks at 7 °C in *F. meleagris*. They found that in the control medium (PGR-free medium), shoots per explant were higher and reached a maximum at the end of the chilling. In the study conducted by Carasso and Mucciarelli [203], seeds of *F. tubiformis* Gren. & Godr were chilled at 4 °C for 30, 60, and 90 days in 1% agar medium. At the end of cold stratification, seeds were disinfected, and immature zygotic embryos were rescued from seed coats. The highest number of somatic embryos occurred in zygotic embryos subjected to 30 days of cold stratification on MS medium supplemented with 8.88 μM BA and 2.68 μM NAA. Successful conversion of somatic embryos into bulblets necessitated transfer to a maturation medium supplemented with 4% (w/v) sucrose. In the presence of 4.92 μM IBA, bulblets sprouted and developed roots, leading to the establishment of newly formed plants suitable for pot cultivation. Çakmak et al. [162] obtained the *F. persica* in vitro bulblets from MS medium containing 20 g/L sucrose. They found major constraints during the acclimatization stage unless a cold treatment (4 °C for 2 months) was applied for dormancy breaking of bulblets.

Bulblets and other storage organs produced in vitro are easily handled, transported, and stored, which makes them preferable propagation plant material. The dormancy period facilitates global commercial handling, and, consequently, the precise regulation of geophyte dormancy and geophyte dormancy release becomes imperative for the efficient management of their production, shipping, and utilization [199].

5.4.6. Stage 5: Ex Vitro Acclimatization and Growth

Transplanting ex vitro plantlets to greenhouse conditions requires specific care, a gradual lowering in air humidity, and a progressive adaptation to high light levels [18]. However, bulblets and other storage organs produced in vitro do not require an extensive acclimatization procedure after transfer to soil. In the last step of micropropagation, in vitro bulblets are produced for many flower bulbs: *Hippeastrum* [157], *Hyacinthus* [204], *Iris* [205], *Lilium* [138], *Muscari* [139], *Narcissus* [206], and *Tulip* [207]. In some cases, such as lily and hyacinth, in vitro bulblets are formed under the normal tissue culture conditions used in the final stage of micropropagation; in other cases, a specific bulb-inducing treatment should be performed (e.g., in tulip and iris). The size and weight of the bulblets produced in vitro greatly can affect their further performance under in vivo conditions [208]. In addition, it was found that sufficiently large lily bulblets were able to sprout with a stem instead of a rosette, enhancing in vivo growth [209]; this was related to a switch in ontogenetic development from juvenile to adult vegetative status. The in vitro-developed bulblets of *F. persica* were transplanted under in vivo conditions according to a two-step procedure. Ex vitro bulblets were first transferred in trays containing compost and placed in growth cabinets under a 16 h light photoperiod at 23 °C and 90% humidity over a 2-week period. Following this, the bulbs were moved to in vivo conditions, where satisfactory survival and noteworthy flowering were observed [162]. In the study by Azeri and Öztürk [180], microbulbs were subjected to rooting media with IBA (0.5 and 1.0 mg/L). Successful plant development with root lengths of 6.4 and 5.9, respectively, was observed in media containing 0.5 and 1 mg/L IBA. All plantlets obtained through tissue culture demonstrated survival when transferred to the soil.

The different substrates and their mixtures used during the acclimatization could affect the success of this stage. However, in the literature, we have several successful examples. Yasemin et al. [95] transferred the *P. maritimum* plantlets into sand, sand:peat (1:1), and peat, and the survival rates were more than 90%. Rafiq et al. [131] transferred the rooted plantlets into media containing perlite and vermiculite (1:1). The survival rates were more than 80%. *Fritillaria ruthenica* was acclimatized to a coconut fiber and sand (3:1) mixture in a greenhouse. The survival rate was found to be 72% [118]. In *Cyclamen*, İzgü et al. [169] obtained embryogenic-like structures (ELSs), which further germinated on PGR-free culture initiation medium (CIM). Developed plantlets were transplanted to pots with peat and sand, and successful acclimatization was achieved, with survival rates of 70%, 63%, 54%, and 25% for *C. mirabile*, *C. pseudibericum*, *C. cilicium*, and *C. parviflorum*, respectively.

6. Somaclonal Variation

In vitro isolation and culture of explants often lead to callus formation, influenced by species and a high auxin-to-cytokinin ratio. Continuous callus culture may result in genetic variation and the loss of regeneration potential over generations. Callus cultures in geophytes have been used for plant regeneration, cell suspension production, and isolating somaclonal variants [18,210]. The evaluation of somaclonal variation is crucial for ensuring the success of in vitro propagation to achieve true-to-type clones. While somaclonal variations may pose challenges in clonal propagation, they can be advantageous in breeding programs [211]. Van Harmelen et al. [212] determined somaclonal variations from callus derived from bulb scales of *L. longliflorum*. The callus was kept for 3 years at 20 °C in the dark, and after this period, regeneration was detected. Their findings showed mutations in the regenerated plants, such as dwarf plants, malformation of the leaves, and male sterility. The use of molecular markers is one of the most effective strategies for monitoring somaclonal variations, and ISSR markers have been used in many studies [213–215]. Memon et al. [216] detected somaclonal variation among in vitro propagated cormels of gladiolus using RAPD and ISSR molecular markers. The observed variations had varying degrees in the mother cormels and were evident across different varieties of gladiolus. Asadi et al. [217] found that indirect embryogenesis resulted in somaclonal variation, while direct embryogenesis produced uniform plants in *Galanthus transcaucasicus*. Higher NAA concentrations increased somaclonal variation, with the highest observed at the maximum NAA concentration. ISSR analysis showed no somaclonal variation in plants at low BA and NAA concentrations, but significant variation occurred at high NAA concentrations. Kritskaya et al. [146] investigated the somaclonal variations in *Tulipa suaveolens* with ISSR markers. They obtained bulblets through direct organogenesis, showing morphological similarities with intact plants. However, ISSR analysis detected a notable level of somaclonal variability, ranging from 13.9% to 15.8%.

7. New Approaches and Future Perspectives for Flower Bulb Micropropagation

As presented in this chapter, all micropropagation stages of the flower bulbs need optimization. Each stage depends on variable factors such as plant genotype, culture medium, different types and concentrations of plant growth regulators (PGRs), etc. The optimization of a tissue culture medium for specific purposes involves numerous components and combinations, requiring considerable time and expertise. Integrating artificial intelligence (AI) into the micropropagation of plants proves to be a promising approach to overcoming the challenges associated with in vitro culture. AI models and optimization algorithms have emerged as effective tools for addressing this complexity [218,219]. AI models started to be applied to increase the efficiency of tissue culture studies such as in vitro sterilization, callus induction, shoot multiplication, organogenesis, somatic embryogenesis, rooting, acclimatization stages, and in vitro haploid production. By categorizing the diverse data derived from plant tissue culture, including binary inputs (e.g., non-embryogenic/embryogenic callus), discrete variables (e.g., the number of roots, shoots, and embryos), continuous variables (e.g., length of shoots or roots, and callus weight), time-series data, temporal data, fuzzy inputs (e.g., the degree of vitrification, callus color, and the developmental stages of embryos), and categorical variables (e.g., the type of reaction, or the type of phytohormones and carbohydrate sources), AI facilitates a comprehensive understanding of the interactive nature of these variables [218]. These computational models have found application across various plant species, and their potential efficacy extends to flower bulbs. Notably, an investigation into the capacity of microshoots to form corms in *Gladiolus hybridus* employed self-organizing mapping and artificial neural network (ANN) models [220]. The utilization of these models underscores the versatility of AI in advancing micropropagation techniques specific to flower bulbs, exemplifying its potential to contribute significantly to the optimization and enhancement of bulbous plant propagation methodologies. However, despite their potential, the use of AI and OA in plant tissue culture processes could be limited due to complex definitions and computational algorithms [218]. Advances in biotechnology should be coupled with advances in AI to reach the goal of new impacts in tissue culture processes.

Like in other plants, gene editing studies are conducted in geophytes as well. In the context of ongoing advancements in genetic engineering, the significance of tissue culture in relation to gene editing is particularly evident [221] Particularly in geophytes of economic importance with high medicinal and aromatic values, we can come across genetic editing studies aimed at increasing the content of important compounds in storage organs or plant parts. Ornamental geophytes, with large genomes, pose research complexities. Kamo et al. [210] emphasized that the application of modern biotechnological approaches to geophytes faces challenges in developing transgenic varieties. Issues include the lack of efficient transformation systems and difficulties in gene integration into specific genome regions. The initial phase in achieving the successful development of genetically modified plants involves creating a straightforward, effective, and practical protocol for transferring and integrating recombinant DNA molecules into host cells. The challenges associated with genetic transformation in plants, such as the resistance of cells to in vitro culture and the genotype-dependent nature of traditional methods, underscore the need for efficient gene delivery systems. The utilization of advanced techniques like pollen transfection, as demonstrated in a study by [222], represents a breakthrough in achieving large-scale, fast, and efficient transfection in lilies. Subsequently, the second step involves regenerating the recombinant cells into mature plants. Researchers have explored various modifications in inoculation, co-cultivation, and regeneration media to devise an efficient gene transformation protocol [223]. The integration of the CRISPR/Cas9 system further amplifies the potential for precise gene editing, allowing for site-directed mutagenesis, sequence insertion, and modulation of gene expression. Protoplasts, an essential component of plant tissue culture, play a crucial role in studying various aspects of plant development, physiology, and genetics. The integration of CRISPR/Cas9 technology into protoplast-based systems highlights the convergence of molecular biology tools with tissue culture methods, offering unprecedented opportunities for the study of molecular mechanisms and the advancement of plant breeding techniques [221]. This approach also touches geophytes, even though geophytes have a lot of challenges. For example, *C. sativus* contains numerous aromatic and volatile compounds. Genetic editing has been targeted to enhance the content of crocin, picrocrocin, and safranal apocarotenoids in this plant, as they are responsible for color, bitter taste, and aroma. Preliminary CRISPR studies have been conducted for this purpose [86]. Biotechnological research on geophytes contributes to expanding our understanding of plant biology. Using the CRISPR method, it is possible to enhance and replicate many important medicinal and aromatical contents of geophytes, overcome difficulties in their multiplication, and increase tolerance to biotic and abiotic stresses.

8. Conclusions

Tissue culture techniques can reply to the requirements of present-day floriculture asking for novelties and good quality of the starting plant material to introduce in the productive flow. Tissue culture has found great application in the flower bulb industry and many new methods of micropropagation have been developed over the last several decades. Many research papers are aimed at describing micropropagation protocols for specific species, but, as deduced from our review, there is a need for a multifunctional approach able to face the important challenges that are still present. Aside from the insights on biochemistry and physiology, molecular biology and artificial intelligence could be of great help in advancing the establishment of valuable tissue culture protocols for new genera/species.

Author Contributions: Conceptualization, S.Y. and M.B.; methodology, S.Y. and M.B.; investigation, S.Y. and M.B.; resources, S.Y. and M.B.; data curation, S.Y. and M.B.; writing—original draft preparation, S.Y. and M.B.; writing—review and editing, S.Y. and M.B.; visualization, S.Y. and M.B.; supervision, M.B. All authors have read and agreed to the published version of the manuscript.

Funding: This research received no external funding.

Data Availability Statement: No new data were created or analyzed in this study. Data sharing is not applicable to this article.

Conflicts of Interest: The authors declare no conflict of interest.

References

1. Benschop, M.; Kamenetsky, R.; Le Nard, M.; Okubo, H.; De Hertogh, A. 1 The global flower bulb industry: Production, utilization research. *Hortic. Rev.* **2010**, *36*, 1–115.
2. Correvon, H.; Massé, H. *Les Iris Dans les Jardins*; Aux Jardins Correvon: Geneva, Switzerland, 1907; p. 214.
3. Reynolds, M.; Meachem, W. *The Garden Bulbs of Spring*; Funk & Wagnalls: Madison, WI, USA, 1967; p. 237.
4. Doerflinger, F. *The Bulb Book*; David & Charles: Newton Abbot, UK, 1973; p. 309.
5. Margaris, N. Flowers in Greek mythology. In Proceedings of the IV International Symposium on New Floricultural Crops 541 Crete, Greece, 22–27 May 1999; pp. 23–29.
6. Warren, P. The fresco of the garlands from Knossos. *Bull. Corresp. Hellénique* **1985**, *11*, 187–208. [CrossRef]
7. Day, J. Crocuses in context: A diachronic survey of the crocus motif in the Aegean Bronze Age. *Hesperia J. Am. Sch. Class. Stud Athens* **2011**, *80*, 337–379.
8. Janick, J.; Kamenetsky, R.; Puttaswamy, S.H. Horticulture of the Taj Mahal: Gardens of the imagination. *Chron. Hortic.* **2010**, *50* 30–33.
9. Bryan, J.E. *Bulbs*; Timber Press, Inc.: Portland, OR, USA, 1989; p. 451.
10. De Hertogh, A.; Schepeen, J.; Kamenetsky, R.; Le Nard, M.; Okubo, H. The Globalization of Flower Bulb Industry. In *Ornamental Geophytes: From Basic Science to Sustainable Production*; Kamenetsky, R., Hiroshi, O., Eds.; CRC Press: Boca Raton, FL, USA, 2013 pp. 1–16.
11. Pavord, A. *The Tulip: The Story of a Flower That Has Made Men Mad*; Bloomsbury Publishing: London, UK, 1999; p. 480.
12. Marasek-Ciolakowska, A.; Sochacki, D.; Marciniak, P. Breeding Aspects of Selected Ornamental Bulbous Crops. *Agronomy* **2021** *11*, 1709. [CrossRef]
13. Orlikowska, T.; Podwyszyńska, M.; Marasek-Ciołakowska, A.; Sochacki, D.; Szymański, R. Tulip. In *Ornamental Crops*; Springer Cham, Switzerland, 2018; pp. 769–802. [CrossRef]
14. Saraç, Y.; Balkaya, A.; Deligöz, İ. Lale Islahı. In *Süs Bitkileri Islahı (Türler)*; Kazaz, S., Mendi, Y.Y., Eds.; Gece Publishing: Ankara, Turkey, 2021; pp. 375–414.
15. Rees, A. The physiology of ornamental bulbous plants. *Bot. Rev.* **1966**, *32*, 1–23. [CrossRef]
16. Bailey, L.H. *Manual of Cultivated Plants—Most Commonly Grown in the Continental United States and Canada*, 2nd ed.; Macmillan Company: New York, NY, USA, 1949; ISBN 0025055208.
17. Gabellini, S.; Scaramuzzi, S. Evolving consumption trends, marketing strategies, and governance settings in ornamental horticulture: A grey literature review. *Horticulturae* **2022**, *8*, 234. [CrossRef]
18. Kim, K.-W.; De Hertogh, A. Tissue culture of ornamental flowering bulbs (geophytes). *Hortic. Rev.* **1997**, *18*, 87–169.
19. Pizano, M. Innovation and sustainability in South American floriculture. In Proceedings of the XXXI International Horticultural Congress (IHC2022): International Symposium on Innovations in Ornamentals: From Breeding to Market, Angers, France, 14–20 August 2022; pp. 77–84. [CrossRef]
20. Sener, B.; Sevim, D. Discovery of bioactive drug candidates from some Turkish medicinal plants-neuroprotective potential of *Iris pseudacorus* L. *Pure Appl. Chem.* **2020**, *92*, 1175–1179. [CrossRef]

21. Janowska, B.; Andrzejak, R. Plant growth regulators for the cultivation and vase life of geophyte flowers and leaves. *Agriculture* **2023**, *13*, 855. [CrossRef]
22. Kocak, M.; Sevindik, B.; Izgu, T.; Tutuncu, M.; Mendi, Y.Y. Synthetic seed production of flower bulbs. In *Synthetic Seeds: Germplasm Regeneration, Preservation and Prospects*; Springer: Cham, Switzerland, 2019; pp. 283–299. [CrossRef]
23. Seyidoglu, N.; Zencirkiran, M.; Ayasligil, Y. Position and application areas of geophytes within landscape design. *Afr. J. Agric. Res.* **2009**, *4*, 1351–1357.
24. Ziv, M.; Lilien-Kipnis, H. Bud regeneration from inflorescence explants for rapid propagation of geophytes in vitro. *Plant Cell Reports* **2000**, *19*, 845–850. [CrossRef]
25. WHO. *Monographs on Selected Medicinal Plants Volume 1*; World Health Organization: Geneva, Switzerland, 1999; 297p, Available online: https://www.who.int/publications/i/item/9241545178 (accessed on 18 February 2024).
26. WHO. *Monographs on Selected Medicinal Plants Volume 2*; World Health Organization: Geneva, Switzerland, 2002; 375p, Available online: https://iris.who.int/bitstream/handle/10665/42052/9241545372.pdf?sequence=2&isAllowed=y (accessed on 18 February 2024).
27. WHO. *Monographs on Selected Medicinal Plants Volume 3*; World Health Organization: Geneva, Switzerland, 2007; 390p, Available online: https://iris.who.int/bitstream/handle/10665/42052/9789241547024_eng.pdf?sequence=3&isAllowed=y (accessed on 18 February 2024).
28. WHO. *Monographs on Selected Medicinal Plants Volume 4*; World Health Organization: Geneva, Switzerland, 2009; 456p, Available online: https://iris.who.int/bitstream/handle/10665/42052/9789241547055_eng.pdf?sequence=4&isAllowed=y (accessed on 18 February 2024).
29. WHO. *Monographs on Medicinal Plants Commonly Used in the Newly Independent States (NIS)*; World Health Organization: Geneva, Switzerland, 2010; 441p, Available online: https://www.e-lactancia.org/media/papers/0-Fitoterapia-Plantas-WHO-O5NIS_2010.pdf (accessed on 18 February 2024).
30. Das, R.; Barman, A.; Ray, S. Traditional knowledge, phytochemistry, and pharmacological properties of the African genus Scadoxus (Amaryllidaceae). *South Afr. J. Bot.* **2022**, *151*, 565–577. [CrossRef]
31. Pautasso, M. Ten simple rules for writing a literature review. *PLoS Comput. Biol.* **2013**, *9*, e1003149. [CrossRef]
32. Rees, A.R. *Ornamental Bulbs, Corms and Tubers*; CABI International: Wallingford, UK, 1992; Volume 1, p. 220.
33. Kamenetsky, R. Biodiversity of geophytes, phytogeography, morphology and survival strategies. In *Ornamental Geophytes: From Basic Science to Sustainable Production*; Kamenetsky, R., Okubo, H., Eds.; CRC Press: Boca Raton, FL, USA, 2013; pp. 57–76.
34. Dafni, A.; Cohen, D.; Noy-Mier, I. Life-cycle variation in geophytes. *Ann. Mo. Bot. Gard.* **1981**, *68*, 652–660. [CrossRef]
35. Woodhead, T.W. Ecology of woodland plants in the neighbourhood of Huddersfield. *Bot. J. Linn. Soc.* **1906**, *37*, 333–406. [CrossRef]
36. De Klerk, G. Micropropagation of bulbous crops: Technology and present state. *Floric. Ornam. Biotechnol.* **2012**, *6*, 1–8.
37. Leeggangers, H.C.F.; Moreno-Pachon, N.; Gude, H.; Immink, R.H. Transfer of knowledge about flowering and vegetative propagation from model species to bulbous plants. *Int. J. Dev. Biol.* **2013**, *57*, 611–620. [CrossRef] [PubMed]
38. Bach, A.; Sochacki, D. Propagation of Ornamental Geophytes:Physiology and Management Systems. In *Ornamental Geophytes from Basic Science to Sustainable Production*; Kamenetsky, R., Okubo, H., Eds.; CRC Press: Boca Raton, FL, USA, 2013; pp. 261–286.
39. Ulus, A.; Seyidoğlu, N. Bazi Doğal Geofitlerin Doku Kültürü İle Üretimi. *J. Fac. For. Istanb. Univ.* **2006**, *56*, 71–80. [CrossRef]
40. Baskin, C.C.; Baskin, J.M. *Seeds: Ecology, Biogeography, and, Evolution of Dormancy and Germination*; Academic Press: London, UK, 1998; p. 666.
41. Anderson, N. Selection tools for reducing generation time of geophytic herbaceous perennials. In Proceedings of the XIII International Symposium on Flower Bulbs and Herbaceous Perennials 1237, Seoul, Republic of Korea, 1–3 May 2019; pp. 53–66. [CrossRef]
42. Langens-Gerrits, M.M.; De Klerk, G.-J.M. Micropropagation of flower bulbs: Lily and Narcissus. In *Plant Cell Culture Protocols*; Humana Press: Totowa, NJ, USA, 1999; pp. 141–147. [CrossRef]
43. Luna, T. Vegetative propagation. In *Nursery Manual for Native Plants: A Guide for Tribal Nurseries*; Dumroese, R.K., Luna, T., Landis, T.D., Eds.; U.S. Department of Agriculture, Forest Service: Washington, DC, USA, 2009; Volume 1, pp. 153–175.
44. Podwyszyńska, M.; Orlikowska, T.; Trojak-Goluch, A.; Wojtania, A. Application and Improvement of In Vitro Culture Systems for Commercial Production of Ornamental, Fruit, and Industrial Plants in Poland. *Acta Soc. Bot. Pol.* **2022**, *91*, 914. [CrossRef]
45. Hartmann, H.T.; Kester, D.E. *Plant Propagation: Principles and Practice*; Soil Science: New York, NY, USA, 1959; p. 662.
46. Debergh, P.; Maene, L. A scheme for commercial propagation of ornamental plants by tissue culture. *Sci. Hortic.* **1981**, *14*, 335–345. [CrossRef]
47. Debergh, P.C.; Read, P. Micropropagation. In *Micropropagation: Technology and Application*; Debergh, P.C., Zimmerman, R.H., Eds.; Springer: Cham, Switzerland, 1991; pp. 1–13. [CrossRef]
48. Ozzambak, M.E.; Senel, U.; Gulden, H.; Zeybekoglu, E.; Baser, S. Mikroçoğaltım teknikleri. In *Süs Bitkileri Islahı (Klasik ve Biyoteknolojik Yöntemler)*; Yalcin Mendi, Y.K.S., Ed.; Gece Publishing: Ankara, Turkey, 2021; Volume 341–360.
49. Bidabadi, S.S.; Jain, S.M. Cellular, molecular, and physiological aspects of in vitro plant regeneration. *Plants* **2020**, *9*, 702. [CrossRef] [PubMed]
50. Ikeuchi, M.; Ogawa, Y.; Iwase, A.; Sugimoto, K. Plant regeneration: Cellular origins and molecular mechanisms. *Development* **2016**, *143*, 1442–1451. [CrossRef] [PubMed]

51. Long, Y.; Yang, Y.; Pan, G.; Shen, Y. New insights into tissue culture plant-regeneration mechanisms. *Front. Plant Sci.* **2022**, *13*, 926752. [CrossRef]
52. Von Arnold, S. Somatic embryogenesis. In *Plant Propagation by Tissue Culture*, 3rd ed.; George, E.F., Hall, M.A., de Klerk, G.J., Eds.; Springer: Dordrecht, The Netherlands, 2008; Volume 1, pp. 335–354.
53. Williams, E.; Maheswaran, G. Somatic embryogenesis: Factors influencing coordinated behaviour of cells as an embryogenic group. *Ann. Bot.* **1986**, *57*, 443–462. [CrossRef]
54. Yang, X.; Zhang, X. Regulation of Somatic Embryogenesis in Higher Plants. *Crit. Rev. Plant Sci.* **2010**, *29*, 36–57. [CrossRef]
55. Zimmerman, J.L. Somatic embryogenesis: A model for early development in higher plants. *Plant Cell* **1993**, *5*, 1411. [CrossRef] [PubMed]
56. Sharp, W.; Sondahl, M.; Caldas, L.; Maraffa, S. The physiology of in vitro asexual embryogenesis. *Hortic. Rev.* **1980**, *2*, 268–310. [CrossRef]
57. Green, P.B. Organogenesis-a biophysical view. *Annu. Rev. Plant Physiol.* **1980**, *31*, 51–82. [CrossRef]
58. Hicks, G.S. Patterns of organ development in plant tissue culture and the problem of organ determination. *Bot. Rev.* **1980**, *46*, 1–23. [CrossRef]
59. Irish, V.F. The Arabidopsis petal: A model for plant organogenesis. *Trends Plant Sci.* **2008**, *13*, 430–436. [CrossRef] [PubMed]
60. Sugiyama, M. Organogenesis in vitro. *Curr. Opin. Plant Biol.* **1999**, *2*, 61–64. [CrossRef]
61. Tutuncu, M.; Basar, S.; Yalcin Mendi, N.Y. Organogenesis. In *Süs Bitkileri Islahı (Klasik ve Biyoteknolojik Yöntemler)*; Yalcin Mendi, Y.K.S., Ed.; Gece Publishing: Ankara, Turkey, 2021; pp. 361–392.
62. Hnatuszko-Konka, K.; Gerszberg, A.; Weremczuk-Jeżyna, I.; Grzegorczyk-Karolak, I. Cytokinin signaling and de novo shoot organogenesis. *Genes* **2021**, *12*, 265. [CrossRef] [PubMed]
63. Schwarz, O.J.; Beaty, R.M. Organogenesis. In *Plant Tissue Culture Concepts and Laboratory Exercises*; Routledge: Abingdon, UK, 2018; pp. 125–138. [CrossRef]
64. Sevindik, B.; Tutuncu, M.; Evci Çürük, P.; Mendi, Y.Y. Somatik embryogenesis. In *Süs Bitkileri Islahı (Klasik ve Biyoteknolojik Yöntemler)*; Yalcin Mendi, Y.K.S., Ed.; Gece Publishing: Ankara, Turkey, 2021; pp. 413–440.
65. Von Arnold, S.; Sabala, I.; Bozhkov, P.; Dyachok, J.; Filonova, L. Developmental pathways of somatic embryogenesis. *Plant Cell Tissue Organ. Cult.* **2002**, *69*, 233–249. [CrossRef]
66. Martini, A.N.; Papafotiou, M.; Vemmos, S.N. Season and explant origin affect phenolic content, browning of explants, and micropropagation of × *Malosorbus florentina* (Zucc.) Browicz. *HortScience* **2013**, *48*, 102–107. [CrossRef]
67. Ascough, G.D.; van Staden, J.; Erwin, J.E. In Vitro Storage Organ Formation of Ornamental Geophytes. *Hortic. Rev.* **2008**, *34*, 417–445.
68. Robb, S.M. The Culture of Excised Tissue 3 Lilium speciosum Thun. *J. Exp. Bot.* **1957**, *8*, 348–352. [CrossRef]
69. Ziv, M. The Contribution Of Biotechnology To Breeding, Propagation And Disease Resistance In Geophytes. In Proceedings of the VII International Symposium on Flowerbulbs, Herzliya, Israel, 1 December 1997. [CrossRef]
70. Hussey, G. *The application of Tissue Culture to the Vegetative Propagation of Plants*; Science Progress: Norwich, UK, 1978; pp. 185–208.
71. Mehbub, H.; Akter, A.; Akter, M.A.; Mandal, M.S.H.; Hoque, M.A.; Tuleja, M.; Mehraj, H. Tissue Culture in Ornamentals: Cultivation Factors, Propagation Techniques, and Its Application. *Plants* **2022**, *11*, 3208. [CrossRef]
72. Dhooghe, E.; Reheul, D.; Van Labeke, M.C. Overcoming Pre-Fertilization Barriers in Intertribal Crosses between Anemone coronaria L. and Ranunculus asiaticus L. *Horticulturae* **2021**, *7*, 529. [CrossRef]
73. Sevindik, B.; İzgü, T.; Tütüncü, M.; Çürük, P.; Sarı, N.; Mendi, Y.Y. Double-haploid plant production through anther and ovule culture of wild *Cyclamen persicum* Mill. and Melody F1 cyclamen cultivar. *Vitr. Cell. Dev. Biol. Plant* **2023**, *59*, 711–723. [CrossRef]
74. Sevindik, B.; İzgü, T.; Tütüncü, M.; Çürük, P.; Söğüt, Z.; Mendİ, Y.Y. Effects of different plant growth regulators on ovule culture of Turkish Cyclamen persicum Mill. and commercial variety "Melody F1". *Alatarım* **2017**, *16*, 1–9.
75. Tütüncü, M.; Mendi, Y.Y. Effect of pollination with gamma irradiated pollen on in vitro regeneration of ovule culture in Cyclamen. *Turk. J. Agric. Food Sci. Technol.* **2022**, *10*, 2415–2420. [CrossRef]
76. Van Creij, M.; Kerckhoffs, D.; De Bruijn, S.; Vreugdenhil, D.; Van Tuyl, J. The effect of medium composition on ovary-slice culture and ovule culture in intraspecific *Tulipa gesneriana* crosses. *Plant Cell Tissue Organ. Cult.* **2000**, *60*, 61–67. [CrossRef]
77. Van Tuyl, J.; Van Creij, M.; Van Dien, M. In vitro pollination and ovary culture as a breeding tool in wide hybridization of Lilium and Nerine. In Proceedings of the VI International Symposium on Flower Bulbs 325, Skiernieweice, Poland, 12 May 1992; pp. 461–466. [CrossRef]
78. Van Tuyl, J.M.; Van Diën, M.P.; Van Creij, M.G.M.; Van Kleinwee, T.C.M.; Franken, J.; Bino, R.J. Application of in vitro pollination, ovary culture, ovule culture and embryo rescue for overcoming incongruity barriers in interspecific Lilium crosses. *Plant Sci.* **1991**, *74*, 115–126. [CrossRef]
79. Wang, J.; Huang, L.; Bao, M.-z.; Liu, G.-f. Production of interspecific hybrids between *Lilium longiflorum* and L. lophophorum var. linearifolium via ovule culture at early stage. *Euphytica* **2009**, *167*, 45–55. [CrossRef]
80. Tütüncü, M.; Mendi, Y. Evaluation of pollen tube growth and fertilization via histological analysis in *Cyclamen persicum*. In Proceedings of the XXVI International Eucarpia Symposium Section Ornamentals: Editing Novelty 1283, Erfurt, Germany, 1–4 September 2019; pp. 21–26. [CrossRef]
81. Karamian, R.; Ebrahimzadeh, H. Plantlet regeneration from protoplast-derived embryogenic calli of *Crocus cancellatus*. *Plant Cell Tissue Organ. Cult.* **2001**, *65*, 115–121. [CrossRef]

82. Koetle, M.; Finnie, J.; Balázs, E.; Van Staden, J. A review on factors affecting the *Agrobacterium*-mediated genetic transformation in ornamental monocotyledonous geophytes. *South Afr. J. Bot.* **2015**, *98*, 37–44. [CrossRef]
83. Babu, P.; Chawla, H. In vitro regeneration and *Agrobacterium* mediated transformation in gladiolus. *J. Hortic. Sci. Biotechnol.* **2000**, *75*, 400–404. [CrossRef]
84. Chib, S.; Thangaraj, A.; Kaul, S.; Dhar, M.K.; Kaul, T. Development of a system for efficient callus production, somatic embryogenesis and gene editing using CRISPR/Cas9 in Saffron (*Crocus sativus* L.). *Plant Methods* **2020**, *16*, 1–10. [CrossRef]
85. Cohen, A.; Meredith, C.P. Agrobacterium-mediated transformation of *Lilium*. In Proceedings of the VI International Symposium on Flower Bulbs 325, Skierniewice, Poland, 12 May 1992; pp. 611–618. [CrossRef]
86. Hoshi, Y.; Kondo, M.; Mori, S.; Adachi, Y.; Nakano, M.; Kobayashi, H. Production of transgenic lily plants by Agrobacterium-mediated transformation. *Plant Cell Rep.* **2004**, *22*, 359–364. [CrossRef]
87. Kamo, K.; Jordan, R.; Guaragna, M.A.; Hsu, H.-t.; Ueng, P. Resistance to Cucumber mosaic virus in *Gladiolus* plants transformed with either a defective replicase or coat protein subgroup II gene from Cucumber mosaic virus. *Plant Cell Rep.* **2010**, *29*, 695–704. [CrossRef]
88. Kondo, T.; Hasegawa, H.; Suzuki, M. Transformation and regeneration of garlic (*Allium sativum* L.) by Agrobacterium-mediated gene transfer. *Plant Cell Rep.* **2000**, *19*, 989–993. [CrossRef]
89. Lu, G.; Zou, Q.; Guo, D.; Zhuang, X.; Yu, X.; Xiang, X.; Cao, J. *Agrobacterium* tumefaciens-mediated transformation of *Narcissus tazzeta* var. *chinensis*. *Plant Cell Rep.* **2007**, *26*, 1585–1593. [CrossRef]
90. Popowich, E.; Firsov, A.; Mitiouchkina, T.; Filipenya, V.; Dolgov, S.; Reshetnikov, V. Agrobacterium-mediated transformation of *Hyacinthus orientalis* with thaumatin II gene to control fungal diseases. *Plant Cell Tissue Organ. Cult.* **2007**, *90*, 237–244. [CrossRef]
91. Suzuki, S.; Supaibulwatana, K.; Mii, M.; Nakano, M. Production of transgenic plants of the Liliaceous ornamental plant *Agapanthus praecox* ssp. *orientalis* (Leighton) Leighton via *Agrobacterium*-mediated transformation of embryogenic calli. *Plant Sci.* **2001**, *161*, 89–97. [CrossRef]
92. Wilmink, A.; Van de Ven, B.; Dons, J. Expression of the GUS-gene in the monocot tulip after introduction by particle bombardment and Agrobacterium. *Plant Cell Rep.* **1992**, *11*, 76–80. [CrossRef]
93. Van Aartrijk, J.; Van der Linde, P. In vitro propagation of flower-bulb crops. In Proceedings of the Tissue culture as a plant production system for horticultural crops: Conference on Tissue Culture as a Plant Production System for Horticultural Crops, Beltsville, MD, USA, 20–23 October 1985; pp. 317–331. [CrossRef]
94. Preece, J. Stock plant physiological factors affecting growth and morphogenesis. In *Plant Propagation by Tissue Culture. Volume 1: The Background*; Springer: Dordrecht, The Netherlands, 2008; pp. 403–422. [CrossRef]
95. Yasemin, S.; Koksal, N.; Buyukalaca, S. Indirect organogenesis and in vitro bulb formation of *Pancratium maritimum*. *Plant Cell Tissue Organ Cult. (PCTOC)* **2023**, *154*, 713–727. [CrossRef]
96. Murashige, T. A revised medium for rapid growth and bioassay with tobacco tissue culture. *Physiol. Plant* **1962**, *15*, 473–497. [CrossRef]
97. Chu, C. Establishment of an efficient medium for another culture of rice through comparative experiments on the nitrogen sources. *Sci. Sin.* **1975**, *18*, 223–231.
98. Lloyd, G.; McCown, B. Commercially-feasible micropropagation of mountain laurel, Kalmia latifolia, by use of shoot-tip culture. *Comb. Proc. Int. Plant Propagators Soc.* **1980**, *30*, 421–427.
99. Gamborg, O.L.c.; Miller, R.A.; Ojima, K. Nutrient requirements of suspension cultures of soybean root cells. *Exp. Cell Res.* **1968**, *50*, 151–158. [CrossRef]
100. George, E.; de Klerk, G. The components of plant tissue culture media I: Macro-and micro-nutrients. In *Plant Propagation by Tissue Culture. Volume 1: The Background*; Springer: Dordrecht, The Netherlands, 2008; pp. 65–113. [CrossRef]
101. Thorpe, T.; Stasolla, C.; Yeung, E.; de Klerk, G.; Roberts, A.; George, E. The components of plant tissue culture media II: Organic additions, osmotic and pH effects, and support systems. In *Plant Propagation by Tissue Culture. Volume 1: The Background*; Springer: Dordrecht, The Netherlands, 2008; pp. 115–173. [CrossRef]
102. Podwyszyńska, M.; Kosson, R.; Treder, J. Polyamines and methyl jasmonate in bulb formation of in vitro propagated tulips. *Plant Cell Tissue Organ Cult. (PCTOC)* **2015**, *123*, 591–605. [CrossRef]
103. Debergh, P.C. Effects of agar brand and concentration on the tissue culture medium. *Physiol. Plant.* **1983**, *59*, 270–276. [CrossRef]
104. Beruto, M.; Curir, P.; Debergh, P. Influence of agar on in vitro cultures: II. Biological performance of *Ranunculus* on media solidified with three different agar brands. *Vitr. Cell. Dev. Biol. Plant* **1999**, *35*, 94–101. [CrossRef]
105. George, E.F.; Hall, M.A.; De Klerk, G.-J. *Plant Propagation by Tissue Culture: The Background*, 3rd ed.; Springer: Dordrecht, The Netherlands, 2008. [CrossRef]
106. Bornwaßer, T.; Tantau, H.-J. Evaluation of LED lighting systems in in vitro cultures. In Proceedings of the VII International Symposium on Light in Horticultural Systems 956, Wageningen, The Netherlands, 15–18 October 2012; pp. 555–562. [CrossRef]
107. Pundir, R.K.; Pathak, A.; Upadhyaya, D.C.; Muthusamy, A.; Upadhyaya, C.P. Red and Blue Light-Emitting Diodes Significantly Improve Tuberization of Potato (L.). *J. Hortic. Res.* **2021**, *29*, 95–108. [CrossRef]
108. Bach, A.; Świderski, A. The effect of light quality on organogenesis of *Hyacinthus orientalis* L. in vitro. *Acta Biol. Cracoviensia Ser. Bot.* **2000**, *42*, 115–120.
109. Bakhshaie, M.; Khosravi, S.; Azadi, P.; Bagheri, H.; van Tuyl, J.M. Biotechnological advances in *Lilium*. *Plant Cell Rep.* **2016**, *35*, 1799–1826. [CrossRef] [PubMed]

110. Lian, M.; Chakrabarty, D.; Paek, K. Growth and uptake of sucrose and mineral ions by bulblets of *Lilium* oriental hybrid 'Casablanca'during bioreactor culture. *J. Hortic. Sci. Biotechnol.* **2002**, *77*, 253–257. [CrossRef]
111. Pałka, P.; Cioć, M.; Hura, K.; Szewczyk-Taranek, B.; Pawłowska, B. Adventitious organogenesis and phytochemical composition of Madonna lily (*Lilium candidum* L.) in vitro modeled by different light quality. *Plant Cell Tissue Organ Cult. (PCTOC)* **2023**, *152*, 99–114. [CrossRef]
112. Van Aartrijk, J.; Blom-Barnhoorn, G.J. Adventitious Bud Formation from Bulb-scale Explants of *Lilium speciosum* Thunb. in vitro Interacting effects of NAA, TIBA, wounding, and temperature. *J. Plant Physiol.* **1984**, *116*, 409–416. [CrossRef]
113. Amaki, W.; Shinohara, Y.; Hayata, Y.; Sano, H.; Suzuki, Y. Effects of bulb desiccation and storage on the in vitro propagation of hyacinth. *Sci. Hortic.* **1984**, *23*, 353–360. [CrossRef]
114. Gavinlertvatana, P.; Read, P.E.; Wilkins, H.; Heins, R. Influence of Photoperiod and Daminozide Stock Plant Pretreatments on Ethylene and CO_2 Levels and Callus Formation from *Dahlia* Leaf Segment Cultures1. *J. Am. Soc. Hortic. Sci.* **1979**, *104*, 849–852. [CrossRef]
115. Hosoki, T.; Sagawa, Y. Clonal Propagation of Ginger (*Zingiber officinale* Roscoe) through Tissue Culture1. *HortScience* **1977**, *12*, 451–452. [CrossRef]
116. Stimart, D.P.; Ascher, P.D. Developmental Responses of *Lilium longiflorum* Bulblets to Constant or Alternating Temperatures in Vitro1. *J. Am. Soc. Hortic. Sci.* **1981**, *106*, 450–454. [CrossRef]
117. Podwyszyńska, M.; Sochacki, D. Micropropagation of tulip: Production of virus-free stock plants. In *Protocols for In Vitro Propagation of Ornamental Plants*; Humana Press: Totowa, NJ, USA, 2010; pp. 243–256. [CrossRef]
118. Muraseva, D.S.; Novikova, T.I. Efficient protocol for in vitro propagation from bulb scale explants of Fritillaria ruthenica Wikst. (Liliaceae), a rare ornamental species. Rendiconti Lincei. *Sci. Fis. Nat.* **2018**, *29*, 491–497. [CrossRef]
119. Kumar, V.; Moyo, M.; Van Staden, J. Enhancing plant regeneration of Lachenalia viridiflora, a critically endangered ornamental geophyte with high floricultural potential. *Sci. Hortic.* **2016**, *211*, 263–268. [CrossRef]
120. Mirici, S.; Parmaksız, İ.; Özcan, S.; Sancak, C.; Uranbey, S.; Sarıhan, E.O.; Gümüşcü, A.; Gürbüz, B.; Arslan, N. Efficient in vitro bulblet regeneration from immature embryos of endangered Sternbergia fischeriana. *Plant Cell Tissue Organ. Cult.* **2005**, *80*, 239–246. [CrossRef]
121. Buckley, P.M.; Reed, B.M. 045 Antibiotic susceptibility of plant-associated bacteria. *HortScience* **1994**, *29*, 434. [CrossRef]
122. George, E.F. *Plant Propagation by Tissue Culture Part 1. The Technology*; CABI: Wallingford, UK, 1993.
123. Silva, J.A.T.d.; Winarto, B.; Dobránszki, J.; Zeng, S. Disinfection procedures for propagation of. *Folia Hortic.* **2015**, *27*, 3–14. [CrossRef]
124. Yasemin, S.; Köksal, N.; Büyükalaca, S. Effects of disinfection conditions and culture media on in vitro germination of sea daffodil (*Pancratium maritimum*). *J. Biol. Environ. Sci.* **2018**, *34*, 13–22.
125. Barrett, C.; Cassells, A.C. An evaluation of antibiotics for the elimination of *Xanthomonas campestris* pv. *pelargonii* (Brown) from *Pelargonium x domesticum* cv.'Grand Slam'explants in vitro. *Plant Cell Tissue Organ. Cult.* **1994**, *36*, 169–175. [CrossRef]
126. Alexopoulos, A.A.; Mavrommati, E.; Kartsonas, E.; Petropoulos, S.A. Effect of Temperature and Sucrose on In Vitro Seed Germination and Bulblet Production of *Pancratium maritimum* L. *Agronomy* **2022**, *12*, 2786. [CrossRef]
127. Nikopoulos, D.; Alexopoulos, A.A. In vitro propagation of an endangered medicinal plant: Pancratium maritimum L. *J. Food Agric. Environ.* **2008**, *6*, 393–398.
128. Lagram, K.; El Merzougui, S.; Boudadi, I.; Ben El Caid, M.; El Boullani, R.; El Mousadik, A.; Serghini, M.A. In vitro shoot formation and enrooted mini-corm production by direct organogenesis in saffron (*Crocus sativus* L.). *Vegetos* **2023**, 1–6. [CrossRef]
129. Sochacki, D.; Orlikowska, T. The obtaining of narcissus plants free from potyviruses via adventitious shoot regeneration in vitro from infected bulbs. *Sci. Hortic.* **2005**, *103*, 219–225. [CrossRef]
130. Redhwan, A.; Acemi, A.; Özen, F. Effects of plant growth regulators on in vitro seed germination, organ development and callogenesis in Pancratium maritimum L. *Plant Cell Tissue Organ Cult. (PCTOC)* **2023**, *154*, 97–110. [CrossRef]
131. Rafiq, S.; Rather, Z.; Bhat, R.A.; Nazki, I.; Al-Harbi, M.S.; Banday, N.; Farooq, I.; Samra, B.N.; Khan, M.; Ahmed, A.F. Standardization of in vitro micropropagation procedure of Oriental Lilium Hybrid Cv.'Ravenna'. *Saudi J. Biol. Sci.* **2021**, *28*, 7581–7587. [CrossRef]
132. Rather, Z.; Nazki, I.; Qadri, Z.; Mir, M.; Bhat, K.; Hussain, G. In vitro propagation of herbaceous peony (*Paeonia lactiflora* Pall.) cv. Sara Bernhardt using shoot tips. *Indian J. Hortic.* **2014**, *71*, 385–389.
133. Farooq, I.; Banday, N.; Nazki, I.; Malik, A.A.; Khan, F.; Mir, M.; Yaseen, T. In vitro sterilization of *Lilium* LA Hybrids "Indian Summerset" and "Nashville" as influenced by different sterilant combinations. *Pharma Innov. J.* **2022**, *11*, 1795–1798.
134. Appleton, M.; Ascough, G.; Van Staden, J. In vitro regeneration of *Hypoxis colchicifolia* plantlets. *South Afr. J. Bot.* **2012**, *80*, 25–35. [CrossRef]
135. Devi, K.; Sharma, M.; Ahuja, P. Direct somatic embryogenesis with high frequency plantlet regeneration and successive cormlet production in saffron (*Crocus sativus* L.). *South Afr. J. Bot.* **2014**, *93*, 207–216. [CrossRef]
136. Lapiz-Culqui, Y.K.; Meléndez-Mori, J.B.; Mállap-Detquizán, G.; Tejada-Alvarado, J.J.; Vilca-Valqui, N.C.; Huaman-Human, E.; Oliva, M.; Goñas, M. In Vitro Bulbification of Five Lily Varieties: An Effective Method to Produce Quality Seeds and Flowers. *Int. J. Agron.* **2022**, *2022*, 8775989. [CrossRef]
137. Patil, A.M.; Gunjal, P.P.; Das, S. In vitro micropropagation of Lilium candidum bulb by application of multiple hormone concentrations using plant tissue culture technique. *Int. J. Res. Appl. Sci. Biotechnol.* **2021**, *8*, 244–253. [CrossRef]

138. Youssef, N.M.; Shaaban, S.A.; Ghareeb, Z.F.; Taha, L.S. In vitro bulb formation of direct and indirect regeneration of *Lilium orientalis* cv. "Starfighter" plants. *Bull. Natl. Res. Cent.* **2019**, *43*, 211. [CrossRef]
139. Ozel, C.A.; Khawar, K.M.; Unal, F. Factors affecting efficient in vitro micropropagation of *Muscari muscarimi* Medikus using twin bulb scale. *Saudi J. Biol. Sci.* **2015**, *22*, 132–138. [CrossRef] [PubMed]
140. Santos, A.; Fidalgo, F.; Santos, I.; Salema, R. In vitro bulb formation of *Narcissus asturiensis*, a threatened species of the Amaryllidaceae. *J. Hortic. Sci. Biotechnol.* **2002**, *77*, 149–152. [CrossRef]
141. Kukulczanka, K.; Kromer, K.; Czastka, B. Propagation of *Fritillaria meleagris* L. through tissue culture. In Proceedings of the III International Symposium on Growth Regulators in Ornamental Horticulture 251, Skierniewice, Poland, 5–10 September 1988; pp. 147–154. [CrossRef]
142. Kumar, P.; Partap, M.; Ashrita; Rana, D.; Kumar, P.; Warghat, A.R. Metabolite and expression profiling of steroidal alkaloids in wild tissues compared to bulb derived in vitro cultures of *Fritillaria roylei*—High value critically endangered Himalayan medicinal herb. *Ind. Crops Prod.* **2020**, *145*, 111945. [CrossRef]
143. Sevindik, B.; Mendi, Y.Y. Somatic embryogenesis in *Crocus sativus* L. In *In Vitro Embryogenesis in Higher Plants*; Humana Press: New York, NY, USA, 2016; pp. 351–357. [CrossRef]
144. Taheri-Dehkordi, A.; Naderi, R.; Martinelli, F.; Salami, S.A. A robust workflow for indirect somatic embryogenesis and cormlet production in saffron (*Crocus sativus* L.) and its wild allies; *C. caspius* and *C. speciosus*. *Heliyon* **2020**, *6*, e05841. [CrossRef] [PubMed]
145. Slimani, C.; El Goumi, Y.; Chaimae, R.; El Ghadraoui, L.; Benjelloun, M.; Lazraq, A. Micropropagation and potential of bioactive compounds of saffron (*Crocus sativus* L.) for nutrition and health. *Not. Sci. Biol.* **2022**, *14*, 11278. [CrossRef]
146. Kritskaya, T.; Kashin, A.; Kasatkin, M.Y. Micropropagation and Somaclonal Variation of *Tulipa suaveolens* (Liliaceae) in vitro. *Russ. J. Dev. Biol.* **2019**, *50*, 209–215. [CrossRef]
147. Pierik, R.; Sprenkels, P.; Van Der Harst, B.; Van Der Meys, Q. Seed germination and further development of plantlets of *Paphiopedilum ciliolare* Pfitz. in vitro. *Sci. Hortic.* **1988**, *34*, 139–153. [CrossRef]
148. Tubić, L.; Savić, J.; Mitić, N.; Milojević, J.; Janošević, D.; Budimir, S.; Zdravković-Korać, S. Cytokinins differentially affect regeneration, plant growth and antioxidative enzymes activity in chive (*Allium schoenoprasum* L.). *Plant Cell Tissue Organ Cult. (PCTOC)* **2015**, *124*, 1–14. [CrossRef]
149. Ziv, M. Morphogenesis of gladiolus buds in bioreactors—Implication for Scaled-up Propagation of Geophytes. In *Progress in Plant Cellular and Molecular Biology. Current Plant Science and Biotechnology in Agriculture*; Nijkamp, H.J.J., Van Der Plas, L.H.W., Van Aartrijk, J., Eds.; Springer: Dordrecht, The Netherlands, 1990; Volume 9. [CrossRef]
150. Pence, V.C. Evaluating costs for the in vitro propagation and preservation of endangered plants. *Vitr. Cell. Dev. Biol. Plant* **2011**, *47*, 176–187. [CrossRef]
151. Backs-Hüsemann, D.; Reinert, J. Embryobildung durch isolierte Einzelzellen aus Gewebekulturen von *Daucus carota*. *Protoplasma* **1970**, *70*, 49–60. [CrossRef]
152. Sochacki, D.; Marciniak, P.; Ciesielska, M.; Zaród, J.; Sutrisno. The Influence of Selected Plant Growth Regulators and Carbohydrates on In Vitro Shoot Multiplication and Bulbing of the Tulip (*Tulipa* L.). *Plants* **2023**, *12*, 1134. [CrossRef] [PubMed]
153. Plessner, O.; Ziv, M.; Negbi, M. In vitro corm production in the saffron crocus (*Crocus sativus* L.). *Plant Cell Tissue Organ. Cult.* **1990**, *20*, 89–94. [CrossRef]
154. Santos, J.; Santos, I.; Salema, R. In vitro production of bulbs of *Narcissus bulbocodium* flowering in the first season of growth. *Sci. Hortic.* **1998**, *76*, 205–217. [CrossRef]
155. Wu, X.Q.; Liao, Z.P.; Luo, P.; Luan, A.Y.; Zeng, L.H. In vitro induction and regeneration of callus from three inflorescence organs of Chinese narcissus (*Narcissus tazetta* L. var. chinensis Roem). *J. Hortic. Sci. Biotechnol.* **2014**, *89*, 619–624. [CrossRef]
156. Bahr, L.R.; Compton, M.E. Competence for in vitro bulblet regeneration among eight Lilium genotypes. *HortScience* **2004**, *39*, 127–129. [CrossRef]
157. Sultana, J.; Sutlana, N.; Siddique, M.; Islam, A.; Hossain, M.; Hossain, T. In vitro bulb production in Hippeastrum (*Hippeastrum hybridum*). *J. Cent. Eur. Agric.* **2010**, *11*, 469–474.
158. Kizil, S.; Khawar, K.M. The effects of plant growth regulators and incubation temperatures on germination and bulb formation of Fritillaria persica L. *Propag. Ornam. Plants* **2014**, *14*, 133–138.
159. Stanišić, M.; Raspor, M.; Ninković, S.; Milošević, S.; Ćalić, D.; Bohanec, B.; Trifunović, M.; Petrić, M.; Subotić, A.; Jevremović, S. Clonal fidelity of Iris sibirica plants regenerated by somatic embryogenesis and organogenesis in leaf-base culture—RAPD and flow cytometer analyses. *South Afr. J. Bot.* **2015**, *96*, 42–52. [CrossRef]
160. Maślanka, M.; Mazur, J.; Kapczyńska, A. In Vitro Organogenesis of Critically Endangered *Lachenalia viridiflora*. *Agronomy* **2022**, *12*, 475. [CrossRef]
161. Doğan, S.; Çağlar, G.; Palaz, E.E. The effect of different applications on in vitro bulb development of an endemic hyacinth plant (*Hyacinthus orientalis* L. subsp. *chionophyllus* Wendelbo) grown in Turkey. *Turk. J. Agric. Food Sci. Technol.* **2020**, *8*, 1713–1719. [CrossRef]
162. Çakmak, D.; Karaoglu, C.; Aasim, M.; Sancak, C.; Özcan, S. Advancement in protocol for in vitro seed germination, regeneration, bulblet maturation, and acclimatization of *Fritillaria persica*. *Turk. J. Biol.* **2016**, *40*, 878–888. [CrossRef]
163. Ebrahimzadeh, H.; Karamian, R.; Nouri, D.M. Somatic Embryogenesis and Regeneration of Plantlet in Saffron, *Crocus sativus* L. 2000. Available online: https://www.sid.ir/paper/536838/en (accessed on 3 January 2024).

164. Sheibani, M.; Nemati, S.H.; Davarinejad, G.H.; Azghandi, A.V.; Habashi, A.A. Induction Of Somatic Embryogenesis In Saffron Using Thidiazuron (TDZ). In Proceedings of the ISHS Acta Horticulturae 739: II International Symposium on Saffron Biology and Technology, Masshad, Iran, 28 October 2006; pp. 259–267. [CrossRef]
165. Marković, M.; Trifunović-Momčilov, M.; Radulović, O.; Paunović, D.M.; Antonić Reljin, D.D.; Uzelac, B.; Subotić, A. The Effects of Different Auxin–Cytokinin Combinations on Morphogenesis of *Fritillaria meleagris* Using Bulb Scale Sections In Vitro. *Horticulturae* **2023**, *9*, 910. [CrossRef]
166. Kocak, M.; Izgu, T.; Sevindik, B.; Tutuncu, M.; Curuk, P.; Simsek, O.; Aka Kacar, Y.; Teixeira da Silva, J.A.; Yalcin Mendi, Y. Somatic embryogenesis of Turkish *Cyclamen persicum* Mill. *Scientia Horticulturae* **2014**, *172*, 26–33. [CrossRef]
167. Bach, A.; Ptak, A. Somatic embryogenesis and plant regeneration from ovaries of *Tulipa gesneriana* L. in in vitro cultures. *Acta Hortic.* **2001**, 391–394. [CrossRef]
168. Podwyszyńska, M.; Marasek-Ciolakowska, A. Micropropagation of tulip via somatic embryogenesis. *Agronomy* **2020**, *10*, 1857. [CrossRef]
169. İzgü, T.; Sevindik, B.; Çürük, P.; Şimşek, Ö.; Aka Kaçar, Y.; Teixeira da Silva, J.A.; Yalçın Mendi, Y. Development of an efficient regeneration protocol for four Cyclamen species endemic to Turkey. *Plant Cell Tissue Organ Cult. (PCTOC)* **2016**, *127*, 95–113. [CrossRef]
170. Kiviharju, E.; Tuominen, U.; Törmälä, T. The effect of explant material on somatic embryogenesis of *Cyclamen persicum* Mill. *Plant Cell Tissue Organ. Cult.* **1992**, *28*, 187–194. [CrossRef]
171. Otani, M.; Shimada, T. Somatic embryogenesis and plant regeneration from *Cyclamen persicum* Mill. leaf cultures. *Plant Tissue Cult. Lett.* **1991**, *8*, 121–123. [CrossRef]
172. Takamura, T.; Miyajima, I.; Matsuo, E. Somatic embryogenesis of *Cyclamen persicum* Mill. *'Anneke'from aseptic seedlings. Plant Cell Rep.* **1995**, *15*, 22–25. [CrossRef] [PubMed]
173. Wicart, G.; Mouras, A.; Lutz, A. Histological study of organogenesis and embryogenesis in *Cyclamen persicum* Mill. tissue cultures: Evidence for a single organogenetic pattern. *Protoplasma* **1984**, *119*, 159–167. [CrossRef]
174. Fersing, J.; Mouras, A.; Lutz, A. First stage in the industrial vegetative propagation of Cyclamen by in vitro culture. *Pepinier. Hortic. Maraichers* **1982**, *224*, 27–30.
175. Kumari, A.; Baskaran, P.; Van Staden, J. In vitro propagation via organogenesis and embryogenesis of *Cyrtanthus mackenii*: A valuable threatened medicinal plant. *Plant Cell Tissue Organ Cult. (PCTOC)* **2017**, *131*, 407–415. [CrossRef]
176. Nautiyal, A.; Ramlal, A.; Agnihotri, A.; Rashid, A. Stress-induced somatic embryogenesis on seedlings of *Azadirachta indica* A. Juss. by thidiazuron and its inhibition by ethylene modulators. *Plant Cell Tissue Organ Cult. (PCTOC)* **2023**, *153*, 357–366. [CrossRef]
177. Yu, Y.; Qin, W.; Li, Y.; Zhang, C.; Wang, Y.; Yang, Z.; Ge, X.; Li, F. Red light promotes cotton embryogenic callus formation by influencing endogenous hormones, polyamines and antioxidative enzyme activities. *Plant Growth Regul.* **2019**, *87*, 187–199. [CrossRef]
178. Mendi, N.Y.Y.; Özdemir, Ş.; Ugur, S.; Sevindik, B.; İzgu, T.; Kaya, E.; Tutuncu, M.; Çuruk, P.; Şimsek, Ö.; Koyuncu, O. In vitro Regeneration and Synthetic Seed Production of *Colchicum cilicicum* Grown Naturally in Turkey. *Alatarım* **2022**, *21*, 18–26.
179. Beruto, M.; Curir, P.; Debergh, P. Callus growth and somatic embryogenesis in thalamus tissue of *Ranunculus asiaticus* L. cultivated in vitro: Cytokinin effect and phenol metabolism. *Vitr. Cell. Dev. Biol. Plant* **1996**, *32*, 154–160. [CrossRef]
180. Azeri, F.N.; Öztürk, G. Microbulb and plantlet formation of a native bulbous flower, *Lilium monodelphum* M. Bieb, var. Armenum, through tissue culture propagation. *Biotechnol. Rep.* **2021**, *32*, e00665. [CrossRef]
181. Murthy, H.N.; Joseph, K.S.; Paek, K.Y.; Park, S.Y. Bioreactor systems for micropropagation of plants: Present scenario and future prospects. *Front. Plant Sci.* **2023**, *14*, 1159588. [CrossRef]
182. Akita, M.; Takayama, S. Induction and development of potato tubers in a jar fermentor. *Plant Cell Tissue Organ. Cult.* **1994**, *36*, 177–182. [CrossRef]
183. Dewir, Y.H.; Alsadon, A.; Al-Aizari, A.A.; Al-Mohidib, M. In vitro floral emergence and improved formation of saffron daughter corms. *Horticulturae* **2022**, *8*, 973. [CrossRef]
184. Gao, R.; Wu, S.-Q.; Piao, X.-C.; Park, S.-Y.; Lian, M.-L. Micropropagation of *Cymbidium sinense* using continuous and temporary airlift bioreactor systems. *Acta Physiol. Plant.* **2014**, *36*, 117–124. [CrossRef]
185. Hao, Z.; Ouyang, F.; Geng, Y.; Deng, X.; Hu, Z.; Chen, Z. Propagation of potato tubers in a nutrient mist bioreactor. *Biotechnol. Tech.* **1998**, *12*, 641–644. [CrossRef]
186. Jiménez, E.; Pérez, N.; de Feria, M.; Barbón, R.; Capote, A.; Chávez, M.; Quiala, E.; Pérez, J.C. Improved production of potato microtubers using a temporary immersion system. *Plant Cell Tissue Organ. Cult.* **1999**, *59*, 19–23. [CrossRef]
187. Jo, U.; Murthy, H.; Hahn, E.; Paek, K. Micropropagation of *Alocasia amazonica* using semisolid and liquid cultures. *Vitr. Cell. Dev. Biol. Plant* **2008**, *44*, 26–32. [CrossRef]
188. Kim, E.; Hahn, E.; Murthy, H.; Paek, K. Enhanced shoot and bulblet proliferation of garlic (*Allium sativum* L.) in bioreactor systems. *J. Hortic. Sci. Biotechnol.* **2004**, *79*, 818–822. [CrossRef]
189. Lian, M.L.; Piao, X.C.; Park, S.Y. Mass Production of Lilium Bulblets in Bioreactors. In *Production of Biomass and Bioactive Compounds Using Bioreactor Technology*; Paek, K.Y., Murthy, H., Zhong, J.J., Eds.; Springer: Dordrecht, The Netherlands, 2014; pp. 389–415. [CrossRef]

90. Lian, M.; Chakrabarty, D.; Paek, K. Bulblet formation from bulbscale segments of Lilium using bioreactor system. *Biol. Plant.* **2003**, *46*, 199–203. [CrossRef]
91. Lim, S.; Seon, J.; Paek, K.; Han, B.; Son, S. Development of pilot scale process for mass production of Lilium bulblets in vitro. In Proceedings of the International Symposium on Biotechnology of Tropical and Subtropical Species Part 2 461, Brisbane, QLD, Australia, 29 September–3 October 1997; pp. 237–242. [CrossRef]
92. Piao, X.C.; Chakrabarty, D.; Hahn, E.J.; Paek, K.Y. A simple method for mass production of potato microtubers using a bioreactor system. *Curr. Sci.* **2003**, *84*, 1129–1132.
93. Teisson, C.; Alvard, D. In vitro production of potato microtubers in liquid medium using temporary immersion. *Potato Res.* **1999**, *42*, 499–504. [CrossRef]
94. Yu, W.-C.; Joyce, P.; Cameron, D.; McCown, B. Sucrose utilization during potato microtuber growth in bioreactors. *Plant Cell Rep.* **2000**, *19*, 407–413. [CrossRef]
95. Raunkiaer, C. *The Life Forms of Plants and Statistical Plant Geography; Being the Collected Papers of C. Raunkiaer*; CABI: Wallingford, UK, 1934.
96. Sheikh, F.R.; Jose-Santhi, J.; Kalia, D.; Singh, K.; Singh, R.K. Sugars as the regulators of dormancy and sprouting in geophytes. *Ind. Crops Prod.* **2022**, *189*, 115817. [CrossRef]
97. Borochov, A.; Spiegelstein, H.; Weiss, D. Dormancy and storage of geophytes. In Proceedings of the VII International Symposium on Flowerbulbs 430, Herzliya, Israel, 10 March 1996; pp. 405–410. [CrossRef]
98. Marković, M.; Trifunović Momčilov, M.; Uzelac, B.; Jevremović, S.; Subotić, A. Bulb Dormancy In Vitro—*Fritillaria meleagris*: Initiation, Release and Physiological Parameters. *Plants* **2021**, *10*, 902. [CrossRef] [PubMed]
99. Zhao, Y.; Liu, C.; Sui, J.; Liang, J.; Ge, J.; Li, J.; Pan, W.; Yi, M.; Du, Y.; Wu, J. A wake-up call: Signaling in regulating ornamental geophytes dormancy. *Ornam. Plant Res.* **2022**, *2*, 8. [CrossRef]
100. Lang, G.A.; Early, J.D.; Martin, G.C.; Darnell, R.L. Endo-, para-, and ecodormancy: Physiological terminology and classification for dormancy research. *HortScience* **1987**, *22*, 371–377. [CrossRef]
101. Takayama, S.; Misawa, M. Differentiation in Lilium bulbscales grown in vitro. Effects of activated charcoal, physiological age of bulbs and sucrose concentration on differentiation and scale leaf formation in vitro. *Physiol. Plant.* **1980**, *48*, 121–125. [CrossRef]
102. Langens-Gerrits, M.M.; Miller, W.B.; Croes, A.F.; De Klerk, G.-J. Effect of low temperature on dormancy breaking and growth after planting in lily bulblets regenerated in vitro. *Plant Growth Regul.* **2003**, *40*, 267–275. [CrossRef]
103. Carasso, V.; Mucciarelli, M. In vitro bulblet production and plant regeneration from immature embryos of *Fritillaria tubiformis* Gren. & Godr. *Propag. Ornam. Plants* **2014**, *14*, 101–111.
104. Santos, A.; Fidalgo, F.; Santos, I. In vitro propagation of *Hyacinthus orientalis* cv. Jan Boss from bulb twin-scale explants. *Floric. Ornam. Plant Biotechnol.* **2006**, *2*, 561–563.
105. Nower, A. Micropropagation and Bulblet Formation In vitro of Dutch *Iris* cv. Blue Magic. *J. Product. Dev.* **2007**, *12*, 329–340. [CrossRef]
106. Staikidou, I.; Watson, S.; Harvey, B.M.; Selby, C. Narcissus bulblet formation in vitro: Effects of carbohydrate type and osmolarity of the culture medium. *Plant Cell Tissue Organ. Cult.* **2005**, *80*, 313–320. [CrossRef]
107. Podwyszyńska, M.; Ross, H. Formation of tulip bulbs in vitro. *Acta Hortic.* **2003**, *616*, 413–419. [CrossRef]
108. Islam, M.S.; Roni, M.Z.K.; Shimasaki, K. Factors affecting bulblet growth of 'Lilium sp. in vitro' and 'in vivo'. *Plant Omics* **2017**, *10*, 263–268. [CrossRef]
109. Langens-Gerrits, M.; De Klerk, G.J.; Croes, A. Phase change in lily bulblets regenerated in vitro. *Physiol. Plant.* **2003**, *119*, 590–597. [CrossRef]
110. Kamo, K.; Roh, M.; Blowers, A.; Smith, E.; Van Eck, J. Transgenic Gladiolus. In *Transgenic Crops III. Biotechnology in Agriculture and Forestry*; Bajaj, Y.P.S., Ed.; Springer: Berlin/Heidelberg, Germany, 2001; Volume 48. [CrossRef]
111. Van Tuyl, J.M.; Arens, P.; Marasek-Ciolakowska, A. Breeding and genetics of ornamental geophytes. In *Ornamental Geophytes: From Basic Science to Sustainable Horticultural Production*, 1st ed.; Kamenetsky, R., Okubo, H., Eds.; CRC Press: Boca Raton, FL, USA, 2013; pp. 131–158.
112. Van Harmelen, M.; Löffler, H.; Van Tuyl, J. Somaclonal variation in lily after in vitro cultivation. In Proceedings of the VII International Symposium on Flowerbulbs 430, Herzliya, Israel, 10 March 1996; pp. 347–350. [CrossRef]
113. Abdolinejad, R.; Shekafandeh, A.; Jowkar, A.; Gharaghani, A.; Alemzadeh, A. Indirect regeneration of Ficus carica by the TCL technique and genetic fidelity evaluation of the regenerated plants using flow cytometry and ISSR. *Plant Cell Tissue Organ Cult. (PCTOC)* **2020**, *143*, 131–144. [CrossRef]
114. Konar, S.; Adhikari, S.; Karmakar, J.; Ray, A.; Bandyopadhyay, T.K. Evaluation of subculture ages on organogenic response from root callus and SPAR based genetic fidelity assessment in the regenerants of Hibiscus sabdariffa L. *Ind. Crops Prod.* **2019**, *135*, 321–329. [CrossRef]
115. Vitamvas, J.; Viehmannova, I.; Cepkova, P.H.; Mrhalova, H.; Eliasova, K. Assessment of somaclonal variation in indirect morphogenesis-derived plants of Arracacia xanthorrhiza. *Pesqui. Agropecuária Bras.* **2019**, *54*, e00301. [CrossRef]
116. Memon, N.; Yasmin, A.; Pahoja, V.; Hussain, Z.; Ahmad, I. In vitro regeneration of gladiolus propagules. *J. Agr. Technol.* **2012**, *8*, 2331–2351.
117. Asadi, N.; Zarei, H.; Hashemi-Petroudi, S.H.; Mousavizadeh, S.J. Micropropagation and assessment of somaclonal variation in *Galanthus transcaucasicus* in vitro plantlets. *Ornam. Hortic.* **2021**, *27*, 505–515. [CrossRef]

218. Hesami, M.; Jones, A.M.P. Application of artificial intelligence models and optimization algorithms in plant cell and tissue cultur *Appl. Microbiol. Biotechnol.* **2020**, *104*, 9449–9485. [CrossRef] [PubMed]
219. Kucukrecep, A.; Tekdal, D. Machine Learning Applications for Plant Biotechnology: Modeling of The Plant Tissue Cultur Procedures with Artificial Neural Networks. *Kadirli Uygulamalı Bilim. Fakültesi Derg.* **2022**, *2*, 306–315.
220. Prasad, V.; Gupta, S.D. Photometric clustering of regenerated plants of gladiolus by neural networks and its biological validatio *Comput. Electron. Agric.* **2008**, *60*, 8–17. [CrossRef]
221. Loyola-Vargas, V.M.; Avilez-Montalvo, R.N. Plant Tissue Culture: A Battle Horse in the Genome Editing Using CRISPR/Cas In *Plant Cell Culture Protocols*; Loyola-Vargas, V.M., Ochoa-Alejo, N., Eds.; Springer: New York, NY, USA, 2018; pp. 131–14 [CrossRef]
222. Zhang, M.; Ma, X.; Jin, G.; Han, D.; Xue, J.; Du, Y.; Chen, X.; Yang, F.; Zhao, C.; Zhang, X. A Modified Method for Transien Transformation via Pollen Magnetofection in Lilium Germplasm. *Int. J. Mol. Sci.* **2023**, *24*, 15304. [CrossRef] [PubMed]
223. Abbasi, H.; Naderi, R.; Kafi, M.; Azadi, P.; Shakh-Asadi, M.; Okazaki, K. Effect of 'Chloroxynil' on Agrobacterium-mediate transformation efficiency of Lilium cv 'Manissa'. *Sci. Hortic.* **2020**, *271*, 109404. [CrossRef]

horticulturae

MDPI

Review

The New Green Challenge in Urban Planning: The Right Genetics in the Right Place

Silvia Farinati †, Angelo Betto †, Fabio Palumbo, Francesco Scariolo, Alessandro Vannozzi and Gianni Barcaccia *

Department of Agronomy, Food, Natural Resources, Animals and Environment (DAFNAE), Campus of Agripolis, University of Padova, Viale dell'Università 16, 35020 Legnaro, PD, Italy
* Correspondence: gianni.barcaccia@unipd.it
† These authors contributed equally to this work.

Abstract: The creation of green areas within urban centers was born as a response to profoundly different problems, such as the demographic increase and the progressive urbanization of landscapes. Moreover, up to date, the genetics of plants has not been considered for urban contexts. The purpose of this review is to discuss the state of the art and the advantages in planning genetic improvement for plants for urban destinations, also providing technical information, that may contribute in a concrete way. Firstly, recent genomic sources and their potential contribution to breeding programs are discussed. Then, dividing the urban scenarios into four macro areas (i.e., urban/metropolitan parks, urban gardens, road verges and roofs/terraces/balconies), we described the role of genetics in the adaptability and sustainability of plants in these different contexts. In addition, we analyzed the genetic traits plants need to provide services for a city environment and population (e.g., pollution reduction, biodiversity conservation, soil stability, and food production).

Keywords: urban area; plant breeding; ideotype; plant genetics; plant genomics urban vegetation; adaptability; sustainability; marker-assisted selection; marker-assisted breeding

Citation: Farinati, S.; Betto, A.; Palumbo, F.; Scariolo, F.; Vannozzi, A.; Barcaccia, G. The New Green Challenge in Urban Planning: The Right Genetics in the Right Place. *Horticulturae* **2022**, *8*, 761. https://doi.org/10.3390/horticulturae8090761

Academic Editors: Margherita Beruto, Emmy Dhooghe and Bruce Dunn

Received: 17 June 2022
Accepted: 20 August 2022
Published: 24 August 2022

1. Introduction

Several examples document the inevitable and continuous interaction between humans and plants, primarily characterized by the human interest in using them for consumption, materials production, medical purposes, decoration, and bioremediation. An example of the most striking human manipulations on the plants' life and environment is the establishment of urban green spaces, for instance, public parks and/or private residential gardens, with surfaces more or less limited, where plants are grown with objectives often different from those of agriculture [1]. In fact, if agriculture has aims that are almost exclusively at production, the creation of green areas within urban centers was born as a response to profoundly different problems, such as the demographic increase and the progressive urbanization of landscapes. The latter is among the dominant anthropogenic factors influencing the natural and adaptive evolution of living populations [2], and for this reason, it is the most important phenomenon affecting the environments in the 21st century [3–6].

Considering the multitude of urban contexts, purposes, and needs for which green spaces in cities are created, it is today very challenging to provide an exhaustive definition of 'urban area' and its relative 'urban vegetation', since the geographic, climatic, and resource-related opportunities, and constraints, are not equally distributed factors across the world and specific for each context. Furthermore, urban vegetation can also include cultural plant typology with agricultural interest related to food production, such as the horticultural species. Starting from these considerations, the present review aims to provide a general overview of how genetics and genomics notions can help in the selection and improvement of plant species and/or suitable varieties for specific urban contexts. In detail, after a brief overview of the genomic tools available and potentially usable in genetic improvement plans, we focused on the evaluation of particular genetic traits for the selection of plants in relation to the urban area in which they are used and the environmental stress to which they are generally subject. Finally, we took into consideration the contribution of genetics in achieving specific goals that are usually set in urban contexts.

Processing Criteria

With a univocal classification of urban green spaces not being present [7–9], in order to allow a better understanding of some of the aspects covered, we judged it appropriate to simplify the classification of these environments by empirically dividing them into four macro areas, according to their surface availability per plant and their intended usage as follows:

(i) *Urban/metropolitan parks:* these are green areas in cities, and other incorporated places, that offer recreation and green spaces to residents of and visitors to the municipality;
(ii) *Urban gardens:* these are areas where urban vegetation is exploited to provide food products, especially employing horticultural species. Urban food production can be carried out by citizens or administrations in private buildings and public spaces for self-consumption, or can be performed by farms with commercial purposes, also using innovative outdoor or indoor growth systems;
(iii) *Road verges:* these are small, vegetated areas composed of grass or plants and sometimes also trees, mainly located between a roadway and a sidewalk or within roundabouts;
(iv) *Roofs/terraces/balconies:* these are small green areas located in private or public buildings. They include both surfaces partially or completely covered with vegetation (e.g., green roofs) and container gardens where plants are maintained in pots.

Each of these four categories is set up according to different purposes (ornamental, urban planning, bioremediation, soil consolidation purposes, etc.), that, in turn, affect the choice of the species/varieties to be used (i.e., woody perennial plants, meadows and turf grasses, horticultural and floricultural ornamental plants), and their adaptive characteristics (resistance to biotic, abiotic stress, hypogeal and epigeal habitus, minimum maintenance requirement, etc.). For this reason, as previously indicated, it is not easy if not impossible to define specific and common breeding programs for urban vegetation, since it includes both agricultural and not species with different genetic and physiological properties, and agronomical needs (Figure 1).

The bibliographic analysis for this work has been performed exploiting Scopus® in order to search for articles and reviews of interest written in English by integrating different combinations of selected keywords for every paragraph, including terms defining:

(a) The environment of interest: "urban areas", "cities", "green areas, "green gardens" "public green", "public parks", "urban agriculture";
(b) The plant typologies: "plant", "ornamental", "flowering", "horticultural", "woody", "trees", "meadows", "turfgrasses";
(c) The genetic subject: "breeding" "molecular markers", "marker-assisted selection" "marker-assisted breeding" "molecular selection" "genomic selection" "genomics" "genetic improvement" "variety" "cultivar";
(d) Specific goals: "abiotic stress", "heat shock", "biotic stress", "pathogen stress", "water stress", "drought", "dwarf" "compacted", "growth habit", "edible flowers" "food production", "leafy vegetation", "baby leaf", "phytoremediation", "air purification", "biodiversity", "soil erosion", "soil stability", "psychosocial", "ecosystem services".

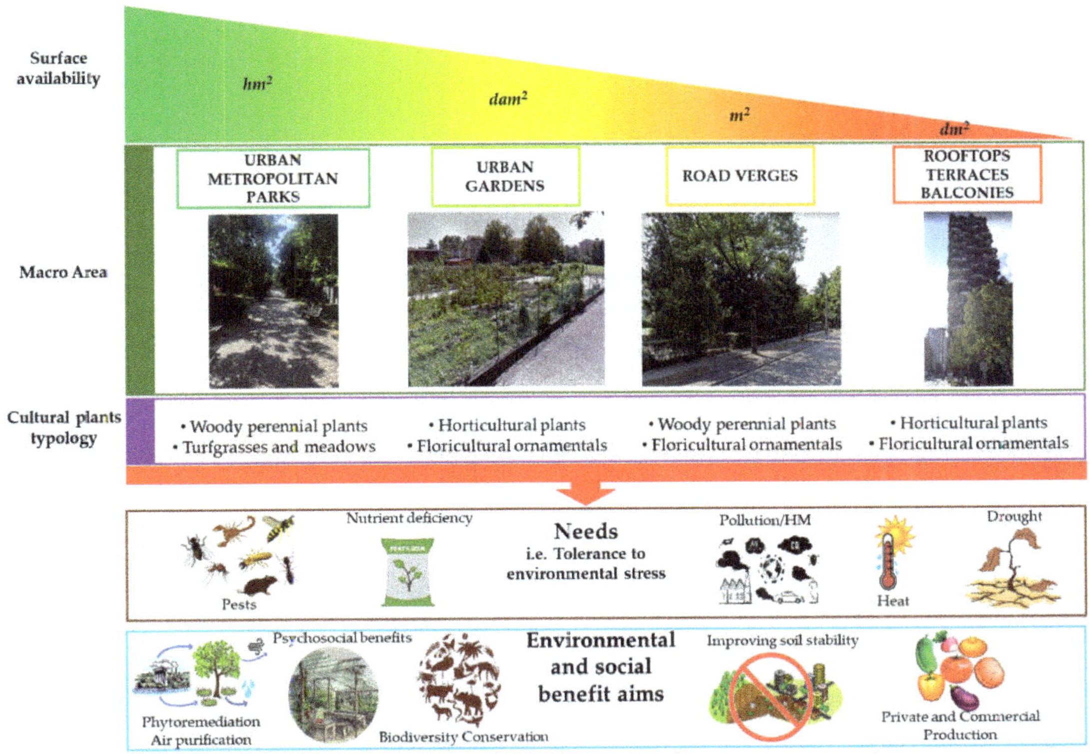

Figure 1. Schematic representation of the multilayered framework of the most common contexts, here defined as Macro Areas, found in an urban environment. In each macro area, the goals to achieve are multiple and specific, depending on the characteristics and needs of every specific zone, and specific vegetation type that is desired to use. However, under all conditions, the need to use specific cultural plant typology able to cope and tolerate environmental stresses is a fundamental requirement. In addition, the potential environmental and social benefits deriving from them are indicated. Details are described in the text.

The lists of documents found have been then manually screened to select the most relevant works. In addition, the research has been deepened with a snowball sampling approach, exploiting citations and references included in the already screened papers.

2. Genetic Information as a Genomic Tool That Is Potentially Helpful in Breeding Approaches for Urban Contexts

Plant breeding is a science-driven creative process that involves the evaluation, selection and propagation of plant populations characterized by a combination of specific desirable traits [10]. Originally, conventional breeding was exclusively based on time-consuming and expensive phenotyping observations: plant selection was accomplished through massive screening of phenotypic parameters mainly related to the morphology of the plant (fruits, flowers, stem), to its tolerance to both biotic (insects, molds and viruses), and abiotic stresses (temperature, drought, salinity, heavy metals) [11].

Genetics made it possible to overcome many of the limitations associated with conventional breeding, facilitating, and speeding up the process of varietal constitution. New opportunities for crop improvement were generated thanks to the increasing availability of whole genome data that, in turn, enabled the discovery of candidate genes related to traits of interest, mutations responsible for phenotypic variability (e.g., single nucleotide polymorphisms, SNP or insertion/deletion, Indel) and molecular markers associated with traits of interest (e.g., simple sequence repeats, SSR) [12]. Overall, these molecular data are powerful predictive tool for the targeted selection of superior genotypes. However, as their discovery is strictly dependent on the availability of sequenced genomes, for many years their application in breeding processes has been limited to a few sequenced species of great agricultural and economic importance, such as rice, barley, tomato, and grapevine [13]. Moreover, before the advent of single molecule real-time (SMRT) sequencing (better known as third-generation sequencing platforms), the complex architecture of a genome represented another major obstacle to the assembly process. The set of sequenced genomes belonging to ornamental species and some tree families still only represents a small fraction of all genomes sequenced thus far for three main and often coexisting reasons [14]. First, in this category of plants, the high level of heterozygosity along with the abundance of GC and repetitive sequences represent a significant source of ambiguity during genome reconstruction. Second, some fruit tree species are often widespread in urban areas (e.g., the *Citrus* genus) [15], and several ornamental plants (e.g., *Rosa*, *Chrysanthemum*, *Lilium*, *Mandevilla*, and *Primula*) [16] are the result of intraspecific and interspecific crosses and ploidy manipulation. These aspects increase the probability of chimeric assemblies because of incorrect connections between scaffolds belonging to homologous chromosomes [17]. Finally, technical problems in the assembly phase are frequently encountered in species characterized by mega-genomes (i.e., genomes > 10 Gb), such as ornamental plants (e.g. peony, 13.79 Gb) [18] and tree genera typical of the urban context (e.g., *Cedrus*, 15–16 Gb and *Pinus*, 31 Gb) [19].

The progressive collapse of sequencing costs and the release of tools able to overcome part of the technical challenges previously mentioned made HTS (High-Throughput Sequencing) platforms accessible also to minor crop species (also known as orphan crops, for review, see for example Simko et al. [20] and Bohra et al. [21]) and ornamental plants (for review, see Zheng et al. [14]). As a matter of example, the number of sequenced genomes completed yearly for ornamental plants significantly increased from 1 in 2012 to 17-20 in the last five years, and currently, the genome sequences of more than 70 ornamental species have been released [14]. Conversely, most of the tree species frequently found in urban green areas of the Mediterranean basin (e.g., *Carpinus betulus*, *Tilia cordata*, *Acer platanoides* and *Pinus pinea*) remain unsequenced today.

Genomics and transcriptomics open a wealth of opportunities not only for large-scale plant breeding but also for small-scale cultivations, such as those in an urban environment. In Table 1 we reported the main molecular approaches currently used to connect phenotype to genotype. In fact, the ultimate goal in plant breeding is to use the genotypic information to predict phenotypes and select improved cultivars [22,23].

Table 1. Overview of the main genomics and transcriptomics techniques useful for breeding purposes.

Techniques		General Description	To Learn More about
Whole Genome Sequencing (WGS)		The genome of a species is assembled for the first time into chromosomes with high coverage and it is functionally annotated to produce a reference assembly and to predict hundreds of loci underlying agronomic traits	[24]
RNA-seq analysis		RNA-seq can be used to examine the RNA sequences that are present in a sample (transcriptome). This is crucial for linking the information contained within the genome with the functional proteins that are expressed. RNA-seq can be used to elucidate which genes are turned on or off within a cell under specific conditions	[25]
Whole genome resequencing (WGR)		The genome is fully sequenced with low or modest coverage and is aligned against the reference genome assembly to predict allelic variants	[26]
Reduced Representation Sequencing (RRS)	GBS	A fraction of the genome is sequenced and aligned against the reference genome assembly to predict allelic variants. For GBS, ddRAD-seq, 2bRAD-seq and ezRAD-seq the regions to be sequenced are randomly chosen using restriction enzymes, for target-seq the regions to be sequenced are selected through PCR	[27]
	ddRAD-seq		[28]
	2bRAD-seq		[29]
	ezRAD-seq		[30]
	target-seq		[31]

Regardless of the methodology used, if the allelic variant(s) is responsible for a favorable trait, it can be subsequently introgressed into an elite variety through a combination of conventional breeding and molecular marker-assisted selection (MAS). MAS is defined as a process where molecular markers associated with a specific trait of interest are used for the selection of the trait itself. In this sense, the use of molecular markers is particularly indicated when the phenotypic evaluation of the trait is hardly doable (e.g., disease resistance) or expensive (e.g., male sterility) [24]. MAS is also effective in selecting desirable gene alleles independently from possible confounding environmental, pleiotropic or epistatic effects, in monitoring the introgression of a desirable allele in backcrossing events, or in identifying and avoiding linkage drag effects [32].

Considering the ratio between extragenic and genic regions, most variants identified through WGR or RRS are unlikely to underlie phenotypic effects. However, the combination of hundreds of noncoding allelic variants produces molecular profiles able to identify uniquely a given genotype. Such information is extensively used for screening populations and selecting parental genotypes in breeding programs (marker-assisted breeding, MAB), testing the purity and uniformity of commercial varieties, evaluating the stability of cultivars through generations, registering new varieties and for assessing their distinctiveness from preexisting varieties [33].

For large-scale crop breeding (open-field and greenhouses), the marker-assisted genetic improvement techniques mentioned above are well documented [34]. The development and incorporation of molecular markers in breeding processes could also be adopted for small-scale cultivation approaches (from urban parks to small private allotments). In the latter case, the goal would be the development of varieties with a genetic background suitable for urban needs. In this context, the contribution of genomics in identifying markers and genes responsible for plant adaptability in urban environments will play a crucial role.

3. The Role of Genetics in the Adaptability and Sustainability of Plants in Different Urban Contexts

In the development of any plant breeding program, it is necessary to define the desired variety's ideotype, i.e., the ideal plant model endowed with every trait of interest for the specific aims and destinations [35]. While some adaptive characteristics of plants are shared by all types of urban vegetation, there are others that must respond to specific context conditions; thus, the ideotype can change according to the macro area considered. For example, the soil requirements of a tree grown on a roadside must necessarily be distinct from those of a tree grown in an urban park. This is both to allow the plant to grow by bypassing the stress of soil deficiency and to prevent it from causing damage to the road surface as it grows. Generally, plants must be able to cope with the environment in which they are introduced, not suffering from abiotic or biotic stresses. In fact, urban areas represent environments characterized by restrictive growing conditions due to a high density of buildings, infrastructures, and heavy traffic, which cause negative effects such as soil compaction, pollution, high temperatures, drought, lack of light, presence of heavy metals, high salinity, and nutrient deficiency [36]. The ability to cope with different kinds of stress has become even more important, considering the increasing impact that environmental changes and global warming have on plants [37]. In past years, it has been reported that abiotic stresses reduced by as much as 50% the yields of most major crops worldwide [37–39]. Providing varieties able to adapt to several different types of environments and resilient to stress factors is in general a major challenge in the immediate future for breeders to guarantee cultivation [40–42]. Moreover, plants must also be resilient to stresses to limit the usage of agronomic inputs. Modern cultivars have been mainly bred for performance under a high supply of factors such as irrigation water, fertilizers, pesticides, and interventions such as tillage and manutention [43]. Hence, the need to provide varieties environmentally and economically sustainable stands with a view of limiting costs and energy exploited for their cultivation, reducing limited resource usage, and avoiding the introduction of polluting chemical compounds into the environment [44,45]. In addition, to guarantee plant sustainability, it is important that they do not present potentially harmful traits for ecosystems and human health, such as invasiveness and high-allergenic compound production [46]. An important strategy to cope with the adaptability and sustainability problem is to recur to wild or traditionally used populations, diffused in areas with environmental conditions similar to those of the destination of interest. This is a key factor because, generally, these accessions are resilient to the biotic and abiotic stress typical of their diffusion area [47,48]. In addition, not or lowly genetically improved populations present a higher genetic variability in comparison to commercial varieties, allowing to have a wider range of alleles to be drawn on for the specific breeding aims. From this point of view, the creation and maintenance of germplasm banks containing wide collections of plant material (seeds or plants conserved in vivo, in vitro or cryo-conserved) derived from wild populations and local varieties is a strategy of great relevance for having a continued availability of genotypes with useful traits [49,50]. With regard to germplasm collections, it is important to point out that presenting a high genetic variability also means an increase in the possibility of having in the selected genotypes alleles linked to undesirable or even harmful traits, e.g., the susceptibility to a pathogen or the production of a toxic compound. These alleles can segregate with traits of interest (linkage drag), therefore, during the breeding process, careful work of selection against them is necessary [51–53].

To analyze more in detail the adaptability of plants to the environment, the subsequent sections provide a specific discussion about two of the most impacting abiotic stresses in urban areas as examples, high temperatures and drought, in addition to the pathogen stress.

3.1. Abiotic Stresses: Heat and Water Stress

In urban areas, the average temperature can be recorded as up to 10 °C warmer than the surrounding rural areas due to impervious surface cover, anthropogenic heat sources, and low vegetation cover. Warmer temperatures increase vapor pressure deficits, creating a greater atmospheric demand for water via transpiration and reducing soil moisture, which limits the amount of water available to roots. In addition, heat stress negatively influences plant physiology as a consequence of protein damage, enzyme inactivation, photosynthesis inhibition, cell membrane deterioration, cell division interruption, and disturbance of reproductive phases [54–56]. Heat stress responses include the involvement of heat shock proteins (HSPs), which have a chaperone function, hence assisting the correct conformational folding of misformed proteins and helping in damaged protein movement and degradation [57]. In recent years, the molecular bases of high-temperature defense have been studied in the species of greatest horticultural interest and potentially usable in an urban context, including tomato (*Solanum lycopersicum* L.) [58,59], lettuce (*Lactuca sativa* L.) [60,61], spinach (*Spinacia oleracea* L.) [62,63], identifying specific transcription factors (heat stress factors—HSFs), HSPs or molecular markers related to them. Considering ornamental species, analogous results were obtained with *Rododendron hainanense* Merr. [64] and carnation (*Dianthus caryophyllus* L.) [65]. However, the use of molecular markers associated with HSPs or HSFs for breeding aims has received little attention outside of cereals. In chili pepper (*Capsicum annuum* L.) a study reported the introgression of two genes encoding HSP70 and HSP24 from a heat-tolerant breeding line to an elite commercial variety using markers and backcrosses with the ladder, without genetic transformation. For the tolerance-related gene selection, a MAS approach was conducted with two markers that were closely linked to the genes; while for the recovery of the recurrent parent line's genome, to obtain the agronomic traits of interest, a MAB strategy was followed with a set of 250 paired SSRs [66].

An increase in global warming leads to enhanced evapotranspiration, reduced rainfall, and consequent water stress [67,68]. The resulting water shock for plants can also be a serious issue for urban cultivation. The solution cannot consist of more frequent irrigation due to a need in reducing water withdrawals for agriculture and in ensuring the supplies of this limited resource all over the world in the future [69]. To select plant species and to develop drought-tolerant varieties, hence requiring minor irrigation, it is important to identify the correct traits of interest, since resorting to the right phenotyping strategies can be onerous, and resistance mechanisms are plentiful and well-differentiated among species through evolution [70,71]. An important drought-resistance plant type is represented by xeromorphic succulent species, such as *Cactaceae, Orchidaceae and Bromeliaceae*, which endure hydric deficiency through water-storing tissues with high vacuoles (*hydrenchyma*), the CAM (*Crassulacean Acid Metabolism*) carbon fixation pathway, or waxes and cuticles on the top page of leaves [72]. These plants are often used in urban areas for ornamental purposes; hence, the introduction of breeding programs for these non-genetically improved species can be an interesting choice for breeders. For other plant types, traits of interest for drought tolerance comprehend high leaf mass per area unit, leaf thickness, unit leaf area, and leaf water potential at the turgor loss point [20,73–76]. In addition to morphological parameters, biochemical analysis of the osmoprotectant content has assumed importance if used as metabolic markers, as well as assays for the enzymes and proteins most involved in signaling drought and other abiotic pathways related to stress response, such as superoxide dismutase (SOD), ascorbate peroxidase (APX) or LEA proteins [77]. Often, morphological and physiological responses to water deficiency are in common with those for other abiotic stresses, mainly derived from heat, salinity, and coldness, but in other cases, they can be in opposition. For example, a high accumulation of proline can protect against wilting but can also lead to toxic effects in the presence of high temperatures [70,78]. Hence, attention to these possible relations must be paid if the goal is plants' genetic improvement for multiple sources of stress. To date, the genetic bases of water deficiency tolerance have been investigated through genomic, transcriptomic, and metabolomic approaches also in species of the greatest horticultural interest but suitable in urban areas [48,79–82], and in sunflower [83–85]. This led to the identification of QTLs (*Quantitative Trait Loci)*, single genes, specific sequences such as miRNAs, or genome-wide marker collections potentially related to drought traits.

3.2. Pathogen Stress

Developing plants able to resist pathogens is a very important goal for breeders since biotic stresses are among the main limiting factors for world agriculture [25,86–88]. In addition, there is a need to reduce the use of harmful pesticides [89–92], an important aspect, especially in urban areas where human presence is considerably high and where people are more likely to come into contact with treatment residues that pollute the environment. Compared to other traits, pathogen resistance was more often found to be controlled by a single gene than by multilocus interactions. Qualitative resistances are mostly inherited dominantly; they concern biotrophic pathogens and are effective only against particular races of pathogen species [93]. This monogenic inheritance has allowed the fixing of the trait in many varieties of species commonly found in urban areas using traditional breeding methods, but the main issue remains that the durability of the resistance is low: fungi, oomycetes, bacteria and viruses can rapidly mutate in the gene coding for the product recognized by the host, making plants return susceptible. In addition, resistant plants exert a selection pressure that favors virulent mutations in pathogens [94–96]. Consequently, the time required for traditional breeding is often too long to provide new resistant varieties before pests adapt to the current ones. Possible methods to overcome this problem include the use of *multiline cultivars*, in which every line presents different resistance genes (R genes) [97], combined with the pyramiding approach, which consists of the transfer of several R genes into the same plant [98]. MAS can be very helpful in facilitating the R gene transfer from wild or local populations, often presenting resistance as showing adaptation to the environment, to the commercial lines and hence reducing breeding times. Many diagnostic markers are available for the horticultural species of highest commercial importance, such as tomato (*Solanum lycopersicum* L.), and they are currently used in private breeding programs (reviewed by Foolad and Panthee [99], Lee et al. [100], Bhardwaj et al. [101], Simko et al. [20]). On the other hand, it is rare for pathogens to adapt in a short time to quantitative resistance, but it is also true that this kind of plant resistance is often difficult to introduce into commercial varieties, even by using MAS. This is because quantitative resistance has been shown to be dependent on a great number of QTLs, each with a small effect [20,97]. A modern tool that could be helpful in future breeding programs for quantitative traits and particularly for pathogen resistances is genomic selection (GS). It refers to the exploitation of high-density molecular marker collections (produced exploiting the modern NGS technologies and mapping approaches such as GBS), with statistical models to predict the breeding value for complex traits of interest [102,103].

3.3. Limited Surface Availability

The surface availability, often limited, may be the most important aspect influencing the subsequent genetic improvement programs of the chosen cultural typology. This aspect is not always transversal for all habitats since it strongly depends on the subarea considered and is specific to the different cultural typologies: for example, the surface can be high in an urban park, medium in an urban garden and/or roadside/traffic islands, and low in the roof/terrace/balcony context. In urban parks, it is possible to plant those species that have high dimensions of aerial parts and roots, such as arboreal ones; however, in roadsides, traffic islands or buildings' terraces and balconies, the choice is more oriented toward plants able to grow in limited surfaces and, consequently, toward the selection of varieties with limited size [104,105].

To reduce the typical large sizes of trees, dwarf rootstocks were selected for the orchard species of greatest interest [106–108], with dwarfism being linked to reduced levels or perception of gibberellins (GA) and brassinosteroids (BR). Hence, genes related to the biosynthesis, and signaling of these phytohormones, in particular, the *Wound-induced Receptor-like protein Kinase* (*WRK*) and *NAC* transcription factors [109,110], are of main interest. An additional aspect that should be considered when developing varieties suitable for tight places is the growth habit. Indeed, also in the case of plants presenting dwarfism great development of lateral shoots would require additional horizontal space, thus limiting the plant density. Studies on shoot architecture have allowed, for example, to characterize the *LAZY1* gene, which is involved in the transport and signaling of auxins, promoting the upward orientation in shoots and the narrowed angles in branches [111,112], or the *Tiller Angle Control 1* (*TAC1*) gene, which induces the outward orientation in branches [113] or genes, e.g., *WEEP*, leading to a weeping growth habit [111,114,115]. In addition, it is important, mainly on roadsides, to select plants in relation to the architecture of the underground parts. This is not only to facilitate nutrient uptake and to limit competition between plants, but also to avoid the risk of ruining and deforming the soil surfaces because of overgrown roots, which is a major issue in many cities' road management [46,116,117].

Because of their generally small dimensions, the herbaceous plants for meadows and turfgrasses can be suitable for several surface availabilities, including buildings. Moreover, given the possibility of walking on them, the use of these species does not prohibit additional uses of the occupied space. The plant varieties that are the most selected for these uses generally belong to species of the *Poaceae* family and can be pure or derived from interspecific hybridizations [118]. Among the most common microthermal species, *Lolium perenne* L., *Poa* spp., *Agrotis* spp., and *Festuca* spp. can be found, while among macro therms, *Buchloe dactiloides* Nutt., *Cynodon* spp., *Eremochloa ophiuroides* Munro, and *Paspalum vaginatum* Sw. are present [119]. Traits of interest for turfgrasses varieties selection regard leaf fineness, rapidity in covering the surface, tolerance to trampling and mowing, slowness in sprout growing, and aesthetic aspects such as greenness and covering uniformity. Breeding programs regarding the species destinated for this use started in the 1970s and until today were carried out with continuous selection schemes, tests of progenies and intercrosses between the best-performing genotypes. For warm-season grasses, selection programs have been dominated by clonal propagation because stolons are easy for most species and interspecific progeny sterility is common [120]. Advances in genetic improvement have been relevantly focused on reducing the turf height increment rate, in finessing leaves, in wear tolerance, in environmental adaptation to different conditions and in crown-rust resistance. In contrast, the improvements that remain to be achieved are related to tolerance to other diseases, the preservation of greenness in winter and summer, and ground cover rapidity, all traits that seem to have high plasticity determined by genotype per environment (G $\times$ E) interactions. Sampoux et al. reported that the selection of plants to lessen the growth rate has probably limited the availability of the genetic resources necessary to improve the ground cover velocity and winter greenness [121].

4. What Are the Achievable Goals with the Help of Genetics?

Anthropogenic activities lead urban environments to be affected by many ecological issues of different natures, starting from pollution, which has an impact on soils as well as on water and air [122–125]. Plants, in addition to withstanding environmental conditions and not contributing to pollution, can directly reduce the contamination already present and in general provide ecosystem services that, consequently, improve the environmental state and human health [126,127]. The ability to offer a beneficial effect on the territory is an aspect of great interest for urban contexts that should be considered in breeding programs and urbanistic management [123]. Indeed, the setting of green areas can improve slope stability reducing erosion [128], it can favor faunal resettlement enhancing biodiversity [129], and can reduce rainfall water surface runoff favoring infiltration into the soil [130]. In addition, it has been demonstrated that plants can limit the phenomenon of urban heat islands by mitigating thermal changes, by increasing evapotranspiration, shadowing, radiation reflection and moisture traps, hence saving energy used for cooling [131,132], keeping lower noise pollution [133] and reducing the CO_2 emissions absorbing the gas through leaves [134]. Another important aspect influencing the choice of a precise cultural species for specific purposes is the ability to provide plant products, particularly food, and their intended use, whether commercial or for private self-consumption. Strongly related to this last factor, surface availability is above the main determinants that can influence the possibility of having an adequate level of productivity. This is in general the largest bottleneck, as space in urban areas is often very limited and indirectly forces specific use choices and selection priorities [104]. Last, although it is not a primary concern for genetic selection plans, the supply of the psychosocial benefits that are derived from the creation of green urban spots assumes an important role in cities, where the high human presence and the reduced availability of cultivated areas are typical [135].

4.1. *Phytoremediation*

Regarding the causes of pollution in urban areas, the major responsibilities are heavy metals, such as cadmium, lead and zinc, mainly found near roads [136], and organic compounds, such as petroleum hydrocarbons (PHCs), polycyclic aromatic hydrocarbons (PAHs), chlorinated solvents, explosives and polychlorinated biphenyls (PCBs) [137–139]. For a long time, it has been known that several woody and herbaceous ornamental species are able to reduce the presence of these molecules in the environment, a process that is called *phytoremediation*, with methods including the accumulation (*phytoextraction*) or conversion to harmless forms (*phytodegradation*) of pollutants [140]. The higher dimensions allow woody plants to absorb larger quantities of pollutants by roots and from deeper soils. In contrast, herbaceous plants can be used in small spaces and can be replaced more frequently, which is a useful aspect to avoid the risk of dispersion of the absorbed pollutants [141]. Instead, more difficulties can be seen in the use of food plants for phytoremediation, unless in the case of total contaminant degradation or if the accumulation occurs in completely separate organs from those related to commercial products. Regarding woody plants, species of the genera *Populus* and *Salix* are the most interesting for phytoremediation. Nevertheless, other species, such as *Ailanthus altissima* (Mill) Swingie, *Betula pendula* Roth, *Carpinus betulus* L., *Platanus × hispanica* Mill. Ex Muenchh. "Acerifolia" and *Robinia pseudoacacia* [142], have been recognized as promising choices for phytoremediation purposes, although in some cases, as for *Ailanthus altissima*, some of their physiological properties make them invasive species. Among herbaceous ornamentals for which the capacity of pollutant presence reduction has been assessed, there are *Dianthus chinensis* L., *Impatiens balsamina* L., *Portulaca grandiflora* Hook., *Portulaca oleracea* L., *Tagetes* spp., and *Vinca rosea* L. [141]. Although many suitable species for phytoremediation in urban areas have been selected, genetic improvement programs, including phytoremediation capacity as a trait of interest, have not yet been carried out with either classical or MAB approaches. Important information on the molecular bases has been obtained in the model species not employed in urban areas *Arabidopsis* and rice, identifying several responsible genes of heavy metal uptake, translocation, vacuolar sequestration, and genic regulation [143]. Using nontransgenic techniques, in rice genotypes with mutations in genes related to this aspect were identified after irradiation with mutagenic agents, and varieties of this cereal were selected with MAB methods only for low accumulation of cadmium or arsenic to avoid the risk of toxic molecules' presence in the edible products [18,144,145]. Nonetheless, these goals highlight the potential of using the same approaches to the opposite objective, hence, to obtain consistent heavy metal accumulation, and to extend this methodology also to plants suitable for urban areas. In addition to high toxic molecule uptake, accumulation, and degradation capacity, other traits of interest for future plant breeding programs for phytoremediation are size, surface and architecture of roots. They ideally would be designed to have the maximum pollutant interception capacity, given space restrictions and without incurring the risk of soil failure, especially along roadways [46]. Another important factor to consider in breeding varieties destinated to phytoremediation is avoiding the return of the absorbed pollutants to the environment. Therefore, for species that accumulate pollutants in their leaves, selecting genotypes without leaf scattering can be relevant [141].

4.2. Air Purification

The removal of CO_2 and gaseous pollutants, such as carbon monoxide, ozone and nitrogen dioxide, is performed by plants through interception by their stomata, while particulate matter (PM10 and PM2.5) adheres to leaf surfaces with the aid of wind currents, where it is then absorbed [146,147]. The effects of plants in reducing atmospheric pollution in cities were studied, demonstrating their potential, particularly that of trees, in improving air conditions [148,149]. It has been reported, for both woody and herbaceous species, that plants with a high presence of leaf hair or wax cover are the most performant in capturing PMs [150,151]. Indeed, wax layers facilitate PM adhesion, while hair greatly extends the surface available to intercept composts, a factor influenced also by leaf morphology, leaf area index (LAI), and, in general, aboveground parts' sizes. Leaf fineness and porosity are also important to favor air movement through the structure without being thrown away. In addition, the maintenance of the leaves for the entire life cycle is relevant to always guarantee pollutant removal services; hence, choosing deciduous trees can be limiting [46,152]. Finally, another aspect to consider is the single species' susceptibility to pollutants. In fact, there are plants, such as some conifers, that are evergreen and very efficient in particulate absorption, due to their complex shoot spatial structures, but they are also particularly susceptible to these pollutants, making them possibly unsuitable for the most polluted urban areas [122,148,150]. The reported traits are very interesting in species selection and genetic improvement for air purification purposes. Among the most performative woody and shrub species, *Albizia julibrissin* Durazz., *Betula pendula* Roth., *Cinnamomum camphora* (L.) J. Presl, *Nerium oleander* L., *Pinus mugo* Turra, *Pinus sylvestris* L., *Stephanandra incisa* (Thunb.) Zabel, *Taxus x Media* Rehder, and *Taxus baccata* L. have been considered for their air pollution tolerance and PM absorption capacity per leaf surface unit [150,153]. As previously reported, despite their small dimensions and their low unit absorption of pollutants, herbaceous plants have anyway *raison d'être*, considering the high plant density for a given surface and the possibility of coupling them with trees, allowing for a synergic effect due to the absorbance of harmful molecules at different height levels and of different types [46]. As potentially interesting herbaceous species for PM removal from the atmosphere, *Sisymbrium loeselii* L., *Polygonum aviculare* L., *Convolvulus arvensis* L., *Chenopodium album* L., *Achillea millefolium* L., *Berteroa incana* (L.) DC., and *Galinsoga parviflora* Cav. have been reported [151].

4.3. Improving Soil Stability

Soil erosion, especially on slope surfaces, is a well-known issue in urban environments and is favored by high compaction and limited drainage capacity. This phenomenon induces hypoxia, water runoff and flooding, and increases the risk of landslides that are potentially dangerous for viability and construction [154,155]. Plants can improve soil stability mainly through the root system, which helps in several ways, such as by anchoring the ground to provide physical support, creating additional pores for oxygen and water infiltration, or supplying organic matter, which induces stable aggregate formation [156–159]. Hence, species must be selected and varieties developed researching high root dimensions, number and stiffness, as useful traits to support the soil because of hardness and tensile strength [160]. Regarding growth habits, roots that grow parallel and close to the surface improve the soil in-plane tensile strength, while roots penetrating deeper and perpendicular to the surface enhance shear strength [161]. However, it is not possible to determine the best habit regardless of soil conditions: in the case of rocky slopes with limited soil availability, plants with superficial root systems must be rewarded; in contrast, if there are adequate soil depths, deeper roots guarantee more stability, and hence, plants having them should be preferred [158]. Attention must also be paid to the canopy, which is important to avoid erosion by rainfall water. High leaf area, elevated branch presence, dense and compact growth habits allowing high plant density can lead to homogeneous soil cover, intercepting more water and gradually reaching the soil. In this context, as already reported, focusing not only on trees but also on turfgrass and shrub selection can be interesting, since different vegetation types, when combined in the same place, can provide synergic benefits, such as multiple levels of canopy to intercept water at different heights [162]. Not including *Poaceae*, the available literature evaluating suitable species for erosion control, such as *Lantana montevidensis* (Spreng.) Brinq., *Lavandula lanata* L., *Origanum vulgare* L., *Rosa abyssinica* R. Br. ex Lindl., and *Rosmarinus officinalis* L., were reviewed by Francini et al. [163].

4.4. Food Production

Plants grown with food productive aims must be subjected to different standards if their production is for commercial purposes, in the context of farms, or if they are intended for self-consumption, as in the case of public spaces and private gardens cultivations. In the first case, it is more important to develop cultivars selecting them for their yield amount and specific qualitative aspects related to products' shelf life and subjection to food processing [164–166]. Many breeding plans focusing on crop productivity have been carried out during the last century by means of conventional methods. Recently, genomic information about coding regions related to production traits and associated molecular markers has become available, particularly in horticultural species [167–169]. This can help breeders enhance the constitution of modern varieties. Compared to productivity traits, factors such as low maintenance requirements, environmental benefits, and decorative aspects are of greater interest in horticultural species to be used in home gardens or public urban spaces than in urban commercial farms [170]. Currently, there is no evidence that horticultural cultivars destinated for self-consumption are genetically different from those employed in farms, hence there is a need in developing the former by rewarding specific traits.

As previously reported, even more in the case of commercial destinations, surface availability can be the most limiting factor in planning an urban plantation. Strategies such as vertical farming, possibly coupled with artificial lights and soilless cultivation systems, could provide solutions to obtain an adequate product quantity for farms while limiting the surface needs [171]. Instead, not having high production level requirements, the possibilities to use horticultural plants are greater when considering private and community gardens, also exploiting simple growth systems such as pots and planters. In any of the mentioned cases, it is important to adopt small-sized plants in terms of roots, shoots, or low lateral growth. For the most important horticultural species, selections for dwarf or compact phenotypes have been made. As an example, in tomato (*Solanum lycopersicum* L.), which is one of the most cultivated horticultural crops worldwide, bush varieties with determinate growth were selected and are some of the most common cultivar types on the market [172–174], even if breeding programs mainly focused on enhancing yield, abiotic and biotic tolerances, product shelf life, organoleptic properties and nutritional values [99,175,176]. A cultivar that combines small-sized mutations, *SP* (*SELF-PRUNING*), *d* (*dwarf*) and *mnt* (*miniaturize*) [177], is represented by Micro-Tom, which has attracted much attention as a model phenotype for plant studies, as it reaches 10–20 cm in height, has a short life cycle with fruits suitable for harvesting 70–90 days after sowing and bears plant densities up to 1357 plants/m^2 [178]. Even if it was originally bred for home garden purposes [178,179], the Micro-Tom variety is now commonly used in laboratory applications for dwarfism studies, as it is particularly suitable to be transformed [180] and to study its possible application for cultivation in space vehicles [181,182]. Crosses between dwarfed varieties and others characterized by high sensory qualities of the products (i.e., sweetness) can represent a possible solution to combine the organoleptic characteristics with little dimensions as demonstrated by Scott et al. [183].

Regarding innovative indoor production systems, such as vertical farming, soilless cultivations and plant factories with artificial lighting (PFALs) [171], the breeding priorities strictly differ from outdoor systems. The controlled environments make plants less affected by pathogens and pedoclimate conditions, allowing stress resistance to be of minor importance [184]. On the contrary, high productive capacity in quantitative and qualitative terms under low light intensities or specific wavelengths can be of primary importance, being the energetic costs a big limit for artificial lighting systems [185,186]. Light-related aspects, such as photoperiod, also influence several phenological stages such as stem elongations and flowering. Hence, developing genotypes responsive to specific manipulation of these factors can lead to advantages such as shorter productive cycles, better harvesting synchronization, and growth habits suitable for the growth system spaces [186,187]. For the obtainment of food products derived from insect-mediated fertilization, the settle of pollinators is often conducted opening the growth environments hence enhancing the risk of pathogen introduction. For this, it can be of interest for the plant species involved, to develop *apomictic* cultivars, i.e., able in producing fruits without requiring fertilization [188,189].

4.4.1. Leafy Vegetables

Aiming at exploiting limited urban spaces for commercial food production or private consumption, high attention should be given to leafy vegetables, which can provide food products needing on average less surfaces and in shorter times compared to fruit species [184,190,191]. A commercial typology of great interest nowadays and for which leafy vegetables are the most popular cultures is fresh-cut production. It consists of fruits or vegetables which have undergone minimal processing after harvest (e.g., trimming washing, decontaminating, packaging), in order to provide ready-to-eat products maintaining their fresh state. The main issue, being the processing made before passage to the distribution chain, is the higher perishability compared to non-fresh-cut products [192]. The leafy vegetables most commonly used for fresh-cut production are spinach (*Spinacia oleracea* L.), kale (*Brassica oleracea* L. var. *acephala* D.C.), salad rocket (*Eruca sativa* Mill.), wild rocket (*Diplotaxis tenuifolia* L.), corn salad (*Valerianella locusta* L.), chicory (*Cichorium intybus* L.), curly endive (*Cichorium endivia* L. var. *crispum*), and lettuce, especially looseleaf varieties (*Lactuca sativa* L. var. *acephala*) [193,194]. To date, lettuce is the most popular species also in vertical farming [195]. In the last decade, breeding for fresh vegetables was mostly based on improving yields and postprocessing performance in terms of shelf life, leaving other aspects to lower priority, such as biotic and abiotic resistance. Moreover, the attention was focused mainly on tolerance to high plant density and hypoxia, post-cutting recovery ability, low core length, solid midrib and good organoleptic and nutritive profiles [192,196,197]. To select cultivars with less browning susceptibility after cutting, the use of browning activity-related enzymes and volatile molecules responsible for off-odors as metabolic markers have been proposed [196]. The market increment of the ready-to-eat products allows farms to harvest leafy vegetables at a very early maturation stage, even further reducing the crop cycle, strongly enhancing the sown density and obtaining higher yields per surface unit [194]. Baby leaf lettuces such as Green Leaf, Red Leaf, and Lollo Red, hence cultivars particularly suitable in providing quality products in a very limited growth time, have been developed and demonstrated to be better than whole-head lettuce in terms of harvest index, simplicity of being processed, oxidation after cutting reduction due to smaller stem diameter and appreciation by consumers [198]. Developing new baby leaf varieties can be very interesting to farms but also to private urban gardens for their cultivation in pots and planters, especially in balconies and other tight spaces.

4.4.2. Edible Flowers

Innovative food production species suitable for urban areas, whose products are having a continuously increasing commercial interest, are edible flower plants [174]. Edible flowers present innovative and original organoleptic characteristics and can be highly nutritive for their nutraceutical values, such as the elevated content of health molecules such as antioxidants, e.g., flavonoids, phenolic acids, and alkaloids [199]. Flowering plants are already one of the first choices for small spaces in cities due to their high aesthetic effect, but species able to provide edible flowers can combine ornamental value with the supply of innovative food products. These species include begonia (*Begonia x tuberhybrida* Voss), rose (*Rosa* spp. L.), chrysanthemum (*Chrysanthemum* spp. L.), pansy (*Viola* × *wittrockiana* Gams), lilac (*Syringa vulgaris* L.), elder (*Sambucus nigra* L.), and Japanese wisteria (*Wisteria floribunda* (Willd.) DC.) [200,201]. Breeding of flowering species has focused almost uniquely on quantitative, morphological, and aesthetic aspects of flowers [45], aside from biotic and abiotic stress tolerance. Good values for these traits were mainly reached using hybridization, also the interspecific trait, followed by vegetative propagation of the most performative individuals [202,203]. This has made the genetic background of varieties highly heterozygous and, in some cases, allopolyploid or aneuploid. The genomic complexity, coupled with a relatively low economic importance for single species in comparison to that of open-field crops, led to severe difficulties in molecular marker development attempts and hence in starting MAB programs [202,204]. However, in recent years, the advent of NGS technologies has made it possible to obtain whole-genome sequences or high-density linkage maps for the most important ornamental species, providing powerful tools for future genomic-based breeding applications [205–208]. Regarding edible flower plants, the biochemical characterization of products was made for the most important species [209–211], but there is no evidence of genetic improvement programs for the obtainment of high organoleptic and health-related molecules content as selection criteria, which can be detected with the use of specific assays or analyzing molecular markers linked to genes controlling these aspects. The improvement of these traits in cultivars, coupled with that of products' shelf life, with flowers being particularly delicate and their appearance preservation being significantly important, is a key factor in promoting edible flower plant diffusion in cities [200].

4.5. Biodiversity Conservation

Urban environments are known to be characterized by low levels of biodiversity, and the increase in urbanization inevitably leads to a reduction in the variability of life in ecosystems [212]. On the contrary, the setting of green areas plays an important role in conserving and also improving biodiversity, favoring the resettlement of wild plants, enhancing their local genetic diversity [213], and providing the right habitat for many animal species through the supply of suitable places for nesting and sheltering or to find food sources [214]. Improving biodiversity can positively influence the ecosystem services provided in the long term, strengthening the resilience to environmental changes [215].

One of the most endangered animal categories in cities without plants is that of pollinating insects, being nectar and pollen primary sources for their sustenance [216,217]. To set plants in green spaces capable of attracting insects, first, it is important to focus on allogamous species with entomophilous pollination. Flower ornamental plants are the most effective, given the attractiveness they have to pollinators due to their flowers. Traits to be considered in plant breeding programs for developing varieties aimed at enhancing insect biodiversity comprehend flower size, morphology, color and scent, pollen and nectar production quantity, and flowering period longevity [129,218,219]. An important aspect on which many studies have focused is the role of native plants in attracting insects in comparison to exotic species, but researchers have not reached a consensus on the matter [220] The traits reported can be good indicators of insect resettlement, but to assess the potential of a variety in improving pollinator biodiversity in a specific environment, it is necessary to count and recognize flowering-visiting individuals for a few years. Garbuzov and Ratnieks [129] evaluated the ability to attract pollinators of 32 garden flowering plant varieties, counting them over two summers in different experimental gardens in Sussex. They noted that *Agastache foeniculum* (Pursh) Kuntze, *Lavandula x intermedia* Emeric ex Loisel. var. Gros Bleau, Edelweiss and Hidcote Giant, and *Nepeta x faassenii* Bergmans ex Stearn var. Six Hills Giant are the most performant plants for the aim. Although works on varietal improvement related to this aspect are not available in the literature yet, providing traits of interest and putative suitable species can favor their insertion into plant breeding plans aimed at enhancing biodiversity.

Despite the setting of green species is in general linked to biodiversity enhancement in an urban area, if it regards the introduction of alien plants, there is a risk of negatively influencing the ecosystem balances due to the risk of invasiveness. In fact, specific traits of the introduced varieties or species can make these plants not only problematic for environmental management as a consequence of their high spread but also predominant in the hoarding of limited resources. This confers them a major fitness advantage in comparison to native plants, reducing the survival possibilities and the reproductive success of these last [221,222]. The competition effect can influence the present biodiversity at the interspecific level, but it may be accentuated at the intraspecific one, likely because of the more probable sharing of an ecologic niche between the introduced varieties and autochthonous populations of the same species [223]. There is a wide range of genetic traits that can make the plant invasive, and they differ depending on species and environment characteristics. However, they generally present a greater capacity for growth under specific environmental conditions (expressible with parameters, e.g., relative growth rate, leaf area index, earliness in blooming, root system development capacity), reproductive and dispersal efficiency (e.g., seed production, flowering duration, seed dispersal distance) [224,225]. These factors often match with traits of interest; hence, it may be not possible to completely avoid the risk of enhancing the invasive potential of selected plants in breeding programs. Another important aspect that can reduce biodiversity in the case of non-native plant introduction is related to intraspecific or interspecific hybridization. In fact, progeny generated through crossing between the introduced plants and native populations of the same species or of compatible species have more chances than the parental lines in presenting competitive traits and become invasive, due to higher genetic diversity, heterosis, trait fixation or trait novelty [226–229]. Therefore, if the fertilization rate within native populations is lower than that between native plants and non-native plants or hybrids, the survival of the former is threatened. Regarding ornamentals or species for which reproductive capacity is not of interest, a possible strategy to counteract this risk is to develop sterile varieties, such as exploiting odd ploidy levels, canceling their possibility of colonizing the surrounding environment [230].

4.6. Psychosocial Benefits

Finally, other important services supplied by green urban areas are the psychosocial benefits derived from their establishment. The sense of being in contact with nature due to the presence of plants in the surrounding environment indeed has several psychological effects on urban communities. For example, making people feel more relaxed and restored, enhancing social cohesion, and improving fitness through the promotion of outdoor physical activities as well as improvement of health through the reinforcement of the immunity system [135]. In general, it is very difficult to define which of the plant traits are useful to provide specific psychosocial benefits to urban communities, but it is known that visual factors such as flower color, morphology, and sizes, as well as the emission of scents, are appreciated not only for their decoration potential but also because of their contribution to the perception of higher psychological well-being. This regards most of all flowering ornamental plants, for which these traits are some of the most researched in current breeding programs [45].

5. Summary and Outlook

The first real difficulty of an urban context lies in the definition and list of diversified typologies of green spaces coexisting often in neighboring areas, created with purposes ranging from niche food production to pure ornamental appearance. For these reasons, a simplification of the system itself, with the definition in macro areas, is of primary importance to focus the attention on the various factors characterizing and influencing such environments. As mentioned in the above sections, genetic/genomic approaches have largely contributed to the development of improved varieties with resilient traits of interest, and the use of molecular markers has amply demonstrated its potential, making it one of the most useful predictive tools in genetic improvement, not only for major plants or in crop science. In fact, in the post-genomics era, there is a striking abundance of genomic resources, such as genome sequence assemblies, germplasm sequencing data and gene expression atlases, for species potentially employed as urban vegetation [14].

This review is intended to evaluate the potential genetic aspects that must be considered in a breeding program for varieties or species to be grown in an urban context. These characteristics must take into account both the adaptability of the plant to the specific area where it must grow and to the purposes for which it is used. As suggested by Henderson and Salt [231], studies on biodiversity, for example, through the genetic characterization of plant germplasm collections and a correct hierarchy of selectable physiological and environmental parameters, can provide a potential genetic resource from which to appeal for targeted planning. Generally, the selection and use of favorable effect alleles in breeding programs are required to enhance genetic variance and to improve the rate of genetic gain in all environmental landscapes, but even more so in an urban scenario, an integrated approach is required to realize genetic gain through the modernization of breeding programs. Thus, in such a context, the appropriate choice of parents and optimized breeding pipelines for the fixation of target alleles present potential ways to enhance breeding efficiency and developing modernized breeding programs will help realize higher genetic gains in urban contexts for developing climate changes-resilient species.

Furthermore, a research improvement in the horti-floricultural sector with new varieties to be entered into the urban market can lead to a boost in socioeconomics that encourages the spread of vegetable, flower, and ornamental species in public and private spaces, with particular reference to environments where the presence of vegetation has so far been limited. In addition, as previously described, it turns out to be an investment in terms of reduction in greenhouse gas emissions and pollution, enhancement of territorial biodiversity, protection of ecosystems, etc. More efficient development of plant varieties can also promote breeding and nursery companies, leading them to increase and diversify the range of products and initiate innovative genetic progress plans. The certification of the products obtained through registration in the official variety registers also guarantees the economic benefit, certifying the intellectual property and safeguarding against fraud.

In conclusion, as summarized in Figure 2, the characterization of the urban vegetation with the best genetic profiles represents the new green challenge in urban planning to have the right genetics in the right place. A simplification and hierarchization of the main aspects that characterize an urban context, such as the surface availability, productive destination, adaptability, sustainability, and benefits contribution (details are reported in the text), play a key role in a breeding program. The identification of more appropriate starting genetic resources, by phenotype collections and physiological traits, is fundamental for the identification of appropriate traits/donors/parents and subsequent crossing programs. The potential application of molecular MAS and MAB and computational tools for performing genetic and/or genomic-assisted selection. Data generated through these trials can be used in the selection of specific leader alleles that can be introduced in species improvement programs for the constitution of new varieties that respond in a more targeted way to different needs in an urban area.

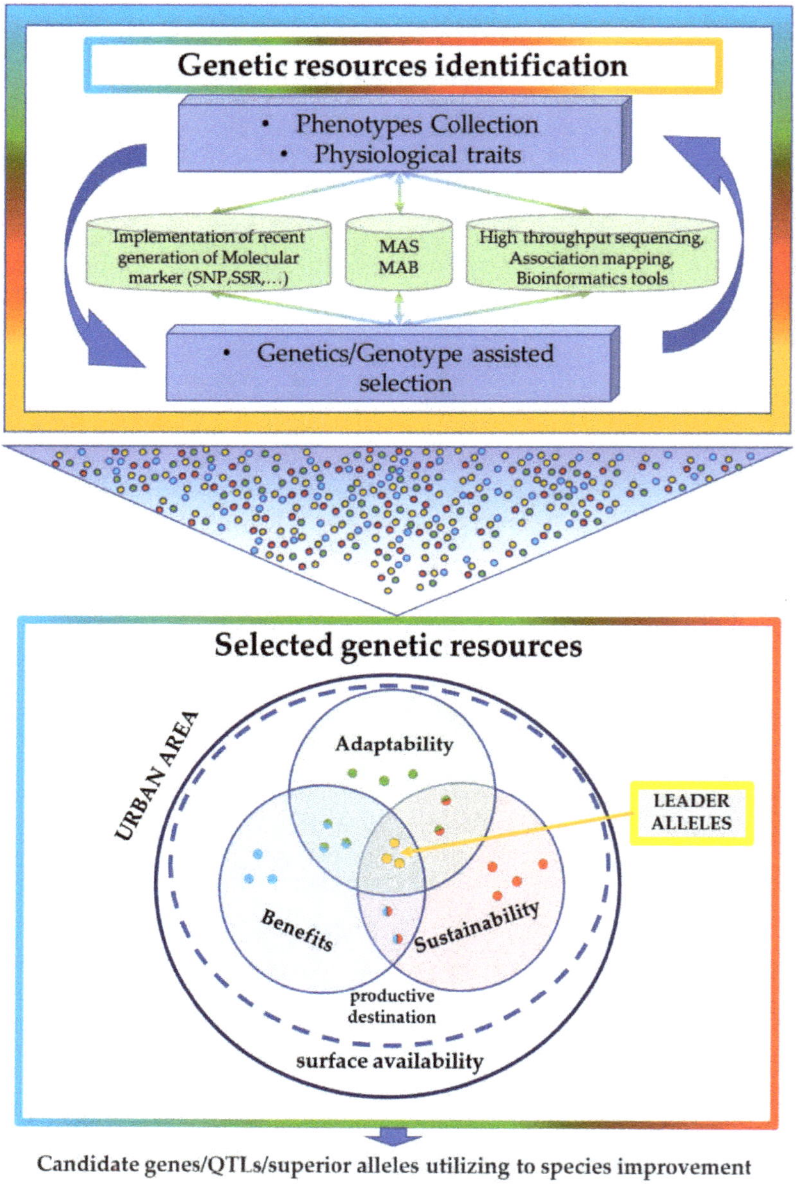

Figure 2. Graphic overview of genetic resource exploitation for plant breeding in urban plans.

Author Contributions: Conceptualization, G.B., S.F. and A.B.; investigation, S.F and A.B.; resources, A.B and F.S.; data curation, S.F., F.P., F.S.; writing—original draft preparation, S.F. and A.B.; writing—review and editing, S.F., F.P. and A.V.; visualization, S.F., A.B., F.P., F.S., A.V. and G.B.; supervision, G.B.; project administration, G.B.; funding Acquisition G.B. All authors have read and agreed to the published version of the manuscript.

Funding: This study was carried out within the research contract signed by the Gruppo Padana S.S. company (Paese, TV, Italy), and Department of Agronomy, Food, Natural resources, Animals and Environment (DAFNAE), University of Padua (Italy), within action IV.5_GREEN, PON 2014–2021.

Institutional Review Board Statement: Not applicable.

Informed Consent Statement: Not applicable.

Data Availability Statement: Not applicable.

Acknowledgments: The authors would like to thank the company Gruppo Padana S.S. (Referent person Marco Gazzola) for funding part of the PhD program of A.B. and the RTDA research project of S.F.

Conflicts of Interest: The authors declare no conflict of interest.

References

1. WHO. *Urban Green Spaces: A Brief for Action*; World Health Organization: Geneva, Switzerland, 2017.
2. Donihue, C.M.; Lambert, M.R. Adaptive evolution in urban ecosystems. *Ambio* **2015**, *44*, 194–203. [CrossRef] [PubMed]
3. McKinney, M.L. Urbanization as a major cause of biotic homogenization. *Biol. Conserv.* **2006**, *127*, 247–260. [CrossRef]
4. Seto, K.C.; Sánchez-Rodríguez, R.; Fragkias, M. The New Geography of Contemporary Urbanization and the Environment. *Annu. Rev. Environ. Resour.* **2010**, *35*, 167–194. [CrossRef]
5. Aronson, M.F.; La Sorte, F.A.; Nilon, C.H.; Katti, M.; Goddard, M.A.; Lepczyk, C.A.; Warren, P.S.; Williams, N.S.; Cilliers, S.; Clarkson, B. A global analysis of the impacts of urbanization on bird and plant diversity reveals key anthropogenic drivers. *Proc. R. Soc. B Biol. Sci.* **2014**, *281*, 20133330. [CrossRef]
6. Zhou, Y.; Smith, S.J.; Zhao, K.; Imhoff, M.; Thomson, A.; Bond-Lamberty, B.; Asrar, G.R.; Zhang, X.; He, C.; Elvidge, C.D. A global map of urban extent from nightlights. *Environ. Res. Lett.* **2015**, *10*, 054011. [CrossRef]
7. Panduro, T.E.; Veie, K.L. Classification and valuation of urban green spaces—A hedonic house price valuation. *Landsc. Urban Plan.* **2013**, *120*, 119–128. [CrossRef]
8. Peña-Salmón, C.; Leyva-Camacho, O.; Rojas-Caldelas, R.; Alonso-Navarrete, A.; Iñiguez-Ayón, P. The identification and classification of green areas for urban planning using multispectral images at Baja California, Mexico. *WIT Trans. Ecol. Environ.* **2014**, *191*, 611–621.
9. Xu, Z.; Zhou, Y.; Wang, S.; Wang, L.; Li, F.; Wang, S.; Wang, Z. A Novel Intelligent Classification Method for Urban Green Space Based on High-Resolution Remote Sensing Images. *Remote Sens.* **2020**, *12*, 3845. [CrossRef]
10. Gerullis, M.K.; Heckelei, T.; Rasch, S. Toward understanding the governance of varietal and genetic diversity. *Ecol. Soc.* **2021**, *26*, 28. [CrossRef]
11. Ahmar, S.; Gill, R.A.; Jung, K.H.; Faheem, A.; Qasim, M.U.; Mubeen, M.; Zhou, W. Conventional and Molecular Techniques from Simple Breeding to Speed Breeding in Crop Plants: Recent Advances and Future Outlook. *Int. J. Mol. Sci.* **2020**, *21*, 2590. [CrossRef]
12. Xu, Y.; Li, P.; Zou, C.; Lu, Y.; Xie, C.; Zhang, X.; Prasanna, B.M.; Olsen, M.S. Enhancing genetic gain in the era of molecular breeding. *J. Exp. Bot.* **2017**, *68*, 2641–2666. [CrossRef] [PubMed]
13. Marks, R.A.; Hotaling, S.; Frandsen, P.B.; VanBuren, R. Representation and participation across 20 years of plant genome sequencing. *Nat. Plants* **2021**, *7*, 1571–1578. [CrossRef] [PubMed]
14. Zheng, T.; Li, P.; Li, L.; Zhang, Q. Research advances in and prospects of ornamental plant genomics. *Hortic. Res.* **2021**, *8*, 65. [CrossRef] [PubMed]
15. Wu, G.A.; Terol, J.; Ibanez, V.; Lopez-Garcia, A.; Perez-Roman, E.; Borreda, C.; Domingo, C.; Tadeo, F.R.; Carbonell-Caballero, J.; Alonso, R.; et al. Genomics of the origin and evolution of Citrus. *Nature* **2018**, *554*, 311–316. [CrossRef] [PubMed]
16. Kato, J.; Mii, M. Production of Interspecific Hybrids in Ornamental Plants. In *Plant Cell Culture Protocols*; Loyola-Vargas, V.M., Ochoa-Alejo, N., Eds.; Humana Press: Totowa, NJ, USA, 2012; pp. 233–245. [CrossRef]
17. Bhattarai, K.; Van Huylenbroeck, J. Breeding, Genetics, and Genomics of Ornamental Plants. *Horticulturae* **2022**, *8*, 148. [CrossRef]
18. Lv, Q.; Li, W.; Sun, Z.; Ouyang, N.; Jing, X.; He, Q.; Wu, J.; Zheng, J.; Zheng, J.; Tang, S.; et al. Resequencing of 1,143 indica rice accessions reveals important genetic variations and different heterosis patterns. *Nat. Commun.* **2020**, *11*, 4778. [CrossRef]
19. Stevens, K.A.; Wegrzyn, J.L.; Zimin, A.; Puiu, D.; Crepeau, M.; Cardeno, C.; Paul, R.; Gonzalez-Ibeas, D.; Koriabine, M.; Holtz-Morris, A.E.; et al. Sequence of the Sugar Pine Megagenome. *Genetics* **2016**, *204*, 1613–1626. [CrossRef]
20. Simko, I.; Jia, M.; Venkatesh, J.; Kang, B.-C.; Weng, Y.; Barcaccia, G.; Lanteri, S.; Bhattarai, G.; Foolad, M.R. Genomics and Marker-Assisted Improvement of Vegetable Crops. *Crit. Rev. Plant Sci.* **2021**, *40*, 303–365. [CrossRef]
21. Bohra, A.; Jha, U.C.; Kishor, P.B.; Pandey, S.; Singh, N.P. Genomics and molecular breeding in lesser explored pulse crops: Current trends and future opportunities. *Biotechnol. Adv.* **2014**, *32*, 1410–1428. [CrossRef]
22. Bhat, J.A.; Ali, S.; Salgotra, R.K.; Mir, Z.A.; Dutta, S.; Jadon, V.; Tyagi, A.; Mushtaq, M.; Jain, N.; Singh, P.K.; et al. Genomic Selection in the Era of Next Generation Sequencing for Complex Traits in Plant Breeding. *Front. Genet.* **2016**, *7*, 221. [CrossRef]
23. Poland, J.A.; Rife, T.W. Genotyping-by-Sequencing for Plant Breeding and Genetics. *Plant Genome* **2012**, *5*. [CrossRef]

24. Edwards, D.; Batley, J. Plant genome sequencing: Applications for crop improvement. *Plant Biotechnol. J.* **2010**, *8*, 2–9. [CrossRef] [PubMed]
25. Wu, Y.; Gong, Z.; Bebber, D.P.; Miao, J.; Zhao, Z.; Jiang, Y.; Xiao, S.; Zhang, G.; Yu, D.; Fang, J.; et al. Phenological matching drives wheat pest range shift under climate change. *bioRxiv* **2019**. [CrossRef]
26. Xu, X.; Bai, G. Whole-genome resequencing: Changing the paradigms of SNP detection, molecular mapping and gene discovery. *Mol. Breed.* **2015**, *35*, 33. [CrossRef]
27. Elshire, R.J.; Glaubitz, J.C.; Sun, Q.; Poland, J.A.; Kawamoto, K.; Buckler, E.S.; Mitchell, S.E. A robust, simple genotyping-by-sequencing (GBS) approach for high diversity species. *PLoS ONE* **2011**, *6*, e19379. [CrossRef] [PubMed]
28. Peterson, B.K.; Weber, J.N.; Kay, E.H.; Fisher, H.S.; Hoekstra, H.E. Double digest RADseq: An inexpensive method for de novo SNP discovery and genotyping in model and non-model species. *PLoS ONE* **2012**, *7*, e37135. [CrossRef]
29. Wang, S.; Meyer, E.; McKay, J.K.; Matz, M.V. 2b-RAD: A simple and flexible method for genome-wide genotyping. *Nat. Methods* **2012**, *9*, 808–810. [CrossRef]
30. Toonen, R.J.; Puritz, J.B.; Forsman, Z.H.; Whitney, J.L.; Fernandez-Silva, I.; Andrews, K.R.; Bird, C.E. ezRAD: A simplified method for genomic genotyping in non-model organisms. *PeerJ* **2013**, *1*, e203. [CrossRef]
31. Ostezan, A.; McDonald, S.C.; Tran, D.T.; Souza, R.S.E.; Li, Z. Target region sequencing and applications in plants. *JCSB* **2020**, *24*, 13–26. [CrossRef]
32. Rajcan, I.; Boersma, J.; Shaw, E. Plant genetic techniques: Plant breeder's toolbox. In *Comprehensive Biotechnology*, 2nd ed.; Moo-Young, M., Ed.; Elsevier: Burlington, NJ, USA, 2011; Volume 4, pp. 133–147.
33. Hasan, N.; Choudhary, S.; Naaz, N.; Sharma, N.; Laskar, R.A. Recent advancements in molecular marker-assisted selection and applications in plant breeding programmes. *J. Genet. Eng. Biotechnol.* **2021**, *19*, 128. [CrossRef]
34. Yang, Y.; Saand, M.A.; Huang, L.; Abdelaal, W.B.; Zhang, J.; Wu, Y.; Li, J.; Sirohi, M.H.; Wang, F. Applications of Multi-Omics Technologies for Crop Improvement. *Front. Plant Sci.* **2021**, *12*, 563953. [CrossRef] [PubMed]
35. Donald, C. The breeding of crop ideotypes. *Euphytica* **1968**, *17*, 385–403. [CrossRef]
36. Czaja, M.; Kołton, A.; Muras, P. The Complex Issue of Urban Trees—Stress Factor Accumulation and Ecological Service Possibilities. *Forests* **2020**, *11*, 932. [CrossRef]
37. Koyro, H.-W.; Ahmad, P.; Geissler, N. Abiotic stress responses in plants: An overview. In *Environmental Adaptations Stress Tolerance of Plants in the Era of Climate Change*; Ahmad, P., Prasad, M.N.V., Eds.; Springer: New York, NY, USA, 2012; pp. 1–28. [CrossRef]
38. Wang, W.; Vinocur, B.; Altman, A. Plant responses to drought, salinity and extreme temperatures: Towards genetic engineering for stress tolerance. *Planta* **2003**, *218*, 1–14. [CrossRef] [PubMed]
39. Cramer, G.R.; Urano, K.; Delrot, S.; Pezzotti, M.; Shinozaki, K. Effects of abiotic stress on plants: A systems biology perspective. *BMC Plant Biol.* **2011**, *11*, 163. [CrossRef]
40. Lobell, D.B.; Burke, M.B.; Tebaldi, C.; Mastrandrea, M.D.; Falcon, W.P.; Naylor, R.L. Prioritizing climate change adaptation needs for food security in 2030. *Science* **2008**, *319*, 607–610. [CrossRef]
41. Renard, D.; Tilman, D. National food production stabilized by crop diversity. *Nature* **2019**, *571*, 257–260. [CrossRef]
42. Husaini, A.M. High-value pleiotropic genes for developing multiple stress-tolerant biofortified crops for 21st-century challenges. *Heredity* **2022**, *128*, 460–472. [CrossRef]
43. Bharadwaj, D.N. Sustainable Agriculture and Plant Breeding. In *Advances in Plant Breeding Strategies: Agronomic, Abiotic and Biotic Stress Traits*; Al-Khayri, J.M., Jain, S.M., Johnson, D.V., Eds.; Springer International Publishing: Cham, Switzerland, 2016; pp. 3–34. [CrossRef]
44. Masclaux-Daubresse, C.; Daniel-Vedele, F.; Dechorgnat, J.; Chardon, F.; Gaufichon, L.; Suzuki, A. Nitrogen uptake, assimilation and remobilization in plants: Challenges for sustainable and productive agriculture. *Ann. Bot.* **2010**, *105*, 1141–1157. [CrossRef]
45. Lutken, H.; Clarke, J.L.; Muller, R. Genetic engineering and sustainable production of ornamentals: Current status and future directions. *Plant Cell Rep.* **2012**, *31*, 1141–1157. [CrossRef]
46. Baldauf, R. Roadside vegetation design characteristics that can improve local, near-road air quality. *Transp. Res.* **2017**, *52*, 354–361. [CrossRef] [PubMed]
47. Jarvis, D.I.; Hodgkin, T.; Sthapit, B.R.; Fadda, C.; Lopez-Noriega, I. An Heuristic Framework for Identifying Multiple Ways of Supporting the Conservation and Use of Traditional Crop Varieties within the Agricultural Production System. *Crit. Rev. Plant Sci.* **2011**, *30*, 125–176. [CrossRef]
48. Kumar, A.; Anju, T.; Kumar, S.; Chhapekar, S.S.; Sreedharan, S.; Singh, S.; Choi, S.R.; Ramchiary, N.; Lim, Y.P. Integrating Omics and Gene Editing Tools for Rapid Improvement of Traditional Food Plants for Diversified and Sustainable Food Security. *Int. J. Mol. Sci.* **2021**, *22*, 8093. [CrossRef]
49. Yang, Y.; Sun, M.; Li, S.; Chen, Q.; Teixeira da Silva, J.A.; Wang, A.; Yu, X.; Wang, L. Germplasm resources and genetic breeding of Paeonia: A systematic review. *Hortic. Res.* **2020**, *7*, 107. [CrossRef]
50. Farida Traore, F.; El-Baouchi, A.; En-Nahli, Y.; Hejjaoui, K.; Metougui, M.L.; Hamwieh, A.; Sohail, Q.; Istanbuli, T.; Boughribil, S.; Amri, M. Exploring the Genetic Variability and Potential Correlations Between Nutritional Quality and Agro-Physiological Traits in Kabuli Chickpea Germplasm Collection (*Cicer arietinum* L.). *Front. Plant Sci.* **2022**, *13*, 905320. [CrossRef]
51. Egorova, A.A.; Chalaya, N.A.; Fomin, I.N.; Barchuk, A.I.; Gerasimova, S.V. De Novo Domestication Concept for Potato Germplasm Enhancement. *Agronomy* **2022**, *12*, 462. [CrossRef]

52. Breider, I.S.; Gaynor, R.C.; Gorjanc, G.; Thorn, S.; Pandey, M.K.; Varshney, R.K.; Hickey, J.M. A multi-part strategy for introgression of exotic germplasm into elite plant breeding programs using genomic selection. *Res. Sq.* **2022**. [CrossRef]
53. Huang, K.; Jahani, M.; Gouzy, J.; Legendre, A.; Carrere, S.; Lázaro-Guevara, J.M.; González Segovia, E.G.; Todesco, M.; Mayjonade, B.; Rodde, N.; et al. The genomics of linkage drag in sunflower. *bioRxiv* **2022**. [CrossRef]
54. Wahid, A.; Farooq, M.; Hussain, I.; Rasheed, R.; Galani, S. Responses and Management of Heat Stress in Plants. In *Environmental Adaptations and Stress Tolerance of Plants in the Era of Climate Change*; Ahmad, P., Prasad, M.N.V., Eds.; Springer: New York, NY, USA, 2012; pp. 135–157. [CrossRef]
55. Mathur, S.; Agrawal, D.; Jajoo, A. Photosynthesis: Response to high temperature stress. *J. Photochem. Photobiol. B* **2014**, *137*, 116–126. [CrossRef]
56. Feller, U.; Vaseva, I.I. Extreme climatic events: Impacts of drought and high temperature on physiological processes in agronomically important plants. *Front. Environ. Sci.* **2014**, *2*, 39. [CrossRef]
57. Young, J.C. Mechanisms of the Hsp70 chaperone system. *Biochem. Cell Biol.* **2010**, *88*, 291–300. [CrossRef] [PubMed]
58. Yang, X.; Zhu, W.; Zhang, H.; Liu, N.; Tian, S. Heat shock factors in tomatoes: Genome-wide identification, phylogenetic analysis and expression profiling under development and heat stress. *PeerJ* **2016**, *4*, e1961. [CrossRef] [PubMed]
59. Aldubai, A.A.; Alsadon, A.A.; Migdadi, H.H.; Alghamdi, S.S.; Al-Faifi, S.A.; Afzal, M. Response of Tomato (*Solanum lycopersicum* L.) Genotypes to Heat Stress Using Morphological and Expression Study. *Plants* **2022**, *11*, 615. [CrossRef] [PubMed]
60. Liu, R.; Su, Z.; Zhou, H.; Huang, Q.; Fan, S.; Liu, C.; Han, Y. LsHSP70 is induced by high temperature to interact with calmodulin, leading to higher bolting resistance in lettuce. *Sci. Rep.* **2020**, *10*, 15155. [CrossRef]
61. Kang, Y.; Jang, S.-W.; Lee, H.J.; Barchenger, D.W.; Jang, S. Expression Profiling of Heat Shock Protein Genes as Putative Early Heat-Responsive Members in Lettuce. *Horticulturae* **2021**, *7*, 312. [CrossRef]
62. Yan, J.; Yu, L.; Xuan, J.; Lu, Y.; Lu, S.; Zhu, W. De novo transcriptome sequencing and gene expression profiling of spinach (*Spinacia oleracea* L.) leaves under heat stress. *Sci. Rep.* **2016**, *6*, 19473. [CrossRef]
63. Li, S.; Yu, J.; Li, Y.; Zhang, H.; Bao, X.; Bian, J.; Xu, C.; Wang, X.; Cai, X.; Wang, Q.; et al. Heat-Responsive Proteomics of a Heat-Sensitive Spinach Variety. *Int. J. Mol. Sci.* **2019**, *20*, 3872. [CrossRef]
64. Li, Z.; Wang, X.; Xia, Y. SMRT sequencing of full-length transcriptome for in-depth understanding of genes under heat stress of Rhododendron hainanense. In Proceedings of the IV International Symposium on Woody Ornamentals of the Temperate Zone, Torino, Italy, 3 March 2021; pp. 261–268.
65. Wan, X.L.; Zhou, Q.; Wang, Y.Y.; Wang, W.E.; Bao, M.Z.; Zhang, J.W. Identification of heat-responsive genes in carnation (*Dianthus caryophyllus* L.) by RNA-seq. *Front. Plant Sci.* **2015**, *6*, 519. [CrossRef] [PubMed]
66. Usman, M.G.; Rafii, M.Y.; Martini, M.Y.; Yusuff, O.A.; Ismail, M.R.; Miah, G. Introgression of heat shock protein (Hsp70 and sHsp) genes into the Malaysian elite chilli variety Kulai (*Capsicum annuum* L.) through the application of marker-assisted backcrossing (MAB). *Cell Stress Chaperones* **2018**, *23*, 223–234. [CrossRef]
67. Sjöman, H.; Hirons, A.D.; Bassuk, N.L. Urban forest resilience through tree selection—Variation in drought tolerance in Acer. *Urban For. Urban Green.* **2015**, *14*, 858–865. [CrossRef]
68. Miller, D.L.; Wetherley, E.B.; Roberts, D.A.; Tague, C.L.; McFadden, J.P. Vegetation cover change during a multi-year drought in Los Angeles. *Urban Clim.* **2022**, *43*, 101157. [CrossRef]
69. FAO & Water World Council. *Towards a Water and Food Secure Future. Critical Perspectives for Policy-Makers*; FAO: Rome, Italy, 2015.
70. Khan, A.; Sovero, V.; Gemenet, D. Genome-assisted Breeding For Drought Resistance. *Curr. Genom.* **2016**, *17*, 330–342. [CrossRef] [PubMed]
71. Kumar, P.; Eriksen, R.L.; Simko, I.; Mou, B. Molecular Mapping of Water-Stress Responsive Genomic Loci in Lettuce (Lactuca spp.) Using Kinetics Chlorophyll Fluorescence, Hyperspectral Imaging and Machine Learning. *Front. Genet.* **2021**, *12*, 634554. [CrossRef]
72. Grace, O.M. Succulent plant diversity as natural capital. *Plants People Planet* **2019**, *1*, 336–345. [CrossRef]
73. Lewis, E.; Phoenix, G.K.; Alexander, P.; David, J.; Cameron, R.W.F. Rewilding in the Garden: Are garden hybrid plants (cultivars) less resilient to the effects of hydrological extremes than their parent species? A case study with Primula. *Urban Ecosyst.* **2019**, *22*, 841–854. [CrossRef]
74. Komar-Tyomnaya, L.D. Drought resistance of ornamental peach cultivars with a different origin. In Proceedings of the I International Conference and X National Horticultural Science Congress of Iran (IrHC2017), Tehran, Iran, 7 September 2017; pp. 353–358.
75. Esfahani, R.E.; Paço, T.A.; Martins, D.; Arsénio, P. Increasing the resistance of Mediterranean extensive green roofs by using native plants from old roofs and walls. *Ecol. Eng.* **2022**, *178*, 106576. [CrossRef]
76. Fuertes, A.; Sixto, H.; González, I.; Pérez-Cruzado, C.; Cañellas, I.; Rodríguez-Soalleiro, R.; Oliveira, N. Time-course foliar dynamics of poplar short rotation plantations under Mediterranean conditions. Responses to different water scenarios. *Biomass Bioenergy* **2022**, *159*, 106391. [CrossRef]
77. Ferrante, A.; Toscano, S.; Romano, D.; Vagge, I. Physiological and morpho-anatomical traits used as markers for the selection of drought tolerance of ornamental plants. In Proceedings of the IV International Symposium on Woody Ornamentals of the Temperate Zone, Torino, Italy, 3 March 2021; pp. 253–260.
78. Salekdeh, G.H.; Reynolds, M.; Bennett, J.; Boyer, J. Conceptual framework for drought phenotyping during molecular breeding. *Trends Plant Sci.* **2009**, *14*, 488–496. [CrossRef]

79. Altunoglu, Y.C.; Keles, M.; Can, T.H.; Baloglu, M.C. Identification of watermelon heat shock protein members and tissue-specific gene expression analysis under combined drought and heat stresses. *Turk. J. Biol.* **2019**, *43*, 404–419. [CrossRef]
80. Diouf, I.; Albert, E.; Duboscq, R.; Santoni, S.; Bitton, F.; Gricourt, J.; Causse, M. Integration of QTL, Transcriptome and Polymorphism Studies Reveals Candidate Genes for Water Stress Response in Tomato. *Genes* **2020**, *11*, 900. [CrossRef]
81. Zhou, R.; Yu, X.; Ottosen, C.O.; Zhang, T.; Wu, Z.; Zhao, T. Unique miRNAs and their targets in tomato leaf responding to combined drought and heat stress. *BMC Plant Biol.* **2020**, *20*, 107. [CrossRef] [PubMed]
82. Taheri, S.; Gantait, S.; Azizi, P.; Mazumdar, P. Drought tolerance improvement in Solanum lycopersicum: An insight into "OMICS" approaches and genome editing. *3 Biotech* **2022**, *12*, 63. [CrossRef] [PubMed]
83. Sarazin, V.; Duclercq, J.; Guillot, X.; Sangwan, B.; Sangwan, R.S. Water-stressed sunflower transcriptome analysis revealed important molecular markers involved in drought stress response and tolerance. *Environ. Exp. Bot.* **2017**, *142*, 45–53. [CrossRef]
84. Darbani, S.P.; Mehrabi, A.A.; Pordad, S.S.; Maleki, A.; Farshadfar, M. Effect of drought stress on agro-morphological traits in sunflower (*Helianthus annuus* L.) genotypes and identification of informative ISSR markers. *Plant Genet. Resour. Characterisation Util.* **2020**, *18*, 49–62. [CrossRef]
85. Muhammad Ahmad, H.; Wang, X.; Fiaz, S.; Mahmood Ur, R.; Azhar Nadeem, M.; Aslam Khan, S.; Ahmar, S.; Azeem, F.; Shaheen, T.; Mora-Poblete, F. Comprehensive genomics and expression analysis of eceriferum (CER) genes in sunflower (*Helianthus annuus*). *Saudi J. Biol. Sci.* **2021**, *28*, 6884–6896. [CrossRef]
86. Yukawa, J.; Kiritani, K.; Kawasawa, T.; Higashiura, Y.; Sawamura, N.; Nakada, K.; Gyotoku, N.; Tanaka, A.; Kamitani, S.; Matsuo, K.; et al. Northward range expansion by Nezara viridula (Hemiptera: Pentatomidae) in Shikoku and Chugoku Districts, Japan, possibly due to global warming. *Appl. Entomol. Zoo* **2009**, *44*, 429–437. [CrossRef]
87. Macfadyen, S.; McDonald, G.; Hill, M.P. From species distributions to climate change adaptation: Knowledge gaps in managing invertebrate pests in broad-acre grain crops. *Agric. Ecosyst. Environ.* **2018**, *253*, 208–219. [CrossRef]
88. Horgan, F.G.; Arida, A.; Ardestani, G.; Almazan, M.L.P. Intraspecific competition counters the effects of elevated and optimal temperatures on phloem-feeding insects in tropical and temperate rice. *PLoS ONE* **2020**, *15*, e0240130. [CrossRef]
89. Bommarco, R.; Kleijn, D.; Potts, S.G. Ecological intensification: Harnessing ecosystem services for food security. *Trends Ecol. Evol.* **2013**, *28*, 230–238. [CrossRef]
90. Forkuoh, F.; Boadi, N.O.; Borquaye, L.S.; Afful, S. Risk Of Human Dietary Exposure To Organochlorine Pesticide Residues In Fruits From Ghana. *Sci. Rep.* **2018**, *8*, 16686. [CrossRef]
91. Jokanovic, M. Neurotoxic effects of organophosphorus pesticides and possible association with neurodegenerative diseases in man: A review. *Toxicology* **2018**, *410*, 125–131. [CrossRef] [PubMed]
92. van der Werf, W.; Bianchi, F. Options for diversifying agricultural systems to reduce pesticide use: Can we learn from nature? *Outlook Agric.* **2022**, *51*, 105–113. [CrossRef]
93. Yin, J.-J.; Xiong, J.; Xu, L.-T.; Chen, X.-W.; Li, W.-T. Recent advances in plant immunity with cell death: A review. *J. Integr. Agric.* **2022**, *21*, 610–620. [CrossRef]
94. Karasov, T.L.; Shirsekar, G.; Schwab, R.; Weigel, D. What natural variation can teach us about resistance durability. *Curr. Opin. Plant Biol.* **2020**, *56*, 89–98. [CrossRef]
95. Sniezko, R.A.; Liu, J.-J. Prospects for developing durable resistance in populations of forest trees. *New For.* **2021**, 1–17. [CrossRef]
96. Tremblay, V.; McLaren, D.L.; Kim, Y.M.; Strelkov, S.E.; Conner, R.L.; Wally, O.; Belanger, R.R. Molecular Assessment of Pathotype Diversity of Phytophthora sojae in Canada Highlights Declining Sources of Resistance in Soybean. *Plant Dis.* **2021**, *105*, 4006–4013. [CrossRef]
97. Miedaner, T. Breeding strategies for improving plant resistance to diseases. In *Advances in Plant Breeding Strategies: Agronomic, Abiotic and Biotic Stress Traits*; Springer: Berlin/Heidelberg, Germany, 2016; pp. 561–599.
98. Peng, T.; Sun, X.; Mumm, R.H. Optimized breeding strategies for multiple trait integration: II. Process efficiency in event pyramiding and trait fixation. *Mol. Breed.* **2014**, *33*, 105–115. [CrossRef]
99. Foolad, M.R.; Panthee, D.R. Marker-Assisted Selection in Tomato Breeding. *Crit. Rev. Plant Sci.* **2012**, *31*, 93–123. [CrossRef]
100. Lee, J.M.; Oh, C.-S.; Yeam, I. Molecular Markers for Selecting Diverse Disease Resistances in Tomato Breeding Programs. *Plant Breed. Biotech.* **2015**, 3, 308–322. [CrossRef]
101. Bhardwaj, V.; Salej, S.; Ashwani, K.; Vanishree, G.; Sanjeev, S.; Sundaresha, S. Efficiency and reliability of marker assisted selection for resistance to major biotic stresses in potato. *Potato J.* **2019**, *46*, 56–66.
102. Meuwissen, T.H.; Hayes, B.J.; Goddard, M.E. Prediction of total genetic value using genome-wide dense marker maps. *Genetics* **2001**, *157*, 1819–1829. [CrossRef]
103. Xu, Y.; Liu, X.; Fu, J.; Wang, H.; Wang, J.; Huang, C.; Prasanna, B.M.; Olsen, M.S.; Wang, G.; Zhang, A. Enhancing Genetic Gain through Genomic Selection: From Livestock to Plants. *Plant Commun.* **2020**, *1*, 100005. [CrossRef]
104. Tan, X.; Shibata, S. Factors influencing street tree health in constrained planting spaces: Evidence from Kyoto City, Japan. *Urban For. Urban Green.* **2022**, *67*, 127416. [CrossRef]
105. Xie, X.; Cheng, H.; Hou, C.; Ren, M. Integration of Light and Auxin Signaling in Shade Plants: From Mechanisms to Opportunities in Urban Agriculture. *Int. J. Mol. Sci.* **2022**, *23*, 3422. [CrossRef] [PubMed]
106. Atkinson, C.; Else, M. Understanding how rootstocks dwarf fruit trees. *Compact Fruit Tree* **2001**, *34*, 46–49.
107. Foster, T.M.; Hooijdonk, B.M.V.; Friend, A.; Seleznyova, A.N.; McLachlan, A.R.G. Apple Rootstock-Induced Dwarfing is Strongly Influenced by Growing Environment. *J. Hortic.* **2016**, 3, 1–8. [CrossRef]

108. Hollender, C.A.; Hadiarto, T.; Srinivasan, C.; Scorza, R.; Dardick, C. A brachytic dwarfism trait (dw) in peach trees is caused by a nonsense mutation within the gibberellic acid receptor PpeGID1c. *New Phytol.* **2016**, *210*, 227–239. [CrossRef] [PubMed]
109. Zheng, X.; Zhao, Y.; Shan, D.; Shi, K.; Wang, L.; Li, Q.; Wang, N.; Zhou, J.; Yao, J.; Xue, Y.; et al. MdWRKY9 overexpression confers intensive dwarfing in the M26 rootstock of apple by directly inhibiting brassinosteroid synthetase MdDWF4 expression. *New Phytol.* **2018**, *217*, 1086–1098. [CrossRef] [PubMed]
110. Jia, D.; Gong, X.; Li, M.; Li, C.; Sun, T.; Ma, F. Overexpression of a Novel Apple NAC Transcription Factor Gene, MdNAC1, Confers the Dwarf Phenotype in Transgenic Apple (Malus domestica). *Genes* **2018**, *9*, 229. [CrossRef]
111. Salojarvi, J.; Smolander, O.P.; Nieminen, K.; Rajaraman, S.; Safronov, O.; Safdari, P.; Lamminmaki, A.; Immanen, J.; Lan, T.; Tanskanen, J.; et al. Genome sequencing and population genomic analyses provide insights into the adaptive landscape of silver birch. *Nat. Genet.* **2017**, *49*, 904–912. [CrossRef]
112. Xu, D.; Qi, X.; Li, J.; Han, X.; Wang, J.; Jiang, Y.; Tian, Y.; Wang, Y. PzTAC and PzLAZY from a narrow-crown poplar contribute to regulation of branch angles. *Plant Physiol. Biochem.* **2017**, *118*, 571–578. [CrossRef] [PubMed]
113. Dardick, C.; Callahan, A.; Horn, R.; Ruiz, K.B.; Zhebentyayeva, T.; Hollender, C.; Whitaker, M.; Abbott, A.; Scorza, R. PpeTAC1 promotes the horizontal growth of branches in peach trees and is a member of a functionally conserved gene family found in diverse plants species. *Plant J.* **2013**, *75*, 618–630. [CrossRef] [PubMed]
114. Zhang, J.; Zhang, Q.; Cheng, T.; Yang, W.; Pan, H.; Zhong, J.; Huang, L.; Liu, E. High-density genetic map construction and identification of a locus controlling weeping trait in an ornamental woody plant (Prunus mume Sieb. et Zucc). *DNA Res.* **2015**, *22*, 183–191. [CrossRef] [PubMed]
115. Hill, J.L., Jr.; Hollender, C.A. Branching out: New insights into the genetic regulation of shoot architecture in trees. *Curr. Opin. Plant Biol.* **2019**, *47*, 73–80. [CrossRef] [PubMed]
116. Guseman, J.M.; Webb, K.; Srinivasan, C.; Dardick, C. DRO1 influences root system architecture in Arabidopsis and Prunus species. *Plant J.* **2017**, *89*, 1093–1105. [CrossRef]
117. Motte, H.; Vanneste, S.; Beeckman, T. Molecular and Environmental Regulation of Root Development. *Annu. Rev. Plant Biol.* **2019**, *70*, 465–488. [CrossRef]
118. Barnes, B.D.; Kopecký, D.; Lukaszewski, A.J.; Baird, J.H. Evaluation of Turf-type Interspecific Hybrids of Meadow Fescue with Perennial Ryegrass for Improved Stress Tolerance. *Crop Sci.* **2014**, *54*, 355–365. [CrossRef]
119. Baldoni, R.; Giardini, L. *Coltivazioni Erbacee*; Pàtron: Bologna, Italy, 2000; pp. 1–409.
120. Casler, M.D. Perennial grasses for turf, sport and amenity uses: Evolution of form, function and fitness for human benefit. *J. Agric. Sci.* **2006**, *144*, 189–203. [CrossRef]
121. Sampoux, J.P.; Baudouin, P.; Bayle, B.; Béguier, V.; Bourdon, P.; Chosson, J.F.; de Bruijn, K.; Deneufbourg, F.; Galbrun, C.; Ghesquière, M.; et al. Breeding perennial ryegrass (*Lolium perenne* L.) for turf usage: An assessment of genetic improvements in cultivars released in Europe, 1974–2004. *Grass Forage Sci.* **2013**, *68*, 33–48. [CrossRef]
122. Beckett, K.P.; Freer-Smith, P.H.; Taylor, G. Urban woodlands: Their role in reducing the effects of particulate pollution. *Environ. Pollut.* **1998**, *99*, 347–360. [CrossRef]
123. Biasioli, M.; Barberis, R.; Ajmone-Marsan, F. The influence of a large city on some soil properties and metals content. *Sci. Total Environ.* **2006**, *356*, 154–164. [CrossRef] [PubMed]
124. Fazekaš, J.; Fazekašova, D.; Chovancová, J. Soil Material Quality and Environmental Potential of Metallically Contaminated Soils. *Key Eng. Mater.* **2020**, *838*, 164–169. [CrossRef]
125. Thomaidi, V.; Petousi, I.; Kotsia, D.; Kalogerakis, N.; Fountoulakis, M.S. Use of green roofs for greywater treatment: Role of substrate, depth, plants, and recirculation. *Sci. Total Environ.* **2022**, *807*, 151004. [CrossRef]
126. O'Sullivan, O.S.; Holt, A.R.; Warren, P.H.; Evans, K.L. Optimising UK urban road verge contributions to biodiversity and ecosystem services with cost-effective management. *J. Environ. Manag.* **2017**, *191*, 162–171. [CrossRef] [PubMed]
127. Evans, D.L.; Falagán, N.; Hardman, C.A.; Kourmpetli, S.; Liu, L.; Mead, B.R.; Davies, J.A.C. Ecosystem service delivery by urban agriculture and green infrastructure—A systematic review. *Ecosyst. Serv.* **2022**, *54*, 101405. [CrossRef]
128. Amoroso, G.; Frangi, P.; Piatti, R.; Fini, A.; Ferrini, F.; Faoro, M. Evaluation of Shrubs for Side Slope Greening and Protection in Urban Landscape. *HortTechnology* **2011**, *21*, 359–366. [CrossRef]
129. Garbuzov, M.; Ratnieks, F.L.W.; Thompson, K. Quantifying variation among garden plants in attractiveness to bees and other flower-visiting insects. *Funct. Ecol.* **2014**, *28*, 364–374. [CrossRef]
130. Armson, D.; Stringer, P.; Ennos, A.R. The effect of street trees and amenity grass on urban surface water runoff in Manchester, UK. *Urban For. Urban Green.* **2013**, *12*, 282–286. [CrossRef]
131. Kumar, R.; Kaushik, S.C. Performance evaluation of green roof and shading for thermal protection of buildings. *Build. Environ.* **2005**, *40*, 1505–1511. [CrossRef]
132. Berardi, U. The outdoor microclimate benefits and energy saving resulting from green roofs retrofits. *Energy Build.* **2016**, *121*, 217–229. [CrossRef]
133. Yang, H.S.; Kang, J.; Choi, M.S. Acoustic effects of green roof systems on a low-profiled structure at street level. *Build. Environ.* **2012**, *50*, 44–55. [CrossRef]
134. Li, J.-f.; Wai, O.W.H.; Li, Y.S.; Zhan, J.-m.; Ho, Y.A.; Li, J.; Lam, E. Effect of green roof on ambient CO2 concentration. *Build. Environ.* **2010**, *45*, 2644–2651. [CrossRef]
135. WHO. *Urban Green Spaces and Health*; World Health Organization, Regional Office for Europe: Copenhagen, Denmark, 2016.

136. Garcia, R.; Millán, E. Assessment of Cd, Pb and Zn contamination in roadside soils and grasses from Gipuzkoa (Spain). *Chemosphere* **1998**, *37*, 1615–1625. [CrossRef]
137. Ferro, A.M.; Sims, R.C.; Bugbee, B. Hycrest crested wheatgrass accelerates the degradation of pentachlorophenol in soil. *J. Environ. Qual.* **1994**, *23*, 272–279. [CrossRef]
138. Chekol, T.; Vough, L.R. A Study of the Use of Alfalfa (*Medicago sativa* L.) for the Phytoremediation of Organic Contaminants in Soil. *Remediation* **2001**, *11*, 89–101. [CrossRef]
139. Alkorta, I.; Garbisu, C. Phytoremediation of organic contaminants in soils. *Bioresour. Technol.* **2001**, *79*, 273–276. [CrossRef]
140. Salt, D.E.; Blaylock, M.; Kumar, N.P.; Dushenkov, V.; Ensley, B.D.; Chet, I.; Raskin, I. Phytoremediation: A novel strategy for the removal of toxic metals from the environment using plants. *Biotechnology* **1995**, *13*, 468–474. [CrossRef]
141. Capuana, M. A review of the performance of woody and herbaceous ornamental plants for phytoremediation in urban areas. *iFor.-Biogeosci. For.* **2020**, *13*, 139–151. [CrossRef]
142. Dadea, C.; Russo, A.; Tagliavini, M.; Mimmo, T.; Zerbe, S.J.A.; Forestry, U. Tree species as tools for biomonitoring and phytoremediation in urban environments: A review with special regard to heavy metals. *Arboric. Urban For.* **2017**, *43*, 155–167. [CrossRef]
143. Zhao, F.J.; Tang, Z.; Song, J.J.; Huang, X.Y.; Wang, P. Toxic metals and metalloids: Uptake, transport, detoxification, phytoremediation, and crop improvement for safer food. *Mol. Plant* **2022**, *15*, 27–44. [CrossRef]
144. Ishikawa, S. Mechanisms of cadmium accumulation in rice grains and molecular breeding for its reduction. *Soil Sci. Plant Nutr.* **2020**, *66*, 28–33. [CrossRef]
145. Hayashi, S.; Kuramata, M.; Abe, T.; Yamaguchi, N.; Takagi, H.; Tanikawa, H.; Iino, M.; Sugimoto, K.; Ishikawa, S. Deficiency in alcohol dehydrogenase 2 reduces arsenic in rice grains by suppressing silicate transporters. *Plant Physiol.* **2021**, *186*, 611–623. [CrossRef] [PubMed]
146. Currie, B.A.; Bass, B. Estimates of air pollution mitigation with green plants and green roofs using the UFORE model. *Urban Ecosyst.* **2008**, *11*, 409–422. [CrossRef]
147. Jayasooriya, V.M.; Ng, A.W.M.; Muthukumaran, S.; Perera, B.J.C. Green infrastructure practices for improvement of urban air quality. *Urban For. Urban Green.* **2017**, *21*, 34–47. [CrossRef]
148. Tallis, M.; Taylor, G.; Sinnett, D.; Freer-Smith, P. Estimating the removal of atmospheric particulate pollution by the urban tree canopy of London, under current and future environments. *Landsc. Urban Plan.* **2011**, *103*, 129–138. [CrossRef]
149. Nowak, D.J.; Greenfield, E.J.; Hoehn, R.E.; Lapoint, E. Carbon storage and sequestration by trees in urban and community areas of the United States. *Environ. Pollut.* **2013**, *178*, 229–236. [CrossRef]
150. Saebo, A.; Popek, R.; Nawrot, B.; Hanslin, H.M.; Gawronska, H.; Gawronski, S.W. Plant species differences in particulate matter accumulation on leaf surfaces. *Sci. Total Environ.* **2012**, *427–428*, 347–354. [CrossRef]
151. Weber, F.; Kowarik, I.; Saumel, I. Herbaceous plants as filters: Immobilization of particulates along urban street corridors. *Environ. Pollut.* **2014**, *186*, 234–240. [CrossRef]
152. Tong, Z.; Baldauf, R.W.; Isakov, V.; Deshmukh, P.; Max Zhang, K. Roadside vegetation barrier designs to mitigate near-road air pollution impacts. *Sci. Total Environ.* **2016**, *541*, 920–927. [CrossRef]
153. Shrestha, S.; Baral, B.; Dhital, N.B.; Yang, H.-H. Assessing air pollution tolerance of plant species in vegetation traffic barriers in Kathmandu Valley, Nepal. *Sustain. Environ. Res.* **2021**, *31*, 3. [CrossRef]
154. Kirk, G.J.D.; Solivas, J.L.; Alberto, M.C. Effects of flooding and redox conditions on solute diffusion in soil. *Eur. J. Soil Sci.* **2003**, *54*, 617–624. [CrossRef]
155. Gallegos Reina, A.; Perles Roselló, M.J. Relationships between Peri-Urbanization Processes and Multi-Hazard Increases: Compared Diachronic Analysis in Basins of the Mediterranean Coast. *ISPRS Int. J. Geo-Inf.* **2021**, *10*, 759. [CrossRef]
156. Bartens, J.; Day, S.D.; Harris, J.R.; Dove, J.E.; Wynn, T.M. Can urban tree roots improve infiltration through compacted subsoils for stormwater management? *J. Environ. Qual.* **2008**, *37*, 2048–2057. [CrossRef] [PubMed]
157. Pohl, M.; Alig, D.; Körner, C.; Rixen, C. Higher plant diversity enhances soil stability in disturbed alpine ecosystems. *Plant Soil* **2009**, *324*, 91–102. [CrossRef]
158. Stokes, A.; Atger, C.; Bengough, A.G.; Fourcaud, T.; Sidle, R.C. Desirable plant root traits for protecting natural and engineered slopes against landslides. *Plant Soil* **2009**, *324*, 1–30. [CrossRef]
159. Hollender, C.A.; Pascal, T.; Tabb, A.; Hadiarto, T.; Srinivasan, C.; Wang, W.; Liu, Z.; Scorza, R.; Dardick, C. Loss of a highly conserved sterile alpha motif domain gene (WEEP) results in pendulous branch growth in peach trees. *Proc. Natl. Acad. Sci. USA* **2018**, *115*, E4690–E4699. [CrossRef]
160. De Baets, S.; Poesen, J.; Reubens, B.; Muys, B.; De Baerdemaeker, J.; Meersmans, J. Methodological framework to select plant species for controlling rill and gully erosion: Application to a Mediterranean ecosystem. *Earth Surf. Process. Landf.* **2009**, *34*, 1374–1392. [CrossRef]
161. Ghestem, M.; Sidle, R.C.; Stokes, A. The Influence of Plant Root Systems on Subsurface Flow: Implications for Slope Stability. *BioScience* **2011**, *61*, 869–879. [CrossRef]
162. Fattet, M.; Fu, Y.; Ghestem, M.; Ma, W.; Foulonneau, M.; Nespoulous, J.; Le Bissonnais, Y.; Stokes, A. Effects of vegetation type on soil resistance to erosion: Relationship between aggregate stability and shear strength. *Catena* **2011**, *87*, 60–69. [CrossRef]
163. Francini, A.; Toscano, S.; Romano, D.; Ferrini, F.; Ferrante, A. Biological Contribution of Ornamental Plants for Improving Slope Stability along Urban and Suburban Areas. *Horticulturae* **2021**, *7*, 310. [CrossRef]

164. Farnham, M.W.; Grusak, M.A. Assessing nutritional changes in a vegetable over time: Issues and considerations. *HortScience* **2014**, *49*, 128–132. [CrossRef]
165. Bumgarner, N.; Dorn, S.; McGinnis, E.; Bennett, P.; Bauske, E.; Krishnan, S.; Bradley, L. Consumer horticulture advancement: Identifying critical research areas and cultivating collaborations. *HortTechnology* **2019**, *29*, 769–776. [CrossRef]
166. Damerum, A.; Chapman, M.A.; Taylor, G. Innovative breeding technologies in lettuce for improved post-harvest quality. *Postharvest Biol. Technol.* **2020**, *168*, 111266. [CrossRef] [PubMed]
167. Diaz, L.M.; Ricaurte, J.; Tovar, E.; Cajiao, C.; Teran, H.; Grajales, M.; Polania, J.; Rao, I.; Beebe, S.; Raatz, B. QTL analyses for tolerance to abiotic stresses in a common bean (*Phaseolus vulgaris* L.) population. *PLoS ONE* **2018**, *13*, e0202342. [CrossRef] [PubMed]
168. Palumbo, F.; Galla, G.; Vitulo, N.; Barcaccia, G. First draft genome sequencing of fennel (Foeniculum vulgare Mill.): Identification of simple sequence repeats and their application in marker-assisted breeding. *Mol. Breed.* **2018**, *38*, 1–17. [CrossRef]
169. Patella, A.; Palumbo, F.; Galla, G.; Barcaccia, G. The Molecular Determination of Hybridity and Homozygosity Estimates in Breeding Populations of Lettuce (*Lactuca sativa* L.). *Genes* **2019**, *10*, 916. [CrossRef] [PubMed]
170. Gabellini, S.; Scaramuzzi, S. Evolving consumption trends, marketing strategies, and governance settings in ornamental horticulture: A grey literature review. *Horticulturae* **2022**, *8*, 234. [CrossRef]
171. Orsini, F.; Pennisi, G.; Michelon, N.; Minelli, A.; Bazzocchi, G.; Sanyé-Mengual, E.; Gianquinto, G. Features and Functions of Multifunctional Urban Agriculture in the Global North: A Review. *Front. Sustain. Food Syst.* **2020**, *4*, 562513. [CrossRef]
172. Bhattarai, K.; Sharma, S.; Panthee, D.R. Diversity among Modern Tomato Genotypes at Different Levels in Fresh-Market Breeding. *Int. J. Agron.* **2018**, *2018*, 1–15. [CrossRef]
173. Burbano-Erazo, E.; Pastrana-Vargas, I.J.; Mejía-Salazar, J.R.; Vallejo-Cabrera, F.A. Selection criteria in tomato lines with determinate growth habit. *Agron. Mesoam.* **2020**, *31*, 1–11. [CrossRef]
174. Nascimento, M.V.; Ávila, M.C.R.; de Abreu-Tarazi, M.F.; Nogueira, A.P.O.; Campos, L.F.C.; dos Reis Nascimento, A. Identification of promising tomato breeding lines with determinate growth by selection index. *Adv. Hortic. Sci.* **2020**, *34*, 337–347.
175. Bergougnoux, V. The history of tomato: From domestication to biopharming. *Biotechnol. Adv.* **2014**, *32*, 170–189. [CrossRef] [PubMed]
176. Rothan, C.; Diouf, I.; Causse, M. Trait discovery and editing in tomato. *Plant J.* **2019**, *97*, 73–90. [CrossRef] [PubMed]
177. Marti, E.; Gisbert, C.; Bishop, G.J.; Dixon, M.S.; Garcia-Martinez, J.L. Genetic and physiological characterization of tomato cv. Micro-Tom. *J. Exp. Bot.* **2006**, *57*, 2037–2047. [CrossRef]
178. Meissner, R.; Jacobson, Y.; Melamed, S.; Levyatuv, S.; Shalev, G.; Ashri, A.; Elkind, Y.; Levy, A. A new model system for tomato genetics. *Plant J.* **1997**, *12*, 1465–1472. [CrossRef]
179. Scott, J.W.; Harbaugh, B.K. Micro-Tom. A miniature dwarf tomato. *Circ. Univ. Fla. Agric. Exp. Stn.* **1989**, *370*, 6.
180. Shikata, M.; Ezura, H. Micro-Tom Tomato as an Alternative Plant Model System: Mutant Collection and Efficient Transformation. In *Plant Signal Transduction: Methods and Protocols*; Botella, J.R., Botella, M.A., Eds.; Springer: New York, NY, USA, 2016; pp. 47–55. [CrossRef]
181. Arena, C.; Vitale, E.; Hay Mele, B.; Cataletto, P.R.; Turano, M.; Simoniello, P.; De Micco, V. Suitability of Solanum lycopersicum L. 'Microtom' for growth in Bioregenerative Life Support Systems: Exploring the effect of high-LET ionising radiation on photosynthesis, leaf structure and fruit traits. *Plant Biol.* **2019**, *21*, 615–626. [CrossRef]
182. Marzioli, P.; Gugliermetti, L.; Santoni, F.; Delfini, A.; Piergentili, F.; Nardi, L.; Metelli, G.; Benvenuto, E.; Massa, S.; Bennici, E. CultCube: Experiments in autonomous in-orbit cultivation on-board a 12-Units CubeSat platform. *Life Sci. Space Res.* **2020**, *25*, 42–52. [CrossRef]
183. Scott, J.; Harbaugh, B.; Baldwin, E. Micro-Tina'andMicro-Gemma'Miniature Dwarf Tomatoes. *HortScience* **2000**, *35*, 774–775. [CrossRef]
184. SharathKumar, M.; Heuvelink, E.; Marcelis, L.F.M. Vertical Farming: Moving from Genetic to Environmental Modification. *Trends Plant Sci.* **2020**, *25*, 724–727. [CrossRef]
185. Al-Kodmany, K. The Vertical Farm: Exploring Applications for Peri-urban Areas. In *Smart Village Technology*; Patnaik, S., Sen, S., Mahmoud, M.S., Eds.; Springer International Publishing: Cham, Switzerland, 2020; pp. 203–232. [CrossRef]
186. Folta, K.M. Breeding new varieties for controlled environments. *Plant Biol.* **2019**, *21* (Suppl. 1), 6–12. [CrossRef] [PubMed]
187. Stagnari, F.; Di Mattia, C.; Galieni, A.; Santarelli, V.; D'Egidio, S.; Pagnani, G.; Pisante, M. Light quantity and quality supplies sharply affect growth, morphological, physiological and quality traits of basil. *Ind. Crops Prod.* **2018**, *122*, 277–289. [CrossRef]
188. Nogler, G.; Johri, B. Embryology of angiosperms. In *Gametophytic Embryogenesis*; Johri, B., Ed.; Springer: Berlin/Heidelberg, Germany; New York, NY, USA, 1984; pp. 475–518.
189. Palumbo, F.; Draga, S.; Vannozzi, A.; Lucchin, M.; Barcaccia, G. Trends in Apomixis Research: The 10 Most Cited Research Articles Published in the Pregenomic and Genomic Eras. *Front. Plant Sci.* **2022**, *13*, 878074. [CrossRef]
190. Kozai, T. Resource use efficiency of closed plant production system with artificial light: Concept, estimation and application to plant factory. *Proc. Jpn. Acad. Ser. B Phys. Biol. Sci.* **2013**, *89*, 447–461. [CrossRef] [PubMed]
191. Touliatos, D.; Dodd, I.C.; McAinsh, M. Vertical farming increases lettuce yield per unit area compared to conventional horizontal hydroponics. *Food Energy Secur.* **2016**, *5*, 184–191. [CrossRef]
192. Nicola, S.; Fontana, E. Fresh-Cut Produce Quality. In *Postharvest Handling*; Florkowski, W.J., Shewfelt, R.L., Brueckner, B., Prussia, S.E., Eds.; Academic Press: San Diego, CA, USA, 2014; pp. 217–273. [CrossRef]

193. Ferrante, A. Ortaggi per la IV gamma [Produce for pre-cut salads]. In *La Concimazione Azotata Degli Ortaggi*; Incrocci, L. Santamaria, P., Pardossi, A., Eds.; Barone e Bella & C: Ragusa, Italy, 2013; pp. 165–180.
194. Saini, R.K.; Ko, E.Y.; Keum, Y.-S. Minimally processed ready-to-eat baby-leaf vegetables: Production, processing, storage, microbial safety, and nutritional potential. *Food Rev. Int.* **2016**, *33*, 644–663. [CrossRef]
195. Halgamuge, M.N.; Bojovschi, A.; Fisher, P.M.J.; Le, T.C.; Adeloju, S.; Murphy, S. Internet of Things and autonomous control for vertical cultivation walls towards smart food growing: A review. *Urban For. Urban Green.* **2021**, *61*, 127094. [CrossRef]
196. Gil, M.I. Preharvest factors and fresh-cut quality of leafy vegetables. In Proceedings of the III International Conference on Fresh-Cut Produce: Maintaining Quality and Safety, Davis, CA, USA, 13 September 2015; pp. 57–64.
197. Jasper, J.; Elmore, J.S.; Wagstaff, C. Determining the quality of leafy salads: Past, present and future. *Postharvest Biol. Technol.* **2021**, *180*, 111630. [CrossRef]
198. Martínez-Sánchez, A.; Luna, M.C.; Selma, M.V.; Tudela, J.A.; Abad, J.; Gil, M.I. Baby-leaf and multi-leaf of green and red lettuces are suitable raw materials for the fresh-cut industry. *Postharvest Biol. Technol.* **2012**, *63*, 1–10. [CrossRef]
199. Pires, E.O., Jr.; Di Gioia, F.; Rouphael, Y.; Ferreira, I.; Caleja, C.; Barros, L.; Petropoulos, S.A. The Compositional Aspects of Edible Flowers as an Emerging Horticultural Product. *Molecules* **2021**, *26*, 6940. [CrossRef]
200. Mlcek, J.; Rop, O. Fresh edible flowers of ornamental plants—A new source of nutraceutical foods. *Trends Food Sci. Technol.* **2011**, *22*, 561–569. [CrossRef]
201. Copetta, A.; Marchioni, I.; Ruffoni, B. The edible flowers from woody ornamental plants. In Proceedings of the IV International Symposium on Woody Ornamentals of the Temperate Zone, Torino, Italy, 3 March 2021; pp. 195–204.
202. Yagi, M. Recent Progress in Genomic Analysis of Ornamental Plants, with a Focus on Carnation. *Hortic. J.* **2015**, *84*, 3–13. [CrossRef]
203. Kuligowska, K.; Lutken, H.; Muller, R. Towards development of new ornamental plants: Status and progress in wide hybridization. *Planta* **2016**, *244*, 1–17. [CrossRef] [PubMed]
204. Debener, T. Molecular Markers for Ornamental Plant Genetics, Genomics and Breeding. In Proceedings of the XXIV International Eucarpia Symposium Section Ornamentals: Ornamental Breeding Worldwide, Warsaw, Poland, 2 September 2012; pp. 193–200.
205. Nowak, M.D.; Russo, G.; Schlapbach, R.; Huu, C.N.; Lenhard, M.; Conti, E. The draft genome of Primula veris yields insights into the molecular basis of heterostyly. *Genome Biol.* **2015**, *16*, 12. [CrossRef]
206. Vukosavljev, M.; Arens, P.; Voorrips, R.E.; van't Westende, W.P.; Esselink, G.D.; Bourke, P.M.; Cox, P.; van de Weg, W.E.; Visser, R.G.; Maliepaard, C.; et al. High-density SNP-based genetic maps for the parents of an outcrossed and a selfed tetraploid garden rose cross, inferred from admixed progeny using the 68k rose SNP array. *Hortic. Res.* **2016**, *3*, 16052. [CrossRef]
207. Yagi, M. Recent progress in whole genome sequencing, high-density linkage maps, and genomic databases of ornamental plants. *Breed Sci.* **2018**, *68*, 62–70. [CrossRef]
208. Scariolo, F.; Palumbo, F.; Vannozzi, A.; Sacilotto, G.B.; Gazzola, M.; Barcaccia, G. Genotyping Analysis by RAD-Seq Reads Is Useful to Assess the Genetic Identity and Relationships of Breeding Lines in Lavender Species Aimed at Managing Plant Variety Protection. *Genes* **2021**, *12*, 1656. [CrossRef]
209. Wan, H.; Yu, C.; Han, Y.; Guo, X.; Ahmad, S.; Tang, A.; Wang, J.; Cheng, T.; Pan, H.; Zhang, Q. Flavonols and Carotenoids in Yellow Petals of Rose Cultivar (Rosa 'Sun City'): A Possible Rich Source of Bioactive Compounds. *J. Agric. Food Chem.* **2018**, *66*, 4171–4181. [CrossRef] [PubMed]
210. Krzyminska, A.; Gasecka, M.; Magdziak, Z. Content of Phenolic Compounds and Organic Acids in the Flowers of Selected Tulipa gesneriana Cultivars. *Molecules* **2020**, *25*, 5627. [CrossRef]
211. Marchioni, I.; Pistelli, L.; Copetta, A.; Dimita, R.; Descamps, S.; Cambournac, L.; Ruffoni, B. Edible roses as novel food with healthy value. In Proceedings of the IV International Symposium on Woody Ornamentals of the Temperate Zone, Torino, Italy, 3 March 2021; pp. 239–244.
212. Goddard, M.A.; Dougill, A.J.; Benton, T.G. Scaling up from gardens: Biodiversity conservation in urban environments. *Trends Ecol. Evol.* **2010**, *25*, 90–98. [CrossRef]
213. Spotswood, E.N.; Beller, E.E.; Grossinger, R.; Grenier, J.L.; Heller, N.E.; Aronson, M.F.J. The Biological Deserts Fallacy: Cities in Their Landscapes Contribute More than We Think to Regional Biodiversity. *Bioscience* **2021**, *71*, 148–160. [CrossRef] [PubMed]
214. Hausmann, S.L.; Petermann, J.S.; Rolff, J. Wild bees as pollinators of city trees. *Insect Conserv. Divers.* **2016**, *9*, 97–107. [CrossRef]
215. Oliver, T.H.; Heard, M.S.; Isaac, N.J.B.; Roy, D.B.; Procter, D.; Eigenbrod, F.; Freckleton, R.; Hector, A.; Orme, C.D.L.; Petchey, O.L.; et al. Biodiversity and Resilience of Ecosystem Functions. *Trends Ecol. Evol.* **2015**, *30*, 673–684. [CrossRef] [PubMed]
216. Tylianakis, J.M. Ecology. The global plight of pollinators. *Science* **2013**, *339*, 1532–1533. [CrossRef] [PubMed]
217. Wenzel, A.; Grass, I.; Belavadi, V.V.; Tscharntke, T. How urbanization is driving pollinator diversity and pollination—A systematic review. *Biol. Conserv.* **2020**, *241*, 108321. [CrossRef]
218. Pellmyr, O. Pollination by animals. In *Plant–Animal Interactions, an Evolutionary Approach*; Herrera, C.M., Pellmyr, O., Eds.; Blackwell Science: Oxford, UK, 2002; pp. 157–184.
219. Payne, A.; Schildroth, D.A.; Starks, P.T. Nest site selection in the European wool-carder bee, Anthidium manicatum, with methods for an emerging model species. *Apidologie* **2011**, *42*, 181–191. [CrossRef]
220. Rahimi, E.; Barghjelveh, S.; Dong, P. A review of diversity of bees, the attractiveness of host plants and the effects of landscape variables on bees in urban gardens. *Agric. Food Secur.* **2022**, *11*, 6. [CrossRef]

221. Slate, M.L.; Matallana-Mejia, N.; Aromin, A.; Callaway, R.M. Nitrogen addition, but not pulse frequency, shifts competitive interactions in favor of exotic invasive plant species. *Biol. Invasions* **2022**. [CrossRef]
222. Rojas-Botero, S.; Kollmann, J.; Teixeira, L.H. Competitive trait hierarchies of native communities and invasive propagule pressure consistently predict invasion success during grassland establishment. *Biol. Invasions* **2021**, *24*, 107–122. [CrossRef]
223. Adler, P.B.; Smull, D.; Beard, K.H.; Choi, R.T.; Furniss, T.; Kulmatiski, A.; Meiners, J.M.; Tredennick, A.T.; Veblen, K.E. Competition and coexistence in plant communities: Intraspecific competition is stronger than interspecific competition. *Ecol. Lett.* **2018**, *21*, 1319–1329. [CrossRef]
224. Drenovsky, R.E.; Grewell, B.J.; D'Antonio, C.M.; Funk, J.L.; James, J.J.; Molinari, N.; Parker, I.M.; Richards, C.L. A functional trait perspective on plant invasion. *Ann. Bot.* **2012**, *110*, 141–153. [CrossRef]
225. Kaushik, P.; Pati, P.K.; Khan, M.L.; Khare, P.K. Plant functional traits best explain invasive species' performance within a dynamic ecosystem—A review. *Trees For. People* **2022**, *8*, 100260. [CrossRef]
226. Ellstrand, N.C.; Schierenbeck, K.A. Hybridization as a stimulus for the evolution of invasiveness in plants? *Proc. Natl. Acad. Sci. USA* **2000**, *97*, 7043–7050. [CrossRef]
227. Takakura, K.; Matsumoto, T.; Nishida, T.; Nishida, S. Effective range of reproductive interference exerted by an alien dandelion, Taraxacum officinale, on a native congener. *J. Plant Res.* **2011**, *124*, 269–276. [CrossRef]
228. Tang, J.; Mao, K.; Zhang, H.; Xu, X.; Xu, X.; Guo, H.; Li, B. Multiple introductions and genetic admixture facilitate the successful invasion of Plantago virginica into China. *Biol. Invasions* **2022**, *24*, 2261–2272. [CrossRef]
229. Uemura, R.; Asakawa, A.; Fujii, S.; Matsuo, A.; Suyama, Y.; Maki, M. Can *Rumex madaio* (Polygonaceae) be threatened by natural hybridization with an invasive species in Japan. *Nord. J. Bot.* **2022**, *2022*, e03543. [CrossRef]
230. Deng, Z.; Wilson, S.B.; Ying, X.; Chen, C.; Freyre, R.; Zayas, V.; Czarnecki, D.M. 'UF-1013-1': An Infertile Cultivar of Lantana camara. *HortScience* **2020**, *55*, 953–958. [CrossRef]
231. Henderson, I.R.; Salt, D.E. Natural genetic variation and hybridization in plants. *J. Exp. Bot.* **2017**, *68*, 5415–5417. [CrossRef]

horticulturae

MDPI

Review

Evolving Consumption Trends, Marketing Strategies, and Governance Settings in Ornamental Horticulture: A Grey Literature Review

Sara Gabellini and Silvia Scaramuzzi *

Department of Agriculture, Food, Environment and Forestry, University of Florence, 50144 Florence, Italy; sara.gabellini@unifi.it
* Correspondence: silvia.scaramuzzi@unifi.it

Abstract: Ornamentals are the most diversified products and fast-changing industry of horticulture. A new flower and ornamental plant market scenario is developing: remarkable opportunities are emerging, but more efforts are required by both public and private stakeholders to seize them and assure a high-value positioning. Our paper aims at filling the gap in the availability of integrated data sources and structured theoretically sound studies on new consumption trends, marketing strategies, and governance settings. Specific objectives are: identifying an innovative ornamental horticulture market data framework; evidencing evolving dynamics of competition in Europe and necessary adaptions of public and private action; defining a new action-research agenda, capable of stimulating the interest of businesses, researchers, and institutions. In terms of methodology, we carry out an innovative integrative review analysis of the wide and most reliable grey literature and statistics, using a comprehensive approach. Results show the emerging consumption dynamics and high-value consumer profiles characterizing the European market, expected to significantly expand and transform, according to the impact of globalization, climate change, urbanization, digitalization, and the affirmation of neo-luxury and sustainability-oriented consumption patterns. The evolution of marketing strategies and governance settings is also highlighted, together with the necessity of developing and integrating public and private initiatives for realizing high-value sustainable and transparent production systems and supply chains. Accordingly, relevant action-research directions are described. These findings are expected to improve the current debate on the competitiveness of the European ornamental industry and contribute to taking a step towards a synergic combination of new differential advantages and wider sustainability goals.

Keywords: market competition; consumer dynamics; market niches; high-value positioning; management approaches; collaborative governance; supply chain; quality products; origin; sustainability transformation

Citation: Gabellini, S.; Scaramuzzi, S. Evolving Consumption Trends, Marketing Strategies, and Governance Settings in Ornamental Horticulture: A Grey Literature Review. *Horticulturae* **2022**, *8*, 234. https://doi.org/10.3390/horticulturae8030234

Academic Editors: Margherita Beruto, Emmy Dhooghe and Bruce Dunn

Received: 4 February 2022
Accepted: 2 March 2022
Published: 8 March 2022

1. Introduction

Ornamentals are the most diversified products and fast-changing industry of horticulture [1]. They include a very big and hugely diverse group of whole plants or parts of plants that are grown usually for decorative purposes [2]. Building on Oxford Economics (2018) [3], we define Ornamental Horticulture as the set of the different sectors engaged in the ornamental horticultural activity: from firms that grow ornamental plants and flower to manufacturers of garden equipment and supplies, to the wholesalers and retailers, as well as floral designers and landscape and arboriculture professionals.

Nowadays, the world ornamental sector is characterized by a significant expansion of both production and consumption, that underpins the growth of international trade and globalization [4–8]. Consequently, major country markets have increased their interdependency and permeability. Accordingly, both flowers' and plants' supply and demand have increased their seasonality and mutability and caused higher volatility of prices [9,10].

From a global perspective, Europe represents one of the main markets for ornamentals, and it is expected to register, over the next decade, one of the highest performances of growth of both production and consumption, in line with the development of other mature producer countries, such as China, Japan, and North America [5,6,9].

Noteworthy, current expectations of the European ornamental demand have evidenced a perspective of unprecedented increase in purchases and price premiums, considering both private and institutional market segments [8,11,12]. As a matter of fact, the affirmation of neo-luxury and sustainability-oriented consumption patterns, together with the expansion of urban upper and middle classes, has generated a new need for both consumers and citizens to green their lives [13–16]. Accordingly, individuals and institutions have shown a higher willingness to buy and pay for flowers and plants and behave as attentive and responsible consumers in choosing sustainability and transparency [11,17,18].

These trends are confirmed by the early insights on the impact of the ongoing COVID-19 pandemic [4,19,20]. Specifically, they highlight the strengthening of new consumer perception of flowers and plants as functional goods, with essential advantages for the environment and human well-being [11,21–23].

In light of the above, we consider the increase in new market opportunities for European domestic producers that show the power to drive processes of requalification and sustainability transformation of the European ornamental sector [24]. The good performance of European demand has also incentivized the growth and higher quality of imported productions, coming from exporting producer countries of Africa and South America [4,5]. This causes new relevant challenges for the competitiveness of high-cost domestic supply systems [6,10,25,26].

The seizing of new opportunities for the sustainable development of the European ornamental sector requires an increase in efforts from both private and public stakeholders. In other words, the enhancement and adaption of existing marketing strategies and governance settings appear essential to face the competition and also favor the realization of win-win approaches among different countries and stakeholders [10,12,27–32]. On this basis, the industry shows a gap in the capacity of prediction, planning, and coordination of relevant actors, thus evidencing a need for improving their knowledge and practices [25,33,34].

In this sense, market research plays a fundamental role [12,35]. Notwithstanding, in the European context, flower and ornamental plants' marketing and governance systems usually receive limited attention from the scientific literature; in addition, the chronic outdatedness and fragmentation of relevant data and information affect the availability of official statistics and institutional sources [36]. As a result, the current economic debate is weak.

Our paper aims at filling the gap in the availability of structured and theoretically sound studies and integrated data sources on new consumption trends, marketing strategies, and governance settings. In particular, the following objectives are pursued: (i) identifying an innovative ornamental horticulture market data framework; (ii) evidencing evolving dynamics of competition in Europe and necessary adaptions of public and private action; (iii) defining a new action-research agenda, capable of stimulating interest from businesses, researchers, and institutions.

As far as methodology is concerned, we carry out an integrative grey literature review. Considering the abovementioned limits of official data and scientific research, the grey literature represents an essential source of information within the sector. Innovatively, the integrative review strategy allows us to overcome the dispersion of grey literature sources and to integrate and synthesize the most reliable and representative data.

We believe our findings give an innovative contribution to the information, enhancement, and coordination of decisions and actions of policymakers, businesses, and industry organizations operating in the European ornamental sector, and sustain the realization and fair remuneration of high-value sustainable ornamental supply chains, with multiple functions for the society and ecosystems [37–39].

Specifically, we illustrate the followed review methodology in Section 2. Results are presented in Sections 3–5. Lastly, Sections 6 and 7 are dedicated to the identification of the new action-research agenda and the conclusions.

2. Methodology

The methodology of this paper is based on an integrative grey literature review [40,41]. We classify the grey literature as a non-systematic aggregate of material and research produced by different organizations outside of the traditional academic or commercial publishing and distribution channels [42].

The panorama of the economic literature on the European ornamental horticulture industry is limited. At the same time, relevant official statistics are often poor and inconsistent. In light of this, we consider grey literature to be the main source of knowledge on the European ornamental horticulture sector structure and dynamics, although characterized by high variability and fragmentation of related institutional sources and documents, as well as dataset types.

In line with this, the adoption of a grey literature approach and integrative review strategy allowed us to select the most reliable and representative institutional sources and providers of grey literature. This led us to identify and review their most recent and comprehensive contributions valuable for the topic [40,43].

The review design does not follow any specific standard [41,44]. The sampling of sources was realized by adopting a purposeful approach, aiming at an inclusive selection of material [40,44].

As a first step, we identified a set of key terms to define the research topic (Table 1). We used the identified terms as keywords to conduct an online search of relevant grey literature. The research was realized by the means of the Google Search engine [45]. The web searching method combined with the key terms identified permits to include in the sample grey literature that is open or easy to access, thus enhancing the study replicability.

Table 1. Key terms for research topic definition and online searching.

Definers	Key Terms
Market and Industry	Floriculture Flower industry Ornamental industry Nursery industry Flower market Ornamental plants market
Product Category	Flowers and ornamental plants Ornamentals Cut flowers and pot(ted) plants Cut flowers and indoor plants
Supply Chain Structure and Characterization	Supply chain structure and dynamics Production Trade Consumption Consumption patterns Sales and spending
Geographical Area	World/Global Europe European Union (EU), i.e., EU (28)/EU (27) + United Kingdom (UK)

Source: our elaboration.

As a preliminary result, we selected 10 institutions that are among the major providers of grey literature concerning the European flower and plant market (i.e., industry organizations, trade fairs, flower auctions, commercial services, governments, and other national and international institutions) and identified their official websites to gather relevant documentary materials and statistics (Table 2).

Table 2. Major institutional sources in the European ornamental sector.

Name of the Institution	Country	Official Websites [1]
International Association of Horticultural Producers (AIPH)	Belgium	http://aiph.org/
International Flower Trade Association (Union Fleurs)	Belgium	https://unionfleurs.org/
Messe Essen GmbH Press Media Centre for IPM Essen	Germany	https://www.ipm-essen.de/world-trade-fair/
Royal Flora Holland (RFH)	The Netherlands	https://www.royalfloraholland.com/en
Association of the German Flower Wholesale and Import Trade (BGI)	Germany	https://bgi-ev.de/en/the-association/
RaboResearch Food and AgriBusiness (Rabobank)	The Netherlands	https://research.rabobank.com/far/en/home/index.html
European Commission Directorate-General for Agriculture and Rural Development (DG AGRI) Unit G2-Wine, spirits, and horticultural products	Belgium	https://ec.europa.eu/info/food-farming-fisheries/plants-and-plant-products/live-plants-and-flowers_en
International Trade Centre (ITC)—Trade Map	n/a	https://www.trademap.org/Index.aspx
Centre for the Promotion of Imports from developing countries of the Netherlands Ministry of Foreign Affairs (CBI)	The Netherlands	https://www.cbi.eu/
Assembly of European Regions producing Fruits, Vegetables and Ornamental Plants (AREFLH)	France	https://www.areflh.org/en/

Source: our elaboration. [1] Accessed on 2 February 2022.

Among the latter, we selected a purposeful sample of 58 secondary data sources. The adopted inclusion criteria are reported in Table 3. We added literature to the sample until achieving the saturation of information [46].

Table 3. Inclusion criteria and sample description.

Inclusion Criteria	Description of the Included Grey Literature
Relevance and reliability	• Published or distributed by the most reliable and representative institutional providers of grey literature for the European ornamental industry • Reporting proprietary elaborations that integrate secondary data-official statistics or other reliable grey information—with primary data—interviews and surveys involving experts and practitioners
Document typology and accessibility	• Including different typologies of documents and datasets, i.e., project reports, periodicals, statistics, yearbooks, press texts, lectures, working documents, opinion statements • Accessible from open-or-easy-to-access online institutional sources, i.e., official websites of the selected institutions
Year of publication	• Published in the timeframe 2015 [1]–2022 (most recent disposable year)
Scope	• Not limited to a single country or product category
Language	• Using English language

Source: our elaboration. [1] We used 2015 as the base year for the gathering of grey literature. Indeed, it represents a turning point in the evolution of European ornamental industry and market dynamics [25,47].

The sampled grey literature was analyzed by the means of a comprehensive qualitative approach [40,41,44]. Accordingly, through the full-text reading of documents and the querying of datasets, we extracted the most relevant text parts and statistics. The extracted information was compared and integrated, based on criteria of logic and conceptual reasoning. The results were elaborated in a narrative synthesis form.

3. The World and European Ornamental Sector

3.1. A Description of the Global Scenario

At the global level, the ornamental sector is expanding in both production and trade, with a consequent increase in market globalization and competition [4–6,15,30,48,49].

On the side of production, flowers and ornamental plants count for a total world value of about EUR 35.5 billion, corresponding to an area of 745,000 Ha [50].

As for trade, Rabobank (2022a) [4] evidences a steady increase in global export, at a compound annual rate (CAGR) of 3.9%. Considering the categories of cut flowers, foliage, and live plants, a twenty-year positive trend in export growth can be identified, which led to an aggregate world value of around EUR 18 billion in 2020 [5,51] (Figure 1).

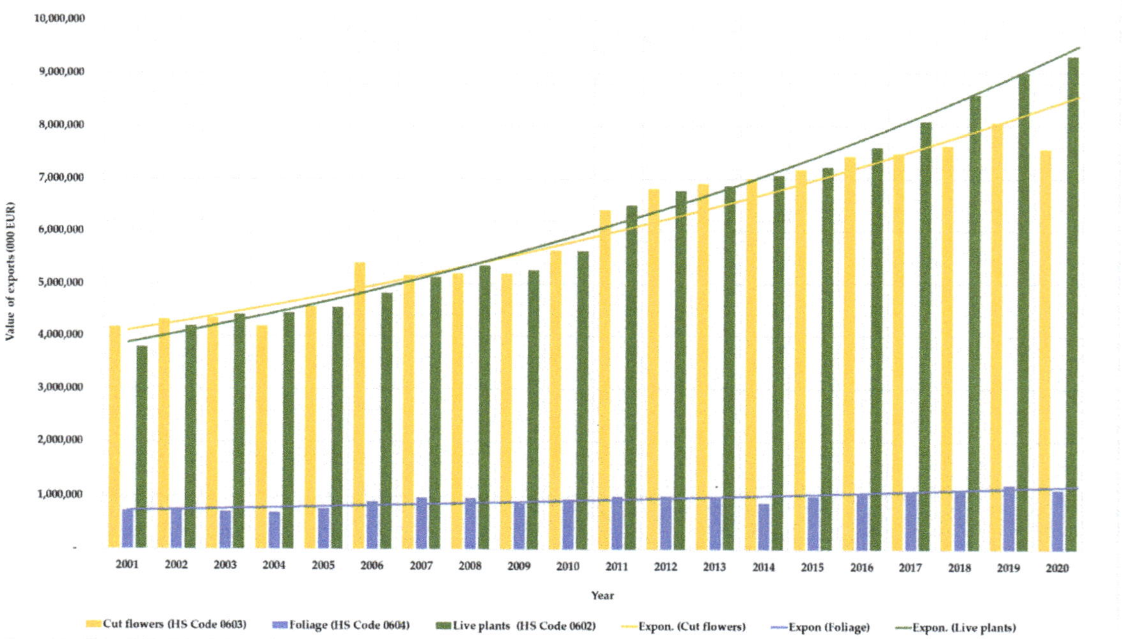

Figure 1. The international trade of flowers and ornamental plants: the trend of world exports (000 EUR, 2001–2020). Source: our elaboration on [51] (data extraction: 24th January 2022). The identified product categories follow the Harmonized System Nomenclature [52], HS Codes: 0603, 0604, 0602.

In order to identify and describe global main flower and ornamental plant markets, we consider a recent classification elaborated by AIPH (2019b) [6]. The latter characterizes four homogenous geographical groups, consisting of the major production and consumption markets at the global level.

Each group includes flowers and plants producer countries that have in common the same market determinants, and report similar conditions of both demand and supply. Building on AIPH (2019b) [6], in Table 4, we present a framework of the current World ornamental market scenario.

Table 4. The World ornamental market scenario: a classification.

Country Group	Included Country/Area [1]	Market Determinants		Characteristics of Demand and Supply
Mature domestic producer countries	Europe Canada United States (US) China Japan	• Good economic performance of included countries • High urbanization rate • High occupation growth • High population aging rate • Interest of individuals in pursuing a higher well-being and life quality • Flowers and plants are core elements of local biocultural heritage • Long-standing tradition of ornamental production • Presence of highly functional logistics networks and hubs for the trade of large volumes of perishables • Subsistence of high flowers and plants production costs, especially due to climate, labor conditions, relevant regulations, and fiscal regimes	High-value domestic demand	• Highest share of world total consumption value • Expected rapid and significant growth of demand, in both volumes and value • High demand fragmentation • Growing attention of consumers to flowers and plants multiple functions
			Strong domestic production base	• Domestic productions cover the highest share of internal demand • China and Japan: expected growth of domestic productions volume to maintain the highest share of consumption growth • Europe and North America: expected growth of domestic productions differentiation to compete with the increase in quality and volume of flower imports from Africa and South America
Emerging domestic producer countries	India Mexico Brazil	• Increase in economic performance of included countries • Growth of per-capita income and purchasing power • High urbanization rate • Upsizing of urban upper-and-middle classes • Favorable climate and soil conditions for flowers and plants cultivation • Low cost of production inputs • Proximity to high-value demand markets pertaining to the group of mature domestic producers	High-growth domestic demand	• Domestic demand exceeds domestic supply • Expected sharp growth in demand volume and value due to the expansion of urban upper and middle classes
			Expanding domestic production base	• Expected increase in domestic production to obtain the major share of growth of internal demand • Domestic production will almost exclusively cover the growth of internal demand • The volume of exports will remain early negligible
Mature exporting producer countries	Colombia Kenya Ecuador	• Unstable economic performance of included countries • High political instability • Low urbanization rate • Low occupation rate • Optimal soil and climate conditions for flower and plant breeding and cultivation • Minimum flower and plant production costs, due to low price of energy and labor, supportive regulations and fiscal regimes • Proximity to fast-growing and high-value markets pertaining to the groups of both mature and emerging domestic producers • Localization of both European and North American production activities that bring with them advanced knowledge and skills, high investment capacity and professional services	Low-growth domestic demand	• Small size of internal demand (Europe and North America currently represent main destination markets) • Low expected growth in consumption (India, Mexico, and Brazil represent new accessible and attractive destination markets)
			Strong domestic production base	• Expected growth of domestic production following a significant expansion of exports • Dominant role of major crops production, i.e., roses, chrysanthemum and carnations • Domestic producers will maintain the role of World cost leaders
Emerging exporting producer countries	Ethiopia Vietnam		Low-growth domestic demand	• Small size of domestic demand • Low expected growth in consumption
			Expanding domestic production base	• Expected increase in domestic production volume and competitiveness to capture a growing share of demand in fast-growing and high-value markets of Europe and Asia • Expected significant increase in the share of domestic productions of world total export

Source: our elaboration on [4–7,9,11,25,53–55]. [1] Countries/areas with either production or consumption above 100 million EUR.

3.2. Relevant Insights from Europe

According to the presented market scenario, the European ornamental industry is transforming, with evolving dynamics of production and trade and the emergence of new strategic orientations [4,5,11,56,57]. Specifically, already since the biennium 2015–2016, the sector entered a period of profound change, determined by market globalization and the evolution of the socio-economic and bio-physical context [25,47,58,59].

In line with the above, in Europe, the flower and plant supply is expanding and structurally mutating, although the effects of climate change contribute to the higher instability and unforeseeability of prices and quantities [15,49]. Dynamics of change are affected by key determinant factors, such as globalization, demand growth, technological development, and favorable commercial policies.

To illustrate, from 2015, significant growth has been characterizing the European flower and plant production [60–62], covering an area of 60.000 Ha for a value of about EUR 11 billion [50]. In terms of value, the major contribution has been made by mature producer countries, i.e., The Netherlands, Italy, Germany, France, Spain, and the UK, which together cover a share of 70% (approx. EUR 7.7 billion) of the European total [50]. At the same time, new domestic producers have emerged, such as the cases of Baltic states, Malta, and Luxembourg, that significantly record percentages of growth between 14% and 15% [60].

Important current dynamics of recovery and qualification of European domestic productions are expected to maintain and even accelerate [4,30], flanked by continuing growth of the flower and plant trade [4,5,30,48,49,54,61–66].

In that regard, Europe is consolidating its position of attractive and accessible target destination for both domestic and exporting producer countries. Hence, intra and extra EU trade is expanding due to: the growth of production, the innovation of information and communication technologies (ICT), the enhancement of logistics, the actualization of free trade agreements (FTA) and preferential trade schemes (GSP), the expansion of mature demand markets, and the emergence of attractive North-Eastern and Middle Eastern destinations (e.g., Denmark, Poland, Russia, Turkey) [4,5,11,12].

The European flower and plant trade is characterized by its dominant role of high quality low-cost imported productions, originating in the Southern countries of Africa (primarily Kenya and Ethiopia) and Latin America (primarily Colombia and Ecuador), mainly distributed by the Dutch wholesale market [4,5,48]. Nonetheless, a stable positive trend can be identified in the improvement of the European trade balance.

To that end, Figure 2 reports our elaboration of ITC trade map data [51] concerning the country group EU (27) + UK in order to evidence the performance of the trade balance in the period 2001–2020 for the categories cut flowers, foliage, and live plants. Figure 2 shows a consistent improvement of the balance during the considered period, leading to a total value of about EUR 191 million in 2020.

In light of the above, European market high internationalization increases its permeability to the political turmoil, economic turbulences, shocks, and crises occurring at the world or country level [54,66], with significant effects on its evolution dynamics and predictability.

To illustrate, among the latest developments affecting the European ornamental market, particular consideration is given to Brexit. On the one hand, the UK maintains its position as one of the major destinations in the European area. As a matter of facts, according to 2021 data, British consumers' demand is very high and Great Britain leads the rankings of Dutch exports growth [11]. On the other hand, current institutional changes (e.g., regulations, fiscal regimes) are partially reducing the attractiveness of the destination (e.g., higher transport and administrative costs) and challenging the European flower and plant exporters and producers [49,65,67,68].

Messe Essen (2022) [11] outlines the high market uncertainty deriving from the forthcoming introduction of a UK policy for the "comprehensive and full monitoring of flowers and plants from The Netherlands". This policy will come into force from the 1 July 2022 and will require flowers and ornamental plants to be physically checked at the border to be imported into the UK.

As a result, higher administrative and transport costs, as well as shipping delays, are expected to affect both the volume and value of EU exports to the UK and influence European traders' and producers' marketing strategies and governance systems. In addition, other similar effects could be generated by the decision of the UK government to look for trade agreements outside the EU, particularly with members of the British Commonwealth, such as Kenya [4].

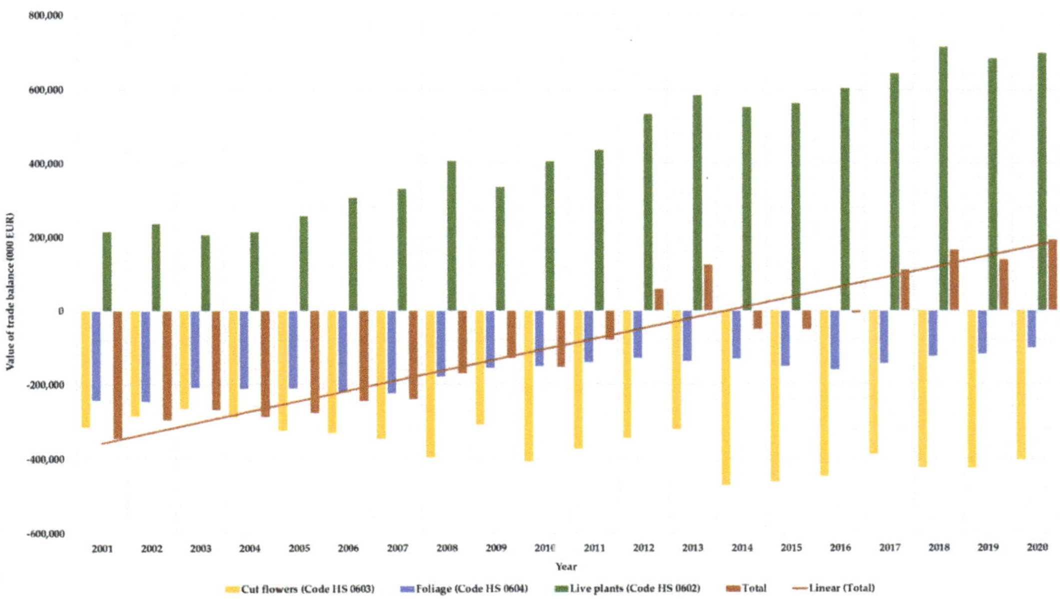

Figure 2. The European market for flowers and ornamental plants: EU (27) + UK trade balance (000 EUR, 2001–2020). Source: our elaboration on [51] (data extraction: 24th January 2022). The identified product categories follow the Harmonized System Nomenclature [52], HS Codes: 0603, 0604, 0602.

Lastly, the ongoing COVID-19 pandemic serves as a game-changer in the European scenario, with breakthrough effects on the flower and plant market. In spite of the limited information available, a few relevant contributions already allow outlining major negative outcomes and positive long-term expectations.

In particular, despite the unpredictable market shock of spring 2020, determined by the sharp brake of trade and sales during the peak season, in the long run, the pandemic is working as an accelerator in the growth of the European flower and plant demand [11,69–72]. A rapid upturn of sales started in the second part of 2020 and accelerated during 2021 affecting a dramatic positive performance of the European flower and plant trade [4,73,74].

As illustrated in detail in Sections 4 and 5, the impact of COVID-19 on ornamentals consumption and competition will drive the emergence of new spaces for growth and profitability of European traders and producers, based on major institutional changes and a with-no-precedent good mood of European consumers [11].

4. Ornamentals Consumption Trends in the European Context

4.1. The New Determinants of Consumption Trends: A Classification

Europe is part of the group of mature domestic producer countries, identified by AIPH (2019b) [6] (Section 3.1). In that regard, in the European context, ornamentals consumption is expanding and changing, due to a selection of key socio-ecological determinants.

In particular, a shift in consumer behavior drives an upturn in the number of consumers and an increase in purchased quantities that both reached record-breaking numbers at the end of 2020 and in 2021 [11]. In addition, it generates a reduction in demand elasticity that makes quality prevail to the detriment of price [68].

In light of that, Table 5 reports a classification and description of the drivers we identified as the most relevant in affecting the evolution of the European flower and plant consumption trends.

Table 5. Main determinants affecting the evolution of flower and plant consumption trends in Europe.

Determinant	Drivers for Change of Consumption Trends
Globalization	• Globalization is the most affecting driver of change in the ornamental sector. The increase in cross-border flows of products, technology, investments, people and information is leading to a strong interdependence of world economies, cultures, and populations. • In the ornamental sector, the high globalization causes a higher permeability among different geographical markets, thus generating a greater variability and mutability of consumer preferences and behavior. • The flower and plant fashion and consumption trends mutually affect each other and rapidly change. As a consequence, the demand for ornamentals appears more fragmented and unpredictable in major consumption markets as Europe. • New high-value demand markets are emerging in growing-income countries of Northern and Eastern Europe, as well as in the areas of Far East and Middle East.
Climate change	• Climate change causes significant alterations in seasonal weather and temperatures and determines a general increase in heat and drought. These changes lead to an upheaval in flower consumption trends compared to regular seasonal dynamics and flower holiday calendars. Flower demand increases its instability and shows sudden unpredictable peaks, thus causing high volatility of exchanged volumes and prices. • Both private and institutional consumers are showing growing attention to the functionality of ornamental plants in dealing with climate issues, thus increasing their demand and willingness to pay.
Urbanization and new city living	• Urbanization is a key driver of flowers' and plants' consumption evolution. AIPH (2019b,c) [6,7] estimates that about 9 billion people will live in cities by 2050. In urban areas, people worry about the adverse effects of cities on the environment and life quality. At the same time, they feel the need to restore direct contact with nature. • By 2030 upper and middle classes are expected to expand to about 2 billion people. The growth of high-income urban consumers is seen to accompany the emergence of a new essential need for greening their lives. This fact increases flower demand volumes and value and reduces its elasticity to price. • Currently, new programmes, regulations, and plans have been realized, intended to requalify urban environments and facilities, thus supporting their capacity to adapt and serve the present and future generations. In that regard, flowers and plants are furtherly considered as essential goods for improving the quality and livability of cities and their surroundings.
Evolution of the socio-demographic context	• The aging rate of populations is rising together with the rate of retirement. Old people are keen to buy and pay for keeping well their houses and gardens and care for their personal health and families. • The segment of urban young workers is expanding. People in this segment show an increase in the caring of their personal image, health, and happiness and in their per-capita spending. • Population aging together with the expansion of the class of young workers are boosting the increase in flowers' and plants' consumption. Old and retired people together with well-off attentive young workers increasingly perceive flowers and plants as lifestyle products, capable to symbolize their identity values and to enhance their well-being • A new biophilia sentiment is emerging that boosts the use of flowers and ornamental plants both as luxury gifts and specialty daily-use goods.
Neo-luxury and sustainability-oriented consumption patterns	• In Europe, a large portion of consumers—even in older age groups—seem to be aware of main global sustainability issues at the social, economic, and environmental level, and are committed to making responsible consumption choices. • With reference to the issue of sustainability, a "neo-ecology" megatrend can be identified, transforming values of individuals, policymakers, and society [11], on Zukunftsinstitut (Future Institute) (p. 18). • Neo-luxury and sustainability-oriented consumption patterns are emerging, thus affecting the evolution of demand and consumption in agri-food markets. • Consumers seem to orient their choices to: valorize hedonistic and ethical principles; value quality instead of price; favor specialty products instead of standardized goods; consider information and transparency as fundamental value attributes; pay a premium for products qualities linked to origin and sustainability; recognize the value of products functionalities for higher life quality. • High value-added goods and services (e.g., compositions, arrangements) are preferred by flower and plant consumers. At the same time, new sustainability values play a dominant role in boosting ornamentals consumption and orienting consumer preferences.
Evolution and spreading of the Internet and ICT	• ICT and Internet are increasingly diffused and accessible in Europe. In line with this, they serve as game-changers in the ornamental sector. • The use of online channels and digital devices is rapidly spreading among flowers and plants consumers of all ages, for both information gathering and product purchasing. The UK is a pathfinder in this area. • The online channel (e.g., web and social media) becomes the primary source of information on ornamental products and brands; at the same time, online sales experience a significant growth which has accelerated due to the current COVID-19 pandemic. • A fundamental role is played by the activities of influencers and bloggers that discovered and promote a new passion for flowers, plants, and gardening. • Multi-channel shopping experiences are increasingly appreciated.

Table 5. *Cont.*

Determinant	Drivers for Change of Consumption Trends
COVID-19 Pandemic	• The ongoing COVID-19 pandemic is expected to positively affect ornamentals consumption in the long run. This is a consequence of the structural changes the pandemic is causing, both socio-cultural and economic. • From a long-term perspective, the pandemic seems to accelerate and amplify the effects of the other identified determinants, thus in particular due to people's reaction to solitude, lockdown, smart working, and fear of new calamities. Consumers came to appreciate flowers and plants during the pandemic and seem to confirm their appreciation and propensity to buy and pay. • Specifically, consumers increase their desire for nature and natural resources (AIPH, Sustainability group) [75]; augment their interest in local products and short supply chains; corroborate their perception of flowers and ornamental plants as everyday products (e.g., to decorate their home offices and gardens); and improve their familiarity with the use of the online channel. • The ornamental plants' segment presents the highest perspective of growth due to the expansion of urban gardening, even including the cases of expanding public green areas and social gardening. In addition, the growing number of people working from home demands rare green plants to qualify home workplaces. At the same time, even cut flowers and foliage are beating the market, based on the recovery and speed up of events and the wedding market. • Flowers and ornamental plants are increasingly recognized by public administrations, businesses, and society as primary goods for enhancing individual well-being and collective life quality. Florists and garden centers have been included in the group of stores rapidly reopened, because of providing essential personal services. In some countries, public campaigns promoted the purchase of ornamental products as a remedy against the negative effects of lockdown solitude and isolation.

Source: our elaboration on [4–7,9–11,13,15,22,23,25,26,30,47,49,54,55,58,64,65,68–70,72,76–85].

Determinants reported in Table 5 show evident effects on the expansion and transformation of ornamentals consumption trends. In perspective, further expansion and qualification of the ornamentals consumption may result from the implementation of the EU Green Deal strategy together with the possible adaptions of the new Common Agricultural Policy (CAP) [12,86–90]. Indeed, flowers and plants seem to be gaining increasing consideration from policymakers as part of the solutions of the overall EU Green Deal objectives (e.g., climate neutrality, natural resources and biodiversity preservation, social inclusion, and well-being) [12,87,90] and the related Biodiversity 2030 strategy [91].

In line with that, expected changes in EU and national (agricultural, environmental, social) policies, flanked by possible transformations of programmes and legislations, and the actualization of pledges and other concerted initiatives concerning life greening, biodiversity, and replanting [92,93], may foster the demand for flowers and plants to mitigate climate change impact, increase social welfare, support green city planning, as well as forestry and agro-forestry activities.

Actually, how and to what extent the new policy orientation will concretely translate in a further and long-lasting expansion of the ornamentals consumption will depend on the capacity of industry and stakeholders to properly organize and effectively sensitize and dynamize both citizens and institutions [89,94].

4.2. *Consumer Profiles and Dominant Consumption Trends*

According to the presented evolution of the socio-ecological context, over the next decade, a positive trend in the growth of flower and ornamental plant consumption is expected to characterize the European market. This trend will concern both private and institutional segments [6,95].

In line with the opinion of the most reliable institutions in the sector, the main engine of this growth is represented by the increase in the number and consumption of high-income urban consumers, which will be profiled in the next paragraph (Section 4.2.1). Consistently, new consumption trends arise and consolidate, concerning both individuals (Section 4.2.2) and institutions (Section 4.2.3). Remarkably, businesses and public administrations are raising their expenditure for flowers and plants to improve the quality of working and living conditions [6,9,11,21,65].

4.2.1. New High-Value Consumer Profiles

In line with the good performance of consumer demand, various attempts are made by the sampled institutions to identify emerging high-value consumer profiles, evidencing dominant consumption trends, and attractive markets.

Among them, we consider the relevant contributions of AIPH (2019b) [6] and Rabobank (2017a) [9]. The latter adopts the criterion of age to identify the most relevant consumer profiles in mature domestic producer countries in Europe, Asia, and North America. Specifically, they highlight two generational segments including the majority of present and early future consumers of flowers and ornamental plants in the considered markets. The first segment is the one of "Millenials", i.e., including people born between 1980 and 2000, while the second is the one of "Baby Boomers", consisting of people born between 1945 and 1964.

In line with these results, Rabobank (2017a) [9] describes the two consumer groups of "Old age people, retired or approaching retirement" and "Young adult people, working or about to get a job" as the most representative ones to be considered in the analysis of prevailing consumption habits in the European market for flowers and plants.

Furtherly, with special consideration of the segment of cut flowers and indoor plants, RFH (2017) [96] uses a multivariable approach to group and classify consumers based on their psychographics and behavioral characteristics, also including flower and plant purchasing frequency and willingness to pay. As a result, three high-value consumer profiles are identified: the "Cultivated performers", the "Cosiness seekers", and the "Individualistic performers" (Table 6). They represent the ones covering the highest share of both the number of consumers and turnover.

Table 6. High-value consumer profiles in the European market for cut flowers and indoor plants: a multivariable approach.

Classification Criterion	Description of Consumer Profile			
	Cultivated Performers	**Cosiness Seekers**	**Individualistic Performers**	
Psychographic	• Pursue high living standards • Are involved in social and cultural life • Are interested in sports, personal care and good nutrition • Perceive creativity as a value • Love to show their knowledge and expertise • Pay for higher quality • Appreciate exclusivity, choice variety and professionalism	• Care of the family • Love travelling and hosting • Have a busy life both at home and outside • Are interested in new technologies • Play sports occasionally • Pursue personal happiness • Are fashion conscious and attentive to seasonal trends • Are careful about price	• Look at appearances • Care of their own image • Consider the importance of style and creativity • Are interested in innovation • Seek and trust experts' advice • Pay for higher quality • Appreciate products sustainability and ecological benefits	
Behavioral	• Buy flowers and plants with a high frequency • Are used to shop through the specialized channel (especially florist shops) • Show a positive attitude towards the purchase of specialty crops and high-value bouquets and arrangements • Pay a premium for products assortment, craftsmanship and experience	• Buy flowers and plants with a high frequency • Are used to shop through supermarkets and garden centers • Buy flowers and plants as a vehicle of happiness • Show a positive attitude towards the purchase of seasonal species and trendy varieties • Consider wide and deep assortment as a choice discriminant	• Buy flowers and plants with a medium frequency • Are used to shop through the specialized channel • Pay a premium for products specialty, creativity and experience	
Market share				**TOT**
% total consumers	14.6	8.5	14.7	37.7
% total turnover	38.2	15	15.5	68.7

Source: our elaboration on [96].

With reference to the profile characterization reported in Table 6, RFH (2017) [96] validates the dominant role of urban young generations and high-income workers, together with old and retired people in boosting the outstanding performance of the European flower market.

4.2.2. Dominant Trends in Flower and Plant Consumption Preferences and Behavior

The evolution of consumer preferences and behavior in major European country markets is permeated by a relevant transformation, based on a new consumer perception of flowers and plants.

In general terms, consumers, especially the younger generations, seem firstly reinforcing their recognition of the unique value of flowers and plants as precious gifts or event arrangements; secondly, developing a new cognition of ornamental products as primary goods and essentials for enhancing their quality of life [11,22,23,55,64].

In more specific terms, new dominant consumption trends characterize the change in consumer beliefs and attitudes in Europe. We categorize these trends according to the following descriptions.

I. Consumers mostly prefer quality over price Consumers ever more valuate and choose flowers and ornamental plants that are qualified as specialty goods and that offer multiple functional and emotional advantages [6,9,11,53,54]. Specifically, they recognize and pay a premium for flowers and plants capable of symbolizing their identity values and tastes, and to provide them with specific benefits, not only in terms of aesthetics but also with reference to their personal realization and well-being. Accordingly, the European market registers an increase in the number and frequency of purchases of high value-added ornamental products. In line with this, refined bouquets and flower arrangements consolidate their positioning in both the segments of luxury gifts and events [4,11,49,81]. In these segments, main differentiation attributes become the following: rareness, creativity, branding, craftsmanship, and personalization. At the same time, flower and plant customized compositions acquire a growing market share in the segment of daily-use goods [5], as either decorations for homes and gardens or horti products [15,81]. In particular, with reference to flower and plant material, consumers choose and remunerate the quality of specialty crops, i.e., using branded seeds, valorizing innovative and trendy species and varieties as well as traditional essences. In particular, good market performance is registered by seasonal off-the-cuff landraces and hardy and heat resistant crops [49,64,65,81]. Besides that, high-value aesthetics and designs reinforce their role in winning the favor of consumers. To that regard, a new ethic-oriented canon is emerging in flower beauty: consumers request and value new organic and fresh-from-the-field styles of flowers and compositions, valorizing effortlessness and naturalness [97–99].

II. Sustainability and transparency play as primary determinants of choice An increasing number of consumers, especially in the group of young educated people, make responsible consumption choices when purchasing flowers and plants. For example, in the representative German market, a share of about 20% of total consumers consider sustainability as a primary determinant of choice [21–23,49,65]. Furthermore, according to a survey carried out by Statista in June 2021, the ongoing pandemic furtherly focuses flowers and plants trade and consumption on the value of sustainability and regionality. In addition, it increases consumer willingness to pay for environmentally friendly and socially sustainable production and distribution processes [11]. The growing consumer desire for sustainability is accompanied by a higher consideration for product transparency. As a consequence, flower and plant consumers are increasingly prepared to listen, learn and appreciate the value of product information [81]. Accordingly, consumers appear to increase their desire and willingness to pay for new quality attributes intended to: disclose products origin; communicate their low environmental impact and high social fairness; promote their contribution to biodiversity preservation; highlight their functionality for a better quality of living [9,11,22,23,55]. Therefore, higher price premiums are obtained by products that strategically use specific signaling tools. For example, a growing number of consumers perceive and remunerate the added value of quality and safety standards as well as origin marks and sustainability certifications schemes (e.g., fair trade, organic, GLOBAL GAP, etc.) [4,26,54,57,65,76,95,96,100]. Likewise, consumers even consider the communicative value of packaging. In particular, they appreciate eco-friendly and plastic-free containers, vases, cartons, and wraps, either recycled or biodegradable [11,101], and appreciate their capacity to claim the product story, identity, and unique benefits. Lastly, consumers show a new attitude for the use of web and social media as the main source of information as well as a channel of dialogue with breeders, producers, and traders [15,55,64].

III. Consumers recognize and remunerate ornamental products for their socio-ecological and therapeutic functionality Ever more consumers are interested in discovering the unique and superior benefits flowers and plants can provide for multiple uses [6,102]. Accordingly, especially in cities, both private ambiances and public spaces are changing their design, focusing on the valorization of the beneficial role of flowers and plants [11,64]. Indeed, in the consumers' intention, the latter is used with multiple functions of: softening and beautifying urban landscapes, home, and commercial spaces; recalling a contact with nature; mitigating temperatures; purifying the air; treating stress disorders, concentration problems, and mental illnesses. In that regard, particular consideration should be given to the new role of plants in the transformation of home environments, such as livings, gardens, balconies, and workspaces. Specifically, concerning the indoor segment, consumers increase their spending on green plants, e.g., split-leaf species, scented, and air-cleaning, as well as on woody plants [49]. In addition, the gardening segment is characterized by an unprecedented positive trend in consumption. As a benchmark, in 2019, it registered a turnover of about EUR 4.4 billion in the representative German market [55]. Moreover, Messe Essen (2022) [11] reports recent estimations by Statista, evidencing that in 2021 around 15 million people aged over 14 spent time working in the garden several times per month, as far as around 9 million people doing it more times per week. Moreover, consumers increasingly perceive home gardens as unique places for happiness, absolving also to the purposes of recreation and food production [11,49]. Specifically, they find satisfaction in the creation and caring of near-natural spaces, providing them with joy and peace of mind, contributing to the surviving of insects and small animals, valorizing regional and native plants, and producing healthy zero-mile food. Consequently, a good market performance is registered by space-saving ornamentals (e.g., flowering perennials, beddings, and balcony plants), trees and shrubs as well as by fruit trees, cacti, vegetable crops, and herbs [11,49,55,65,81]. For example, a new trend for "nibbles gardens" is emerging [49,65]. In that regard, consumers research and pay for the specialty of snack and dwarf vegetable crops, fruit nibbles, aromatics, and officinal plants as well as easy-care and insect-friendly ornamental plants [11,49,55,64,81]. Lastly, the expansion of the gardening segment boosts the growth of complementary markets, such as Do-It-Yourself (DIY) and hobby gardening, e.g., to buy integrated pests, near-natural fertilizers, recyclable materials, and innovative protection stuff [11,49,64]. At the same time, a fast-growing "smart gardening" segment is emerging, combining consumer interest in gardening with their desire to experiment with the high functionality of new smart devices and home automation technologies (e.g., robotic lawn mowers, digitally controlled irrigation systems, drones, etc.) [11,49,55,65].

IV. Consumers valuate ornamental products origin and show a preference for locally-grown and seasonal flowers Consumers appreciate the origin of ornamental products as a distinctive quality attribute, thus valuing their territorial linkage as a determinant of choice. In this sense, they consider not only the geographical provenance of the product but also the typicality of the used species and varieties and the adoption of traditional production and processing techniques [32]. In light of that, a large part of consumers favors the purchase of both locally-and-nationally-grown flowers and plants. In addition, the market shows a new consumer interest in buying native species and varieties, even valorizing specialty crops and landraces that are typical of other countries [11]. In light of the above, consumers are willing to pay a premium for ornamental products qualified by specific signs or storytelling, identifying their local or national provenance.

V. Consumers attribute a growing added value to customized services In line with the abovementioned trends, flower and plant consumers are changing their perception and remuneration of services (e.g., assistance, information, advice, composition, etc.). Accordingly, the flower market shows a shift in the composition of total consumer spending, characterized by a higher share for the remuneration of services, which become the main determinant of consumer choices and willingness to pay, while flowers and plants are considered components or "ingredients" [9,26,30,54,55,57,96]. In light of that, new high-value niches are emerging in the specific market of ornamental services. Among them, we consider the growth potential of floral and garden design segments [32], together with landscaping, that registers a positive trend in both sales volume and turnover [6].

VI. Consumers use alternative shopping channels and favor multi-channel experiences Specialized trade still maintains the largest share of the market. Nonetheless, consumers are increasingly hybrid in alternatively using specialized and non-specialized shopping channels [4,5,11]. As a matter of fact, consumers increase channel switching frequency, on the basis of the purchase occasion and the wanted product category [49,54,63,66]. Accordingly, consumers are more likely to combine the use of specialized shops to purchase premium price products (e.g., rare essences, personalized compositions, or arrangements), with the use of non-specialized channels to buy standardized products (e.g., mono or mixed bouquets, ordinary houseplants, seeds, and gardening material) [9,54]. Specifically, on the one hand, florists, kiosks, and street stalls still keep the highest share of the market, especially due to the expansion of luxury gifts and event segments. On the other hand, super-/hypermarkets and garden centers, followed by discounts and DIY, are rising their share, because of the higher sales of bouquets, houseplants (outdoor), and gardening products [4,5,49,50]. To illustrate, in the representative German market, in 2021, specialized trade accounted for a 60% share of private customers' total expenditure for flowers and plants, covering 30% of the total purchased quantity, while large-scale retailers reached a share of 40% of total expenditure, and covered 70% of the total purchased quantity [11]. Besides that, consumers are increasing their use and appreciation of online shopping channels. Remarkably, in consumer perception, online trade is complementing and integrating the role of stationary trade, but not replacing it. Indeed, while the market shows a swipe up of consumer spending on the online channel, the turnover of offline trade remains largely stable [9,65]. In line with this, consumers–and particularly "Baby Boomers"-, show a preference for multi-channel shopping experiences [9,15,26,54,57,96]. In that regard, on the one hand, they choose and value the higher convenience, entertainment, and personalization of online searching and purchasing processes [9,11,65]. On the other hand, they still prefer to visit offline shops to experience and valuate products, picking up orders, enjoy moments of leisure, and relate with local growers and retailers [54,63]. The positive trend in online sales started in the UK and Dutch markets, which represent the initiators of the trend in the European context, with a good performance in both the handicrafts and gardening segments [25,49]. Nowadays, online sales are experiencing a steady growth in major European markets (The Netherlands, France, Germany, and the UK), with the best performances of indoor and outdoor potted plants [4,11]. The good mood of online purchases in the gardening and DIY segments is flanked by the increase in online sales in the markets for flower and plant gifts and floral design arrangements [63]. On these bases, during the next decade, online trade is expected to cover a share of about 30% of the entire European ornamental market [9]. Significantly, we retain this estimate should even be revised upwards, to include the effect of acceleration generated by the COVID-19 pandemic [4,11].

4.2.3. A Focus on the Institutional Demand Segment

The institutional segment consists of both the demand of the private sector and public administration [100]. In accordance with the new determinants of consumption trends (Section 4.1), this segment is characterized by the growth of consumption and the change of institutional consumers' preferences and attitudes.

Concerning the private sector, flowers and ornamental plants are ever more integrated as a resource in the planning of business models [6,7]. Indeed, businesses and organizations demonstrate to recognize and research the high functionality of flowers and plants for the caring of the health and happiness of workers, the enhancement of their creativity and productivity, and the qualification of work and commercial spaces [11,21].

Likewise, the public administration is increasing the demand for ornamentals for their role in mitigating the adverse effects of climate change, urbanization, and modernization. Currently, new programmes, regulations, and initiatives have been realized, at both EU and national level, fostering the realization of incentive measures and territorial plans including support to the flowering and greening of urban areas, the (re-)foresting of woodlands, and the diffusion of agroforestry [49,64,81,103].

The strong increase in public spending plays as an unprecedented accelerator for the expansion of demand. In accordance with that, a good performance is registered in both sales and prices by woody plants, with specific reference to native species and landraces, temperature mitigator and air purifier trees and shrubs, and low maintenance species. Similarly, even seeds and propagation materials expand their share of the institutional market, thus considering, in particular, the case of wildflowers and bee-friendly species [15,49,64,65].

Forward looking, the ongoing political transformations, particularly the evolution of the discussion concerning the implementation of the EU Green Deal strategy, seem to open new opportunities for further growth in institutional demand [12,75,90].

Significantly, the new policy orientation is also encouraged by the effects of the current COVID-19 pandemic [104]. Actually, the latter boosted the recognition of flowers' and plants' unique social and ecological functions, thus stimulating the expansion and qualification of both market and social demand, and attracting and directing the attention and commitment of the EU and national institutions [11,15].

4.3. Consumer Spending and Consumption Value

The revised literature evidences a lack in the availability of quantitative data, referring to the value of flower and plants consumption and consumer spending. Despite that, some valuable estimations are made in relation to the trends registered in consumers' per-capita income and purchasing power, e.g., per-capita gross domestic product (GDP) or gross national income (GNI) ratios.

We consider the valuable contribution of AIPH (2020a) [50] in aggregating the most recent available data referring to both per-capita and total consumer spending on flowers and plants (and related goods), in main world consumption markets excluding China.

On the one hand, for each of the considered countries, the value of per-capita spending related to the level of GNI per capita is reported, with the highest registered performances by Switzerland (EUR 127), Denmark (EUR 114), Germany (EUR 108), and the UK (EUR 104) (Figure 3). According to the latest contribution of Rabobank (2022b) [5], Norway is also included in this group of high-value spending countries.

Messe Essen (2022) [11] highlights a perspective increase in per-capita spending, with reference to the major German market. This is attributed to the significant increase in the number of first-time consumers together with consumers' willingness to accept higher prices for plants (e.g., the average purchase amount per receipt increased by at least 10% from 2020 to 2021), and their higher propensity to pay for high-quality flowers as unique gifts.

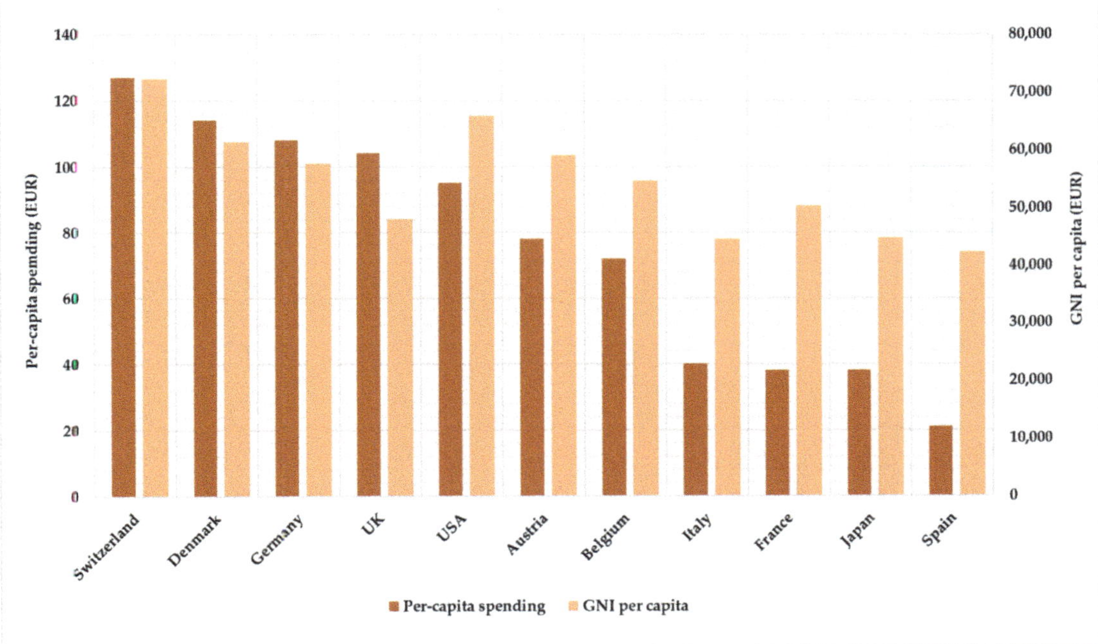

Figure 3. Consumption of flowers and plants (and related goods): per-capita spending and GNI in major world country markets (excl. China) (EUR). Source: our elaboration on [50]. The reported data do not include the share of spending concerning the market for services, e.g., floral and garden design, landscaping, green urban planning, green urban maintenance, etc. The data of per-capita spending refer to the latest available year in the timeframe 2015–2019. GNI = gross national income (at PPP = purchasing power parity) in 2016.

Besides that, regarding the estimations of total consumer spending, the highest values are registered in the US (EUR 31.1 billion), followed by Germany (EUR 8.9 billion), the UK (EUR 6.9 billion), and Japan (EUR 4.8 billion) (Figure 4).

In order to give a long-term perspective of the trend of the world and European consumption value, Rabobank (2017a) [9] provides an estimation of the global major ornamental markets, with reference to the decade 2017–2027. The estimation is referred to the aggregate flowers and pot plants.

In this regard, Rabobank [9] reports that by 2027 the value of flowers and plants consumption will grow of a percentage of about 20% in both Europe and North America, while Asia will be characterized by a percentage of growth standing between 60% and 80%. According to these estimations, the total world consumption value is expected to reach about USD 100 billion in 2027. More specifically, Europe and Asia are expected to count, respectively, for approx. USD 37 billion, while North America is expected to register a value of approx. USD 20 billion.

Lastly, in light of the ongoing COVID-19 pandemic, Rabobank [9] estimations could be partially changed to include the effects of the relevant socio-economic phenomena [11,84]. Nonetheless, according to the most recently available information, since the end of 2020, flower and plant sales, after a relevant but short-time decrease, recovered in both volume and value and soared at a rate equal or even higher than the one of the pre-pandemic period [4,11,22,23,25,73,74].

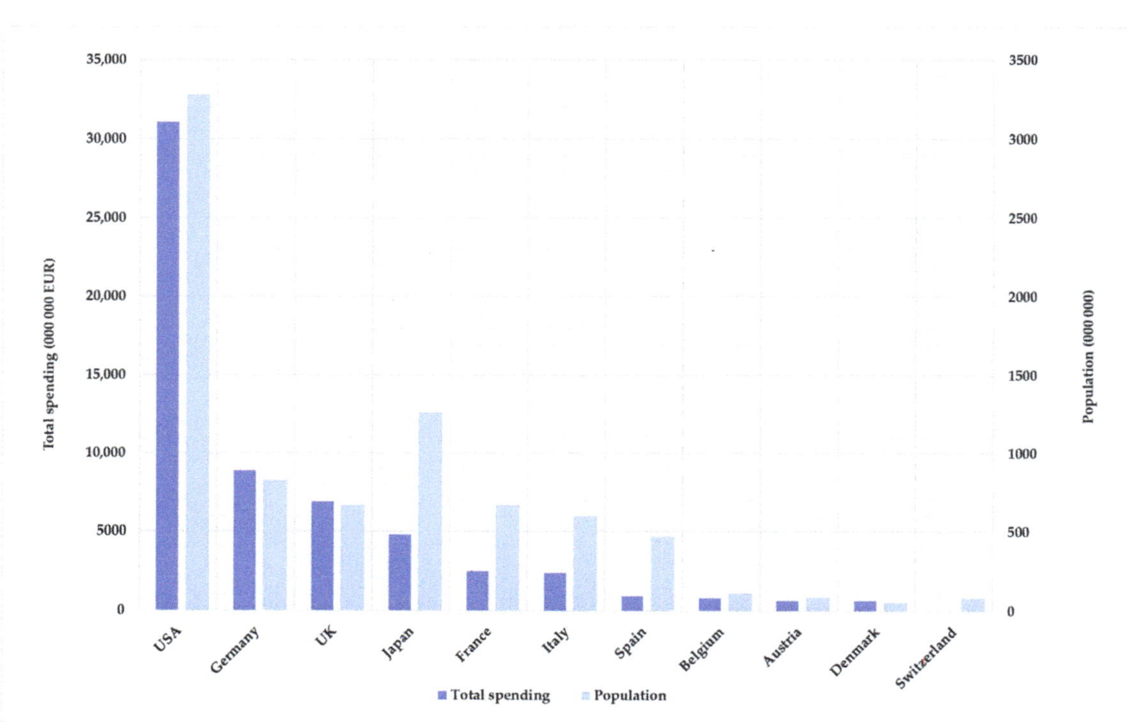

Figure 4. Consumption of flowers and plants (and related goods): total spending (EUR 000 000) over population (000 000) in major world country markets (excl. China). Source: our elaboration on [50]. [1] The data of total spending refer to the latest available year in the timeframe 2015–2019. The reported data do not include the share of spending concerning the market for services, e.g., floral and garden design, landscaping, green urban planning, green urban maintenance, etc.

5. Emerging Competitive Dynamics in the European Ornamental Sector

5.1. The European Competitive Environment and Evolving Marketing Strategies

The presented evolving dynamics determine a situation of hyper-competition in the European ornamental sector, characterized by a mutable competitive environment, dominated by rapid or even unpredictable changes [57]. As a consequence, the whole flower and plant industry is required to increase its efforts for the adaption of its marketing strategies to pursue higher competitiveness and sustainability.

Specifically, involved business operators aim at increasing their sharing of resources, reducing their costs and risks, enhancing their predictive and planning capability, and developing sustainable competitive advantages [12,55,57].

The problem of costs reduction is particularly sensitive, also considering the significant increase in the costs of production and logistics in the European context [11]. As a matter of fact, a low availability of qualified personnel and a steep increase in land prices are contributing to a higher cost of inputs limiting the upscaling potential of localized systems and supply chains. In addition, the ongoing pandemic is furtherly worsening producers' and traders' costs, due to the shortage of labor and higher prices of raw materials (e.g., energy, potting soils and peat, and construction materials). As a consequence, the unprecedented good performance of flowers and plants purchases, in both quantity and price, demonstrates to be not always capable to improve the profit of the European ornamental businesses, thus requiring a change of their marketing strategies and organization.

In order to sustain the competitiveness of high-cost domestic producers and supply systems, businesses and institutions are enhancing their efforts to pursue sustainable differential advantages. In that way, they intend to: on the one hand, address the problem of small scale and increasing costs; on the other hand, afford the challenge deriving from imported productions, outstanding for quality standardization and convenience.

What comes of it is a shift in the competitive orientation of the European ornamental industry, passing from cost minimization to quality maximization [11,25,33,105].

In other words, the current evolution of competition in the European sector makes the adoption of differentiation strategies an increasingly valuable solution for the effective positioning of domestic productions and supply systems [12]. Hence, European operators are growingly developing differentiation strategies, combining them with the adoption of focalized approaches. Specifically, they often decide for one or a few high-potential markets to serve and specialize their activities to target the satisfaction of specific consumers' needs.

In that regard, the development of effective marketing strategies appears an essential factor to target new high-value consumer profiles and emerging market niches. Specifically, we consider the selection of marketing levers should be intended to: (i) identify and qualify flowers and plants locality and territorial linkage; (ii) enhance and promote their specific functionalities for life quality and well-being, even by the integration of smart technologies and DIY; (iii) strengthen and communicate the commitment of businesses to the protection of the environment (e.g., reduction in CO_2 emissions, circular economy, preservation of biodiversity and ecosystems; renewable energy, etc.) and the pursuit of higher social fairness (e.g., protection of intellectual property rights and varietal innovation; promotion of human and labor rights; fair distribution of value, etc.) [106].

As an essential aspect, product policies should be enhanced by considering the potential advantage of the qualification of flowers and plants as specialty goods and services.

To that end, specific quality attributes can be valorized, as, for example: the selection of native species and traditional varieties (also with reference to the valorization of underutilized or neglected landraces); the adoption of organic production methods and sustainable processing techniques; the valorization of both multifunctional species, e.g., bee-friendly wildflowers, edible flowers, dwarf vegetable, herbs, and principles of flower and plant design and arrangement [11,15,28,55,81,107].

To give evidence of the above, we consider the representative case of Germany, where since 2018, a large part of the ornamental production has been converted from major crops to seasonal and traditional species and varieties [81]. As a matter of fact, the increase in consumer recognition and remuneration of the value of flowers diversity and sustainability is driving a structural change in the orientations of both florists and floral designers, thus making them innovate their assortments and differentiate. Consequently, the latter have increased their need and demand for specialty crops and locally and sustainably produced flowers and plants, thus stimulating the expansion and diversification of the domestic production market [11,49,65].

In light of the foregoing, a growing number of quality standards are also adopted, together with the implementation of higher requirements of safety and sustainability [4,56,108]. In that regard, the creation of private and collective marks (e.g., geographical indications), the participation in voluntary certification schemes (e.g., GLOBALGAP, MPS-ABC, ISO, IFS), and the creation of eco-friendly talking packs can furtherly contribute to the enhancement of flowers and plants market value.

The described innovation in product policies is accompanied by the adoption of premium price policies and the adaptation of promotion and distribution strategies.

To illustrate, particular attention is given to the development of multi-channel distribution systems and the realization of seamless shopping experiences. Indeed, ever more traditional retailers use to improve their offer by combining the advantages of stationary trade (e.g., physical experience of goods and services, sociality, proximity, etc.) with the ones obtainable by online channels, such as higher convenience and personalization [54,80,81].

Besides that, producers' and traders' promotion strategies are increasingly valorizing the role and efficacy of new media marketing tools, sustaining direct marketing approaches and storytelling. Indeed, web and social media serve as trendsetters and main sources of consumer information in the European market. At the same time, according to some relevant studies, referred to the comparable US context, web marketing strategies (i.e., the use of web and social media, forum, blogs, and newsletters) are demonstrating both lower costs and higher returns, in terms of product and brand image and customer loyalty, with respect to more traditional approaches [109–112].

Lastly, public and collective promotion campaigns are demonstrating their potential in enhancing consumers' awareness and feeding their desire. As a matter of fact, during the ongoing pandemic period, a growing number of initiatives have been taken, intended to stimulate flowers and plants consumption as a remedy against isolation and social distancing as well as essential goods for well-being and green living [11].

5.2. The European Competitive Environment: The Role of Governance

The described evolution of marketing strategies in the European ornamental sector goes hand in hand with the development of adequate organizational strategies and governance settings [4,9,56,57,63,65,77,81,112,113].

New threats and opportunities have emerged. European producers and supply chains are requested to improve their flexibility by the means of higher collaboration and investments. In light of that, flowers and plants growers, together with breeders and traders are either creating or consolidating stable cooperative networks, and integrating supply chains, both at a horizontal and vertical level. To rapidly respond to the acceleration and diversification of demand, they must enhance their logistics efficiency [4,55,64].

In this regard, the case of European floral supply chains is of particular relevance. As a matter of fact, the latter are changing their structure [26,77,114,115], thus with consideration of processes of: (a) shortening, with the emerging role of new facilitators, aggregating different functions, and sustaining the connection of producers with final consumers (b) decentralization, with reference to the emergence of diffused logistic hubs and virtualized networks at the international level [5]; and (c) specialization, with a view to the targeting of new high-value consumer profiles and emerging market niches.

Concerning these points, Rabobank (2017e) [77] describes the major changes that will affect the organization of the European floral supply chains during the next few years and classifies the chain typologies that are expected to obtain the highest share of the market According to the study, by 2027, three different supply chains will consolidate in Europe and cover each a 30% market share:

i. the specialist: targeting consumers who buy flowers and plants as a gift or for special occasions. It is centered on the role of specialist shops (florists, garden centers) auctions, wholesale markets, and growers, that are focused on the enhancement of specialty products and differential quality attributes;
ii. the big-box: specialized in serving large retailer outlets. This chain includes growers indeed large growing companies and associations, service providers, that play a dominant role in dealing with sourcing, logistics, payments, and quality control and large-scale retailers, as super/hypermarkets, DIY, and discount stores. The focus is on the realization of sustainable cost and operational advantages, valorizing responsive logistics and economies-of-scale;
iii. the e-commerce: targeting consumers buying flowers and plants online. It is characterized by short flexible connections between growers, digital marketplaces, and online retailers, committed to the pursuit of higher logistics efficiency, for assuring the satisfaction of a great number of small client-specific orders. Particular consideration should be also given to the entering into the market of new online retailers such as Amazon, or retailers with subscription models, such as Bloomon, or new logistical players, such as Post.nl.

Accordingly, with a view to improving their organization, business operators increase their investments and collaboration strategies [9,10,116].

In that way, they can easily specialize their activities, and synergically plan and act for the targeting of selected markets and the realization of high-value positionings, based on higher chains traceability, product quality and sustainability, and efficient distribution [114].

To that end, a central role is even played by digitalization and ICT [4,54,80–82,84,85]. What emerges is the relevant contribution that the use of new software and devices (e.g., systems for track-and-trace, big data management, etc.) can give to the enhancement of supply chain coordination and transparency, as well as the increase in the management of logistics.

In light of the above, the evolution of private actors' organizational strategies should also be supported by the efforts of public administrations and policymakers, at both the EU and national level [6,54,65]. Increasing public actors' attention and commitment should favor the adaption and innovation of mechanisms of support for the overall supply chains and the existing market structures, favoring the growth of private investments and the collaboration in the sector [12,89,90].

To illustrate, specific consideration should be paid to the importance of an effective adaption of the new CAP policy and related national programmes. As a matter of fact, until now, the sector benefited from little or no financial assistance within this framework, being eligible to receive sustain only through a few instruments linked to national rural development programmes (RDPs) [12,90]. AREFLH (2020) [12] affirms the opportunity to include ornamental horticulture among the mandatory sectoral interventions for EU Member States (Art. 40 of the draft regulation 2018/392 [117]), considering at least the countries that possess a sizeable production of flowers and plants.

Besides the abovementioned mechanisms, new EU and national policy measures, plans, and regulations can work, within or outside the new CAP framework, to give further support to flowers and plants producers and traders, with reference to the possible: (a) unification and standardization of certification schemes across Europe; (b) harmonization of labor, transport, environmental and plant products protection legislations among the EU Member States, as well as raising of social and environmental EU market requirements; (c) registration and marketing of flowers and plants origin labels and quality schemes; (d) definition of multi-actor action-research programs on key topics as pest management, climate change, innovative breeding, etc.; (e) promotion of education and professional training.

With a view to favoring these transformations, industry stakeholders should further increase their commitment and collaboration in order to sensitize and direct the ongoing political debate and obtain full recognition of the strategic role of the sector [12,89,90,94].

Significantly, the latter demonstrates a huge potential in contributing to the processes of social and ecological transition involving European countries, also related to the EU Green Deal implementation [86,88]. Accordingly, the emergence of high-impact concerted initiatives (e.g., the EU Action Plan Towards Zero Pollution for Air, Water and Soil, the EU Pollinators Initiatives, etc.) [93], as well as relevant pledges (e.g., 3 billion Trees Pledge) [92], give significant evidence of the growing public attention and social recognition flowers and ornamental plants are acquiring for their role in dealing with major social and environmental challenges (e.g., climate change, resources erosion, urbanization, margination, etc.). Moreover, the consequences of the current COVID-19 pandemic have been shown to further strengthen this trend [104].

Despite that, the political debate is still ongoing and new institutional arrangements have not yet been finalized. Noteworthy, programmes and legislations aiming at meeting the new EU goals of sustainability could determine the necessity of balancing the necessary expansion of ornamentals production, to respond to the growing demand, with the required reduction of environmental negative externalities (e.g., lowering greenhouse emissions) [12,32,37,38,79]. As a result, an effective innovation of governance systems, favoring public and private action coordination, will be fundamental for the qualification and sustainability transformation of European ornamental supply chains.

6. A New Action-Research Agenda for the European Ornamental Horticulture Industry Development and Sustainability

The planning and implementation of new effective marketing strategies and governance settings are based on the improvement of knowledge and capabilities within the industry and also at the institutional level.

In view of the above, we consider the advancement of scientific research as a critical condition. Accordingly, we identify a new action-research agenda. We designed the latter on the basis of the review analysis results, in order to improve the current debate and support the real needs of the industry. The agenda includes the following action-research directions.

- Product innovation and multifunctionality Action research should favor the specification and implementation of new high-value quality and related product attributes capable of obtaining a price premium for the remuneration of producers and supply chains. In particular, the creation of innovative products with specific reference to the local provenance of flower and plant material, underutilized and neglected landraces, traditional crops, the sustainability of production methods, the socio-ecological functionality of varieties, and arrangement techniques should be further investigated.
- Consumer analysis Advances in consumer analysis should sustain the action of high-cost producers and localized supply systems in identifying consumers' attitudes and evaluating their willingness to pay in both private and institutional segments. In this regard, specific attention should be paid to the description analysis of new consumer profiles.
- Quality-oriented marketing strategies Action research should sustain new valuable approaches to market segmentation and sustainable differentiation, favoring the identification and targeting of emerging niche markets, recognizing and remunerating specific quality attributes. Accordingly, future goals should evaluate the potential of creating product brands, adopting origin signs and certification schemes, enhancing products transparency, and consumer engagement. The institutional market segment should be also considered for the valorization of high-value products and services, with specific attention to floral design, landscaping, and urban greening. The pursuit of higher competitiveness of high-cost domestic producers and localized supply systems asks for in-depth research on the potential of direct or short distribution channels, also paying attention to the role of online trade and digitalization.
- Collaborative Governance settings Research advances are needed to favor the innovation and reinforcement of governance settings, both public and private. On the public side, research should support the improvement and harmonization of policies, standards, and legislations, at both the EU and national level. To that end, particular consideration should be given to foster the recognition and remuneration of the strategic role of the ornamental sector in sustaining the realization of the EU Green Deal strategy goals and of the related EU and national agricultural, social and environmental policies, programmes, and regulations. On the private side, new forms of coordination, cooperation, and collaboration, at both the horizontal and vertical level, should be studied, discussed, and validated, for the enhancement of supply chains competitiveness, guaranteeing not only a generation but also a fair distribution of benefits, towards higher social, economic, and environmental sustainability.
- Dedicated research observatories The construction of dedicated research observatories at the national or European level, committed to improving the availability of harmonized, updated, and reliable quantitative and qualitative data, is fundamental to support the new positioning of the ornamental sector and the implementation of effective marketing strategies and multi-actor governance models and the realization of participatory action-research. This could support the development of academic and institutional research, according to the hypotheses identified by this work, and, on the other hand, promote a more widespread ability to forecast and strategic planning among the various actors for the realization of new competitive objectives.

7. Conclusions

Our paper seeks to innovatively contribute to the advancement of market research in ornamental horticulture and to the improvement of public and private action and coordination for the enhancement of the industry-specific potential in economic, social, and environmental terms. As a matter of fact, whereas the relevant economic debate is limited and discontinued, we retain the development of scientific research as an essential lever, providing a reconnection of the academy and research institutes with the real necessities of the sector.

Therefore, our study contributes to filling the gap in the availability of structured and theoretically sound studies and integrated data sources on ornamentals' new consumption trends, marketing strategies, and governance settings. To that end, this paper proposes an innovative data framework, presenting major changes occurring in the world and European market, and emerging big threats and opportunities, affecting the transformation of European competitive dynamics.

As a result, this framework can be fundamental support for policymakers, business operators, and industry organizations for the planning and combination of effective informed public policies and private strategies.

Furthermore, new supporting action-research directions are identified, capable of stimulating the interest of businesses, researchers, and institutions. Noteworthy, our study outlines the necessity of advancing research to sustain an increase in the efforts and collaboration of public and private stakeholders, towards a synergic combination of new differentiation advantages and wider social and environmental goals. To this respect, a prompt adaption and effective innovation of production and marketing strategies as well as governance settings are seen as unavoidable.

Specifically, on the public side, EU and national institutions should increase their attention and commitment towards the investigation and realization of collaborative governance systems, and the identification of a common strategic orientation. In that way, an effective adaption, integration, and harmonization of EU and national policies, programmes, and legislations should be favored, and new supporting measures and regulations should be provided to the sector (e.g., financial and technical assistance, quality and safety requirements, certification and quality schemes, etc.).

In that regard, we believe particular attention should be given to the ongoing political debate, concerning the implementation of the EU Green Deal strategy, and related Biodiversity 2030 directions, fostering a new strategic role for the ornamental industry. Accordingly, favorable adaptions of EU and national agricultural, environmental, and social policies, can boost ornamentals demand and support an adequate reorganization of production and trade. To that end, policymakers, together with citizens, and consumers, should be properly informed and sensitized, to recognize ornamental industry concrete multiple values and sustainability potential.

In line with that, on the private side, businesses and supply chains need to improve their planning capacity and investment policies. They should develop more collaborative strategic and governance approaches for the realization of win–win production and marketing strategies and effective communication initiatives.

As a result, we expect the European ornamental industry can increase its competitiveness and resilience, as well as affirm its unique role in the greening of the life of present and future generations.

Author Contributions: Conceptualization, S.S. and S.G.; methodology, S.G.; validation, S.S.; investigation, S.G.; resources, S.G.; data curation, S.G.; writing—original draft preparation, S.S. and S.G.; writing—review and editing, S.S. and S.G.; visualization, S.S. and S.G.; supervision, S.S.; project administration, S.S. All authors have read and agreed to the published version of the manuscript.

Funding: This research received no external funding.

Institutional Review Board Statement: Not applicable.

Informed Consent Statement: Not applicable.

Data Availability Statement: The data presented in this study are available on request from the corresponding author.

Conflicts of Interest: The authors declare no conflict of interest.

References

1. Volckaert, E.; Gobin, B. "Ornamental Plants and Floriculture" Soils, Plant Growth and Crop Production. Encyclopedia of Life Support System (EOLLSS). UNESCO-EOLSS Sample Chapters C 10. 2014. Available online: https://www.eolss.net/Sample Chapters/C10/E1-05A-51.pdf (accessed on 2 February 2022).
2. Yahia, E.M. Chapter 3—Classification of Horticultural Commodities. In *Postharvest Technology of Perishable Horticultural Commodities*; Yahia, E.M., Ed.; Woodhead Publishing: Cambridge, UK, 2019; pp. 71–97. ISBN 9780128132760.
3. Oxford Economics. The Economic Impact of Ornamental Horticulture and Landscaping in the UK. A Report for the Ornamental Horticulture Round Table Group. October 2018. Available online: https://www.rhs.org.uk/science/pdf/The-economic-impact-of-ornamental-horticulture-and.pdf (accessed on 3 February 2022).
4. Van Horen, L. A Mixed Bouquet of Development in Floriculture-World Floriculture Map 2021. RaboResearch Food&Agribusiness Rabobank, January 2022. Available online: https://research.rabobank.com/far/en/documents/179560_Rabobank_A-Mixed Bouquet-of-Developments-World-Floriculture-Map-2021_vanHoren_January2022.pdf (accessed on 31 January 2022).
5. RaboResearch Food&Agribusiness World Floriculture Map 2021. Rabobank: Utrecht, The Netherlands, January 2022. Available online: https://research.rabobank.com/far/en/documents/175926_Rabobank_Flower-Map-2021_20211230.pdf (accessed on 31 January 2022).
6. Hendricks, J.; Briercliffe, T.; Oosterom, B.; Treer, A.; Kok, G.; Edwards, T.; Kong, H. Ornamental Horticulture, A Growing Industry? International Vision Project Reports. AIPH Horticulture House: Chilton Didcot, Oxfordshire, UK, September 2019. Available online: https://aiph.org/giic/international-vision-project-reports/ (accessed on 25 January 2022).
7. Hendricks, J.; Briercliffe, T.; Oosterom, B.; Treer, A.; Kok, G.; Edwards, T.; Kong, H. Production and Markets, the Future of Ornamentals. International Vision Project Reports. AIPH Horticulture House: Chilton Didcot, Oxfordshire, UK, July 2019. Available online: https://aiph.org/giic/international-vision-project-reports/ (accessed on 25 January 2022).
8. Coherent Market Insights (CMI) Floriculture Market. Global Industry Insights, Trends, Outlook, and Opportunity Analysis 2019–2027. CMI Floriculture Market Report. August 2019. Available online: https://www.coherentmarketinsights.com/market-insight/floriculture-market-1586 (accessed on 4 February 2022).
9. Van Horen, L. Flourishing Flowers, Promising Plants: Changes in Consumer Behaviour. RaboResearch Food&Agribusiness Rabobank, December 2017. Available online: https://research.rabobank.com/far/en/sectors/regional-food-agri/flourishing_flowers_promising_plants_changes_in_consumer_behaviour.html (accessed on 5 June 2021).
10. Centre for the Promotion of Imports from Developing Countries of The Netherlands Ministry of Foreign Affairs (CBI). What Competition Do You Face on the European Cut Flowers and Foliage Market? CBI Ministry of Foreing Affairs Market Information. 3 May 2017. Available online: https://www.cbi.eu/market-information/cut-flowers-foliage/competition (accessed on 3 February 2022).
11. Löbke, A. Record Sales for the Flower and Plant Market. MESSE ESSEN Press Text. 25 January 2022. Available online: https://www.ipm-essen.de/press/press-texts/ (accessed on 27 January 2022).
12. Assembly of European Horticultural Regions (AREFLH). Position Paper on the European Ornamental Plant Sector. AREFLH Position Statements. April 2020. Available online: https://www.areflh.org/en/aboutus/positions/position-statement-plants (accessed on 25 January 2022).
13. Mattia, G. *Il Neo-Lusso. Marketing e Consumi di Qualità in Tempi di Crisi*; FrancoAngeli: Milan, Italy, 2013; ISBN 8820451727.
14. Borsellino, V.; Schimmenti, E.; El Bilali, H. Agri-food markets towards sustainable patterns. *Sustainability* **2020**, *12*, 2193. [CrossRef]
15. MESSE ESSEN GmbH Press Media Center for IPM Essen. Green, Great, Gorgeous! Excellent Mood at IPM ESSEN 2020 IPM ESSEN 2020 Final Report. Essen, Germany, 31 January 2020. Available online: https://www.ipm-essen.de/news-en/ (accessed on 30 October 2021).
16. Sharathkumar, M.; Peter, K.V.; Rajeevan, P.K. Ornamentals for greening. *Acta Hortic.* **2017**, *1165*, 45–56. [CrossRef]
17. Yue, C.; Dennis, J.H.; Behe, B.K.; Hall, C.R.; Campbell, B.L.; Lopez, R.G. Investigating Consumer Preference for Organic, Local, or Sustainable Plants. *HortScience* **2011**, *46*, 610–615. [CrossRef]
18. Isaak, M.; Wolfgang, L. Consumer Preferences for Sustainability in Food and Non-Food Horticulture Production. *Sustainability* **2020**, *12*, 7004. [CrossRef]
19. Bulgari, R.; Petrini, A.; Cocetta, G.; Nicoletto, C.; Ertani, A.; Sambo, P.; Ferrante, A.; Nicola, S. The Impact of COVID-19 on Horticulture: Critical Issues and Opportunities Derived from an Unexpected Occurrence. *Horticulturae* **2021**, *7*, 124. [CrossRef]
20. McBain, J. Post COVID-19 Consumer Landscape. Finding Opportunities Amid Upheaval. Lecture Presented at the Conference "AIPH Virtual International Conference. Recovery from Crisis—The Future for Ornamentals". 15 September 2020. Available online: https://aiph.org/event/recovery-from-crisis/ (accessed on 4 February 2022).
21. Hall, C.R.; Knuth, M.J. An Update of the Literature Supporting the Well-Being Benefits of Plants: Part 4—Available Resources and Usage of Plant Benefits Information. *J. Environ. Hortic.* **2020**, *38*, 68–72. [CrossRef]

22. International Association of Horticultural Producers (AIPH). How Gardening Will Keep You Well during the Coronavirus Pandemic. Factsheets. March 2020. Available online: https://aiph.org/latest-news/gardening-will-keep-you-well-during-the-coronavirus-pandemic/ (accessed on 27 January 2022).
23. International Association of Horticultural Producers (AIPH). How Flowers Can Help during the Coronavirus Pandemic. Factsheets. March 2020. Available online: https://aiph.org/latest-news/how-flowers-will-help-keep-you-well-during-quarantine/ (accessed on 27 January 2022).
24. Havardi-Burger, N.; Mempel, H.; Bitsch, V. Sustainability Challenges and Innovations in the Value Chain of Flowering Potted Plants for the German Market. *Sustainability* **2020**, *12*, 1905. [CrossRef]
25. Van Rijswick, C. World Floriculture Map 2016: Equator Countries Gathering Speed. RaboResearch Food&Agribusiness Rabobank, November 2016. Available online: https://research.rabobank.com/far/en/sectors/regional-food-agri/world_floriculture_map_2016.html (accessed on 23 May 2021).
26. van Horen, L. Towards Smarter Floriculture Chains. Lecture Presented at the Conference "CROP Innovation & Business", Amsterdam, The Netherlands. 4 April 2017. Available online: https://www.cropib.com/storage/app/media/Programme_2017/Presentations/4b.%20Ornamentals%20-%20Lambert%20van%20Horen.pdf (accessed on 5 June 2021).
27. Zarbà, A.S.; Di Vita, G.; Allegra, V. Strategy Development for Mediterranean Pot Plants: A Stakeholder Analysis. *Qual.-Access Success* **2013**, *14*, 52–58.
28. Darras, A. Overview of the Dynamic Role of Specialty Cut Flowers in the International Cut Flower Market. *Horticulturae* **2021**, *7*, 51. [CrossRef]
29. Karpun, O. Conceptual model of floriculture supply chain management. *Intellect. Logist. Supply Chain Manag.* **2020**, *4*, 41–52. [CrossRef]
30. van Horen, L. Flourishing Flowers, Promising Plants: Internationalisation Strategy. RaboResearch Food&Agribusiness Rabobank, December 2017. Available online: https://research.rabobank.com/far/en/sectors/regional-food-agri/Flourishing_flowers_promising_plants_Internationalisation_strategy.html (accessed on 6 June 2021).
31. Lambrecht, E.; Taragola, N.; Kühne, B.; Crivits, M.; Gellynck, X. Networking and innovation within the ornamental plant sector. *Agric. Food Econ.* **2015**, *3*, 10. [CrossRef]
32. Gabellini, S.; Scaramuzzi, S. Luxury strategies for agricultural products: A new sustainable governance model for the valorisation of the Tuscan flower supply chain. In *Green Metamorphoses: Agriculture, Food, Ecology. Proceedings of the LV Conference of SIDEA Studies*; Torquati, B., Marchini, A., Eds.; Wageningen Academic Publishers: Wageningen, The Netherlands, 2020; pp. 413–423. ISBN 978-90-8686-347-1.
33. Hall, C.R. Business Decisions to Help Me Outshine the Competition. The Association of Horticulture Professionals (OFA) Bullettin No. 932 March/April 2012. Available online: https://cdn.coverstand.com/4456/102937/102937.1.pdf (accessed on 2 February 2022).
34. Di Vita, G.; Allegra, V.; Zarbà, A.S. Building scenarios: A qualitative approach to forecasting market developments for ornamental plants. *Int. J. Bus. Glob.* **2015**, *15*, 130–151. [CrossRef]
35. Owen, J.S., Jr.; LeBude, A.V.; Calabro, J.; Boldt, J.K.; Gray, J.; Altland, J.E. Research Priorities of the Environmental Horticultural Industry Founded through Consensus. *J. Environ. Hortic.* **2019**, *37*, 120–126. [CrossRef]
36. Cardoso, B.F.; Rasetti, M.; Giampietri, E.; Finco, A.; Shikida, P.F.A. Trade Dynamics in the Italian Floriculture Sector within EU Borders: A Gravity Model Analysis. *AGRIS-Line Pap. Econ. Inform.* **2017**, *9*, 23–32. [CrossRef]
37. Wani, M.A.; Nazki, I.T.; Din, A.; Iqbal, S.; Wani, S.A.; Khan, F.U. Floriculture Sustainability Initiative: The Dawn of New Era. In *Sustainable Agriculture Reviews 27. Sustainable Agriculture Reviews*; Lichtfouse, E., Ed.; Springer: Cham, Switzerland, 2018; Volume 27, pp. 91–127. ISBN 978-3-319-75190-0.
38. Dominguez, G.B.; Mibus-Schoppe, H.; Sparke, K. Evaluation of Existing Research Concerning Sustainability in the Value Chain of Ornamental Plants. *Eur. J. Sustain. Dev.* **2017**, *6*, 11. [CrossRef]
39. Floriculture Sustainability Initiative (FSI) 2025. FSI 2025 Strategy Summary for Approval by the FSI General Assembly. FSI 2025 Summary Strategy Paper. 2 February 2021. Available online: https://www.fsi2025.com/wp-content/uploads/2021/02/FSI-2025-SUMMARY.pdf (accessed on 3 February 2022).
40. Snyder, H. Literature review as a research methodology: An overview and guidelines. *J. Bus. Res.* **2019**, *104*, 333–339. [CrossRef]
41. Torraco, R.J. Writing Integrative Literature Reviews: Using the Past and Present to Explore the Future. *Hum. Resour. Dev. Rev.* **2016**, *15*, 404–428. [CrossRef]
42. Farace, D.; Schöpfel, J. Grey Literature. In *Encyclopedia of Library and Information Sciences*; Bates, M.J., Maack, M.N., Eds.; CRC Press: Boca Raton, FL, USA, 2015; pp. 2029–2039. ISBN 9780203757635.
43. Da Silva, R.N.; Brandão, M.A.G.; Ferreira, M.D.A. Integrative Review as a Method to Generate or to Test Nursing Theory. *Nurs. Sci. Q.* **2020**, *33*, 258–263. [CrossRef] [PubMed]
44. Whittemore, R.; Knalf, K. The integrative review: Updated methodology. *J. Adv. Nurs.* **2005**, *52*, 546–553. [CrossRef] [PubMed]
45. Bonato, S. *Searching the Grey Literature. A Handbook for Searching Reports, Working Papers, and Other Unpublished Research*; Rowman & Littlefield Publishers: Lanham, MD, USA, 2018; ISBN 978-1-5381-0063-9.
46. Saunders, B.; Sim, J.; Kingstone, T.; Baker, S.; Waterfield, J.; Bartlam, B.; Burroughs, H.; Jinks, C. Saturation in qualitative research: Exploring its conceptualization and operationalization. *Qual. Quant.* **2018**, *52*, 1893–1907. [CrossRef] [PubMed]

47. van Rijswick, C. World Floriculture Map 2015: Gearing Up For Stronger Competition. RaboResearch Food&Agribusiness Rabobank Industry Note #475. January 2015. Available online: https://research.rabobank.com/far/en/sectors/regional-food-agri/world_floriculture_map_2015.html (accessed on 12 March 2021).
48. Hübner, S. *International Statistics Flowers and Plants 2021*; International Association of Horticultural Producers (AIPH) and International Flower Trade Association (Union Fleurs); AIPH Horticulture House: Oxfordshire, UK, 2021; Volume 69. ISBN 978-1-9164807-9-7.
49. Kirchhoff, A. BGI Markt + Trend. Ausbage IPM 2020. BGI Service UG: Straelen-Herongen, Germany, 2020. Available online: https://bgi-ev.de/data/2020/01/BGI_Folder_MarktTrend_A4_2020_WEB.pdf (accessed on 6 June 2021).
50. Hübner, S. *International Statistics Flowers and Plants 2020*; International Association of Horticultural Producers (AIPH) and International Flower Trade Association (Union Fleurs); Horticulture House: Oxfordshire, UK, 2020; Volume 68.
51. International Trade Center (ITC) Trade Map. Available online: https://www.trademap.org/Index.aspx (accessed on 31 January 2022).
52. European Commission Taxation and Customs Union Harmonized System-General Information. Available online: https://ec.europa.eu/taxation_customs/business/calculation-customs-duties/customs-tariff/harmonized-system-general-information_en (accessed on 1 February 2022).
53. Altmann, M. Developments and Trends in the Flower and Plant Market for 2015/2016, Stability Is Not Enough: New Markets Are Important-IPM ESSEN 2016. MESSE ESSEN Press Text. Essen, Germany, October 2015. Available online: https://www.ipm-essen.de/press/press-texts/ (accessed on 27 May 2021).
54. Kirchhoff, A. BGI Markt + Trend. Ausbage IPM 2019. BGI Service UG: Straelen-Herongen, Germany, 2019. Available online: https://bgi-ev.de/data/2019/01/BGI_Folder_MarktTrend_IPM19_WEB.pdf (accessed on 6 June 2021).
55. MESSE ESSEN GmbH Press Media Center for IPM Essen. "We Gardeners Can Do Climate!": Sustainability and Climate Change Were Defining Subjects at the World's Leading Fair for Horticulture. IPM ESSEN 2019 Final Report. Essen, Germany, 25 January 2019. Available online: https://www.ipm-essen.de/news-en/ (accessed on 16 September 2021).
56. Mamias, S. The Floriculture Supply-Chain: Characteristics & Prospects. Lecture Presented at the Seminar "Supply-Chains in the Agri-Food Sector as the UK Leaves the EU", Amsterdam, The Netherlands. 8 February 2018. Available online: https://unionfleurs.org/industry/ (accessed on 25 January 2022).
57. Mamias, S. Opportunities for Market Diversification. Lecture Presented at the "Kenya Flower Industry Sustainability Conference", Nairobi, Kenya. 6 June 2017. Available online: https://unionfleurs.org/industry/ (accessed on 25 January 2022).
58. RaboResearch Food&Agribusiness World Floriculture Map 2016. Rabobank: Utrecht, The Netherlands, November 2016. Available online: https://research.rabobank.com/far/en/sectors/regional-food-agri/world_floriculture_map_2016.html (accessed on 23 May 2021).
59. RaboResearch Food&Agribusiness World Floriculture Map 2015. Rabobank: Utrecht, The Netherlands, January 2015. Available online: https://research.rabobank.com/far/en/sectors/regional-food-agri/world_floriculture_map_2015.html (accessed on 12 March 2021).
60. European Commission Directorate-General for Agriculture and Rural Development (DG AGRI). Unit G.2—Wine, Spirits, and Horticultural Products Working Document. Horticultural Products. Flowers and Ornamental Plants-Production Statistics 2010–2019. DGAGRI-G2. 10 February 2020. Available online: https://ec.europa.eu/info/food-farming-fisheries/plants-and-plant-products/live-plants-and-flowers_en (accessed on 3 January 2022).
61. European Commission Directorate-General for Agriculture and Rural Development (DG AGRI). Unit G.2—Wine, Spirits, and Horticultural Products Working Document. Horticultural Products. Flowers and Ornamental Plants Statistics 2006–2016. DGAGRI-G2. 23 November 2017. Available online: https://ec.europa.eu/info/food-farming-fisheries/plants-and-plant-products/live-plants-and-flowers_en (accessed on 28 October 2020).
62. Lariviere, V. Live Plants and Products of Floriculture Sector in the EU. Lecture Presented at the Parliament's Committee on Agriculture and Rural Development (AGRI Committee), Brussels, Belgium. 12 December 2017. Available online: https://ec.europa.eu/info/food-farming-fisheries/plants-and-plant-products/live-plants-and-flowers_en (accessed on 28 October 2020).
63. Löbke, A. The Flower and Plant Market in 2019 (IPM Essen 2020). MESSE ESSEN Press Text. 30 October 2019. Available online: https://www.ipm-essen.de/press/press-texts/ (accessed on 30 October 2021).
64. Altmann, M.; Löbke, A. IPM Market Description of the Flower and Plant Markets, Part 1. Best Economic Prerequisites for IPM ESSEN 2018. MESSE ESSEN Press Text. Essen, Germany, 7 December 2017. Available online: https://www.ipm-essen.de/press/press-texts/ (accessed on 12 June 2021).
65. Altmann, M.; Löbke, A. IPM Market Description of the Flower and Plant Markets, Part 2. IPM ESSEN 2018 Focuses on Individualisation and Digitalisation. MESSE ESSEN Press Text. Essen, Germany, 14 December 2017. Available online: https://www.ipm-essen.de/press/press-texts/ (accessed on 12 June 2021).
66. MESSE ESSEN GmbH Press Media Center for IPM Essen. Heat, Water Shortage and Rising Ecological Awareness: IPM ESSEN 2020 Shows Trends and New Products. IPM ESSEN 2020 News. Essen, Germany, 27 January 2020. Available online: https://www.ipm-essen.de/news-en/ (accessed on 30 October 2021).
67. Wakefield, R. Growers and Traders Discuss the Impacts of Brexit on the Global Ornamental Horticulture Industry. AIPH News. 3 December 2020. Available online: https://aiph.org/latest-news/growers-and-traders-discuss-the-impacts-of-brexit-on-the-global-ornamental-horticulture-industry/ (accessed on 27 January 2022).

68. Altmann, M.; Löbke, A. The Climate Influences the Turnover in the International Green Sector-IPM ESSEN 2019. MESSE ESSEN Press Text. Essen, Germany, 21 November 2018. Available online: https://www.ipm-essen.de/press/press-texts/ (accessed on 13 June 2021).
69. International Flower Trade Association (UNION FLEURS). EU-Wide Survey Provides a First Estimate of the Brutal Impact of COVID-19 Pandemic on the European Flower & Live Plants Sector (March–April 2020). Union Fleurs News. 16 June 2020. Available online: https://unionfleurs.org/news_events/eu-wide-survey-provides-a-first-estimate-of-the-brutal-impact-of-covid-19-pandemic-on-the-european-flower-live-plants-sector-march-april-2020/ (accessed on 25 January 2022).
70. FloraCulture International-International Association of Horticultural Producers (AIPH). Coronavirus Global Impact Survey-Datasheet Revised Part 1. March 2020. Available online: https://aiph.org/latest-news/horticultural-industry-looks-to-the-future-in-the-latest-covid-19-global-impact-survey/ (accessed on 26 January 2022).
71. FloraCulture International-International Association of Horticultural Producers (AIPH). Coronavirus Global Impact Survey-Datasheet Revised Part 2. May 2020. Available online: https://aiph.org/latest-news/horticultural-industry-looks-to-the-future-in-the-latest-covid-19-global-impact-survey/ (accessed on 26 January 2022).
72. Van Horen, L.; van Rijswick, C. Floriculture Demand Collapses Dramatically Under Coronavirus Pressure. RaboResearch Food&Agribusiness Rabobank, March 2020. Available online: https://research.rabobank.com/far/en/sectors/fresh-produce/floriculture-demand-collapses-under-coronavirus.html (accessed on 31 October 2021).
73. Van Tol, F. FCI—Reflecting on the Long-Term Impact of COVID-19. FloraCulture International-AIPH Issue: July-August 2020. Available online: https://aiph.org/covid-19/long-term-impact/ (accessed on 26 January 2022).
74. Wakefield, R. Resilience and Positivity in the Face of Adversity at the AIPH Recovery from Crisis Conference. AIPH News. 17 September 2020. Available online: https://aiph.org/latest-news/resilience-and-positivity-in-the-face-of-adversity-at-the-aiph-recovery-from-crisis-conference/ (accessed on 27 January 2022).
75. International Association of Horticultural Producers (AIPH). Sustainability. The Growing Global Population Places Increasing Demands on Our Natural Resources. Available online: https://aiph.org/ornamentals-production/sustainability/ (accessed on 3 February 2022).
76. Van Horen, L. Flourishing Flowers, Promising Plants: Embracing Sustainability. RaboResearch Food&Agribusiness Rabobank, December 2017. Available online: https://research.rabobank.com/far/en/sectors/regional-food-agri/Flourishing_flowers_promising_plants_Embracing_sustainability.html (accessed on 5 June 2021).
77. Van Horen, L. Flourishing Flowers, Promising Plants: Chain Organisation in European Floriculture. RaboResearch Food&Agribusiness Rabobank, November 2017. Available online: https://research.rabobank.com/far/en/sectors/regional-food-agri/flourishing_flowers_promising_plants_chain_organisation_in_european_floriculture.html (accessed on 5 June 2021).
78. Corbellini, E.; Saviolo, S. *L'Esperienza del Lusso. Mondo, Mercati, Marchi*; Rizzoli: Milan, Italy, 2007; ISBN 8817095273.
79. International Association of Horticultural Producers (AIPH). AIPH International Conference: The Path to Sustainability in Ornamental Horticulture. Available online: https://aiph.org/event/sustainability-conference-2021/ (accessed on 30 January 2022).
80. Kirchhoff, A. BGI Trade Center IPM 2018. BGI Service UG: Straelen-Herongen, Germany, 2018. Available online: https://bgi-ev.de/data/2018/01/BGI_TradeCenter_2018_WEB-final.pdf (accessed on 5 June 2021).
81. Altmann, M.; Löbke, A. The Green Sector is Characterised by these Currents-IPM ESSEN 2019. MESSE ESSEN Press Text. Essen, Germany, 21 November 2018. Available online: https://www.ipm-essen.de/press/press-texts/ (accessed on 13 June 2021).
82. Wakefield, R. Horticultural Industry Looks to the Future in the Latest COVID-19 Global Impact Survey. FloraCulture International-AIPH Press Release. 29 May 2020. Available online: https://aiph.org/latest-news/horticultural-industry-looks-to-the-future-in-the-latest-covid-19-global-impact-survey/ (accessed on 26 January 2022).
83. International Association of Horticultural Producers (AIPH). Global Impact of Coronavirus Pandemic on Garden Centres. April 2020. Available online: https://aiph.org/latest-news/global-impact-of-coronavirus-pandemic-on-garden-centres/ (accessed on 27 January 2022).
84. Van Rijswick, C.; Fumasi, R.; van Horen, L.; Higgins, H.; Magaña, D. Coronavirus Concerns in the Global Fresh Produce Sector: Different Every Day. RaboResearch Food&Agribusiness Rabobank, March 2020. Available online: https://research.rabobank.com/far/en/sectors/fresh-produce/corona-concerns-in-the-global-fresh-produce-sector.html (accessed on 31 October 2021).
85. International Association of Horticultural Producers (AIPH). Coronavirus Global Impact Survey on the Ornamental Horticultural Industry (Part 2). 2020. Available online: https://aiph.org/wp-content/uploads/2020/11/Coronavirus-Global-Impact-Survey-Particpant-Comments.pdf (accessed on 27 January 2022).
86. European Commission. A European Green Deal. Striving to be the First Climate-Neutral Continent. EU Commission Strategy, Priorities 2019–2024. Available online: https://ec.europa.eu/info/strategy/priorities-2019-2024/european-green-deal_en#thebenefitsoftheeuropeangreendeal (accessed on 21 February 2022).
87. Moonen, G. (Ed.) *The New CAP Creating New Horizons. Journal n. 2/2021*; European Court of Auditors: Luxembourg, 2021; QJ-AD-21 002-2A-N.
88. European Commission. The CAP Reform's Compatibility with the Green Deal's Ambition. EU Commission News. Agriculture and Rural Development: Brussels, Belgium, 20 May 2020. Available online: https://ec.europa.eu/info/news/cap-reforms-compatibility-green-deals-ambition-2020-may-20_it (accessed on 21 February 2022).

89. Bruns, J.D. Let Us Plant and Tree Nurseries Be Part of the European Green Deal. FloraCulture International-AIPH EU, 31 May 202 Available online: https://aiph.org/floraculture/news/let-us-plant-and-tree-nurseries-be-part-of-the-european-green-deal (accessed on 21 February 2022).
90. International Flower Trade Association (UNION FLEURS). STATEMENT: Situation of the Ornamental Sector—COVID-1 Crisis. Union Fleurs News. 24 March 2020. Available online: https://unionfleurs.org/news_events/statement-situation-of-the ornamental-sector-covid-19-crisis/ (accessed on 21 February 2022).
91. European Commission. Biodiversity Strategy for 2030. EU Commission Strategy. Available online: https://ec.europa.eu environment/strategy/biodiversity-strategy-2030_en (accessed on 21 February 2022).
92. European Commission. 3 Billion Trees Pledge. EU Commission Environment. Available online: https://ec.europa.eu environment/3-billion-trees_it (accessed on 21 February 2022).
93. European Union Assembly of Regional and Local Representatives. Concerted Action Needed to Green the EU's Communitie European Committee of the Regions Press Release. 27 January 2022. Available online: https://cor.europa.eu/en/news/Pages concerted-action-needed-green-communities.aspx (accessed on 21 February 2022).
94. Baccino, F. Le Affinità (elettive) tra New Green Deal e Florovivaismo. Terra e Vita, 15 September 2021. Available online: https://terraevita edagricole.it/featured/le-affinita-elettive-tra-new-green-deal-e-florovivaismo/ (accessed on 21 February 2022).
95. Centre for the Promotion of Imports from Developing Countries of The Netherlands Ministry of Foreign Affairs (CBI). Through What Channels Can You Get Cut Flowers or Foliage onto the European Market? CBI Ministry of Foreign Affairs Marke Information. 9 May 2017. Available online: https://www.cbi.eu/market-information/cut-flowers-foliage/channels-segments (accessed on 3 February 2022).
96. Betjes, J.; Vallen, J.; Luca, E.R.; Tufano, G. Con RoyalFloraHolland Verso il Futuro Grandi Opportunita' di Sviluppo per I Produttor Agricoli. Lecture Presented at the RoyalFloraHolland and Veiling Rhein-Maas Meeting "I Fiori di Roma", Oasi di Kufra, Sabaudia Latina, Italy. 25 November 2017. Available online: https://www.royalfloraholland.com/en (accessed on 25 May 2021).
97. Byczynski, L.; Benzakein, E. *Fresh from the Field Wedding Flowers*, 1st ed.; Fairplain Publications Incorporated: Lawrance, KS, USA 2014; ISBN 0977978133.
98. Needleman, D. What Happened to Traditional Floral Bouquets? The New York Times Style Magazine. 20 March 2017. Available online https://www.nytimes.com/2017/03/20/t-magazine/traditional-floral-bouquets.html (accessed on 3 February 2022).
99. Prinzing, D. *Slow Flowers: Four Seasons of Locally Grown Bouquets from the Garden, Meadow and Farm*, 1st ed.; St. Lynn's Press Pittsburgh, PA, USA, 2013; ISBN 0983272689.
100. Centre for the Promotion of Imports from Developing Countries of The Netherlands Ministry of Foreign Affairs (CBI). What Requirements Should Your Cut Flowers and Foliage Comply with to Be Allowed on the European Market? CBI Ministry of Foreing Affairs Market Information. 4 May 2017. Available online: https://www.cbi.eu/market-information/cut-flowers-foliage/buyer-requirements (accessed on 3 February 2022).
101. Hall, C.R.; Campbell, B.L.; Behe, B.K.; Yue, C.; Lopez, R.G.; Dennis, J.H. The appeal of biodegradable packaging to floral consumers. *HortScience* **2010**, *45*, 583–591. [CrossRef]
102. Hall, C.R.; Dickson, M.W. Economic, Environmental, and Health/Well-Being Benefits Associated with Green Industry Products and Services: A Review. *J. Environ. Hortic.* **2011**, *29*, 96–103. [CrossRef]
103. European Network for Rural Development (ENRD). *Green Economy Opportunities for Rural Europe. EU Rural Review No. 23* Thorpe, E., Ed.; Publications Office of the European Union: Luxembourg, 2017.
104. Schouten, M. EU Action Plan: Towards Zero Pollution for Air, Water and Soil. European Committee of the Regions, ENVE Commission, Opinion No. CDR 3178/2021, Adopted. 27 January 2022. Available online: https://cor.europa.eu/en/our-work/Pages/OpinionTimeline.aspx?opId=CDR-3178-2021 (accessed on 21 February 2022).
105. Ronco, R. La Filiera Florovivaistica nel Veneto. Veneto Agricoltura: Legnaro, Padova, Italy, December 2002. Available online https://www.venetoagricoltura.org/upload/pubblicazioni/PDF%20Economia/SC36.pdf (accessed on 3 February 2022).
106. Lufkin, B. Why Are Flowers so Expensive? BBC Worklife Economics. 8 May 2019. Available online: https://www.bbc.com/worklife/article/20190507-why-are-flowers-so-expensive (accessed on 28 January 2022).
107. Joyce, D.C.; Turner, C. Developing a Commercial Floriculture Activity in a Research Environment and a Supply Chain Context *Acta Hortic.* **2007**, *755*, 45–54. [CrossRef]
108. International Trade Center (ITC) Standards Map App. Available online: https://standardsmap.org/en/identify (accessed on 3 February 2022).
109. Stebner, S.; Baker, L.M.; Peterson, H.H.; Boyer, C.R. Marketing with More: An In-depth Look at Relationship Marketing with New Media in the Green Industry. *J. Appl. Commun.* **2017**, *101*, 7–18. [CrossRef]
110. Paniagua, J.; Sapena, J. Business performance and social media: Love or hate? *Bus. Horiz.* **2014**, *57*, 719–728. [CrossRef]
111. Weinberg, B.D.; Pehlivan, E. Social spending: Managing the social media mix. *Bus. Horiz.* **2011**, *54*, 275–282. [CrossRef]
112. Yao, B.; Shanoyan, A.; Peterson, H.H.; Boyer, C.; Baker, L. The use of new-media marketing in the green industry: Analysis of social media use and impact on sales. *Agribusiness* **2018**, *35*, 281–297. [CrossRef]
113. Hall, C.R. How to Market Yourself in a Questionable Economy. The Association of Horticulture Professionals (OFA) Bullettin No. 929 September/October 2011. Available online: https://cdn.coverstand.com/4456/81472/81472.1.pdf (accessed on 2 February 2022).

114. Malindretos, G.; Moschuris, S.; Folinas, D. Cut-Flowers Supply Chain and Logistics. The Case of Greece. *Int. J. Res. Manag. Bus. Stud.* **2015**, *2*, 15–25.
115. Allegra, V.; Bellia, C.; Zarbà, A.S. Direct Sales as a Tool for Competitiveness for Smes in the EU. The Case of Farms "Ornamental Floriculture and Nursery Products". *Qual.-Access Success* **2014**, *15*, 19–24.
116. Serra, G. La Filiera Della Qualità nel Florovivaismo: Qualità-Valore-Servizio-Convenienza-Scelta. Lecture Presented at the Conference "La Qualità Totale nel Florovivaismo". Baveno (Verbano-Cusio-Ossola), Italy. 2009; *To be submitted*.
117. European Commission. *Proposal for a Regulation of the European Parliament and of the Council COM(2018) 392 Final 2018/0216 (COD)*; European Commission: Brussels, Belgium, 1 June 2018; Available online: https://eur-lex.europa.eu/legal-content/EN/TXT/?uri=COM%3A2018%3A392%3AFIN (accessed on 21 February 2022).

MDPI AG
Grosspeteranlage 5
4052 Basel
Switzerland
Tel.: +41 61 683 77 34

Horticulturae Editorial Office
E-mail: horticulturae@mdpi.com
www.mdpi.com/journal/horticulturae

www.ingramcontent.com/pod-product-compliance
Lightning Source LLC
Chambersburg PA
CBHW061559220926
51850CB00005B/377
9783725818112